Clusters and Colloids

From Theory to Applications

Edited by Günter Schmid

Distribution:
VCH, P.O. Box 10 1161, D-69451 Weinheim, Federal Republic of Germany
Switzerland: VCH, P.O. Box, CH-4020 Basel, Switzerland
United Kingdom and Ireland: VCH, 8 Wellington Court, Cambridge CB1 1HZ, United Kingdom
USA and Canada: VCH, 220 East 23rd Street, New York, NY 10010–4606, USA
Japan: VCH, Eikow Building, 10-9 Hongo 1-chome, Bunkyo-ku, Tokyo 113, Japan

ISBN 3-527-29043-5 (VCH, Weinheim) ISBN 1-56081-753-4 (VCH, New York)

Clusters and Colloids

From Theory to Applications

Edited by Günter Schmid

Weinheim · New York
Basel · Cambridge · Tokyo

Prof. Dr. Günter Schmid
Institut für Anorganische Chemie
Universität GH Essen
Universitätsstraße 5–7
D-45117 Essen
Federal Republic of Germany

Published jointly by
VCH Verlagsgesellschaft mbH, Weinheim (Federal Republic of Germany)
VCH Publishers, Inc., New York, NY (USA)

Editorial Directors: Dr. Thomas Mager and Dr. Thomas Kellersohn
Production Manager: Elke Littmann

The cover shows a high resolution microscopic image of a single gold colloid (about 11 x 13 nm), protected by a shell of P(m-$C_6H_4SO_3Na$) ligands. The picture has kindly been provided by Prof. J. O. Bovin and A. Carlsson, University of Lund, which is gratefully acknowledged.

Library of Congress Card No. applied for.

A catalogue record for this book is available from the British Library.

Deutsche Bibliothek Cataloguing-in-Publication Data:
Clusters and colloids : from theory to applications / ed.
by Günter Schmid. – Weinheim ; New York ; Basel ; Cambridge ;
Tokyo : VCH, 1994
ISBN 3-527-29043-5 (Weinheim ...)
ISBN 1-56081-753-4 (New York)
NE: Schmid, Günter [Hrsg.]

Printed on acid-free and chlorine-free paper.

Composition: Hagedornsatz GmbH, D-68519 Viernheim. Printing: Druckhaus Diesbach, D-69442 Weinheim. Bookbinding: J. Schäffer GmbH, D-67269 Grünstadt.
Printed in the Federal Republic of Germany.

Contents

List of Contributors

Dr. John S. Bradley
Exxon Research & Engineering Company
Corporate Research
Materials Science
Route 22 East, Clinton Township
Annandale, NJ 08801
USA

Prof. Dr. Dieter Fenske
Inst. für Anorganische Chemie
der Universität Karlsruhe
Engesserstraße
D-76128 Karlsruhe
Germany

Prof. Dr. Bruce C. Gates
Dept. of Chemical Engineering
and Materials Science
University of California
Davis, CA 95616
USA

Dr. Maria Carmela Iapalucci
Dipartimento di Chimica Fisica
e Inorganica
Università di Bologna
Viale de Risorgimento 4
I-40136 Bologna
Italy

Dr. Sibudjing Kawi
Dept. of Chemical Engineering
and Materials Science
University of California
Davis, CA 95616
USA

Prof. Dr. Giuliano Longoni
Dipartimento di Chimica Fisica
e Inorganica
Università di Bologna
Viale de Risorgimento 4
I-40136 Bologna
Italy

Prof. Dr. Gianfranco Pacchioni
Dipartimento di Chimica Inorganica,
Metallorganica e Analitica
Università di Milano
Via Venezian 21
I-20133 Milano
Italy

Prof. Dr. Notker Rösch
Lehrstuhl für
Theoretische Chemie
der Technischen Universität München
Lichtenbergstraße 4
D-85748 Garching
Germany

Prof. Dr. Günter Schmid
Inst. für Anorganische Chemie
der Universität GH Essen
Universitätsstr. 5–7
D-45117 Essen
Germany

Prof. Dr. Arndt Simon
Max-Planck-Institut
für Festkörperforschung
Heisenbergstr. 1
D-70569 Stuttgart
Germany

Color Plates

Figure 3-34. Scanning tunnel microscopic image of five shell Pd cluster molecules $[Pd_{561}phen_{36}O_{\sim 200}]$. The ball like molecules are probably imaged together with their ligand shell.

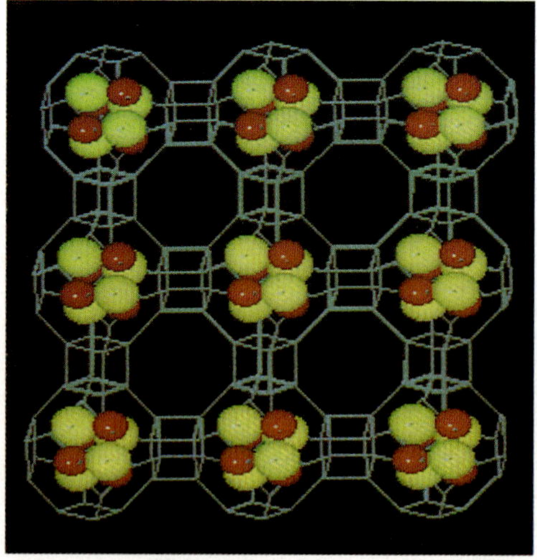

Figure 4-14. Structures proposed for $(CdS)_4$ clusters in the sodalite cages of zeolite A. [217] Reproduced from *Science* with permission of the American Association for the Advancement of Science.

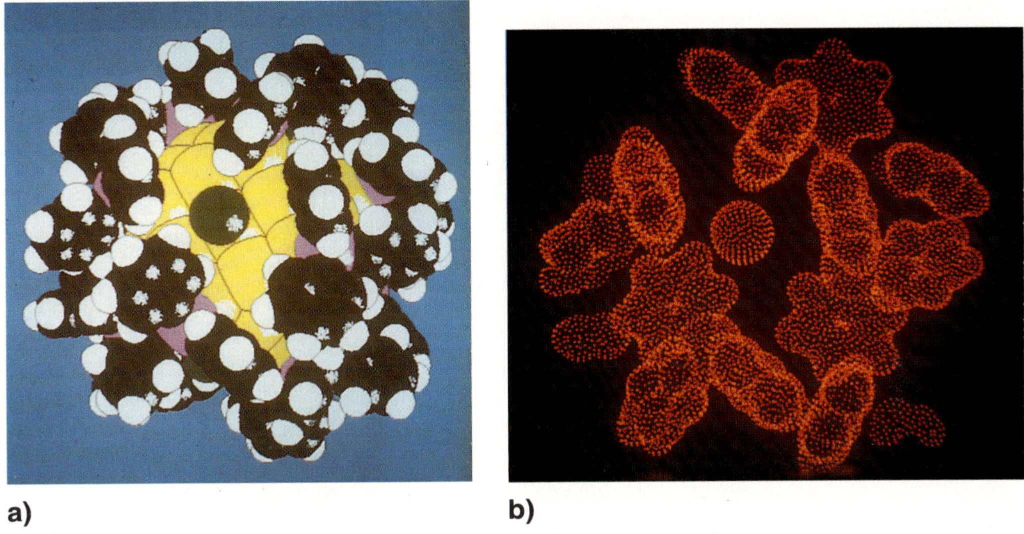

Figure 3-35. a) A computer simulated space filling model of $[Au_{55}(PPh_3)_{12}Cl_6]$. b) A computer simulated "two-dimensional" electron density image of $[Au_{55}(PPh_3)_{12}Cl_6]$ for comparison with c) the STM image of the same cluster molecule in probably the same direction. The similarities between both images are evident. A chlorine atom is positioned in the center of the images.

1 General Introduction

Günter Schmid

The continuous reduction in size of a solid finally leads to a situation where the original solid state properties can be only partially observed or may be even completely lost, as these properties are exclusively the result of the cooperation between an infinite number of building blocks. Further reduction of size finally leads to typical molecular behavior. On the other hand, even here are structural relations to the bulk occasionally detectable. For instance, the arrangements of the sp^3 hybridized carbon atoms in cyclohexane or in adamantane can easily be traced back to the diamond lattice, whereas benzene or phenanthrene represent derivatives of the graphite lattice. However, neither cyclohexane, benzene, nor phenanthrene have chemical properties which are comparable with those of the carbon modifications they originate from. The existence of the above mentioned C_6, C_{10} or C_{14} units is only made possible by the saturation of the free valencies by hydrogen atoms. Comparable well known examples for other elements are numerous, for instance the elements boron, silicon, and phosphorous. Figure **1-1** illustrates some of the relations between elementary and molecular structures.

Carbon atoms with sp^2 hybridization offer a fascinating example for the transition from the infinite crystal lattice to the molecular state. In this case, not 6, 10, or 14, but 60 carbon atoms are used as cutouts of the lattice, and the free valencies are not saturated by hydrogen atoms: such nano sized cutouts are too small to exist as a stable graphitic structure and consequently they create a spheric shape consisting of five- and six membered rings with altogether 60 vertices, the famous soccer-like so-called fullerene, C_{60}.

If a piece of metal is reduced to a size of a few thousand atoms we enter the world of metal colloids, unique particles which were already handled by Michael Faraday in the last century. Smaller units of a few hundred or dozen atoms are usually called 'clusters'. This term is also well tried for small molecular species consisting of only a few metal atoms. Metals, especially transition metals, offer an exceptional opportunity to study the pathway which leads from the bulk to the molecular state and finally to mononuclear complexes (Figure **1-2**).

The present book aims for a general overview of our present knowledge in the field of cluster and colloid science, without calling on completeness. On the contrary, it is intended to elucidate developments, highlights, and the actual situation. The broadness of this field is documented by chapters with a mainly theoretical background, sections where the physics of small metal particles dominate,

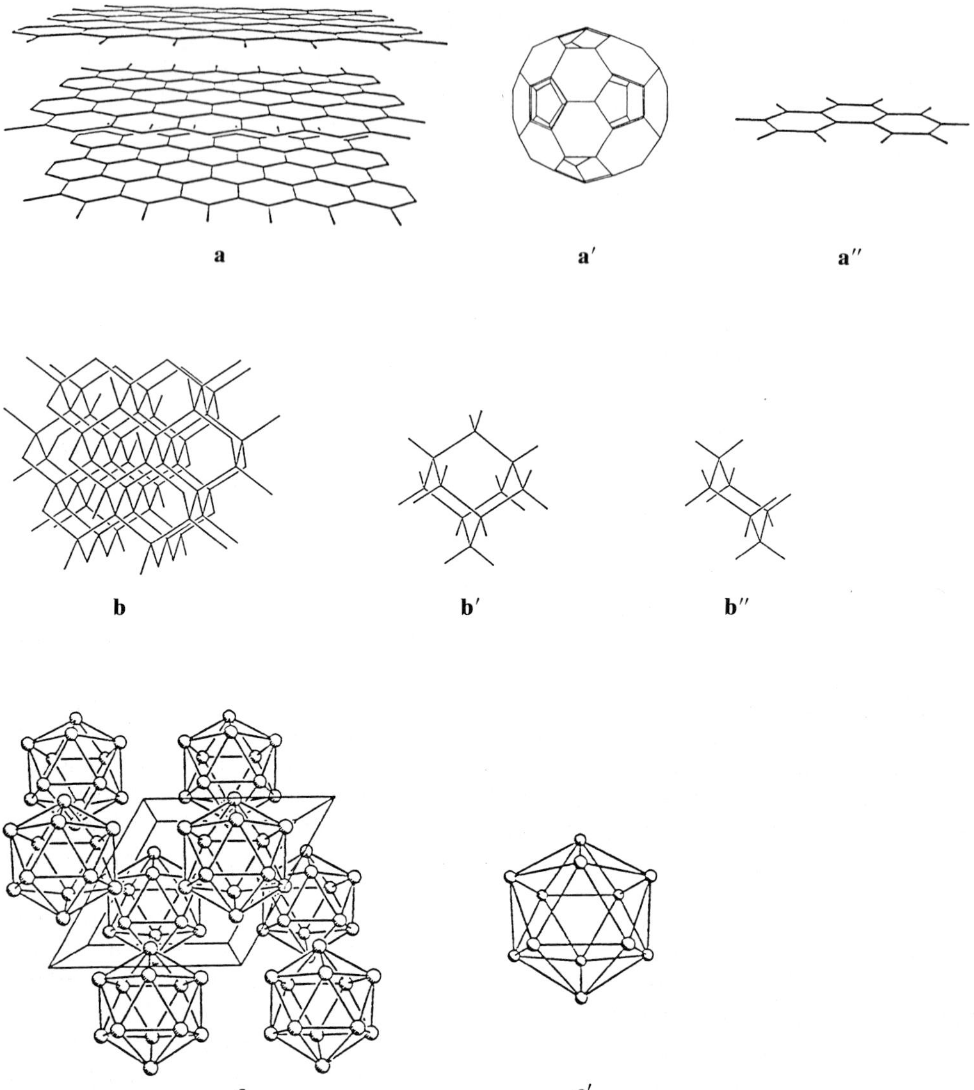

Figure 1-1. Structural relations between the bulk and the molecular state of some elements.
1-1 a) The graphite lattice, a′) the fullerene molecule C_{60}, formally derived from a C_{60} cutout of a graphite layer, and a″) the skeleton of the phenanthrene molecule as a representative of aromatic systems.
1-1 b) The diamond lattice and the molecular structures of the molecules, b′) adamantane, and b″) cyclohexane. The chair configuration of its C_6-skeleton can be easily recognized in b′) and b).
1-1 c) The crystal structure of the rhombohedral boron modification consisting of linked icosahedra. c′) Most boranes derive from the icosahedral building block. $[B_{12}H_{12}]^{2-}$ consists of a complete icosahedron. (The hydrogen atoms are omitted)

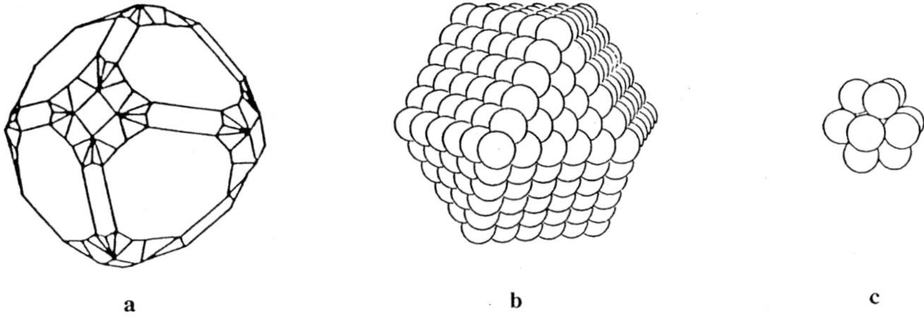

Figure 1-2. Illustration of the transition from a) bulk metal *via* b) colloids and clusters to c) small molecular cluster species.

and others in which the syntheses of clusters and colloids are described. The syntheses, physical and chemical properties, and first applications as well as the structures of these nano sized particles are treated. The generation and the properties of 'isolated clusters' ('naked clusters'), which are only available in so-called cluster beams, are not intended to be described in this book. They are of special interest in physics, however, they can never be isolated as stable materials. The description of the various and complicated cluster beam techniques as well as the discussion of the physical properties of clusters in beams would exceed the frame of this book considerably.

Where does the scientific importance and the fascination for clusters and colloids originate from? There are several answers to this question, not the least of which is due to subjectively different views. For many scientists, the synthesis of isolable metal clusters is a big challenge, whereas others are interested in structural details or physical properties. A possible application in catalysis is a further motive to employ oneself with small and reactive metal particles. All these different interests are finally based on a mutuality to which the known and the expected properties are based on: that is, the dramatic reduction of freely mobile electrons on the path from the bulk to the cluster. However, such a miniaturization also affects numerous other properties, for instance the melting point. This can be demonstrated with the element gold: the melting point of bulk gold is 1064 °C, whereas that of a 1.5 nm gold particle is decreased to about 500–600 °C. Of much more importance is the change in the electronic properties on the way from the bulk to the nano sized species. This can be followed best in an imaginary experiment. If a threedimensional piece of metal is reduced to a layer of only a few atoms in thickness, the original electrons, which were freely mobile in three dimensions, now can only move in two dimensions: a quantum wall has been generated. Further reduction to a quantum wire leads to the one dimensional case. If finally a quantum wire was to be cut into a piece which was as short as it was thick, then an electronically zero dimensional quantum dot would result. In such a quantum dot, the last few 'metallic' electrons are fenced and, due to quantum size effects, they behave like electrons in a box and differ generally from electrons in the bulk. These electrons determine the physical and chemical characteristics of metal clusters and colloids to a significant extent.

Another factor deserves consideration if we are to discuss the properties and applications of metal clusters and colloids. Most of the investigated species do not exist without being influenced by a surrounding media. In practice, most of the clusters and colloids described in this book are protected by a shell of ligand molecules or they are embedded in cages or matrices like polymers or solids, in order to separate them from each other to prevent coalescence. Ligand molecules, as well as cages, chemically interact with the surface atoms of the metal particles and so have a remarkable influence on their electronic character. Just as the electronic states of single metal atoms or ions in simple complexes are determined by the ligand field, the surface atoms of a cluster or colloid will be affected by their environment as well. Consequently, we have to realize that the properties of 'naked' clusters must be considerably different from those of ligated or somehow fenced metal particles.

This book tries to regard most of these aspects. Internationally recognized scientists describe those fields of cluster and colloid research in which they have been working for many years and therefore are endowed with fundamental knowledge. The book is intended for those scientists working in research as well as in practice who wish to gain a fundamental insight into one or more areas of the world of small metal particles. However, it is addressed to advanced students in physics, chemistry, or materials sciences as well.

2 Electronic Structures of Metal Clusters and Cluster Compounds

Notker Rösch and Gianfranco Pacchioni

2.1 Introduction

Inorganic chemists understand the term "cluster" differently than physical chemists or physicists do. It is no wonder then that the objects designated in these fields as clusters have very different characteristics and that their investigations give rise to quite different concepts and require many diverse methods, both experimental and theoretical. In the following, we will focus our attention on the electronic structures of those metal clusters and metal cluster compounds which represent the most important systems in each area. It will become clear that despite their often very different characteristics, the same, or at least rather similar, concepts and methods in theoretical chemistry may be profitably applied to both types of metal clusters. By highlighting both their similarities and their differences, the juxtaposition of these two classes of metal clusters will provide a more lucid view on each of them.

In the newly emerging field of cluster science, as an offspring of physical chemistry and physics, the term "cluster" refers to a new form in the aggregation of matter which lies intermediate between the molecular and the solid state. Thus, cluster science will provide information on how the properties of a solid gradually evolve as atoms are brought together to form increasingly larger units. Given this general definition, it might be difficult to distinguish between a molecule and a small cluster. For instance, one may wish to exclude a tetrahedron of four phosphorus atoms or a ring of eight sulphur atoms from the class of clusters. In fact, the P_4 and S_8 molecular units exist in various aggregation states (solid, liquid, and vapour) and may be more properly considered as homonuclear molecules. In this contribution, the term "cluster" will be used to designate aggregates of atoms, not necessarily of the same element, which do not exist in measurable quantities in an equilibrium vapor. In this respect, the fullerenes would be considered as large molecules while Na_n aggregates certainly belong to the category of clusters.

Even within the field of inorganic chemistry, the term "cluster" is used to designate a wide variety of molecular entities. The boranes (e. g. $[B_{12}H_{12}]^{2-}$), the basic Fe_4S_4 unit contained in the iron-sulphur proteins, the transition metal carbonyls (e. g. $[Ni_5(CO)_{12}]^{2-}$), the "metal-only" clusters like Bi_9^{5+}, Ge_9^{2-}, and

Pb_5^{2-}, the ternary chalcides of general formula $[M_xMo_6X_8]$ (Chevrel phases), and such cage molecules as P_4 (the basic component of white phosphorus) are all examples of "cluster cages" but with completely different characteristics.

Undoubtedly, the metal clusters form a unique and exciting subgroup of inorganic cluster compounds. More than a thousand examples of ligated metal clusters have been reported in the literature since the first examples of polynuclear complexes containing metal-metal bonds were discovered about 30 years ago. This number gives an idea of the exceptional growth which has taken place in this area of inorganic chemistry. It is useful to classify an inorganic cluster as "a compound containing a finite group of metal atoms which are held together entirely, mainly, or at least to a significant extent, by bonds directly between metal atoms". [1] This definition, originally proposed by Cotton in 1966, is valuable from a conceptual point of view, although it is difficult to apply since there is generally no simple way to establish the existence or to measure the strength of a metal–metal bond within a cluster compound. Thus, the distinction between a metal cluster and other metal containing inorganic compounds where metal–metal bonds are completely absent is not always straightforward. Actually, characterizing the nature and the extent of the metal–metal interactions in metal clusters is one of the most challenging problems for theoreticians and will be discussed at length below.

Metal clusters are the most interesting from several points of view, although a great deal of attention has also been given to clusters of semiconducting materials [2–6] and to Van der Waals clusters [7, 8] over the past decade. In the following, we will discuss the electronic structures of both gas phase (or "naked") metal clusters and inorganic metal cluster compounds. Rather than attempting to give an exhaustive review, we will highlight the key concepts and methods and then discuss the theoretical results, mostly from a quantum chemical point of view, for important examples in each class of metal cluster. This implies two obvious restrictions for our presentation. First, when one is interested in the transition to the bulk limit, it is natural to discuss clusters using concepts and methods derived from condensed matter theory. Although we will occasionally mention such methods, we will not explore them in any great detail. Furthermore, gas phase clusters (and in several cases also inorganic clusters) have interesting and novel dynamic properties as a consequence of their unusual geometric and electronic properties. These aspects of clusters are beyond the scope of this presentation.

2.2 The Description of the Clusters Electronic Structure

The role of theory in cluster research is twofold. On the one hand, theory must be able to provide a basis for understanding the chemical and physical properties of small metal aggregates, whether naked or ligated, and to rationalize any observed trends. On the other hand, theory is also expected to furnish quantitative answers and to have a definite predictive power. An obvious question among the many that may be posed to theory is the one already mentioned concerning the amount of metal–metal bonding. Another important one is about the geometry of a cluster and its relation to the other characteristics of the cluster. It is clear from these considerations that one would like to apply highly accurate methods as well as qualitative schemes. Also, methods which are able to describe not only small but also large clusters as well, and even extended systems will have a special appeal. It should be remembered that the quantitative description of metal compounds, and especially those of the transition metals, is still quite a challenge for all computational methods, despite the substantial progress which has been made in the recent past, at least for mononuclear complexes. [9, 10]

Given the complexity of the systems and the diversity of the questions still open in the field of metal clusters, it is no wonder that essentially all the methods available from the ample arsenal of quantum chemistry have been applied to cluster problems. We will not give an extensive overview of the many different methods (let alone aim for completeness) and leave aside most technical aspects. This information can be found in specialized publications (e. g. [11–15]), from which some are even devoted to the electronic structures of clusters. [16, 17] Instead, we will summarize the basic features of the methods and comment on their applicability to the description of both naked and ligated metal clusters. We will start the discussion with wave function based methods and then proceed to density functional methods. Although the latter have only recently gained a broader acceptance for chemical applications, they have a rich tradition in the metal cluster field, particularly due to their solid state heritage. We will also briefly mention simplified approaches to the electronic structure of metal clusters.

First principle quantum chemical methods, whether wave function based (*"ab initio"*) or density based, are aimed at solving the electronic Schrödinger equation without any reference to adjustable parameters or empirical data. In their standard form, they invoke the Born-Oppenheimer separation of electronic and nuclear motion and employ a nonrelativistic Hamiltonian which does not include any explicit reference to spin-dependent terms. Many quantum chemical methods are based on the variational principle which, for computational convenience, is implemented in algebraic form *via* either one-electron functions built from linear combinations of atomic orbitals or n-electron functions constructed from Slater determinants. [11, 12]

2.2.1 Wave Function Based Methods

The basis for all wave function based *ab initio* methods is the Hartree-Fock (HF) approach. [11, 12] It makes use of a single-determinant ansatz constructed from one-electron spin orbitals. These orbitals describe the motion of each electron within the field of the nuclei and the mean field of the remaining n–1 electrons. The mean field is not known *a priori*, but depends on the orbitals which are determined self-consistently from the eigenvalue problem of the Fock operator.

[12] Therefore, the resulting iterative procedure is referred to as a self-consistent field (SCF) technique. In the case of metal clusters, one is often faced with an open shell system in which at least one set of degenerate spin orbitals is not fully occupied. The familiar form of the restricted HF (RHF) theory must then be replaced by a more complicated formalism. Often, one resorts to unrestricted HF (UHF) theory which allows different spatial orbitals for different spins at the expense of employing a wave function which is not an eigenfunction of the total spin operator. [12, 17] For computational efficiency, the molecular orbitals are usually constructed as a linear combination of atomic Gaussian type basis functions (GTO). In this way, the accuracy of the description of a metal cluster is very dependent on the choice of the basis set (see [18] for a comprehensive description of *ab initio* basis sets). The number of integrals which have to be computed in the HF method formally scales with N^4 where N is the number of basis functions. This means that the treatment of clusters having more than about 20 atoms becomes difficult, even with the computational facilities available today. This problem can be partially overcome by using a "direct" SCF approach. Here, the storage requirement is significantly reduced by following a strategy whereby the integrals are evaluated upon demand as required for constructing the change in the Fock operator during the iteration process. [19, 20]

The HF method provides a transparent interpretation of the n-electron wave function. According to Koopmans' theorem, the one-electron energies are directly related to the ionization potentials. [11, 12] In general, the method yields acceptable results for the properties of clusters near the equilibrium configuration, provided the HOMO-LUMO gap is not too small. Unfortunately, this is not the case for many naked metal clusters. Although it may not be a problem for ligated clusters, they tend to be too large for this level of theory. A severe limitation to the HF mean field approach is that it ignores the spatial correlation of the electrons. Thus, the absolute values of observable properties, in particluar the binding energies and vibrational frequencies, deviate considerably from their experimental values. [12] Another disadvantage with respect to metal clusters is the well known fact that the HF method incorrectly describes dissociation when it is accompanied by a change in spin multiplicity. [12]

In order to improve the mean field description of the electronic structure one has to go beyond the single-configuration approach. [12, 13] Two main strategies have been developed to introduce correlation effects. In the first case, one employs methods based on many-body perturbation theory (MBPT). [12, 21] They allow the treatment of so-called dynamical correlation effects in cases where the HF method already provides a reasonable description of the ground state. However, these perturbation theoretical methods are not variational, that is the calculated value for the energy does not provide an upper bound to the true energy of the system.

An alternative is represented by methods where one mixes one-electron configurations to obtain a many-determinant wave function. [12] These configurations are generated by distributing the electrons among the mean field spin orbitals. If one takes all the possible "substitutions" (single, double, triple, etc.) into account, one obtains in principle the exact solution to the Schrödinger equation.

This approach is called full configuration interaction (full CI). The energy difference between the full CI and the SCF solutions is defined as the correlation energy.

In practice, however, such calculations are hardly feasible for systems containing more than 10–15 electrons because post-HF methods scale as at least N^5. Thus, one has to resort to limited multi-determinant expansions of the *n*-electron wave function. This "truncated" CI introduces only part of the correlation energy with the main consequence being that the method is not size consistent: the energy of a system and its separated components are not described on an equal footing. This represents a serious drawback when studying the change in cluster stability as a function of cluster size and ultimately precludes the investigation of large clusters and thus the convergence to the bulk cohesive energy. [14] Another limitation is that the interpretation of the CI wave function is often less facile. The problems of size consistency can be removed by employing multi-configuration SCF (MCSCF) techniques. [22] The MCSCF wave function is a truncated CI expansion in which both the coefficients of the atomic orbitals in the one-electron wave functions and the coefficients of the determinants in the CI expansion are simultaneously optimized. The underlying equations are considerably more complicated than those of either the HF or the CI methods. For this reason, MCSCF and its variant GVB [23] and CASSCF [4] techniques have so far been applied only to clusters of relatively small size. [17]

Nevertheless, these methods provide the only viable alternative for naked transition metal clusters since they can treat the nondynamical correlation effects of the near-degeneracy problems which typically occur in these systems. For clusters of simple metals, the previously mentioned MBPT methods are applicable when one configuration dominates the multi-determinant expansion. Post-HF techniques yield much better values for the various measureable properties of a cluster. The allowed optical transition energies play a particularly important role among these because they often permit the indirect determination of the geometry of a gas phase metal cluster. [25] In this context, the only computational approaches which allow for an accurate description of the excited states are the CI or MCSCF methods. Since the electronic properties of small metal clusters depend strongly on the details of the geometric structure, a geometry optimization should only be performed with a method that includes a description of the correlation effects. However, due to the availability of analytical gradient techniques, most of the geometry optimization studies on clusters have so far been performed at only the HF level of theory. [17, 25]

Since the computational effort required for post-HF *ab initio* methods is quite substantial, one often treats only the valence electrons explicitly and replaces the atomic cores by analytical effective core potentials (ECP). [17, 26] This technique is also referred to as a pseudopotential approach. A variety of strategies for the design of ECP's has been suggested. [27–30] Although these techniques have allowed the description of relatively large naked metal clusters, their application to ligated clusters is still scarce and restricted to low nuclearity complexes. For certain properties the ECP technique represents a considerable restriction, in particular when the polarization of the core is not negligible; however, more sophisticated formalisms are being developed to include appropriate corrections. [17, 31] ECP's also provide a convenient way for treating heavier atoms by incorporating relativistic effects. [29] These are important for clusters of the heavy elements in general and in particular for clusters of gold and mercury. [32] Relativistic corrections may be as large as correlation effects in gold cluster compounds. [33]

2.2.2 Density Functional Methods

Density functional methods for finite electronic systems were suggested some time ago, [15, 34, 35] but only in the last decade have these techniques found wider acceptance for chemical applications. [36, 37] Density functional theory (DFT) starts from the assertion that the ground state energy of an electronic system can be expressed as a unique functional of the density ϱ and that it fulfills a variational principle. [38, 39] A convenient technique for solving this minimization problem is provided by the Kohn-Sham (KS) formalism [40] which results in a one-electron Schrödinger equation with a density dependent effective local potential. A very appealing aspect of this formalism is its simplicity; yet it incorporates exchange and correlation effects on an equal footing. Limitations of DFT are that the fundamental form of the energy functional is known only approximately and that, in contrast to wave function based methods, there is no hierarchy in the approximations which can provide for systematic improvements.

The most common choice for the energy functional starts from its separation into three terms: a kinetic energy contribution of a "noninteracting" reference system, the classical Coulomb interaction of the charge distribution under study, and a remainder which comprises the exchange and correlation effects. [15] Various approximations have been suggested to treat the latter term, [15, 34, 37] whereby a popular choice is to assume the same functional form as in a weakly inhomogeneous electron gas. This approach is called the local density approximation (LDA) and several parametrizations have been suggested. If one takes only the exchange interaction into account, then the famous $\varrho^{1/3}$ dependence of the "exchange-correlation" potential, well known from the Xα formalism, [41] is obtained. It should be noted that the "correlation energy" in DFT is defined differently than in *ab initio* methods. The relationship of the exchange-only approximation in LDA to the HF formalism has been the subject of an intense and controversial debate which has not reached a definite conclusion. [42] At this level, the method provides good results for bond lengths and vibrational frequencies. [37] However, the values for binding energies may be in serious error; in many cases, they are too large. This deficiency is related to the LD approximation. Improvements can be made by chosing a more sophisticated form for the exchange-correlation functional in that it also depends on the gradients of the electronic density. These so-called nonlocal corrections do not affect the local character of the effective one-electron potentials. This aspect of the theory is currently undergoing intense development. [34, 37, 43, 44]

In the LDA methods, the one-electron functions are usually expanded into atomic basis sets whereby numerical orbitals, GTO's, Slater-type orbitals (STO), and the very special linearized muffin tin orbitals (LMTO) are used. [37] The use of GTO's or STO's facilitates a direct comparison between the one-electron functions in the LDA to those in HF theory. In contrast to HF, however, Koopmans' theorem is not valid here. [15] In order to relate the one-electron energies in DFT to ionization potentials or core level binding energies one must resort to Slater's transition state procedure. [41]

The local spin density (LSD) approximation is an extension of the above method to spin-polarized cases in which different densities are defined for electrons with up and down spins. [15, 34] This is particularly important for the study of transition metal clusters with magnetic ground states, such as clusters of Co, Fe, Ni, etc. In the following, the term 'local density functional (LDF) methods' will be used as a joint designation of DFT methods that employ either the LD or the LSD approximation. Density functional based methods have been further augmented by including pseudopotentials and relativistic effects, as well as energy gradients for geometry optimization. [37, 45] The molecular dynamics (MD) approach of Car and Parrinello [46] provides a useful tool for determining the global energy minimum of a system by simultaneously solving the KS equations as the nuclear positions vary.

The various computational schemes based on DFT are attractive alternatives to conventional *ab initio* methods and particularly for the study of large clusters since the computational effort increases with the number of basis functions as roughly N^3. They allow an accurate treatment of transition metal clusters where the standard HF technique is not easily applicable. Furthermore, they provide a natural way for describing the transition from the molecular to the metallic regime since DFT theory underlies most of the first principle methods for solid state band structure calculations. [47] Although the method is still restricted to ground state properties, possible extensions for the treatment of excited states are under discussion. [34]

2.2.3 Simplified Methods

Over the past 25 years, a large variety of approximations have been introduced to treat the electronic structure problem for large systems. Many of these methods try to simplify the HF-SCF formalism by restricting themselves to the valence electrons and by drastically reducing the number of integrals which have to be explicitly calculated. [48, 49] In order to compensate for these crude approximations, parameters are introduced and adjusted to fit various experimental quantities. Several strategies have been followed and an important distinction between these is in their treatment of the electron-electron interaction. This interaction either is considered only implicitly as in Hückel and extended Hückel (EH) methods [11] or is treated similar to that in HF theory within the zero differential overlap (ZDO) methods. [48, 49] The EH method has been widely used in the study of ligated metal clusters, [50] and it also provides a framework to rationalize electron counting rules. [51] These very simplified techniques have a special merit when one investigates those aspects of the electronic structure which are governed by topological factors. Details of this will be discussed in later sections. It should be pointed out, however, that the reliability of these methods for quantitative predictions is in general very limited, especially when applied to transition metal clusters.

Simplified methods based on DFT have also been suggested. A rather simple, but very effective one used for the description of some cluster properties, like the "magic numbers" for gas phase clusters, is the jellium model. [52] This approximation does not provide any insight into the nature of the chemical bonding between the atoms and, in particular, it neglects all details of the nuclear framework. The jellium model can be used to extract information on only the topological aspects of the cluster geometry. [51] Other approximate DFT based schemes have been developed and applied to the investigation of cluster stability and geometry and include the embedded atom [53] and the effective medium theory. [54] These approximate methods will be discussed further below in context with special aspects of the cluster electronic structure.

2.3 Structure and Properties of Naked Clusters

We will discuss metal clusters of both the main group and transition metal elements, although such a distinction is not really necessary; in fact, many of the features typical of metal clusters and also their size effects have been observed for both transition metal as well as simple metal clusters. However, an obvious, but important difference is that the latter are much easier to treat theoretically and, indeed, have been studied at a much higher level of accuracy. Thus, a coherent understanding of size effects in main group clusters will also be very helpful in the description of transition metal clusters.

It has become quite apparent that the properties of small naked clusters are dominated by their very different average atomic coordination compared to that in the bulk. The low coordination in the clusters is the origin of several peculiar properties, including the large oscillations and discontinuities observed as function of the cluster size. The rationalization of these discontinuities represents one of the main goals of cluster theory. The second basic question that has attracted the attention of both chemists and physicists is that concerning the size which is required for a cluster to exhibit bulk-like character. There seems to be general agreement [17, 55, 56] that this question does not have a unique answer. As the cluster size increases, some physical properties converge to the crystalline value more rapidly than others. Many chemically interesting properties reach values within a few percent of their bulk limit for clusters containing about 100 atoms. On the other hand, pronounced cluster size dependent variations are observed for clusters having less than 100 atoms. [56]

2.3.1 The Theoretical Description of Metal Clusters

The need for theory in cluster research is quite evident since there are several properties and characteristics of bare clusters that are not easily determined experimentally, such as the geometry of stable isomers and the energy barriers which separate different structures on a potential energy hypersurface. The most serious problem to constructing a theory capable of giving an adequate description of a cluster's electronic structure is the requirement that it treats a very small cluster of only a few atoms at a comparable level of accuracy as a large metallic aggregate with typical bulk-like properties. In other words, one would like to have a theory that can extrapolate correctly to both the molecular (or atomic) and the bulk limits. Clearly, this is not a simple task and, so far, most of the theoretical approaches used in cluster theory have been derived from theories which were developed to describe one or the other extreme.

Band structure theories used for the treatment of bulk metals are based on the concept of translational symmetry and on the itinerant electron model and are therefore not applicable in the case of finite clusters where structural anisotropy, small dimensions, and anomalies in the electronic structure are common. At the molecular level, clusters are better described with the methods of quantum chemistry at either the semiempirical, the *ab initio*, or the density functional level. [17, 57–59] In some cases, even simple considerations based on the symmetry and topology of a cluster provide useful qualitative information. The problem with these simplified methods is that they can, at best, fulfill the first requirement for a cluster theory of qualitative understanding, yet in no way can they satisfy the second requirement of quantitative prediction. Both types of theory, qualitative and quantitative (or at least semiquantitative), have been and continue to be used in the study of naked clusters, and their success largely depends on the property or on the problem under investigation. There are few areas in modern science where the role of theory is so pivotal as it is in cluster research. Experiments need to be substantiated by theory and theory is continuously being challenged by new experimental findings. The combined use of theory and experiment is one of the reasons for the impressive advances which have been achieved in this field over the past 15 years. It is not a mere coincidence that in several cases theoretical predictions on cluster structures and properties have anticipated the experimental findings.

2.3.2 Structure, Bonding, and Stability

2.3.2.1 Geometrical Structures

Without any doubt, the question concerning the structure of metal clusters is one of the most intriguing. Unfortunately, as mentioned previously, it is also one of the most difficult to probe directly by experiment. The only available techniques are electron differaction from gas phase clusters or information from low temperature matrix studies (*e. g.* absorption, Raman, ESR, EXAFS). [55, 56] From the very beginning then, the main goal of the theoretical study of clusters has been the prediction of their structures.

The first semiquantitative *ab initio* investigations on cluster structures were performed on Li clusters, whereby the computational simplicity of these systems was exploited. [60–63] If one surmises that metal clusters are the "seed" for crystal growth, it is logical to expect that they will assume the most compact structure for any given nuclearity, that is, a tetrahedron for the tetramer or an octahedron for the hexamer. In these cases, the driving force which determines the shape of a small metal aggregate would be the tendency to maximize the coordination. Therefore, it came as some suprise when one found that the most stable structures for low nuclearity Li clusters, [61, 63] and of alkali metal clusters in general, [64–66] were not the most compact ones. The tetramers Li_4, Na_4, K_4, etc., assume, in fact, a nearly planar rhombic structure. This theoretical prediction was confirmed in recent years by thermodynamic data on matrix isolated Li_4 [67] and by photodetachment experiments on gas phase Li_4 (see Section 2.3.5). Even more astonishing was the finding that Li_6 is not an octahedron but rather an edge-bridged planar triangle, with a pentagonal pyramid being very close in energy. [68, 69] The theoretical prediction that Li_7 should have a bipyramidal pentagonal form was confirmed by ESR spectroscopy on matrix isolated Li clusters. [70] The appearance of cluster structures having pentagonal symmetry has opened new horizons for the understanding of cluster growth. Based on *ab initio*, density functional, and Car-Parrinello calculations, stable isomers with pentagonal or icosahedral symmetry have also been proposed for larger size clusters (e. g. Li_{13} [71]) and clusters of other elements (e. g. Be_7, [72] Mg_7, [73] Ni_7, [74] Nb_7, [75] and Cu_6, Ag_6 as well as Au_6 [76] all exhibit pyramidal or bipyramidal pentagonal arrangements). Given the difficulty of determining the structure of gas phase clusters, it is not surprising that discrepancies between theory and experiment still exist. For instance, based on vibrational autodetachement studies on Au_6, [77] it was suggested that the neutral cluster has a planar ring structure.

Using the large amount of theoretical data available for the alkali metal clusters, an "aufbau" algorithm for cluster growth was proposed but met with little success. [17] The idea that the growth sequence starts from some given "seed" structure, like a tetrahedron or a rhombus, does not seem valid. Alkali metal clusters derived from condensed tetrahedra do not correspond to the absolute minima on the potential energy surface. While heptamers prefer a pentagonal bipyramid structure, the octamer has a tetrahedral shape which cannot be derived by simply adding an atom to the heptamer. Substantial rearrangement of the atoms must take place as a cluster grows, and the dynamics of cluster growth can certainly be quite complicated when several channels compete. Similar qualitative conclusions about the most stable structures of alkali metal clusters have been obtained at very different levels of theory (graph theory, Extended Hückel, jellium, semiempirical, Hartree-Fock, valence bond, density functional, etc.). [17] Thus, it seems that the leading interaction mechanism is the overlap between the outer s orbitals of the alkali atoms with little (but not negligible) mixing of the empty p orbitals (hybridization). This causes the bonding to be nondirectional and so the geometry may be predicted on the basis of topological arguments.

Even more complex are the clusters formed from metal atoms which have more than one electron in the valence shell, that is, those from the majority of the elements of the periodic table. Since these atoms undergo substantial rehybridization in their cluster bond formation, the geometrical shape cannot be predicted by simple topological schemes. An illustrative example is provided by the alkaline earth metal clusters [72, 73, 78–86] whose atoms have a $(ns)^2$ $(np)^0$ configuration. Only closed shell interactions in the form of van der Waals forces would occur if a change in configuration did not take place. True chemical bonds can only be formed after promoting one electron from the valence s to the valence p shell. This is what actually happens in small alkaline earth metal clusters like Be_4 and Be_5. The potential energy curves for the interactions between the Be atoms in forming Be_4 and Be_5 clusters exhibit an interesting double minimum feature (Fig. **2–1**). [81, 83]

The Be atoms are in their atomic $(2s)^2$ $(2p)^0$ configuration at longer interatomic distances and the interactions are very weak. As the separation decreases,

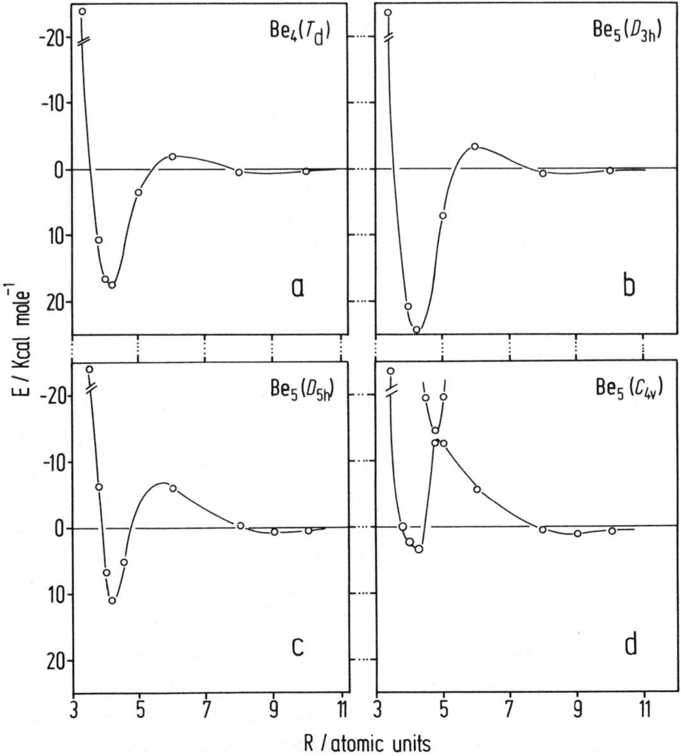

Figure 2-1. HF-CI potential energy curves for the ground states of a) a Be_4 tetrahedron, b) a Be_5 trigonal bipyramid, c) a Be_5 pentagonal planar, and d) a Be_5 square pyramidal. Distances are in atomic units. Reproduced with permission from [83].

the atoms begin to modify their electronic configuration and the 2p levels become populated. This corresponds to an electron promotion and the related energetic cost shows up as an increase in the total energy. As the distances decrease further, the price paid to change the Be configuration is overcompensated for by the formation of strong directional bonds. Thus, an energy barrier results at an interatomic distance of about 2.5–3 Å (Fig. **2-1**) and the potential surface exhibits a deep minimum at a Be–Be separation close to the bulk value of 2.29 Å. [81, 83] This behavior is typical for all small alkaline earth clusters, but the balance between the energy necessary to promote electrons from the s to the p shell and the energy gain due to the bond formation depends strongly on the cluster geometry. The interaction of four Be atoms to form a rhombic cluster is much less favorable than that for a tetrahedron, as shown by an analysis of the wavefunction. The formation of the sp hybrids requires the average coordination of the Be atoms to be as high as possible. Thus, completely different bonding mechanisms have been identified in the formation of Li_4 and Be_4. This is also generally true for clusters of different elements and accounts for the different geometrical structures observed for tetramers of different elements. [17]

When one goes from Be to heavier alkaline earth elements like Mg and Ca, the situation becomes even more complex since the tendency to form hybrid sp orbitals decreases as a group is decended. Indeed, it has been shown [80, 82] that in order to correctly describe the bonding in clusters of the heavier alkaline earth metals, one has to employ large basis sets which include not only p but also d polarization functions. Furthermore, correlation has to be taken into account simultaneously.

The special situation with clusters formed from atoms having a fully occupied s valence shell is of particular interest since in these cases, the non-metal to metal transition may occur more suddenly than with other metals. To use band structure terminology, closed shell atoms can form stable aggregates only via an overlap of the filled valence s-band with a low lying empty band (most likely of p character). The energy gap between the s- and the p-like level manifolds will decrease as the cluster size increases until a certain particle size is reached at which point the overlap will occur. Here, measurable changes in the electronic structure should take place and possibly be accompanied by a sudden change in some properties, such as the cohesive energy and the ionization potential. Whereas this transition has been theoretically predicted to take place with very small sized Be clusters, [81, 83] it seems to require 50 or more atoms in the case of Hg. [87]

Various techniques have been employed to theoretically determine metal cluster structures. In general, structural optimization is performed by standard minimization techniques which make use of first and second order derivatives of the total energy with respect to the nuclear displacements. This procedure may be used in connection with semiempirical, [88] *ab initio*, or DF-type electronic structure calculations (e. g. [57]) and is well suited for finding local minima which lie close to the starting configuration. An alternative approach is that of a molecular dynamics (MD) simulation based either on empirical many-body potentials or on the density functional MD method as proposed by Car and Parrinello. [46]

The problem with empirical many-body potentials lies in the choice of the parameters used to define the interatomic force fields. [89] Since empirical potentials are usually derived by fitting bulk properties, their application to finite systems can lead to incorrect answers. However, successful examples of empirical potential functions which incorporate cluster size dependent effects, such as the sp hybridization in Be clusters, [90] have been reported.

Much more appealing for the purpose of structure determination is the computationally expensive Car-Parrinello method, [46] where the relaxations of both the electronic and the nuclear degrees of freedom are treated simultaneously. The applications of this procedure to Na [91] and Mg [73] clusters containing up to 20 atoms have confirmed the earlier predictions based on standard *ab initio* methods for the existence of stable pentagonal forms, the oscillatory behavior of many properties (in particular for low nuclearity clusters), and the abundance of structural isomers with similar binding energies in clusters whose sizes do not correspond to specially stable situations ("magic numbers"). Finally, simulated annealing calculations have further demonstrated that the cluster geometry is not determined by a tendency for denser packing, but rather by the requirement for a more favorable electronic configuration.

Since the latter procedure is quite expensive, approximate treatments have been designed for transition metal clusters based on the embedded atom method (EAM). [53] In EAM, the dominant contribution to the energy of the metal is viewed as the energy of an atom embedded in the local electron density of its environment, supplemented by short range two-body interactions and possibly other terms. The basic idea underlying this method is similar to the effective medium theory [54, 92, 93] and both procedures find their roots in density functional theory. Applications of these techniques to transition metal clusters of medium to very large size (up to 5000 atoms) have been reported. It has been suggested that low nuclearity Ni and Pd clusters (up to about 20 atoms) tend to maximize the minimum coordination at any atom. [94] On the other hand, clusters containing from a few hundred to a few thousand atoms exhibit transitions from icosahedral to decahedral and finally to face-centered cubic polyhedra. [95]

2.3.2.2 The Jellium Model

The idea that clusters which exhibit high stability correspond to particular "closed shell" electronic configurations is basic for the use of the jellium model. It is well known that the mass spectra of Na_n clusters in molecular beams show a higher abundance, indicating greater stability, of masses corresponding to $n = 2, 8, 20, 40, 58, 92$, etc. [52] This sequence of "magic numbers" has been rationalized by Knight and coworkers by means of the jellium model. [52] The jellium model describes the cluster electronic structure by considering only the valence electrons which are assumed to move "freely" in a smooth attractive mean field potential. A crucial approximation is that the detailed positions of nuclei do not play a significant role. An important characteristic of the jellium approach is its separation of the Schrödinger equation into radial and angular parts, a situation familiar from the electronic structure theory of an atom. For the case of a spherical potential, the angular solutions are the spherical harmonics. However, the radial behavior of the jellium potential is nonsingular at the center of the cluster and thus quite different from the hydrogen-like Coulombic potential in an atom. The calculated order of the energy levels is 1s < 1p < 1d < 2s < 1f < 2p < 1g < 2d < 3s < 1h ... (note the different conventions for the notation of the radial quantum numbers in the jellium model and in atomic structure theory.) The peaks observed in the mass spectra of the Na clusters correspond to the filling of the 1s, 1p, 1d, 2s, 1f, 2p, 1g, etc. levels (Fig. **2-2**). [52] Formally, the cluster shell model is rather similar to the nuclear shell model, but the latter also takes a spin-orbit type interaction into account. [96] The order and degeneracy of the levels and the set of "magic numbers" thus predicted by the model by filling these orbitals depend on the shape of the potential assumed. [97] Although initially suggested for the alkali metal clusters, the jellium model has also been successfully applied to the interpretation of the cluster ion distributions observed with other s-bonded metal clusters, such as those of Cu, Ag, and Au. [98] Attempts to apply the jellium theory to clusters characterized by bonds having more directional properties, for example in Al_n and Pb_n were less successful. [99, 100] For this type of clusters, the electronic structure is strongly dependent on the geometric structure of the nuclear framework and on the resulting average atomic coordination.

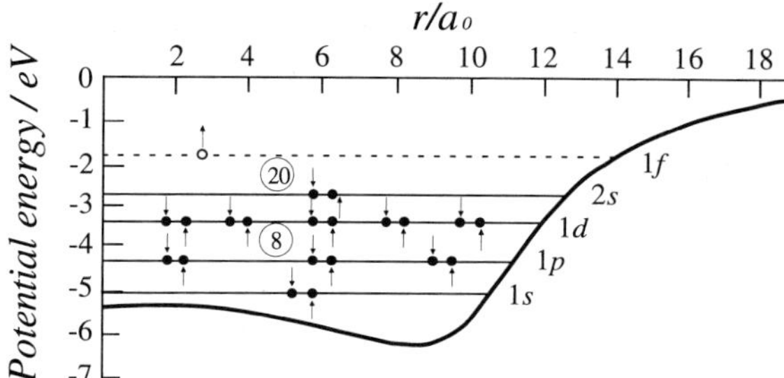

Figure 2-2. Jellium potential for a spherical Na_{20} cluster. Filled circles indicate electrons occupying the lowest levels; the open circle marks the level where an extra electron should go. Adapted from [96].

Although the simplicity of the jellium model is quite attractive, its validity has been the subject of considerable controversy. The problem can be formulated in a simple question. Is it necessary to explicitly take into account the detailed geometric structure of a cluster or is it sufficient to study the electronic properties assuming a general cluster shape without worrying about the localization of the nuclei and the corresponding singularities in the potential? The considerable success of the jellium model in predicting the proper sequence of the most stable alkali clusters suggests that the shell model reflects some basic and generally valid properties of the electronic structure of a cluster; these include the topological (nodal) properties and the concomitant degeneracies of the molecular orbitals. On the other hand, the shape of the effective potential in the jellium model is always assumed to be highly symmetric, either spherical [52] or spheroidal, [101] yet accurate quantum mechanical calculations have shown that this is not always the case. [91] Apart from the fact that some highly symmetrical structures are Jahn-Teller unstable, [17] it turns out that the ground state structures of alkali clusters are not necessarily those with the highest symmetry compatible with the corresponding electronic configuration. [91] In conclusion, the jellium model is mainly appropriate for large s orbital bonded clusters, while methods which explicitly take into account the delicate interplay between the nuclear and electronic positions are required for a proper description of low nuclearity clusters, and most probably for those comprised of elements outside the groups Ia and Ib.

2.3.2.3 Fluxionality

In general, the determination of cluster structures is complicated by the fact that a large number of isomers may have very similar stabilities. [102] The possibility for an interconversion among these structures requires a consideration of the fluxional nature of these systems. Metal clusters may be considered rigid at the zero-point vibrational level and they may become fluxional as their temperatures increase. Unfortunately, the cluster temperature is often not known experimentally or is known only with a large degree of uncertainty. Most of the techniques used to produce clusters, like laser vaporization, generate a large number of isomeric forms which are far from the thermodynamic equilibrium. [56] This complicates any comparison between the predicted stability and the observed abundance of a given cluster ion (or of its neutral parent form). This is particularly problematic for clusters having many atoms since the so-called "isomer density" increases exponentially with the size of the particle. [103]

2.3.2.4 Stability and Fragmentation

Beside its structure, the most interesting ground state property of a M_n cluster is its stability. Using the total energy $E(M_n)$ of a cluster M_n, this property can be easily converted into an average binding energy per atom, $D_e/n = (n \times E(M_1) - E(M_n))/n$. This quantity generally increases with the cluster size n and, for very large metallic particles, levels off asymptotically to the bulk cohesive energy. It also provides a convenient measure for deviations from an overall monotonic trend in the cluster dissociation energy. Theoretical studies on cluster stability have revealed large oscillations for small clusters of less than 20 atoms. [25] These oscillations present themselves as peaks of particular abundance in the mass spectrum of size selected clusters ("magic numbers").

The theoretical determination of cluster stability requires some care. The HF method severely underestimates the cluster binding energy. Correlation accounts for about 50 % of the stability in Li clusters and 90 % of the stability in Na and K clusters. [25] This makes it difficult, if not impossible, to treat larger aggregates (especially of transition metals) by means of wavefunction based methods. Alternatives are provided by density functional methods or, with some caution, by well parametrized semiempirical techniques, although both types of methods also have inherent problems. In the local density approximation, which is the most widely used variant of density functional theory, the binding energy is often overestimated and "nonlocal" corrections are required in order to obtain an accurate value for the stabilization energy. [15, 37] In semiempirical methods, it is often difficult to control the quality of the results. These methods have been used to study the evolution of cluster properties, including cluster stability, from small to large aggregates of typically up to 50–70 atoms. The stabilities of Cu_n and Al_n clusters as sections of the corresponding crystals have been determined from $X\alpha$ [104] and semiempirical [105] calculations respectively. It was found that clusters which contain less than 100 atoms exhibit only about 70 % of the bulk stability. If one assigns an equal volume for each atom of a spherical cluster, then the surface-to-volume ratio can be represented by the parameter $n^{-1/3}$. It is close to 1 for small clusters in which all the atoms are on the surface, and decreases to zero in the limit of an extended particle. The quantity D_e/n correlates linearly with $n^{-1/3}$ for both Cu_n and Al_n clusters and extrapolates quite well to the bulk cohesive energy. [104, 105] For aluminum, this linear relationship reproduces the experimentally determined stability of both the dimer and the bulk metal with only a very small error (Fig. **2-3**). From this curve, it was predicted that a cluster of about 1000 Al atoms is required in order to achieve 90 % of the bulk stability. [105] This roughly corresponds to a particle with a diameter of 3 nm.

Gas phase clusters undergo fragmentation when their internal temperature is high enough. The theoretical study of such fragmentation processes, $M_n \rightarrow M_{n-m} + M_m$, is very important for the interpretation of a variety of phenomena. Here a convenient quantity to analyze is

$$\Delta E(n,m) = E(M_{n-m}) + E(M_m) - E(M_n).$$

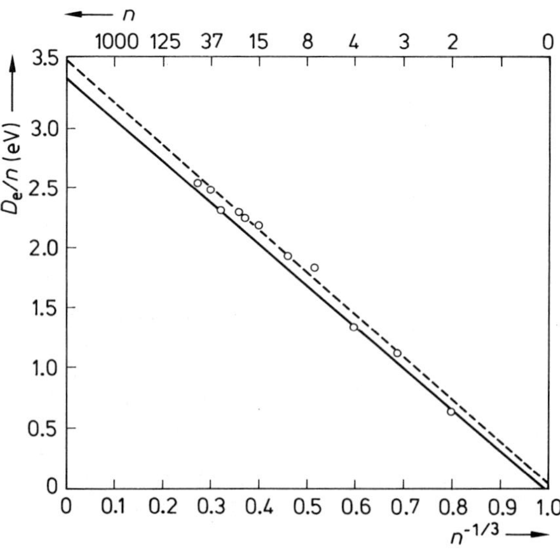

Figure 2-3. Binding energy per atom (D_e/n) for Al_n clusters versus cluster size as measured by the parameter $n^{-1/3}$. The solid line connects the atomic with the bulk limits. The dash-dotted line represents a least squares fit of the computed D_e/n values. Adapted from [105].

For the special case where the dissociation involves a single metal atom, M_1, one obtains

$$\Delta E(n,1) = E(M_{n-1}) + E(M_1) - E(M_n).$$

This quantity is also used to define the second difference

$$\Delta^2 E(n) = \Delta E(n,1) - \Delta E(n+1,1) = E(M_{n-1}) + E(M_{n+1}) - 2E(M_n).$$

Whereas a negative value for $\Delta^2 E(n)$ indicates that the process $2\ M_n \rightarrow M_{n-1} + M_{n+1}$ is energetically feasible, a positive value signals that the process $M_n \rightarrow M_{n-1} + M$ is less favorable than the reaction $M_{n+1} \rightarrow M_n + M$. When one plots the quantity $\Delta^2 E(n)$ against the cluster size n, the odd-even oscillations in the cluster stability become quite obvious (Fig. **2-4**). For alkali metal clusters, the sharp maxima found at $n = 2$, 4, 6 and 8 indicate that clusters having odd numbers of atoms undergo fragmentation more easily than clusters with even nuclearity. [25, 91]

2.3.2.5 Bond Lengths

Bond lengths play an important role in the structural characterization of micro-clusters. The majority of the experimental studies have shown that the nearest neighbor distances contract as the cluster size decreases. [55] The determination of the bond lengths in supported clusters or clusters in a matrix have been based, in general, on the EXAFS technique. The bond lengths in a cluster are easily obtained computationally, although it is well known that some computational techniques, such as the HF method, substantially overestimate the equilibrium

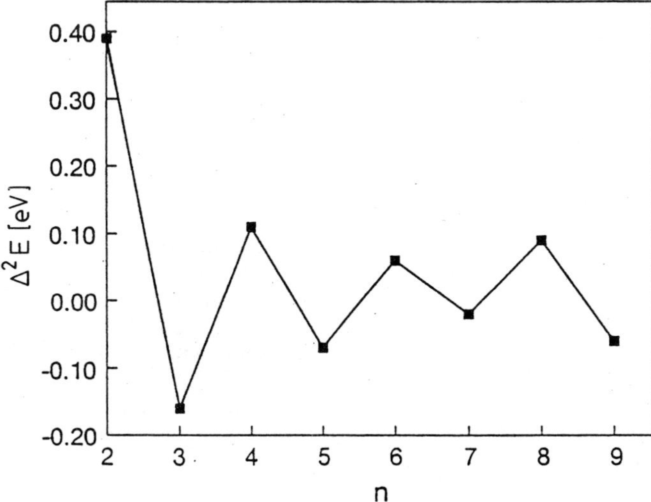

Figure 2-4. Second energy difference for Na_n clusters ($n = 2$–9). The maxima at $n = 2$, 4, 6, and 8 indicate the particularly high stability of clusters with even nuclearity. Reproduced with permission from [91].

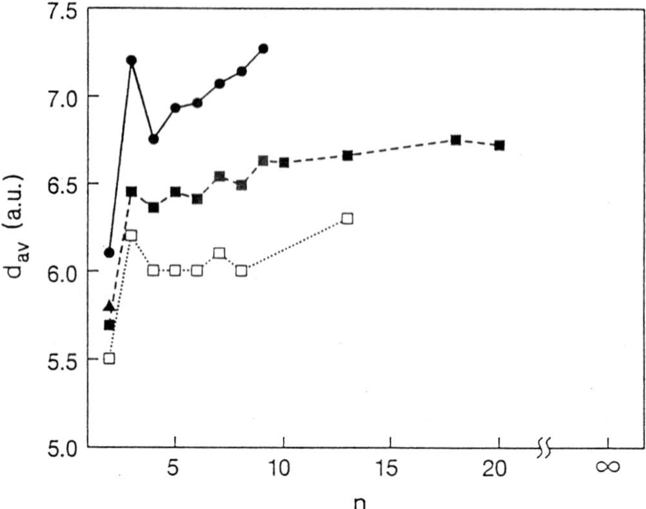

Figure 2-5. Average nearest neighbor distance for the most stable isomers of Na_n versus cluster size. – – – LDF-MD (Car-Parrinello); · · · · · · LDA; ––––– SCF-CI. The filled triangle shows the experimental value for the dimer. Reproduced with permission from [91].

bond lengths (Fig. **2-5**). If one introduces an extensive treatment of correlation effects or uses density functional methods, then bond lengths are obtained which are in much closer agreement with experiment. Among the numerous cluster properties, the mean bond length is one of the parameters that converges most rapidly to its bulk value. Even clusters of relatively few atoms have metal-metal distances which are rather close to the bulk value, although large deviations are found for clusters with less than 10 atoms (Fig. **2-5**). This may be rationalized if the metal-metal separation is assumed to depend largely on the local characteristics of the bond, that is, the equilibrium bond length represents a balance of several attractive and repulsive contributions, all of which act in a region close to the atom. Therefore, any similarity between the metal-metal distances in a small cluster and those in the bulk can not be taken as a criterion for the convergence of other properties of the cluster to those of the extended metal.

2.3.2.6 Electron Delocalization in Clusters

All MO treatments of the electronic structure of metal clusters give a delocalized description for the valence s and p electrons. Such a charge distribution is also one of the crucial assumptions underlying the jellium model. Clearly, a more detailed analysis of the cluster electron density is essential for understanding the metal-metal interactions in small aggregates. The interpretation of the wave functions has been limited in most cases to the gross atomic and overlap populations from the classical Mulliken population analysis. [106] Although this procedure serves as a very useful tool for identifying the degree of hybridization, its limits are well known, especially when large basis sets with very small exponents are used. [11] A more sophisticated analysis of the CI wave functions in terms of valence bond structures has shown that ionic configurations contribute to a non-negligible degree. [107] A very careful study of the nature of the bonding in small alkali metal clusters, based on a topological analysis of the charge density as proposed by Bader, [108] established the unambiguous existence of electron density maxima at the center of the triangular faces of the Li clusters. [109] This finding is completely different than classical chemical bonds where electron density maxima are found only at the nuclear positions. These "non-nuclear" maxima contribute significantly to the cluster stability. Thus, the bonding in small Li and Na clusters must be viewed as manifestations of delocalized three-center bonds. A similar conclusion about the existence of non-localized electrons in alkali metal and aluminum clusters was reached from an anlysis of GVB-type wave functions. [110]

2.3.3 Ionization Potentials and Electron Affinities

The energy released from a bulk metal upon addition of one electron is equal to the energy required to remove an electron. Therefore, the experimental and theoretical determination of ionization potentials (IP) or electron affinities (EA) represents a powerful criterion for the degree of convergence between a metal cluster and a metallic particle. [111] Moreover, both the IP and the EA are very important properties for determining, in a rather direct way, the chemical reactivity of a cluster towards adsorbed or interacting molecules. Accurate experimental determinations of cluster IP's have been recently reported. For metals, the (first) IP varies dramatically with the cluster size and typically decreases by 2–3 eV (or more) as one goes from the atom to the bulk. [55]

Cluster IP's can be derived theoretically in various ways. At the simplest theoretical level (in a wave function based method), one may use Koopmans' theorem in HF theory and take the negative of a cluster orbital energy, $-\varepsilon_i$, as a crude measure of the energy required to remove one electron from this orbital. This procedure neglects the electronic relaxation which will accompany the ionization process. Such relaxation effects can be accounted for by computing the energy difference between a neutral cluster and its cation,

$$\mathrm{IP} = E(\mathrm{M}_n^+; \mathbf{R}^+) - E(\mathrm{M}_n; \mathbf{R}).$$

If the nuclear positions, $\mathbf{R}^+$, in the M_n^+ cluster are kept fixed as in M_n *(i. e.* $\mathbf{R} = \mathbf{R}^+$), then one obtains the vertical ionization energy, IP_v. Adiabatic IP's can be obtained by using the energy of the optimized M_n^+ structure as a reference. Here, not only the electronic, but also the geometric relaxation which follows the ionization is accounted for.

The IP's of several low nuclearity clusters have been computed by various methods with different degrees of success. As shown by the following examples, the trend in the IP's as a function of the cluster size strongly depends on the metal considered. For the alkali metal clusters Li_n and Na_n, odd-even oscillations have been observed, [112] whereby the higher IP's correspond to those clusters having closed shells and higher stability (n even). In general, however, the IP decreases with increasing nuclearity (Fig. **2-6**). A reverse trend is found for small Al clusters such that the IP's of Al_n with $n = 2–6$ are higher than for the Al atom. [113] This has been rationalized as being due to a low degree of 3s-3p hybridization and a lowering of the cluster Fermi level compared to the single atom in small Al clusters. No odd-even oscillations and no discontinuities that might be correlated with shell closing have been found in the IP's of Ni clusters which contain less than 9 atoms. [74] The absence of odd-even oscillations in the Ni clusters compared to the alkali metal clusters is not related to any differences in their geometries which, on the contrary, are rather similar. Rather, it has been explained as being due to differences in the electronic structure and, in particular, by the direct involvement of the 3d electrons of Ni in the ionization process. [74] It is worth noting that all these computed trends are in very good agreement with the experimental observations. [114, 115]

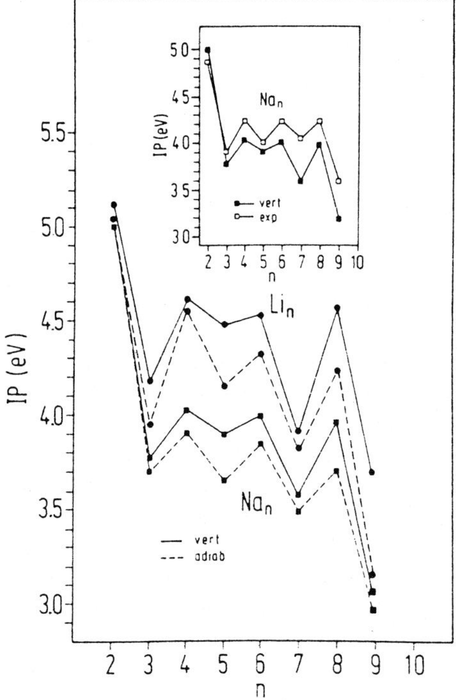

Figure 2-6. Calculated vertical (———) and adiabatic (– – –) IP's of Li (●) and Na (■) clusters. The experimentally determined vertical IP's for Na clusters are given in the box for comparison. Reproduced with permission from [25]. Copyright 1991 American Chemical Society.

The IP's and EA's for larger particles can be interpreted in terms of classical electrostatics. In the classical droplet model, [116, 117] one assumes that a metal cluster can be viewed as a conducting sphere and thus relates the energy required to extract (or add) one electron from a sphere of radius R to the IP (or EA) of a cluster. Furthermore, one correlates the value for a sphere of infinite size to the corresponding quantity for the bulk metal, the work function W, and arrives at the following relations: [116, 117]

$$\mathrm{IP}(R) = \mathrm{W} + (3/8\, e^2)/R$$
$$\mathrm{EA}(R) = \mathrm{W} - (5/8\, e^2)/R$$

These relations clearly indicate that the IP will decrease and the EA will increase with the cluster size. Thus, it will be easier to remove or add an electron to a larger cluster than it will to a smaller one. The origin and significance of the factor (3/8) (e^2/R) has been critically discussed. [118] The measured IP's do not always agree with the predictions from the spherical droplet model. Significant deviations were found for Hg clusters [87] (attributed to the occurence of a non-metal to metal transition) and for Fe_n and Ni_n clusters. [119, 120] The discrepancy between the metallic droplet model and the IP's of transition metal clusters suggests that the ionization process may directly involve the d electrons and that the d electron energies may be independent of the clusters size. Indeed, CASSCF [74] and density functional [121] calculations have shown that the first ionization in Ni clusters is more properly described as a 3d rather than a 4sp ionization. On the other hand, when the droplet model closely describes the trend in the cluster IP's this may be taken as an indicator for an almost spherical cluster symmetry and for a high degree of delocalization of the valence electrons. In general, these conditions are fulfilled only for very large clusters of 100 atoms and more. Below this size, typical quantum size effects lead to considerable deviations in the IP's compared to the classical electrostatic behavior. [122] These deviations are usually related to the character of the cluster's electronic structure and, in particular, to the presence of singly occupied orbitals of bonding or anti-bonding character.

The determination of cluster EA's is less straightforward than the calculation of the IP's. Two important methodological aspects must be pointed out. First, it is important to use basis sets which are large and flexible enough to adequately describe the diffuse electron distribution of a negatively charged particle, particularly when the cluster is small and the localization is high. Second, a correct prediction of the EA requires a more extended treatment of correlation effects than in neutral clusters. It is for these reasons that theoretical studies addressing the dependence of the EA on cluster size have been restricted to group Ia [123] and Ib [124, 125] clusters since they are computationally more amenable than other metals. The theoretical EA's show an oscillatory behavior similar to that found for the IP's. In the case of Cu, a good correlation with the measured electron detachment energies has been found. [126]

2.3.4 Electronic States, "Band Structure", and Band Gap

A cluster becomes metallic when its valence electrons are fully delocalized and/or when there is a high density of states (DOS), both occupied and unoccupied, in the immediate vicinity of the Fermi level, E_F, whose average spacing is small enough so that the unoccupied levels become thermally accessible. Very crucial measures of the evolution to metallic properties are thus the number of excited electronic states close to the ground state and their corresponding excitation energies. Although the gap between the filled and unfilled levels (the HOMO-LUMO gap) in a cluster becomes smaller as the size increases, this general trend may not follow a monotonic course. For instance, the gap in Mg clusters, as determined from density functional calculations, gradually decreases from 2 eV in Mg_2 to 0.5 eV in Mg_{12} with local maxima appearing in correspondence of shell closing (Fig. **2-7**). [73] In some cases, several excited states were found to lie within 0.5 eV of the ground state level even for clusters of less than 10 metal atoms. [17] Competition between closed shell singlet states and open shell triplet states have been predicted even for clusters of closed shell atoms of rather small size. [72, 127]

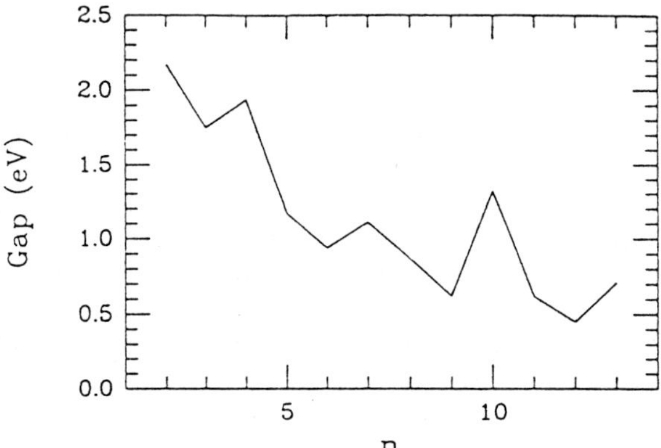

Figure 2-7. Energy gap between the highest occupied and the lowest unoccupied levels of Mg_n clusters (n = 2–13). Reproduced with permission from [73].

In all these cases, the theoretical determination of the cluster ground state may be rather difficult and could require extensive inclusion of correlation effects. This is particularly true for transition metal clusters. Accurate correlated calculations have shown a near degeneracy of states already for such very small aggregates as Ni_3 [128] and Pt_3. [129] CASSCF calculations on the octahedral Ni_6 cluster [130] identified six very low lying electronic states with an average energy separation of about 60 meV, roughly twice the thermal energy at room temperature (kT = 26 meV). These electronic states arise from a redistribution of the electrons among the partially filled narrow 3d manifold as well as from different spin coupling schemes. However, they do not affect the delocalized 4sp conduction band which is responsible for the metal bonding.

Electronic transitions involving the 4sp electrons have been calculated for large Ni clusters of up to 181 Ni atoms by performing large scale one-electron ECP calculations (3d electrons were not explicitly included). [131] The excitation energies show only a gradual decrease with cluster size, such that even in Ni_{181} the energy separation between the lower excited states is 100–200 meV. [131] Clearly, the calculation of the total energy of each electronic state and the determination of the cluster "ground" state for systems containing a few tens of transition metal atoms is simply not feasible with such wavefunction based techniques as HF-CI or MCSCF methods. In this respect, spin polarized density functional calculations, most often carried out in the LSD approximation, provide a useful and computationally much simpler alternative.

The spin polarized approach removes the restriction that up and down spin electrons be energetically degenerate in a molecular orbital (similar to the UHF approach). This induces a stabilization of the majority (spin-up) levels with respect to their minority (spin-down) partners and the resulting energy difference can be compared to the experimental exchange splitting of the metal. The problem of the dense manifold of states in large metal clusters or in clusters of transition metal atoms can be further simplified within the density functional method by means of the fractional occupation number (FON) technique. [132] This method assigns fractional electronic charges to the numerous close lying levels near the Fermi energy and the resulting cluster electronic structure can be considered, in the spirit of a mean-field approach, as an average over several single configurations (and spin states) of very similar energy.

Density functional calculations have been used to determine the "band structure" of a cluster and its evolution towards the bulk. [104, 121, 133–135] This is usually done by a Gaussian broadening of the discrete spectrum of one-electron levels of the finite cluster. The density of states, DOS, profile for a large octahedral cluster of 44 Ni atoms [121] (Fig. **2-8**) shows remarkable analogies to the corresponding total DOS curve of bulk Ni (Fig. **2-9**). [47] The width of the 3d band in Ni_{44} of about 4.5 eV is quite close to that determined from band structure calculations for the fcc metal. [136] The DOS profile and the 3d band width are not the only characteristics of Ni_{44} which have almost converged to the bulk limit. The cohesive energy and the atomic magnetic moments are also similar to the metallic regime. On the other hand, other characteristics like Ni–Ni bond lengths and the cluster IP are still different than the bulk metal distances and work function respectively. [121] This is a typical example illustrating the different rates at which cluster properties can evolve toward the metallic limit.

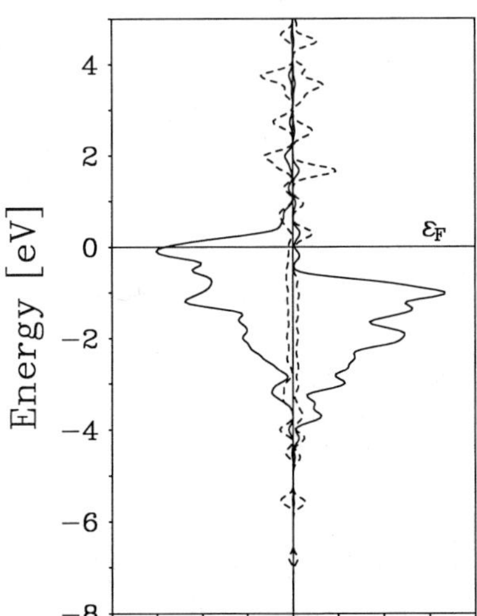

Figure 2-8. Density of states (DOS) curves obtained by Gaussian broadening (0.1 eV) of the discrete LDF one-electron energy spectrum of an actahedral Ni_{44} cluster. Left side: minority spin; right side: majority spin.

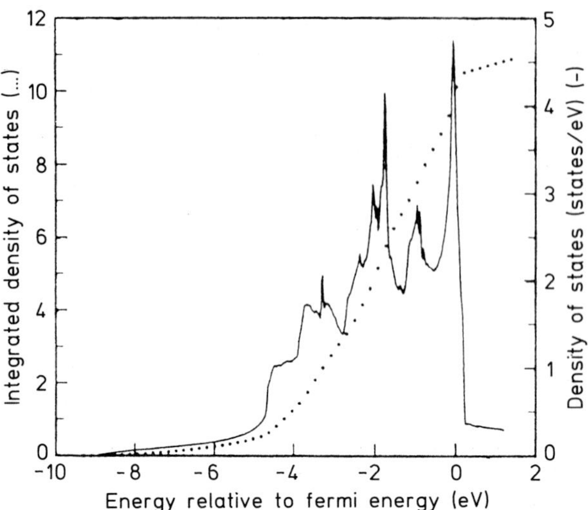

Figure 2-9. DOS curve and integrated density for bulk Ni from a LDF band structure calculation. Reproduced with permission from [47].

2.3.5 Optical Responses

In the last years there has been a rapid development in experimental techniques designed to study the optical response of clusters to electromagnetic radiation. Optical spectroscopy on gas phase clusters is a very useful tool for determining the electronic and geometric features of clusters. In particular, photodetachment and photodissociation spectroscopies, as applied to negative ions and neutral or positively charged cluster beams, have made the spectrum of excited states in clusters of various sizes accessible. [25] Among the clusters most studied with optical spectroscopies are those of the group Ia and Ib elements, whereas most of the theoretical work has been done on Li, Na, and K clusters. [25]

Metal clusters can absorb light through the excitation of a collective mode produced by the motion of electrons in the field of the nuclear charges. This is similar to the free oscillations of the electron gas in metals which are called plasmons and cannot be described in terms of one-electron excitations. Plasmon excitations have been observed in very large clusters and they have recently been reported for low nuclearity clusters as well. [25] In large clusters or in metals, the surface plasmons are described in terms of the classical Mie theory, [137] whereby one single surface plasmon peak carries 100 % of the oscillator strength. Experimentally, it has been found that the width of this resonance varies as $1/R$, where R is the radius of the cluster. Small clusters containing only 40–50 atoms or less, exhibit a much more complex lineshape and a high degree of fragmentation into several peaks of various intensities. For small clusters, the $1/R$ dependence of the peak broadening also breaks down. This is a typical example of a quantum size effect, and the evolution of the absorption spectra of alkali metal clusters towards the classical Mie behavior has stimulated considerable theoretical interest.

Accurate *ab initio* CI calculations have been reported for small Li and Na clusters. [123, 138] These calculations have successfully described the optical response of Na clusters with less than 8 atoms. This has also permitted the unambiguous identification of the structures of these small alkali metal clusters. For instance, the excellent agreement between the photodepletion spectrum and the optically allowed transitions in Li_4 and Na_4 (Fig. **2-10**) provides compelling evidence for the rhombic shape of the alkali tetramers. In some cases, the existence of cluster isomers with very similar stabilities and not too different absorption spectra makes the structure assignment less transparent. It should also be noted that *ab initio* CI calculations have failed to describe the double line in the photodepletion spectrum of Na_{20}, [138] suggesting that the application of these methods to large clusters is not only computationally very expensive, but also less accurate because of the difficulty in correlating a large number of electrons.

An alternative approach to the description of collective excitations in clusters is represented by the random-phase approximation (RPA) [139] within the jellium model. [140, 141] Calculations of transition energies and oscillator strengths *via* the RPA-jellium model are rather similar to those performed in the investigations of giant dipole resonances in nuclei. The RPA-jellium model predicts for Na_8 the existence of one dominant line carrying 75 % of the total intensity whereas, for Na_{20}, about the same intensity is equally divided among two lines. [140] Thus, although the fragmentation of the experimental spectra into more than one line is qualitatively reproduced, the positions and intensities of the lines are described less satisfactorily. Moreover, *ab initio* RPA calculations based on HF one-electron wave functions [25] have yielded quantitatively different results, mainly due to the difference between the detailed forms of the nuclear potential in the jellium and standard quantum chemical methods where the nuclear positions are explicitly taken into account.

Sophisticated *ab initio* techniques are clearly superior for calculating the excited states in very small clusters. On the other hand, the RPA-jellium approach can describe the entire range of excitations, from small to large clusters, although quite large discrepancies occur for low nuclearity clusters. Indeed, RPA theory predicts two regimes for the optical properties of alkali metal clusters. The first applies to smaller sized clusters and is characterized by response functions exhibiting quantum size effects which lead to a fragmentation of the total oscillator strength into several lines. The second regime applies to larger sized clusters and is characterized by a smooth response yielding a single peak whose $1/R$ width dependence follows the classical Mie theory. The transition between the two regimes has been predicted to occur around Na_{58}. [142]

In conclusion, the optical response of gas phase metal clusters in conjunction with their theoretical interpretation represents a powerful method for characterizing these systems. In the low nuclearity regimes, accurate *ab initio* CI calculations can be applied, at least to clusters of simple metals having small numbers of valence electrons. They have provided a basis for the characterization of the geometric structures of smaller aggregates. On the other hand, the RPA-jellium approach yields a simple and sufficiently accurate picture for the evolution of the optical properties of metal clusters towards the bulk. The principle underlying this method is the assumption of an average potential which governs the motion of the electrons but is independent of the detailed positions of the nuclei. Changes in the photoabsorption cross sections can thus be related to changes in the potential which result from the increase in the cluster size.

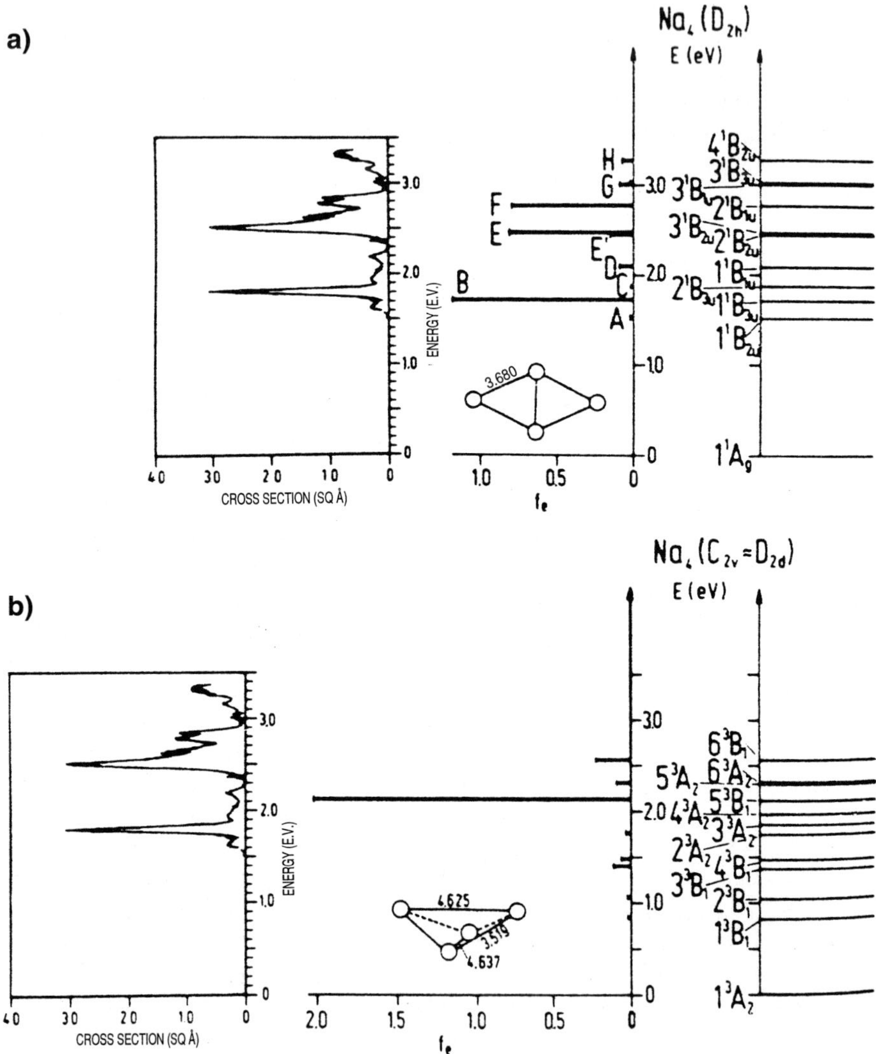

Figure 2-10. Comparison for the depletion spectrum of Na_4 and SCF-CI energies for optically allowed states and the oscillator strengths, f_e, for a) rhombic Na_4 (singlet state) and b) deformed tetradron Na_4 (triplet state). The good agreement with the rhombic structure lends confidence to the assignment of this geometrical structure. Reproduced with permission from [25]. Copyright 1991 American Chemical Society.

2.3.6 Clusters in External Fields

Another set of fundamental properties of metal clusters involves their response to static external electric and magnetic fields. Transition metal clusters embedded in matrices have been extensively studied with these techniques. [143] Unfortunately, the size distribution of the particles is broad in these experiments and interactions with the matrix can introduce changes in the properties of the metal aggregates. This is also true for metal clusters stabilized by ligands, as will be discussed in Section 2.4.5.3. The study of clusters in molecular beams overcomes these difficulties. This powerful approach, when combined with mass spectrometric detection, allows the investigation of mass selected free clusters. The main disadvantage is that since the particle densities are quite low, most of the standard spectroscopic techniques cannot be used.

2.3.6.1 Magnetic Behavior

One of the most intriguing questions concerns the magnetic properties of metal clusters: at which size does an iron or nickel cluster become paramagnetic or even ferromagnetic? The magnetic moments of free K, Al, Fe, Co, etc. clusters have been determined using the Stern-Gerlach depletion method. [55] The theoretical description of the magnetic properties of metal clusters has only been attempted in a few cases. We have already mentioned that clusters often exhibit several low lying excited states with different spin multiplicities and that the determination of the ground state (diamagnetic or paramagnetic) may require a considerable computational effort. This task becomes even more difficult with transition metal clusters of ferromagnetic materials due to the complex spin coupling of the localized d electrons. Thus, theoretical investigations addressing the magnetic behavior of metal clusters have been based almost exclusively on density functional methods.

Recently, the magnetic behavior of Ni clusters which contain from 6 to 44 Ni atoms has been studied. [121, 144] The total number of unpaired electrons in the cluster, n_s, was defined as the difference between the majority (up) and minority (down) spins, $n_s = n_u - n_d$, and found to be 7.0, 7.9, 15.1, and 31.1 in Ni_6, Ni_8, Ni_{19}, and Ni_{44} respectively. These results correspond to an average of about 0.7 unpaired electrons per atom in Ni_{44} and they show that the magnetization in Ni_{44} is not too different than that in the bulk metal of 0.56 μ_B. [145] Ni_2 has a triplet ground state corresponding to one unpaired electron per Ni atom which is localized in the 3d shell. Thus, the average magnetic moment per atom decreases as one goes from very small clusters to large particles, a trend confirmed by experiment. [146] In addition, a more detailed analysis of the spin distribution in Ni_{44} showed that the Ni atoms on the cluster "surface" carry a slightly larger magnetic moment than the atoms with a "bulk" coordination. [121, 144] Both these effects, the cluster size dependence and the different behavior of "surface" and "bulk" atoms in a cluster, are related to the average coordination. There is a well known tendency for the local magnetic moments to be larger for atoms with fewer nearest neighbors. This effect is responsible for the enhanced magnetic moment in the first layer of Ni (100) compared to a bulk layer. [147, 148] This has also been found from density functional calculations on Ni_{13} clusters with various structures. [149]

Finally, the magnetization also depends on the metal–metal distances. In transition metal clusters, the paramagnetic behavior arises from the weak coupling of the partially filled d levels. As the metal–metal distances decrease, the d–d overlap increases in conjunction with the amount of spin coupling. In general, the results indicate that Ni clusters [121] (but also Fe [134] and Co [150]) are clearly precursors for bulk magnetism. Magnetic ordering of the spins is already found in quite small clusters because it is strictly related to the localized nature of the metal d electrons.

2.3.6.2 Electric Polarizability

The response of metal clusters to a static external electric field is also of interest. The static dipole polarizability is a physical quantity directly related to the electronic charge distribution in the ground state of the cluster. Static electric polarizabilities have been measured for Na and K clusters with less than 40 atoms. [151] Computational studies based on HF-CI [25] or DF [152] methods and aimed at determining the components of the polarizability tensor have been reported. Calculations of alkali metal polarizability have also been performed within the jellium model. [153] An interesting association between theory and experiment is found in the identification of cluster isomers through the comparison of computed and measured polarizabilities. In fact, theory gives values which only follow the experimental trend for those clusters having the ground state geo-

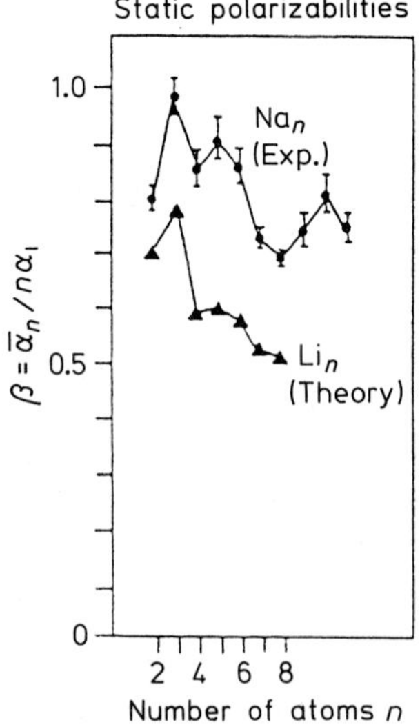

Figure 2-11. Comparison of experimental static dipole polarizabilities of Na_n clusters with theoretical SCF-CI values for Li clusters as a function of the cluster size. Reproduced with permission from [25]. Copyright 1991 American Chemical Society.

metry or their energetically close lying isomers. Whereas odd-even oscillations have been observed for small alkali metal clusters with less than 10 atoms (Fig. **2-11**), [25] the averaged polarizability per atom becomes a smooth function as the cluster nuclearity increases. The fact that larger aggregates respond to an external electric field in a similar way supports the idea which forms the basis for the cluster jellium model of electrons moving in a central field.

2.4 Structures and Properties of Ligated Clusters

In the previous section we discussed gas phase metal clusters. We called these units of matter "naked" or elemental metal clusters in order to emphasize that they are most often composed of atoms of a single element and are not stabilized and surrounded by ligand molecules or fragments. Here, we discuss the electronic characteristics of ligated clusters, where a metal core is stabilized by an outer shell of ligands. These clusters possess chemical and physical properties completely different from those of the naked clusters and are sometimes referred to as molecular or organometallic clusters. Of course, there is no definition without exceptions, and examples of gas phase ligated metal clusters or of solid state "naked" metal clusters are also known.

Nevertheless, there is no doubt that the ligands play a unique role in the stabilization and thus the chemistry of molecular clusters. It is their presence which prevents the clusters from condensing to form a bulk metal. Thus, any attempt to understand and rationalize the electronic structures and the properties of molecular clusters must take into account the role and the nature of the metal–ligand interactions.

2.4.1 The Metal–Ligand Interactions

There are two main groups of ligands in molecular clusters: the so-called π-donor ligands, such as Cl^-, S^{2-}, O^{2-}, and OR^-, which stabilize clusters of the early transition metals with high formal oxidation states, and the π-acceptor ligands, such as CO, phosphines, and isocyanides, which are particularly effective in removing excesses of electronic charge and thus in stabilizing clusters of the late transition metals with low oxidation states. Classic examples of clusters stabilized by these different classes of ligands are $[Mo_6Cl_8]^{4+}$ and $[Ir_4(CO)_{12}]$ respectively. The cluster chemistry of elements from the left side of the d-block is dominated by low valency halide clusters where the metal atoms have oxidation numbers +2 or +3. These clusters have the general formulas $[M_6X_8]$, $[M_6X_{12}]$, and $[M_3X_9]$ (Fig. **2-12**) in which the metal atoms typically assume an octahedral or a triangular geometry with the halogen atoms, X, acting as bridging units. They are characterized by strong metal–metal bonds and are often linked to form extended structures as in

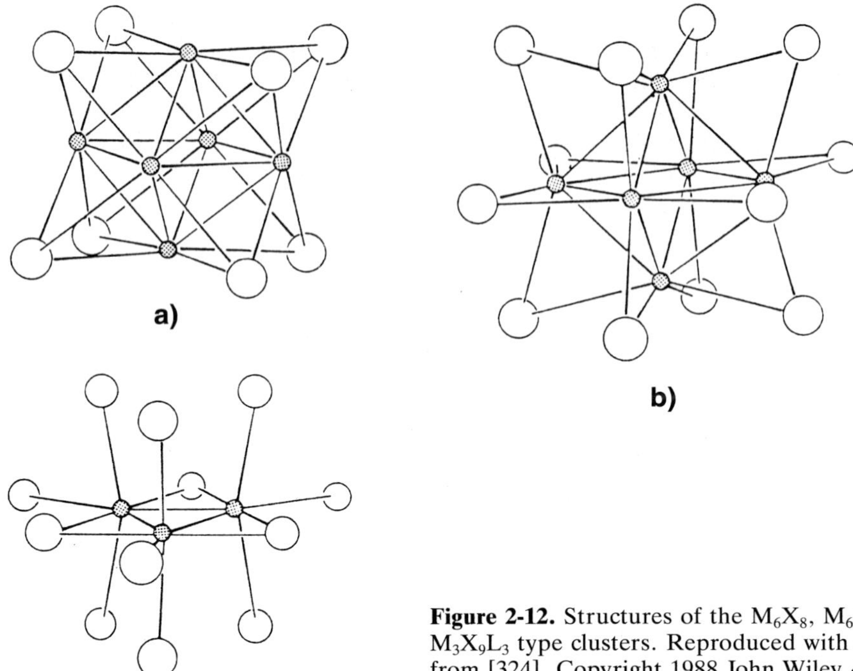

Figure 2-12. Structures of the M_6X_8, M_6X_{12}, and $M_3X_9L_3$ type clusters. Reproduced with permission from [324]. Copyright 1988 John Wiley & Sons.

the Chevrel phases [154] where the partial filling of the d band leads to interesting properties. [154, 155] The properties of these clusters will be discussed in more detail in chapter 5 of this book and so we will not address their bonding and electronic structures here.

Late transition metal, TM, atoms preferentially form clusters which are stabilized by strong acceptor ligands. In these clusters, the metal atoms are formally in oxidation state zero. Since clusters stabilized by π-accepting ligands represent by far the most numerous group of molecular clusters, the character of their M–L bonding, where L is either CO, PR_3, NO, etc., deserves special attention. The generally accepted picture for the interaction of π-acceptor ligands with a metal atom is that originally proposed by Dewar, Chatt and Duncanson of a σ-donation and a π-back donation. [156, 157] According to this scheme, the ligand (CO, PR_3, olefins, etc.) donates charge to the metal through a lone pair, and the metal d orbitals donate charge back into low lying empty ligand orbitals of the proper symmetry. This description, which has the value of being both simple and intuitive, has been revised in recent years by recognizing the importance of yet another type of interaction besides the σ-donation and π-back donation. This additional contribution is the Pauli repulsion, also called steric repulsion or frozen orbital repulsion, which occurs at the initial stage of the interaction between two molecular fragments before the chemical bonds are formed. [158, 159] The Pauli repulsion has a clear quantum mechanical origin in its antisymmetry requirement for the wavefunction and reflects the fact that the probability of finding two electrons with the same spin in close proximity to each other is quite small. The Pauli principle thus forces electrons of the same spin to remain separated from one another. To illustrate this effect, consider the interaction between a single TM atom and CO. Transition metal atoms have the general configuration $(n)\mathrm{d}^{m-l}(n+1)\mathrm{s}^{l}$, with $l = 1$ or 2 (Pd is the only TM atom where $l = 0$ in the ground state: ^{1}S [$5\mathrm{d}^{10}$]). [160] The (n)d, $(n+1)$s, and $(n+1)$p orbitals of a TM atom have very different radial extensions, the d-shell being much more spatially contracted than the s-p valence shell (Fig. **2-13**).

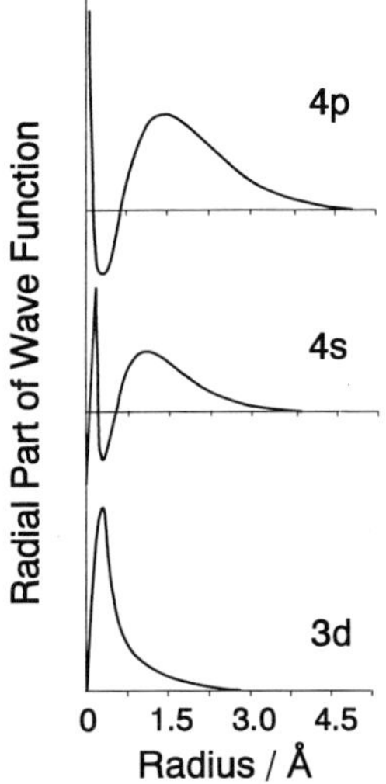

Figure 2-13. Radial extension of 3d, 4s, and 4p orbitals for a transition metal atom.

When a ligand approaches the metal atom with its lone pair directed towards the metal, the interaction between an occupied, diffuse $(n+1)$s metal orbital and the ligand lone pair will have a repulsive character. In order to form a bond through the π-back donation mechanism, the ligand must get close enough to the metal atom so that a significant bonding overlap can occur between the filled metal d(π) orbitals and the acceptor orbitals of the ligand. This requirement for a short distance contrasts with the presence of the diffuse, occupied $(n+1)$s orbital of the TM atom. The bond can be formed only if the TM atom undergoes a change in its electronic configuration, in particular, a promotion from $(n)\mathrm{d}^{m-l}$ $(n+1)\mathrm{s}^{l}$ to $(n)\mathrm{d}^{m-l+1}(n+1)\mathrm{s}^{l-1}$. This step is energetically much more important than the σ-donation. While the interaction between a $(n)\mathrm{d}^{m}(n+1)\mathrm{s}^{0}$ TM atom and the ligand is purely attractive, a repulsion occurs when the interaction involves a TM atom in a $(n)\mathrm{d}^{m-l}(\mathrm{n}+1)\mathrm{s}^{l}$ state, and especially when the (n + 1)s shell is filled ($l = 2$; Fig. **2-14**). [161–165] The rearrangement and configurational change in the metal is a very important consequence of its interaction with the ligand, and this also has an effect on the electronic structure of the metal core of a ligated cluster.

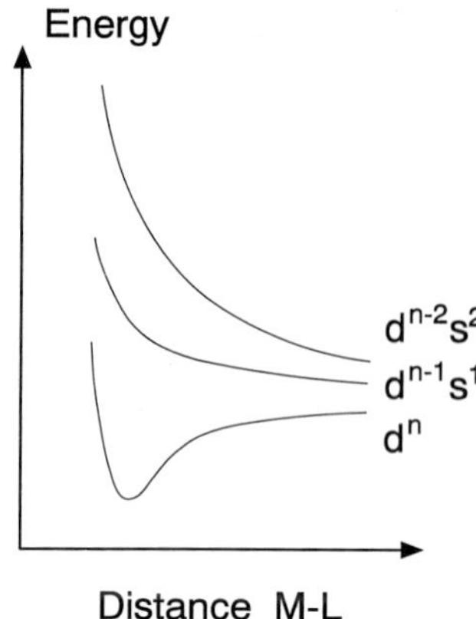

Figure 2-14. Potential energy curves for the interaction of a single CO molecule with a transition metal atom in different configurations.

In summary, the bonding of a π-acceptor ligand, L, to a TM atom can be ideally divided into three steps: a) initial repulsion between the L lone pair and the metal $(n+1)$s charge distribution, b) rearrangement of the metal configuration whereby an s to d electron promotion takes place and thus an increase in the d^m character at the expense of the $d^{m-l}s^l$ contribution, and c) formation of a dative covalent bond through the σ-donation and the π-back donation mechanisms. The relative importance of the two charge transfer channels depends on several factors, although, for a TM in a low or a zero oxidation state, there is strong evidence that the π-back donation mechanism dominates. [166] Nevertheless, CO, PR_3, NO, olefines, etc. are usually considered as either two (CO, PR_3) or three (NO) electron donors when counting the cluster valence electrons, despite the fact that these ligands give rise to little, if any, σ-donation when interacting with TM in low oxidation states. [167] It is worth noting that also the phosphine ligands possess a considerable π-acidity, although they are not as strong π-acceptors as carbonyls. [168] The capability of alkyl phosphines to accept charge from metals is not due to the direct involvement of the empty P 3d orbitals with the d(π) orbitals of the metal, as it had been assumed for a long time. Rather, it is P–R antibonding orbitals of proper symmetry and energy on the PR_3 group which accept the charge from the metal d(π) orbitals. [168]

2.4.2 Structures and Bond Lengths

2.4.2.1 Geometrical Structures

Although molecular metal clusters exhibit a large variety of structures and nuclearities, the vast majority of the structures reported in the literature correspond to small clusters containing less then ten atoms. [169] High nuclearity metal clusters (with 10–50 atoms) whose structures have been well characterized by

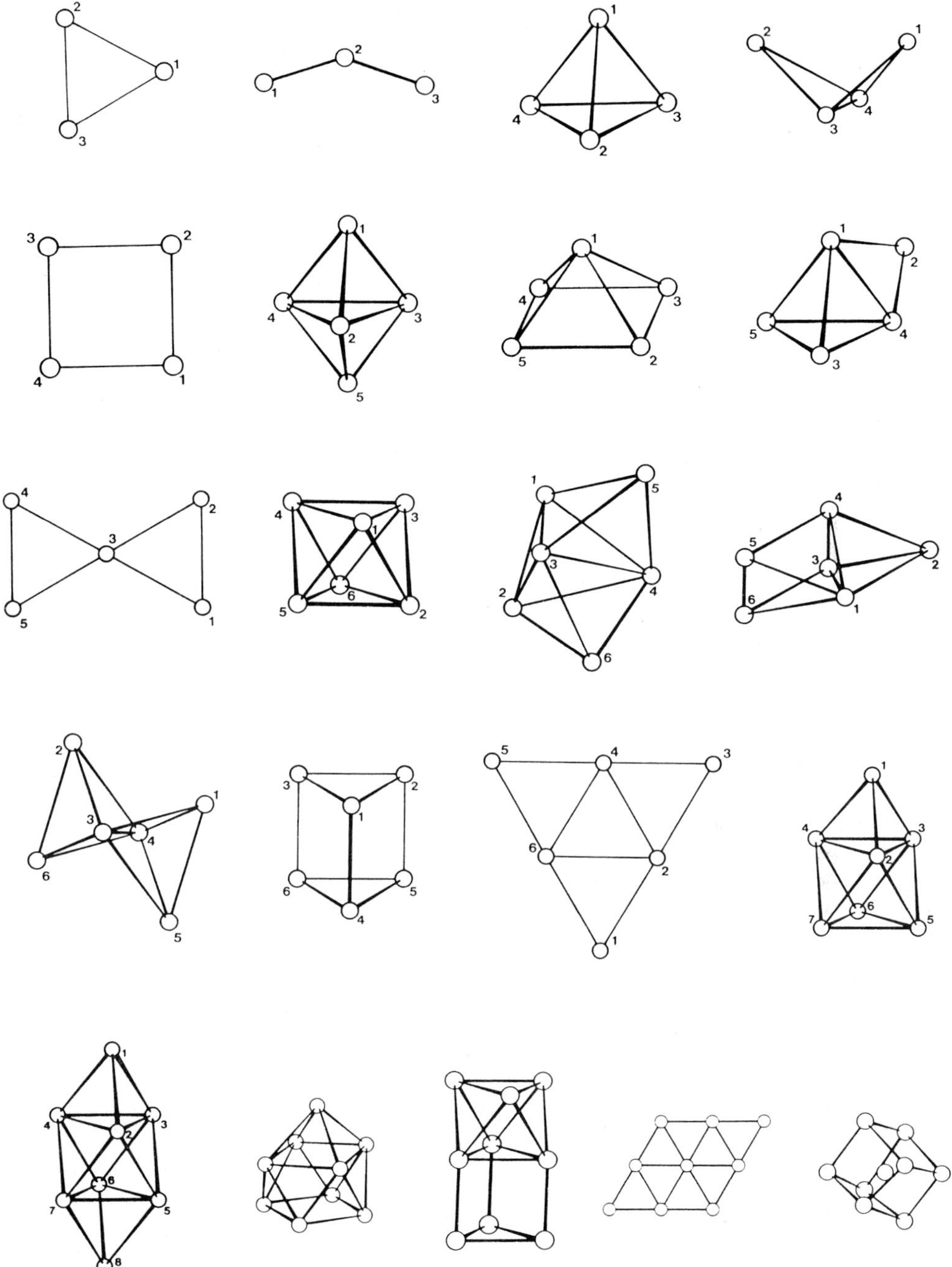

Figure 2-15. Geometrical structures of low nuclearity organometallic clusters. Adapted from [169]. See also Table **2-1**.

Table 2-1. Examples of low nuclearity organometallic clusters with different geometries and electron counts. See also Figure **2-15**.

Cluster	Structure	CVE	Ref.
$[Fe_3(CO)_{12}]$	triangular	48	a
$[Au_2Os(CO)_4(PPh_3)_2]$	triangular	42	b
$[Ru_3(CO)_8(Cp)_2]$	chain	50	c
$[Co_2Pt(CO)_8(C_5H_5N)_2]$	chain	48	d
$[Pd_4(CO)_5(PR_3)_4]$	tetrahedral	58	e
$[Re_4H_4(CO)_{12}]$	tetrahedral	56	f
$[Re_4(CO)_{16}]^{2-}$	butterfly	62	g
$[Ru_4(CO)_{11}(PPh)_2]$	square	62	h
$[Os_5(CO)_{16}]$	trigonal bipyramidal	72	i
$[Ni_5(CO)_{12}]^{2-}$	trigonal bipyramidal	76	j
$[Fe_5C(CO)_{15}]$	square pyramidal	74	k
$[Os_5H_2(CO)_{16}]$	edge-capped tetrahedron	74	l
$[Pt_5(CO)_6(PPh_3)_4]$	edge-capped tetrahedron	70	m
$[Fe_4Pd(CO)_{16}]^{2-}$	atom-sharing triangle	76	n
$[Co_6(CO)_{16}]$	(pseudo) octahedral	86	o
$[Au_6(PPh_3)_6C]^{2+}$	octahedral	80	p
$[Os_6(CO)_{18}]$	bicapped tetrahedron	84	q
$[Ni_3Os_3(CO)_9(Cp)_3]$	bicapped tetrahedron	90	r
$[Os_6H_2(CO)_{18}]$	capped square pyramidal	86	s
$[Au_6(PPh_3)_6]^{2+}$	edge-sharing tetrahedron	76	t
$[Co_6C(CO)_{15}]^{2-}$	trigonal prismatic	90	u
$[Pt_6(CO)_{12}]^{2-}$	trigonal prismatic	86	v
$[Fe_3Pt_3(CO)_{15}]^{2-}$	triangular planar	86	w
$[Fe_3Cu_3(CO)_{12}]^{3-}$	triangular planar	84	y
$[Os_7(CO)_{21}]$	capped octahedron	98	z
$[Os_8(CO)_{22}]^{2-}$	bicapped octahedron	110	aa
$[Co_9Si(CO)_{21}]^{2-}$	capped square antiprism	129	ab
$[Ni_9C(CO)_{17}]^{2-}$	capped square antiprism	130	ac
$[Ni_9(CO)_{18}]^{2-}$	trigonal prismatic	128	ad
$[Cu_5Fe_4(CO)_{16}]^{3-}$	planar rhombus	122	ae
$[Au_9(PPh_3)_8]^{+}$	centered cube	114	af

X-ray diffraction methods are still quite scarce and only a few tens of examples are known to date.

To give an idea of the diversity of the geometrical arrangements assumed by low nuclearity clusters, we have listed examples of the more frequently encountered cluster structures in Table **2-1** and illustrated them in Figure **2-15**. Also reported in Table **2-1** are the formal number of cluster valence electrons, CVE, associated with these structures, whereby in determining these, each ligand is considered as a donor of one (H), two (CO, PR_3), three (NO), four (C, Si), or five (N, Cp) electrons. It is obvious from even a cursory inspection of Table **2-1** that the structures of these TM clusters cannot be simply interpreted and rationalized on the basis of their number of valence electrons since for almost every cluster shape there are examples with different electron counts. More complex electron counting rules, largely based on the topological properties of the metal cage, have been developed in order to account for the observed structures and these rules will be discussed in Section 2.4.3. Although the majority of these low

---

Table 2-1. (Continued).

(a) F. A. Cotton and J. M. Troup, *J. Am. Chem. Soc.* **1974**, *96*, 4155. (b) B. F. G. Johnson, J. Lewis, P. R. Raithby and A. Sanders, *J. Organomet. Chem.* **1984**, *260*, C29. (c) N. Cook, L. E. Smart, P. Woodward and J. D. Cotton, *J. Chem. Soc. Dalton Trans.* **1979**, 1032. (d) D. Moras, J. Dehand and R. Weiss, *C. R. Acad. Sci., Ser. C.* **1968**, *267*, 1471. (e) J. Dubrawski, J. C. Kriege-Simonsen and R. D. Feltham, *J. Am. Chem. Soc.* **1980**, *102*, 2089. (f) R. D. Wilson and R. Bau, *J. Am. Chem. Soc.* **1976**, *98*, 4687. (g) M. R. Churchill and R. Bau, *Inorg. Chem.* **1968,** *7*, 2606. (h) J. S. Field, R. J. Haines, and D. N. Smit, *J. Organomet. Chem.* **1982**, *224*, C49. (i) C. R. Eady, B. F. G. Johnson, J. Lewis, B. E. Reichert and G. M. Sheldrick, *J. Chem. Soc. Chem. Commun.* **1976**, 271. (j) G. Longoni, P. Chini, L. D. Lower and L. F. Dahl, *J. Am. Chem. Soc.* **1975**, *97*, 5034. (k) E. H. Braye, L. F. Dahl, W. Hubel and D. S. Wampter, *J. Am. Chem. Soc.* **1962**, *84*, 4633. (l) J. J. Guy and G. M. Sheldrick, *Acta Cryst.* **1978**, *B34*, 1725. (m) J. P. Barbier, R. Bender. P. Braunstein, J. Fischer and L. Ricard, *J. Chem. Res.* **1978**, *230*, 2913. (n) G. Longoni, M. Manassero and M. Sansoni, *J. Am. Chem. Soc.* **1980**, *102*, 3242. (o) V. G. Albano, P. Chini and V. J. Scatturin, *J. Chem. Soc. Chem. Commun.* **1968**, *163*. (p) F. Scherbaum, A. Grohmann, B. Huber, H. Schmidbaur, *Angew. Chem. Int. Ed. Engl.* **1988**, *27*, 1544. (q) R. Mason, K. M. Thomas and D. M. P. Mingos, *J. Am. Chem. Soc.* **1973**, *95*, 3802. (r) E. Sappa, M. Lanfranchi, A. Tiripicchio and M. T. Camellini, *J. Chem. Soc. Chem. Commun.* **1981**, 995. (s) A. G. Orpen, *J. Organomet. Chem.* **1978**, *159*, C1. (t) C. E. Briant, K. P. Hall and D. M. P. Mingos, *J. Organomet. Chem.* **1983**, *254*, C18. (u) S. Martinengo, D. Strumolo, P. Chini, V. G. Albano and D. Braga, *J. Chem. Soc. Dalton Trans.* **1985**, 35. (v) J. C. Calabrese, L. F. Dahl, P. Chini, G. Longoni and S. Martinengo, *J. Am. Chem. Soc.* **1974**, *96*, 2614. (w) G. Longoni, M. Manassero and M. Sansoni, *J. Am. Chem. Soc.* **1980**, *102*, 3242. (y) G. Doyle, K. A. Eriksen and D. Van Engen, *J. Am. Chem. Soc.* **1986**, *108*, 445. (z) C. R. Eady, B. F. G. Johnson, J. Lewis, R. Mason, P. B. Hitchcock and K. M. Thomas, *J. Chem. Soc. Chem. Commun.* **1977**, 385. (aa) P. F. Jackson, B. F. G. Johnson, J. Lewis and P. R. Raithby, *J. Chem. Soc. Chem. Commun.* **1980**, 60. (ab) K. M. Mackay, B. K. Nicholson, W. T. Robinson and A. W. Sims, *J. Chem. Soc. Chem. Commun.* **1984**, 1276. (ac) A. Ceriotti, G. Longoni, M. Manassero, M. Perego and M. Sansoni, *Inorg. Chem.* **1985**, *24*, 117. (ad) R. A. Nagaki, L. D. Lower, G. Longoni, P. Chini and L. F. Dahl, *Organometallics* **1986**, *5*, 1764. (ae) G. Doyle, K. A. Eriksen and D. Van Engen, *J. Am. Chem. Soc.* **1986**, *108*, 445. (af) J. G. M. Van der Linden, M. L. H. Paulissen and J. E. J. Schmitz, *J. Am. Chem. Soc.* **1983**, *105*, 1903.

nuclearity molecular clusters adopt deltahedral geometries (tetrahedron, trigonal bipyramid, octahedron, etc.) or their mono and bicapped structural derivatives, several clusters exhibit less common geometrical arrangements, such as the large planar aggregates in $[Fe_3Pt_3(CO)_{15}]^{7-}$ [170] and $[Fe_4Cu_5(CO)_{16}]^{3-}$ (Fig. **2-16**). [171]

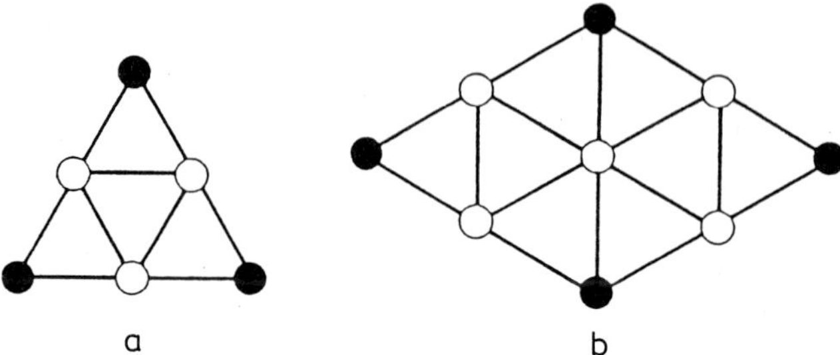

Figure 2-16. Planar packing in the molecular metal clusters a) $[Fe_3Pt_3(CO)_{15}]^{7-}$ and b) $[Fe_4Cu_5(CO)_{16}]^{3-}$.

As the nuclearity increases, three types of high nuclearity structures are distinguished as occurring most often: layered or columnar clusters (Fig. **2-17**), closed packed clusters (Fig. **2-18**), and clusters with icosahedral or pentagonal symmetry (Fig. **2-19**). Most of the columnar clusters are based on trinuclear, triangular building block layers of general formula M_3 $(\mu_2\text{-L})_3(\text{L}')_3$. Two layers in an eclipsed conformation lead to a trigonal prism and, if staggered, to an octahedron of metal atoms. The extension to oligomers and to one-dimensional polymers is obvious. Beautiful examples for this construction principle are provided by the Pt carbonyl clusters of general formula $[Pt_3(\mu_2\text{-CO})_3(CO)_3]_n^{2-}$ [172] where $n = 2, 3, 4, 5$, and 6. Analogous layered Ni carbonyl clusters are also known. [173, 174] Although the stacking arrangement for the members of this series ideally conform to D_{3h} symmetry, the actual crystal structures may show significant deviations from regularity.

The largest cluster structures resolved by X-ray crystallography show hcp or fcc metal atom arrangements. Typical examples of this class of clusters are $[Rh_{22}(CO)_{37}]^{4-}$, [175] $[Pt_{38}(CO)_{44(?)}]^{2-}$, [176] and $[Ni_{38}Pt_6(CO)_{44}]^{n-}$, [177] (Fig. **2-18**). The structure of the latter cluster is particularly remarkable considering the highly symmetrical arrangement of the metal atoms which form an almost undistorted octahedron. Pentagonal or icosahedral packings have also been reported for a

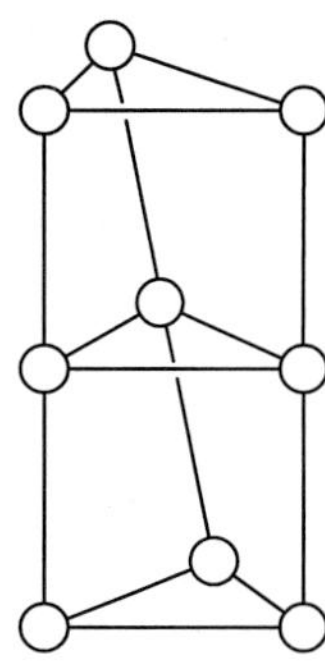

Figure 2-17. The structure of the layered clusters $Ni_9(CO)_{18}$.

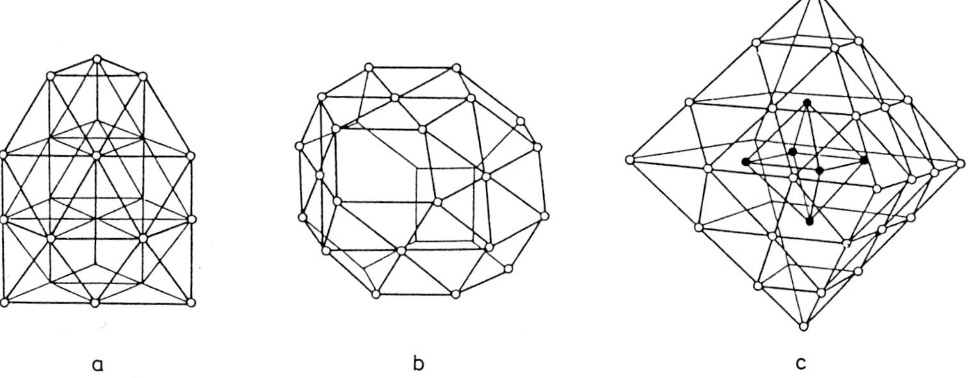

Figure 2-18. Compact packing in high nuclearity molecular metal clusters: a) $[Rh_{22}(CO)_{37}]^{4-}$, fcc-hcp; b) $[Pt_{38}(CO)_{44}]^{2-}$, fcc; c) $[Ni_{38}Pt_6(CO)_{44}]^{6-}$, fcc.

series of clusters (Fig. **2-19**). The icosahedron and the cuboctahedron are closely related. However, there is still no satisfactory answer to the question of why M_{13} clusters can be either icosahedral or cuboctahedral. Among the homonuclear systems, the icosahedral clusters $[Au_{13}(PR_3)_{10}Cl_2]^{3+}$ [178] (Fig. **2-19a**) and $[Rh_{12}Sb(CO)_{27}]^{3-}$ [179] have particularly simple structures. The structures of the clusters $[Pt_{19}(CO)_{22}]^{4-}$ [180] and $[Ni_{34}Se_{22}(PR_3)_{10}]$ [181] are more complicated. The Pt system can be viewed as resulting from the condensation of three pentagonal bipyramids (Fig. **2-19b**). The $[Ni_{34}Se_{22}(PR_3)_{10}]$ cluster contains a pentagonal antiprismatic Ni_{10} core which is surrounded by five Ni_4 "butterfly" units, the remaining four Ni atoms are located between the two pentagonal Ni_5 rings of the antiprism. Another example of a cluster with five-fold symmetry is the mixed metal cluster $[Au_{13}Ag_{12}(PPh_3)_{12}Cl_6]^{m+}$ [182] in which the metal cage consists of three condensed icosahedra (Fig. **2-19c**).

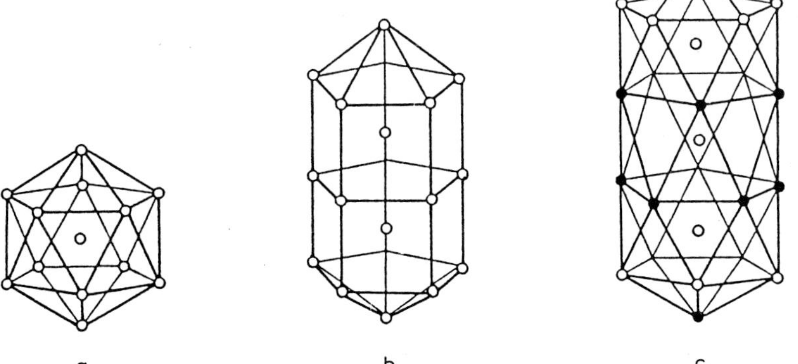

Figure 2-19. Icosahedral or pentagonal packing in molecular metal clusters: a) $[Au_{13}(PR_3)_{10}Cl_2]^{3+}$; b) $[Pt_{19}(CO)_{22}]^{4-}$; c) $[Au_{13}Ag_{12}(PR_3)_{12}Cl_6]^{m+}$.

$[Ni_{34}Se_{22}(PR_3)_{10}]$ [181] is only one of several examples of TM clusters stabilized by such main group bridging atoms as As, Sb, Bi, S, and Se, [183] or groups like NH, PR, or SR. [184] Some of the clusters formed by these stabilizing ligands have very complicated structures as observed in $[Ni_{15}Se_{15}(PPh_3)_6]$, [183] $[Cu_{20}Se_{13}(PEt_3)_{12}]$, [185] or the very large $[Cu_{70}Se_{35}(PEt_3)_{22}]$. [185]

In molecular clusters, the symmetry of the metal cage depends on a variety of factors which include the number, the type, and the coordination mode of the ligands, the number of valence electrons, the total molecular charge, etc. Often, an interchange from one kind of packing to another has been observed in clusters of the same nuclearity. Systems with very similar chemical formulae may exhibit different cluster geometries, as illustrated by the clusters $[Os_6Pt_2(CO)_{16}(C_8H_{12})]$ (**1**) and $[Os_6Pt_2(CO)_{17}(C_8H_{12})]$ (**2**) (Fig. **2-20**). [186] The metal core in **1** can be described as a Pt-bicapped Os-tetrahedron condensed to a second Os tetrahedron through a common edge. The metal cage in **2** can be regarded as a regular octahedron of Os atoms with two face capping $[Pt(C_8H_{12})]$ units. It has been found that clusters with the same formal number of skeletal electrons assume similar metal polyhedron geometries (see Section 2.4.3), although several exceptions to this rule are known. Small changes in this number (e. g. by adding a CO ligand as in the example described above) can result in substantial rearrangements of the metal cage and can thus be considered as a sign of the facile deformability of the systems. It has been proposed [187] that the growth sequence of simple metal aggregates proceeds initially by capping the triangular faces of the core polyhedron. The tendency to cap triangular faces seems to be supported by the large number of five-atom clusters with the trigonal bipyramidal structure and by the fact that several seven-atom clusters have a structure corresponding to a capped octahedron. On the other hand, there are not too many examples of seven-atom clusters with a genuine pentagonal arrangement as observed in the pentagonal bipyramidal cluster $[Au_7(PPh_3)_7]^+$. [188]

Many clusters incorporate heteroatoms such as H, C, N, or even As, Sb, P, S, Ge, etc. inside their metallic cores (see Chapter 3.2). [189] The metal core structure usually forms a cavity in which the heteroatom is accomodated. The metallic core can be fundamentally cubic or pentagonal. The number of interstitial atoms embedded within the metal core depends on the geometry and the dimensions of the cluster. In some cases a small molecule is encapsulated in the metal cluster, for example, in the clusters $[Co_6Ni_2C_2(CO)_{16}]^{2-}$ [190] and $[Ni_{16}(C_2)_2(CO)_{23}]^{4-}$ [191]

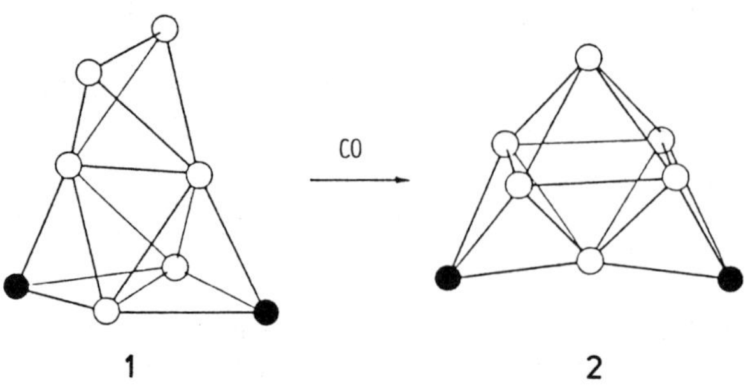

Figure 2-20. Metal core structures of $[Os_6Pt_2(CO)_{16}(C_8H_{12})_2]$ (**1**) and $[Os_6Pt_2(CO)_{17}(C_8H_{12})_2]$ (**2**).

which contain one and two C_2 molecules respectively. The heteroatoms do not always occupy interstitial positions. A classical example is the complex $[Fe_4C(CO)_{12}]^{2-}$ [192] which has an exposed C atom. Many other elements besides carbon can be included in metal clusters. One such system is the cluster $[Co_9Si(CO)_{21}]^{2-}$ in which a Si atom sits in the center of a capped square antiprism of cobalt atoms and represents a rare example of a kinetically stable paramagnetic cluster. [193] The great variety of geometrical arrangements reported in the literature suggests that interstitial compounds may be structurally complex since they do not follow simple rules. They may provide examples for studying building blocks of technologically relevant materials like carbides and nitrides.

Mixed metal clusters with two or more types of metal atoms are also of particular interest and we have already mentioned the clusters $Ni_{38}Pt_6$ and $Au_{13}Ag_{12}$. In both cases, there is an aggregate of one metal surrounded by a shell of the other metal. This observation may be relevant for understanding the formation of metal alloys. These cluster structures seem to indicate that alloy formation may proceed through similar homogeneous blocks such that the final effect would be to dilute the two metals without seriously changing the microscopic properties of the individual metals. This would support the “minimum polarity model” of alloys. [194, 195] According to this model, the electronic structure of each component of an alloy remains similar to that of its pure phase. It is also worth mentioning that many mixed clusters can be easily transformed either thermally or photochemically into mixtures of simple homonuclear clusters. This might suggest that, despite the complexity of the molecular structures of these mixed metal clusters, there is a propensity for total dissociation into homonuclear metallic structures.

2.4.2.2 Metal–Metal Bond Lengths

The analysis of the bonding in organometallic clusters has often been based on the measured lengths of the metal–metal bonds. In particular, attempts have been made to correlate the metal–metal distances in ligated clusters with the corresponding internuclear separations in bulk metals. [196] In molecular metal clusters, the M–M bond is relatively deformable as a result of both steric and electronic effects. Generalizations and trends are thus difficult to establish because of the great variety of electronic and geometric situations. For instance, bridging ligands are able to induce a considerable contraction or elongation of the M–M distances, an effect which is not easily related in a simple way to an electronic change in the metal–metal interaction.

The trinuclear metal carbonyl clusters $[Fe_3(CO)_{11}]^{2-}$, [197] $[Ru_3(CO)_{12}]$, [198] and $[Os_3(CO)_{12}]$ [199] do not contain bridging ligands and the M–M distances are 4–8 % longer than in the corresponding bulk metals. On the other hand, tetranuclear and hexanuclear metal clusters often show M–M distances which are very close to those of the metal crystals, but significant deviations have also been observed (Table **2-2**).

For high nuclearity clusters, the similarity of the M–M distances to the bulk values becomes more evident. In $[Ni_{38}Pt_6(CO)_{44}]^{6-}$, [177] the average distance of

Table 2-2. Metal-metal bond lengths (in Å) in polynuclear metal clusters and their respective bulk metals.

	M–M (av.)	bulk value	difference
$[Co_4(CO)_{12}]$	2.49	2.51	−1 %
$[Co_6(CO)_{15}]^{2-}$	2.51	2.51	0
$[Rh_4(CO)_{12}]$	2.73	2.69	+1 %
$[Rh_6(CO)_{16}]$	2.78	2.69	+3 %
$[Rh_{22}(CO)_{37}]^{4-}$	2.80	2.69	+4 %
$[Fe_4(CO)_{13}]^{2-}$	2.55	2.48	+3 %
$[Os_6(CO)_{18}]^{2-}$	2.86	2.67	+7 %
$[Ni_6(CO)_{12}]^{2-}$	2.38, 2.77	2.49	−4 %, +11 %
$[Pt_6(CO)_{12}]^{2-}$	2.66, 3.04	2.75	−3 %, +11 %

the 108 surface Ni–Ni bonds (2.58 Å) is about 4 % larger than that in bulk Ni, while the 2.72 Å mean value for the 12 Pt–Pt distances is 1 % shorter than in bulk Pt.

The relative similarity between the M–M bond distances in clusters and their bulk metals has been taken as a sign for the ligated clusters having a similar nature as the extended metal bonding regimes. [196] However, this similarity is not supported by a more accurate analysis of the cluster electronic structures, as will be discussed below. Here, it is important to mention that a direct comparison between the metal–metal bond lengths in clusters and those in the bulk metals is not fully justified. In fact, clusters are coordinatively saturated compounds with a very high ligand to metal ratio. A more meaningful comparison would be with the distances found at metal surfaces which are fully covered by adsorbed molecules or atoms. Even in these cases, the ratio between the number of ligands and the number of metal atoms is not found to exceed 1, while in coordination compounds it may be as high as 6 (Table **2-3**). It is well known that clean, and in particular covered metal surfaces undergo reconstruction, whereby a significant departure from the geometrical parameters characteristic of the bulk occurs. [200]

Despite these limitations, some general trends can be observed for the metal–metal distances in TM clusters. The Ni–Ni bond lengths obtained from X-ray crystallographic data for a series of Ni_8 clusters with cubic cores are given in Table **2-4** (see also Fig. **2-21**). These experimental values have been compared with the computed optimal Ni–Ni bond lengths for the "hypothetical" naked or low coordinated Ni_8 cubic clusters. [201] In this way, it was possible to establish a relationship between the total number of ligands on the cluster and the intermetallic separation. The computed values, being obtained from density functional calculations and accurate to within 1–2 % (a typical error for bond lengths with this computational technique), are directly comparable to the experimental ones. The level of accuracy achieved by theory can be deduced from a comparison of

Table 2-3. The CO to metal atom ratios in a number of mono- and polynuclear transition metal complexes.

System	CO/metal atoms ratio
$[Ni(CO)_4]$	4
$[Fe(CO)_5]$	5
$[Cr(CO)_6]$	6
$[Fe_3(CO)_{12}]$	4
$[Fe_4(CO)_{13}C]$	3.2
$[Ni_5(CO)_{12}]^{2-}$	2.4
$[Ni_6(CO)_{12}]^{2-}$	2
$[Ni_9(CO)_{18}]^{2-}$	2
$[Ni_{32}C_6(CO)_{36}]^{2-}$	1.1
$[Ni_{38}Pt_6(CO)_{48}]^{n-}$	1.1
metal surface (full coverage)	1

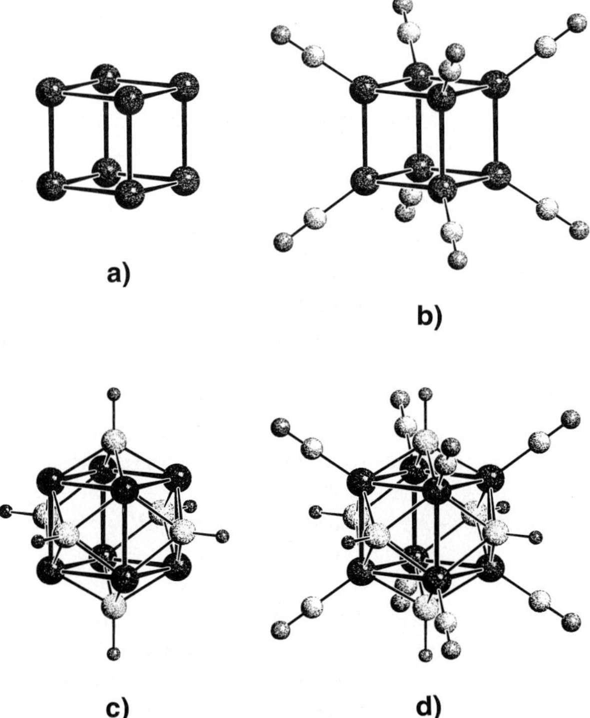

Figure 2-21. Structures of cubic Ni_8 clusters with complete and incomplete ligand shells. a) Ni_8 (**3**), b) $[Ni_8(CO)_8]$ (**4**), c) $[Ni_8(\mu_4\text{-PH})_6]$ (**5**), d) $[Ni_8(\mu_4\text{-PH})_6(CO)_8]$ (**6**) Reproduced with permission from [201]. Copyright 1993 American Chemical Society.

Table 2-4. Computed and experimentally determined Ni-Ni bond distances (in Å) for a series of Ni_8 clusters having the cubic cage structure.

Cluster		CVE's	r(Ni–Ni)	Ref.
Computed				
3	Ni_8	80	2.19	a
4	$[Ni_8(CO)_8]$	96	2.27	a
5	$[Ni_8(\mu_4\text{-PH})_6]$	104	2.37	a
6	$[Ni_8(\mu_4\text{-PH})_6(CO)_8]$	120	2.64	a
Measured				
7	$[Ni_8(\mu_4\text{-PPh})_6(CO)_8]$	120	2.65	b
8	$[Ni_8(\mu_4\text{-PPh})_6(PPh_3)_4(CO)_4]$	120	2.67	c
9	$[Ni_8(\mu_4\text{-PPh})_6(PPh_3)_4]$	112	2.53	d
10	$[Ni_8(\mu_4\text{-S})_6(PPh_3)_6Cl_2]$	118	2.68	e
11	$[Ni_8(\mu_4\text{-PPh})_6(PPh_3)_4Cl_4]$	116	2.61	c
12	$[Ni_9(\mu_4\text{-GeEt})_6(CO)_8]$	124	2.67	f
13	$[Ni_9(\mu_4\text{-Te})_6(PEt_3)_8]$	130	2.86	g
14	$[Ni_9(\mu_4\text{-As})_6(PPh_3)_5Cl_3]$	121	2.81	h
	bulk Ni		2.49	k

(a) N. Rösch, L. Ackermann and G. Pacchioni, *Inorg. Chem.* **1993**, *32*, 2963. (b) L. D. Lower and L. F. Dahl, *J. Am. Chem. Soc.* **1976**, *98*, 5046. (c) D. Fenske, R. Basoglu, J. Hachgenei, and F. Rogel, *Angew. Chem. Int. Ed. Engl.* **1984**, *23*, 160. (d) D. Fenske, J. Hachgenei and F. Rogel, *Angew. Chem. Int. Ed. Engl.* **1984**, *23*, 982. (e) D. Fenske, J. Hachgenei and J. Ohmer, *Angew. Chem. Int. Ed. Engl.* **1985**, *24*, 706. (f) J. P. Zebrowski, R. K. Hayashi, A. Bjarnason and L. F. Dahl, *J. Am. Chem. Soc.* **1992**, *114*, 3121. (g) J. G. Brennan, T. Siegrist, S. M. Stuczynski and M. L. Steigerwald, *J. Am. Chem. Soc.* **1989**, *111*, 9240. (h) D. Fenske, J. Ohmer and K. Merzweiler, *Angew. Chem. Int. Ed. Engl.* **1988**, *27*, 1512. (k) N. W. Ashcroft, N. D. Mermin: *Solid State Physics*, Saunders, Philadelphia, **1988**.

the computed with the measured Ni–Ni distances in $[Ni_8(\mu_4\text{-PR})_6(CO)_8]$ for R = H and Ph respectively.

A feature common to all the ligated Ni_8 clusters that have been isolated experimentally (compounds **7–14** of Table **2-4**) is that the Ni–Ni distances are larger than in the bulk. However, a simple relationship between the average Ni–Ni bond lengths and either the steric parameters or the number of cluster valence electrons does not exist. This can be seen, for instance, by comparing the compounds **7** and **12** where the addition of a Ni atom inside the Ni_8 cage does not induce a significant change in the dimensions of the metal frame. The chemical nature of the ligands also does not alter the metal–metal distances significantly since replacing the CO ligands in **8** by the electronegative Cl atoms in **11** has only a moderate effect on r(Ni–Ni). Steric effects seem play a more important role; the capping Te atom in **13** induces a considerable elongation of r(Ni–Ni). This effect is probably due to the fact that the covalent radius of Te (1.43 Å) is about 0.3–0.4 Å larger than that of either P (1.09 Å) or S (1.03 Å). On the other hand, Ge has a covalent radius of 1.22 Å, which is larger than those of S or P, but the Ni–Ni distances are nevertheless similar in the clusters **7**, **8**, **10**, and **12**.

A clear difference is found between low coordinated or even naked clusters and the high coordinated clusters described above. The computed Ni–Ni distances in the hypothetical low coordinated Ni_8 clusters **3–6** of Table **2-4** are considerably shorter (about 0.1–0.3 Å) than the bulk value. [201] This result might suggest the existence of a direct correlation between the number of ligands and the metal–metal distances such that the larger the number of ligands, the longer the Ni–Ni bonds. The geometry of the cluster $[Ni_8(\mu_4\text{-}PPh)_6(PPh_3)_4]$ [202] (see compound **9** of Table **2-4** and Fig. **3-48**) provides support for this view. The cluster **9** is the only known case of a cubic Ni_8 cluster where the metal frame is ligated by only four terminal ligands, instead of the usual eight in the other members of this series, and so four of the Ni atoms can be considered as low coordinated. Indeed, the measured metal–metal distances in **9** are, on the average, 0.14 Å shorter than in the other members of the group. [202] It has been suggested that the elongation of the metal–metal distances in the high coordinated clusters is more electronic than steric in origin. [201]

2.4.2.3 The Ligand Polyhedron

In the previous discussion, we focused attention on the geometric structure of the metal frame. However, the arrangement of the ligands in molecular metal clusters is no less significant than that of the metal atoms. On the contrary, it is important to realize that in coordinatively saturated cluster compounds the non-bonding interactions between the ligands play a non-negligible role in determining the geometric shape of the metal atoms. To account for the effect of the steric repulsion between the ligands, Johnson and Benfield [203, 204] proposed a model which may be considered as an extension of the valence shell electron pair repulsion model, VSEPR. [205] They suggested that electronic bonding mechanisms may be less important than the number and position of the ligands in determining the geometry of a metal cluster. The hard sphere packing model [203, 204] assumes that the ligands, and in particular the CO groups, are disposed around the metal frame in such a way that their mutual repulsion is minimized and the M_n metal core is then positioned symmetrically inside the ligand shell. For a $[M_4(CO)_{12}]$ cluster, the most stable polyhedron for the 12 CO ligands is the icosahedron. If the cavity inside the ligand polyhedron is large enough to accomodate the M_n metal unit, then the cluster will assume this geometry. This is the case, for instance, in $[Co_4(CO)_{12}]$ and $[Fe_3(CO)_{12}]$ where the CO groups do indeed form icosahedra. As the metal atomic radius increases (*e. g.* with Ir or Os) the icosahedral cavity becomes too small to accomodate a tetrahedral Ir_4 or a triangular Os_3 unit and the ligand shell rearranges to form the second best polyhedron, namely the cuboctahedron. In fact, the CO ligands in $[Ir_4(CO)_{12}]$ and $[Os_3(CO)_{12}]$ show a cuboctahedral arrangement, even though the icosahedron is more favorable. Thus, a simple analogy based on the repulsive interaction of equal point charges moving on a sphere accounts for the observed behavior.

The model does not always work so nicely, and particularly not for larger clusters. In fact, as the size of the ligand polyhedron increases the energy differences between the various geometries become very small. For large polyhedra, it is virtually impossible to predict whether the tendency to minimize the nonbonding interactions of the ligands will prevail over other competing factors, such as electronic bonding mechanisms (which are certainly not negligible), packing effects in the crystal, etc. Nevertheless, it is a merit of the hard sphere packing model to focus attention on the fact that the ligand shell constitutes more than simply an "envelope" of innocent spectator molecules. On the contrary, it suggests that the geometry of the metal cluster may be governed, at least in part, by that of the ligands in the coordination sphere and not only by the electronic interactions among the metal atoms. This is opposite to the case of adsorption at metal surfaces, where, to a large extent, the adsorption site of the ligand is determined by the orbital properties of the substrate. However, in such cases the repulsive interactions between the adsorbates can also lead to changes in the adsorption geometry as the coverage increases. An example for this effect is provided by the chemisorption of C_2N_2 on the 110 surface of Ni or Pd. It was demonstrated by angular resolved photoemission and supported by force field calculations that the molecular axis of C_2N_2 is oriented across the troughs of these surfaces. Whereas the molecules lie flat on Pd, they are tilted away from the surface in the case of Ni where the closer spacing of the adsorption sites induces strong repulsive interactions between neighboring adsorbates. [206, 207]

An important aspect of metal clusters which is intimately related to the ligand shell repulsion model is the mobility of the ligands on the cluster "surface". This mobility has been demonstrated many times, mainly through the use of NMR spectroscopy [208–210] (e.g. by the temperature dependent ^{13}C NMR). It is often found that the CO ligands are highly fluxional and undergo a continous change between terminal and bridging positions as long as enough sites are available on the cluster surface for a concerted motion of several ligands. Fluxionality in clusters is not a peculiar property of the ligands. Cases where the motion involves the metal atoms have also been reported. [210] These small motions of the metal atoms can lead to structural changes in the metal frame. This phenomenon is another, indirect, indication for the "special" nature of the metal–metal interaction in organometallic clusters compared to those in the bulk metal. In this context, it seems worth pointing out that the diffusion of atoms over a metal surface has also been described as a concerted motion of several atoms. [211]

2.4.3 Topological Relationships and Simplified Bonding Models

The nature of the bonding in molecular metal clusters represents a central question in organometallic chemistry and a real challenge to theoreticians. The similarity of both the metal framework structure and the ligands' stereochemistry in clusters with those at surfaces has led to the formulation of the so-called "cluster-surface" analogy. This analogy was first suggested in the pioneering work of Muetterties. [196] Indeed, the language of organometallic chemistry is largely the same as that used in surface chemistry. [212, 213] For instance, the bonding of CO to a metal surface or to a metal cluster is described in the same terms of σ-donation and π-back donation originally proposed in the Dewar-Chatt-Duncanson model for metal complexes [156, 157] and then extended to metal surfaces by Blyholder. [214] Although the metal–ligand interactions in mono- or polynuclear complexes can often be formulated and discussed in similar terms as those on extended surfaces, the nature of the metal–metal bonding is much more complex and has been the subject of considerable theoretical controversy.

Since organometallic clusters are well defined molecular objects of finite size (although perhaps a bit large), standard molecular quantum mechanical methods like *ab initio* or density functional techniques are the ideal tools for the description of their electronic structures. Unfortunately, the dimensions of these "molecules", combined with the fact that they contain transition metal atoms, places this class of compounds well beyond the range of practical *ab initio* calculations. Only recently, due to improvements in quantum mechanical computer codes and to the extremely rapid increase in computing power, has it become possible to successfully tackle the electronic structure problem of ligated clusters on a quantitative or semiquantitative basis. So far, most of the theoretical analyses have been based on approximate computational schemes, topological considerations, or even simple empirical rules.

The first empirical correlations between cluster structures and the number of electrons involved in their skeletal bonding were formulated about 20 years ago thanks to the contributions of Williams, [215] Wade, [216] and Mingos. [217] Since then, other schemes have been suggested to account for the relationships between the geometric structures and the cluster valence electrons, CVE, and include the graph theoretical method proposed by King, [218] the topological electron counting procedure formulated by Teo, [219] and the extended Hückel based scheme suggested by Lauher. [220, 221] Here, we give only a brief account of some of these procedures, but refer the reader to the specialized literature for a more extensive treatment.

2.4.3.1 Effective Atomic Number (EAN) Rule

The effective atomic number rule is an extension of the 18 electron rule for transition metal complexes. [222] One assumes that each CO ligand contributes two electrons and that the polyhedral edges represent two-electron/two-center bonds.

According to this rule, $[Os_5(CO)_{16}]$ [223] has an electron count of [(5 x 8) + (16 x 2)] = 72 electrons, yet in order for all five Os atoms to attain a noble gas configuration, 90 (*i.e.* 5 x 18) electrons are required. The formation of nine metal–metal bonds in the trigonal bipyramidal structure of $[Os_5(CO)_{16}]$ thus accounts for the missing nine electron pairs. Many exceptions to this rule are known. For instance, $[Ni_5(CO)_{12}]^{2-}$ [224] has an electron count of 76 valence electrons, but nevertheless has the same structure as $[Os_5(CO)_{16}]$.

2.4.3.2 Polyhedral Skeletal Electron Pair (PSEP) Model

The PSEP method [217, 225–227] is derived from rules that had originally been developed to rationalize the structures of borane compounds. [215, 216] The borane anions $B_nH_n^{2-}$ are so-called *closo* compounds and exhibit deltahedral structures which maximize the number of nearest neighbors. Each B–H bond involves one boron orbital and two electrons, thus leaving three orbitals and two electrons to be used in the B_n skeleton. Two of these orbitals lie on the "surface" (e.g. p_x or p_y) and one (s, p_z, or an sp_z hybrid) is directed towards the center of the polyhedron. This scheme yields $2n$ "surface" or tangential orbitals, n bonding and n antibonding, and n radially bonding orbitals of which one is strongly bonding radially and $n-1$ are either weakly bonding, nonbonding, or weakly antibonding radial orbitals. Since the total number of bonding orbitals is therefore $n+1$, the number of cage electrons is $2n+2$ and a closed shell electronic structure results from a total number of $4n+2$ valence electrons. By removing a boron atom from a vertex of a B_{n+1} *closo* skeleton one obtains a B_nH_{n+4} *nido* structure with an electron count of $4n+4$. If two vertex boron atoms are removed from a B_{n+2} *closo* structure, then a B_nH_{n+6} *arachno* structure results which obeys the $4n+6$ electronic rule. [215, 216]

The extension of these rules to transition metal clusters is based on the isolobal analogy. [50] Two molecular fragments are said to be isolobal if they have the same number of electrons and the same number and symmetry of valence orbitals at nearly the same energy. In principle, a wide range of atoms or groups of atoms can substitute for the BH fragments in the boranes. Neutral analogues of the BH moiety include $[Fe(CO)_3]$, $[Ni(PR_3)_2]$, AlR, Sn, Pb, etc. In fact, all these fragments contribute two electrons and three orbitals to the cluster bonding. In the $[Fe(CO)_3]$ fragment, the Fe atom contributes 9 valence orbitals and 8 electrons. Three empty σ orbitals and three filled π orbitals interact with the CO orbitals, leaving two electrons in three orbitals available for the bonding with other metal atoms. On the basis of these analogies it is possible to conclude that isolobal fragments will produce isostructural clusters.

Transition metal analogues of *closo* boranes are known. They are characterized by a number of valence electrons equal to $14n+2$ as a consequence of the filling of the additional five d orbitals. Also, transition metal polyhedra exhibiting *nido* and *arachno* structures (with electron counts of $14n+4$ and $14n+6$ respectively) have been identified. [227] Thus, the PSEP model represents a very simple, yet powerful method to rationalize or predict the structures of cluster compounds and it has been succesfully applied to a wide range of systems. [226, 227] The original rules have been generalized in order to explain the structures of high nuclearity organometallic clusters which may be considered as derived by condensing several polyhedra. [228]

A topological foundation for the PSEP rules has been formulated by Stone. [229–231] The basic idea relies on a classification of the cluster MO's according to their nodal characteristics, irrespective of their detailed form. Stone's tensor surface harmonic (TSH) theory constructs a set of appropriate MO's from the nodal properties of the spherical harmonics which are the solutions to the Schrödinger equation for an electron moving in a spherical potential shell. For a TM cluster M_n, the TSH theory identifies $2n$–1 orbitals which are likely to be vacant, thus leaving $7n+1$ MO's which can take on the cluster electrons in agreement with the $14n+2$ rule. An extensive account of the PSEP and TSH theories and their application to a wide class of cluster compounds has been given recently by Mingos and Wales. [232]

2.4.3.3 Topological Electron Counting (TEC)

Although the PSEP theory has the advantage of being easily applicable to metal clusters derived from triangulated polyhedra, it becomes less easy to apply to TM clusters with unusual polyhedral frameworks having no borane analogues. Moreover, examples of metal clusters which do not fit the PSEP model are known. To overcome this problem, Teo [219, 233] proposed an electron count which is based on Euler's theorem. This theorem states that for a polyhedron with V vertices, F faces, and E edges, the reationship $E = V + F - 2$ holds. Assuming that each atom on the surface of the metal skeleton attains an 18 electron noble gas configuration and that each edge may be considered as a two-center/two-electron metal–metal bond, the total electron count, N, for the cluster will be $N = 18V - 2E$ and the number of occupied cluster valence molecular orbitals, CVMO, is CVMO $= 9V - E$. By substituting E from Euler's theorem, one obtains CVMO $= 8V - F + 2$. Indeed, it is well known that the bonding in metal clusters is better described in terms of multicenter bonds (*vide infra*). In order to take into account this important correction, Teo proposed a new formula, CVMO $= 8V - F + 2 + X$, where X is a parameter dependent on the shape of the polyhedron. The model successfully describes polyhedra which follow the PSEP rule (e.g. $[Os_5(CO)_{16}]$: observed $N = 72$, PSEP = 72, TEC = 72) as well those which violate it (e.g. $[Ni_5(CO)_{12}]^{2-}$: observed $N = 76$, PSEP = 72, TEC = 76). Exceptions to the TEC model are also knwon (e.g. $[Pt_6(CO)_{12}]^{2-}$: observed $N = 86$, TEC = 90).

2.4.3.4 Electron Counting Based on the Extended Hückel Approach

Lauher [220, 221] proposed that the metal derived MO's of a carbonylated cluster can be derived from an extended Hückel calculation on the bare metal cluster and then divided into sets of filled cluster valence molecular orbitals, CVMO, and high lying antibonding orbitals, HLAO (Fig. **2-22**). The energy of the atomic p level of the metal atom was chosen as an empirical upper limit to separate the occupied from the unoccupied MO's. Only the CVMO's are energetically suitable for ligand bonding and the occupation by metal electrons and the total number of CVMO's is a function of the cluster geometry. According to this model, the number of these MO's fully determines the stoichiometry, structure, and chemistry of a cluster compound. [220, 221, 234]

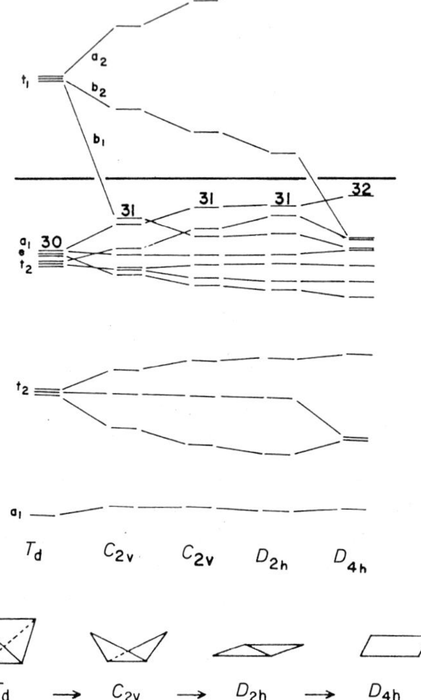

Figure 2-22. Correlation diagram of the molecular orbitals of various M_4 clusters. The horizontal line separates the CVMO's from the HLAO's. The number of CVMO's is indicated for each geometry. Reproduced with permission from [220]. Copyright 1978 American Chemical Society.

The method has been applied to a series of low nuclearity organometallic clusters, like $[Os_6(CO)_{18}]$, $[Fe_5(CO)_{15}C]$, and $[Rh_4(CO)_{12}]$. [220, 221] For larger clusters, it has been shown [234] that any compact M_n compound should have $6n+7$ CVMO's or at least a number very close to this. The factor $6n$ has been interpreted to represent a situation where, for a large cluster, each metal atom contributes a maximum of 6 orbitals (one s and five d) simliar to that in the bulk metal. Despite the fact that this method seems apparently more grounded on quantum mechanical principles than other electron counting rules which are essentially based on topological arguments, it suffers from a number of limitations. The most serious one is that the energy levels of the bare cluster are considerably modified by their interactions with the ligands and therefore do not reflect the MO energy ordering of the ligated cluster (see below). Moreover, for large systems, the MO scheme tends to become a quasi-continuum whereby it is virtually impossible to establish a clear cut separation between bonding and antibonding orbitals.

2.4.3.5 Clusters Stabilized by s–s Interactions: A Unified View

Mingos has recently stressed the analogies between the TSH theory and the jellium model and has proposed a general bonding scheme to account for the stability of clusters stabilized by interactions between the metal s orbitals. [51, 235] This bonding model has the value of being valid for a number of very different classes of clusters such as naked, gas phase, alkali metal clusters or ligated, solid state gold-phosphine clusters. The requirement, however, is that the cluster structures are held together primarily by interactions between the outer s orbitals of the metal atoms. Indeed, this is the case for the clusters formed from the group Ia and Ib elements (Li, Na, K, Cu, Ag, Au, etc.), both with and without surrounding ligands.

In these clusters, the MO's are delocalized over the metal skeleton and one is reminded of the assumptions underlying the jellium model, namely that all electrons are moving freely over the whole cluster (see Section 2.3.2.2). Indeed, there is a close similarity between the jellium model and the TSH theory. [51, 235] In the TSH method, linear combinations of atomic orbitals, LCAO, are formed for clusters with a single shell of atoms where the signs and magnitudes of the coefficients are determined by the amplitudes of the spherical harmonics (*i.e.* by the wave functions of a free electron moving on a spherical surface). From a set of s atomic orbitals, all having local σ character, one can construct a set of cluster orbitals with different nodal properties as illustrated in Figure **2-23** for the special case of an octahedral cluster. These orbitals are denoted as S^σ, P^σ, and D^σ according to their nodal structures (Fig. **2-23**). In cases where only s orbitals are important for the bonding, which is true to a first approximation with alkali metal and gold-phosphine clusters, the TSH theory and the jellium model give the same qualitative energy spectrum.

The S^σ, P^σ, and D^σ subhells are filled by 2, 6, and 10 electrons respectively. Following an aufbau principle, one obtains closed shells of 2, 8, and 18 electrons. For these electron counts, in correspondence to the shell closing, pseudo-spherical clusters characterized by high stability and high ionization potentials are expected. Indeed, accurate *ab initio* calculations have shown that both Li_8 and Na_8 are pseudo-spherical tetracapped tetrahedra [236] whose S^σ (a_1) and P^σ (t_2) MO's accomodate the eight valence electrons of the clusters. The Li_8 and Na_8 clusters have greater stabilities than the adjacent members of the series containing 7 or 9 atoms. [236, 237]

Clusters with an insufficient number of electrons to completely fill the S^σ, P^σ, or D^σ shells are expected to adopt prolate (like a rugby ball) or oblate (like a discus) geometries. This is the case for Li_6 and Na_6 clusters, [69, 238] each with six valence electrons, which assume flat (planar triangle) or nearly flat (pentagonal pyramid) oblate structures instead of "spherical" octahedral geometries (Fig.

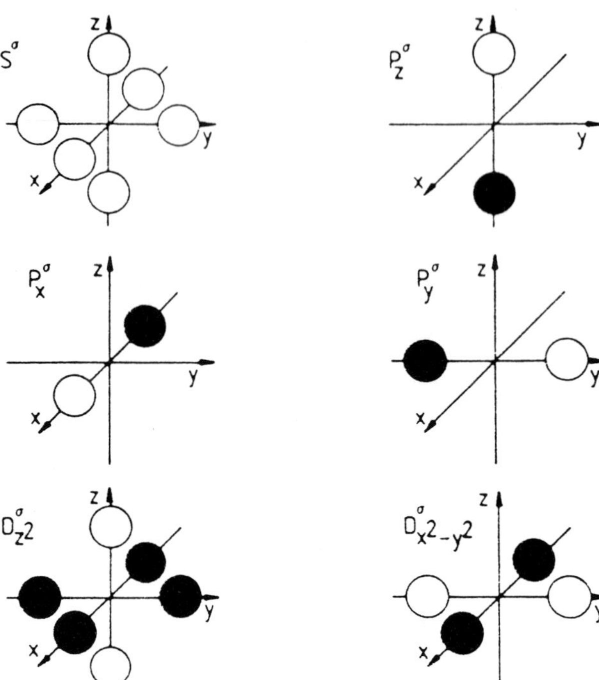

Figure 2-23. Nodal structures of cluster orbitals constructed from s atomic orbitals for an octahedral cluster. Reproduced with permission from [235]. Copyright 1990 American Chemical Society.

2-24). The change in geometry is accompanied by a splitting of the partially filled P^{σ} levels, an effect reminiscent of the Jahn-Teller distortion in molecular compounds. Mingos has extended this picture to organometallic $(Au{-}PR_3)_n$ clusters. [51] According to this model, the $(Au{-}PR_3)_n$ bonding is assumed to be dative in character and resulting from a donation of the lone pair on P to an empty sp hybrid on Au, whereby the single valence electron of Au is left in another sp hybrid of largely s character. The model ignores bonding contributions from the Au 5d electrons (not entirely negligible, according to recent relativistic calculations, *vide infra* [239]). The $[Au_6(PPh_3)_6]^{2+}$ cluster, with only four Au 6s valence electrons, has a skeleton based on two tetrahedra sharing a common edge (Fig. **2-25**). This roughly corresponds to a prolate geometry and a $(S^{\sigma})^2(P^{\sigma})^2$ configuration. In contrast, $[Au_7(PPh_3)_7]^+$, with six Au 6s valence electrons, assumes an oblate structure (a distorted pentagonal bipyramid, Fig. **2-25**) consistent with the $(S^{\sigma})^2(P^{\sigma})^4$ configuration. The same structure has been predicted for the isoelectronic Li_7^+ and Na_7^+ clusters, [17] giving support to the general nature of the model.

Thus, there is a close resemblance in the results for different types of clusters which are mainly stabilized by a s–s bonding. A rationalization for the observed geometric structures can be derived from either the TSH or the jellium model. This is a typical example for the isolobal analogy whereby the $[Au(PPh_3)]$ fragment is isolobal to an alkali metal atom. The electronic structure becomes much more complicated when the valence p (or d) orbitals are involved in the skeletal

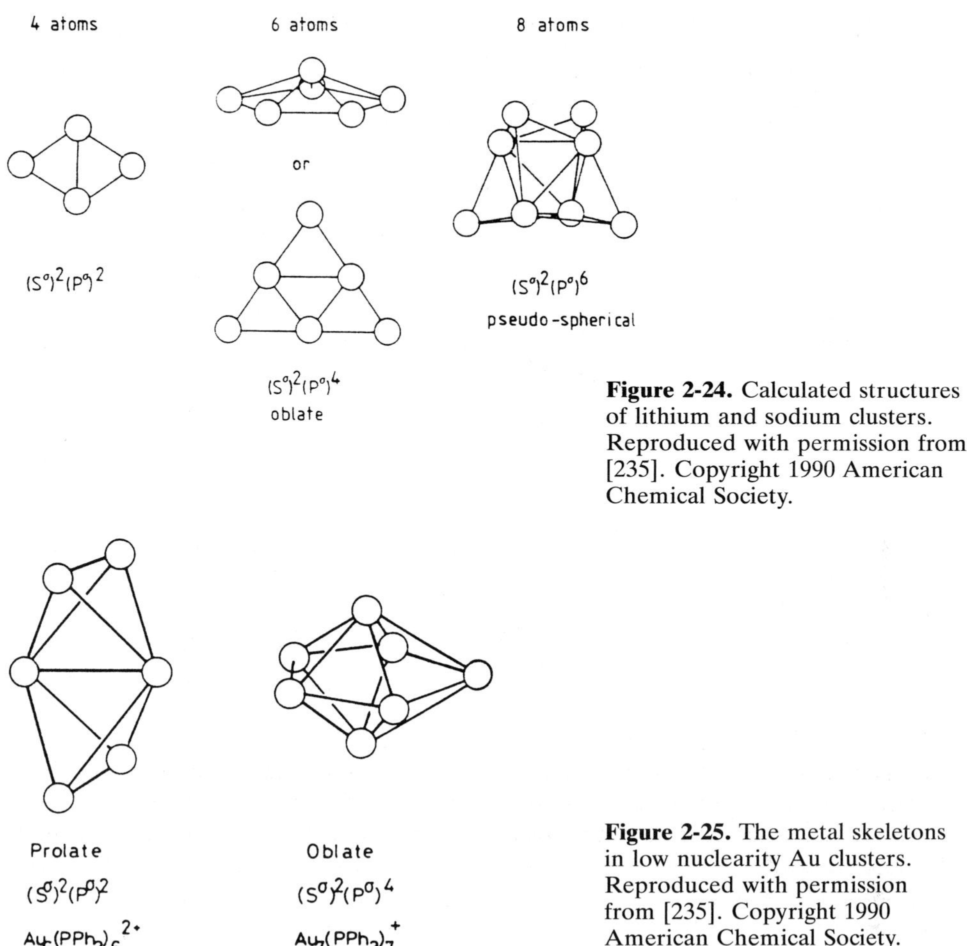

Figure 2-24. Calculated structures of lithium and sodium clusters. Reproduced with permission from [235]. Copyright 1990 American Chemical Society.

Figure 2-25. The metal skeletons in low nuclearity Au clusters. Reproduced with permission from [235]. Copyright 1990 American Chemical Society.

bonding. No simple free-electron type electronic shell structure has been observed for gas phase main group or transition metal clusters where the p and d orbitals participate directly in the bonding. The mixing, or hybridization, of s, p, and d orbitals under formation of more directional bonds greatly changes the nature of the potential and makes the details of the electronic structure more important than the shell closing requirement. An additional complication in organometallic clusters of transition metal atoms comes from the fact that the ligands are not "innocent" and may induce substantional changes in the position of the cluster energy levels and in the configuratiojn of the metal atoms. In these cases, a more rigorous theoretical treatment which explicitly takes into account the detailed nature of the interaction becomes necessary.

2.4.4 Quantum Chemistry of Organometallic Clusters: A Deeper Look into the Bonding

2.4.4.1 Semiempirical Calculations

Simple models are quite valuable as tools for rationalizing the cluster structures and eventually as an aid for designing new synthetic routes to cluster species. On the other hand, they provide only a rather limited insight into the real nature of the chemical bonding in transition metal clusters. For this reason, several approximate quantum chemical calculations have been reported since the pioneering work of Brown on $Fe_3(CO)_{12}$. [240] Most of these studies have been based on the LCAO approach within the extended Hückel (EH), [241–250] CNDO or INDO, [69, 251–253] Fenske-Hall, [254] etc. approximations (see also [255] and [256]). These investigations provide a more grounded analysis of the bonding in terms of MO energy levels and, in some cases, of energy considerations. In particular, qualitative correlations have been established between bond orders and experimental bond distances, electronic transitions and optical spectra, one-electron energy levels and photoelectron spectra, and, in a few cases, spin or charge distributions and EPR data and redox potentials. These studies have shown that, in general, although the s and p character in the bonding is not large, it is nonetheless essential since the d orbitals are better suited for metal–metal and metal–ligand bonding if s and p orbitals are mixed in. Along a series, the strength of the metal–metal bond is found to be correlated with the atomic number due to the larger extension of the d–d overlap in compounds of heavier TM elements. [257] Semiempirical methods, and the EH approach in particular, have been extremely valuable in clarifying the bonding mechanisms in several classes of compounds such as columnar clusters, [245] clusters with interstitial atoms, [248, 250] and clusters stabilized by π-acceptor ligands [69, 252] or bonded by main group atoms. [249, 258] In addition, they have also provided a link between the molecular and the solid state views. [213, 245, 249]

2.4.4.2 Limitations of Simplified Approaches

The combination of semiempirical molecular orbital calculations and symmetry arguments has led to some important generalizations which relate the observed structures to elementary rules. However, semiempirical approaches suffer from the typical deficiencies well known for these methods and, in particular, the dependence of their results on the choice of parameters. While semiempirical methods can be very successful in reproducing a particular cluster property, they usually fail in giving a simultaneous correct description for the full range of aspects of the cluster electronic structure (energetics, geometry, vibrations, position and width of the metal bands, etc.). Moreover, a serious drawback to the simplified methods is that most of them are based on the nodal, or symmetry, properties of the individual molecular orbitals of a cluster and, hence, to their bonding, nonbonding, or antibonding character. Bonding in clusters, however, is of a rather delocalized nature and can hardly be described in terms of the properties of single orbitals or of HOMO-LUMO concepts. This is particularly true for high nuclearity clusters where the bonding is no longer molecular and starts to resemble that of the bulk metal. Thus, when one aimes at understanding the electronic structures, charge distributions, and reactivities of high nuclearity clusters, the use of a "first principle" method becomes mandatory. To this end, more well grounded theoretical approaches, based on the chemical pseudopotential, [259–261] the Xα, [150, 248, 262–265] or the local density approximations [168, 206, 266, 267] have been employed.

Although high nuclearity molecular clusters containing a few tens of metal atoms are still molecular objects exhibiting a discrete energy spectrum, the energy spacings between the levels and the number of orbital degeneracies, or near degeneracies, become so high that it is very cumbersome to specify the position and the symmetry character of each individual orbital. It is more practical to develop a common language which can be used when describing small as well as very large clusters or even small colloidal particles. In this way, the non-metal to metal transition can be followed by the change in the cluster electronic structure. This language makes use of the concept of density of states, DOS, which gives the number of states within a given energy interval. DOS curves are commonly used when describing the electronic structures of solids [268] where, in case of perfect translational symmetry, they are commonly derived from the band structure. One may also construct them in an approximate fashion by convoluting the discrete one-electron energy levels of a cluster with a Gaussian function of fixed width, that is, by "broadening" each level and by summing up all the resulting Gaussians. This "computational" broadening underlies the fractional occupation number [FON] technique [132] used in DFT calculations and ensures convergence in cases of near degeneracy, a quite common situation in metal clusters. Whereas in those cases the broadening is a computationally motivated device, it is used here to generate a convenient representation of the cluster level spectrum. Thus, it provides a natural transition to a quasi-continuum as the cluster size increases and thereby approaches the metallic regime. As an illustrative example, the concept of DOS will be used in the next section to describe the bonding in $[Ni_5(CO)_{12}]$.

2.4.4.3 Bare Versus Ligated Clusters: The Effect of the Ligands

It is useful to start the description of the electronic structure of a ligated metal cluster by separately considering the bare metal atom cluster, with the exact same geometry as the metal core in the ligated cluster, and the ligand polyhedron, as an empty spherical cavity, and then finally allowing the two parts to interact (Fig. **2-26**).

The schematic representation of the interaction of a naked metal cluster with its ligand shell allows one to analyze in detail the electronic interaction between the two fragments in terms of a) the relative positions of the metal and the ligand

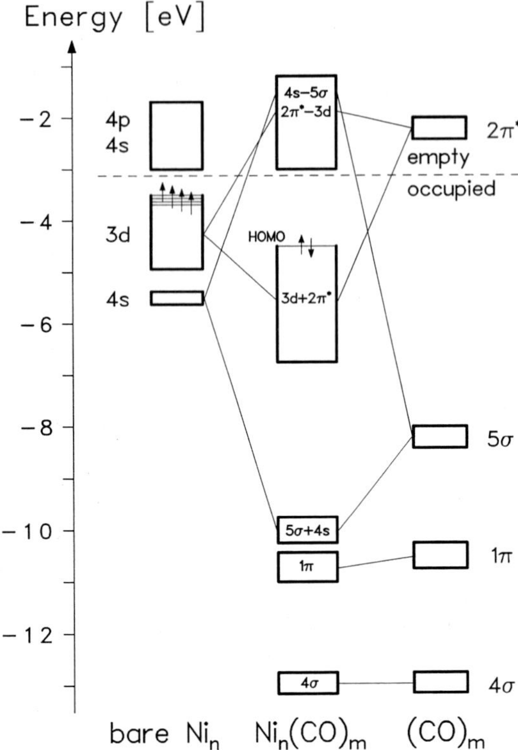

Figure 2-26. Schematic orbital interaction diagram for the interaction of a Ni_5 core with a $[(CO)_{12}]$ ligand polyhedron to form the $[Ni_5(CO)_{12}]$ cluster. Reproduced from [269].

energy levels, b) the extent and direction of the charge transfer, and c) the configuration change and rehybridization of the metal atoms. In principle, the analysis can be carried out on a qualitative basis by simply considering the symmetry properties of the fragments, or on a more quantitative basis by also taking into account the relative importance of the various bonding mechanisms. This latter approach is a necessary prerequisite for the interpretation of most of the physical measurements performed on cluster compounds (see Section 2.4.5).

For small clusters, which still must be considered as molecular aggregates, the nodal characteristics of the individual orbitals play a relatively important role in determining the stability of any given structure. This has been shown for the $[Ni_5(CO)_{12}]$ cluster (Fig. **2-27**) [266, 269] by considering the interaction of the naked trigonal bipyramidal Ni_5 cluster with its ligand shell. The DOS curves for this cluster were determined by all-electron LDF calculations and are displayed in Figure **2-28**. [266]

In the bare Ni_5 cluster, the 4s orbitals of the Ni atoms form bonding, nonbonding, and antibonding combinations with a slight mixing in of the 4p orbitals. The total symmetric bonding combination of the 4s atomic orbitals, also called the S^{σ} orbital in TSH theory (Section 2.4.3.5), is a very stable MO lying below the 3d band in energy (Fig. **2-28**). Strictly speaking, the term "band" refers to the energy bands generated from perfectly periodic structures. In cluster theory,

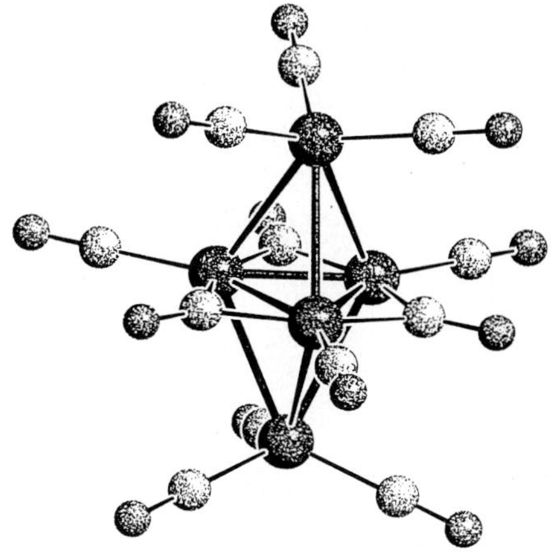

Figure 2-27. Structure of the $[Ni_5(CO)_{12}]$ cluster.

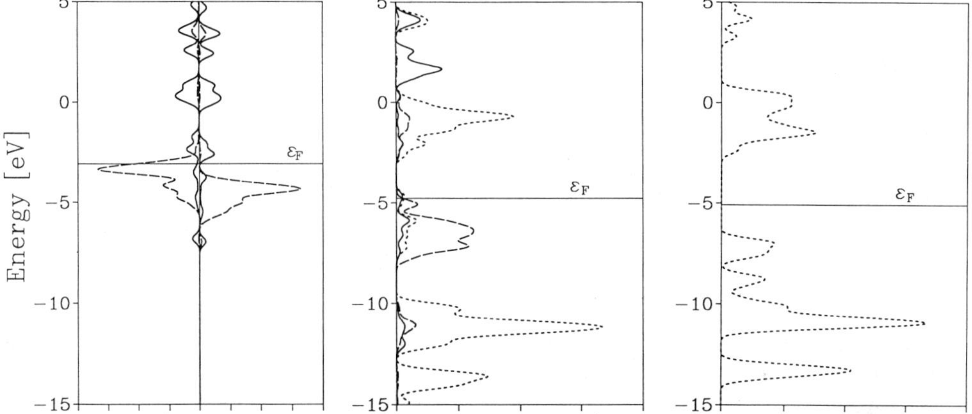

Figure 2-28. Density of states for Ni_5 (left side), $[Ni_5(CO)_{12}]$ (center), and $[(CO)_{12}]$ (right side); – – – Ni 3d contribution; ——— Ni 4s contribution; · · · · · · CO contribution. For Ni_5 both majority and minority spin-orbital manifolds are given. Reproduced from [269].

however, it has become common practice to also use this term to designate submanifolds of a DOS, and so we will occasionally employ this loose terminology when there is no danger of confusion. The S^{σ} orbital makes a large contribution to the stability of the bare Ni_5 cluster. The 3d band in Ni_5 extends about 2 eV below the Fermi level. While the majority spin component of its DOS curve is completely filled and well below the Fermi level, the minority spin manifold is only partially filled and gives rise to a magnetic ground state for the Ni_5 cluster. In other words, there is a different number of α and β spins in the 3d shell. This is a consequence of the fact that in Ni_5, as well as in other small Ni clusters, the bonding occurs with Ni atoms in a $3d^94s^1$ configuration, rather similar to that of the bulk metal. The magnetic behavior of both bulk Ni and small Ni clusters is due to the fact that the 3d orbitals are not completely filled. In the metal, this results in about 0.6 unpaired electrons per Ni atom, [270] while in Ni_5 there is a total of five unpaired electrons, one unpaired electron per Ni atom. [266] The difference in magnetization between Ni_5 and bulk Ni is related to the different average coordination and to different Ni–Ni distances in the two systems (see also Section 2.3.6.1).

As the group of 12 CO ligands is brought together to form the polyhedral shell found in $[Ni_5(CO)_{12}]$, the individual valence orbitals of the isolated CO molecule (1π, 5σ, 2π, etc.) interact with one another and broaden out into narrow "bands" of levels. This is shown in the DOS profile of the CO_{12} polyhedreon (Fig. **2-28**). The 5σ orbitals, which point toward the center of the structure, and the 2π orbitals, which are tangential to the cluster "surface", form bands of about 1–2 eV in width. Whereas the center of gravity of the ligands shell's 5σ band is well below both the 3d band and the 4s levels of the Ni_5 cluster, the 2π levels lie just above the Fermi level of Ni_5. When the Ni_5 and $(CO)_{12}$ fragments are allowed to interact, the most significant changes occur in the DOS profile of the metal core (Fig. **2-28**). The structure and position of the CO 5s and 1π bands in fact does not change dramatically, while the position and character of the metal orbitals undergo a substantial change. The 3d band in $Ni_5(CO)_{12}$ is stabilized with respect to that of Ni_5 but, much more important, the 4s derived MO's are destabilized by their interaction with the CO ligands and pushed above the Fermi level (see also Fig. **2-26**). As a consequence, the 4s electrons are promoted into the 3d shell which becomes fully occupied. This results in a complete quenching of the Ni_5 magnetic moment: the majority and minority spin components of the 3d band are equally populated and the fully carbonylated cluster is diamagnetic, in contrast to the naked Ni_5 cluster which has a high spin ground state. This mechanism is schematically illustrated in Figure **2-26**. The basic concept here is that the interaction of the CO 5σ orbitals with the diffuse 4s derived MO's of the Ni core is purely repulsive as already described in Section 2.4.1.

Energetically, it is more convenient to formally change the average Ni configuration from $3d^94s^1$ to $3d^{10}$ (in the free atom the two configurations are separated by about 1.82 eV [160]) since the $3d^{10}$ Ni atoms are then able to bind strongly with the CO ligands through the classical π-back donation mechanism. A Mulliken analysis [269] showed that the actual population of the Ni 3d orbitals is less than 10 electrons as a consequence of this charge transfer and that there is a mixing in of the 4s and 4p orbitals. The important feature here is that the 4s orbital bonding combination is no longer occupied in the ligated cluster and therefore cannot contribute to the stability of the metal frame in $[Ni_5(CO)_{12}]$.

2.4.4.4 Ligand-Field Effects in Clusters

The destabilization of the 4s derived MO in small Ni clusters caused by the interaction with the CO ligands, and the consequent change in magnetization, is reminiscent of the high-spin to low-spin transition in TM complexes which results from the "field" created by the surrounding ligands. In ligand field theory, if the d orbital splitting, Δ, exceeds the energy required for electron pairing, then a low-spin complex results. [271] The analogy with molecular metal clusters is apparent. If the number and the nature of the ligands destabilize the 4sp derived orbitals to such an extent that a 4s $\rightarrow$ 3d transition occurs, then the resulting molecular cluster will have a diamagnetic or only weakly magnetic ground state. It is difficult to establish a scale for the strenght of the field generated by the ligands analogous to the one known for TM complexes [272] since only a small variety of ligands bond to zero valent metal clusters (CO, PR_3, CNR, and olefins). However, as a general rule, "strong field" ligands (*i.e.* ligands which are most effective in quenching the magnetic moment of the bare metal cluster) are those which give rise to a strong repulsive interaction in the σ space and generally include the π-acceptor molecules. [158] Finally, not only the nature of the ligand, but also its position (terminal, bridge bonded, face bonded, etc.) can have an effect on the splitting of the levels and on the magnetic quenching. It has been shown that bridging CO ligands are more efficient than terminal ones in reducing the magnetic moment of the Ni atoms. [266]

Another important feature regarding the presence of the ligand shell is the appearance of a sizable energy gap at the Fermi level E_F (or the HOMO-LUMO gap in MO language). Whereas in small metal clusters the DOS curves cross the Fermi level so that there is virtually no gap (a typical sign of metallic character), in the carbonylated form the gap is about 1–2 eV. [265, 266] In an extended system, this change would correspond to a metal to semiconductor transition. The size of the gap is also a function of the cluster nuclearity, and decreases as the cluster gets larger. This aspect will be discussed further below in connection with the interpretation of some experimental data (see Section 2.4.5.3)

2.4.4.5 The Strength of Metal–Metal bonds

The bonding mechanism described above is quite general and, indeed, similar effects have also been found with Co clusters. [265] Furthermore, its generality is indirectly supported by the fact that virtually all smaller sized carbonylated clusters do not exhibit the magnetic properties which are typical of the corresponding bulk metals (*e.g.* Fe, Co, Ni). Clearly, this bonding mechanism requires the presence of holes in the d shell of the metal atoms in order to accomodate the electrons from the $(n + 1)$ s orbitals, which thus entails the corresponding change in the metal configuration.

The situation is quite different, in this respect, for clusters of the coin metals Cu, Ag, or Au (*e.g.* in $[Au_6(PR_3)_6]^{2+}$), where the metal atomic valence configuration is $5d^{10}6s^1$ and the d shell is complete. In these clusters, a substantial contribution to the metal–metal bonding arises from the 6s electron interactions, as already discussed in Section 2.4.3.5. Recently, however, it was demostrated that the 5d orbitals also contribute to the metal–metal bonding, [239] a result which is related to the relativistic contraction of the 6s shell and the expansion of the 5d shell, which consequently enables an increased s-d hybridization. This leads to an opening of the formally closed 5d shell and results in an enhancement of the tangential Au–Au bonding. An important contribution to this rehybridization comes from the interactions with the phosphine ligands. [273, 274]

This brings up the questions regarding the metal–metal bond strengths in ligated metal clusters. These have been one of the main concerns of theoreticians ever since the first theoretical investigations on metal cluster compounds were performed. [246, 275–279] For a long time it was assumed that the cluster-surface analogy is a direct consequence of the similarity between the nature of the metal–metal bonding in organometallic clusters and on metal surfaces. [196] Qualitative theoretical analysis in favor of this argument have been presented. However, it is evident from recent, more well grounded theoretical investigations that the large changes induced by the ligand shell on the metal cluster also involve deep modifications in the metal–metal interactions and thus in the metal–metal bond strengths. [266, 267, 273] For instance, accurate *ab initio* calculations have shown that a CO molecule bridge-bonded to a Ni_2 molecule causes the Ni–Ni bond to break. [280] In an analogous way, it was concluded that there is no direct M–M bonding in either $Fe_2(CO)_9$ [281] nor $Co_2(CO)_8$. [282]

Although it is very difficult to determine the strength of a bond within such a complex system as a cluster, some attempts have been reported which were based on overlap populations, [245] on the partitioning of the total energy into two-body contributions, [279] or on electron density differences. [266] However, since the $(n+1)$ s metal bonding orbitals, which play a fundamental role in the stabilization of the naked metal clusters, are unoccupied due to their destabilization from the interactions with the ligands, some weakening in the metal–metal bonding may occur. The (n) d shell is, in fact, rather contracted and the d–d overlap may not be sufficient to restore the bond strength originally found for the naked metal particle. Clearly, the nature of the metal–metal interaction may differ significantly as one goes from the left to the right side of the periodic table and from the first to the third transition metal series. However, all metal atoms are on the cluster "surface" in small clusters, and hence interact directly with the ligand shell. Therefore, the idea seems plausible that the metal electronic structure may differ significantly from that of the metal atoms in a bulk crystal. This hypothesis is supported by a large number of experimental facts (see Section 2.4.5).

2.4.4.6 Clusters with Interstitial Atoms

While many organometallic clusters with interstitial atoms are known, the detailed nature of the bonding of these interstitial atoms, usually H, B, C, N, P, Si, Ge, etc., with the metal atoms of the cluster has been addressed in very few theoretical studies. [239, 248, 250, 283] In a comparative relativistic DV-Xα investigation [239, 284] on a series of octahedral cluster ions $\{[(R_3P)Au]_6X_m\}^{m+}$, where X_1 = B, X_2 = C, and X_3 = N, it was shown that the central atom formally takes up electrons that would otherwise reside in energetically unfavorable molecular orbitals. In this way, the central atom contributes to the stabilization of the cluster by forming radial bonds to the Au atoms. This has also been found in quasi-relativistic pseudopotential calculations on ligand free, element centered gold clusters. [285]

The experimentally observed coordination number of carbon in $\{[(R_3P)Au]_6C\}^{2+}$ is unexpected, but may be rationalized by the attractive Au–Au interactions. The term "aurophilicity" has been coined for this phenomenon. [286] Recent calculations [283] using a scalar-relativistic extension of the LCGTO-LDF method [287] have shed some light on the question of why only the carbon centered cluster has so far been synthesized and not the analogous compounds with interstitial boron and nitrogen atoms. Three mechanisms have to be taken into account. i) The favorable radial interaction between the gold cage and the central atom decreases along the series B, C, N due to the reduced Au–X overlap. Thus, for the ligand free case, the boron centered cluster cation is the most stable one. ii) The ligand shell has a pronounced stabilizing effect [274] which increases with the positive charge of the cluster. iii) The destabilizing effect of a net positive charge on the cluster, even stronger in the ligand free models, is largest for the N-centered compound. The global effect of these mechanisms is a stability maximum for the C-centered cluster. [283]

In a LCGTO-LDF study on the high nuclearity $[Ni_{32}(C)_6(CO)_{36}]^{n-}$ cluster model (Fig. **2-29**) [267] it was found that the interstitial C atoms give a large contribution to the stability of the structure, but also induce a partial quenching of the total magnetic moment of the bare Ni_{32} cluster. The 26 unpaired electrons in Ni_{32} are reduced to only 10 in the $Ni_{32}C_6$ unit from the coupling of the C 2sp electrons with the metals' unpaired electrons. When the CO ligands are finally added, the $[Ni_{32}(C)_6(CO)_{36}]^{n-}$ cluster becomes diamagnetic. On the other hand, when the CO ligands are first added to the Ni_{32} unit to form $[Ni_{32}(CO)_{36}]^{n-}$ (*i. e.* without interstitial C atoms), a small magnetic moment remains. This suggests that interstitial atoms may significantly contribute to changing the magnetic behavior of very large clusters or colloidal particles.

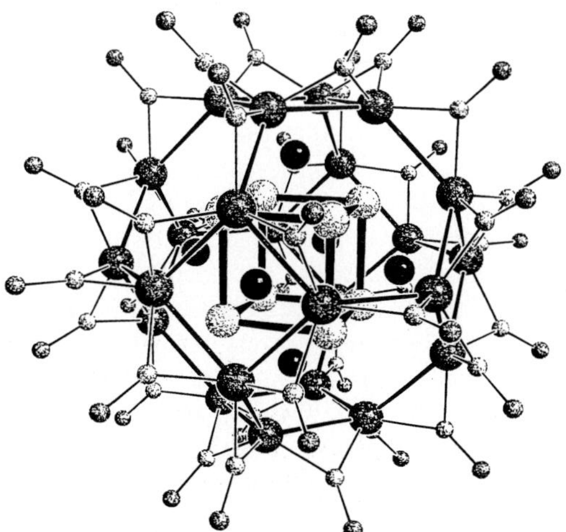

Figure 2-29.
The structure of $[Ni_{32}(C)_6(CO)_{36}]^{n-}$.

2.4.5 Physical Measurements and Chemical Bonding

In recent years, a large number of chemical and physical techniques have been used to better characterize the electronic properties of ligated metal clusters. Information from X-ray diffraction, still the most widely used and probably one of the most powerful methods to investigate the cluster structures, has been supplemented by electron microscopy, optical spectroscopies (infrared, Raman, ultraviolet), nuclear magnetic and electron spin resonances, magnetic susceptibility and electrical conductivity measurements, direct and inverse photoemission, X-ray absorption (EXAFS), redox measurements, Mössbauer spectroscopy, etc. [288, 289] In this section, we juxtapose the main features which can be identified by each of these techniques with the theoretical models for cluster bonding outlined in the previous sections. A general, unified picture of the electronic structure and bonding in organometallic clusters will emerge.

2.4.5.1 Photoelectron Spectroscopy

Photoelectron spectroscopy provides a rather direct way to characterize the electronic structure of metal cluster compounds. [290] Depending on whether the wavelength of the incident photon is in the X-ray (XPS) or ultra-violet (UPS) regimes, either core or valence electron binding energies, BE's, can be measured. The shift of the core (or valence) level BE of a given atom is influenced by the surroundings of this atom and provides a local probe of the electronic structure of the compound, thus allowing to distinguish the presence of atoms in different chemical environments. For instance, the O 1s spectrum of $Co_4(CO)_{12}$ shows a broad peak resulting from the presence of both terminal and bridging CO ligands in the cluster. [291] The ratio between the two groups of ligands is 3:1 and the experimental spectrum can be deconvoluted and resolved into two components which integrate to approximately this ratio. The fact that the BE's of the terminal CO ligands are about 1 eV higher than those of the bridging carbonyl groups is in accord with the expectation that a stronger back donation occurs with the bridging CO ligands which induces a better screening of the O 1s core levels.

Both conventional and synchrotron radiation core level photoemission spectroscopy have been used to study the evolution from molecular clusters to metallic particles or the interaction of the clusters with inorganic supports. [292, 293] The 4f and 4d core level BE's for a series of layered Pt cluster anions, $[Pt_3(CO)_6]_n^{2-}$ (n = 2, 3, 4, 5, 6, and about 10), increase regularly with increasing cluster size and approach the asymptotic value at about n = 5 (Fig. **2-30**). [292] The 4f and 4d BE's for Pt clusters, however, are always about 2 eV larger than the corresponding values for Pt metal (Fig. **2-30**). This has been rationalized by the acceptor properties of the CO ligands which reduce the electron density on the metal, and by the different extra-atomic relaxation of the cluster compared to the metal. This reasoning also holds for the high nuclearity Pt clusters with n = 10 since the individual $[Pt_3(CO)_6]_n^{2-}$ units interact only weakly. It is worth noting that an analogous increase (about 1 eV) in the core level BE's of the surface

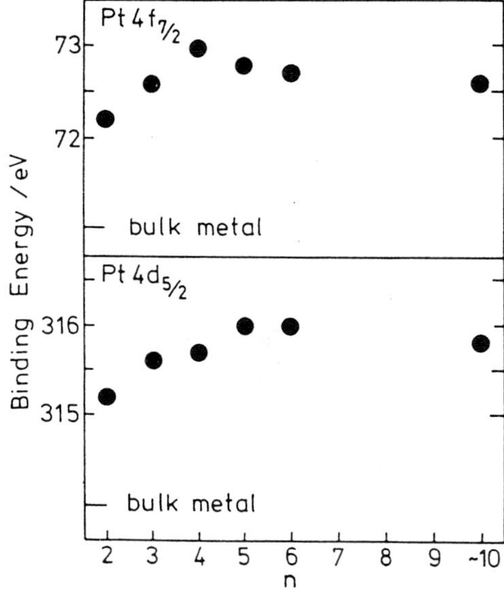

Figure 2-30. Pt 4f and 4d core level binding energies as a function of the cluster size for Pt clusters of the general formula $[Pt_3(CO)_6]_n^{2-}$. Adapted from [292].

metal atoms of a covered Pd(110) or Pd(100) surface have been measured by synchrotron radiation, [294] further supporting the notion for the local nature of the metal–CO interaction. Thus, XPS provides clear proof that a substantial charge donation to the ligands occurs in metal clusters.

More valuable for the understanding of the electronic structure in metal cluster compounds are the UPS spectra. It has been shown that the first ionization for the $[Pt_3(CO)_6]_n^{2-}$ cluster occurs from the d band. [295] Moreover, the split in the ionization spectra is clearly indicative of two features below the Fermi level (Fig. **2-31**) and suggests that there should be a non-uniform distribution of the d levels which roughly corresponds to a bonding/antibonding division. [295]

The UPS spectra for these clusters reveal the general appearance of a 3–4 eV wide metal d "band" lying immediately below the Fermi level which is separated by 2–3 eV from an intense peak due to the CO 5σ and 1π orbitals. A second intense peak at 3–4 eV below the 5σ and 1π features is due to the CO 4σ levels. These spectra exhibit a remarkable qualitative similarity to the corresponding spectra of CO adsorbed on metal surfaces. More important for the present discussion is that they can be directly compared with the ionization energies obtained from theoretical calculations. For instance, the calculated spectra for a series of tetracobalt carbonyl clusters using the DV-Xα method have been compared with the He(I) UPS spectrum of $[Co_4(CO)_{12}]$. [150] The transition state procedure [41] used in these calculations also accounted for the final state effects, that is, for the effects of orbital relaxation following the ionization process. This relaxation is typically about 2 eV for the valence orbitals of transition metal carbonyls and the calculated IP's compare well with the experimental ones. In a similar way, ruthenium, osmium, and rhenium carbonyl cluster UPS spectra have been calculated and reproduce the general trends and positions of the

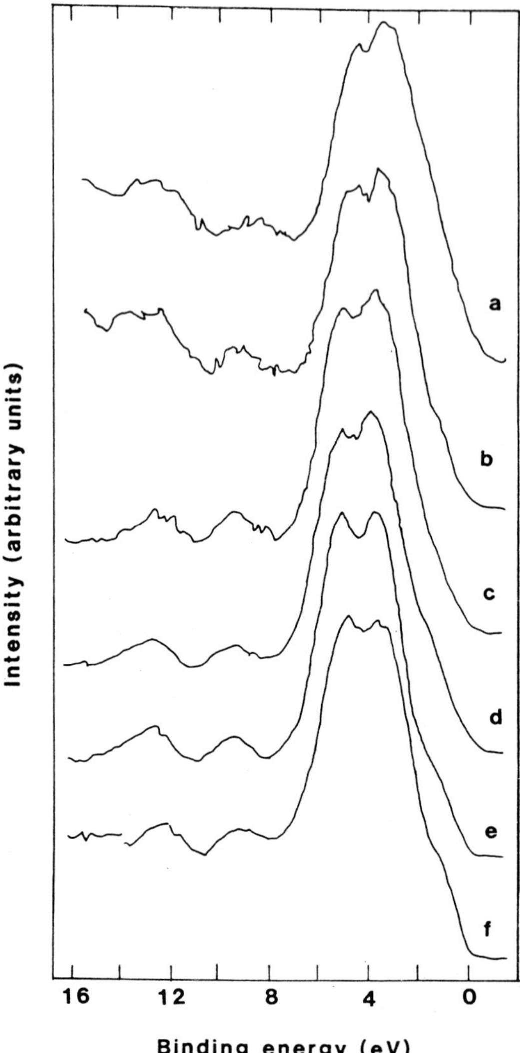

Figure 2-31. Valence photoelectron spectra of $[Pt_3(CO)_6]_n^{2-}$ clusters. a) $n = 2$, b) $n = 3$, c) $n = 4$, d) $n = 5$, e) $n = 6$, and f) $n = 10$. Reproduced with permission from [295].

experimental peaks. [255, 262] In particular, the broad spectral features found for the d ionization band are in line with the delocalized nature of the cluster bonding.

A direct comparison of the computed DOS curves with the UPS spectrum of a cluster has no theoretical justification since the former does not take into account orbital relaxation and many-body effects. However, this is not too serious a limitation given the relatively small relaxation and the qualitative nature of the comparison. With this in mind, the DOS curves and the UPS spectra of osmium carbonyl clusters have been compared. [261, 296] The positions of the computed energy levels correlate well with the structure of the UPS spectrum (Fig. **2-32**).

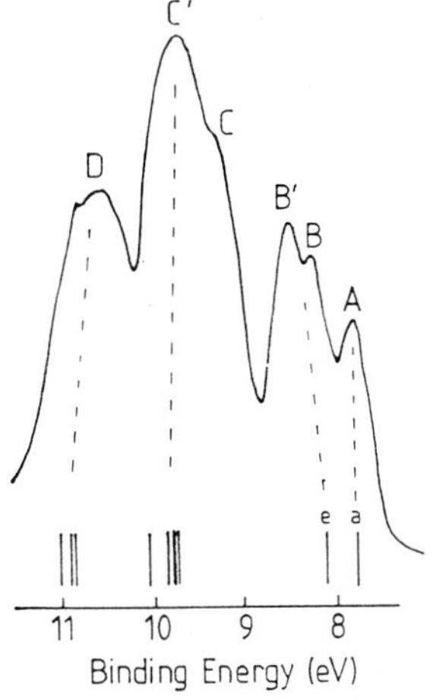

Figure 2-32. Comparison of the valence photoelectron spectrum with the computed energy levels for the $[Os_3(CO)_{12}]$ clusters. Reproduced with permission from [261].

Given the inherent limitations of the method, the computed DOS curves reproduce the features of the valence structure of the clusters to an acceptable accuracy. For instance, the split in the d levels observed in layered Pt clusters (Fig. **2-31**) can also be seen in the theoretical DOS curves of analogous layered Ni clusters (Fig. **2-33**). [266] It should be noted, however, that this good agreement between theory and experiment has only been obtained with rather sophisticated computational methods like the local density approximation or its variants. Other methods, in particular the EH approach, may fail to reproduce the width, and sometimes the position, of the metal s and d bands.

Valence photoelectron spectroscopy has also been used to follow the evolution of the electronic structure as the size of the metal cluster increases. For instance, the Au 5d binding energies of $[Au_{11}(PPh_3)_7Cl_3]$, $[Au_{55}(PPh_3)_{12}Cl_6]$ and Au metal

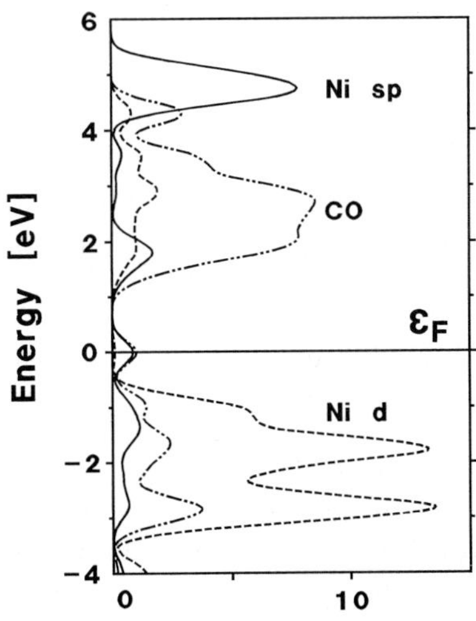

Figure 2-33. Calculated DOS curves for the $[Ni_6(CO)_{12}]^{2-}$ cluster. Reproduced with permission from [266]. Copyright 1990 American Chemical Society.

exhibit a surprising similarity (see Chapter 3, Fig. **3-41**). [297, 298] The major difference is a shift toward higher binding energies and a modest narrowing of the 5d band as the size of the cluster decreases. Even in Au_{11}, however, the 5d band shows a substantial broadening. This suggests that a considerable d–d overlap occurs between neighboring Au atoms, in agreement with recent DV-Xα relativistic calculations on octahedral Au_6 clusters. [239]

2.4.5.2 Optical Spectroscopy

Infrared, IR, [299] and ultraviolet, UV, [300] spectroscopic experiments have been widely used to characterize organometallic clusters. In particular, because of the extreme experimental simplicity and the high sensitivity in the carbonyl as well as in other ligand regions, IR spectroscopy plays a very important role in the study of these compounds. Only a few reports of low frequency Raman spectra dealing with metal–metal vibrations have appeared. [301] In $[Os_3(CO)_{12}]$ and $[Ru_3(CO)_{12}]$, Raman bands were observed at frequencies lower than 200 cm^{-1} [262, 302] and attributed to metal–metal stretching vibrations. In a detailed analysis of this low frequency region, the cluster was described as a "plastic" unit in which a change in one metal–metal bond length has only a small effect on the energy required to change another metal–metal bond. [303, 304]

Most molecular metal clusters are colored and show rich UV-visible spectra. Although these spectra usually exhibit pronounced peaks, they are also very broad. This is the result of the large number of dipole allowed and vibronically allowed electronic transitions occuring when light interacts with a cluster of even moderate size. This makes any theoretical attempt to interprete the spectrum extremely complex. In principle, excitation energies must be computed as the difference between the total energies of two electronic states, the initial one and the final one, whereby the change in the electron–electron interactions or, in a one-electron framework, the orbital relaxation effects which occur upon excitation are accounted for. This would require a huge number of calculations, even larger than that necessary for the interpretation of photoemission spectra. Nevertheless, electronic transitions in molecular clusters have been computed and compared to the corresponding optical spectra. [266, 305] The method usually followed for determining the excitation energies consists of taking the differences in the ground state one-electron energies involved in the electronic transitions. Although this procedure is certainly a very gross approximation, it has been applied with some success when combined with the local density approximation in the interpretation of the optical spectra of other inorganic compounds. Experience with Slater's transition state method for excitation energies shows that orbital energy differences provide a rough first approximation if the spatial characteristics of the two orbitals involved are similar.

Following this procedure, the UV spectra of carbonylated Ni (Fig. **2-34**), Pt, and Os clusters of various sizes have been interpreted with acceptable accuracy, considering the limitations of the theoretical approach. [261, 266, 295] The lowest dipole allowed transitions were calculated to be around 2–3 eV (about

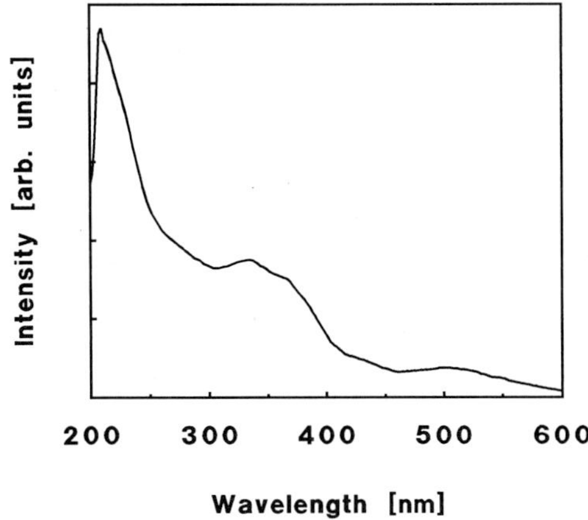

Figure 2-34. Experimental UV-visible spectrum of $[Ni_6(CO)_{12}]^{2-}$. Reproduced with permission from [266]. Copyright 1990 American Chemical Society.

400–600 nm), where the compound just starts to adsorb. These transitions involve the cluster metal orbitals around the top of the d band, near the Fermi level, and the empty CO 2π MO's. A second group of intense transitions occurs at higher energy, around 5 eV (about 200 nm, Fig. **2-34**) and involves metal orbitals at the bottom of the d band. Accordingly, it is possible to assume a rough division of the d band into two components, one with metal–metal bonding (M–M) and one with metal–metal antibonding (M–M*) character, in agreement with both the photoemission experiments (Fig. **2-31**), [292] and the theoretical DOS curves (Fig. **2-33**). [266] The region of the optical spectrum between 2 and 4 eV is assigned mainly to excitations from the top of the d band into the CO 2π MO's, while the transitions involving the M–M bonding orbitals in the lower part of the d band and the empty 2π levels may be responsible for the strong features occuring between 4 and 6 eV. Finally, it is worth noting that a significant number of transitions from the top of the (n)d band into the empty $(n+1)$s and $(n+1)$p levels should occur immediately above this region of the spectrum. These transitions, however, are not expected to carry much intensity since they involve much smaller changes in the electronic distribution, and hence in the dipole moment, compared to the d $\rightarrow$ CO 2π excitations.

2.4.5.3 Magnetic Susceptibility Measurements

The temperature dependence of the magnetic susceptibility of large carbonyl clusters like $[Pt_{38}(CO)_{44}H_2]^{2-}$, [306] $[Ni_{34}(C)_4H(CO)_{38}]^{5-}$, [307] $[Ni_{38}Pt_6(CO)_{44}]^{6-}$, [307] and $[Au_{55}(PPh_3)_{12}Cl_6]$, [308] as well as small carbonyl clusters like $[Ru_6(CO)_{18}]^{2-}$, [309] $[Pt_6(CO)_{12}]^{2-}$, [306] $[Ni_9C(CO)_{17}]^{2-}$, [307] and $[Os_{10}H_2C(CO)_{24}]$, [310] have been determined. In all cases, it is possible to describe the temperature dependence of the susceptibility with the equation:

$$\chi(T) = \chi_0 + C_0/(T-\theta).$$

Here, χ_0 is the temperature independent negative diamagnetic susceptibility, and the second term accounts for the Curie-Weiss law which dominates at low temperatures, typically < 100 K. The values found for the Curie-Weiss temperature θ are generally rather low, < 10 K. The effective magnetic moments deduced from the Curie constant, C_0, are very small (0.1–0.5 μ_B per cluster) for clusters composed of nonmagnetic metals (Pt, Au, Rh, Ru, etc.) and for Ni clusters containing up to 44 metal atoms (about 1 μ_B/clusters or less). This magnetization is much smaller than that found in bulk Ni where the magnetic moment is about 0.6 μ_B/atom. [270] Assuming a similar magnetization for the large Ni clusters, one should expect a total of about 20 μ_B/cluster for the Ni_{34} and $Ni_{38}Pt_6$ clusters. Similar results have been obtained for other metal clusters having the general formula $[M_{55}L_{12}Cl_x]$.

It has only recently been possible to formulate a general theory to explain the observed behavior. This theoretical model is fully consistent with the bonding mechanism described in the previous section, and in particular with the perturbation induced by the ligands on the metal atoms of a cluster and the resultant modification in the cluster electronic structure. We already mentioned (Section 2.4.4) that the ligand interactions split the metal orbitals in a somewhat analogous way as the crystal field effects in mononuclear transition metal complexes. The ligand shell induces a strong charge redistribution inside the cluster core which leads to a splitting of the levels and to a pairing of the spins. The overall result is that low nuclearity clusters composed of magnetic metals are expected to have low-spin ground states, in contrast to their naked counterparts. This has been theoretically demonstrated for small Co and Ni clusters. [253, 265, 266] When the cluster size is small enough so that all the metal atoms are on the "surface" and the cluster is coordinatively saturated (*i.e.* the ligand polyhedron is complete), then there is no magnetic moment (Table **2-5**). However, if the Ni or Co cluster has some metal atoms in a low coordinative situation (*i.e.* if some ligands have been removed), then a residual magnetization is found and the unpaired spin density is entirely localized on these "exposed" atoms (Fig. **2-35**). [253, 265]

Table 2-5. Distribution of unpaired electrons in free and carbonylated Ni clusters as derived from local spin density calculations [266, 267].

	total number of unpaired electrons	unpaired electrons/atom		ground state
		surface atoms	bulk atoms	
Ni_5	4.2	0.84	–	magnetic
$[Ni_5(CO)_{12}]$	0.0	0.00	–	diamagnetic
Ni_{44}	33.7	0.74	0.72	magnetic
$[Ni_{44}(CO)_{48}]$	3.4	0.00	0.56	magnetic

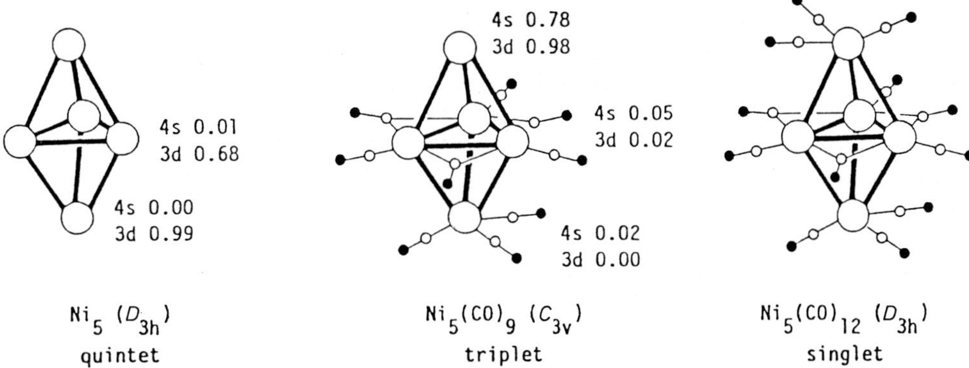

Figure 2-35. Geometric structures of $[Ni_5(CO)_{12}]^{2-}$ (singlet ground state), the hypothetical coordinatively unsaturated $[Ni_5(CO)_9]^{2-}$ (triplet ground state), and the bare Ni_5 (quintet ground state) clusters. The localization of the unpaired spins in the latter two clusters is indicated. Adapted from [253].

Thus, for small, coordinatively saturated clusters in which all the metal atoms interact directly with the ligand shell there is no magnetization expected, in agreement with the experimental observations. As the clusters become larger, however, two or more shells of metal atoms can be present and it becomes possible to distinguish between "surface" metal atoms which are directly bound to the ligands and "bulk" metal atoms which interact only with neighboring metal atoms. This is the case for instance, in the cluster $[Ni_{38}Pt_6(CO)_{48}]$, whose electronic structure has been modeled with a $[Ni_{44}(CO)_{48}]$ cluster in a series of local spin density calculations (Fig. **2-36**). [144, 267, 273]

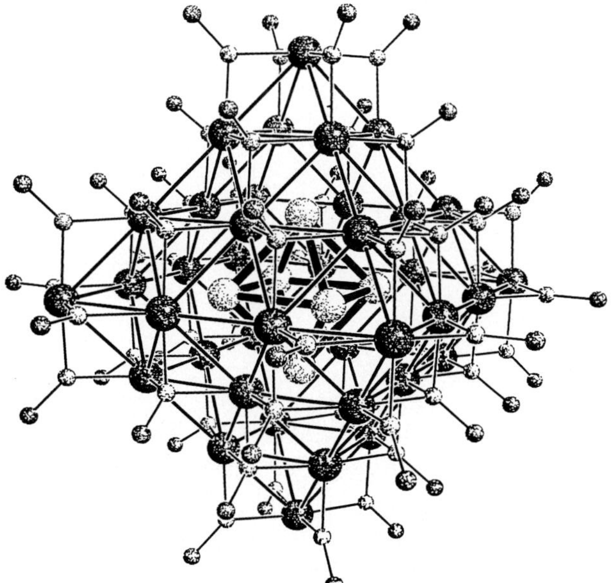

Figure 2-36. Geometric structure of the $[Ni_{44}(CO)_{48}]^{n-}$ cluster used to model the magnetic behavior of $[Ni_{38}Pt_6(CO)_{48}]^{n-}$.

It was found that while all the surface Ni atoms are substantially perturbed by the ligands, the bulk atoms are only indirectly affected by this interaction. As a consequence, the d band of the cluster is split into two components, a broad one well below the Fermi level due to the surface atoms and one at the Fermi level due to the bulk atoms (Fig. **2-37**). [267, 273] The surface component is much broader since the cluster contains 38 surface atoms but only 6 bulk Ni atoms. While the majority and minority spins compensate each other in the surface component and provide for a complete spin pairing, in the bulk component a residual magnetic moment remains. In other words, the magnetic moments of the surface atoms, about 0.7 unpaired electrons per Ni atom in the free Ni_{44} cluster (Table **2-5**), are completely quenched in the carbonylated cluster, while there remain about 0.6 unpaired electrons per atom on the 6 bulk Ni atoms, thus resulting in a total of 3–4 unpaired electrons for the whole cluster. This is another case where clusters and metal surfaces show a surprising similarity. In fact, the same mechanism described above is responsible for the observed local quenching of the magnetic moment induced by CO when adsorbed on Ni surfaces. [132, 311, 312]

2.4.5.4 ESR Spectra

Although the vast majority of small organometallic clusters are diamagnetic, a few examples of paramagnetic compounds have been reported. [209] In these cases, ESR spectroscopy provides a useful tool for studying the symmetry and localization of the (possibly degenerate) orbitals carrying the unpaired electron(s). One of the first reported cases of a paramagnetic molecular cluster was the triangular $[Co_3(CO)_9S]$. [313] The unpaired electron in this cluster was found

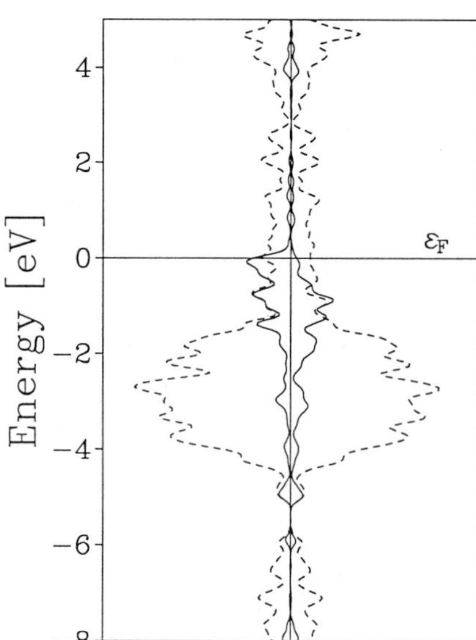

Figure 2-37. Density of states for $[Ni_{44}(CO)_{48}]^{4-}$ determined by LDF calculations. Left side: minority spin, right side: majority spin. Solid line indicates the contribution from the six bulk Ni atoms and the dashed line the contribution from the 38 surface Ni atoms. Reproduced with permission from [267]. Copyright 1992 American Chemical Society.

to occupy a Co–Co antibonding orbital of a_2 symmetry resulting from the combination of cobalt d_{xz} orbitals. Analogous investigations on other cobalt clusters have shown that the unpaired electron has a high degree of delocalization. [209] ESR measurements thus provide direct evidence for the delocalized nature of the metal–metal bonding in ligated clusters.

Quite interestingly, some even electron clusters composed of the nonmagnetic metal Os exhibit an intrinsic paramagnetism. [314, 315] A series of 21 Os clusters with nuclearities ranging from $[Os_3(CO)_{12}]$ to $[Os_{40}Hg_3(C)_4(CO)_{96}]^{2-}$ have been studied by ESR spectroscopy. [316] While all the Os clusters containing less than 10 metal atoms are ESR silent, as expected, a single ESR line was observed for the higher nuclearity clusters. This has been explained in terms of a reduced HOMO-LUMO energy gap as the size of the cluster increases. [316] In the following, let us designate the HOMO and the LUMO of the cluster by a and b, respectively. The energy of one electron in orbital a, $I_{eff}(a)$, will be a function of its kinetic energy, its attraction by the nuclei, and its Coulomb and exchange interactions with all the other electrons. If a second electron is placed in orbital a, its repulsion from the first electron will be characterized by the Coulomb integral J_{aa}. The energy of the corresponding configuration a^2 will thus be $2I_{eff}(a) + J_{aa}$. On the other hand, if the second electron is placed in orbital b, then the resulting a^1b^1 configuration will be a triplet with an energy $I_{eff}(a) + I_{eff}(b) + J_{ab} - K_{ab}$, where the repulsion between the two electrons is measured by the difference in the Coulomb integral J_{ab} and the exchange integral K_{ab}. The a^1b^1 triplet configuration will be the ground state if the difference in the repulsion energies, $J_{aa} - J_{ab} + K_{ab}$, is greater than the absolute difference in the effective one-electron energies, $|I_{eff}(a) - I_{eff}(b)|$. This is probably what happens for the medium nuclearity Os clusters which have a triplet ground state and thus exhibit an ESR signal. In small Os clusters, $I_{eff}(a) - I_{eff}(b)$ is too large and the ground state is a singlet. As more metal atoms are added to the cluster, such that a small metallic particle is eventually formed, the energy separation between the a and b levels becomes very small. At the same time, one would expect that J_{aa} and J_{ab} become very similar and that K_{ab} becomes rather small as the orbitals a and b get delocalized over the entire system. Under these conditions, an open shell ground state is not forthcoming and the metal particle assumes a diamagnetic ground state with an a^2 configuration. Finally, the one-electron energy levels in the bulk metal can be grouped into bands which form a quasi-continuum. At absolute zero, and in the absence of external magnetic or electric fields, the closed shell a^2 configuration is the electronic ground state and the metal is nonmagnetic. This model provides a convincing explanation for the observed paramagnetic behavior of medium to high nuclearity Os clusters. [316]

The above model is not an alternative to that used to explain the temperature dependent magnetic susceptibility of clusters of magnetic metals presented in the previous sections. [266, 267] It merely provides an additional view of how the cluster properties may evolve as a function of the density of states and of the band gap at the Fermi level. Indeed, a significant decrease in the HOMO-LUMO energy separation (or band gap) as a function of the cluster size has been predicted for both Os [296] (Fig. **2-38**) and Ni clusters. [121, 266]

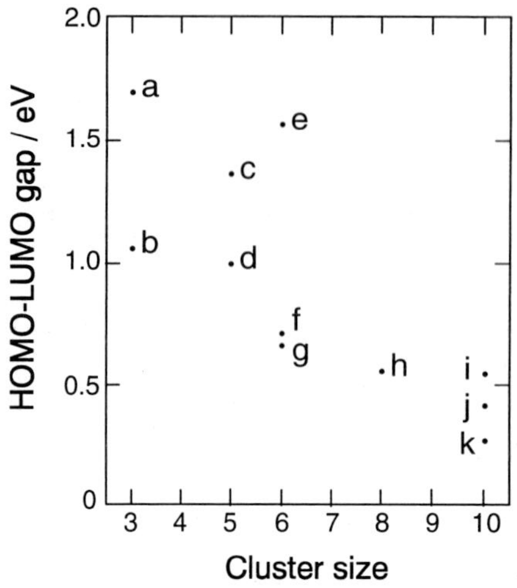

Figure 2-38. Calculated values of the HOMO-LUMO energy gaps as function of the cluster size. a) $[Os_3(CO)_{12}]$, b) $[Os_3(CO)_{12}I_2]$, c) $[Os_5(CO)_{19}]$, d) $[Os_5(CO)_{16}]$, e) $[Os_6(CO)_{18}]^{2-}$, f) $[Os_6(CO)_{18}]$, g) $[Os_6O(CO)_{19}]$, h) $[Os_8(CO)_{22}]^{2-}$, i) $[Os_{10}C(CO)_{24}]^{2-}$, j) $[Os_{10}C(CO)_{24}I_2]$, and k) $[Os_{10}S_2(CO)_{23}]$. Adapted from [296].

2.4.5.5 NMR Spectra

Experiments designed to determine the structures and the temperature dependent fluxional behavior of clusters in solution are routinely performed using ^{1}H, ^{13}C, ^{15}N, and ^{31}P NMR. [208, 209] As such, NMR represents a very powerful technique for the characterization of organometallic clusters. ^{59}Co, ^{103}Rh, ^{183}W, or ^{195}Pt NMR experiments can provide information about the electronic structure of large metallic clusters or colloidal particles. In the previous sections we drew attention to the distinction between "surface" and "bulk" metal atoms. It is therefore expected that these two types of metal atoms will exhibit somewhat different NMR characteristics. Recently, the spin-lattice relaxation time and the metallic Knight shift of a large $[Pt_{309}Phen_{36}O_{30}]$ cluster [317] have been measured. Chemical shifts in insulating Pt clusters are usually small (0.1–1 %). For metals, on the other hand, an additional shift known as the Knight shift is introduced by the Pauli spin susceptibility associated with the conduction electrons. For Pt metal, this additional shift is –3.4 %. A second measure for the "metallic" character of the metal atoms in a ligated cluster comes from the nuclear spin-lattice relaxation time, T_1. In insulators, T_1 is very long, while it is relatively short for metals because the conduction electrons provide an efficient relaxation channel. Both these criteria, when applied to the Pt_{309} cluster, show the clear existence of two kinds of metal atoms in the cluster. [317] The ^{195}Pt NMR signal from $[Pt_{309}Phen_{36}O_{30}]$ can be interpreted in terms of two contributions with different chemical shifts and relaxation times (see Chapter 3, Fig. **3-38**). While the first peak has a long T_1 and a chemical shift typical of nonconducting Pt compounds, the second one has a small value for T_1 and a resonance typical of metals. The signal intensities are consistent with a "normal" contribution from the 162 surface Pt atoms and a "metallic" term arising from the 147 bulk atoms. [317]

2.4.5.6 Specific Heat

Other physical measurements which depend on the effective DOS near the Fermi energy have also indicated the presence of atoms with "metallic" character in large ligated clusters. The electronic contribution to the specific heat of a cluster is $C = \gamma T$, where γ is a coefficient directly proportional to the DOS around the Fermi level. Thus, the appearance of an electronic contribution to the specific heat is considered as a strong test for metallic character and a gauge for the evolution of the energy levels towards a continuous manifold. For the cluster $[Pd_{561}Phen_{36}O_{200}]$, γ was found to be about one third of the bulk metal value. [297] By taking into account the 252 (about 50%) surface atoms in the almost spherical Pd_{561} core, this low value could be explained by assuming virtually no contribution from the surface atoms to the DOS near the Fermi level and a significantly reduced contribution from the next inner shell of bulk atoms. Both these assumptions are fully supported by the general features of the LDF results for the $[Ni_{44}(CO)_{48}]$ cluster [267, 273] previously described (Fig. **2-37**).

It is interesting at this point to analyze the average energy spacings of the levels in the vicinity of the Fermi level as one goes from a very small cluster to a medium or high nuclearity cluster. This can be taken as a measure of the evolution of the cluster one-electron energy levels toward the quasi-continuum expected for large metallic particles. It has been found, based on LDF calculations, [144] that the average energy spacing around E_F decreases from about 170 meV in $[Ni_5(CO)_{12}]$ to about 40 meV in $[Ni_{44}(CO)_{48}]$. This latter value is still larger than the thermal energy kT of about 24 meV at room temperature. In order to reach a quasi-continuum, M_n clusters containing a few hundreds of metal atoms are probably required since the level spacings decrease approximately as n^{-1}.

2.4.5.7 Redox Properties

The electrochemical behavior of organometallic clusters has been studied by cyclic voltammetry and controlled potential coulometry. [318, 319] Since a wide range of clusters have been found to undergo reversible one-electron reactions, they may be considered as sources of or sinks for electrons, or in other words, electron "reservoirs". However, it is interesting to note that electrochemically generated radicals are often unstable. In trinuclear clusters, the frontier orbital involved in the reduction was found to perhaps have some metal–metal antibonding character, whereas the HOMO orbital involved in the oxidation process is usually metal–metal bonding. The consequence of this is that regardless of whether an oxidation or a reduction takes place, the result is a weakening of the metal–metal interactions and an associated increase in the metal–metal distances. [318, 319]

Such simple HOMO-LUMO arguments, however, do not seem applicable to even slightly larger metal clusters. The redox potentials of a series of tetracobalt carbonyl clusters with different ligands were found to correlate well with the total charge on the metal core, but not with the HOMO or LUMO energies. [150] This result may seem surprising since a linear relationship between HOMO energies and oxidation potentials has been reported for mononuclear carbonyl complexes. [320] Apparently, the energies of individual orbitals in medium size cluster compounds do not reflect the gross molecular properties. Indeed, an excellent correlation was found between the global charges on the metal core, as computed by Mulliken population analyses, and the redox properties of the clusters. [150] These redox experiments, together with their theoretical interpretation, provide strong evidence for the delocalized nature of the metal electrons in these metal clusters. Individual orbitals may thus be localized on particular atoms, or even ligands, and not reflect the gross cluster properties. Frontier orbital arguments, although extremely useful for mononuclear complexes in which the HOMO is metal localized, do not seem to be appropriate for the description of the redox properties, and in general the electronic structures, of organometallic clusters.

2.5 Conclusions: from Clusters to Surfaces

2.5.1 The Role of Theory in Clusters Research

Cluster science is a relatively new area of research and lies at the borderline between physics and chemistry. Its interdisciplinary character follows simply from the fact that clusters are units of matter with properties intermediate between those of molecules and solids. This is also the main reason why the rationalization of the chemical and physical properties of clusters represents a formidable challenge for theory. In principle, one would like to be able to describe the transition from the molecular state to the bulk solid state within a single theoretical framework, using the same language and, possibly, the same degree of theoretical accuracy.

There is no doubt that theory plays a central role in elucidating many of the features of metal clusters. The possible recourse to theoretical schemes at various levels of accuracy, from highly sophisticated quantum mechanical calculations to simplified modeling approaches, has permitted the description of a large variety of phenomena, and in particular for those of gas phase clusters. This goes from the detailed interpretation of cluster optical spectra to the reactivity of clusters in molecular beams or the study of collective properties of very large metallic aggregates like the plasmon resonances. Through the combined use of theory and experiment, it has also been possible to identify the key electronic mechanisms which contribute to cluster stability (hybridization, electron delocalization, etc.) and to draw up some general systematic schemes for the bonding in clusters from various groups in the periodic table.

One important aspect of clusters is the peculiar nature of these "molecular" entities. Although several cluster properties exhibit a rapid convergence to the corresponding bulk values, there are some unique features of metal clusters which have no analogy in the massive metals. Another intriguing question, from a theoretical point of view, is the interplay between the "optimal" geometry of a metal cluster and its electronic structure. The special relationship between the geometry of a cluster and its electronic properties may be the reason for the marked size-dependent variations in the rates and selectivities of cluster reactions.

It is a well known fact that the cluster stability increases with cluster size and that this is related to the increasing average coordination number. However, maximum coordination is not the most important "ingredient" for stabilizing low nuclearity clusters. Indeed, the preferred geometry for naked tetrameric aggregations of several elements is a rhombus and not the more compact tetrahedron. The cluster symmetry, this meaning the detailed geometric arrangement of the nuclei, may produce degeneracies and near degeneracies in the electronic energy levels. These orbital degeneracies can be predicted by simple topological arguments. The existence of particularly stable clusters with pseudo-spherical shapes corresponds to the complete filling, or shell closing, of such degenerate energy levels ("magic numbers") and represents one of the most fascinating aspects of cluster physics. [96] It also has a surprising analogy to nuclear structures. In addition, orbital degeneracies also determine the spin multiplicity of the cluster ground state and thus govern the occurrence of Jahn-Teller distorsions.

Attempts to define a theoretical basis for the growth sequence of small clusters and thus to predict how they evolve into large metallic particles and finally form an extended system have met with little success. This is primarily due to the existence of competing growth mechanisms and of isomers with similar stability. Furthermore, clusters whose structures form a section of a fcc or a bcc lattice are known, as are clusters with pentagonal symmetry. Clearly, the two channels lead to mutually exclusive growth paths, either to cubic or icosahedral structures.

The theoretical description of organometallic clusters also presents formidable problems. Here, the great variety of metal atoms, ligands, nuclearities, geometric structures, and bonding modes makes it quite difficult to formulate a general theory that is capable of explaining the very large number of observed phenomena. Furthermore, only in recent years has it been possible to combine physical measurements with accurate quantum mechanical calculations in order to investigate the electronic structures of these systems. These studies have opened new horizons whereby this fascinating class of inorganic compounds are used as tools for the understanding of more fundamental and general questions like the gradual transition from the molecular to the colloidal and finally to the metallic state.

The apparent completely different nature of the bonding between clusters like $[Mo_6Cl_{12}]$ and $[Os_3(CO)_{12}]$ makes the application of similar concepts or of a unifying view quite difficult. However, it is certainly possible to draw some general conclusions about the electronic character of the late transition metal clusters, that is, for clusters of metal atoms in low oxidation states. These systems also represent the largest fraction of the clusters reported so far.

2.5.2 On the Analogy between Metal Clusters and Surfaces

A good analogy can be drawn between the bonding modes of the ligands in organometallic clusters and of small molecules adsorbed on metal surfaces. It can be deduced from both structural and spectroscopic data [196] and its origin can be traced to the close similarity of the bonding of small molecules like CO, C_2H_4, or PR_3 to a metal center. This interaction is very local, directly involves only the neighboring atoms, and, most importantly, has the same electronic origin. Thus, the analogy between organometallic clusters and surfaces is largely due to the similarity in the nature of their chemical bonding to ligands either attached to a finite metal cluster or adsorbed on an extended metal surface.

The ligands not only simply bind to the metal atoms, they also induce a perturbation on the local electronic structure of the metal cluster core in a way that is reminiscent of the ligand field splitting of the metal levels in a mononuclear complex. Such an energetic splitting also occurs with the levels of a metal cluster when surrounded by a complete ligand shell. The most remarkable, and also observable, consequence of this splitting is spin pairing. Thus, there is a complete suppression of the magnetic moments in low nuclearity ligated clusters composed of magnetic materials since they are always diamagnetic, in contrast to their naked counterparts which often exhibit high-spin magnetic ground states. It is likely that this ligand induced change in the electronic structure of the metal core will also affect the strength of the metal–metal interactions to some extent. Taking the length of the intermetallic bonds, which increases upon coordination, as a crude measure, it can be concluded that the ligands induce a weakening, and not a strengthening, of the metal–metal bonds in organometallic clusters. It also indicates that the ligand polyhedron plays a very important role in the stabilization of cluster compounds as a whole. Consequently, the metal–ligand bond strength is comparable to or larger than the metal–metal bond strength and organometallic clusters undergo structural reorganization quite easily. Thus, one has to be cautious in extending the cluster-surface analogy beyond the apparent similarity between their metal–ligand bonding modes since most of their physical properties have very little in common.

The modifications induced by the ligands on the metal atoms of a cluster are a direct consequence of the high ligand-to-metal ratio in organometallic clusters as compared to metal surfaces. Of course, this ratio decreases as the size of the cluster increases; for large colloidal particles, stabilized by ligands and containing a few hundreds of metal atoms, this perturbation is certainly less significant for the global physical properties of the particle than for a small cluster containing 20 atoms or less. In fact, all metal atoms in a small cluster are on the "surface", while for larger particles the fraction of "surface" atoms which interact directly with the ligands decreases compared to the number of "bulk" or internal metal atoms. However, each of these two types of atoms have their own characteristic properties and each responds differently to the various physical measurements which are used to test the cluster electronic structure. For medium sized clusters or colloids with sizes ranging from 20 to a few hundreds of metal atoms, the approximate ratio between "bulk" and "surface" atoms is between 0.2 and 0.5.

This means that there are enough "bulk" metal atoms to be detected by experimental techniques of sufficient resolution. In this sense, large organometallic clusters and colloidal particles are ideal systems for studying the transition from the molecular to the metallic state.

The similarity in the bonding in a small molecule to that in a metal cluster or on a metal surface provides a strong justification for representing and describing the chemisorption phenomena on surfaces with the cluster model approach. [321] In this theoretical model, one represents the surface as a cluster composed of a finite number of substrate atoms such that the geometry of the cluster model chosen is usually equivalent to a section of the surface under consideration. The cluster model approach has been applied successfully to basically all types of surface–gas interactions (metals, oxides, semiconductors, from weak physisorption to strong chemisorption, from covalent to ionic bonds, etc.). Embedding techniques have been developed in order to take into account the boundary conditions of a chemisorption cluster. The cluster model technique provides a simple, yet effective way to gain information on the interactions of small molecules with extended surfaces. The cluster model approach has two great advantages: the same language can be used to describe completely different physical situations, such as the bonding in a transition metal complex or on a metal surface, and the same sophisticated techniques of modern quantum chemistry can be employed to both types of systems.

Basically, the same phenomena which have been observed in organometallic clusters also occur on metal surfaces. For instance, the configurational change in the atoms on the surface which is induced by the adsorbed molecules has exactly the same electronic origin as that in a metal cluster. The magnetic quenching which leads to a diamagnetic ground state in a coordinatively staturated cluster causes a local reduction of the magnetic moment at the surface. The fluxional nature and the easy structural reorganization of organometallic clusters has its analogy in the considerable reconstruction of metal surfaces induced by the chemisorbed species, although here the rest of the crystal has a stabilizing effect which is obviously not present in clusters. [200] The addition of ligands to an ensemble of metal atoms and the subsequent formation of metal–ligand bonds occurs at the expense of the metal–metal interactions which are weakened in a ligated cluster compared to its naked counterpart. In a similar way, the metal–adsorbate bonding may lengthen and weaken the bonds of the surface atoms to the rest of the metal, a mechanism which has been described as bond order conservation. [322]

Besides these apparent similarities, there are, of course, also substantial differences. Long range effects on metal surfaces are completely absent in those cluster models (and clusters) which contain only a few surface atoms. Some characteristics of the cluster model which have a large impact on the chemisorption behavior (*e.g.* the ionization potential, the electron affinity, the position of the Fermi level, the static polarizability, etc.) may be considerably different than the corresponding bulk values. These differences obviously reflect the truncated nature of the cluster and the atypical coordination of its constituent atoms. The so-called "bond preparation" approach has been proposed to overcome, at least in part, this problem. [323] Here, by chosing special excited states of the cluster which are more closely related to the electronic structure of the surface, one arrives at an improved representation of the interaction with an adsorbed molecular fragment.

Despite these limitations, the use of local cluster models for the description of the interactions at surfaces has been very successful. The main reason is that, as mentioned before, the same "ingredients" govern the formation and the breaking of the chemical bonds between the "ligands" and a metal atom, a cluster, or at a surface. As suggested by Hoffmann, [213] the metal–ligand bonding is a delicate balance of the very same interactions; the metal–ligand bond is accomplished at the expense of the bonding within both the metal and the ligand. In this respect, dissociative chemisorption and surface reconstruction are just two extremes of the same phenomenon. To the extent that this bonding is local and involves, to a first order approximation, only the neighboring metal atoms, the same chemical picture applies to both clusters and surfaces.

Acknowledgments

We would like to thank L. Ackermann, T. Fox, O. Häberlen, S. Köstlmeier, and S. Krüger for their scientific efforts and for their help during the preparation of this review. Our work has been generously supported over the years by the Deutsche Forschungsgemeinschaft, by the Fonds der Chemischen Industrie and by the European Community Science Program.

References

[1] F. A. Cotton, *Q. Rev. Chem. Soc.* **1966**, 416.

[2] E. Bleisten-Barojas, D. Lavesque, *Phys. Rev. B* **1986**, *34*, 3910.

[3] K. Raghavachari, *J. Chem. Phys.* **1986**, *84*, 5672.

[4] G. Pacchioni, J. Koutecky, *J. Chem. Phys.* **1986**, *84*, 3301.

[5] D. Tomanek, M. Schlüter, *Phys. Rev. Lett.* **1986**, *56*, 1055.

[6] W. L. Brown, R. R. Freeman, K. Raghavachari, M. Schlüter, *Science* **1987**, *255*, 860.

[7] H. Haberlandt, *Clusters of Atoms and Molecules*, Springer, Berlin-Heidelberg, **1992**.

[8] N. Ben-Horin, U. Even, J. Jortner, *J. Chem. Phys.* **1992**, *97*, 5296, 5988.

[9] *Quantum Chemistry: The Challenge of Transition Metals and Coordination Compounds* (Ed.: A. Veillard), NATO ASI Series C 176. Reidel, Dordrecht, **1986**.

[10] *The Challenge of d and f Electrons: Theory and Computation* (Eds.: D. R. Salahub, M. C. Zerner), Symp. Series 394, American Chemical Society, Washington D. C., **1989**.

[11] I. N. Levine, *Quantum Chemistry*, Prentice Hall, Englewood Cliffs, **1991**.

[12] A. Szabo, N. S. Ostlund, *Modern Quantum Chemistry: Introduction to Advanced Electronic Structure Theory*, first revised ed., McGraw-Hill, New York, **1989**.

[13] *Ab Initio Methods in Quantum Chemistry I and II* (Ed.: K. P. Lawley) Wiley, New York, **1987**.

[14] J. Simons, *J. Phys. Chem.* **1991**, *95*, 1017.
[15] R. G. Parr, W. Yang, *Density Functional Theory of Atoms and Molecules*, Oxford University Press, New York, **1988**.
[16] J. Sauer, *Chem. Rev.* **1989**, *89*, 199.
[17] J. Koutecky, P. Fantucci, *Chem. Rev.* **1986**, *86*, 539.
[18] D. Feller, E. R. Davidson in *Reviews in Computational Chemistry* (Eds.: K. B. Lipkowitz, D. B. Boyd), VCH, New York, **1990**, p. 1.
[19] J. Almlöf, K. Fægri, K. Korsell, *J. Comput. Chem.* **1982**, *3*, 385.
[20] R. Ahlrichs, M. Bär, M. Häser, H. Horn, C. Kölmel, *Chem. Phys. Lett.* **1989**, *162*, 165.
[21] J. A. Pople, R. Krishnan, H. B. Schlegel, J. S. Binley, *Int. J. Quantum Chem.* **1978**, *14*, 545.
[22] A. C. Wahl, G. Das in *Methods of Electronic Structure Theory* (Ed.: H. F. Schaefer III), Plenum, New York, **1977**, p. 51.
[23] W. A. Goddard III, T. H. Dunning, W. G. Hunt, P. J. Hay, *Acc. Chem. Res.* **1973**, *6*, 368.
[24] P. E. M. Siegbahn, J. Almlöf, A. Heiberg, B. O. Roos, *J. Chem. Phys.* **1981**, *74*, 2384.
[25] V. Bonacic-Koutecky, P. Fantucci, J. Koutecky, *Chem. Rev.* **1991**, *91*, 1035.
[26] L. Szasz, *Pseudopotential Theory of Atoms and Molecules*, Wiley, New York, **1985**.
[27] S. Huzinaga, M. Klobukowski, Y. Sakai, *J. Chem. Phys.* **1984**, *88*, 4880.
[28] J. C. Barthelat, P. Durand, A. Serafini, *Mol. Phys.* **1977**, *33*, 159.
[29] P. J. Hay, W. R. Wadt, *J. Chem. Phys.* **1985**, *82*, 270, 284, 299.
[30] R. B. Ross, J. M. Powers, T. Atashroo, W. C. Ermler, L. A. LaJohn, P. A. Christiansen, *J. Chem. Phys.* **1990,** *93*, 6654.
[31] H. Stoll, P. Fuentealba, M. Dolg, J. Flad, L. von Szentpaly, H. Preuss, *J. Chem. Phys.* **1983**, *78*, 5532.
[32] P. Pyykkö, *Chem. Rev.* **1988**, *88*, 563.
[33] P. Schwerdtfeger, *Chem. Phys. Lett.* **1991**, *183*, 457.
[34] R. O. Jones, O. Gunnarsson, *Rev. Mod. Phys.* **1989**, *61*, 689.
[35] *Density Functional Theory of Many-Fermion Systems* (Ed.: S. B. Trickey), vol. 21 of *Adv. Quantum Chem.* **1990**.
[36] *Density Funktional Methods in Chemistry* (Eds.: J. Labanowski, J. Andzelm), Springer, New York, **1991**.
[37] T. Ziegler, *Chem. Rev.* **1991**, *91*, 651.
[38] P. Hohenberg, W. Kohn, *Phys. Rev. B* **1964**, *136*, 864.
[39] M. Levy, *Proc. Natl. Acad. Sci.* **1979**, *76*, 6062.
[40] W. Kohn, C. J. Sham, *Phys. Rev. B* **1965**, *140*, 1133.
[41] J. C. Slater, *The Self-Consistent Field for Molecules and Solids*, McGraw-Hill, New York, **1974**.
[42] J. R. Sabin, S. B. Trickey in *Local Density Approximations in Quantum Chemistry and Solid State Physics* (Eds.: J. P. Dahl, J. Avery), Plenum, New York, **1984**, p. 333.
[43] A. D. Becke, *Phys. Rev. A* **1988**, *38*, 3098.
[44] J. P. Perdew, J. A. Chewary, S. H. Vosko, K. A. Jackson, M. R. Pederson, D. J. Singh, C. Fiolhais, *Phys. Rev. B* **1992**, *46*, 6671.
[45] B. I. Dunlap, J. Andzelm, J. W. Mintmire, *Phys. Rev. A* **1990**, *42*, 6354.
[46] R. Car, M. Parrinello, *Phys. Rev. Lett.* **1985**, *55*, 2471.
[47] V. L. Moruzzi, J. F. Janak, A. R. Williams, *Calculated Properties of Metals*, Pergamon, London, **1978**.
[48] J. J. P. Steward in *Reviews in Computational Chemistry* (Eds.: K. Lipkowitz, D. B. Boyd), VCH, New York, **1990**, p. 45.
[49] M. C. Zerner in *Reviews in Computational Chemistry Vol. 2* (Eds.: K. Lipkowitz, D. B. Boyd), VCH, New York, **1992**, p. 313.
[50] R. Hoffmann, *Angew. Chem. Int. Ed. Engl.* **1982**, *21*, 711.
[51] D. M. P. Mingos, T. Slee, L. Zhenyang, *Chem. Rev.* **1990**, *90*, 383.

[52] W. D. Knight, K. Clemenger, W. A. de Heer, W. Saunders, M. Y. Chau, M. L. Cohen, *Phys. Rev. Lett.* **1984**, *52*, 2124.
[53] M. S. Daw, M. I. Baskes, *Phys. Rev. B* **1984**, *29*, 6443.
[54] J. K. Norskov, N. D. Lang, *Phys. Rev. B* **1980**, *21*, 2136.
[55] M. D. Morse, *Chem. Rev.* **1986**, *86*, 1049.
[56] M. M. Kappes, *Chem. Rev.* **1988**, *88*, 369.
[57] *Physics and Chemistry of Small Clusters* (Eds.: P. Jena, B. Rao, K. Khanna), Plenum Press, New York, **1987**.
[58] *Elemental and Molecular Clusters* (Eds.: G. Benedek, T. P. Martin, G. Pacchioni), Springer, Berlin-Heidelberg, **1988**.
[59] T. Halicioglu, C. W. Bauschlicher, *Rep. Prog. Phys.* **1988**, *51*, 883.
[60] R. F. Marshall, R. J. Blint, B. Kunz, *Phys. Rev. B* **1976**, *13*, 3333.
[61] P. Fantucci, P. Balzarini, *J. Mol. Catal.* **1978**, *4*, 337.
[62] H. O. Beckmann, J. Koutecky, P. Botschwina, W. Meyer, *Chem. Phys. Lett.* **1979**, *67*, 119.
[63] H. O. Beckmann, J. Koutecky, V. Bonacic-Koutecky, *J. Chem. Phys.* **1980**, *73*, 5182.
[64] J. Flad, H. Stoll, H. Preuss, *J. Chem. Phys.* **1979**, *71*, 3042.
[65] R. Car, J. Martins, *Surf. Sci.* **1981**, *106*, 280.
[66] G. Pacchioni, H. O. Beckmann, J. Koutecky, *Chem. Phys. Lett.* **1982**, *87*, 151.
[67] C. H. Wu, *J. Phys. Chem.* **1983**, *87*, 1534.
[68] D. Plavsic, J. Koutecky, G. Pacchioni, V. Bonacic-Koutecky, *J. Phys. Chem.* **1983**, *87*, 1096.
[69] P. Fantucci, J. Koutecky, G. Pacchioni, *J. Chem. Phys.* **1984**, *80*, 325.
[70] G. A. Thompson, F. Tischler, D. M. Lindsay, *J. Chem. Phys.* **1983**, *78*, 5946.
[71] G. Pacchioni, J. Koutecky, *J. Chem. Phys.* **1984**, *81*, 3588.
[72] M. M. Marino, W. C. Ermler, *J. Chem. Phys.* **1987**, *86*, 6283.
[73] V. Kumar, R. Car, *Phys. Rev. B* **1991**, *44*, 8243.
[74] M. A. Nygren, P. E. M. Siegbahn, U. Wahlgren, A. Akeby, *J. Phys. Chem.* **1992**, *96*, 3633.
[75] L. Goodwin, D. R. Salahub, *Phys. Rev. A* **1993**, *47*, R774.
[76] D.-W. Liao, K. Balasubramanian, *J. Chem. Phys.* **1992**, *97*, 2548.
[77] K. J. Taylor, C. Lin, J. Conceicao, L.-S. Wang, O. Cheshnowsky, B. R. Johnson, P. J. Nordlander, R. E. Smalley, *J. Chem. Phys.* **1990**, *93*, 7515.
[78] R. B. Brewington, C. F. Bender, H. F. Schaefer III, *J. Chem. Phys.* **1976**, *64*, 905.
[79] H. Stoll, J. Flad, E. Golka, T. Krüger, *Surf. Sci.* **1981**, *106*, 251.
[80] C. W. Bauschlicher, P. S. Bagus, B. N. Cox, *J. Chem. Phys.* **1982**, *77*, 4032.
[81] R. A. Whiteside, R. Krishnan, J. A. Pople, M. B. Krogh-Jespersen, P. von Rague Schleyer, G. Wenke, *J. Comp. Chem.* **1980**, *1*, 307.
[82] G. Pacchioni, J. Koutecky, *J. Chem. Phys.* **1982**, *77*, 5850.
[83] G. Pacchioni, J. Koutecky, *Chem. Phys.* **1982**, *71*, 181.
[84] W. C. Ermler, R. B. Ross, C. W. Kern, R. M. Pitzer, N. W. Winter, *J. Phys. Chem.* **1988**, *92*, 3042.
[85] B. K. Rao, S. N. Khanna, P. Jena, *Phys. Rev. B* **1987**, *36*, 953.
[86] T. J. Lee, A. P. Randell, P. R. Taylor, *J. Chem. Phys.* **1990**, *92*, 489.
[87] K. Rademann, *Ber. Bunsenges. Phys. Chem.* **1989**, *93*, 653.
[88] H. J. Nolte, K. Jug, *J. Chem. Phys.* **1993**, *93*, 2584.
[89] J. Jellinek, I. L. Garzon, *Z. Phys. D* **1991**, *20*, 239.
[90] E. Blaisten-Barojas, S. Khanna, *Phys. Rev. Lett.* **1988**, *61*, 1477.
[91] U. Röthlisberger, W. Andreoni, *J. Chem. Phys.* **1991**, *94*, 8129.
[92] J. K. Norskov, *Phys. Rev. B* **1982**, *26*, 2875.
[93] J. D. Kress, A. E. DePristo, *J. Chem. Phys.* **1988**, *88*, 2596.
[94] M. S. Stave, A. E. DePristo, *J. Chem. Phys.* **1992**, *97*, 3386.
[95] C. L. Cleveland, U. Landman, *J. Chem. Phys.* **1991**, *94*, 7376.
[96] S. Bjornholm, *Contemporary Phys.* **1990**, *31*, 309.

[97] W. Knight, W. de Heer, W. Saunders, *Z. Phys. D* **1986**, *3*, 109.
[98] I. Katakuse, I. Ichihara, Y. Fujita, T. Matsuo, T. Sukurai, H. Matsuda, *Int. J. Mass. Spectron. Ion Phys.* **1986**, *74*, 33.
[99] W. Begemann, K. Meiwes-Broer, H. Lutz, *Phys. Rev. Lett.* **1986**, *56*, 2248.
[100] J. Phillips, *Chem. Rev.* **1986**, *86*, 619.
[101] K. Clemenger, *Phys. Rev. B* **1985**, *32*, 1359.
[102] T. Halicioglu, *Surf. Sci.* **1988**, *197*, L233.
[103] M. Hoare, *Adv. Chem. Phys.* **1979**, *40*, 49.
[104] B. Delley D. E. Ellis, A. J. Freeman, E. J. Baerends, D. Post, *Phys. Rev. B* **1983**, *27*, 2132.
[105] P. Fantucci, G. Pacchioni, *Phys. Stat. Sol. (b)* **1989**, *153*, 193.
[106] R. J. Mulliken, *J. Chem. Phys.* **1955**, *23*, 1833.
[107] D. Maynau, J. P. Malrieu, *J. Chem. Phys.* **1988**, *88*, 3163.
[108] R. F. W. Bader, T. T. Nguyen-Dang, *Adv. Quantum Chem.* **1981**, *14*, 63.
[109] C. Gatti, P. Fantucci, G. Pacchioni, *Theoret. Chim. Acta* **1987**, *72*, 433.
[110] M. H. McAdon, W. A. Goddard III, *Phys. Rev. Lett.* **1985**, *55*, 2563.
[111] J. A. Alonso, L. C. Balbas, *Struct. Bond.* **1993**, *80*, 229.
[112] I. Boustani, W. Pewestorf, P. Fantucci, V. Bonacic-Koutecky, J. Koutecky, *Phys. Rev. B* **1987**, *35*, 9437.
[113] T. H. Upton, *Phys. Rev. Lett.* **1986**, *56*, 2168.
[114] W. D. Knight, W. A. de Heer, W. Saunders, *Lecture Notes in Physics* **1987**, *269*, 15.
[115] K. J. Taylor, C. L. Pettiette, M. J. Craycraft, O. Cheshnowsky, R. E. Smalley, *Chem. Phys. Lett.* **1988**, *152*, 347.
[116] J. M. Smith, *AIAA J.* **1965**, *3*, 648.
[117] D. Wood, *Phys. Rev. Lett.* **1981**, *46*, 749.
[118] G. Makov, A. Nitzan, L. E. Brus, *J. Chem. Phys.* **1988**, *88*, 5076.
[119] E. Rohlfing, D. Cox, A. Kaldor, K. Johnson, *J. Chem. Phys.* **1984**, *81*, 3846.
[120] M. B. Knickelbein, S. Yang, S. Railey, *J. Chem. Phys.* **1990**, *93*, 94.
[121] N. Rösch, L. Ackermann, G. Pacchioni, *Chem. Phys. Lett.* **1992**, *199*, 275.
[122] W. P. Halperin, *Rev. Mod. Phys.* **1986**, *58*, 533.
[123] I. Boustani, J. Koutecky, *J. Chem. Phys.* **1988**, *88*, 5657.
[124] H. Akeby, I. Panas, L. G. M. Pettersson, P. E. M. Siegbahn, U. Wahlgren, *J. Phys. Chem.* **1990**, *94*, 5471.
[125] C. W. Bauschlicher, S. R. Langhoff, H. Partridge, *J. Chem. Phys.* **1990**, *93*, 8133.
[126] K. M. McHugh, J. C. Eaton, G. H. Lee, H. W. Sarkas, L. H. Kidder, J. T. Snodgrass, M. R. Manaa, K. H. Bowen, *J. Chem. Phys.* **1989**, *91*, 3792.
[127] G. Pacchioni, W. Pewestorf, J. Koutecky, *Chem. Phys.* **1984**, *83*, 201.
[128] S. P. Walch, *J. Chem. Phys.* **1987**, *86*, 5082.
[129] H. Wang, E. A. Carter, *J. Phys. Chem.* **1992**, *96*, 1197.
[130] O. Gropen, J. Almlöf, *Chem. Phys. Lett.* **1992**, *191*, 306.
[131] L. G. M. Pettersson, T. Faxen, *Theoret. Chim. Acta* **1993**, *85*, 345.
[132] N. Rösch, P. Knappe, P. Sandl, A. Görling, B. I. Dunlap in *The Challenge of d and f Electrons: Theory and Computation*, (Eds.: D. R. Salahub, M. C. Zerner), Symp. Series 394, Am. Chem. Soc., Washington D.C. **1989**, p. 180.
[133] D. R. Salahub R. P. Messmer, *Phys. Rev. B* **1977**, *16*, 2526.
[134] K. Lee, J. Callaway, K. Kwong, R. Tang, A. Ziegler, *Phys. Rev. B* **1985**, *31*, 1796.
[135] A. L. Benze, P. Lamande, R. Lissillour, H. Chermette, *Phys. Rev. B* **1985**, *31*, 5094.
[136] C. S. Wang, J. Callaway, *Phys. Rev. B* **1977**, *15*, 298.
[137] G. Mie, *Ann. Physik (Leipzig)* **1908**, *25*, 377.
[138] V. Bonacic-Koutecky, P. Fantucci, J. Koutecky, *J. Chem. Phys.* **1990**, *93*, 3802.
[139] A. A. Lushnikov, A. J. Simonov, *Z. Physik* **1977**, *270*, 17.
[140] C. Yannouleas, R. A. Broglia, M. Brack, P. F. Bortignon, *Phys. Rev. Lett.* **1989**, *63*, 255.
[141] C. Yannouleas, R. A. Broglia, *Phys. Rev. A* **1991**, *44*, 5793.

[142] C. Yannouleas, E. Vigezzi, R. A. Broglia, *Phys. Rev. B* **1993**, *47*, 9849.
[143] M. Moskovitz in *Metal Clusters* (Ed.: M. Moskovitz), Wiley, New York, **1986**, p. 185.
[144] L. Ackermann, N. Rösch, B. I. Dunlap, G. Pacchioni, *Int. J. Quant. Chem. Quant. Chem. Symp.* **1992**, *26*, 605.
[145] H. Danan, A. Herr, A. J. P. Meyer, *J. Appl. Phys.* **1968**, *39*, 669.
[146] W. A. de Heer, P. Milani, A. Chatelain, *Phys. Rev. Lett.* **1990**, *65*, 488.
[147] H. Krakauer, A. J. Freeman, E. Wimmer, *Phys. Rev. B* **1983**, *28*, 610.
[148] F. Liu, M. R. Press, S. N. Khanna, P. Jena, *Phys. Rev. B* **1988**, *38*, 5760.
[149] B. I. Dunlap, N. Rösch, *J. Phys. Chim. Physico-Chim. Biol.* **1989**, *86*, 671.
[150] G. F. Holland, D. Ellis, W. C. Trogler, *J. Am. Chem. Soc.* **1986**, *108*, 1884.
[151] W. D. Knight, K. Klemenger, W. A. de Heer, W. A. Saunders, *Phys. Rev. B* **1985**, *31*, 2539.
[152] I. Mullet, J. L. Martins, F. Reuse, J. Buttet, *Phys. Rev. Lett.* **1990**, *65*, 476.
[153] M. Manninen, *Phys. Rev. B* **1986**, *34*, 6886.
[154] H. Chevrel, *J. Solid State Chem.* **1971**, *3*, 515.
[155] A. Simon, *Angew. Chem. Int. Ed. Engl.* **1981**, *20*, 1.
[156] M. J. S. Dewar, *Bull. Soc. Chem. Franc.* **1951**, *18*, C71.
[157] J. Chatt, L. A. Duncanson, *J. Chem. Soc.* **1953**, 2939.
[158] P. S. Bagus, C. J. Nelin, C. W. Bauschlicher, *Phys. Rev. B* **1983**, *28*, 5423.
[159] D. Post, E. J. Baerends, *J. Chem. Phys.* **1983**, *78*, 5663.
[160] C. E. Moore, *Circular N. 467*, Natl. Bur. Std., Washington D.C., **1952**.
[161] S. P. Walch, W. A. Goddard III, *J. Am. Chem. Soc.* **1976**, *98*, 7908.
[162] A. B. Rives, R. F. Fenske, *J. Chem. Phys.* **1981**, *75*, 1293.
[163] H. Basch, D. Cohen, *J. Am. Chem. Soc.* **1983**, *105*, 3856.
[164] J. Koutecky, G. Pacchioni, P. Fantucci, *Chem. Phys.* **1985**, *99*, 87.
[165] N. Rösch, H. Jörg, B. I. Dunlap in *Quantum Chemistry: The Challenge of Transition Metals and Coordination Compounds* (Ed.: A. Veillard), NATO ASI Series C 176, Reidel, Dordrecht, **1986**, p. 179.
[166] C. W. Bauschlicher, P. S. Bagus, C. J. Nelin, B. O. Roos, *J. Chem. Phys.* **1986**, *85*, 354.
[167] C. W. Bauschlicher, P. S. Bagus, *J. Chem. Phys.* **1984**, *81*, 5889.
[168] G. Pacchioni, P. S. Bagus, *Inorg. Chem.* **1992**, *31*, 4391.
[169] D. M. P. Mingos, A. S. May in *The Chemistry of Metal Cluster Complexes* (Eds.: D. F. Shriver, H. D. Kaesz, R. D. Adams), VCH, Weinheim, **1990**, p. 22.
[170] G. Longoni, M. Manassero, M. Sansoni, *J. Am. Chem. Soc.* **1980**, *102*, 7974.
[171] G. Doyle, K. A. Eriksen, D. van Engen, *J. Am. Chem. Soc.* **1986**, *108*, 445.
[172] J. C. Calabrese, L. F. Dahl, P. Chini, G. Longoni, S. Martinengo, *J. Am. Chem. Soc.* **1974**, *96*, 2614.
[173] J. C. Calabrese, L. F. Dahl, A. Cavalieri, P. Chini, G. Longoni, S. Martinengo, *J. Am. Chem. Soc.* **1974**, *96*, 2616.
[174] R. A. Nagaki, L. D. Lower, G. Longoni, P. Chini, L. F. Dahl, *Organometallics* **1986**, *5*, 1764.
[175] S. Martinengo, G. Ciani, A. Sironi, *J. Am. Chem. Soc.* **1980**, *102*, 7564.
[176] P. Chini, *J. Organomet. Chem.* **1980**, *200*, 37.
[177] A. Ceriotti, F. Demartin, G. Longoni, M. Manassero, M. Marchionna, G. Piva, M. Sansoni, *Angew. Chem. Int. Ed. Engl.* **1985**, *24*, 697.
[178] C. E. Briant, B. R. C. Theobald, J. W. White, L. K. Bell, D. M. P. Mingos, A. J. Welsh, *J. Chem. Soc. Chem. Commun.* **1981**, 201.
[179] J. L. Vidal, J. M. Troup, *J. Organomet. Chem.* **1981**, *213*, 351.
[180] D. M. Washecheck, E. J. Wucherer, L. F. Dahl, A. Ceriotti, G. Longoni, M. Manassero, M. Sansoni, P. Chini, *J. Am. Chem. Soc.* **1979**, *101*, 6110.
[181] D. Fenske, J. Ohmer, J. Hachgenei, *Angew. Chem. Int. Ed. Engl.* **1985**, *24*, 993.
[182] B. K. Teo, K. Keating, *J. Am. Chem. Soc.* **1984**, *106*, 2224.
[183] D. Fenske, J. Ohmer, J. Hachgenei, K. Merzweiler, *Angew. Chem. Int. Ed. Engl.* **1988**, *27*, 1277.

[184] G. Huttner, K. Knoll, *Angew. Chem. Int. Ed. Engl.* **1987**, *26*, 743.
[185] D. Fenske, H. Krautscheid, *Angew. Chem. Int. Ed. Engl.* **1990**, *29*, 1452.
[186] C. Couture, D. H. Farraw, *J. Chem. Soc. Chem. Commun.* **1985**, 197.
[187] J. M. Basset, R. Ugo in *Aspects of Homogeneous Catalysis* (Ed.: R. Ugo), Reidel, Dordrecht, **1977**, p. 137.
[188] J. W. A. Van der Velden, P. T. Beurskens, J. J. Bour, W. P. Bosman, J. H. Noordik, M. Kolenbrander, J. A. K. M. Buskes, *Inorg. Chem.* **1984**, *23*, 497.
[189] V. Albano, S. Martinengo, *Nach. Chem. Tech. Lab.* **1980**, *28*, 654.
[190] A. Arrigoni, A. Ceriotti, R. Della Pergola, G. Longoni, M. Manassero, M. Masciocchi, M. Sansoni, *Angew. Chem. Int. Ed. Engl.* **1984**, *23*, 322.
[191] A. Ceriotti, G. Longoni, M. Manassero, M. Masciocchi, G. Piro, L. Resconi, M. Sansoni, *J. Chem. Soc. Chem. Commun.* **1985**, 1402.
[192] M. Tachicawa, E. L. Muetterties, *J. Am. Chem. Soc.* **1980**, *102*, 4541.
[193] V. M. Mackay, B. K. Nicholson, W. T. Robinson, A. W. Sims, *J. Chem. Soc. Chem. Commun.* **1984**, 1276.
[194] N. D. Lang, H. Ehrenreich, *Phys. Rev.* **1968**, *168*, 605.
[195] K. Christmann, G. Ertl, *Surf. Sci.* **1972**, *33*, 254.
[196] E. L. Muetterties, T. N. Rhodin, E. Band, C. F. Brucker, W. R. Pretzer, *Chem. Rev.* **1979**, *79*, 91.
[197] F. Y. K. Lo, G. Longoni, P. Chini, L. D. Lower, L. F. Dahl, *J. Am. Chem. Soc.* **1980**, *102*, 7691.
[198] M. R. Churchill, F. J. Hollander, J. P. Hutchinson, *Inorg. Chem.* **1977**, *16*, 2655.
[199] M. R. Churchill, B. G. D. Boer, *Inorg. Chem.* **1977**, *16*, 878.
[200] G. A. Somorjai, M. A. Van Hove, *Prog. Surf. Sci.* **1989**, *30*, 201.
[201] N. Rösch, L. Ackermann, G. Pacchioni, *Inorg. Chem.* **1993**, *32*, 2963.
[202] D. Fenske, J. Hachgenei, F. Rogel, *Angew. Chem. Int. Ed. Engl.* **1984**, *23*, 982.
[203] B. F. G. Johnson, R. E. Benfield, *J. Chem. Soc. Dalton Trans.* **1978**, 1554.
[204] R. E. Benfield, B. F. G. Johnson, *J. Chem. Soc. Dalton Trans.* **1980**, 1743.
[205] R. J. Gillespie, *J. Chem. Educ.* **1963**, *40*, 295.
[206] N. Rösch, T. Fox, F. P. Netzer, M. G. Ramsey, D. Steinmüller, *J. Chem. Phys.* **1991**, *94*, 3276.
[207] T. Fox, N. Rösch, *Surf. Sci.* **1991**, *256*, 159.
[208] B. T. Heaton, *Phil. Trans. R. Soc. London* **1982**, *A308*, 95.
[209] H. Knözinger in *Metal Clusters in Catalysis* (Eds.: B. C. Gates, L. Guczi, H. Knözinger), Elsevier, Amsterdam, **1986**, p. 187.
[210] G. Schmid, *Struct. Bond.* **1985**, *62*, 51, and references therein.
[211] P. J. Feibelman, *Phys. Rev. Lett.* **1990**, *65*, 729.
[212] M. R. Albert, T. J. Yates, *The Surface Scientist's Guide to Organometallic Chemistry*, American Chemical Society, Washington D.C., **1987**.
[213] R. Hoffmann, *Solids and Surfaces: A Chemist's View of Bonding in Extended Structures*, VCH, New York, **1988**.
[214] G. Blyholder, *J. Phys. Chem.* **1964**, *68*, 2772.
[215] R. E. Williams, *Inorg. Chem.* **1971**, *10*, 210.
[216] K. Wade, *J. Chem. Soc. Chem. Commun.* **1971**, 792.
[217] D. M. P. Mingos, *Nature* **1972**, *236*, 99.
[218] *Chemical Applications of Topological and Graph Theory* (Ed.: R. B. King), Elsevier, Amsterdam, **1983**.
[219] B. K. Teo, *Inorg. Chem.* **1984**, *23*, 1251.
[220] J. W. Lauher, *J. Am. Chem. Soc.* **1978**, *100*, 5305.
[221] J. W. Lauher, *J. Am. Chem. Soc.* **1979**, *101*, 2604.
[222] B. F. G. Johnson, J. Lewis, *Advan. Inorg. Chem. Radiochem.* **1981**, *24*, 215.
[223] C. R. Eady, B. F. G. Johnson, J. Lewis, B. E. Reichert, G. M. Sheldrick, *J. Chem. Soc. Chem. Commun.* **1976**, 271.
[224] G. Longoni, P. Chini, L. D. Lower, L. F. Dahl, *J. Am. Chem. Soc.* **1975**, *97*, 5034.

[225] R. Mason, K. M. Thomas, D. M. P. Mingos, *J. Am. Chem. Soc.* **1973**, *95*, 3802.
[226] D. M. P. Mingos, *Acc. Chem. Res.* **1984**, *17*, 311.
[227] D. M. P. Mingos, *Chem. Soc. Rev.* **1986**, *15*, 31.
[228] D. M. P. Mingos, *J. Chem. Soc. Chem. Commun.* **1985**, 1352.
[229] A. J. Stone, *Mol. Phys.* **1980**, *41*, 1339.
[230] A. J. Stone, *Inorg. Chem.* **1981**, *20*, 563.
[231] A. J. Stone, *Polyhedron* **1984**, *3*, 1299.
[232] D. M. P. Mingos, D. J. Wales, *Introduction to Cluster Chemistry*, Prentice Hall, London, **1990**.
[233] B. K. Teo, *Inorg. Chem.* **1985**, *24*, 4209.
[234] G. Ciani, A. Sironi, *J. Organomet. Chem.* **1980**, *197*, 233.
[235] D. J. Wales, D. M. P. Mingos, T. Slee, L. Zhenyang, *Acc. Chem. Res.* **1990**, *23*, 17.
[236] J. Koutecky, P. Fantucci, *Z. Phys. D* **1986**, *3*, 147.
[237] P. Fantucci, J. Koutecky in *Elemental and Molecular Clusters* (Eds.: G. Benedek, T. P. Martin, G. Pacchioni), Springer, Berlin, **1988**, p. 125.
[238] G. Pacchioni, J. Koutecky, *Ber. Bunsenges. Phys.Chem.* **1984**, *88*, 242.
[239] A. Görling, N. Rösch, D. E. Ellis, H. Schmidbaur, *Inorg. Chem.* **1991**, *30*, 3986.
[240] D. A. Brown, *J. Inorg. Nucl. Chem.* **1958**, *5*, 289.
[241] J. Evans, *J. Chem. Soc. Dalton Trans.* **1980**, 1005.
[242] A. Dedieu, R. J. Hoffmann, *J. Am. Chem. Soc.* **1978**, *100*, 2074.
[243] J. F. Halet, R. Hoffmann, J. Y. Saillard, *Inorg. Chem.* **1985**, *24*, 1695.
[244] D. G. Evans, D. M. P. Mingos, *Organometallics* **1983**, *2*, 435.
[245] D. J. Underwood, R. Hoffmann, K. Tatsumi, A. Nakamura, Y. Yamamoto, *J. Am. Chem. Soc.* **1985**, *107*, 5968.
[246] D. G. Evans, *Inorg. Chem.* **1986**, *25*, 4602.
[247] C. Mealli, D. M. Proserpio, *J. Am. Chem. Soc.* **1990**, *112*, 5484.
[248] H. Müller, U. Fuhr, C. Opitz, H. F. Fritsche, *J. Less. Common Met.* **1988**, *137*, 195.
[249] J. Burdett, G. J. Miller, *J. Am. Chem. Soc.* **1987**, *109*, 4081.
[250] R. A. Wheeler, *J. Am. Chem. Soc.* **1990**, *112*, 8737.
[251] H. J. Freund, G. Hohlneicher, *Theoret. Chim. Acta* **1979**, *51*, 145.
[252] G. Pacchioni, P. Fantucci, V. Valenti, *J. Organomet. Chem.* **1982**, *224*, 89.
[253] G. Pacchioni, P. Fantucci, *Chem. Phys. Lett.* **1987**, *134*, 407.
[254] A. B. Rives, Y. Xiao-Zeng, R. F. Fenske, *Inorg. Chem.* **1982**, *21*, 2286.
[255] M. C. Manning, W. C. Trogler, *Coord. Chem. Rev.* **1981**, *38*, 89.
[256] G. P. Kostikova, D. V. Korolkov, *Russ. Chem. Rev.* **1985**, *54*, 344.
[257] P. A. Cox, *The Electronic Structure and Chemistry of Solids*, Oxford University Press, Oxford, **1987**.
[258] T. A. Albright, K. A. Yee, J.-Y. Saillard, S. Kahlal, J. F. Halet, J. S. Leigh, K. H. Whitmire, *Inorg. Chem.* **1991**, *30*, 1179.
[259] K. W. Chan, R. G. Woolley, *J. Phys. C* **1979**, *12*, 2745.
[260] D. W. Bullet, *Chem. Phys. Lett.* **1985**, *115*, 450.
[261] D. W. Bullet, *Chem. Phys. Lett.* **1987**, *135*, 373.
[262] W. C. Trogler, *Acc. Chem. Res.* **1990**, *23*, 239.
[263] G. G. Hoffmann, J. K. Bashkin, M. Karplus, *J. Am. Chem. Soc.* **1990**, *112*, 8705.
[264] D. Delley, M. C. Manning, J. Berkowitz, D. Ellis, W. C. Trogler, *Inorg. Chem.* **1982**, *21*, 2247.
[265] G. F. Holland, D. Ellis, W. C. Trogler, *J. Chem. Phys.* **1985**, *83*, 3507.
[266] G. Pacchioni, N. Rösch, *Inorg. Chem.* **1990**, *29*, 2901.
[267] N. Rösch, L. Ackermann, G. Pacchioni, *J. Am. Chem. Soc.* **1992**, *114*, 3549.
[268] N. W. Ashcroft, N. D. Mermin, *Solid State Physics*, Saunders, Philadelphia, **1988**.
[269] G. Pacchioni, L. Ackermann, N. Rösch, *Gazz. Chim. Ital.* **1992**, *122*, 205.
[270] H. Darmon, R. Heer, J. P. Meyer, *J. Appl. Phys.* **1968**, *39*, 669.
[271] C. J. Ballhausen, *Ligand Field Theory*, McGraw-Hill, New York, **1962**.
[272] C. K. Jorgensen, *Absorption Spectra and Chemical Bonding in Complexes*, Pergamon, Elmsford, New York, **1962**.

[273] N. Rösch, L. Ackermann, G. Pacchioni, B. I. Dunlap, *J. Chem. Phys.* **1990**, *95*, 7004.
[274] D. M. P. Mingos, *J. Chem. Soc. Dalton Trans.* **1976**, 1163.
[275] R. G. Woolley in *Transition Metal Chemistry* (Ed.: B. F. G. Johnson), Wiley, New York, **1980**, p. 607.
[276] R. G. Woolley, *Inorg. Chem.* **1985**, *24*, 3519.
[277] R. G. Woolley, *Inorg. Chem.* **1985**, *24*, 3525.
[278] R. G. Woolley, *Inorg. Chem.* **1988**, *27*, 430.
[279] P. Fantucci, G. Pacchioni, V. Valenti, *Inorg. Chem.* **1984**, *23*, 247.
[280] M. R. A. Blomberg, C. B. Lebrilla, P. E. M. Siegbahn, *Chem. Phys. Lett.* **1988**, *150*, 522.
[281] C. W. Bauschlicher, *J. Chem. Phys.* **1986**, *84*, 872.
[282] A. A. Low, K. L. Kunze, P. J. MacDougall, M. B. Hall, *Inorg. Chem.* **1991**, *30*, 1079.
[283] O. D. Häberlen, N. Rösch, unpublished.
[284] N. Rösch, A. Görling, D. E. Ellis, H. Schmidbaur, *Angew. Chem. Int. Ed. Engl.* **1989**, *28*, 1357.
[285] P. Pyykkö, Y. Zhao, *Chem. Phys. Lett.* **1991**, *177*, 103.
[286] F. Scherbaum, A. Grohmann, B. Huber, C. Krüger, H. Schmidbaur, *Angew. Chem. Int. Ed. Engl.* **1988**, *27*, 1544.
[287] O. D. Häberlen, N. Rösch, *Chem. Phys. Lett.* **1992**, *199*, 491.
[288] L. J. de Jongh, J. Baak, H. B. Brom, D. van der Putten, J. M. van Ruitenbeek, R. C. Thiel in *Physics and Chemistry of Finite systems: from Clusters to Crystals* (Eds.: P. Jena, S. N. Khanna, B. K. Rao), Kluwer, Dordrecht, **1992**, p. 839.
[289] R. E. Benfield, *J. Organomet. Chem.* **1989**, *372*, 163.
[290] S. C. Avanzino, W. L. Jolly, *J. Am. Chem. Soc.* **1976**, *98*, 6505.
[291] W. L. Jolly, *Top. Curr. Chem.* **1977**, *71*, 149.
[292] G. Apai, S. T. Lee, M. G. Mason, L. J. Gerenser, S. A. Gardner, *J. Am. Chem. Soc.* **1979**, *101*, 6880.
[293] R. Zanoni in *Cluster Models for Surface and Bulk Phenomena* (Eds.: G. Pacchioni, P. S. Bagus, F. Parmigiani), NATO ASI Series B 283, Plenum, New York, **1992**, p. 169.
[294] G. Comelli, M. Sastry, G. Paolucci, K. C. Prince, L. Olivi, *Phys. Rev. B* **1991**, *43*, 14385.
[295] R. G. Woolley, *Chem. Phys. Lett.* **1987**, *143*, 148.
[296] D. W. Bullet, *Surf. Sci.* **1987**, *189/90*, 583.
[297] L. J. de Jongh, H. B. Brom, J. M. van Ruitenbeek, R. C. Thiel, G. Schmid, G. Longoni, A. Ceriotti, R. E. Benfield, R. Zanoni in *Cluster Models for Surface and Bulk Phenomena* (Eds.: G. Pacchioni, P. S. Bagus, F. Parmigiani), NATO ASI Series B 283, Plenum, New York, **1992**, p. 151.
[298] G. K. Wertheim, J. Kwo, B. K. Theo, K. A. Keating, *Solid State Commun.* **1985**, *55*, 357.
[299] I. A. Oxton, *Rev. Inorg. Chem.* **1982**, *4*, 1.
[300] H. Knözinger in *Metal Clusters in Catalysis* (Eds.: B. C. Gates, L. Guczi, H. Knözinger), Elsevier, Amsterdam, **1986**, p. 173.
[301] T. G. Spiro, *Prog. Inorg. Chem.* **1970**, *11*, 1.
[302] H. Knözinger in *Metal Clusters in Catalysis* (Eds.: B. C. Gates, L. Guczi, H. Knözinger), Elsevier, Amsterdam, **1986**, p. 129.
[303] C. O. Quicksall, T. G. Spiro, *Inorg. Chem.* **1968**, *7*, 2365.
[304] S. F. A. Kettle, P. L. Stanghellini, *Inorg. Chem.* **1979**, *18*, 2749.
[305] S. R. Drake, B. F. G. Johnson, J. Lewis, R. G. Woolley, *Inorg. Chem.* **1987**, *26*, 3952.
[306] B. K. Teo, F. J. D. Salvo, J. V. Waszczak, G. Longoni, A. Ceriotti, *Inorg. Chem.* **1986**, *25*, 2262.
[307] B. J. Pronk, H. B. Brom, L. J. de Jongh, G. Longoni, A. Ceriotti, *Solid State Commun.* **1986**, *59*, 349.
[308] L. J. de Jongh, J. A. O. de Aguiar, H. B. Brom, G. Longoni, J. M. van Ruitenbeek, G. Schmid, H. H. Smit, M. P. J. van Steveren, R. C. Thiel, *Z. Phys. D* **1989**, *12*, 445.

[309] J. A. O. de Aguiar, A. Mees, J. Darriet, L. J. de Jongh, S. R. Drake, P. P. Edwards, B. F. G. Johnson, J. Lewis, *Solid State Commun.* **1988**, *66*, 913.
[310] D. C. Johnson, R. E. Benfield, P. P. Edwards, W. H. J. Nelson, M. D. Vargas, *Nature* **1985**, *314*, 231.
[311] F. Raatz, D. R. Salahub, *Surf. Sci.* **1986**, *176*, 219.
[312] C. W. Bauschlicher, C. D. Nelin, *Chem. Phys.* **1986**, *108*, 275.
[313] E. Klumpp, L. Marko, G. Bor, *Chem. Ber.* **1964**, *97*, 926.
[314] R. E. Benfield, P. P. Edwards, A. M. Stacy, *J. Chem. Soc. Chem. Commun.* **1982**, 525.
[315] R. E. Benfield, *J. Phys. Chem.* **1987**, *91*, 2712.
[316] S. R. Drake, P. P. Edwards, B. F. G. Johnson, J. Lewis, E. A. Marseglia, S. D. Orbetelli, N. C. Pyper, *Chem. Phys. Lett.* **1987**, *139*, 336.
[317] D. van der Putten, H. B. Brom, L. J. de Jongh, G. Schmid in *Physics and Chemistry of Finite Systems: from Clusters to Crystals* (Eds.: P. Jena, S. N. Khanna, B. K. Rao), Kluwer, Dordrecht, **1992**, p. 1007.
[318] P. Lemoine, *Coord. Chem. Rev.* **1982**, *47*, 55.
[319] P. Lemoine, *Coord. Chem. Rev.* **1988**, *83*, 169.
[320] R. E. Bursten, *J. Am. Chem. Soc.* **1982**, *104*, 1299.
[321] *Cluster Models for Surface and Bulk Phenomena* (Eds.: G. Pacchioni, P. S. Bagus, F. Parmigiani), NATO ASI Series B 283, Plenum, New York, **1992**.
[322] E. Schustorovich, *Surf. Sci. Reports*, **1986**, *6*, 1.
[323] I. Panas, J. Schule, P. Siegbahn, and U. Wahlgren, *Chem. Phys. Lett.* **1988**, *149*, 265.
[324] F. A. Cotton, G. Wilkinson, *Inorganic Chemistry*, 5th ed. Wiley, New York, **1988**.

3 Clusters in Ligand Shells

3.1 Introduction

Günter Schmid, Giuliano Longoni and Dieter Fenske

The extensive coverage in chapter 3 is dedicated to clusters in protecting ligand shells. Just as single metal atoms or ions have to be coordinated by ligands so as to prevent reactions between each other or with any other materials, metal clusters must be protected by a ligand shell if they are to be preparatively handled as stable compounds. Such a 'chemical treatment' of metal clusters has considerable consequences: due to the interaction of the ligand molecules with the metal atoms of the cluster, the latter will be heavily influenced in their electronic behavior. Analogous to the way the electronic states of a single atom in a simple complex change depending on the sort, number, and arrangement of the ligands, the electronic properties of a metal cluster are affected by its protecting ligands. With increasing numbers of metal atoms in the cluster this influence decreases, however. The metal atoms of main concern are those which are directly bonded to ligands. In addition to the surface atoms, metal rich clusters also contain inner core atoms which are coordinated only by other metal atoms. These will be much less involved in the electronic interactions with the ligand molecules. From all this it follows that ligand stabilized clusters should differ considerably from free clusters in their electronic properties.

The enormous variety of ligand stabilized metal clusters makes it advisable to divide them into three categories. Chapter 3.2 will deal with the absolutely largest class of cluster compounds, namely those carrying organic ligands. Organometallic low-valent clusters of transition metals are almost as old as their mononuclear congeners. Indeed, the first reports of a polynuclear carbonyl derivative, *viz.* $[Co_4(CO)_{12}]$, go back to the early days of the chemistry of carbonyl compounds. Although progress in this particular area which lies at the interface of coordination and colloid chemistry was initially slow due to the complexity of the matter. Clever experimentalists like W. Hieber and coworkers were nevertheless able to develop this field without the help of the plethora of spectroscopic techniques which are nowadays taken for granted. Several iron and nickel polynuclear carbonyl species, as well as transition metal clusters containing main group elements, were first obtained between the thirties and the fifties, e.g. $[Fe_4(CO)_{13}]^{2-}$, $[Rh_4(CO)_{12}]$, and $[Rh_6(CO)_{16}]$, and their formulas and structures were often correctly assigned on the basis of the chemical composition.

In the sixties and the seventies, X-ray crystallography, computer science and NMR spectroscopy had advances in a major impact on cluster chemistry. Discoveries of unprecedented structural and spectroscopic features became an almost routine occurence, and cluster chemistry started to blossom, while the frontiers of the area were continously moved ahead.

Intriguing structural features, such as a variety of metal frame shapes and unsuspected bonding modes for a given ligand, only possible because of the presence of polyatomic metal sites, provided models for highly dispersed metal particles, and for chemisorbed atoms, ligands or molecular fragments on a metal surface. Sometimes, their stereochemistries seem to represent possible snapshots of the transition states that these molecules could presumably traverse along the reaction path of their miscellaneous catalytic reactions. The isolation and characterization of species of increasing size eventually captured the attention of solid state chemists and physicists, and cluster properties began to be investigated as particulate metals in a diamagnetic shell.

It is not the aim here to completely review all the past findings, but rather to concentrate on a few leading examples and to restate the present state-of-the-art through a concise, if often rough, summary encompassing the miscellaneous facets of the chemistry of low-valent organometallic clusters.

At present, the maximum size organometallic clusters reach is about 40–50 metal atoms. Giant clusters will be discussed in chapter 3.3 and are of special interest to the investigation of various fundamental issues. For instance, how large does a cluster particle have to be in order to start showing metallic behavior or, coming from the opposite direction, how small does it have to be so that typical bulk properties like magnetism, electrical conductivity etc. begin to disappear, that is, the point at which the quasi-continuous density of states changes into a series of discrete energy levels? Such quantum size effects are of very actual interest. To realize them, the synthesis of ligand stabilized clusters with an increasing number of metal atoms is necessary to observe the transition from the typical molecular behavior to the metallic state. In contrast to the development of organometallic clusters, the synthesis of giant 1–3 nm clusters succeeds if metal atoms are generated in solution and are allowed to grow to larger units. Here, the cluster growth is mainly determined by the nature of the ligands.

A fascinating class of ligand stabilized transition metal clusters is described in chapter 3.4. A relatively simple synthetic principle leads to systems where metal atoms are bridged by different group 15 and 16 elements. Silyl derivatives of these elements are reacted with ligated metal halides and clusters containing P, As, Sb, S, Se, and Te atoms are formed. In some cases, for instance with S and Se, the resulting clusters can be regarded as excerpts of metal sulfide and selenide lattices. They open the door to ligand stabilized clusters of semiconducting materials which are of enormous interest with respect to their electronic properties. Thus chapter 3.4 leads to the far-reaching conclusion that miniaturization of solid materials to nanometer scale is not restricted to metals, but can be extended to semiconducting materials. It can be predicted that the miniaturization of solids to cluster size is just at the beginning.

3.2 Low-valent Organometallic Clusters

Giuliano Longoni and Maria Carmela Iapalucci

3.2.1 Interplay Between Electronic and Steric Factors in the Growth of Transition Metal Molecular Clusters in Ligand Shells

Several ligands are in principle suitable for the stabilization of homoleptic low-valent transition-metal molecular clusters[1], all of them being σ-donors and π-acceptors. Carbon monoxide is by far the most important, as documented by the isolation and characterization of thousands of carbonyl and carbonyl-substituted clusters. Although the usual cluster growth in a ligand shell is limited to only a few metal atoms (2–13) for most ligands, clusters containing several dozen metal atoms can be stabilized in a carbon monoxide shell. The reason for this predominant role of CO is a combination of electronic (attractive) and steric (repulsive) factors, some of them less obvious than others. Initially, however, it is probably convenient to gain some rough idea about how much a cluster in a ligand shell might be expected to grow as a function of the electronic and steric character of the miscellaneous ligands.

Experiments and theory agree in the finding that the number of cluster valence electrons (CVE; *viz.* the sum of the number of valence electrons of the metals, the electrons donated by the ligand shell and any free charges present) necessary to support a stable compound with π-acceptor ligands is $12\,n + x$, where n is the number of metal atoms making up the cluster. [1–6] For most transition metals the value of x falls in the range between 12 and 26, and is usually a function of cluster shape. Furthermore, those clusters which have a metallic core representing a fragment of a close-packed metal lattice generally display an x value close to the above lowest limit, almost independent of n and the position of the transition metal in the periodic table. As a first approximation, we may

[1] The adjective "molecular" is intended to specify that all the compounds taken into consideration in this chapter are well defined molecules or molecular ions of constant composition and structure.

therefore assume that an M_n metallic core with a close-packed structure requires $[(12-y)n + 14]/2$ bielectronic ligands (L) in order to approximately attain the required CVE, where y is the number of the metal valence electrons.

On the other hand, the importance of a suitable packing of the ligands around the metal core in order to stabilize a given cluster was recognized as early as the first results of the structural characterization of binary carbonyl clusters became available. For instance, L. F. Dahl, analyzed the non bonding molecular parameters of $[Fe_4(CO)_{13}]^{2-}$ and suggested that the existence of a tetrahedral isoelectronic $[Fe_4(CO)_{14}]$ species should be very unlikely. [7] Later, both qualitative and quantitative theoretical studies were carried out in an attempt to understand the relative importance of steric effects, and to rationalize the miscellaneous carbonyl stereochemical behaviour experimentally observed in binary carbonyl clusters with up to six metal atoms. [8–13] Relevant to the present discussion is the expectation that, simply on the basis of steric considerations, it should be possible to accommodate up to 16 and 20 carbonyl groups respectively around a metal tetrahedron and octahedron having edges of ca. 2.90 Å. Such expectations have not yet been fully realized by actual compounds, but are partially supported by the isolation of $[MFe_3(CO)_{14}]^{2-}$ (M = Cr, Mo,W) [14] and $[Re_6C(CO)_{19}]^{2-}$ [15] respectively. Since related limiting steric numbers for ligands other than carbon monoxide are not available (and indeed for CO also for higher nuclearity clusters) we must, in these cases, rely on very approximate guesses. Taking for instance octahedral clusters of increasing frequency ν_x (where x is the number of M–M bond units along each edge), [16] as being representative of cubic close-packed metal fragments, the total number of required mono-hapto bielectronic ligands (N_L) and their corresponding number per peripheral metal atom (N_L/n_p) may be readily calculated. A few exemplary results are shown in Table **3-1** and are compared with the maximum number of sterically allowed carbonyl groups (last row). These numbers were calculated on the assumption that the peripheral metal atoms should approximately adhere to the following limiting upper coordination ratios: $N_{CO}/n_v = 3$, $N_{CO}/n_e = 2$ and $N_{CO}/n_f = 1$ (n_v, n_e and n_f are the numbers of metal atoms at vertex sites, edge sites, and embedded in a face respectively).

It turns out that the pseudo-spherical cluster growth examined here should soon face insurmountable difficulties. As shown in Table **3-1**, the existence of such octahedral carbonyl clusters of frequency 2, 3 and 4 is progressively hindered for metals belonging to group 7, 8 and 9 respectively, because of the above combination of obvious electronic and steric factors. Passing to other considerations, the required N_L/n_p ratio for metals belonging to group 11 remains less than 1 up to frequency 7 and might formally allow the existence of octahedral clusters containing more than 344 metal atoms in a shell of mono-hapto bielectronic ligands which are isosteric with CO. In contrast, sterically more demanding mono-hapto ligands should become entangled at an earlier stage following the above limitations. The same applies to poly-hapto ligands even though the increased hapticity corresponds to a greater number of electrons formally donated to the cluster. For instance, it has been calculated that a MCp (Cp = η^5-C_5H_5) moiety has a cone angle amounting to ca. 83–85 % of that of a corresponding

Table 3-1. Comparison between the total number of required and allowed carbonyl groups around a metal octahedron of increasing frequency.[a)]

Frequency		ν_1	ν_2	ν_3	ν_4	ν_5
$n = n_p + n_i$		6	19	44	85	146
	n_i	0	1	6	19	44
	n_v	6	6	6	6	6
	n_e	0	12	24	36	48
	n_f	0	0	8	24	48
Group 7	N_L	22	54.5	117	219.5	372
	N_L/n_p	3.67	**3.03**	3.08	3.32	3.64
Group 8	N_L	19	45	95	177	299
	N_L/n_p	3.17	**2.5**	**2.5**	2.68	2.93
Group 9	N_L	16	35.5	73	134.5	226
	N_L/n_p	2.67	1.97	**1.92**	2.04	2.21
Group 10	N_L	13	26	51	92	153
	N_L/n_p	2.16	1.44	**1.34**	1.39	1.50
Group 11	N_L	10	16.5	29	49.5	80
	N_L/n_p	1.66	0.92	0.76	**0.75**	0.78
Max. N_{CO}		18[b)]	42	74	114	162

a) n_i represents the number of interstitial atoms. For definition of other symbols see the text. The minimum N_L/n_p ratio encountered on increasing n is indicated by bold characters

b) See text for more reliable limits

$M(CO)_3$ group [11]. Therefore, the possible existence of such homoleptic octahedral M_nCp_x clusters should be limited to clusters of frequency 1 composed of metals at the end of the transition series. Accordingly, $[Ni_6Cp_6]^{n+}$ ($n = 0,1$) species have been isolated and characterized. [17]

Although the above reasoning is very rough, and several other factors should be taken into account for more reliable predictions, [13] there are experimental indications that such a wide range of existence could in fact be realized with particular combinations of metal and ligand. For example, platinum carbonyl anionic clusters with platinum/cation ratios of up to 80 (from elemental analyses) have metal core diameters of up to ca. 21 Å according to their transmission electron micrographs. [18] Although obtained in a macroscopic crystalline state, their poor X-ray diffraction pattern excluded a crystal structure determination. The presence of a threedimensional crystalline network, reproduced by translation of constant arrangements of molecular ions of identical size, composition and structure cannot be excluded, however. Indeed, lack of sufficient diffraction may have several alternative explanations, such as loss of clathrated co-crystallized solvent molecules, or excessive thermal motion probably arising from insufficiently tight interlocking between the cluster ion units.

Going back to cluster growing, one should also consider the fact that at least a partial relief of the steric pressure within the ligand shell, and a consequent possibility of increasing the size of the inner metal core, may be obtained in several ways. For instance, using interstitial main group elements such as C or N may help in fulfilling the clusters' electronic requirements internally by filling the cavities of the metal lattice.

As is well illustrated in standard textbooks, [19–22] the linear bonding of CO to metals occurs through a transfer of electron density from the carbonyl 5σ orbital (carbon lone pair) to an empty metal orbital, and a concurrent electron density backdonation from filled d_π metal orbitals into the carbonyl 2π orbitals in a synergetic fashion. Edge- and face-bridging bonding modes are thought to occur in a similar way through polycentric interactions. [23] The M–M distances spanned by bridging CO ligands are generally shorter than those unsupported by bridging carbonyls. This shortening indicates a strengthening of the M–M bond, which may be affected by both the increased electron density between the metal nuclei owing to the above polycentric interactions, as well as the partial release of repulsive interaction between the ligands. The latter can be a result of the fact that terminal and bridging ligands bound to the same pseudospherical surface reside on concentric spheres of different radii. It should be noted, however, that the above M–M bond shortening effect of bridging carbonyls may only be restricted to first and second row transition metals, as shown by the comparative analysis of the structures of several tetra-iridium derivatives. [24]

Many other π-acceptor ligands such as NO, CN^-, CNR, alkynyl, ene and enyl derivatives and phosphines are thought to behave similarly because they have suitable empty π* or dπ orbitals. However, the relative effectiveness of the σ and π bonding components differs greatly in each case. [25] An increasing inbalance between σ-donor and π-acceptor behavior, as demonstrated by several experimental and theoretical studies, can bring about Pauling's electroneutrality principle in cluster compounds, as it does in homoleptic mononuclear complexes. On the other hand, a lower energy and contraction of the metal d orbitals will disfavor back-donation of electron density.

As a trivial consequence, although a ligand such as NO is a 3-electron donor of comparable size to CO and should be potentially suitable for cluster growth exceeding the above limits, it has, in fact, not been very effective [26]. Little sterically demanding alkyl isocyanides, alkynyl ($R\text{-}C{\equiv}C^-$) and CN^- ligands appear to be good candidates on the basis of their σ and π bonding behaviour. [27, 28] Accordingly, homoleptic clusters such as $[Pt_3(CN^tBu)_{6},]$, [29] $[Pt_7(2,6\text{-}Me_2C_6H_3NC)_{12}]$, [30], $[M(C{\equiv}CR)]_n$, $[LM(C{\equiv}C\text{-}R)]_n$ (M = Cu,Ag,Au; L = phosphine and amine), and $[Ag_6Cu_7(C{\equiv}C\text{-}Ph)_{14}]^{2-}$ are known. [31–34] With the former ligands, cluster growth should be enhanced by the use of less bulky alkyl substituents, whereas the negative charge of the cyanide ligand may greatly limit the range of existence of this type of cluster.

All the above considerations indicate that carbon monoxide is a very effective ligand for clusters composed of metals from group 7 to 10 in which their average formal oxidation state is zero or even slightly negative. In contrast, ligands with predominently σ-donor character, are more adequate when the metals have a fractional positive formal oxidation state. Thus, these ligands are more effective for the heavier elements of groups 10 and 11 owing to the progressive contraction of the d orbitals of these metals. In this portion of the periodic table, carbon monoxide is replaced as prefered ligand by the phosphines as illustrated by the homoleptic $[Au_{13}(dppm)_6]^{4+}$ cluster [35] and several other higher nuclearity heteroleptic compounds (see chapter 3.3).

3.2.2 Bonding and Spectroscopic Behavior of Carbon Monoxide

Another property, also shared by several other ligands, which further contributes toward making carbon monoxide the most successful ligand in low-valent clusters is its bonding versatility. [36, 37] In addition to the most common mono-hapto terminal and symmetrical μ_2- and μ_3-bridging modes, carbon monoxide has recently been documented in a monohapto μ_4-bridging mode. [38] The variety of mono- and dihapto bonding modes displayed by carbon monoxide are illustrated in Figure **3-1**, where it is postulated to contribute up to three electron pairs to the CVE. Moreover, the sketches in Figure **3-1** are not the only possible bonding modes of carbon monoxide in clusters. Indeed, the degree of asymmetry exhibited by the carbonyl groups in the crystal structures of these type clusters encompasses a quasi continuum of modes between the linear terminal and the symmetrical μ_2- and μ_3-bridging limits. The significance of the presence of unsymmetrical bridges in these solid state structures is potentially relevant in order to gain a more detailed picture to explain the predominant role of carbon monoxide in cluster chemistry.

The occurrence of linear, bent (cyclic forms excluded), and prone semibridges (such as those represented in sketches A–C of Figure **3-1** and their related range of semibridging modes on a three metal atom array beginning from both the terminal and symmetrical μ_2-bridging extremes) has been explained by various, and sometimes controversial, concepts; these have focused on either attractive electronic or repulsive intra- and intermolecular interactions. [37, 39–42] In the electronic explanation, it is the presence of electronically inhomogeneous metal sites which supplies the driving force to bend the carbonyl. This is an attractive force arising either from a shift in electron density from the filled d orbital of a metal into a π* orbital of an adjacent metal atom's CO group, or from an opposite flow of electron density from a filled π orbital of a CO into an empty d orbital of an adjacent electronically unsaturated metal. These interactions have the effect of either opposing charge separation, or eliminating the otherwise unavoidable electronic unsaturation of a metal atom within the molecule.

In the steric explanation, it is assumed that the attractive interactions with the second (or third) metal atom are negligible, and the steric requirements of the whole ligand shell are thought to be responsible for the unsymmetrical bridging in the solid state structure.

The electronic explanation is likely to be valid for low nuclearity clusters in at least two cases: (i) when the carbonyl group leans toward a metal atom which is itself sterically crowded but would have to bear a more negative charge in the absence of such an interaction; (ii) when, in the absence of an electron density shift from the carbonyl π orbitals, the second metal atom would be formally electron deficient. It is worth noting that case (i) also implies several potentially significant consequences for higher nuclearity clusters. On the one hand, such an interaction may provide a mechanism for mitigating an uneven charge distribution because of uneven M–L and M–M connectivities or that the nature of the peripheral metal atoms vary. On the other hand, leveling off the charge separations which have been either induced on homometallic sites through nonequivalent coordination, or arise from the presence of heterometallic sites, may result in concomitant loss in site-selective reactivity. Additionally, heterometallic sites may also loose their own individual bonding and structural features. Both metal types may show identical stereochemistry by the random occupation of unequivalent polyhedral sites. Such random distribution of metals is often found in Co-Ni systems (e.g.$[Co_3Ni_9C(CO)_{20}]^{n-}$) [43], Fe-Rh systems (e.g. the isostructural $[Fe_xRh_{6-x}(CO)_{12}(\mu_3\text{-}CO)_4]^{x-}$ series of clusters), [44] and is also indicated in the partial metal disorder observed in some heterometallic clusters composed of metals from non adjacent groups (e. g. $[Pt_6Ru_9(CO)_{28}]^{2-}$). [45]

As stated earlier, the electron density transfer from a filled CO π orbital into an unfilled metal orbital alleviates the otherwise unavoidable electronic unsatura-

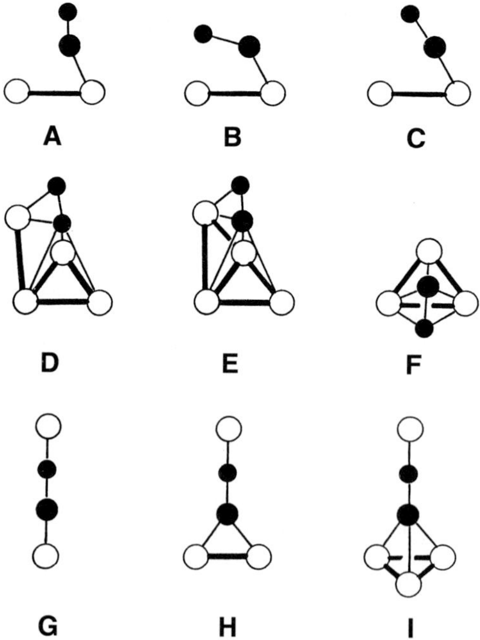

Figure 3-1. Schematic representation of some less usual bonding stereochemistries of carbon monoxide (blackened spheres).

tion at the metal site, and is a consequence of the experimental reaction conditions and/or is triggered by steric factors. The dihapto bonding modes of CO, as implied in diagrams B and C and unambiguously illustrated in diagrams D–F of Figure **3-1**, can occur not only intra- but also intermolecularly; the known dihapto bonding modes have been classified into Σ and Π modes in order to differentiate between those in which the oxygen lone pair is involved and those which use the carbonyl 1π electrons in their bonding. [36] In the bonding modes represented in diagrams G–I, the CO uses both the C and O lone pair electron for bonding, whereas in diagrams D and E a π carbonyl orbital is employed as the second electron pair. In F the unusual mode represented in diagram both the π electron pairs of the carbonyl are formally donated to the cluster. At present, there is no documented case in which all four electron pairs are formally available for donation. A few leading examples of the less usual or unusual modes of coordination for CO are collected in Table **3-2** together with their corresponding CO stretching frequencies. These frequencies are significantly lower than those observed for typical linear, edge- and face-bridging carbonyl ligands and indicate a significant weakening of the C–O bond. Bonding sites with similar or related features are also present on metal and metal oxide surfaces. The binding of CO at these sites is thought to promote an effective activation of the molecule towards nucleophilic attack, as well as its cleavage into C and O adsorbed atoms. [54–56]

In several compounds which have unsymmetrical bridging carbonyls, there are no apparent electronic reasons to explain the observed asymmetry. Intra- and/or

Table 3-2. Selected compounds containing carbon monoxide groups displaying less usual bonding modes.

Compound	Bonding mode of CO	ν_{CO} (cm^{-1})	Reference
	Mono-hapto		
$[Mo_2Ni_2S_2Cp_4(CO)]$	η^1–μ_4	1654	[38]
	Dihapto of Σ type		
$[Mo(CO)_2Cp(CO)]_2MgPy_4$	η^2	1667	[46]
$[Fe_2(CO)_2Cp_2(CO\text{-}AlEt_3)_2]$	η^2–μ_2	1460	[47]
$[Co_3(CO)_9(CO\text{-}TiClCp_2]$	η^2–μ_3	1401	[48]
	Dihapto of Π type		
$[Mn_2(CO)_5(dppm)_2]$	η^2–μ_2		[49]
$[MoNb(CO)_3Cp_3]$	η^2–μ_2	1560	[50]
$[Nb_3(CO)_7Cp_3]$	η^2–μ_3	1330	[51]
$[HFe_4(CO)_{13}]^-$	η^2–μ_4		[52]
$[Co_2Mo_2(CO)_4Cp_2Cp^*_2]$	η^2–μ_4	1633	[53]

intermolecular steric interactions are thought to be primarily responsible for these semibridges even though, once established, the unsymmetrical bridge may experience an attractive electronic interaction. It is worth mentioning that a concerted distortion of a set of carbonyl groups from their terminal positions to edge bridging positions, in which any degree of unsymmetrical bridging is attained, can self-compensate for any eventual inbalance of charge within a cluster and thereby preserve the equivalence of the metal atoms involved in the ligand rearrangement. As pointed out by F. A. Cotton, a compensating set of semibridging carbonyls is "a game any number can play". [19, 39]

A related controversial topic was the relative importance of electronic *vs.* steric factors in explaining the occurrence of symmetrically bridged (e.g. $M_4(CO)_9(\mu_2\text{-}CO)_3$ where M = Co,Rh) and unbridged (e.g. $Ir_4(CO)_{12}$) structures of congener carbonyl species and also to the induction of bridges in unbridged structures up on substitution of one or more carbonyl groups by more bulky and less π-acidic ligands or by negatively charged substituents. [9, 10, 23, 57–59] Recently, molecular mechanics studies on hexanuclear carbonyl clusters have suggested that intramolecular steric interactions are not the overriding factor in determining the ligand stereogeometry around the metal core, but can only justify small distortions within a given geometry. [13]

Intermolecular effects are much more difficult to unravel, both experimentally and theoretically. Clearly, their relevance can be seen by the isolation of a given species exhibiting different ligand stereochemistries in the solid state, as for example with the isomers of $[Rh_3(CO)_3Cp_3]$ (C_{3v} *vs.* C_s), [60] $[H_3Ru_4(CO)_{12}]^-$ (C_2 *vs.* C_{3v}), [61] $[Ir_6(CO)_{16}]$ (T_d *vs.* C_{2v}) [62] and $[Rh_{11}(CO)_{23}]^{3-}$. [63] In particular, the molecular ions making up the crystalline network of an ionic lattice experience rather large packing forces together with the constraints arising from the crystallographic symmetry. As a result, a rather soft molecular anion such as a cluster may be expected to show significant structural changes upon substitution of the counterion. Investigations in which differences in both electronic effects and intramolecular steric crowding are negated, as for example, the determination of the structures of the $[Fe_5N(CO)_{14}]^-$ anion with different counterions and the Mössbauer spectra of its salts, have shown the importance of intermolecular steric effects to the presence of bridging carbonyls in clusters. [64] In contrast, the results from the structure determinations of the $[Rh_6C(CO)_{15}]^{2-}$ and $[Fe_4S_3(NO)_7]^-$ anions with different tetrasubstituted ammonium or phosphonium cations indicate significant variations in the M–M interatomic distances. [65, 66] It is interesting to note that the carbonyl groups in the C_{3v} isomer of $[Rh_3(CO)_3Cp_3]$ are fluxional on the NMR time scale in solution, however, as shown in Scheme **3-1**, the carbonyl scrambling does not imply the interconversion of the C_{3v} into the C_s isomer. [67] Furthermore, results from various spectroscopic techniques have shown that the ligand stereogeometry observed in the solid state by X-ray diffraction is often retained in solution or is obtained at the slow-exchange limit. Therefore, although the importance of intermolecular packing forces has probably been generally under estimated, care should be taken that it not be overemphasized in many cases.

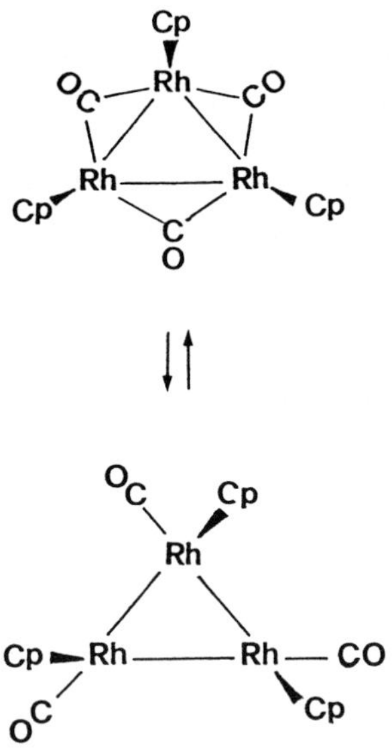

Scheme 3-1.

3.2.3 Stereochemical Non-rigidity of Clusters in Ligand Shells

The occurrence of isomers, as discussed in the previous section, is related to a well documented phenomena observed for the carbonyl ligands bonded to these metal clusters though not unique to them which is called fluxionality. As ligand-unsaturated metal moieties condense into clusters, the increased number of suitable orbitals, available for ligand interactions, combined with the bonding versatility of the carbonyl ligand, is expected to provide low-barrier kinetic paths for intramolecular carbonyl exchange. Furthermore, there is some indication from the little thermodynamic date available and such experimental evidence as the isolation of isomers, that the binding energies of terminal and bridging carbonyls are similar. [68] The flatness of the potential energy surfaces [13] and the results from structure correlation calculations [37] concur in the suggestion that, in many cases, very small differences in energy exist between the different ligand stereogeometries possible around a given metal core. It is not our intention to review the enormous amount of experimental work carried out in recent years on the subject of the stereochemical non-rigidity in clusters. It seems more appropriate here to address the general experimental observations and provide some guidelines for further, more specialized articles.

Stereochemical non-rigidity has been observed in all the various types of clusters described in the following sections and has been reported for the ligand shell, the metal core, or both. Furthermore, this phenomenon is not limited to clusters in solution but, with some obvious differences, can also be present in the solid state. This behavior has mostly been observed from multinuclear NMR studies in solution, [69–71] or by ^{1}H spin-lattice relaxation time measurements, ^{13}C cross-polarization magic-angle spinning (CPMAS) NMR spectroscopy, and X-ray diffraction studies in the solid state. [72] Whereas infrared spectroscopy, owing to its fast time-scale (ca. 10^{-13} s for carbonyl compounds), only gives information on the ground state structure of a given species and affords its "fingerprint", NMR techniques, because of their much lower radiation frequency and limited spectral field, have a time-scale of a few orders of magnitude longer and may, therefore identify exchange processes. Depending on whether the site exchange process is slower or faster than the time-scale of the experimental method, stereochemically inequivalent atoms may or may not be distinguishable and, if not, the observed lineshape will reflect the ground state structure or an averaged situation.

A different case is represented by X-ray diffraction studies. Although the frequencies of the X rays employed in these studies imply a time-scale of ca. 10^{-18} s, it must be realized that this value has no meaning in a diffraction experiment because the measurement of a single reflection may take up to several minutes. As a result, several types of molecular motions may take place during this time and can include small and large amplitude oscillations of the metal core with respect to the ligand shell, or *vice versa.* The relevant point then becomes how many of these occur during a measurement. [72]

By far the most widely studied site exchange processes are those involving carbonyl ligands. Several mechanisms which account for the observed carbonyl migration processes in solution and/or in the solid state have been put forward. All of them assume that the intramolecular carbonyl exchange reaction takes place in a concerted fashion so as to keep the electron count at each metal constant. Differences between the theories arise from the presumption as to whether it is the metal or the carbonyl polyhedron which is fixed, whilst the other is free to rotate or indeed, that only a part of one of the polyhedra has freedom of motion. Unambiguous experimental evidence supporting the proposal that carbonyl scrambling could arise from extensive motion of the metal cluster core within a rigid carbonyl polyhedron has yet to be reported from either solution or solid state studies. [70, 72] Small amplitude oscillations of both the metal cluster core and its outer carbonyl polyhedron around the equilibrium position are possibly responsible for most of the documented examples of stereochemical nonrigidity in binary carbonyls whether in the solid state or in solution. These small amplitude oscillations, however, are not sufficient to account for all the different kinds of intramolecular carbonyl exchange processes which have been experimentally observed (see, for example, the one depicted in Figure **3-2**). Furthermore, variable temperature NMR experiments often reveal the coexistence of distinct migration processes within a molecule which differ only slightly in their energy barriers and cooperate within the overall intramolecular carbonyl exchange process. These distinct migration paths may take a variety of forms, such as the interconversion of terminal and bridging carbonyls which can be either localized across a M–M bond or delocalized over a particular metal ring (so called planar and conical merry-go-round processes), rocking or rotational motion of particular ligand groups, more complicated pairwise interconversions which nevertheless

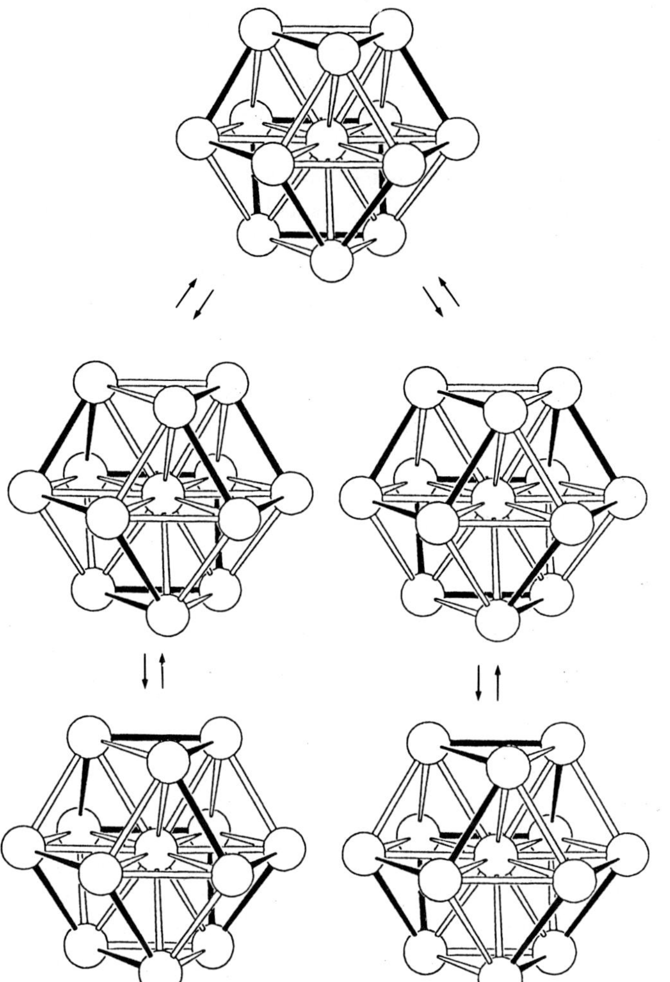

Figure 3-2. Proposed mechanism for the carbonyl intra-exchange process in the $[H_3Rh_{13}(CO)_{12}(\mu_2\text{-}CO)_{12}]^{2-}$ dianion (blackened Rh–Rh bonds are those spanned by the three stereochemically rigid and the exchanging carbonyl bridges).

regenerate a structure identical to the original one at each step, or completely delocalized scrambling. [69, 70] As an example, the carbonyl migration process observed in $[H_3Rh_{13}(CO)_{24}]^{2-}$ equilibrates only 21 of the 24 carbonyl groups whereby the three carbonyls spanning three alternate edges of the central Rh_6 ring are not involved in the exchange process. Such behaviour has been interpreted as proceeding through the pairwise interexchange of terminal and bridging CO's and is depicted schematically in Figure **3-2**. [73]

In contrast, intermolecular carbonyl exchange processes seem to play only a secondary role and just a few examples, which take place through either dissociative or associative mechanisms, have been reported. [70]

Processes which involve the reorientation of ene ligands (e.g. $\eta^2\text{-}C_2H_4$, $\eta^5\text{-}C_5H_5$, $\eta^6\text{-}C_6H_6$) are well documented in both mononuclear and cluster complexes. For the latter, μ_3-(η^2:η^2:η^2)-coordinated benzene ligands have been found to undergo reorientation motion in both solution and the solid state. For example, the variable temperature CPMAS spectra of $[Os_3(CO)_8(\eta^2\text{-}C_2H_4)(\mu_3\text{-}[\eta^2:\eta^2:\eta^2]\text{-}C_2H_6)]$ show only one carbon resonance for each type of coordinated ligand at temperatures above 196 K, whereas several resolved lines can be observed below 245 K. [74] The two reorientation processes for ethene and benzene show identical activation energies of 55 $kJmol^{-1}$. The rotation of these ligands are probably related and show that the term "helicopter molecule" is quite appropriate. Another spectacular example for stereochemical nonrigidity of arene ligands is provided by the compound $[Ru_6(\mu_6\text{-}C)(CO)_{11}(\eta^6\text{-}C_6H_6)(\mu_3\text{-}[\eta^2:\eta^2:\eta^2]\text{-}C_6H_6)]$, [75] whose structure [76] is shown in Figure **3-3**. The high fluxional behavior of this molecule in solution can be seen from the single broad resonance of the benzene ring protons in the ^1H NMR spectrum at room temperature; the averaging process can be slowed down only at temperatures below 180 K. [74,77]

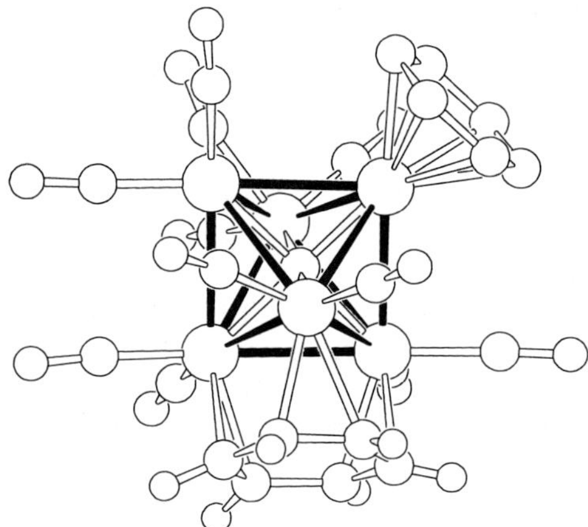

Figure 3-3. The molecular structure of $[Ru_6C(CO)_{11}(\eta^6\text{-}C_6H_6)(\mu_3\text{-}C_6H_6)]$ [76]

The stereochemical non-rigidity of metal fragments in a supposedly rigid metal cluster core has been documented in both lower and higher nuclearity mixed-metal clusters from various multinuclear NMR experiments. A recent report on a variable temperature ^{13}C NMR study of the $[Os_{10}C(CO)_{24}(HgX)]^-$ cluster anion unambiguously demonstrated that the HgX (X = Cl,Br,I,CF_3 and $[Mo(CO)_3Cp]$) fragment is highly mobile over the three μ_3 sites of the capping tetrahedron of the $[Os_{10}C(CO)_{24}]$ moiety. [78] In a analogous study, the analyses of the variable temperature ^{31}P NMR solution spectra of $[Ru_6Au_2C(CO)_{16}(PEt_3)_2]$ and $[Ru_5Au_2MC(CO)_{17}(PEt_3)_2]$ (M = Cr,Mo,W) are consistent with a stereochemical nonrigidity of the $[Au(PEt_3)]$ fragments over the Ru_6C and Ru_5MC cores. [79] In these clusters, several mechanisms can account for the equivalency of the phosphine ligands, e.g. a see-saw motion of the $[(Et_3P)Au\text{-}Au(PEt_3)]$ moiety across a Ru–Ru bond axis, nuclear permutation through a triangulated dodecahedron transition state, or free rollover of either the AuL or the Au_2L_2 fragments. The latter mechanisms would seem to be the most appropriate to explain the spectral behavior of $[Rh_6Au_2(\mu_6\text{-}C)(CO)_{13}(PEt_3)_2]$, whose solid state structure reveals an arrangement of metals intermediate between a distorted triangulated dodecahedron and a *cis*-bicapped octahedron. [80] The room temperature ^{13}C and ^{31}P NMR spectra of this remarkable compound show peaks split into septets (J(C–Rh) = 13.1 and J(P-Rh) = 3.81 Hz) from the coupling to either six magnetically equivalent or fortuitously coincident Rh atoms. Unfortunately, variable temperature spectra down to 183 K show only a broadening of the resonances and do not allow the differentiation between the two possibilities. [80]

Metal cluster core rearrangements are fairly common. For instance, the multinuclear NMR spectra of the butterfly cluster $[Pt_4(CO)_5L_4]$ (L = phosphine) are consistent with a rapid interconversion of the wingtip and backbone platinum atoms. [81] The equivalence of the rhodium atoms in $[Rh_9(\mu_8\text{-}P)(CO)_{21}]^{2-}$, whose solid state structure consists of a D_{4d} monocapped square antiprism, can be explained either by a pairwise concurrent shortening and lengthening of two Rh–Rh bonds, which preserves the capped square antiprism geometry throughout the nuclear permutation mechanism, or with a tricapped trigonal prismatic transition state. [82, 83] The $[Pt_3(CO)_3(\mu\text{-}_2CO)_3]_n{}^{2-}$ (n = 3,4) clusters, whose metal gemetries are based on fused trigonal prisms, were the first recognized examples of stereochemically non-rigid metal cluster skeletons. [84] Thus, the ^{195}Pt NMR of $[Pt_9(CO)_{18}]^{2-}$ is consistent with a rapid free rotation of the rigid $[Pt_3(CO)_6]$ building units. A snapshot of a possible transition state for this rearrangement is represented by the solid state structure of the $[Ni_9(CO)_{18}]^{2-}$ congener, which can be approximately described as deriving from the fusion of a trigonal prismatic unit with a trigonal antiprismatic unit. [85] The analogy between outer shapes of a $[Pt_3(CO)_3(\mu\text{-}_2CO)_3]$ fragment bound to a prismatic $[Pt_6(CO)_{12}]$ moiety and the $\mu_3\text{-}[\eta^2\text{:}\eta^2\text{:}\eta^2]\text{-}C_6H_6$ ligand bound to the octahedral Ru_6 metal core shown in Figure **3-3**, in conjunction with the poor or complete absence of X-ray scattering from most of the tetrasubstituted ammonium salts of the $[Pt_3(CO)_3(\mu\text{-}_2CO)_3]_n{}^{2-}$ dianion, suggests that these clusters might also be stereochemically nonrigid in the solid state.

^{31}P NMR studies have shown that fluxional behaviour in gold phosphine cluster cations is also rather common. These often show only one phosphorus resonance, whereas X-ray crystallography studies have shown that the ligands are not equivalent. [86–88] More recently, the interconversion of the hydrido-alkynyl cluster $[Ru_3Pt(\mu\text{-}H)(\mu_4\text{-}C_2{}^tBu)(CO)_9(dppe)]$ and its butterfly vinylidene $[Ru_3Pt(\mu_4\text{-}C{=}CH^tBu)(CO)_9(dppe)]$ isomer has been followed by EXAFS in solution. [89]

Finally, it is worth mentioning that the detection of magnetic nuclei and unpaired electron resonances becomes more difficult, if not impossible, as the nuclearity of the cluster increases. For example, the largest platinum cluster for which solution ^{195}Pt and ^{13}C NMR resonances are believed to have been observed is $[Pt_{19}(CO)_{22}]^{4-}$ and, in spite of several attempts under different experimental conditions, neither of the above nor proton resonances have ever been detected for either the ^{13}C enriched or natural abundance clusters $[Pt_{26}(CO)_{32}]^{2-}$ and $[HNi_{38}Pt_6(CO)_{48}]^{5-}$ (Figure **3-4**). [18] The negative results from these experiments may be due to several, perhaps synergetic, effects: exceedingly small molar concentrations, high spin multiplicity, wide resonance field, very long relaxation

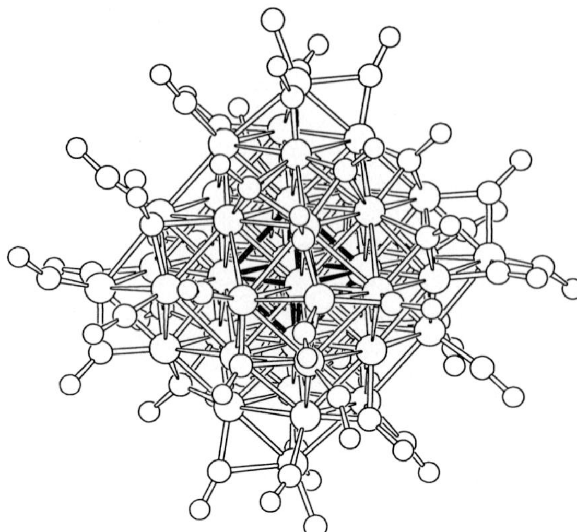

Figure 3-4. The structure of the $[HNi_{38}Pt_6(CO)_{48}]^{5-}$ ion: blackened bonds illustrate the fully interstitial Pt_6 octahedron. [155]

times,[2] foreign magnetic impurities, etc. Furthermore, the full characterization of the highly stable $[H_{2-x}Os_{10}C(CO)_{24}]^{x-}$ (x = 0,1,2) clusters [90–92] and the detection of the related unprotonated $[Os_{10}C(CO)_{24}]^{x-}$ (x = 0,1) cluster, which is stable on the cyclovoltammetry timescale, [93] suggests that clusters with equal charges but even and odd numbers of electrons might coexist to some extent in both solution and the solid state. In such a situation, these species may escape detection by both chemical means and infrared spectroscopy, and even the investigation of a diamagnetic species by NMR could be hampered by the presence of minor paramagnetic congeners. On the other hand, detection of the latter by ESR techniques might be likewise difficult. For example, the odd-electron cluster $[Ni_{11}Bi_2(CO)_{18}]^{3-}$ exhibits an ESR signal with an enormous half-height width of 1500 G. [94] It is easy to imagine that in the case of a low concentration of unpaired electrons, a correspondingly broad signal could easily render its detection impossible.

As a first conclusion to these three introductory sections, the often claimed amenability of molecular clusters to a plethora of solution and solid state spectroscopic techniques, which allow detailed structural investigations, seems to fade away as the metal nuclearity increases beyond a certain limit. For some metals this limit seems to correspond to only about 20–30 atoms. It is, however, to be predicted that the chemistry of high nuclearity clusters should profit in the near future from the development of new experimental techniques which were originally designed for colloid and solid state chemistry investigations.

---

[2] For example, the carbide atoms lodged with interstitially in the square antiprismatic cavities of some nickel carbonyl clusters display ^{13}C relaxation times in the 100–400 s range (B. T. Heaton, personal communication).

3.2.4 Homo- and Heterometallic Transition Metal Clusters

Beginning with this section, we will describe the synthetic methods which have been used to induce the growth of metal clusters in ligand shells, and discuss their chemical reactivities and structural features. Since the syntheses and reactivities of lower nuclearity clusters have been reviewed several times in the past, [95–97] we will focus mainly on higher nuclearity compounds. Since synthesis and reactivity are only different points of view to the same chemical phenomena, at the present stage it would seem more justified to place the main emphasis on the former. In order to avoid repetitions and to allow better exemplification, in the present section, which is formally dedicated to homo- and heterometallic clusters, we will describe the most general synthetic methods, without bothering as to whether the clusters are homo- or heteroleptic, or whether they contain main group elements or any molecular fragments derived thereof. In subsequent sections, however, only the more specific methods for each particular category of clusters will be examined.

The type and range of metal carbonyl clusters which have been isolated and characterized for each metal, or combination of metals, are illustrated in Table **3-3** and **3-4** respectively, and for each entry the most prominent examples have been included.

It must be emphasized at this point that a “tailored synthesis expressly devised for a given cluster” is well beyond the present state-of-the-art in the synthesis of high nuclearity clusters. At best, there are a few guidelines to follow. Each transition metal has its own chemical characteristics which descend from its particular nuclear composition and electronic configuration, and the steric and electronic factors influencing cluster growth mentioned in the previous section roughly indicate the general frame of principles. Within this frame there is ample room in which more subtle electronic factors may have an influence which can significantly affect the stereochemical behaviour of each metal, the relative strength of M–M and M–L bonds, the charge distribution and more. A simple inspection of the available data shows that although isoelectronic clusters can be isostructural, and *vice versa,* this is not necessarily always the case.

Moreover, only occasionally does a given product represent a relatively deep potential well along the reaction coordinate, and thus only supposedly minor changes in the reagents or in the experimental conditions can result in completely different products. However, even if the deliberate synthesis of a particular cluster having the desired composition and structural features is in most cases fortuitous, and there is much yet to be learned before being able to plan in detail the synthesis of a particular cluster molecule, there are at least a few general methods available to induce the growth of metal clusters. Although the individual tactics which have been adopted so far may appear to vary from case to case, the essential strategy which lies behind each successful experiment is one or a combination of the following three:

1) Ligand elimination
2) Elimination of metal fragments
3) Redox processes

Table 3-3. Some representative high nuclearity homometallic clusters.

Compound	Reference
Group 8	
$[H_4Os_{10}(CO)_{24}]^{2-}$	[98, 99]
$[Os_{17}(CO)_{36}]^{2-}$	[100]
$[Os_{20}(CO)_{40}]^{2-}$	[101]
Group 9	
$[Rh_{12}(CO)_{30}]^{2-}$	[102, 103]
$[H_2Rh_{12}(CO)_{25}]$	[104]
$[HRh_{13}(CO)_{24}]^{4-}$	[105]
$[H_2Rh_{13}(CO)_{24}]^{3-}$	[106]
$[H_3Rh_{13}(CO)_{24}]^{2-}$	[107]
$[Rh_{14}(CO)_{25}]^{4-}$	[108, 109]
$[HRh_{14}(CO)_{25}]^{3-}$	[110]
$[Rh_{14}(CO)_{26}]^{2-}$	[111, 112]
$[Rh_{15}(CO)_{27}]^{3-}$	[113, 114]
$[Rh_{15}(CO)_{30}]^{3-}$	[114]
$[Rh_{17}(CO)_{30}]^{3-}$	[115]
$[Rh_{22}(CO)_{37}]^{4-}$	[116]
$[HRh_{22}(CO)_{35}]^{5-}$	[117]
$[H_2Rh_{22}(CO)_{35}]^{4-}$	[117]
$[Ir_{12}(CO)_{24}]^{2-}$	[118]
$[Ir_{12}(CO)_{26}]^{2-}$	[119]
$[Ir_{14}(CO)_{27}]^{-}$	[120]
Group 10	
$[Ni_{12}(CO)_{21}]^{4-}$	[121, 122]
$[HNi_{12}(CO)_{21}]^{3-}$	[121, 123]
$[H_2Ni_{12}(CO)_{21}]^{2-}$	[121, 123]
$[Pd_{16}(CO)_{13}(PEt_3)_9]$	[124]
$[Pd_{23}(CO)_{22}(PEt_3)_{10}]$	[125]
$[Pd_{23}(CO)_{20}(PEt_3)_8]$	[126]
$[Pd_{38}(CO)_{28}(PEt_3)_{12}]$	[127]
$[Pt_{12}(CO)_{24}]^{2-}$	[128, 129]
$[Pt_{15}(CO)_{30}]^{2-}$	[128, 130]
$[H_xPt_{15}(CO)_8(P^tBu_3)_6]$	[131]

Table 3-3. (Continued).

Compound	Reference
$Pt_{17}(CO)_{12}(PEt_3)_8$	[132]
$[Pt_{18}(CO)_{36}]^{2-}$	[128, 129]
$[Pt_{19}(CO)_{22}]^{4-}$	[133]
$[Pt_{24}(CO)_{30}]^{2-}$	[134]
$[Pt_{26}(CO)_{32}]^{2-}$	[135]
$[Pt_{38}(CO)_x]^{2-}$	[135]

It should be clear, however, that a certain amount of overlap between these three classes is unavoidable, not only because the categorization of chemistry often leads to oversimplification, but also because the above categories have been set up without any support from mechanistic studies. Often partition has been made only on the basis of the small amount of miscellaneous information which is presently known about the experimental conditions and the kind of reagents employed in each case.

3.2.4.1 Synthesis of High Nuclearity Clusters by Ligand Elimination

Elimination of one or more ligands from the surface of a cluster with a given geometry temporarily generates sites of open coordination and electronic unsaturation. Although there are known examples of clusters which can withstand the change in the number of cluster valence electrons with only minor changes in the overall geometry, the creation of one or more "holes" in the initial ligand shell very seldom takes place without causing major consequences. Since both the metal core and the ligand shell are very soft they will tend to readapt to one another so as to give rise to a species which is electronically and/or coordinatively saturated. The possibilities which these ready rearrangements of both ligand and metal polyhedra offer has probably been underestimated in the past, and are a direct result of the ligand and metal mobilities described in the previous section. Incidentally, may also be the basic reason for the meager achievements of clusters in homogeneous catalysis.

There are four distinct ways to cause cluster condensation through the removal of ligands. Whereby three of the four are intramolecular processes and very rare, the fourth is intermolecular and by far the most common. The first intramolecular route consists of a simple readaptation of the ligand shell as exemplified by the $[Rh_{10}Au_4C_2(CO)_{20}(PPh_3)_4]$–$[Rh_{10}Au_4C_2(CO)_{18}(PPh_3)_4]$ pair of cluster compounds. [156] The second process results in only the permutation of two different metals in their positions, whereby the shape of the metal polyhedron is preserved, e.g. the $[PtRh_4(CO)_{14}]^{2-}$–$[PtRh_4(CO)_{12}]^{2-}$ pair of clusters. [157] The third way, which we may define as intramolecular condensation, is more common than the previous two and involves a change of the shape of the metal polyhedron. A pertinent example of this is the equilibrium reaction (3.1). [158]

$$[Rh_6C(CO)_{15}]^{2-} \rightleftharpoons [Rh_6C(CO)_{13}]^{2-} + 2\ CO \tag{3.1}$$

Table 3-4. Some representative heterometallic transition metal clusters.

Compound		Reference
	Group 6	
$[Mo_8Hg_4(CO)_{12}Cp_4]$		[136]
	Group 7	
$[MnAu_6(CO)_3(PPh_3)_6]^+$		[137]
$[Re_{14}Ag_2C_2(CO)_{42}Br]^{5-}$		[138]
	Group 8	
$[HFe_6Pd_6(CO)_{24}]^{3-}$		[139]
$[Fe_4Pt_6(CO)_{22}]^{2-}$		[140]
$[Fe_4Cu_5(CO)_{16}]^{3-}$		[141]
$[Fe_4Cu_6(CO)_{16}]^{2-}$		[142]
$[Fe_8Ag_{13}(CO)_{32}]^{4-}$		[143]
$[Ru_6Pd_6(CO)_{24}]^{2-}$		[144]
$[Ru_9Pt_6(CO)_{28}]^{4-}$		[45]
$[Os_6Pt_7(CO)_{21}(cod)_2]$		[145]
$[Os_9Hg_3(CO)_{33}]$		[146]
	Group 9	
$[Co_6Hg_9(CO)_{18}]$		[147]
$[H_xRh_5Ni_6(CO)_{21}]^{3-}$		[148]
$[Rh_{11}Pt_2(CO)_{24}]^{3-}$		[149]
$[Rh_{12}Pt(CO)_{24}]^{4-}$		[149]
$[Rh_{13}Pt(CO)_{25}]^{3-}$		[150]
$[Rh_{18}Pt_4(CO)_{35}]^{4-}$		[151]
	Group 10	
$[Ni_9Pt_3(CO)_{21}]^{4-}$		[152]
$[HNi_9Pt_3(CO)_{21}]^{3-}$		[152]
$[Ni_{12}Au_6(CO)_{24}]^{2-}$		[153]
$[Ni_{36}Pt_4(CO)_x]^{6-}$		[154]
$[HNi_{38}Pt_6(CO)_{48}]^{5-}$		[155]

In reaction (3.1), the thermally induced loss of two carbon monoxide ligands from the ligand shell and the resulting decrease in the electron count are compensated for by an intramolecular increase in the metal connectivity, whereby the metal polyhedron rearranges from an original trigonal prism into an octahedron. There are several other examples of what we have called here intramolecular condensation, and also of its reverse reaction, i.e. reactions in which additional ligands coordinate to the cluster without inducing cluster fragmentation but only requiring the sacrifice of one or two M-M bonds. We mention here only a few other scattered examples of such as the sequence of intramolecular condensation products, $[H_2Os_7(CO)_{22}]$, $[H_2Os_7(CO)_{21}]$ and $[H_2Os_7(CO)_{20}]$, [159] and the $[Ir_{12}(CO)_{26}]^{2-}$– $[Ir_{12}(CO)_{24}]^{2-}$, [118, 119] and $[Rh_{15}(CO)_{30}]^{3-}$–$[Rh_{15}(CO)_{27}]^{3-}$pairs of compounds. [113, 114]

More often, however, the elimination of ligands from the cluster surface results in cluster growth through an intermolecular condensation process and hereafter we will focus on the methods used to achieve this. In principle, there are several ways to promote the elimination of one or more ligands and these include, photochemical, thermal and chemical methods by the use of suitable reagents. Photochemistry, however, has had very little relevance so far in high nuclearity clusters. The only reported application of this techniques is shown in equation (3.2) [160, 161] and proceeds through the elimination of a mercury atom from the metal core of the cluster, rather than ligands from its surface.

$$[Os_{18}Hg_3C_2(CO)_{42}]^{2-} \xrightarrow{h\nu} [Os_{18}Hg_2C_2(CO)_{42}]^{2-} + Hg \tag{3.2}$$

In contrast, the thermal activation or pyrolysis of a lower nuclearity cluster having volatile ligands, carried out either in solution or in the solid state and under an inert atmosphere or in vacuum, is probably the most successful method for the preparation of higher nuclearity clusters of a variety of second and third row metals. Most of the highest nuclearity clusters of Re, Ru, Os, Rh, Ir, Pd and Pt have been obtained more or less directly in this way. Thermal activation, however, is very seldom selective and almost always give rise to very complex mixtures containing several products. Normally, these can only be separated into their separate components by employing time consuming techniques such as thin layer chromatography (TLC), high pressure liquid chromatography (HPLC), differential solubility in organic solvents, or the more or less selective precipitation with alkali and ammonium counterions in water. Nevertheless, the results have been very rewarding in several cases. Furthermore, once a given compound has been successfully separated in a pure state and its spectroscopic characteristics are known, the method can then be optimized to achieve yields in the 60–90 % range, which often eases the separation procedure. Thus, several products, which were first obtained on only a few milligrams scale, subsequently become available on a much larger scale and, in turn, have been used as the starting material for further condensation processes. As partially shown in Scheme **3-2**, the best example for this is provided by the variety of products obtained from the pyrolysis by J. Lewis, B. F. G. Johnson and coworkers of $Os_3(CO)_{12}$ under various conditions, and the recognition that throughout the years several of these have been

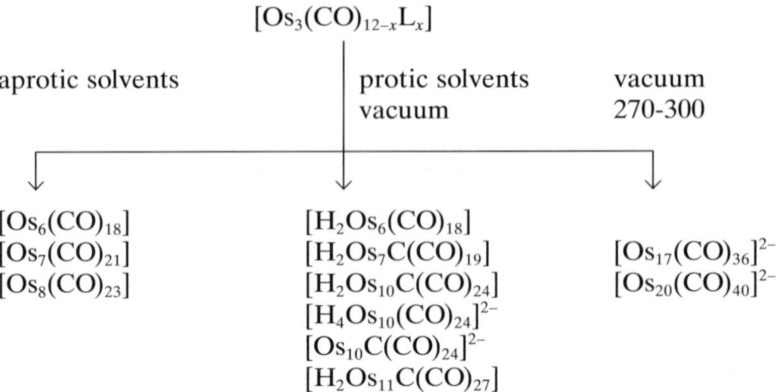

Scheme 3-2. Some selected osmium clusters obtained from the pyrolysis of $[Os_3(CO)_{12-x}L_x]$ (L = CO, MeCN, C_5H_5N, OR). [95, 100, 101]

employed as parent compounds for a wide variety of related derivatives. [92] Carbon monoxide and weakly bound nitrogen or oxygen donor ligands are not the only examples which can be thermally removed. Indeed, the thermal activation of $[Pt_5(CO)_6(PEt_3)_4]$ gave rise to $[Pt_{17}(CO)_{12}(PEt_3)_8]$ via the contemporary elimination of carbon monoxide and triethylphosphine. [132] It is reasonable to assume that these methods can probably be extended to a wider range of clusters stabilized in assorted ligand shells.

High selectivity can be achieved even in pyrolysis if the starting metal clusters carry a charge and the experimental conditions are milder. A relevant example is the pyrolysis of $[Pt_9(CO)_{18}]^{2-}$ in refluxing acetonitrile outlined in equation (3.3). Here, if the proper amount of $[Pt_{12}(CO)_{24}]^{2-}$ is used to match the metal/charge ratio in the final product then $[Pt_{19}(CO)_{22}]^{4-}$ can be obtained in greater than 80 % yield. [133] Obviously, such a delicate balance between the reagents can only be adopted when the exact composition of the product has been ascertained.

$$5\,[Pt_9(CO)_{18}]^{2-} + [Pt_{12}(CO)_{24}]^{2-} \rightarrow 3\,[Pt_{19}(CO)_{22}]^{4-} + 48\,CO \qquad (3.3)$$

In those cases where either a different transition metal or a main group heteroatom is present, the composition and the metal/charge ratio of the final product can be very close to that of the starting material. For example, the thermal decomposition of $[Rh_6N(CO)_{15}]^-$ has been reported to give rise to the formation of $[Rh_{23}N_4(CO)_{38}]^{3-}$, [162] whose composition does not exactly match that of a tetramer of the parent compound because of the loss of one rhodium atom and one negative charge per molecule of product. In general, however, either the precipitation of metal (e.g. equation 3.4), [163] the loss of charge from the decomposition of the counter cation, or the occurrence of disproportionation reactions, do not allow the complete preservation of the initial metal/charge ratios.

$$3\,[Co_6N(CO)_{15}]^- \xrightarrow{\text{diglyme},150^\circ} [Co_{14}N_3(CO)_{26}]^{3-} + 4\,Co + 19\,CO \qquad (3.4)$$

The use of either an anionic carbonyl precursor or a suitable mixture of two anions, so as to set the metal/charge ratio from the outset, is not compulsory because the final metal/charge ratio can also be controlled or achieved by addition of a carefully calibrated amount of oxidizing or reducing agent to the reaction mixture. A recent example in which reductive conditions were introduced during thermal activation of a carbonyl anion precursor is shown in equation (3.5), [164]

$$2[Rh_6N(CO)_{15}]^- + OH^- \xrightarrow{\text{refluxing MeOH}} [HRh_{12}N_2(CO)_{23}]^{3-} + 6\,CO + CO_2 \quad (3.5)$$

A more general picture of the potential of this method, whether starting from a neutral cluster or a mononuclear complex, is shown in Scheme **3-3**. In the reaction equation (3.5) and in several of the reactions summarized in Scheme **3-3**, the reductant also contributes to the loss of some moles of ligands through their chemical conversion to CO_2 or HCO_3^-.

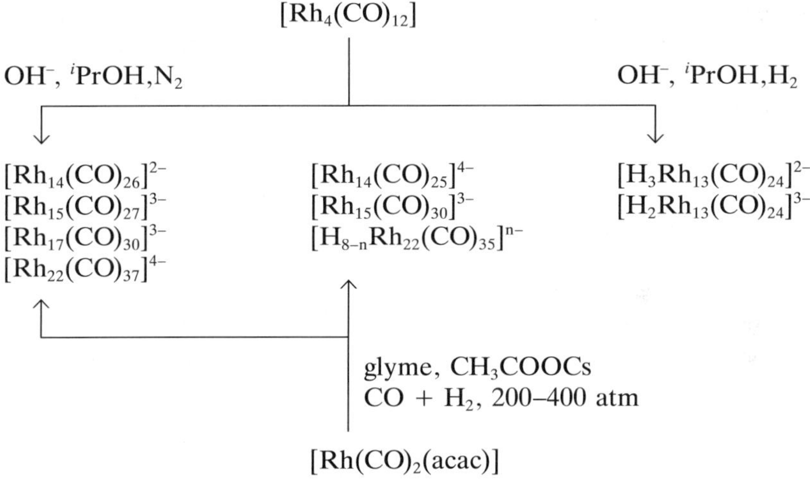

Scheme 3-3. Some selected rhodium carbonyl clusters obtained from the reductive thermolysis of $[Rh_4(CO)_{12}]$ under N_2 or H_2 at atmospheric pressure, and $[Rh(CO)_2(acac)]$ under superatmospheric pressure of carbon monoxide (acac = acetylacetonate). [95]

The elimination of coordinated ligands through their chemical transformation is a classical method for inducing the exchange of strongly coordinated ligands with weaker ones, *e.g.* CO can be substituted by nitrogen donor ligands via oxidation of the coordinated carbon monoxide with Me_3NO in the presence of the nitrogen donor. In the absence of any added donor, the above procedure can give either trimethylamine complexes or result in condensation to higher nuclearity clusters. The synthesis of $[Pd_{16}(CO)_{13}(PEt_3)_9]$ from $[Pd_{10}(CO)_{12}(PEt_3)_6]$, [124] as well as that of $[Os_6Pt_7(CO)_{21}(cod)_2]$ (cod = cyclooctadiene) from $[Os_6Pt_4(CO)_{22}$

(cod)] and Pt(cod)$_2$, [145] through the reaction of the starting materials with Me_3NO illustrates the potential of this method. Careful selection of the relative amounts of the carbonyl cluster and Me_3NO, even under pyrolytic conditions, might be a method of general applicability for presetting the CO/M ratio in the products. Unfortunately, this is not so straightforward. The trimethylamine liberated during the oxidation step can enter the coordination sphere of the cluster and induce a variety of transformations. For instance, it has been reported that $[Os_6(CO)_{18}]$ is reduced to $[HOs_6(CO)_{17}]^-$ by trialkylamineoxides via formation of an $[Os_6(CO)_{17}(NR_3)]$ intermediate, which then gives rise to the final product by transfer of a hydrogen atom from the amine to the cluster with concomitant elimination of a $[R_2N{=}CHR']^+$ iminium cation. [165] In contrast, the reaction of trimethylamineoxide with anionic species results in the preferential oxidation of the whole cluster rather than the coordinated ligands and often leads to decomposition.

A few syntheses have been reported which can be classified as ligand abstraction methods, and all of them involve the use of palladium salts. For example, treatment of $[Os_3(CO)_{10}(MeCN)_2]$ with $PdCl_2$ in dichloromethane resulted in the formation of $[Os_6(CO)_{21-n}(MeCN)_n]$, [166] whereas the reaction of $[Pd_{10}(CO)_{12}(PEt_3)_6]$ with palladium acetate afforded a mixture of $[Pd_{23}(CO)_{22}(PEt_3)_{10}]$, $[Pd_{23}(CO)_{20}(PEt_3)_8]$ and $[Pd_{38}(CO)_{28}(PEt_3)_{12}]$. [125–127] It should, however, be mentioned for the latter case that it is not clear whether palladium(II) is reduced and incorporated into the clusters, or whether it simply behaves as a phosphine scavenger; the increasing CO/PEt_3 ratios observed in the products might eventually support it functioning as a scavenger.

3.2.4.2 Synthesis of Higher Nuclearity Clusters by Elimination of Metal Fragments

The abstraction of metal atoms from clusters via their complexation by suitable ligands is a well documented method for the preparation of lower nuclearity clusters. Some more recent reports on the syntheses of lower nuclearity clusters through the degradation of higher nuclearity parent compounds are represented in equations (3.6)–(3.8). [167–169]

$$[Ni_{12}Ge(CO)_{22}]^{2-} + 6\ CO \rightleftharpoons [Ni_{10}Ge(CO)_{20}]^{2-} + 2\ Ni(CO)_4 \tag{3.6}$$

$$[Rh_{15}(CO)_{27}]^{3-} + 2\ I^- \rightarrow [Rh_{14}(CO)_{25}]^{3-} + [Rh(CO)_2I_2]^- \tag{3.7}$$

$$[Ni_9C(CO)_{17}]^{2-} + 3\ PPh_3 \rightarrow [Ni_7C(CO)_{12}]^{2-} + [Ni(CO)_2(PPh_3)_2] + [Ni(CO)_3(PPh_3)] \tag{3.8}$$

Degradation under mild conditions in the presence of carbon monoxide is generally limited to compounds of the first row metals and, in particular, to nickel derivatives. This is probably due to the combination of a greater difference between the Ni–Ni and Ni–CO bond energies and the presence of “holes” in the cluster's ligand shell which allow a more facile approach of attacking ligands to the metal atoms of the cluster.

Apart from a few relevant exceptions, e.g. $[Rh_5(CO)_{14}(PPh_3)]^-$, [170] and $[Co_6Ni_2C_2(CO)_{14}(MeCN)_2]^{2-}$, [171] ligands which are less π-acidic than CO, e.g. nitrogen or phosphorus donors, are rarely found coordinated to an anionic metal cluster. This presents an opportunity to revert the above degradation trend and instead promote cluster condensation, rather than cluster degradation. The necessary condition is that the ligand whose function is to abstract the metal atom from the cluster has as its most stable carbonyl substituted derivative, a complex which has a CO/M ratio greater than the corresponding average ratio for the metals in the anionic cluster. In such a case the removal of metal atoms from the anionic carbonyl cluster as complexes with larger CO/M ratios as the cluster, will generate progressively more coordinatively unsaturated and unstable cluster units. These may then condense with one another to form stabil compounds. An enlightening example of this type of reaction is provided by the almost quantitative yield of $[Ni_{16}C_4(CO)_{23}]^{4-}$ from the reaction of $[Ni_{10}C_2(CO)_{16}]^{2-}$ with triphenylphosphine. [172] As shown in equation (3.9), this reaction is thought to occur *via* the formation of a transient "$[Ni_8C_2(CO)_{12}]^{2-}$" cluster complex, which readily dimerizes into the final product.

$$2\ [Ni_{10}C_2(CO)_{16}]^{2-} + 8\ L \rightarrow 2\ \text{“}[Ni_8C_2(CO)_{12}]^{2-}\text{”} + 4\ Ni(CO)_2L_2$$
$$\downarrow$$
$$[Ni_{16}(C_2)_2(CO)_{23}]^{4-} + CO \qquad (3.9)$$

Other higher nuclearity clusters obtained by the related, although much less selective, abstraction of $[Ni(CO)_2]$ and $[Ni(CO)_3]$ moieties from $[Ni_9C(CO)_{17}]^{2-}$ with phosphine ligands include the $[Ni_{12}C_2(CO)_{16}]^{2-}$ and $[Ni_{11}C_2(CO)_{15}]^{2-}$ dicarbide compounds. These clusters have been obtained in mixtures containing the lower nuclearity $[Ni_7C(CO)_{12}]^{2-}$ derivative. [169]

3.2.4.3 Synthesis of Higher Nuclearity Clusters by Redox Processes

Only a few clusters have been shown to withstand redox changes in their formal oxidation states without major structural modifications. Some leading examples which have been substantiated by X-ray structural investigations are $[Fe_3Pt_3(CO)_{15}]^{n-}$ (n = 0,1,2), [140, 173] $[Rh_{12}C_2(CO)_{23}]^{n-}$ (n = 3,4), [174, 175] $[Co_3Ni_9C(CO)_{20}]^{n-}$ (n = 2,3), [43, 176] $[Co_{13}C_2(CO)_{24}]^{n-}$ (n = 3,4), [177, 178] and $[Ni_{13}Sb_2(CO)_{24}]^{n-}$ (n = 2,3). [179] Their occurrence arises from the fact that the frontier orbitals of these clusters are delocalized over several nuclei and are only weakly antibonding, nonbonding, or weakly bonding. As a result, very small changes are found in the structural parameters and the cohesive energies of the M–M and M–L bonds, as well as intraligand interactions. As shown in Table **3-5**, the major change observed in the solid state structures of the $[Fe_3Pt_3(CO)_{15}]^{n-}$ (n = 0,1,2) series is a slight and progressive lengthening of all the Pt–Pt and Fe–Pt interatomic distances as the charge increases from n = 0 to n = 2, due to the progressive population of a weakly antibonding Pt–Pt and Fe–Pt LUMO (Table **3-5**). Accordingly, the unpaired electron present in the $[Fe_3Pt_3(CO)_{15}]^-$ radical anion is weakly coupled to three equivalent platinum nuclei. [180] Incidentally, the combination of structural, spectroscopic and electrochemical information available for these clusters provides a sound experimental basis for the theoretical investigation of the bonding characteristics within these clusters and can lend support to the purely theoretical results obtained on other related systems. [181] In the above cases, the minor structural changes correspond to complete electrochemically reversible redox changes. The latter can often extend over an even wider range of oxidation states. [182–184]

In other cases, the attempt to introduce additional electrons into a cluster results in the elimination of ligands or the disruption of some M–M interactions, and these can often be chemically reversible. The former process generally keeps the cluster valence electron count unchanged through the substitution of ligands with negative charges, whereas the latter may lead to either an intramolecular rearrangement of both the metal and ligand polyhedron, or cluster fragmentation. In the first event, the severe structural changes undergone by the metal polyhedron are probably triggered by the change in the ligand arrangement and by the necessity to find a new equilibrium position which allows the most effective distribution of ligands and minimizes charge separation among the metal sites. In all the above events, the significant structural changes brought about by the change in oxidation state prevent the observation of electrochemically reversible redox processes. [182–184] A few illustrative examples of the various results obtained when reduction is accompanied by the elimination of ligands or disruption of M–M interactions are shown pictorially in Schemes **3-4** and **3-5**. It should be noted that fragmentation processes resulting from the reduction of a cluster may be complicated by comproportionation reactions like those outlined in equations (3.10) and (3.11).

$$[Pt_{12}(CO)_{24}]^{2-} + 2\,e^- \rightarrow 2\,[Pt_6(CO)_{12}]^{2-} \qquad (3.10)$$

$$[Pt_{12}(CO)_{24}]^{2-} + [Pt_6(CO)_{12}]^{2-} \rightarrow 2\,[Pt_9(CO)_{18}]^{2-} \qquad (3.11)$$

Conversely, the oxidation of a cluster with "innocent" oxidizing agents[3], probably because it is mainly performed in the absence of free ligands, generally results in intra- or intermolecular condensations not involving concomitant loss of coordinated ligands (e.g. the reverse processes in Scheme **3-5**), or into ever increasingly complicated condensation processes with or without concomitant elimination of some ligands. Probably the most spectacular example of dimerization reaction, possibly triggered by the simple elimination of electrons[4], is the collapse of two

[3] By the term "innocent" we mean here an oxidizing agent which is fully shielded by its own ligands and capable of transfering electrons with the cluster through an outer sphere mechanism, without being involved in condensation processes, e.g. $FeCp_2^+$.

[4] The caution here is due to the fact that the exact number of carbonyl groups surrounding the Pt_{38} metal core could not be determined by the X-ray structure investigation carried out so far.

Table 3-5. Comparison among some molecular parameters in the $[Fe_3Pt_3(CO)_{15}]^{n-}$ series clusters with variable oxidation states.

Compound	Pt–Pt (Å)	Fe–Pt (Å)	Ref
$[Fe_3Pt_3(CO)_{15}]$	2.590	2.578	[173]
$[Fe_3Pt_3(CO)_{15}]^-$	2.656	2.587	[140]
$[Fe_3Pt_3(CO)_{15}]^{2-}$	2.750	2.596	[140]

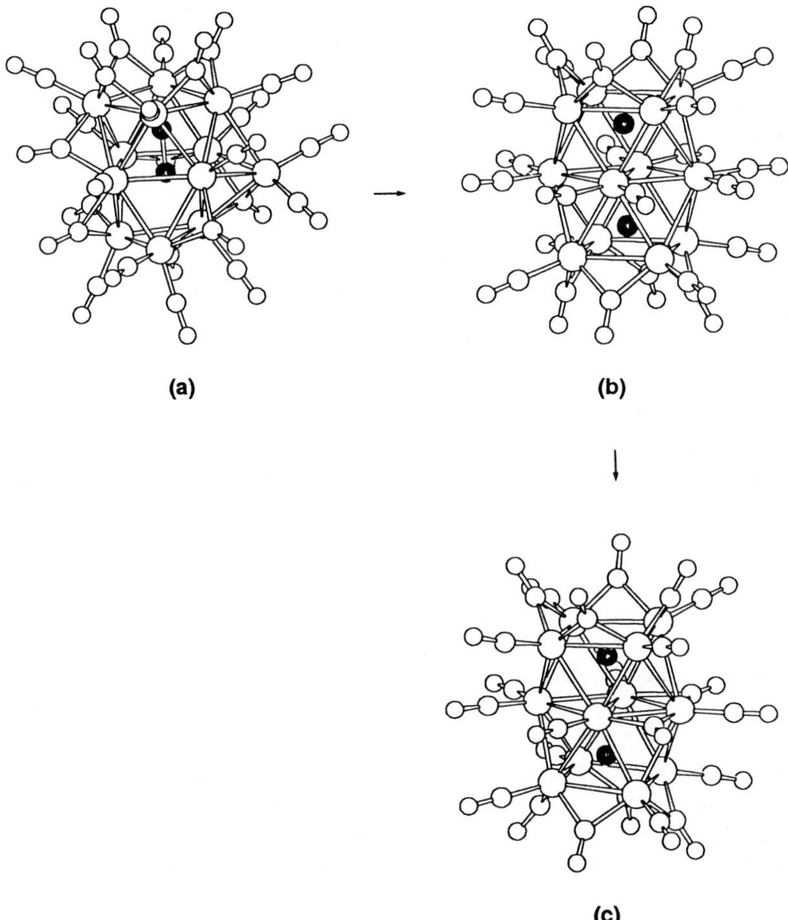

Scheme 3-4. Pictorial representation of the structural changes caused by the substitution of a carbonyl group with a pair of electrons in the series of $[Rh_{12}C_2(CO)_{25-n}]^{2n-}$ (n = 0, 1, 2) [174, 185, 186] compounds.[a]

[a] Lacking experimental verification, the first step should only be considered as formal.

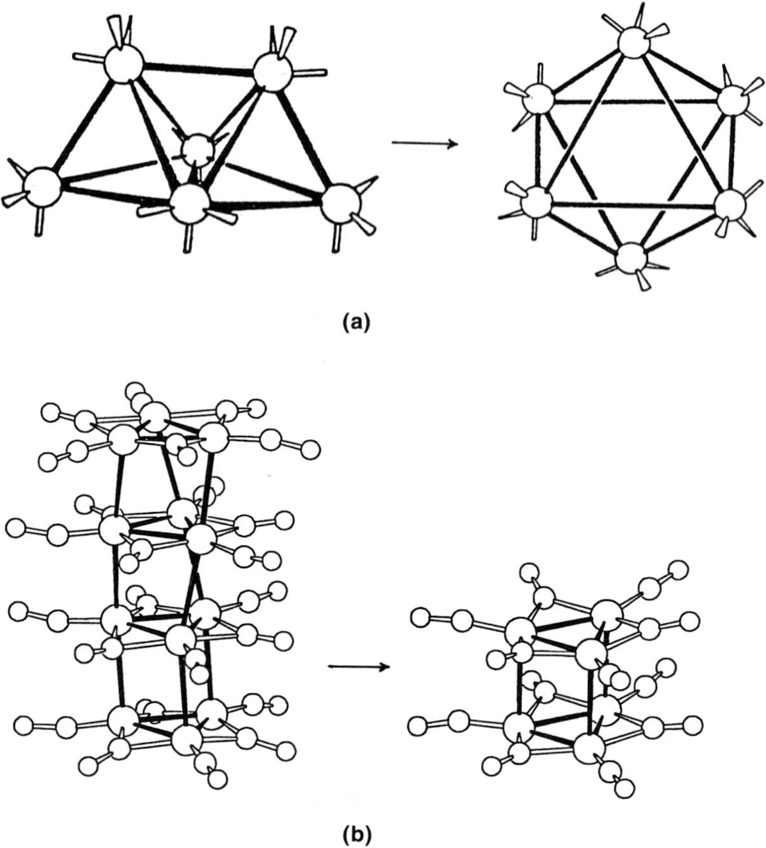

Scheme 3-5. Pictorial representation of the structural changes caused by the addition of electrons: (a) intramolecular loss of metal cohesion upon a change in the number of CVE in the $[Os_6(CO)_{18}]$-$[Os_6(CO)_{18}]^{2-}$ pair of compounds, [187, 188] (b) fragmentation of $[Pt_{12}(CO)_{24}]^{2-}$ into two $[Pt_6(CO)_{12}]^{2-}$ ions.

molecules of $[Pt_{19}(CO)_{22}]^{4-}$ into one molecules of $[Pt_{38}(CO)_{44}]^{2-}$ through the reaction with protic acids. [135, 189] On the other hand, the oxidation of $[Ir_6(CO)_{15}]^{2-}$ gave $[Ir_{12}(CO)_{26}]^{2-}$ through the concomitant loss of two carbon monoxide groups per molecule of reagent. [119] The structure of $[Ir_{12}(CO)_{26}]^{2-}$ is represented schematically in Figure **3-5a**.

Potential oxidizing agents can also enter the coordination sphere of the cluster and be associated in a stable or a labile fashion. An impressive number of products derived from the simple addition of one or more CuL, AgL, AuL, or HgR moieties (where L is a phosphine or acetonitrile ligand and R is either an alkyl group or a halogen atom) at the surface of a cluster are examples of stable adducts. [190] The few compounds collected in Table **3-6** illustrate the enormous potential variety of these adducts, and some of their very interesting properties

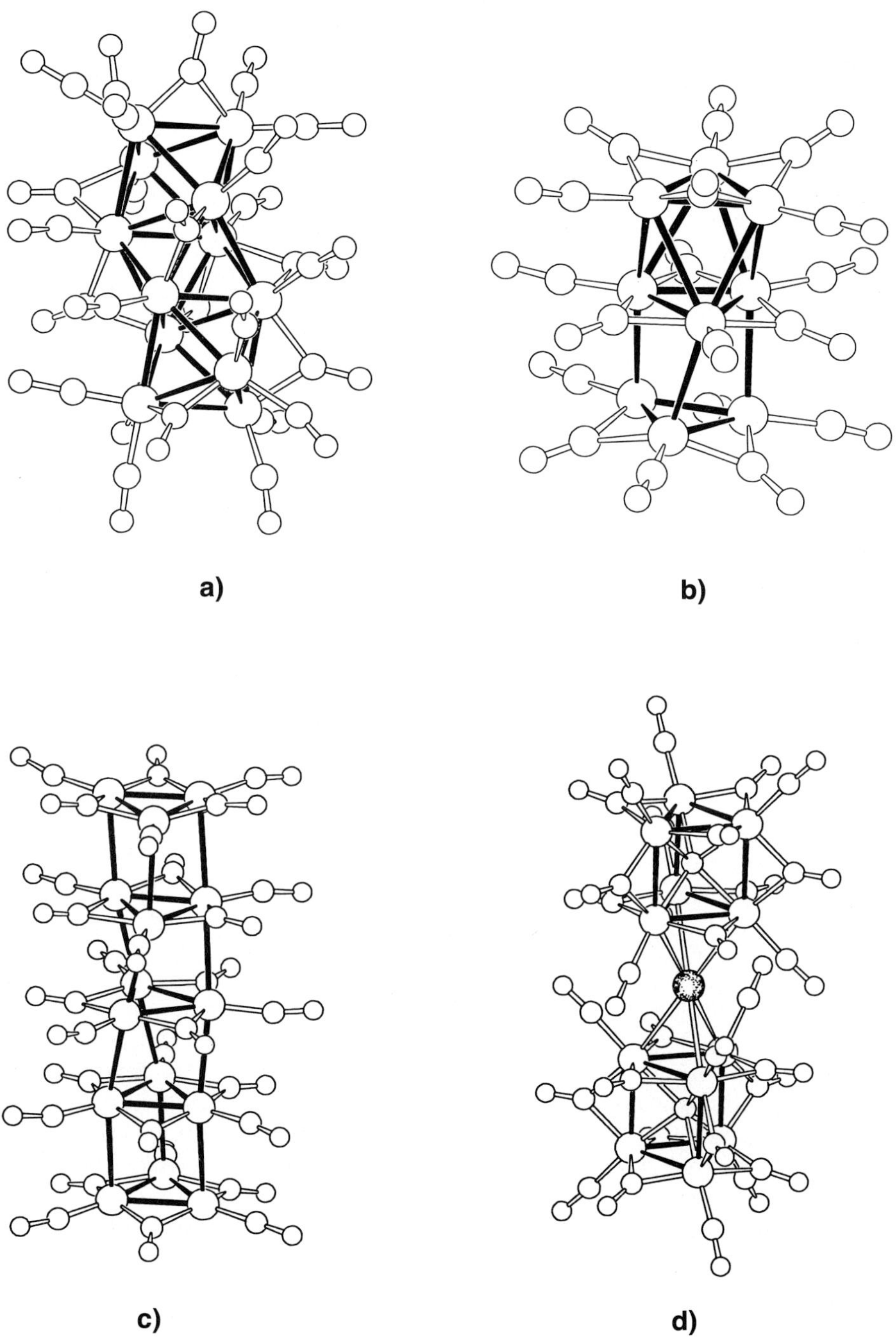

Figure 3-5. Some selected examples of C_3 pseudo one dimensional growth: a) $[Ir_{12}(CO)_{26}]^{2-}$, [119] b) $[Ni_9(CO)_{18}]^{2-}$, [85] c) $[Pt_{15}(CO)_{30}]^{2-}$, [130] and d) $[Rh_{12}AgC_2(CO)_{30}]^{3-}$. [213]

Table 3-6. Selected products from the condensation of CuL, AgL, AuL and HgR fragments onto clusters of various metals.

Compound		Reference
	Cu	
$[Rh_6C(CO)_{15}\{Cu(MeCN)\}_2]$		[192]
$[Os_{10}C(CO)_{24}\{Cu(MeCN)\}]^-$		[193]
$[Os_{11}C(CO)_{27}\{Cu(MeCN)\}]^-$		[194]
$[Os_{11}C(CO)_{27}\{Cu(MeCN)\}_2]$		[195]
	Ag	
$[H_2Ru_4(CO)_{12}\{Ag(PPh_3)\}_2]$		[196]
	Au	
$[Os_{10}C(CO)_{24}\{Au(AuPPh_3)_3\}]$		[197]
$[Os_{11}C(CO)_{27}\{Au(PPh_3)\}]^-$		[194]
$[Rh_{10}C_2(CO)_{18}\{Au(PPh_3)\}_4]$		[156]
$[Rh_{10}C_2(CO)_{20}\{Au(PPh_3)\}_4]$		[156]
$[Rh_{12}C_2(CO)_{23}\{Au(PPh_3)\}]^-$		[198]
$[Os_{10}C(CO)_{24}(AuBr)]^-$		[199]
	Hg	
$[Os_{10}C(CO)_{24}(HgR)]^-$		[199, 200]
$[Ru_{10}C_2(CO)_{28}Cl_2(HgCl)_2]$		[201]

have already been mentioned in the previous section. Since a ML (M = Cu,Ag,Au) group and a hydrogen atom are isolobal, [191] we will discuss at this stage those derivatives having peripheral hydride ligands (derivatives with interstitial hydride ions will be covered in the next section if only for the sake of structural homogeneity even though no real peculiar chemical feature differentiates the two coordination types). Despite the possibility of a more or less rapid exchange via peripheral mobility (or even by plunging inside the metal core and re-emerging at the surface in a different site) the hydrogen must find a well-defined coordination site on the surface of a cluster. Consequently, even the simple action of adding a proton to a cluster may not only cause a change in the ligand distribution, but also reversible rearrangements of the metal frame. As examples consider the conversion of the octahedral $[HOs_6(CO)_{18}]^-$ into the μ_3-capped square pyramidal $[H_2Os_6(CO)_{18}]$ upon protonation [202] (in contrast to their ruthenium congeners) [203, 204] and the structural differences between the clusters $[HOs_7C(CO)_{19}]^-$ and $[H_2Os_7C(CO)_{19}]$. [205] Such differences can be attributed to the fact that different types of metal polyhedra may display the same number of CVE and their overall energies are probably not very different. Often, the hydride cluster resulting from the protonation reaction is labile and can readily lose hydrogen and/or carbon monoxide and undergos dimerization reactions or more complex condensation processes (Equations 3.12 through 3.14). [206, 186, 112]

$$[Rh_6(CO)_{15}]^{2-} \rightarrow [HRh_6(CO)_{15}]^{-} \rightarrow [Rh_{12}(CO)_{30}]^{2-} \tag{3.12}$$

$$[Rh_6C(CO)_{15}]^{2-} \rightarrow [HRh_6C(CO)_{15}]^{-} \rightarrow [Rh_{12}C_2(CO)_{24}]^{2-} \tag{3.13}$$

$$[Rh_7(CO)_{16}]^{3-} \rightarrow \text{“}[HRh_7(CO)_{16}]^{2-}\text{”} \rightarrow [Rh_{14}(CO)_{26}]^{2-} \tag{3.14}$$

The use of a metal ion with weakly coordinated ligands as a potential oxidizing agent can result in oxidation from which the eventual product may or may not incorporate the metal in the cluster. Some relevant examples are represented in equations (3.15)–(3.17). [205, 152–155]

$$3\,[Ni_6(CO)_{12}]^{2-} + [Ni(EtOH)_x]^{2+} \rightarrow 2\,[Ni_9(CO)_{18}]^{2-} + Ni + x\ EtOH \tag{3.15}$$

$$[Ni_6(CO)_{12}]^{2-} + [PtL_2Cl_2] \rightarrow [H_{6-n}Ni_{38}Pt_6(CO)_{48}]^{n-},\ [Ni_{36}Pt_4(CO)_x]^{6-} \text{ and } [H_{4-n}Ni_9Pt_3(CO)_{21}]^{n-}\ (L = SEt_2, PhCN, Cl^-) \tag{3.16}$$

$$[Ni_6(CO)_{12}]^{2-} + [AuLCl] \rightarrow [Au_6Ni_{12}(CO)_{24}]^{2-}\ (L = PPh3) \tag{3.17}$$

The reaction outlined in equation (3.16) ultimately affords $[Pt_{12}(CO)_{24}]^{2-}$ and Ni^{2+} ions.

The corresponding reactions of iron carbonyl anions with Pd,Pt, Cu, and Ag halides have been used to synthesize several bimetallic clusters such as $[Fe_3Pt_3(CO)_{15}]^{n-}$, [140] $[Fe_3Cu_3(CO)_{12}]^{3-}$, [141] $[Fe_4Ag_4(CO)_{16}]^{4-}$, [208] $[Fe_4M_5(CO)_{16}]^{3-}$ (M = Cu,Ag), [141, 208] $[Fe_4Cu_6(CO)_{16}]^{2-}$, [142] $[Fe_4Pt_6(CO)_{22}]^{2-}$, [140] $[H_{4-n}Fe_6Pd_6(CO)_{24}]^{n-}$, [139] and $[Fe_8Ag_{13}(CO)_{32}]^{4-}$. [143] A recent attempt to extend these reactions to ruthenium carbonyl anions resulted in the synthesis of $[Ru_6Pd_6(CO)_{24}]^{2-}$ [144] and $[Ru_9Pt_6(CO)_{28}]^{2-}$ derivatives. [45] The reduced tendency of one of the two metals for binding carbon monoxide in several of the above compounds favors the application of the isolobal analogies between organic, inorganic and organometallic fragments. [191] As already used with great success in smaller clusters, isolobal concepts can not only aid in the rationalization of the structural features and the electronic configuration of larger metal clusters, but may also have some fallout in the design of new syntheses and products.

To close this section let us consider the more classical redox condensation processes involving the comproportionation reactions between two carbonyl species having different oxidation states. The great importance and general applicability of these processes was first recognized by P. Chini, [57] and gave a vigorous impulse to cluster chemistry during the seventies and triggered the synthesis of a wide range of new carbonyl anions of Co, Rh, Ir, Ni and Pt of increasing nuclearity within just a few years. Several of these are included in the Tables of this chapter for they are still the most representative and cited examples of their general class or in their own specific fields.

Redox condensation can occur by the simple addition of a mononuclear carbonyl cationic species to an anionic cluster as shown in equation (3.18). Such condensation reactions allow for a stepwise and controlled construction of higher nuclearity clusters whereby the regeneration of a sufficiently high negative charge on the cluster can be accomplished by appropriate reduction or deprotonation steps in the condensation sequence. [168]

$$[Rh_{14}(CO)_{25}]^{4-} + [Rh(CO)_2(MeCN)_2]^{+} \rightarrow [Rh_{15}(CO)_{27}]^{3-} + 2\ MeCN \quad (3.18)$$

The most striking class of redox condensation reactions is that in which two otherwise stable clusters readily and quantitatively collapse into one cluster upon being brought in contact with each other. Some remarkable more recent examples of homo- and heterometallic clusters formed by this process are shown in equations (3.19) and (3.20). [209, 210]

$$[Rh_5(CO)_{15}]^{-} + [Rh_4(CO)_{11}]^{2-} \rightarrow [Rh_9(CO)_{19}]^{3-} + 7\ CO \quad (3.19)$$

$$[Rh_5Pt(CO)_{15}]^{-} + [Rh_4Pt(CO)_{12}]^{2-} \rightarrow [Rh_9Pt_2(CO)_{22}]^{3-} + 5\ CO \quad (3.20)$$

The course of the reaction is often more complicated, however, and one or more side-products, of variable size and charge, may accompany the expected comproportionation product. It is conceivable that this situation arises from a competition between disproportionation steps and ligand elimination steps resulting from one or more of the transition states which the comproportionation process should traverse. For the sake of clarity, the above presumption is illustrated in Scheme **3-6** with an interpretation for the formation of $[Co_6Ni_2C_2(CO)_{16}]^{2-}$ from the condensation of two molecules of $[Co_3(CO)_9CCl]$ with $[Ni_6(CO)_{12}]^{2-}$. [169, 171] Thus, $[Ni(CO)_4]$ is lost during the condensation process and $[Co_6Ni_2C_2(CO)_{16}]^{2-}$ is formed rather than the "expected" $Co_6Ni_6C_2(CO)_{21}$ species. Such a cluster should be either similar to the structure of the isoelectronic $[Ni_{12}C_2(CO)_{16}]^{4-}$ species or display a 3,3,3,3 stacking arrangement of Co,Ni,Ni and Co atoms.

$$2[Co_3(CO)_9CCl] + [Ni_6(CO)_{12}]^{2-}$$

$$\downarrow$$

$$\text{“}[Co_6Ni_6C_2(CO)_{30}]\text{“} + 2\ Cl^{-}$$

$$\swarrow \qquad \searrow$$

$$\text{“}[Co_6Ni_6C_2(CO)_{21}]\text{“} + 9\ CO \qquad [Co_6Ni_2C_2(CO)_{16}]^{2-} + Ni^{2+} + 3[Ni(CO)_4] + 2\ CO$$

Scheme 3-6. The stoichiometry of the product of a redox condensation reaction may greatly differ from the one expected for a simple condensation by elimination of ligands.

3.2.4.4 Structural Features of Homo- and Heterometallic Clusters

As in the previous sections, rather than give a complete and systematic description of the structural features of transition metal clusters, we will confine ourselves to some trends in cluster growth, which can be roughly delineated from the enormous plethora of metal cluster structures stabilized in ligand shells known to date. The obvious fee which must be paid in doing this is a certain degree of approximation. Only a few key structures will be highlighted which support some particularly interesting structural aspects or broad concepts which effect cluster growth.

a) The Pseudo One-dimensional Growth Path

The pseudo one dimensional growth of metal clusters, besides assembling metal atoms in straight or zig-zag chains (see Section 3.2.6), occurs primarily through the stacking of triangular, square or pentagonal arrays of metal atoms. Consequently, these have been classified by this symmetry into the C_3, C_4, or C_5 paths illustrated schematically in Schemes **3-7** to **3-9**. As can be seen in Scheme **3-7**, triangular arrays of three metal atoms can be stacked in a prismatic fashion (path ***a***) an antiprismatic fashion (path ***c***), or a combination of the two (path ***b***).

Real examples of clusters formed by path ***a*** include the $[Pt_3(CO)_3(\mu_2\text{-}CO)_3]_n^{2-}$ derivatives which, by virtue of the progressively longer stacks of inner Pt_3 triangles and outer $(CO)_6$ hexagons, are composed of pseudo-cylindrical metal cores surrounded by concentric pseudo-cylindrical shells of ligands and may be considered as fragments of a molecular cable. All compounds in which $2 \leq n \leq 6$ have been structurally characterized [129, 130] and, there is experimental evidence for species with much higher values of *n*. [128] In principle, there are no steric limits to the infinite growth of such a "cable" along path ***a***. The formation of two concentric metal and carbon monoxide pseudo-cylinders is perfectly allowed as long as the CO/M ratio of 2 and the interlayer spacing of *ca.* 3 Å, which prevents the occurrence of strained non-bonding repulsions among the carbonyl layers, are preserved. In contrast, the progressive growth of clusters via the idealized paths ***b*** and ***c*** will be hampered because of the shorter interlayer spacing possible with an antiprismatic orientation at equivalent interlayer M–M contacts. The validity of such a limitation is supported by the observation that clusters from paths ***b*** and ***c*** have been limited to only $n = 3$ and $n = 4$ respectively. Indeed, the only known example of a path ***b*** cluster is $[Ni_9(CO)_{18}]^{2-}$ (Figure **3-5b**), [85] whereas path ***c*** is more well documented with the structures of $[Rh_9(CO)_{19}]^{3-}$, $[H_2Rh_{12}(CO)_{25}]$, and $[Ir_{12}(CO)_{26}]^{2-}$ (Figure **3-5a**). [209, 104, 119] It is worth noting that the latter two clusters display the same metal frame yet are not isoelectronic. It may be argued that the ligand shells of the two are different, however, the recent characterization of a $[Ru_6Pd_6(CO)_{12}(\mu_2\text{-}CO)_6(\mu_3\text{-}CO)_6]^{2-}$ cluster, [144] which is isostructural with the corresponding $[HFe_6Pd_6(CO)_{24}]^{3-}$ congener, [139] supports the view that small variations in electron count may be allowed under the same cluster core symmetry as a result of the different transition metal orbital energies and orbital overlaps on descending a group.

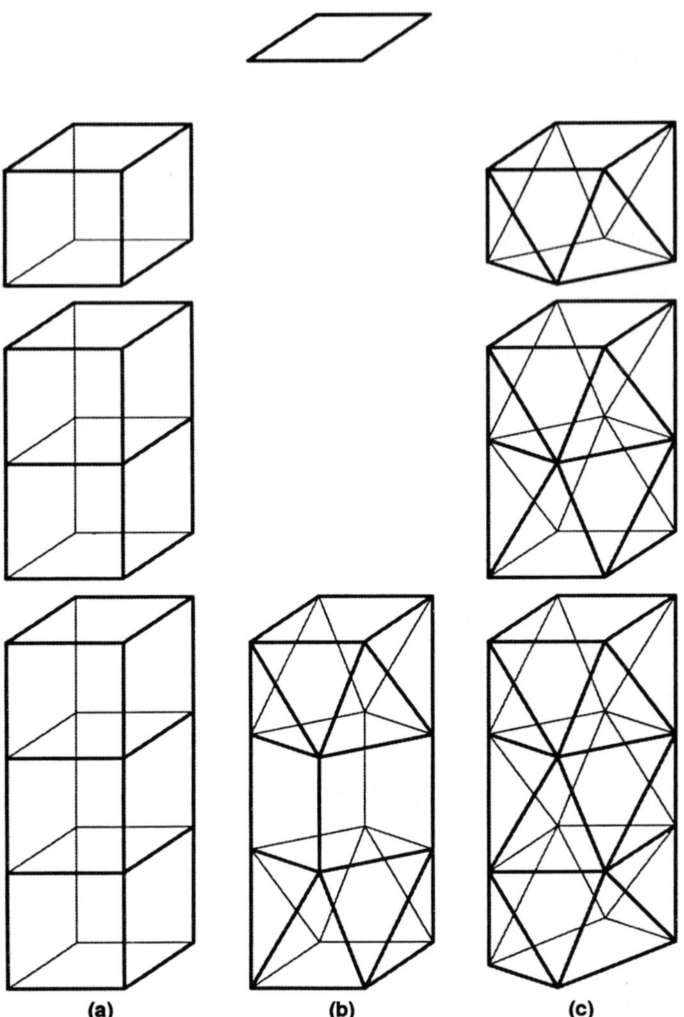

Scheme 3-7. The pseudo one-dimensional growth along a C_3 axis.

As an alternate option to the homogeneous stacking of metal triangles, the pseudo one dimensional C_3 symmetry growth can be sustained by μ_3 capping of the trigonal prism or the octahedron on opposite ends with single metal atoms, or by the formal insertion in any possible combination of triangles and/or single atoms. This alternative growth path is, for instance, chosen by $[Pt_6Hg(2,6\text{-}Me_2C_6H_3NC)_{12}]$, $[Pt_6Hg_2(2,6\text{-}Me_2C_6H_3NC)_6(dpph)_3]$ (dpph = bis-diphenylphosphine hexane), and $[Pt_6Hg_2(CO)_6(PR3)_6]$, [211, 212] whereby the former has a 3,1,3 and the latter two a 3,1,1,3 stacking sequence, and may be described as sandwich Pt_3L_6 complexes of a Hg atom and a Hg_2 molecule respectively. Other related examples are the 3,3,1,3,3 stacking sequence in $[AgRh_{12}C_2(CO)_{30}]^{3-}$

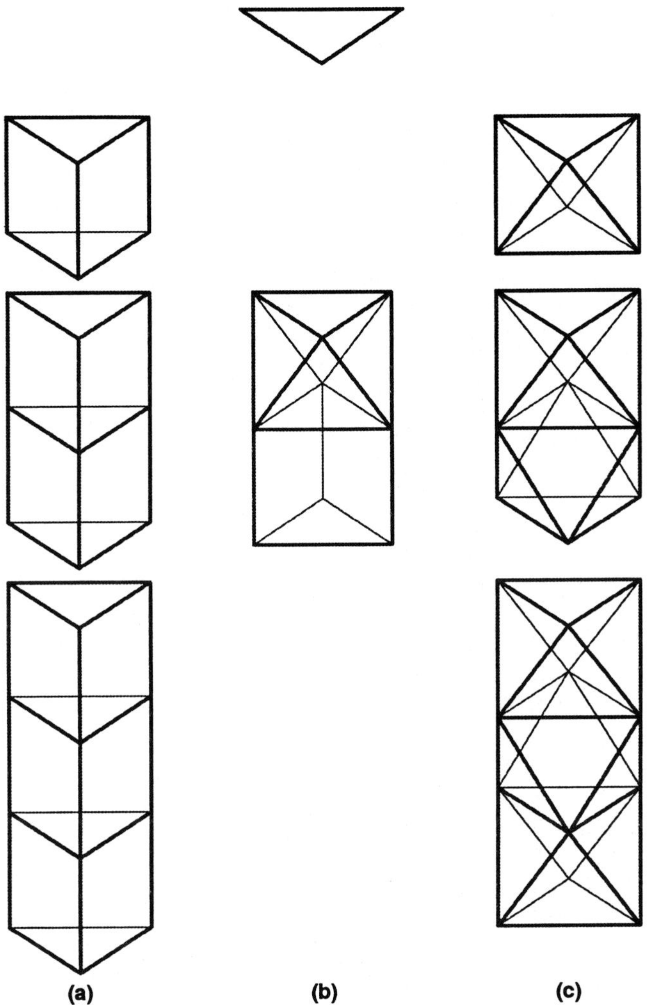

Scheme 3-8. The pseudo one-dimensional growth along a C_4 axis.

(Figure **3-5d**), [213] and the 3,3,1,1,3,3 stacking sequence displayed by $[Rh_{14}C_2(CO)_{33}]^{2-}$. [214]

For the sake of exemplification, the last two examples necessitated the mention of clusters which contain interstitial main group elements. This necessity will become even more pressing on examining the C_4 and C_5 growth paths and arises from the fact that transition metals, in contrast to main group elements, possess valence d electrons which may be used to maximize the cohesive energy. As a result, the formation of polyhedra with cavities larger than simple tetrahedral or octahedral sites is disfavored unless they can be sustained by the presence of a heteroatom. There are only very few exceptions to this rule; one is represented by the above mentioned $[Pt_3(CO)_6]_n^{2-}$ platinum clusters, which in the solid state show empty trigonal prismatic cages. At this stage, however, we only wish to highlight the metal frame character as a whole, rather than focus on the specific features brought about by the presence of an interstitial main group element or molecular fragment derived thereof. These aspects and the presentation of other exceptions will be discussed in the next section.

The C_4 pseudo one dimensional growth path is based on the correspondingly stacking of squares, either in eclipsed or staggered conformations.

The staggered 4,4′ as well the related 1,4,4′ and 1,4,4′,1 stacking forms[5] (i.e. uncapped, monocapped, and bicapped) are displayed by several square antiprismatic clusters, and their cavities are filled by miscellaneous main group atoms. Some representative compounds are $[Ni_8C(CO)_{16}]^{2-}$ [215] and $[Rh_9P(CO)_{23}]^{2-}$. [82] Furthermore, several other compounds also have a metal frame based on a staggered 4,4′,4 stacking sequence, such as the previously shown $[Rh_{12}C_2(CO)_{25-n}]^{2n-}$ (see Scheme **3-4**). These may only be approximately included in this path, owing to the presence of rhombohedral distortions of the central Rh_4 layer which probably arise from the necessity to provide a better lodging for the interstitial carbide atoms. In our rough presentation here we have arbitrarily decided to underestimate the relevance of these distortions and to refer those readers who are interested in more fine structural details to the original papers. [174, 185, 186] Finally, an example of a staggered 4,4′,4,4′ stack is found in the structure of $[Rh_{17}S_2(CO)_{32}]^{3-}$. [216] In this compound, the outer square antiprismatic cages are filled by two S atoms, whereas the central cavity is rather swelled by the presence of the 17-th rhodium atom. Alternatively, if the rhodium atom is counted as a distinct layer, then the stacking sequence can be written as 4,4′,1,4,4′. However, in order to avoid possible conflicting definitions, we shall adopt the convention of dismissing all metallic or non-metallic interstitial atoms from the stacking sequence, even when the interstice is occupied by a main group element whose size is similar to that of the surrounding metal atoms.

The first step along the eclipsed C_4 path is represented by the 4,4 stack, and the only relevant examples are the non-centered $[Ni_8(CO)_8(\mu_4\text{-}PPh)_6]$ and the nickel centered $[Ni_8(\mu_8\text{-}Ni)(CO)_8(\mu_4\text{-}GePh)_6]$ clusters. [217, 218] The next step, that is the 4,4,4 stack, can be recognized in the $[H_{8-n}Rh_{22}(CO)_{35}]^{n-}$ ($n = 4{,}5$) and $[Rh_{18}Pt_4(CO)_{35}]^{4-}$ clusters; [117, 151] the 10 additional metal atoms in these clusters are found at two interstitial and eight μ_4-capping sites. The suggested presence of three or four hydride atoms in the homometallic rhodium clusters stems from the equivalence and interchangeability of platinum atoms for RhH moieties, which has been established in several structures.

As shown in Scheme **3-9**, the C_5 path starts from a pentagon whereby, this geometry has only recently been established in a transition metal cluster with the synthesis of $[Mn_5(\mu_5\text{-}In)(CO)_{20}]^{2-}$. [219] Transition metal clusters with a pentagonal bipyramidal geometry are likewise very rare, e.g. $[Au_7(PPh_3)_7]^+$ and $[Au_6Mn(CO)_3(PPh_3)_6]^+$. [220, 137] As for the C_3 and C_4 paths, the successive stacking of

[5] In this and the following pages, the primed layer has a different orientation with respect to the pertinent rotation axis as the previous layer.

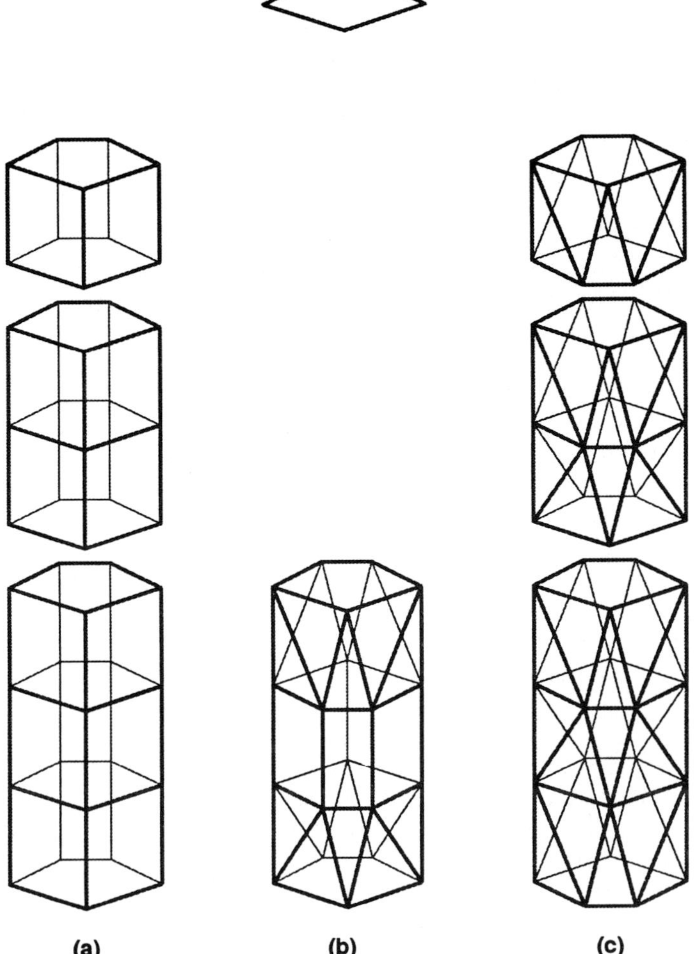

Scheme 3-9. The pseudo one-dimensional growth along a C_5 axis.

pentagons can give rise to pseudo one dimensional clusters which can show systematically eclipsed, staggered, or mixed conformations. The only known example of an eclipsed 5,5 stack is represented by $[Ni_{10}(\mu_5\text{-PMe})_2(\mu_4\text{-PMe})_5(CO)_{10}]^{2-}$, [240] whereas the $[Ni_{10}(\mu_{10}\text{-Ge})(CO)_{20}]^{2-}$ cluster adopts the staggered 5,5′ stacking (Figure **3-6a**). [167]

As can be seen from the above two compounds, each of these metal geometries requires a different number of valence electrons, as in the eclipsed and staggered conformations of the triangle and square stacks.

An eclipsed 1,5,5,1 stack, however, can be identified in the structure of the $[Rh_{15}C_2(CO)_{28}]^-$ dicarbide cluster (Figure **3-6b**). [221] The next step along path ***a*** is exemplified by the 1,5,5,5,1 sequence in $[Pt_{19}(CO)_{22}]^{4-}$ [133] (Figure **3-6c**)

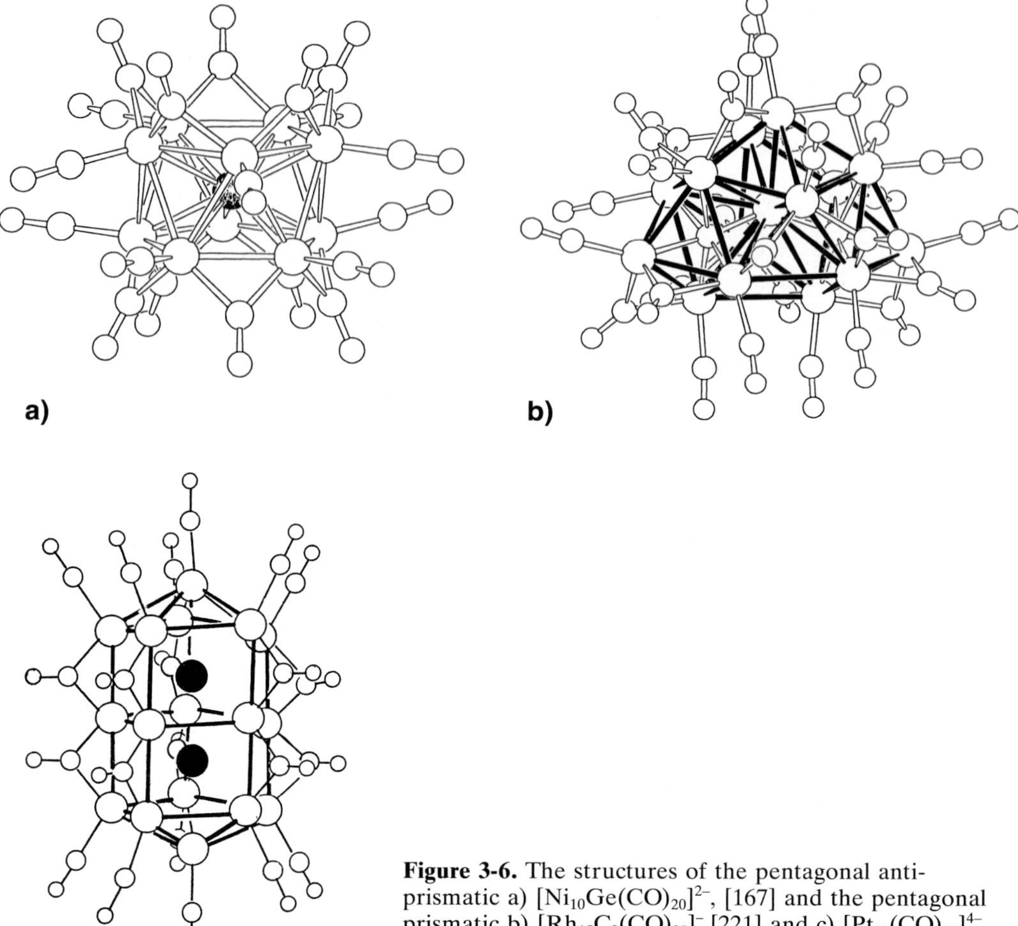

Figure 3-6. The structures of the pentagonal antiprismatic a) $[Ni_{10}Ge(CO)_{20}]^{2-}$, [167] and the pentagonal prismatic b) $[Rh_{15}C_2(CO)_{28}]^-$ [221] and c) $[Pt_{19}(CO)_{22}]^{4-}$ clusters. [133]

where the two extra platinum atoms occupy the interstitial sites of the two pentagonal prisms. The recently characterized $[Ag_{12}Au_{13}Br_8(PPh_3)_{10}]^+Br^-$ and $[Ag_{12}Au_{13}Br_8\ (PPh_3)_{10}]^+SbF_6^-$ salts exhibit 1,5,5′,5,5′,1 and 1,5,5′,5′,5,1 stacking sequences respectively, and very nicely illustrate paths ***b*** and ***c*** of Scheme **3-9**. [222, 223] It is worth noting that these two rotamers have been obtained selectively by the simple crystallization with different counter-anions. This observation supports the previous suggestion that packing forces may play a key role in determining the actual metal geometry of the cluster. Evidently, the two geometries exhibited with this combination of metals require the same number of valence electrons, and the eclipsed and staggered conformations differ very little in energy so that they are probably separated by a low rotational energy barrier. This would be in accord with the results from Extended Hückel Molecular Orbital (EHMO) calculations which suggest that tangential bonds in gold clusters (and probably in silver) are less energetic than radial interactions. [88] The above situation formally resembles that found with the eclipsed and staggered conformations of the metallocenes. [72]

b) Two-dimensional Growth Path

A few clusters adopt a two dimensional metal frame based on almost perfect tiling of triangular, or mixed triangular and square, building units on a plane. Tiling of triangles is the most documented. The most relevant examples are the ν_2-triangular array exhibited by the previously discussed $[Fe_3Pt_3(CO)_{15}]^{n-}$, [140] and $[Fe_3Cu_3(CO)_{12}]^{3-}$ clusters (Figure **3-7a**), [141] $[Os_6(CO)_{21-n}L_n]$ (L = CO, MeCN,P(OR)$_3$), [224] $[HRu_6(CO)_{18}(OCNMe_2)_2]^{-}$, [225] $[Os_6(\mu_3\text{-}O)(CO)_{19}]$ [226] and $[HRu_6(\mu_3\text{-}S)_3(CO)_{14}]^{-}$. [227] Interestingly, the hetero- and homometallic clusters are not isoelectronic and have either 86 (eventually 88 in $[Fe_3Pt_3(CO)_{15}]^{2-}$) or 90 valence electrons. This event parallels the 42 (eventually 44) or 48 valence electron configurations of ν_1-triangular clusters (e.g.$[Pt_3(CNR)_6$ and $[Pt_3(CO)_3L_4]$ *vs.* $[Os_3(CO)_{12}]$). Recognition that a $[Fe(CO)_4]$ fragment is isolobal to carbon monoxide (as well as carbene) provides an elegant rationalization of the above differences on the basis of the corresponding behavior of simpler compounds. [191] Furthermore, the structure of all these clusters represents a chunk of a (111) metal surface stabilized in a molecular species. Consequently, the structure of $[Os_6(\mu_3\text{-}O)(\mu_3\text{-}CO)(CO)_{18}]$ provides a very nice snapshot of an oxygen atom chemisorbed onto a metal surface.

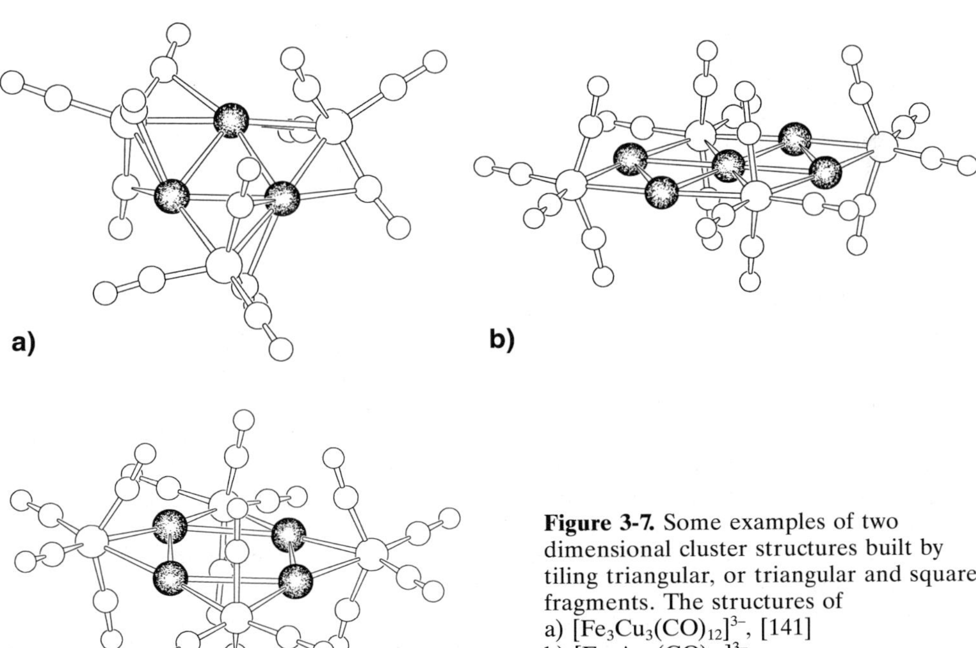

Figure 3-7. Some examples of two dimensional cluster structures built by tiling triangular, or triangular and square fragments. The structures of
a) $[Fe_3Cu_3(CO)_{12}]^{3-}$, [141]
b) $[Fe_4Ag_5(CO)_{16}]^{3-}$,
and c) $[Fe_4Ag_4(CO)_{16}]^{4-}$. [208]

The progressive tiling of triangles gives rise to the ν_2-rhombic structures found in $[Fe_4M_5(CO)_{16}]^{3-}$ (M = Cu,Ag; Figure **3-7b**) [141, 208] and $[Os_9Hg_3(CO)_{30}]$ [228] Tiling of triangles and a square is at the core of the structure of the clusters $[Co_4M_4(CO)_{16}]$ (M = Cu,Ag), [229, 230] $[Fe_4Ag_4(CO)_{16}]^{4-}$ (Figure **3-7c**), [208] $[Fe_4Cd_4(CO)_{16}]$, [231] $[Hg_4Mn_4(CO)_8Cp_4]$ [232] and $[Pt_4W_4(CO)_8Cp_4(CC_6H_4Me_4)_4]$. [233]

The occurrence of the above planar structures is due to both the high CO/M ratios of the outer coordinated metal sites and/or the tendency of some metals for digonal or square planar coordination (e.g. Cu, Ag, Hg and Pt).

c) Three-dimensional Growth Around a Center-of-mass

The vast majority of clusters grow more or less spherically around a center-of-mass. This can be either a real metal atom or simply a geometrical point and may correspond to the idealized center of symmetry of the metal frame of the molecule. Transition metal atoms lodged interstitially within a swelled cube or square antiprisms of metal atoms have already been cited in the previous subsections. A surrounding cubic array of metal atoms represents the unit cell of a body-centered metal lattice. Several clusters are known, which grow around a central metal atom and conform to pieces of this metal lattice, e.g. $[Ni_9(CO)_8(\mu_4\text{-GeEt})_6]^{2-}$ (Figure **3-8a**), [218] $[Rh_{14}(CO)_{25}]^{4-}$, [108, 109] $[Rh_{15}(CO)_{30}]^{3-}$ (Figure **3-8b**), [120] $[Ru_9Pt_6(CO)_{28}]^{4-}$ [45] and $[Pd_{23}(CO)_{20}(PEt_3)_8]$. [126] The two pentadecanuclear clusters differ slightly in their M–M contacts, whereby the former displays a metal array consisting of an almost regular centered rhombic dodecahedron and the latter conforms better to a centered hexa-capped cube. In the palladium clus-

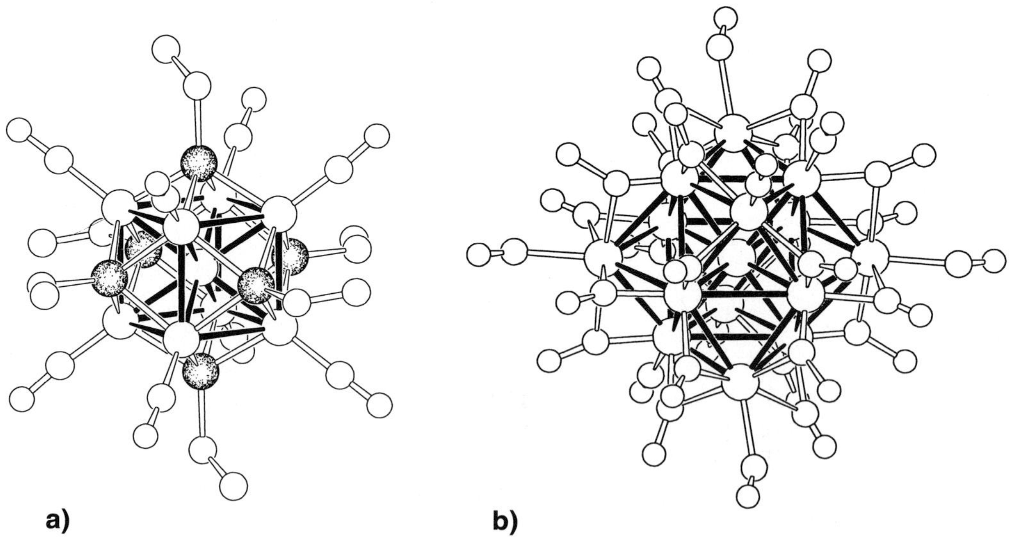

Figure 3-8. Structures of a) $[Ni_8(\mu_8\text{-Ni})(\mu_4\text{-GeEt})_6(CO)_8]$ [218] and b) $[Rh_{15}(CO)_{30}]^{3-}$. [114]

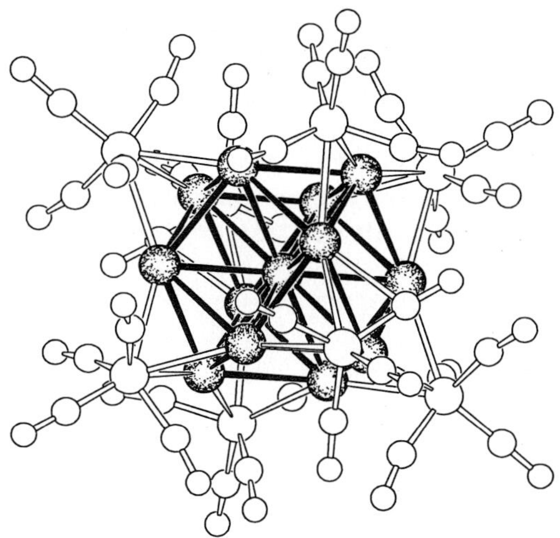

Figure 3-9. The structure of the octa-capped cuboctahedral $[Ag_{13}Fe_8(CO)_{32}]^{4-}$ cluster. [143]

ter, eight of the twelve edges of the cube are spanned by exo-polyhedral Pd(CO)-(PEt_3) moieties.

In the penta-capped body-centered cubic $[Rh_{14}(CO)_{25}]^{4-}$ cluster, the formal substitution of two negative charges for a carbonyl group or the addition of a proton or a $Rh(CO)_2^+$ group, results in the cluster $[Rh_{14}(CO)_{26}]^{2-}$, $[HRh_{14}(CO)_{25}]^{3-}$ and $[Rh_{15}(CO)_{27}]^{3-}$ respectively. Their structures reveal that their metal frames have been slightly, yet perhaps significantly, distorted and can now be considered as intermediate between body-centered cubic and cubic-close-packed. [110–114]

The smallest fragment of a cubic close packed lattice containing a twelve coordinated metal atom is the cuboctahedron. The structure of the bimetallic $[Ag_{13}Fe_8(CO)_{32}]^{4-}$ derivative, [143] as shown in Figure **3-9**, consists of a centered cuboctahedron of silver atoms capped at the eight triangular faces by eight C_{3v} $Fe(CO)_4$ groups. These may be considered as four electron donors and would be equivalent to μ_3-S capping atoms. Interestingly, both a noncentered $[Cu_{12}(\mu_3\text{-}S)_8]^{4-}$ [234] and a Na-centered $[Au_{12}NaSe_8]^{3-}$ derivative are known. [235] The latter has been described as a Na^+ kryptate of the $[Au_{12}Se_8]^{4-}$ inorganic kryptand. This description can also be used for $[Ag_{13}Fe_8(CO)_{32}]^{4-}$, which may be thought of as a diamagnetic $[Ag_{12}Fe_8(CO)_{32}]^{4-}$ kryptand of a paramagnetic silver atom. The ESR spectrum of this compound partially supports this interpretation and consists of two well separated pairs of partially overlapping multiplets due to the coupling of the unpaired electron with the unique interstitial nucleus. Each multiplet in turn consists of thirteen closely spaced lines arising from coupling with the twelve equivalent peripheral silver atoms. [143] A centered cuboctahedral array of metal atoms also constitutes the central units in $[Rh_{14}N_2(CO)_{25}]^{2-}$, [236] $[H_xPt_{15}(CO)_8\text{-}(P^tBu_3)_6]$ [131] and $[Pd_{23}(CO)_{22}(PEt_3)_{10}]$. [125]

A centered anticuboctahedral arrangement of metal atoms, although not matching the elementary cell, is at the heart of the hexagonal close packed metal lattice. The first cluster which was shown from X-ray studies to contain a fully interstitial metal atom, namely $[H_3Rh_{13}(CO)_{24}]^{2-}$, [107] adopts this metal geometry, as do all the derivatives in the $[H_{5-n}Rh_{13}(CO)_{24}]^{n-}$ (n = 1–4) series, [105–107] and also the bimetallic $[Rh_{11}Pt_2(CO)_{24}]^{3-}$ and $[Rh_{12}Pt(CO)_{24}]^{4-}$ clusters. [149] Capping four of the six square faces on the above polyhedron with $Rh(CO)_2$ fragments gives rise to the structure found for $[Rh_{17}(CO)_{30}]^{3-}$. [115]

Alternative stereogeometries for which the central metal atom has a coordination number of twelve are bicapped pentagonal prisms and antiprisms. As shown in Table **3-7**, both the regular bicapped pentagonal prism and the icosahedron allow the interstitial occupation of only those atoms having a radius which is slightly smaller than that of the peripheral atoms in a rigid sphere model. These geometrical limitations, however, should not be taken too seriously for the occurrence of such metal arrangements in molecular compounds. Minor distortions of the polyhedron, for instance a localized elongation along a C_5 axis or a general loosening of the peripheral contacts, are sufficient to overcome this restriction. Although it is possible to pack spheres in concentric pentagonal prisms of increasing frequency and obtain a density approaching that of close packing, [346] so far there are no reported examples of pentagonal prismatic three dimensional growth around a metal other than the previously cited $[Rh_{15}C_2(CO)_{28}]^-$ (Figure

Table 3-7. Radius of the cavity for some selected regular polyhedra with edge length, *e*, of 2.5 and 2.8 Å.[a)]

N_v	Polyhedron	R_C	range of values of R_C for e = 2.8–2.5 Å	
4	Tetrahedron	0.114*e*	0.32–0.29	
6	Octahedron	0.207*e*	0.58–0.52	H(0.37),B(0.92), C(0.77),N(0.75)
6	Trigonal prism	0.264*e*	0.74–0.66	B,C,N,P(1.20)
7[b)]	Pentagonal Bipyramid	0.026*e*		
8	Square Antiprism	0.321*e*	0.90–0.80	C,Si(1.11),P,S(1.02), As(1.19)
8	Triangular Dodecahedron	0.350*e*	0.98–0.88	C
8	Cube	0.364*e*	1.02–0.91	
12	Icosahedron	0.451*e*	1.26–1.13	Ge(1.22),Sn(1.40), Sb(1.38)
12	Bicapped (5,5) Pentagonal Prism	0.486*e*	1.36–1.22	
12	Cuboctahedron	0.500*e*	1.40–1.25	
12	Anticuboctahedron	0.500*e*	1.40–1.25	

a) N_v = number of vertices, R_C = radius of the cavity for a rigid sphere model, e = edge length, the covalent radius of a few selected main group elements is reported in parenthesis.
b) The cavity of the pentagonal bipyramid is discoidal and only the limiting value of the height is given.

3-6b), [221] $[Pt_{19}(CO)_{22}]^{4-}$ (Figure **3-6c**), [133] and the mixed $[Ag_{13}Au_{12}Br_8(PEt_3)]^+$ rotamer. [222, 223] With regard to the pentagonal antiprism the situation is somewhat better and such a geometry is observed in the structures of $[Ni_{10}Ge(CO)_{20}]^{2-}$ [167] (Figure **3-6a**) and the centered icosahedral $[Au_{13}Cl_2(PR_3)_{10}]^{3+}$ and $[Au_{13}(dppm)_6]^{4+}$ clusters. [35, 88, 237] Two slightly larger carbonyl-substituted clusters, i.e. $[Pd_{16}(CO)_{13}(PEt_3)_9]$ and $[Pt_{17}(CO)_{12}(PEt_3)_8]$, [124, 132] likewise contain an inner centered icosahedral unit. Furthermore, both the centered and non-centered icosahedral geometry is well documented in transition metal clusters containing main group elements or molecular fragments derived thereof (see next section).

As shown in Figure **3-10**, the structures of $[H_4Os_{10}(CO)_{24}]^{2-}$ and $[Os_{20}(CO)_{40}]^{2-}$ conform very nicely to tetrahedra of frequencies 2 and 3. [98, 99, 101] Correspondingly, the metal framework of the not fully characterized bimetallic $[Ni_{36}Pt_4(CO)_x]^{6-}$ cluster is derived from a tetrahedron of frequency 5, which would require 56 metal atoms and a central, completely encapsulated, tetrahedron of Pt atoms. The larger tetrahedron is reduced to the actual metal framework by cutting away three full rows of metal atoms along three edges. [154]

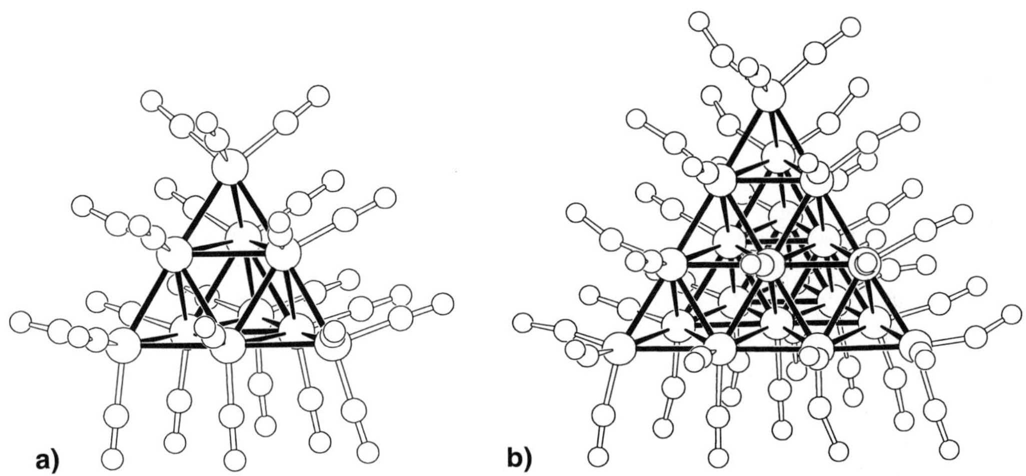

Figure 3-10. The frequency 2 and 3 tetrahedra exhibited by the a) $[H_4Os_{10}(CO)_{24}]^{2-}$ [99] and b) $[Os_{20}(CO)_{40}]^{2-}$ clusters. [101]

Analogously, the recently characterized $[Ir_{14}(CO)_{27}]^-$ cluster represents a trigonal bipyramid of frequency 2. [120] Incomplete trigonal bipyramids of frequency 2 are observed in $[Ir_{12}(CO)_{24}]^{2-}$ [118] and the $[H_{4-n}Ni_{12-x}Pt_x(CO)_{21}]^{n-}$ (n = 2,3,4; x = 0,3) bimetallic clusters [121–123, 152]; however, while the former lacks the two axial vertices, the latter has one axial and one equatorial vertex missing. The second largest homometallic osmium cluster known to date, *viz.* $[Os_{17}(CO)_{36}]^{2-}$, [100] also contains a trigonal bipyramid of frequency 2 in which the three additional osmium atoms may be considered as μ_3-capping one of the axial tetrahedral cap. The largest fully characterized platinum cluster known is $[Pt_{26}(CO)_{32}]^{2-}$; [135, 189] and analogous to the previously discussed $[Ni_{36}Pt_4(CO)_x]^{6-}$ cluster, it may be formally thought of as being derived from a much larger trigonal bipyramid of frequency 5 through the loss of 6 rows of metal atoms along the six axial-equatorial edges and the resulting top and bottom triangular layers.

Finally, the previously discussed $[Pd_{23}(CO)_{22}(PEt_3)_8]$ cluster may be alternatively described as an octahedron of frequency 2 with four additional exopolyhedral $[Pd(CO)(PEt_3)]$ groups. Likewise, the partially characterized $[Pt_{38}(CO)_{44}]^{2-}$ cluster [135, 189] has a metal frame composed of an incomplete octahedron of frequency 3, whereas metal frameworks based on complete ν_3-octahedra have been identified in all the structures of the $[H_{6-n}Ni_{38}Pt_6(CO)_{48}]^{n-}$ ($n = 6,5,4$) series of bimetallic clusters. [155] The full structure of the $[HNi_{38}Pt_6(CO)_{48}]^{5-}$ pentaanion has been unraveled and is shown in Figure **3-4**.

3.2.5 Transition Metal Clusters Containing Main Group Elements

Main group elements, or molecular fragments derived thereof, may combine with transition metals to form clusters with a wide range of interesting and often unique characteristics. The main group elements themselves show a wide cluster chemistry which, however, is beyond the purpose of this chapter. It will suffice here to simply recall the chemistry of the boranes and carboranes as well as the clusters of heavier elements, either bare or stabilized by σ and π-donor ligands, such as those of Ge, Sn, Pb, P, As, Sb, Bi, Se, and Te. [238, 239] A first reason for the interest in combining main group elements and transition metals in a cluster may be the search for new coordinations and geometries. Elements with valence d electrons tend to limit themselves to deltahedral metal geometries derived from the tetrahedron, the octahedron, or to a combination of the two. As the nuclearity increases, the geometries of the metal cluster cores will begin to represent fragments of a typical metallic lattice, and is probably due to the gain in cohesive energy arising from the presence of valence d-orbitals. In contrast, the main group elements tend to exhibit either chain, ring, or cage structures, which may often be described with two-center two-electron localized bonds. Not surprisingly, as the connectivity increases, the necessity for a delocalized bond description becomes more pronounced, and those cages based on spherical deltahedra, such as the pentagonal bipyramid, the triangulated dodecahedron, the bicapped trigonal and square antiprism and the icosahedron, are preferentially adopted. As already discussed in the previous section, a further difference between the cage compounds of these two types of clusters rests in their ability to interstitially lodge a heteroatom. Thus, while transition metal clusters will only very rarely have frames with unfilled cavities larger than octahedral (the only reported examples being $[Pt_3(CO)_6]_n^{2-}$, [130] $[Ni_8(CO)_8(\mu_4\text{-PPh})_6]$, [217] $[Ni_8Te_4(CO)_{12}]^{2-}$, [240] and $[Ni_{12-x}(\mu_5\text{-ER})_x(CO)_{24-3x}]^{2-}$ where E = P, As, Sb and $x = 2,3,4$), [241–243] there are no reports to our knowledge of main group element clusters having either homo- or heteroatoms (main group or transition metal) interstitially lodged in their cavities. It is possible that this difference is artificial, however because for certain combinations there should be no insurmountable electronic or steric reasons which would hinder the formation of main group clusters containing interstitial atoms. The $[Pb_6(\mu_4\text{-}O)(OH)_6]^{4+}$ cation [245] and the proposed endohedral structures of the fullerene-noble gas and fullerene-M (M = La, Ni, Li, Na) clusters [244] are a few examples which support this conjecture.

A second line of interest in combining transition metals and main group elements in clusters stems from the possibility that these compounds could be useful precursors for the synthesis of new materials (see Section 2.6 for some relevant literature references).

For the purpose of classification, the clusters reported so far, which combine transition metal atoms and either main group elements or elemental organic fragments in their frames, can be divided into the following categories on the basis of their structural features:

A. transition metal clusters containing interstitial main group elements (Table **3-8**), [246–250]
B. transition metal clusters containing exposed main-group elements or elemental organic moieties (Table **3-9**), [246–250]
C. mixed transition metal and main-group element clusters (Table **3-10**), [248–250]
D. main group element clusters containing transition metal atoms (Table **3-10**). [248–250]

The clusters in category **A** exhibit only M–M and M–E interactions, but as one progresses down to category **D** the presence of E–E interactions become increasingly more important, and eventually become as predominant as M–E bonds at the end.

The assignment of a given compound to one of the above categories is not clear cut in some cases. For instance, clusters such as the previously cited $[Ni_8(CO)_8(\mu_4\text{-}PPh)_6]$ and $[Ni_8(\mu_8\text{-}Ni)(CO)_8(\mu_4\text{-}GeEt)_6]$ compounds [217, 218] can be described as noncentered or centered Ni_8 hexacapped cubes belonging to category **B**. Alternatively, they can be considered as noncentered or centered examples of a Ni_8E_6 rhombic dodecahedral mixed cluster, stabilized by carbonyl and phenyl or ethyl ligands, and thus may formally belong to category **C**. Since these clusters derive from the interpenetration of a M–M bonded nickel cube and an E···E nonbonded phosphorus or germanium octahedron, they are classified as **B** types and are listed in Table **3-9**. Their description in category **C** and **D** would, however, gain more significance upon a progressive loosening of the Ni–Ni bonds and a tightening of the E–E contacts, so to obtain an octa-capped E_6 octahedron.

Of the above four categories, the first two are by far the most documented in high nuclearity clusters and share common synthetic methods. Therefore, in the following brief overview of the synthetic methods for transition metal clusters containing main group elements, they will be treated together. The remaining two (**C** and **D**) present some peculiar features and will be described separately.

Table 3-8. Selected metal clusters containing interstitial main group elements (Category **A**).[a)]

Compound		Reference
	Group 13	
$[Fe_4Rh_2(\mu_6\text{-}B)(CO)_{16}]^-$	e	[251]
$[HRu_6(\mu_6\text{-}B)(CO)_{17}]$	e	[252]
$[H_2Ru_6(\mu_6\text{-}B)(CO)_{18}]^-$	f	[253]
$[H_2Ru_8Cu_4(\mu_6\text{-}B)_2(CO)_{24}(\mu_2\text{-}Cl)]$		[254]
	Group 14	
$[Re_6(\mu_6\text{-}C)(CO)_{19}]^{2-}$	e	[15]
$[H_2Re_6(\mu_6\text{-}C)(CO)_{18}]^{2-}$	e	[255]
$[Re_8(\mu_6\text{-}C)(CO)_{24}]^{2-}$	e	[256]
$[\{Re_7Ag(\mu_6\text{-}C)(CO)_{21}\}_2Br]^{4-}$	e	[257]
$[Ru_6(\mu_6\text{-}C)(CO)_{14}(\text{arene})]$	e	[258]
$[M_{10}(\mu_6\text{-}C)(CO)_{24}]^{2-}$ (M = Ru,Os)	e	[259, 90]
$[Ru_{10}(\mu_6\text{-}C)_2(CO)_{24}]^{2-}$	e	[260]
$[M_6(\mu_6\text{-}C)(CO)_{13}]^{2-}$ (M = Co,Rh)	e	[261, 158]
$[M_6(\mu_6\text{-}C)(CO)_{15}]^{2-}$ (M = Co,Rh)	f	[262, 263]
$[Co_8(\mu_8\text{-}C)(CO)_{18}]^{2-}$	h	[215]
$[Co_{13}(\mu_6\text{-}C)_2(CO)_{24}]^{n-}$ (n = 3,4)		[178, 179]
$[Rh_{12}(\mu_6\text{-}C)_2(CO)_{25-n}]^{2-}$ (n = 0,1,2)		[174, 175, 185, 186]
$[Rh_{15}(\mu_6\text{-}C)_2(CO)_{28}]^-$	e	[221]
$[Ni_7(\mu_7\text{-}C)(CO)_{12}]^{2-}$	g	[169]
$[Ni_8(\mu_8\text{-}C)(CO)_{16}]^{2-}$	h	[215]
$[Ni_{10}(C_2)(CO)_{16}]^{2-}$		[264]
$[Ni_{16}(C_2)_2(CO)_{23}]^{4-}$		[172]
$[Ni_{35}(\mu_7\text{-}C)_4(CO)_{39}]^{6-}$	g	[265]
$[HNi_{38}(\mu_8\text{-}C)_6(CO)_{42}]^{5-}$	h	[266]
$[Co_9(\mu_8\text{-}Si)(CO)_{21}]^{n-}$	h	[267]
$[Ni_{10}(\mu_{10}\text{-}Ge)(CO)_{20}]^{2-}$		[167]
$[Ni_{12}(\mu_{12}\text{-}Ge)(CO)_{22}]^{2-}$		[167]
$[Ni_{12}(\mu_{12}\text{-}Sn)(CO)_{22}]^{2-}$		[167]
	Group 15	
$[Ru_6(\mu_6\text{-}N)(CO)_{16}]^-$		[268]
$[Ru_8(\mu_6\text{-}P)(\mu_4\text{-}PPh)(\mu_2\text{-}PPh_2)(CO)_{21}]$		[269]
$[Ru_8(\mu_8\text{-}P)(CO)_{19}(\mu_2\text{-}\eta^1\text{-}\eta^6\text{-}CH_2Ph)]$		[270]
$[Os_6(\mu_6\text{-}P)(CO)_{18}]^-$	f	[271]

Table 3-8. (Continued).

Compound		Reference
$[M_6(\mu_6\text{-}N)(CO)_{15}]^-$ (M = Co,Rh)	f	[272, 273]
$[Co_{14}N_3(CO)_{26}]^{3-}$	f	[274]
$[HRh_{12}N_2(CO)_{23}]^{3-}$		[275]
$[Rh_{14}N_2(CO)_{25}]^{2-}$		[236]
$[Rh_{23}N_4(CO)_{38}]^{3-}$		[276]
$[Rh_{10}PtN(CO)_{21}]^{3-}$		[277]
$[Co_{14}P_2(CO)_{27}]^{4-}$		[278]
$[Rh_9(\mu_8\text{-}P)(CO)_{21}]^{2-}$	h	[82]
$[Rh_{10}E(CO)_{22}]^{3-}$ (E = P,As)	h	[279, 280]
$[Rh_{12}(\mu_{12}\text{-}Sb)(CO)_{27}]^{3-}$		[281]
	Group 16	
$[Rh_{10}S(CO)_{22}]^{2-}$	h	[282]
$[Rh_{17}S_2(CO)_{32}]^{3-}$	h	[210]

a) The letter appearing in the second column refers to the stereochemical coordination of the interstitial atom as pictorially represented in Fig. 3-13.

3.2.5.1 Overview of the Synthetic Methods and Spectroscopic Characteristics of Transition Metal Clusters Containing Interstitial or Exposed Main Group Elements

Clusters with interstitial atoms potentially encompass an enormous variety of compounds differing in the kind of metal and the interstitial atom involved, as well as in their respective numbers. So far, cluster chemistry has provided examples of interstitial clusters which contain one or more interstitial atoms of H, B, C, N, Si, P, S, Ge, As, Se, In, Sn, and Sb, however, for most of these interstitial atoms only one or very few examples are known. Probably due to their obvious relevance to interstitial alloys, the most highly documented and widely investigated classes of interstitial clusters are those incorporating hydrides, borides, carbides and nitrides. If transition metal clusters which have exposed main group elements or elemental organic fragments are also taken into consideration, practically all the elements of groups 13 to 17 have been documented.

The synthetic approaches taken for the preparation of interstitial clusters are based on the activation of E–E, E–H, E–C, E–Si, E–O, E–S and E–Cl bonds of suitable molecules. The methods to obtain clusters having exposed maingroup elements or elemental organic fragments are either exactly the same or differ only in their use of a source molecule which contains two kinds of E–X bonds

Table 3-9. Selected metal clusters containing exposed main group elements or elemental organic fragments (Category B).

Compound		Reference
	Group 13	
$[Mn_5(\mu_5\text{-}In)(CO)_{20}]^{2-}$		[219]
$[Ni_6(\mu_3\text{-}InBr_3)(\eta^2\text{-}\mu_6\text{-}In_2Br_5)(CO)_{11}]^{3-}$		[283]
$[Ni_6(\eta^2\text{-}\mu_6\text{-}In_2Br_5)_2(CO)_{10}]^{4-}$		[283]
$[Ni_{12}(\mu_6\text{-}In)(\eta^2\text{-}\mu_6\text{-}In_2Br_4OH)(CO)_{21}]^{4-}$		[283]
	Group 14	
$[Re_4(\mu_4\text{-}C)(CO)_{15}I]^-$	a	[284]
$[Fe_4(\mu_4\text{-}C)(CO)_{12}]^{2-}$	b	[285]
$[Ru_4(\eta^2\text{-}\mu_4\text{-}C_2Me_2)(CO)_{12}]$		[286]
$[M_5(\mu_5\text{-}C)(CO)_{15}]$(M = Fe,Ru,Os)	c	[287–289]
$[Os_5(\mu_5\text{-}C)(CO)_{16}]$	d	[290]
$[Ru_5(\eta^2\text{-}\mu_5\text{-}C_2)(\mu_2\text{-}PPh_2)_2(\mu_2\text{-}SMe)_2(CO)_{11}]$		[291]
$[Co_6(\eta^2\text{-}\mu_6\text{-}C_2)(\mu_4\text{-}S)(CO)_{14}]$		[292]
$[Co_8Ge_3(CO)_{26}]$		[293]
$[Ni_8(\mu_8\text{-}Ni)(\mu_4\text{-}GeEt)_6(CO)_8]$		[218]
$[Ni_{11}(\mu_6\text{-}SnR)_2(CO)_{18}]^{2-}$		[294]
	Group 15	
$[M_5(\mu_5\text{-}N)(CO)_{14}]^-$ (M = Fe,Ru)	c	[295, 296, 64]
$[Co_6(\mu_6\text{-}P)(CO)_{16}]^-$		[297]
$[Fe_4Cr(\mu_4\text{-}P)_2(CO)_{18}]$		[298]
$[Ni_8(\mu_4\text{-}PPh)_6(CO)_8]$		[217]
$[Ni_{10}(\mu_5\text{-}PMe)_2(\mu_4\text{-}PMe)_5(CO)_{10}]^{2-}$		[241]
$[Ni_{12-x}(ER)_x(CO)_{24\text{-}3x}]^{2-}$ (E = P,As,Sb)		[241–243]
$[Fe_4\{\mu_4\text{-}SbFe(CO)_4\}_2(CO)_{12}]^{2-}$		[299]
$[Ni_{11}\{\mu_6\text{-}SbNi(CO)_3\}_2(CO)_{18}]^{n-}$ $(n = 2,3,4)$		[300]
$[Ni_{11}Bi_2(CO)_{18}]^{n-}$ $(n = 2,3,4)$		[94]
	Group 16	
$[Os_6(\mu_3\text{-}O)(CO)_{19}]$		[226]
$[Fe_3Mn(\mu_4\text{-}O)(CO)_{12}]^-$	b	[301]
$[Fe_2Ru_3(\mu_4\text{-}O)(CO)_{15}]$		[302]
$[HRu_6(\mu_3\text{-}S)_3(CO)_{15}]^-$		[227]
$[Os_6(\mu_4\text{-}S)_n(CO)_{17}]$$(n = 1,2)$		[303]
$[Os_6(\mu_4\text{-}S)(\mu_3\text{-}S)(CO)_{16}]$		[304]
$[Os_7(\mu_4\text{-}S)_2(CO)_{20}]$		[305]
$[Ni_8Te_4(CO)_{12}]^{2-}$		[240]
$[Ni_{13-x}E_x(CO)_{24-3x}]^{2-}$(E = Se,Te; x = 2,3)		[240]

a) The letter appearing in the second column refers to the stereochemical coordination of the interstitial atom as pictorially represented in Fig. 3.13.

Table 3-10. Selected examples of mixed clusters of transition metal and main group elements and clusters of main group elements containing transition metal atoms (Categories C and D).

Compound	Reference
Group 13	
$[Au_6(PEt_3)_4(B_{10}H_{12})_2]$	[307]
$[H_2Mo_2Cu_2(CO)_6(C_2B_9H_{10})]$	[308]
$[Tl_4Fe_8(CO)_{30}]^{4-}$	[309]
$[Tl_6Fe_{10}(CO)_{36}]^{6-}$	[309]
Group 14	
$[Sn_6Cr_6(CO)_{30}]^{2-}$	[306]
Group 15	
$[P_6Mo_2Cp^*_2]$	[310]
$[As_5Mo_2Cp_2]$	[311]
$[Bi_4Fe_4(CO)_{13}]^{2-}$	[312]
$[Co_{16}(\mu_6\text{-}As)_2(\mu_4\text{-}As)_2(\mu_4\text{-}AsPh_2)_4(CO)_{32}]$	[313]
$[Co_4E_2(CO)_{10}]^{n-}$(E = Sb,Bi; n = 1,2)	[314, 315]
Group 16	
$[S_8Co_6(CO)_6.3S_8]$	[316]
$[Se_4W_2(CO)_{10}]^{2+}$	[317]
$[Se_{10}Ni_{15}(CO)_3Cp^*_8]$	[318]
$[Se_{10}Ni_{15}(CO)Cl_2Cp^*_8]$	[318]
$[Te_8W_6(CO)_{18}]^{2-}$	[319]
$[Te_4Fe_5(CO)_{14}]^{2-}$	[320]
$[Te_{10}Fe_8(CO)_{20}]^{2-}$	[320]
$[Te_{14}Ru_6(CO)_{12}]^{2-}$	[321]

each requiring different experimental conditions for activation. For example, although PCl_3, PH_3, or even PPh_3 can be used to introduce an interstitial phosphorus atom into a cluster, these three phosphorus sources require increasingly severe experimental conditions to be activated because of their difference in polarity and in the strengths of the bonds which must be broken and subsequently formed. It is therefore conceivable that a $PCl_{3-x}Ph_x$ (x = 1,2) mixed bond species would be more effective in the formation of a cluster containing exposed PPh_x fragments. The stabilization of a bare main group element at an exposed or semi-interstitial site seems to be based on a variety of factors such as the reluctance of the heteroatom to share its valence electrons, as in the case of oxygen and the halogens, the unsuitable size of the cavity made available by the self-assembling ML_x fragments, the reaction procedure, and/or the reducing strength of the metal complex *(vide supra)*.

Generally speaking, very little is known about the mechanism of most of these reactions and often the only available tool to selectively drive the reaction toward the formation of a sterically and electronically feasible compound are suitable experimental conditions and the careful control of the reagents' stoichiometries. However, a Lewis acid-base interaction may be envisioned in those cases where the main group elements possess empty low energy orbitals, whereas an opposite type interaction could occur when the main group elements possess unshared valence electron pairs. The latter interaction can also be represented by a ligand substitution reaction. The subsequent steps are probably a concerted sequence of intra- or intermolecular oxidative additions of E–X bonds to metal centers or across M–M bonds with alternating reductive elimination steps. A representative example of such a pathway is represented with the step-by-step synthesis of $[Os_6(\mu_6\text{-P})(CO)_{18}]^-$ [271] as illustrated in Scheme **3-10**.

In other cases, due to the lack of suitable filled or unfilled orbitals on the main group elements, the suggested Lewis acid-base interaction cannot occur as the initial step. These orbitals can be made available through an initial redox reaction, or from either homolytic or heterolytic cleavage of the E–X bond. A direct oxidative addition also cannot be ruled out. It is worth noting that the use of molecules which contain the potential interstitial or exposed main-group atom in a more or less formal high oxidation states as reagents and ligated clusters or complexes in low oxidation states is often a very effective combination. Since the source molecule is reduced under these circumstances at the expense of cluster charge, or by sacrificial oxidation of some of the metal atoms or ligands, the previously suggested requirements may also be readily satisfied under very mild conditions. The main group atom becomes progressively less coordinated and is captured in a self-assembling process by the unreacted starting cluster or complex, or a derivative thereof.

We shall discuss in more detail only the synthesis of the exposed and interstitial carbides, since this class of clusters is one of the most investigated and well developed. Moreover, their syntheses require methods which have been, or can likely be, extended to other main group elements.

Previous to this, however, it seems appropriate to shortly describe the synthesis of clusters with interstitial hydride and deuteride ions not only due to their intrinsic interest but also because of their obvious relationship to metal hydrides. The interstitial and semi-interstitial hydrides known to date include $[Co_6(\mu_6\text{-H})(CO)_{15}]^-$, [322] $[Ru_6(\mu_6\text{-H})(CO)_{18}]^-$, [203, 323, 324] $[Ni_{12}(\mu_6\text{-H})_{4-n}(CO)_{21}]^{n-}$ (n = 2,3), [121] $[Ni_9Pt_3H_{4-n}(CO)_{21}]^{n-}$ (n = 2,3), [152] $[Rh_{13}H_{5-n}(CO)_{24}]^{n-}$ (n = 2,3,4) [105–107] and $[Rh_{14}H(CO)_{25}]^{3-}$. [110] Several other carbonyl species, e.g. $[HRu_{11}(CO)_{27}]^{3-}$, [325] $[HRu_{10}(\mu_6\text{-C})(CO)_{24}]^-$, [326] and $[HOs_{10}(\mu_6\text{-C})(CO)_{24}]^-$, [327] are thought to contain interstitial hydride atoms based on the results from NMR solution studies and/or their solid state structural characteristics (swelling of a particular cage and packing of the surface bound carbonyls). It has been demonstrated that conclusions drawn from the carbonyl packing may be misleading; for instance, the surface bound hydride atoms on $[H_2Os_{10}(\mu_6\text{-C})(CO)_{24}]$ [92] and $[H_4Os_{10}(CO)_{24}]^{2-}$ [98, 99] do not significantly disturb the carbonyl distribution as it is very similar to that of the $[Os_{10}(\mu_6\text{-C})(CO)_{24}]^{2-}$ dianion. [90]

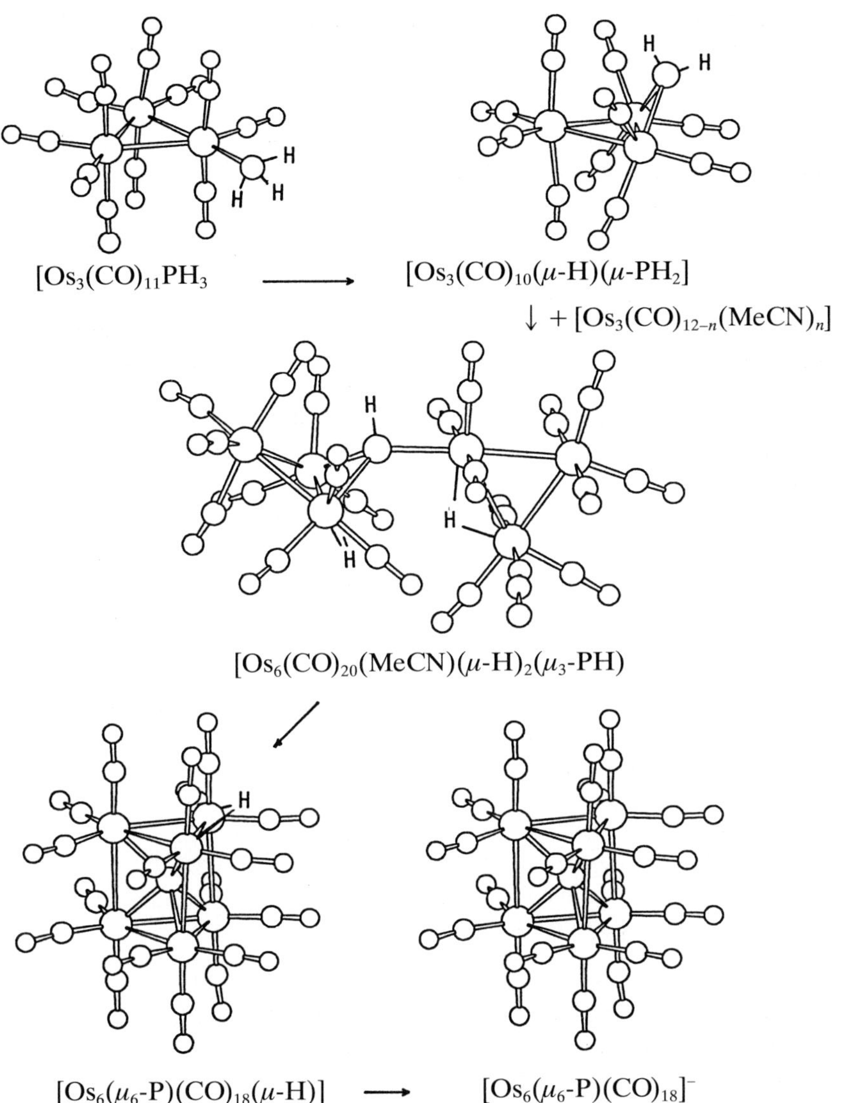

Scheme 3-10. Step-by-step construction of the $[Os_6(\mu_6\text{-}P)(CO)_{18}]^-$ cluster.

All the above hydrides, regardless of their stereochemistry, have been obtained either by direct protonation of the preformed parent anion, e.g. Equation (3.21), [324] or by protonation of a precursor anion which subsequently undergoes a more or less straightforward redox condensation, as schematically indicated in Equation (3.22). [121] At least one example for the synthesis of interstitial hydride clusters through the activation of molecular hydrogen is known and is represented in Equation (3.23). [328, 329]

$$[Ru_6(CO)_{18}]^{2-} + H^+ \rightarrow [Ru_6(\mu_6\text{-}H)(CO)_{18}]^- \quad (3.21)$$

$$2\ [Ni_6(CO)_{12}]^{2-} + H_2O \rightarrow [Ni_{12}(\mu_6\text{-}H)(CO)_{21}]^{3-} + OH^- + 3\ CO \quad (3.22)$$

$$[Rh_{12}(CO)_{30}]^{2-} + H_2 \xrightarrow{70\,^\circ C} [Rh_{13}H_{5-n}(CO)_{24}]^{n-} + \text{by-products} \quad (3.23)$$

In this latter reaction, unambiguous spectroscopic evidence for the formation of a $[HRh_6(CO)_{15}]^-$ intermediate containing a terminal hydride has been obtained at lower temperature, and the formation of the final $[Rh_{13}H_{5-n}(CO)_{24}]^{n-}$ product containing interstitial hydrides arises from the thermal decomposition of this compound. Although evidence for the existence of a $[HNi_6(CO)_{12}]^-$ species is lacking, formation of such a labile species by hydrolysis of $[Ni_6(CO)_{12}]^{2-}$ and its subsequent redox condensation with the latter is a conceivable route to $[Ni_{12}(\mu_6\text{-}H)(CO)_{21}]^{3-}$. The corresponding deuteride analogs of the above hydrides have been similarly obtained by employing deuterated acids or water in the respective synthesis, or by statistical exchange of the preformed hydride with D_2O or molecular deuterium, as shown in Equation 3.24. [121, 329]

$$[Ni_{12}(\mu_6\text{-}H)_2(CO)_{21}]^{2-} + 2\ D_2 \rightarrow [Ni_{12}(\mu_6\text{-}D)_2(CO)_{21}]^{2-} + 2\ HD \quad (3.24)$$

The reasons why hydrogen is interstitially bonded in the above compounds rather than taking up a surface coordination site on the cluster are not straightforward. In the first member of each of the $[Co_6(\mu_6\text{-}H)(CO)_{10}(\mu_2\text{-}CO)_5]^-$-$[HRh_6(CO)_{11}(\mu_3\text{-}CO)_4]^-$ [322, 330] and $[Ru_6(\mu_6\text{-}H)(CO)_{18}]^-$-$[Os_6(\mu_3\text{-}H)(CO)_{18}]^-$ [324] congener pairs, the interstitial lodging of hydrogen is probably due to a very subtle compromise between steric and electronic effects, both inside and outside the metal cage. It has been shown that the hydrogen atom in the Co_6 cage has enough room to rattle around and therefore, the expansion of the metal lattice which normally accompanies the occupation of interstial sites by hydrogen atoms in both molecular clusters and metal lattices appears to be electronic in nature. It may be speculated that the incorporation of an interstitial hydrogen, and perhaps the slight swelling observed for the octahedral metal frames in $[Co_6(\mu_6\text{-}H)(CO)_{15}]^-$ and $[Ru_6(\mu_6\text{-}H)(CO)_{18}]^-$ upon occupation of the hole with a hydrogen atom, might help in alleviating the steric strain among the outer carbonyl ligands and the eventual surface bound hydride. The increased size of the metal atoms upon descending a group makes interstitial lodging less favourable in each of the above pairs.

It is worth mentioning that, at least in the case of the $[Co_6(\mu_6\text{-}H)(CO)_{15}]^-$ salts, the counterions also seem to play a significant role on the position of the hydride. Thus, IR and inelastic neutron scattering spectra recorded on $[HCo_6(CO)_{15}]^-$ salts with different counterions, or in various matrices, are not uniformly consistent and show the presence of both μ_3- and μ_6-H atoms. [331–333]

The chemical shifts observed in the spectra NMR of these octahedral single cavity interstitial hydride clusters, *viz.* $[Co_6(\mu_6\text{-}H)(CO)_{15}]^-$, $[Ru_6(\mu_6\text{-}H)(CO)_{18}]^-$, appeared to be anomalous to those shown by μ_2 and μ_3 surface bound hydrides, as well as by most multi cavity interstitial hydride clusters. The anomaly is that of a downfield rather than upfield shift of the ^1H resonance whose interpretation has caused a controversy over the years. Since these anomalous chemical shifts were observed in solution studies (unfortunately, no solid state NMR studies on these compounds are yet available) and interstitial lodging of the hydride atom was ascertained from solid state studies, it was at first suggested that the downfield resonance could arise from their possible interaction with either the carbon or oxygen atoms of the surface bound carbonyls if the hydride atoms could move out of the metal cage once the cluster salts are dissolved. [334] This migration in solution would also have been in keeping with the ready deprotonation of $[Co_6(\mu_6\text{-}H)(CO)_{15}]^-$ in most organic solvents and in water. An alternative explanation offered for these anomalous downfield shifts was based on the symmetrical surrounding of hydrogen in the octahedral $[Co_6(\mu_6\text{-}H)(CO)_{15}]^-$ and $[Ru_6(\mu_6\text{-}H)(CO)_{18}]^-$ clusters versus the nonsymmetrical coordination or the nonstatic behavior of the interstitial hydride atoms exhibiting upfield shifts, *viz.* those present in $[Ni_{12}(\mu_6\text{-}H)_{4-n}(CO)_{21}]^{n-}$ (n = 2,3) and $[Rh_{13}H_{5-n}(CO)_{24}]^{n-}$ (n = 1,2,3,4). [335] As shown in Table **3-11**, the chemical shifts of the interstitial hydride atoms are spread over a wide range and so neither of the above interpretations seem sufficiently convincing to explain such behaviour.

The surface bound hydride atoms of $[H_4Os_{10}(CO)_{24}]^{2-}$ were computed to be partially submerged in the van der Waals spheres of the carbon atoms of the surrounding CO ligands whereby, an incipient H···C interaction of the hydride atoms with the nearest three CO's is suggested by the short C···H (2.23–2.35 Å) contacts and the concomitant outward bending of the oxygen atoms. Nevertheless, the chemical shifts (δ = –14.7 and –19.08 ppm) of these μ_2 and μ_3 hydrides are perfectly comparable with those of other edge and face bridging hydrides which do not display this structural peculiarity and are far away from typical formyl resonances. [98, 99] In addition, the recently isolated $[Ru_{11}H(CO)_{27}]^{3-}$ trianion [325] is reported to have a hydride chemical shift similar to that of $[Ru_6(\mu_6\text{-}H)(CO)_{18}]^-$ and, as shown in Figure **3-11**, has a metal framework closely related to that in the $[H_{4-n}Ni_{12-x}Pt_x(CO)_{21}]^{n-}$ (x = 0,3; n = 2,3) interstitial

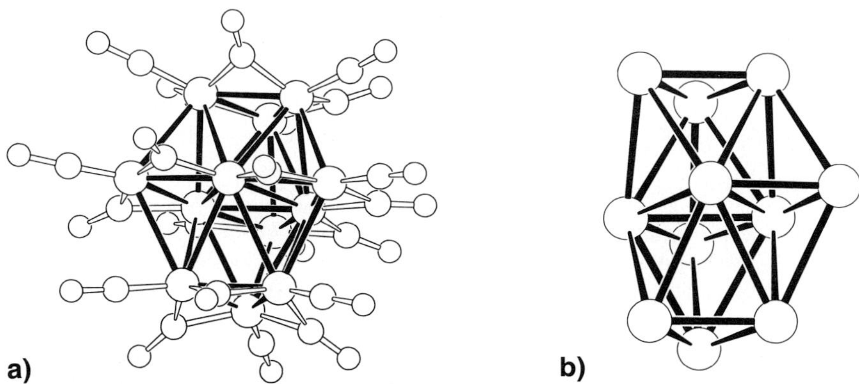

Figure 3-11. Schematic structures representations of the a) $[H_{4-n}Ni_{12-x}Pt_x(CO)_{21}]^{n-}$ (x = 0,3; n = 2,3) [123, 152] and b) the metal skeleton of the $[Ru_{11}H(CO)_{27}]^{3-}$ trianion. [325]

Table 3-11. Spectroscopic features of some selected interstitial or semi-interstitial atoms (δ^1H and δ^{13}C are relative to Me_4Si, δ^{11}B to BF_3, δ^{15}N to NH_3 and δ^{31}P to H_3PO_4).[a)]

Compound	ν_{M-E}[cm^{-1}]	δ_E[ppm]	Reference
$[Co_6(\mu_6\text{-}H)(CO)_{15}]^-$ (e)	1086, 949	+ 23.2	[322, 331]
$[Ru_6(\mu_6\text{-}H)(CO)_{18}]^-$ (e)	845, 806	+ 16.4	[203, 323, 337]
$[Ru_{11}H(CO)_{27}]^{3-}$		− 5.47	[325]
$[Ni_{12}(\mu_6\text{-}H)_2(CO)_{21}]^{2-}$ (e)	667	− 18.0	[121, 331]
$[Rh_{13}H_2(CO)_{24}]^{3-}$ (c)		− 26.7	[335]
$[HRu_6(\mu_6\text{-}B)(CO)_{17}]$ (e)		+193.8	[252]
$[H_2Ru_6(\mu_6\text{-}B)(CO)_{18}]^-$ (f)		+206	[253]
$[Fe_4(\mu_4\text{-}C)(CO)_{12}]^{2-}$ (b)		+478	[336]
$[Fe_5(\mu_5\text{-}C)(CO)_{15}]$ (c)	805, 775, 766	+486	[336, 338]
$[Ru_5(\mu_5\text{-}C)(CO)_{15}]$ (c)	757, 738, 730		[338]
$[Os_5(\mu_5\text{-}C)(CO)_{15}]$ (c)	793, 769, 757		[338]
$[Fe_6(\mu_6\text{-}C)(CO)_{16}]^{2-}$ (e)		+484	[336]
$[Rh_6(\mu_6\text{-}C)(CO)_{13}]^{2-}$ (e)		+335.8	[344]
$[Au_6(\mu_6\text{-}C)(P(p\text{-}C_6H_4NMe_2)Ph_2)_6]^{2+}$ (e)		+137	[341]
$[Co_6(\mu_6\text{–}C)(CO)_{15}]^{2-}$ (f)	772, 719	+332.8	[342, 343]
$[Rh_6(\mu_6\text{-}C)(CO)_{15}]^{2-}$ (f)	689, 653	+264.7	[340, 342, 344]
$[Ni_8(\mu_8\text{-}C)(CO)_{16}]^{2-}$ (h)	580, 525	+278.0	[345, 18]
$[Ru_4(\mu_4\text{-}N)(CO)_{12}]^-$ (b)		+519.1	[296]
$[Ru_5(\mu_5\text{-}N)(CO)_{14}]^-$ (c)		+464.9	[296]
$[Ru_6(\mu_6\text{-}N)(CO)_{16}]^-$ (e)		+559.8	[296]
$[Co_6(\mu_6\text{-}N)(CO)_{15}]^-$ (f)	722, 698	−184[b)]	[272, 342]
$[Rh_6(\mu_6\text{-}N)(CO)_{15}]^-$ (f)	645, 622	−272.6[b)]	[272, 342]
$[Os_6(\mu_6\text{-}P)(CO)_{18}]^-$ (f)		+523	[271]
$[Rh_9(\mu_8\text{-}P)(CO)_{21}]^{2-}$ (h)		+282.3	[82]
$[Rh_{10}(\mu_8\text{-}P)(CO)_{22}]^{3-}$ (h)		+369.3	[279]

a) The letters appearing in parenthesis refer to the stereochemical coordination of the interstitial atom as pictorially represented in Fig. 3-13.
b) Referenced to external $CD_3{}^{15}NO_2$

hydride clusters, the only difference being the absence of one metal atom in the central triangle of frequency 2.

Despite the exceptions cited above, interstitial lodging of hydride atoms appears to be more favourable in polycavity clusters and, as previously discussed, at least a dozen higher nuclearity hydride clusters have been unambiguously shown, or are reasonably thought, to have interstitial hydrides. Of those, $[H_{5-n}Rh_{13}(CO)_{24}]^{n-}$ (n = 1,2,3,4) and $[HRh_{14}(CO)_{27}]^{3-}$ deserve particular mention for they represent spectacular examples of the migration of hydride atoms within the metal skeleton of a cluster and provide an excellent model for hydrogen diffusion in metal lattices, e.g. the permeability of palladium towards hydrogen. As shown in Figure **3-12**, the high field hydride resonance of $[H_3Rh_{13}(CO)_{24}]^{2-}$ at room temperature consists of a doublet of septets of septets and is due to coupling of the hydrogen atoms with the unique interstitial rhodium atom, the six equatorial rhodium atoms, and the six rhodium atoms belonging to the top and bottom triangles. [335] Since the coupling with the unique interstitial rhodium atom is 3–4 times larger than the coupling with the peripheral rhodium atoms, migration must occur within the metal frame. The hydrogen's migration can be stopped by lowering the temperature to –90 °C and ^{1}H- ^{103}Rh-NMR measurements indicate that the hydrides then occupy the square-pyramidal sites, in accordance with the situation in the solid state as deduced from X-ray diffraction studies. [117, 350]

An unprecedented stereochemistry, for a hydride atom although not fully interstitial, is to be seen in $[HRu_6(\mu\text{-}O{=}CNMe_2)_2(CO)_{18}]^-$ (δ = –17.08 ppm). The hydrogen atom is located within the plane of the central nonbonded ruthenium triangle at distances of 1.50 Å from one ruthenium and 2.07 Å from the remaining two. [225]

The spread of the NMR chemical shifts over a more or less wide range is a common feature for all the interstitial main group atoms. A comparison of the

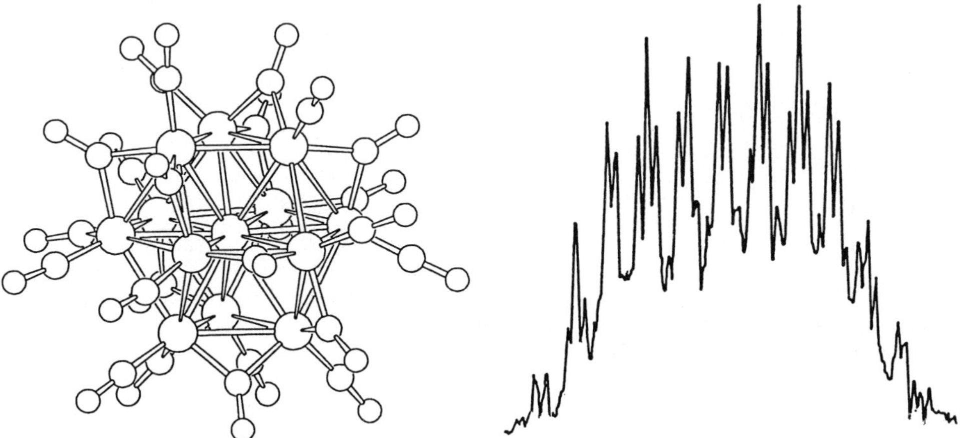

Figure 3-12. Molecular structure [107] and ^{1}H NMR spectrum [335] of the $[H_3Rh_{12}(\mu_{12}\text{-}Rh)(CO)_{24}]^{2-}$ ion.

$\delta_{carbide}$ values from a number of homometallic molecules having closely related geometries shows that the charge on the molecule has very little or no influence on the chemical shifts. For example, in the series $[Fe_4(\mu_4\text{-}C)(CO)_{13}]$, $[Fe_4(\mu_4\text{-}C)(\mu\text{-}H)(CO)_{12}]^-$, and $[Fe_4(\mu_4\text{-}C)(CO)_{12}]^{2-}$ the carbide resonances were found at 469.8, 464.2, and 478.0 ppm respectively. [336] Whereas the ^{11}B NMR chemical shifts of boron atoms residing in Ru_6 octahedral or trigonal prismatic cages are very close, a large variation was found in the ^{13}C and ^{15}N shifts of related Co_6 and Rh_6 clusters. The expected increase in shielding of a given interstitial atom in isostructural clusters on descending a group in the transition metal series is generally observed. The strong low field shifts observed might be interpreted in terms of deshielded positively charged interstitial main group atoms, however, a correlation between the chemical shifts and the charge associated with the interstitial atom based on the increase or decrease of its radius with respect to its accepted covalent size is often ambiguous (compare for instance the data in Tables **3-11** and **3-7** and is probably because size is not a decisive factor.

Clusters with interstitial carbides have been obtained from noncarbide precursors by activation of C–O, C–S, C–B, and C–Cl bonds, or from carbide cluster precursors through the synthetic methods already presented in Section 3.2.4 (thermal activation, oxidation, reduction, redox condensation, etc.). Therefore, we shall only examine here the established procedure to introduce carbide atoms into noncarbide precursors in more detail.

A very successful method for synthesizing interstitial carbide clusters is based on the metal assisted disproportionation of carbon monoxide into a carbide atom and carbon dioxide, which is related to the Bouduard reaction. Thus, most of the Re, Fe, Ru, and Os carbide clusters have been obtained from the thermal decomposition of a carbonyl precursor and the evolution of CO_2 has been confirmed in one case. [95]

Suitable carbonyl precursors can be both neutral and anionic mono- or polynuclear carbonyl complexes. Leading examples are represented in Equations (3.25)–(3.27). [347–349]

$$2[Ru_3(CO)_{12}] \rightarrow [Ru_6(\mu_6\text{-}C)(CO)_{17}] + CO_2 + 5\ CO \tag{3.25}$$

$$[Os_3(CO)_{12}] \rightarrow [Os_{11}(\mu_6\text{-}C)(CO)_{27}]^{2-} \tag{3.26}$$

$$[H_3Re_3(CO)_{10}]^{2-} \xrightarrow{n\text{-tetradecane},250°} [Re_7(\mu_6\text{-}C)(CO)_{21}]^{3-} \tag{3.27}$$

Running the above reactions in a π-donor solvent can lead to the synthesis of carbonyl substituted carbide clusters, such as the reaction shown in Equation (3.28). [350]

$$[Ru_3(CO)_{12}] \xrightarrow{\text{mesitylene}} Ru_6(\mu_6\text{-}C)(CO)_{14}(\eta^6\text{-}C_6H_3Me_3) \tag{3.28}$$

In Equation (3.27), as well as in related reactions which afford anionic carbide clusters, the anionic precursor can also be generated directly *in situ* by the addition of a reducing agent such as an alkali metal (Equation 3.29) [260] or a mononuclear carbonyl anion (Equation 3.30). [351]

$$[Ru_3(CO)_{12}] \xrightarrow{Na,glyme} [Ru_{10}(\mu_6\text{-}C)_2(CO)_{24}]^{2-} \tag{3.29}$$

$$[Fe(CO)_5] \xrightarrow{[Mn(CO)_5]^-,diglyme,160^\circ} [Fe_6(\mu_6\text{-}C)(CO)_{16}]^{2-} \tag{3.30}$$

The latter reaction has been shown to occur with several other reducing agents such as $[V(CO)_6]^-$, $[Mo(CO)_3Cp]^-$, $[Fe(CO)_4]^{2-}$ and $[Co(CO)_4]^-$. Furthermore, as a result of the well known disproportionation of $Fe(CO)_5$ in basic solvents, salts of the $[Fe(solvent)_x][Fe_6(\mu_6\text{-}C)(CO)_{16}]$ complex cluster have also been obtained on refluxing $Fe(CO)_5$ in diglyme or quinoline. [352–354]

One can postulate that the cleavage of carbon monoxide in all the above syntheses occurs *via* activation by dual coordination. In principle, this may occur either (a) by the intra- or intermolecular nucleophilic attack of the oxygen atom of a coordinated carbon monoxide group at the carbon atom of a second CO group bonded to either the same or a different metal atom, or (b) by the η^2-coordination of a CO group in ways schematically shown in Section 3.2.1. This suggestion arises from the consideration that the dual coordination of carbon monoxide should correspond to a significant lengthening of the C–O bond, whereby the stabilization by electron delocalization over the resulting O–C or O–M bond may favor the splitting of CO. Formation of CO_2 has occasionally been observed experimentally and formally supports mechanism (a), however, this observation is not sufficient to truly distinguish between the two. Mechanism (b) parallels the irreversible chemisorption of carbon monoxide on a metal surface and although the formation of metal oxides has never been detected in the above reactions, this mechanism cannot be ruled out. In this respect, it is worth noting that of all the metal carbide clusters known, those obtained by the thermally activated splitting of carbon monoxide are composed of the most oxophilic metals. That CO_2 would be detected as by products rather than metal oxide derivatives, could be a result of intra- or intermolecular deoxygenation reactions via elimination of CO_2. Consistent with this suggestion is the observation that the chemisorbed oxygen atom on $[Fe_3(\mu_3\text{-}O)(CO)_9]^{2-}$ [355] is readily eliminated intermolecularly in several ways when exposured to a carbon monoxide atmosphere. [301]

The cleavage of carbon monoxide has been accomplished under mild conditions by using strong protonic acids, CH_3COCl or $SiCl_4$ as deoxygenating agents. Interstitial carbide clusters such as $[Co_6C(CO)_{14}]^-$, $[Ni_8C(CO)_{16}]^{2-}$, and $[Rh_6C(CO)_{15}]^{2-}$, semi-interstitial carbides such as $[H_{2-n}Fe_4C(CO)_{12}]^{n-}$ (n = 0,1,2) and $[HFe_4AuC(CO)_{12}(PPh_3)]$, [54, 356–358] or ketenylidene derivatives such as $[H_{2-n}M_3(CO)_9(CCO)]^{n-}$ (M = Fe,Ru,Os) [359, 360] have also been obtained by this alternative route. For iron derivatives, it has been shown that the CO cleavage can be carried out in the two distinct steps shown in Equations (3.31) and (3.32) from which the carbide clusters are obtained almost quantitatively. [356, 357]

$$[Fe_4(CO)_{12}(\mu_3\text{-}CO)]^{2-} + RCOCl \rightarrow [Fe_4(CO)_{12}(\mu_3\text{-}CO\text{-}C(=O)\text{-}R]^- + Cl^- \tag{3.31}$$

$$[Fe_4(CO)_{12}(\mu_3\text{-}CO\text{-}C(=O)\text{-}R]^- + 2Na \rightarrow [Fe_4(\mu_4\text{-}C)(CO)_{12}]^{2-} + RCOO^- + 2\,Na^+ \tag{3.32}$$

In contrast, $[Co_6C(CO)_{14}]^-$ can be obtained in high yields by the simple addition of CH_3COCl to $[Co_6(CO)_{15}]^{2-}$. Here, the dual coordinated carbon monoxide group of the $[Co_6(CO)_{14}(CO\text{-}C(=O)\text{-}R]^-$ intermediate can be cleaved through a sacrificial redox reaction with the starting cobalt carbonyl anion, or a derivative thereof, and the resulting carbide atom can then be adsorbed into the metal frame. [356] Analogously, in the related reaction of $[HFe_4(CO)_{13}]^-$ with strong protonic acids, protonation of its most basic oxygen atom of the bridging carbonyl group followed by cleavage of the dual coordinated CO group through sacrificial oxidation of some iron is at the basis for the formation of these carbide clusters. [361]

Although conclusive evidence is lacking, it has been suggested that either dihapto or dual coordination of a nitrosyl group could also be at the heart of the synthesis of nitride derivatives like $[M_6N(CO)_{15}]^-$ (M = Co,Rh), $[M_5(\mu_5\text{-}N)(CO)_{14}]^-$ (M = Fe,Ru), and $[HFe_3Ru(\mu_4\text{-}N)(CO)_{12}]$ through the reaction of metal carbonyl anions with nitrosonium salts or of nitrosyl substituted carbonyl clusters with protonic acids. [268]

Cleavage of less energetic C–S, C–Cl, C–B and C–Si bonds represents an alternative way to introduce carbon atoms into a cluster and has been successfully applied in the syntheses of several Co, Rh, Ni, and Au clusters, e.g. $[Co_6(\mu_6\text{-}C)(CO)_{12}(\mu_3\text{-}S)_2]$, [362] $[Rh_6(\mu_6\text{-}C)(CO)_{15}]^{2-}$, [340] $[Ni_9(\mu_8\text{-}C)(CO)_{17}]^{2-}$ [215] and $[Au_6(\mu_6\text{-}C)(PPh_3)_6]^{2+}$. [341] This method, in addition to forming monocarbide clusters by employing molecules such as CS_2, $CHCl_3$, CCl_4, and $C\{B(OMe)_2\}_4$ as the source of the carbide atom, also offers the possibility of directly introducing up to five additional carbide atoms into a cluster by using perhalocarbons such as tetrachloroethene or hexachloropropene. [264–266]

A direct derivation of the above approach made the synthesis of most of the clusters containing B, In, Si, Ge, Sn, P, As, Sb, S, Se, and Te interstitial heteroatoms or elemental organic fragments possible by the cleavage of E–H, E–Cl and E–C bonds under suitable experimental conditions.

At this point, it only remains to report on an alternative synthesis for nitride clusters Equation (3.33), which is based on a different approach as the previously described cleavage of NO. A nucleophilic attack by N_3^- on coordinated CO followed by the elimination of dinitrogen will generate a coordinated isocyanate ligand. The subsequent step to form the nitride is based on the observation that the preformed $[Ru_4(CO)_{13}(NCO)]^-$ isocyanate substituted carbonyl cluster slowly converts into the $[Ru_4(\mu_4\text{-}N)(CO)_{12}]^-$ nitride derivative on standing in solution. [268, 363]

$$2[Ru_3(CO)_{12}] + N_3^- \rightarrow [Ru_6(\mu_6\text{-}N)(CO)_{16}]^- + N_2 + 8\ CO \qquad (3.33)$$

The spectroscopic features of some selected interstitial atoms are listed in Table **3-11**.

3.2.5.2 Overview of the Synthetic Methods of Mixed Clusters and Main Group Element Clusters Containing Transition Metal Atoms

A few significant and recent examples of mixed clusters containing M–M, M–E and E–E bonds, also a few others where only M–E and E–E bonds are present are collected in Table **3-10**. These are clusters from categories **C** and **D** as opposed to the clusters belonging to categories **A** and **B**, which only contain M–M and M–E bonds. To be more comprehensive, examples of polynuclear compounds having only M-E interactions should also be included, since these are often closely related to the former, however, the absence of homoatomic interactions put them beyond the purposes of this brief review. Mixed clusters of main group and metal atoms have been known for a long time, e.g. $[Co_{4-n}As_n(CO)_{12-3n}]$ (n = 1,2,3), [364, 365] but only in recent years has there been an upsurge of interest, probably stimulated by the spectacular results obtained in phosphine stabilized clusters (see chapters 3.3 and 3.4). Another reason for the increased interest probably stems from the possibility of intercepting excised fragments of lattices of solid state homologues with various ligands and to thereby generate entirely new types of compounds.

A few of the synthetic methods adopted so far are closely related to the previously examined methods used for the synthesis of clusters containing interstitial or exposed main group elements or elemental organic fragments. For example, as shown in Scheme **3-11**, the very peculiar $[Bi_4Fe_4(CO)_{13}]^{2-}$ cluster (see Section 3.2.6) has been obtained in two steps through classical redox condensation reactions or by the clockwise condensation induced by fragment elimination. [312] Related syntheses have also been used for the E–Co (E = Sb,Bi) [314, 315] and Tl–Fe [309] mixed clusters listed in Table **3-11**.

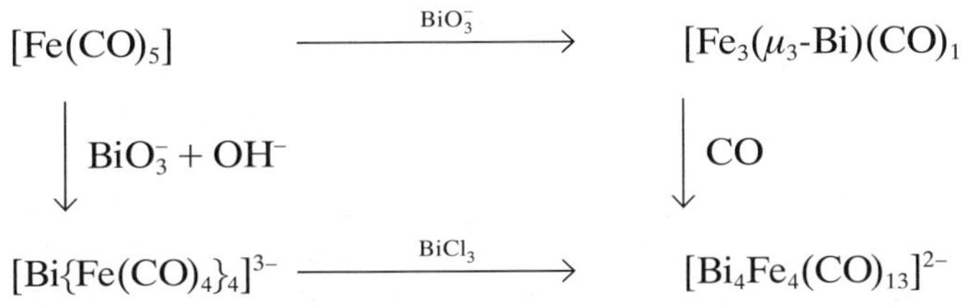

$$
\begin{array}{ccc}
[Fe(CO)_5] & \xrightarrow{BiO_3^-} & [Fe_3(\mu_3\text{-}Bi)(CO)_{10}]^- \\
\downarrow {\scriptstyle BiO_3^- + OH^-} & & \downarrow {\scriptstyle CO} \\
[Bi\{Fe(CO)_4\}_4]^{3-} & \xrightarrow{BiCl_3} & [Bi_4Fe_4(CO)_{13}]^{2-}
\end{array}
$$

Scheme 3-11.

Main group elements form a wide variety of neutral, cationic, and anionic compounds having chain, ring, and cage structures (e.g. Zintl anions). [366] All of these may be useful starting materials to be combined with mono- and polynuclear transition metal moieties to generate molecules with novel and exciting structural features. This alternative type of synthetic approach has been used for the synthesis of all the other derivatives included in Table **3-11**. For example, as shown in Scheme **3-12**, the reaction of either Se_4^{2+} or Te_4^{2-} with such mononuclear metal carbonyls as $[Fe(CO)_5]$ and $[W(CO)_6]$ can yield cluster compounds

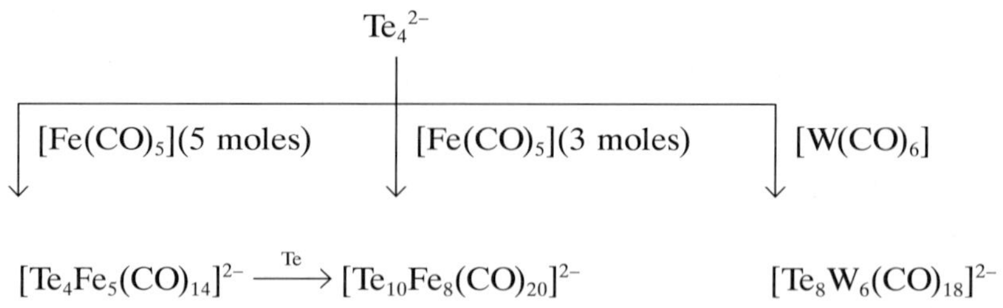

$$Te_4^{2-}$$

$$\xrightarrow{[Fe(CO)_5](5\ moles)} [Te_4Fe_5(CO)_{14}]^{2-} \xrightarrow{Te} [Te_{10}Fe_8(CO)_{20}]^{2-}$$

$$\xrightarrow{[Fe(CO)_5](3\ moles)} [Te_{10}Fe_8(CO)_{20}]^{2-}$$

$$\xrightarrow{[W(CO)_6]} [Te_8W_6(CO)_{18}]^{2-}$$

$$Se_4^{2+} + 2[W(CO)_6] \longrightarrow [Se_4W_2(CO)_{10}]^{2+} + 2\ CO$$

Scheme 3-12.

with various proportions of M–M, M–E and E–E bonds, as well as polynuclear derivatives containing essentially only M–E interactions. [317, 319, 320]

Reactions of the neutral ring compounds As_5Me_5 and As_6Ph_6 with $[Mo_2Cp_2(CO)_4]$ and $[Co_2(CO)_8]$ respectively, are at the origin of mixed metal clusters as different as $[As_5Mo_2Cp_2]$ and $[Co_{16}(\mu_6\text{-}As)_2(\mu_4\text{-}As)_2(\mu_4\text{-}AsPh)_4(CO)_{32}]$. [311, 313] The former compound can be viewed as a triple decker sandwich derivative in which the inner As_5 ring formally replaces an isolobal C_5H_5 ring. A related triple decker $[P_6Mo_2Cp^*_2]$, compound in which the inner P_6 ring is the counterpart of a C_6H_6 ring, has been obtained by the reaction of P_4 with $[Mo_2Cp^*_2(CO)_4]$. [310]

The close electronical and structural relationships between the cyclopentadienide ion and the $[B_9C_2H_{11}]^{2-}$ nido carborane were recognized from early on. Accordingly, $[M(B_9C_2H_{11})_2]^{n-}$ (e.g. M = Fe,Co, and Cu) derivatives related to the corresponding MCp_2 sandwich compounds have been synthesized and characterized. The extensive chemistry of the metalloborane and metallocarbaborane clusters has been developed to a large extent by M. F. Hawthorne, R. N. Grimes and N. N. Greenwood. [238]

3.2.5.3 Structural characteristics of Transition Metal Clusters Containing Interstitial or Exposed Main Group Elements or Elemental Organic Fragments

In both metal lattices and closo polyhedral metal clusters there are cavities present, whose dimensions are a function of the number of vertices, the shape of the polyhedral moiety, and the interatomic separation. Partial occupation of these sites within the metal lattice by main group elements results in the formation of interstitial alloys such as, for example, metal hydrides, carbides, and nitrides. Occupation of the cavity within a molecular metal cluster gives rise to interstitial clusters. Although close topological relationships are sometimes found between interstitial alloys and interstitial clusters, the greater degree of freedom of a molecular assembly of metals, in comparison to a three dimensional infinite array of metals, increases the possible number of interstitially lodged elements and can give rise to peculiar structural features. These will be highlighted in this chapter together with the most striking similarities between such clusters and metal lattices.

Under a rigid sphere model, the radius of a cavity is easily evaluated from geometrical considerations and, as a first approximation, may be taken as the limiting factor for determining the size of the potential interstitial atom. The calculated radii of the cavities of a few selected regular polyhedra with edge *e* are listed in Table **3-7**. For comparison, numerical values for edge lengths, *e*, of 2.5 and 2.8 Å, which may be assumed to represent typical M–M bond lengths of first and second or third row group VIII metals respectively, are given in the fourth column of Table **3-7**. The fifth column summarizes the main group elements (accepted covalent radii in parentheses) which have been reported for a given cavity. The apparent radius of a given element (E) which is located within the interstitial cavity of a metal (M) cluster is usually calculated as $r_E = d_{M-E}$-$1/2d_{M-M}$ (where *d* is the interatomic separation). In real compounds, the radius of the interstitial atom depends greatly on the shape of the host polyhedron, the kind of metal making up the cavity, and probably, the ligands bonding the metals. Therefore, the above steric limit should not be overemphasized. It must always be kept in mind that both the metal framework and the interstitial atoms may be rather soft. For instance, interstitial carbide atoms have been squeezed into cavities with radii as small as 0.54 Å or effectively swelled to fill cavities with radii of up to 0.87 Å depending on the shape of the polyhedral frame and/or the type of metal. In a rigid sphere model, the covalent radius of carbon may be thought to give a satisfactory fit to only the square antiprismatic cavities of first row metals and the trigonal prismatic cavities of second and third row metals. On the contrary, as shown in Figure **3-13** and by inspection of the second column in Tables **3-8** and **3-9**, carbon atoms have also been found to take up interstitial sites having octahedral (μ_6), trigonal prismatic (μ_6), distorted capped trigonal prismatic (μ_7) and square antiprismatic (μ_8) symmetry, virtually without regard to the particular metal. In addition, there are several examples of carbon in other semi-interstitial site stereochemistries, including a slightly distorted square planar coordination. Similar observations have also been made with other lighter elements such as boron and nitrogen. Although an example is known in which phosphorus takes up a trigonal prismatic interstitial site, the third row main group elements, as well as arsenic, display a marked tendency to lodge in square antiprismatic cages. Upon further descending the groups, the increasing size of the elements allows for an increase in the formal coordination number to twelve, whereby this has only been realized so far by interstitial lodging in an icosahedral cage. This represents a major stereochemical difference from transition metal atoms which have been found to be twelve coordinate with almost all the possible suitable stereochemistries including the icosahedron. Examples of icosahedral metal clusters are the $[Ni_{12}Sn(CO)_{22}]^{2-}$ dianion and its germanium congener. The structure and space filling model of the former are illustrated in Figure **3-14**. [167] As shown in the top part of the Figure **3-14**, the 22 carbonyl groups are divided into three sets with the first set comprising twelve terminally bonded carbonyl groups, one each per nickel atom. The second and third sets are composed

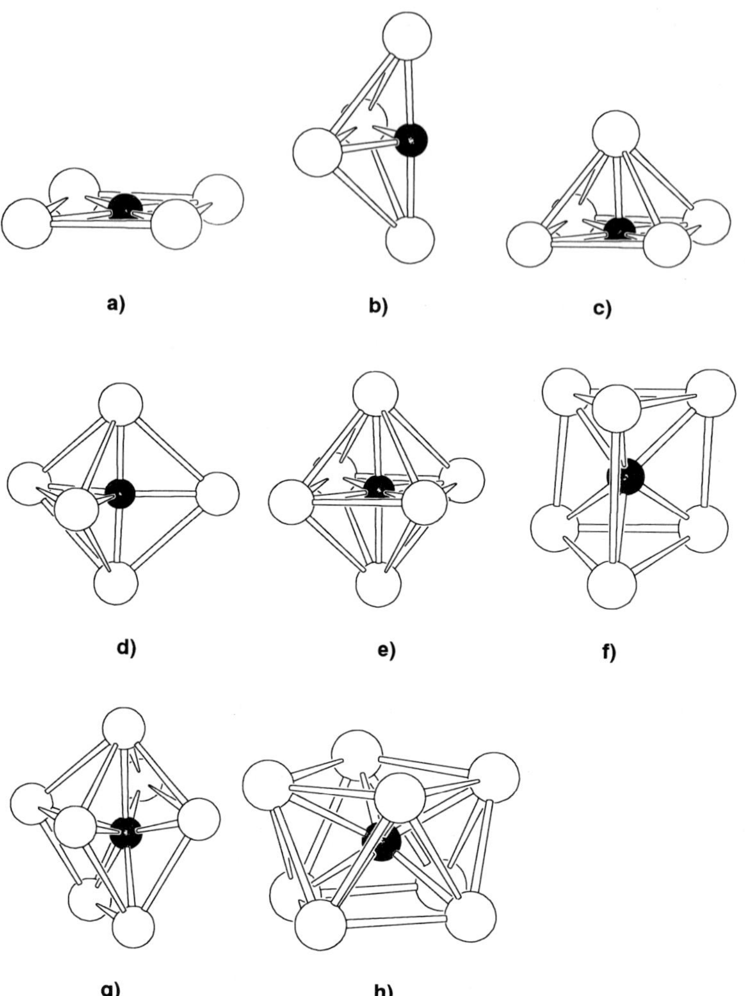

Figure 3-13. Stereochemical behavior of the metal-coordinated carbon atom.

of three edge and two face bridging carbonyls respectively, which are tilted outward with respect to the central pentagonal antiprismatic moiety. This particular stereochemistry of the carbonyl groups allows all the nickel atoms to be coordinated to three carbon monoxides and most likely ensures equalization of the charge at all the metal sites. Inspection of the bottom part of Figure **3-14** shows that the metal core is almost perfectly shielded by the carbonyl groups which are themselves arranged as a 1,5,5′,5,5′,1 stacked polyhedron approximating two fused icosahedra and related to the one illustrated in Scheme **3-9** of Section 3.2.4. Notably, the only other known icosahedral cluster lodging a main group

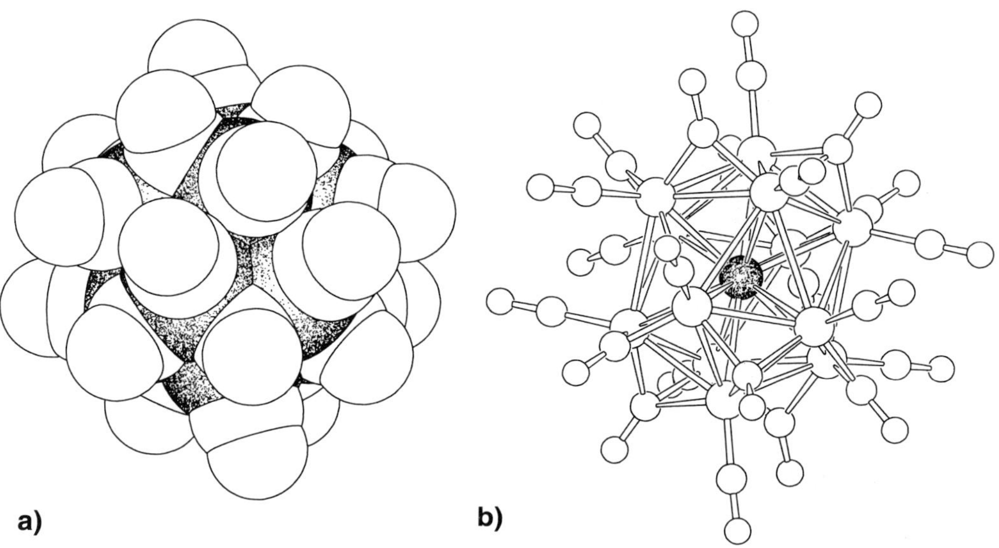

Figure 3-14. The $[Ni_{12}Sn(CO)_{22}]^{2-}$ dianions a) molecular structure (tin atom is shaded) and b) space filling sketch (nickel atoms are shaded). [167]

element interstitially, *viz.* $[Rh_{12}Sb(CO)_{27}]^{3-}$, [281] has a bulkier metal core and coordinates five additional carbonyl ligands. These ligands appear to span the interlayer edges of the central pentagonal antiprism and, as a result, the carbonyl polyhedron can be described as three fused icosahedra. Such arrangements of carbonyls as in the above, which reproduce the C_5 symmetry of the inner metal core, underline the importance of a good match between the metal and the carbonyl polyhedra and may ultimately affect the shape of the cluster and the stereochemistry of the interstitial atom.

The possible relevance of factors other than the covalent radius of the interstitial atom to the formation of interstitial clusters is suggested by the results of the attempted synthesis of an icosahedral $Ni_{12}(\mu_{12}\text{-}Sb)$ carbonyl cluster. Despite of the similarity between the covalent radii of tin and antimony, the corresponding reaction of $[Ni_6(CO)_{12}]^{2-}$ with $SbCl_3$ results in the formation of a $[Ni_{13}Sb_2(CO)_{24}]^{n-}$ species, in which an interstitial nickel atom is lodged within a $[Ni_{10}Sb_2]$ icosahedral moiety (Fig. **3-15**). [300] This difference has been attributed to the greater electronegativity of antimony which opposes the necessary shrinking of the atom required in order for it to fit into the Ni_{12} interstitial cavity.

It is not intended here to give a full account of all the variety of structures which can be stabilized by the incorporation of interstitial atoms or the peripheral coordination of main group elements. However, a few more examples seem appropriate so as to give some indication about the potential vastness of this area of cluster chemistry.

A prerogative of carbon, according to the data available so far, is to give rise to interstitially lodged C_2 units displaying interatomic separations in the 1.48–

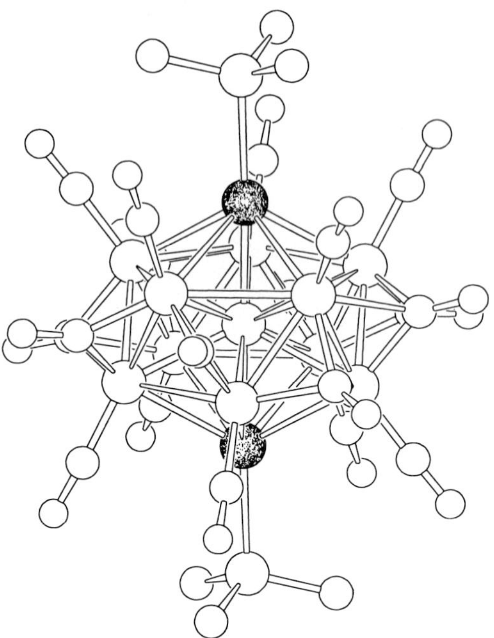

Figure 3-15. The molecular structure of $[Ni_{13}(\mu_7\text{-}Sb)_2(CO)_{24}]^{2-}$. [300]

1.36 Å range. In addition, there are also several examples known of exposed C_2 fragments which show C–C interatomic separations as short as 1.30 Å, e.g. $[Ru_5C_2(PPh_2)_2(SMe)_2(CO)_{11}]$. [291] The bonding interaction of these C_2 moieties with the metal cage has been shown by EHMO calculations to be related to the interaction of ethylene or acetylene with a metal atom and suggests the occurrence of a significant $d_\pi-\pi^*$ back-donation. The first reported example of this type was $[Rh_{12}C_2(CO)_{25}]$, [185] whose structure is shown in Scheme **3-4**. The most striking example, however, is provided by the structure of $[Ni_{16}(C_2)_2(CO)_{23}]^{4-}$. [172] As shown in Figure **3-16**, the nickel atoms form a truncated octahedron of frequency 2, in which an interstitial nickel atom is replaced by two pairs of bonded carbon atoms.

There are several interstitial polycarbides and polynitrides known for both carbon and nitrogen. These represent increasingly larger chunks from the structure of bulk carbide and nitride binary alloys stabilized in a molecular compound. For instance, the recently reported $[Co_{14}(\mu_6\text{-}N)_3(CO)_{26}]^{3-}$ shows a metal frame consisting of a distorted hexagonal prism formally deriving from the fusion along one interlayer edge of three N-centered $[Co_6(\mu_6\text{-}N)]$ trigonal prismatic moieties. [274] As shown in Figure **3-17**, the molecular structure of this anion represents a well defined fragment of the hexagonal packing lattice adopted by several carbides and nitrides of the earlier transition metals (e.g. WC).

In a molecular compound, the steric and electronic constraints which disfavor or hinder the occurrence of certain M_xC_y and M_xN_y binary alloys, and arise from the requirement for a unit cell which repeats periodically in three dimensions, are no longer operative. The smaller late transition metals in particular may give rise

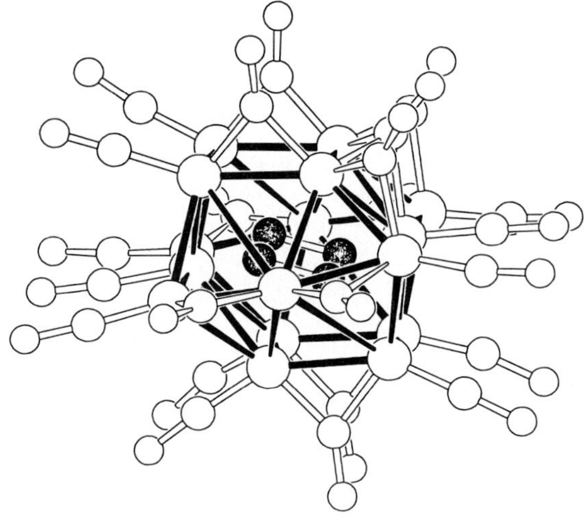

Figure 3-16. The molecular structure of the $[Ni_{16}(C_2)_2(CO)_{23}]^{4-}$ tetraanion [172], based on a truncated octahedron of frequency 2, which lacks the interstitial nickel atom (the two interstitial C_2 units caged in the unique cavity are highlighted by shading).

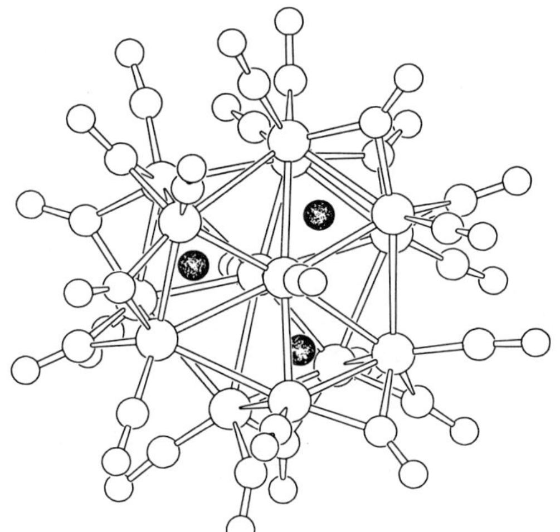

Figure 3-17. The molecular structure of the $[Co_{14}(\mu_6\text{-}N)_3(CO)_{26}]^{3-}$ anion [274] as an example of stabilization of a fragment of the hexagonal lattice in a molecular cluster.

to interstitial carbides and nitrides, eventually approximating fragments of early transition metal binary alloys. [367] The extensive chemistry of nickel carbonyl carbide clusters, compared to the reported existence of an ill-defined Ni_3C phase, highlights this difference. The largest chunk of a bulk metal carbide which has been stabilized so far in a molecular compound can be recognized in the structure of the $[HNi_{38}C_6(CO)_{42}]^{5-}$ [266] as the truncated octahedral $M_{32}C_6$ building block of the $Cr_{23}C_6$ phase. [368] As shown in Figure **3-18**, the metal frame of this compound comprises an inner Ni_8 cube capped on its six faces by 6 carbide atoms; these are lodged in regular square antiprismatic cavities arising from further condensation of 6 square Ni_4 units on top in a staggered conformation. The resulting truncated octahedron displays eight concave hexagonal faces. The

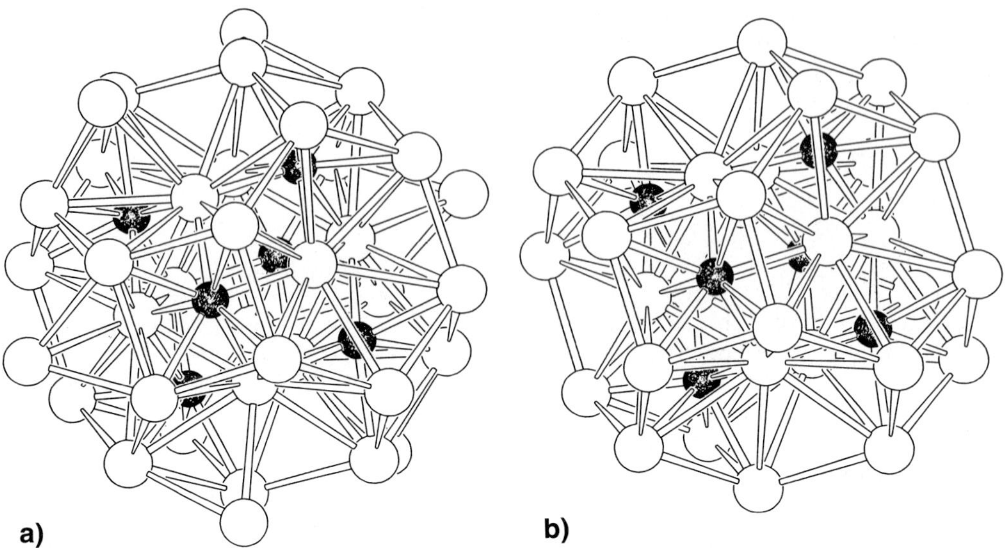

Figure 3-18. a) The molecular structure of the $[HNi_{38}C_6(CO)_{42}]^{5-}$ anion, [266] and b) its relationship with an extended chunk of the structure of bulk $Cr_{23}C_6$. The truncated octahedral building unit of b) shows an intimate structure identical to that of the $Ni_{32}C_6$ central unit of a).

remaining six nickel atoms condense in a μ_3-capping fashion onto six of the above hexagonal faces. This close relationship between a molecular compound and a bulk phase suggests the possibility of stepwise construction of the architecture of the latter and, perhaps, of the eventual synthesis of unknown binary phases starting from molecular compounds under mild conditions.

3.2.6 Clusters of Clusters and Beyond

In addition to mobile electrons and mixed valency, low dimensionality, resulting from the presence of infinite chains, ribbons, layers, etc., or stacks of molecules in the crystal network, is a common structural feature of organic, organometallic and, inorganic conductors and superconductors. [369, 370] As shown in the previous chapters, a variable oxidation state, although not general, is a characteristic displayed by several clusters. Furthermore, it is predictable that the occurrence of variable valency should become more frequent as the nuclearity of the cluster increases, owing to a decrease in the HOMO-LUMO separation and electron pairing energy as a result of increasing delocalization and a progressive change from discrete to quasi-continuous energy levels. In keeping with the above suggestion, electrochemically reversible redox behavior and the occurrence of several stable oxidation states for a given species are not prerogatives of the metal clusters stabilized by main group atoms or fragments derived thereof (e.g. cubane derivatives such as $[Fe_4(\mu_3\text{-}S)_4Cp_4]^{n-}$ and the pseudo-cubane $[Fe_4(\mu_3\text{-}CO)_4Cp_4]$ [182–184], or those having "ad hoc" favorable conditions such as low lying weakly antibonding orbitals (e.g. $[Fe_3Pt_3(CO)_{15}]^{n-}$). [181] Clusters supported by only M–M bonds (e.g. Ni_6Cp_6, [17] $[Pt_{24}(CO)_{30}]^{2-}$, [189, 371] $[Pt_{26}(CO)_{32}]^{2-}$, [189] and $[Pt_{38}(CO)_{44}]^{2-}$ [189]) may also demonstrate such behavior. A few selected examples are listed in Table **3-12**. When multiple reversible redox changes occur, alternately larger and smaller ΔE_f are commonly observed. This general trend suggests that the electrons are added or removed in pairs since the energy gap between the frontier orbitals is larger than the pairing energy. A rough estimate of the HOMO-LUMO gap, as well as the energy separations between orbitals surrounding it, and the pairing energy in the corresponding orbitals, can sometimes be made and compared with the results from molecular orbital calculations. For example, under the crude assumption that the solvation free energies associated with the various ions and neutral compounds are invariant, energy separations of ca. 0.7–1.0 eV and pairing energies of ca. 0.15–0.30 eV may be calculated for the frontier orbitals of $[Pt_{24}(CO)_{30}]^{n-}$. These may be compared, for instance, with the HOMO-LUMO gap of *ca.* 1.4 eV, which is close to the value of 1.3 eV found from EHMO calculations, and the pairing energy of *ca.* 0.5 eV estimated for the lower nuclearity cluster $[Fe_3Pt_3(CO)_{15}]^{n-}$. [181]

It should also be possible to fullfil the low-dimensionality requisite cited at the beginning of this section, by assembling individual molecular cluster entities in pseudo one or two dimensional arrays within three dimensional networks. Cluster chemistry has already provided several examples of compounds, namely clusters of clusters, which suggest that such a possibility is feasible. In the following few pages we shall describe some known compounds which may be seen as relevant examples of expressly devised or auto-assembling syntheses for the way to build up both low dimensional as well as three dimensional clusters of clusters. In principle, assemblage of molecular cluster entities can be obtained in several ways. A first possibility is given by the formation of a homometallic M–M bond

Table 3-12. Formal redox potentials (E_f in V *vs.* $FeCp_2^{+1/0}$) in acetonitrile solution of some selected clusters.[a)]

Starting	redox couples								
material	+2/+1	+1/0	0/−1	−1/−2	−2/−3	−3/−4	−4/−5	−5/−6	−6/−7
$[Co_4Sb_2(CO)_{11}]^-$			−0.54[b]						
$[Os_4Au_2(CO)_{12}L_2]$	0.78	0.60	−0.61	−0.89					
$[Fe_3Pt_3(CO)_{15}]^{2-}$			0.19	−0.31					
$[Ni_{11}Bi_2(CO)_{18}]^{2-}$					−0.67	−1.42			
$[Ni_{13}Sb_2(CO)_{24}]^{2-}$					−0.56	−1.28			
$[Pt_{24}(CO)_{30}]^{2-}$			0.40[b]	0.15[b]	−0.60	−0.90	−1.35	−1.50	−2.15

a) The data have been taken from references 314, 339, 190, 94, 300, 189 respectively.
b) Dichloromethane as solvent.

between two or more units (Category **A**). Since several clusters may be viewed as deriving from a more or less formal condensation of two or more subunits, we will take into consideration only those compounds where the assembling subunits can be unambiguously identified at a glance. Probably the oldest examples are $[Rh_{12}\text{-}(CO)_{30}]^{2-}$, [102, 103] $[Ir_8(CO)_{22}]^{2-}$ [372] and $[Pt_{15}(CO)_{30}]^{2-}$. [130] As shown in Figure **3-19**, the first two examples derive from two octahedral $[Rh_6(CO)_{15}]^-$ and two tetrahedral $[Ir_4(CO)_{11}]^-$ units joined together by a single Rh–Rh or Ir–Ir bond respectively. The $[Pt_{15}(CO)_{30}]^{2-}$ cluster, taken as a representative of the $[Pt_3(CO)_6]_n^{2-}$ series of oligomers, has already been discussed in Section 3.2.4 and, as shown in Figure **3-5b**, is made up of a stack of five $[Pt_3(CO)_6]$ subunits in a quasi-eclipsed conformation. Other members of the above series include the structurally characterized oligomers with $n = 2, 3, 4$, and 6 and those which have

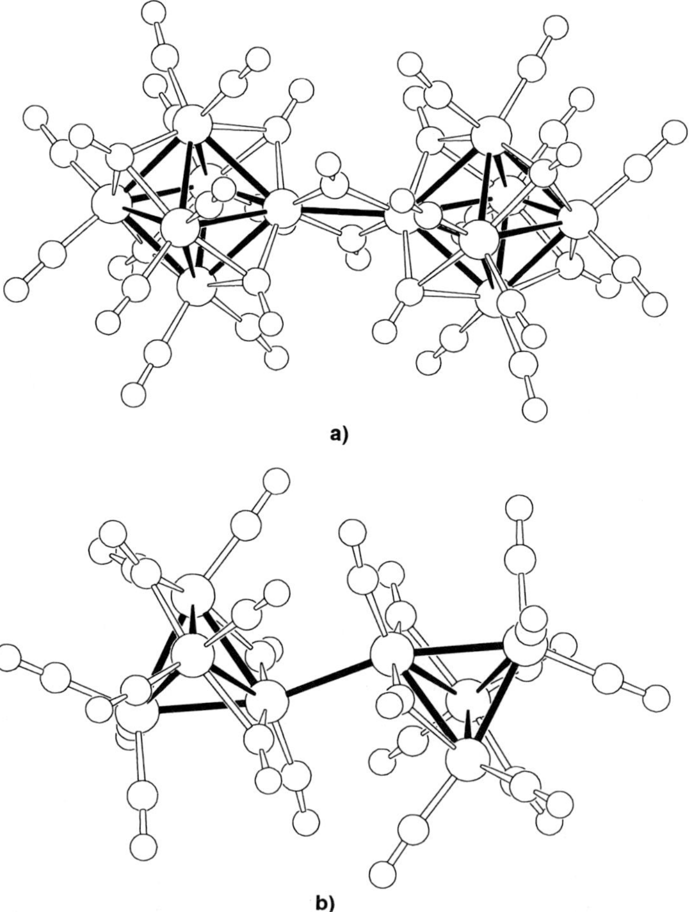

Figure 3-19. Some examples of cluster of clusters: $[Rh_{12}(CO)_{30}]^{2-}$ (a), [103] $Ir_8(CO_{22}]^{2-}$ (b), [372]

only been spectroscopically and analytically characterized due to their decreasing solubility in all organic solvents with $n = 7$, 8, 9, 10, and beyond. [128]

In contrast to the stacking characteristics of the $[Pt_3(CO)_6]$ units, the stacking of analogous $[Ni_3(CO)_6]$ units only gives rise to $[Ni_3(CO)_6]_n^{2-}$ oligomers with $n = 2$ and 3, [373, 85] and is probably due to a combination of both electronic and steric factors. In particular, the nonbonding repulsions between the piled carbonyl ligands of consecutive $[Ni_3(CO)_6]$ units are thought to limit the growth of these pseudo one dimensional oligomers to $n = 3$. However, once attention is focused on the collective behavior of the crystal, rather than the individual properties of the molecular entity, it turns out that the crystalline structure of $[NMe_4]_2[Ni_6(CO)_{12}]$ derives from an infinite one dimensional stacking of $[Ni_6(CO)_{12}]^{2-}$ molecular ions held together by weak van der Waals forces rather than Ni–Ni bonds. In other words, the crystal network may be seen as bundle of pillared $[Ni_3(CO)_6]^-$ units assembled by alternating covalent and van der Waals interactions. The contiguous sets of $[Ni_3(CO)_6]$ units are alternatively separated by 2.40 and 4.64 Å within and between each dianion moiety respectively; the individual pillars are separated from each other by the $[NMe_4]^+$ counterions. [374] This structural feature is reminiscent of the structural characteristics of several inorganic and organometallic materials. For example, gold(I) dithiocarbamates $[R_2NCS_2Au]$ (R = *n*-propyl or *n*-butyl), [375, 376] and gold(I) dithiophosphate $[(i\text{-}C_3H_7O)_2PS_2Au]$ complexes [377] form dimers and are assembled in the crystal to give infinite Au-Au$\cdots$Au-Au$\cdots$ chains. Moreover, several square planar complexes of groups 9, 10 and 11 which contain less sterically demanding ligands, e.g. $K_2[Pt(CN)_4].3H_2O$, [31, 369] are arranged in their crystalline structure into infinite stacks in which linear chains of nonbonded metal atoms are present. The stoichiometric compounds are essentially colorless and nonconducting. However, partial oxidation of these compounds often gives rise to bronze or gold colored conductors owing to the generation of holes in the d band which strengthen the M–M interactions and results in the manifestation of electrical conductivity (e.g. $K_{0.6}[Ir(CO)_2Cl_2]$ and $K_2[Pt(CN)_4]Cl_{0.3} \cdot 3H_2O$ [31, 369, 378, 379]).

At this stage it is worth mentioning that several carbonyl and carbonyl-arene clusters have been recently observed to form *cluster cables* within their three dimensional network via a tight interlocking of the carbonyl and arene ligands, regardless of the cluster nuclearity (Figure **3-20**). [392, 398] The intermolecular nonbonding contacts do not appear to be significantly affected by the ionic charge. [374, 380] Since some of the clusters arranged in the solid state as molecular cables are capable of existing in different oxidation states in solution, a more detailed investigation of both their solution and solid state physical behavior seems desirable. In this respect, electron exchange processes in clusters has received very little attention so far.

Assemblage of anionic cluster units can be readily obtained by the use of a suitable metal cation of groups 11 to 13 which gives rise to a continuum of intermediate situations between the two limiting extremes of simple tight ion pairing and hetero metallic bonding (Category **B**). This second method of polymerization may have the nontrivial advantage of easy handling and allow the crystallization of compounds with exceedingly high molecular weights by reversible Lewis base

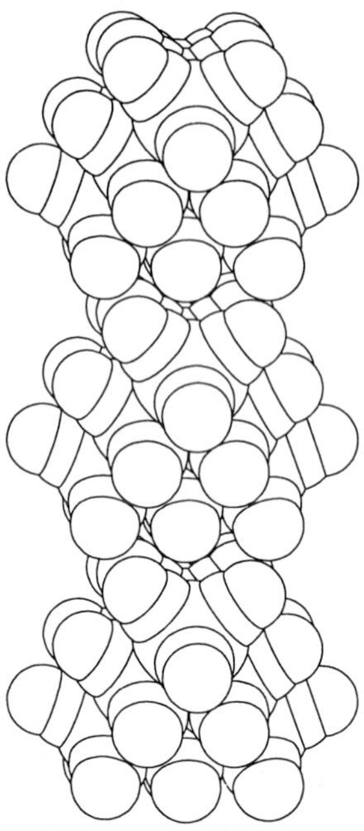

Figure 3-20. Carbonyl ligand interlocking of $[Ru_{10}(\mu_6\text{-}C)(CO)_{24}]^{2-}$ molecular ions. [398]

cleavage of an eventual intractable polymer into soluble oligomeric building blocks. These can be subsequently purified and repolymerized by slow removal of the Lewis base or dilution in a less solvating solvent.

Relevant examples of increasing complexity for this second category are $[Cd\{Fe_3(CO)_{11}\}_2]^{2-}$, [381] $[Hg_3\{Os_3(CO)_{11}\}_3]$, [146] $[Hg_3\{Co_3Hg_3(CO)_9\}_2]$, [147] $[M\{Ru_6C(CO)_{16}\}_2]^{n-}$ (M = Hg, n = 2; M = Tl, n = 1), [382] $[Hg\{Os_{10}C(CO)_{24}\}_2]^{2-}$ (Fig. **3-21**), [383] $[Hg_2\{Os_9C(CO)_{21}\}_2]^{2-}$, [160, 161] $[Hg_3\{M_9C(CO)_{21}\}_2]^{2-}$ (M = Ru, Os), [201, 383, 384] $(HgCl)_2\{Ru_5C(CO)_{14}Cl\}_2]$, [201] and $[Ag_2Br\{Re_7C(CO)_{21}\}_2]^{5-}$. [138] Probably the most striking example in this context is the co-polymerization of $[Rh_6C(CO)_{15}]^{2-}$ with Ag^+ ions to form oligomers of the general formula $[Ag_n\{Rh_6C(CO)_{15}\}_{n+1}]^{(n+2)-}$ which can be followed by depolymerization to obtain $[Ag_{n+1}\{Rh_6C(CO)_{15}\}_n]^{(n-1)-}$ oligomers. The final valve of n is a function of the stoichiometric ratio between the $[Rh_6C(CO)_{15}]^{2-}$ and the Ag^+ ions. Several of the above steps have been characterized by multinuclear NMR studies and the $[Ag\{Rh_6C(CO)_{15}\}_2]^{3-}$ oligomer has been shown by X-ray studies to have trigonal prismatic $[Rh_6C(CO)_{15}]^{2-}$ units assembled in a staggered conformation and joined through a silver ion (see Figure **3-5d**). [213]

It has long been thought that materials with a $[MFe(CO)_4]$ (M = Zn,Cd, and Hg) composition might have polymeric structures consisting of infinite –M–Fe– chains. However, a single crystal X-ray study of $[Cd_4Fe_4(CO)_{16} \cdot 2Me_2CO]$, showed that they seem rather to consist of oligomeric octametal planar rings. [231] A closely related metal structure has been found more recently in several other octametal mixed clusters, among which the isoelectronic $[MCo(CO)_4]_4$ (M = Cu and Ag) compounds, [229, 230] and this structure has also been ascertained by

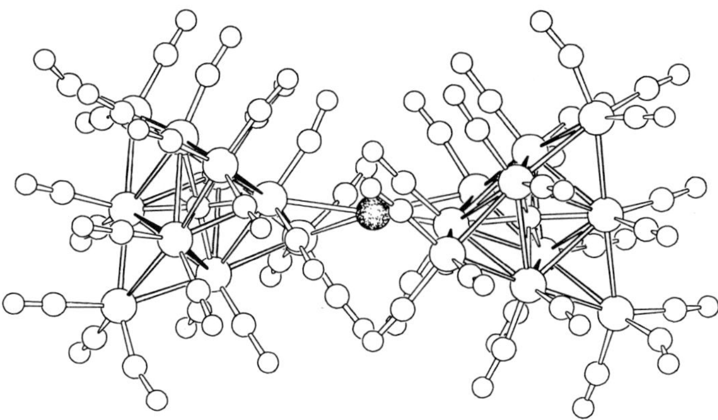

Figure 3-21. The molecular structure of $[Hg\{Os_{10}C(CO)_{24}\}_2]^{2-}$. [383]

X-ray powder diffraction studies on the closely related insoluble $[Hg_4Ru_4(CO)_{16}]$ compound. [385] Although the occurrence of a chain polymer, rather than a ring, should be entropically unfavorable, at least $[CuCo(CO)_4]$ crystallizes dimorphically and the second crystalline modification consists of wavy zig-zag –Cu–Co– infinite chains. [229] In addition to Cu–Co bonds, Cu–Cu interatomic contacts of ca. 2.70 Å are also present along the chains and might be indicative of a weak Cu–Cu interactions. As shown schematically in Figure **3-22**, consideration of these additional bonds makes $[CuCo(CO)_4]_n$ topologically reminiscent of the polymeric alkynyl compound $[Cu(C{\equiv}C\text{-}Ph)]_n$. [31–34] However, the clear lemon yellow color of this crystalline modification indicates that charge delocalization along the chain is unlikely. Nevertheless, it is surprising that, to the best of our knowledge, materials with the above, or previously described structural features have not yet been subjected to more detailed physical studies, such as before and after doping with redox agents.

A cluster containing exposed main group elements (E) can give rise to oligomers or polymers through E–E homoatomic interactions (Category **C**). As in the previously discussed homoatomic M–M linkage, the E–E interactions can vary from well defined covalent E–E bonds to much weaker van der Waals interactions and so self assemble the individual molecular entities into a giant supercluster in the crystalline network. The earlier reported $[(CO)_9Co_3C\text{–}(=O)\text{–}CCo_3\text{-}(CO)_9]$ [386] and $[(CO)_9Co_3(\mu_3\text{-}C\text{–}S\text{–}S\text{-}\mu_3\text{-}C)(Co_3(CO)_9]$ clusters, [387] the more recent $[Bi_4Fe_4(CO)_{13}]^{2-}$ [312] cluster exemplify these two limiting situations. The latter compound has been described as a hybrid Zintl-iron carbonyl ion since the four bismuth atoms define a tetrahedron which is capped on three of the four triangular faces by three μ_3-$Fe(CO)_3$ moieties. The remaining $[Fe(CO)_4]$ group is terminally bound to the unique Bi atom along the pseudo C_3 axis (Figure **3-23**). The two sets of intramolecular Bi–Bi contacts (3.46 and 3.16 Å, displayed by the capped and uncapped faces respectively) are similar to the closest contacts found in crystalline bismuth metal. A third set of Bi–Bi contacts (3.98 Å) is intermolecular and extends in three dimensions throughout the crystalline network. These

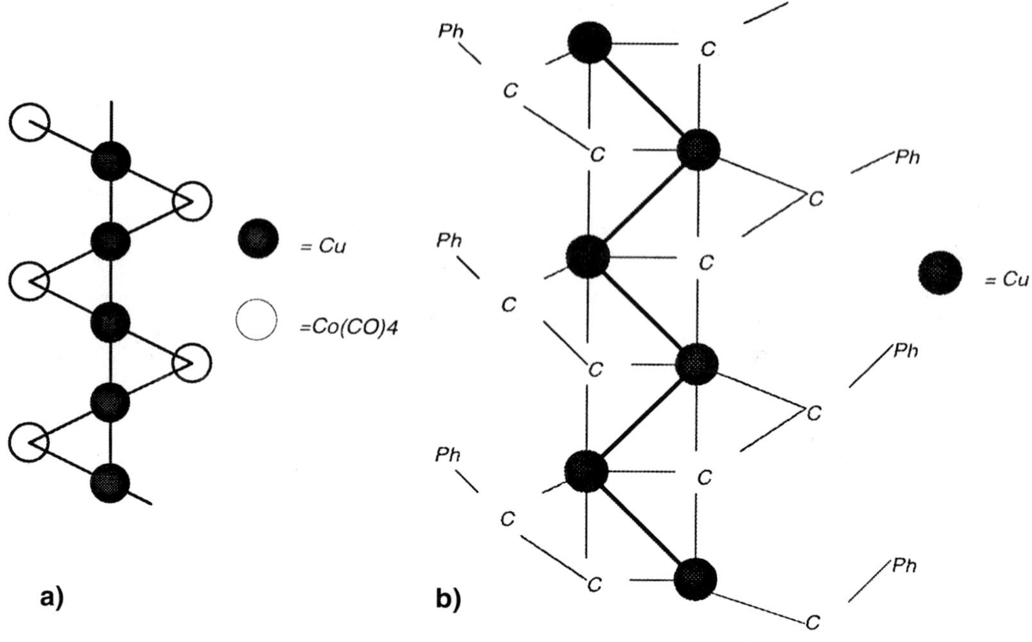

Figure 3-22. Schematic representations of the zig-zag chain structures of **a**) $[CuCo(CO)_4]_n$ ($Cu{-}Co_{av}$ = 2.37, $Cu{-}Cu_{av}$ = 2.71 Å) [229] and **b**) $[Cu(C{\equiv}C\text{-}Ph)]_n$ ($Cu{-}Cu_{av}$ = 2.45 Å). [32]

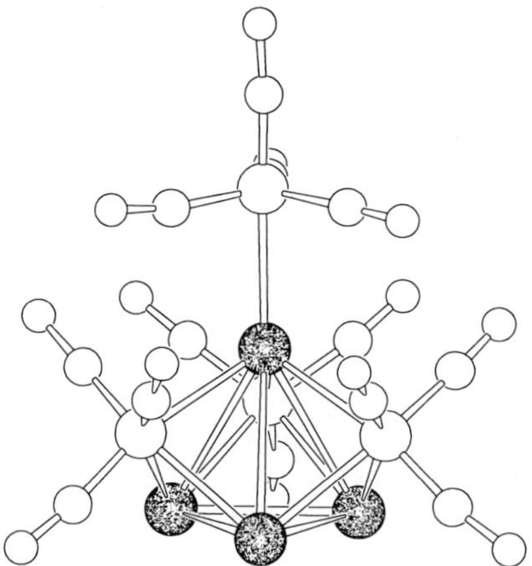

Figure 3-23. The molecular structure of the $[Bi_4Fe_4(CO)_{13}]^{2-}$ dianion. [312]

intermolecular interactions are comparable to those exhibited in the solid state by some Zintl ion salts, [239] e.g. $Ca_{11}Bi_{10}$, [388] and give even greater support to the description of the $[Bi_4Fe_4(CO)_{13}]^{2-}$ dianion as a hybrid Zintl-metal carbonyl ion.

Related thallium-iron carbonylates have been obtained by treating alkali methanol solutions of $[Fe(CO)_5]$ with thallium trichloride and include $[Tl_2Fe_4(CO)_{16}]^{2-}$, $[Tl_4Fe_8(CO)_{30}]^{4-}$ and $[Tl_6Fe_{10}(CO)_{36}]^{6-}$. [248, 309] However, in all of these ions the Tl–Tl contacts are rather long (*ca.* 3.7 Å), and thus may be considered more similar to intramolecular van der Waals interactions. There are no significant Fe–Fe interactions in either the former or the latter ion and, as a result, these compounds, as well as the corresponding $[Te_4Fe_5(CO)_{14}]^{2-}$ and $[Te_{10}Fe_8(CO)_{20}]^{2-}$ dianions presented in Scheme **3-12** and Table **3-10** of the previous chapter, [320] are on the borderline between clusters and the polynuclear compounds whose frame is only supported by M–E bonds. Thus, the structure of $[Te_{10}Fe_8(CO)_{20}]^{2-}$ may be thought to derive from the condensation of two neutral $[Fe_4(\mu_3\text{-}Te)_4(CO)_{12}]$ cubane moieties with Te_2^{2-} upon loss of carbon monoxide. The Te_2^{2-} unit links two iron atoms from each of the two cubane moieties in an η^2 fashion by donation of the lone pair electrons. The structure of the chemically related $[Te_4Fe_5(CO)_{14}]^{2-}$ dianion formally derives from two butterfly $[Fe_2Te_2(CO)_6]^{2-}$ units acting as chelating ligands on an unsaturated $[Fe(CO)_2]^+$ moiety; the Fe–Fe and Te–Te contacts in each butterfly moiety are 2.59 and 3.15 Å respectively. [320] Ditelluride ions are at the heart of the remarkable $[Te_{14}Ru_6(CO)_{12}]^{2-}$ compound [321], whose structure is based on a nonbonded Ru_6 octahedron held together by a central ditelluride moiety. The remaining twelve tellurium atoms are present as six Te_2 units bound in an $\eta^2\text{-}\mu_3$ fashion at the periphery of the nonbonded Ru_6 octahedron. The most notable structural feature of this compound is that the $Te_{14}Ru_6$ core is closely related to an intercepted fragment of the pyrite like $RuTe_2$ lattice. [321]

Although most of the Fe–Fe and Te–Te interactions in the above compounds are clearly nonbonding, the inclusion of these at this stage seems appropriate since the distinction between clusters and polynuclear compounds is often not clear cut and transformations from one type to the other often readily occur with redox reactions. Moreover, these compounds anticipate a behavior which is becoming increasingly documented for clusters containing exposed main group elements having lone electron pairs in their frame: *viz.* the possibility to act as *cluster ligands* by formation of dative E→M bonds with unsaturated metal atoms belonging to mono- or polynuclear complexes or cluster moieties. Some selected examples are listed in Table **3-13** where their donor and acceptor moieties are also indicated. The first part of Table **3-13** contains some leading examples of triangular metal clusters bearing a μ_3-capping main group element of group 15 or 16; these behave as monodentate ligands through the lone pairs on the heteroatoms which donate electron density to unsaturated carbonyl or carbonyl substituted mononuclear groups. For instance, $[(CO)_9Co_3(\mu_3\text{-}As{\rightarrow}Cr(CO)_5)]$ can be considered as a chromium pentacarbonyl arsine complex in which the ligand is a trimetallasubstituted arsine. [390] Likewise, a carbonyl substituted $[Co_4(CO)_{11}L]$ species has already been characterized in which L is the above *cluster ligand*. [391]

Table 3-13. Selected examples of *cluster ligand* complexes.

Compound	Monodentate donor moiety	acceptor moiety	Reference
$[Fe_4(\mu_4\text{-}P)(CO)_{14}]^-$	$[Fe_3(\mu_3\text{-}P)(CO)_{10}]^-$	$[Fe(CO)_4]$	[389]
$[Co_3Mn(\mu_4\text{-}P)(CO)_{11}Cp]$	$[Co_3(\mu_3\text{-}P)(CO)_9]$	$[Mn(CO)_2Cp]$	[390]
$[Co_3Cr(\mu_4\text{-}As)(CO)_{14}]$	$[Co_3(\mu_3\text{-}As)(CO)_9]$	$[Cr(CO)_5]$	[390]
$[Co_7(\mu_4\text{-}As)(CO)_{20}]$	$[Co_3(\mu_3\text{-}As)(CO)_9]$	$[Co_4(CO)_{11}]$	[391]
$[Co_9(\mu_4\text{-}As)_3(CO)_{24}]$	$[Co_3(\mu_3\text{-}As)(CO)_8]$		[392]
$[Co_{16}(\mu_6\text{-}As)_2(\mu_4\text{-}As)_2(\mu_4\text{-}AsPh)_4(CO)_{32}]$	$[Co_8(\mu_6\text{-}As)(\mu_3\text{-}As)(\mu_4\text{-}AsPh)_2(CO)_{16}]$		[313]
$[Fe_4(\mu_4\text{-}Sb)(CO)_{14}]^-$	$[Fe_3(\mu_3\text{-}Sb)(CO)_{10}]^-$	$[Fe(CO)_4]$	[393]
$[Fe_4(\mu_3\text{-}Bi)(\mu_4\text{-}Bi)(CO)_{13}]^{2-}$	$[Fe_3(\mu_3\text{-}Bi)_2(CO)_9]^{2-}$	$[Fe(CO)_4]$	[394]
$[CrCo_2Fe(\mu_4\text{-}S)(CO)_{14}]$	$[Co_2Fe(\mu_3\text{-}S)(CO)_9]$	$[Cr(CO)_5]$	[395]
$[Ru_9(\mu\text{-}H)_6(\mu_4\text{-}S)_3(CO)_{24}]$	$[H_2Ru_3(\mu_3\text{-}S)(CO)_8]$		[396]
	Bidentate donor moiety		
$[Fe_3Mn_2(P)_2(CO)_{11}Cp_2]$	$[Fe_3(\mu_3\text{-}P)_2(CO)_9]$	$[Mn(CO)_2Cp]$	[298]
$[Cr_3Co_2(\mu_4\text{-}As)_2(CO)_{20}]$	$[CrCo_2(\mu_3\text{-}As)(CO)_{10}]$	$[Cr(CO)_5]$	[390]
$[Fe_6(\mu_5\text{-}Sb)_2(CO)_{20}]^{2-}$	$[Fe_4(\mu_4\text{-}Sb)_2(CO)_{12}]^{2-}$	$[Fe(CO)_4]$	[299]
$[Ni_{13}(\mu_7\text{-}Sb)_2(CO)_{24}]^{2-}$	$[Ni_{11}(\mu_6\text{-}Sb)_2(CO)_{18}]^{2-}$	$[Ni(CO)_3]$	[300]
$[Co_{18}Zn_4(CO)_{54}(\mu_3\text{-}CCO_2)_6(\mu_4\text{-}O)]$	$[Co_3(CO)_9(\mu_3\text{-}CCO_2]^-$	$[Zn_4(\mu_4\text{-}O)]^{6+}$	[397]
$[Co_{22}(CO)_{54}(\mu_3\text{-}CCO_2)_6(\mu_4\text{-}O)]$	$[Co_3(CO)_9(\mu_3\text{-}CCO_2]^-$	$[Co_4(\mu_4\text{-}O)]^{6+}$	[398]
$[HOs_9(CO)_{27}(CO_2)]^-$	$[Os_6(CO)_{17}(CO_2)]^-$	$[HOs_3(CO)_{10}]$	[399]

Often the free trimetallasubstituted ligands, e.g. $[Co_3(CO)_9(\mu_3\text{-}E)]$ (E = P,As), $[H_2Ru_3(CO)_9(\mu_3\text{-}S)]$, and $[Co_8(\mu_6\text{-}As)(\mu_4\text{-}AsPh)_2(CO)_{17}(\mu_3\text{-}As)]$, are rather difficult to isolate in a pure crystalline state because of their instability under the experimental conditions required for their synthesis. In these cases, such clusters of clusters as $[Co_9(\mu_4\text{-}E)_3(CO)_{24}]$ (Fig. **3-24a**), [392] $[H_6Ru_9(\mu_4\text{-}S)_3(CO)_{24}]$, [396] and $[Co_{16}(\mu_6\text{-}As)_2(\mu_4\text{-}As)_2(\mu_4\text{-}AsPh)_4(CO)_{32}]$ (Figure **3-24b**) [313] can be formed from their intermolecular dimerization or trimerization reactions via carbon monoxide substitution and directly isolated.

Within this class of *cluster ligands* or *complexes,* those containing two basic heteroatoms on opposite sites have additional potential interest. As shown in Table **3-13**, there are several known examples of carbonylsubstituted complexes bridged by trimetallasubstituted bidentate ligands such as $[Fe_3(CO)_9(\mu_3\text{-}P)_2]$. [298] Furthermore, tetra- and hexametallasubstituted heteroatoms, *e.g.* the antimony atoms of the $[Fe_4(CO)_{12}(\mu_4\text{-}Sb)_2]^{2-}$ and $[Ni_{10}(\mu_{12}\text{-}Ni)(CO)_{18}(\mu_6\text{-}Sb)_2]^{n-}$ bidentate ligands, may also retain sufficient basicity to give rise to similar complexes (see

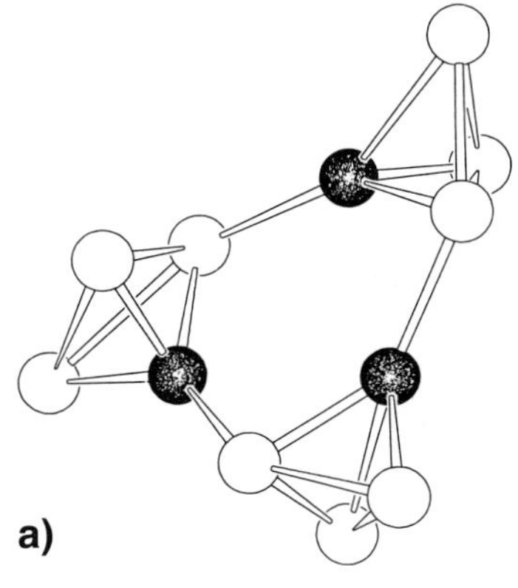

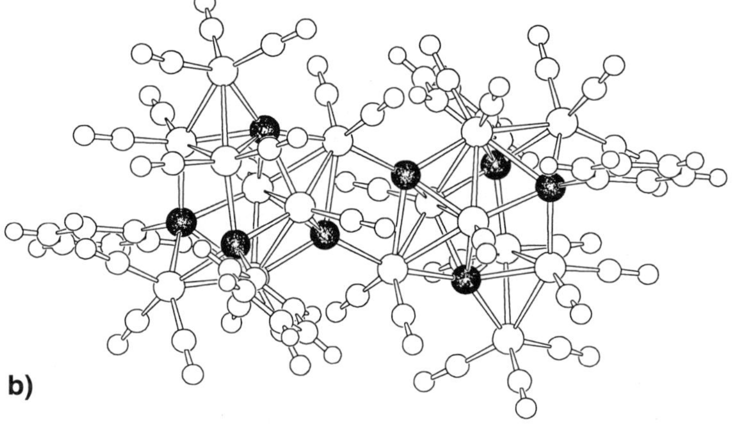

Figure 3-24. a) Schematic representation of the structure of $[Co_9(\mu_4\text{-}As)_3(CO)_{24}]$, [392] and b) molecular structure of $[Co_{16}(\mu_6\text{-}As)_2(\mu_4\text{-}As)_2(\mu_4\text{-}AsPh)_4(CO)_{32}]$. [313]

Figures **3-15** and **3-25**). [299, 300] The potential application of these metalla-substituted ligands for the synthesis of new metal coordination complexes, clusters of clusters, and, perhaps, low dimensional intercalation compounds is worthy of investigation. As stated earlier, the handling and crystallization of co-oligomers with increasingly higher molecular weights soon becomes insurmountably difficult, however, reversible ligand substitution equilibria might ease such a task. An appealing characteristic of these latter materials as potential building blocks for intercalation compounds of low dimensionality is their variable free negative charge so that mixed valence may be deliberately introduced along the chain.

A completely different approach to *cluster ligands* is the one exemplified by the use of $[Co_3(CO)_9(\mu_3\text{-}C\text{-}COOH)]$. This cluster may be considered as a trimetallasubstituted acetic acid and, as such, has been used to synthesize trimetallasubstituted acetate complexes with Mo–Mo multiple bonded dinuclear moieties and with $[M_4(\mu_4\text{-}O)]^+$ polynuclear oxo-cations supporting their description as so-called basic acetates. [398, 400] One of these remarkable salts is, for instance, constituted by a tetrahedral $[Co_4(\mu_4\text{-}O)]^{6+}$ oxo-cation edge-bridged by six $[Co_3(CO)_9(\mu_3\text{-}C\text{-}COO)]^-$ tricobaltasubstituted acetate anions. [398]

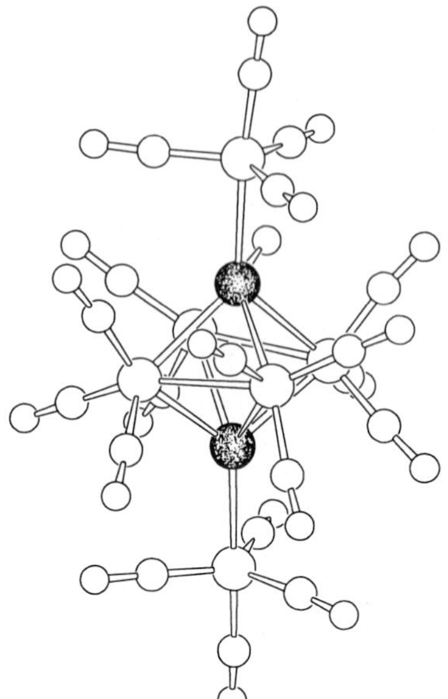

Figure 3-25. The molecular structures of $[Fe_6(\mu_5\text{-}Sb)_2(CO)_{20}]^{2-}$. [299]

This section would not be complete if some apparently unrelated experiments were not cited. First, Analytical Electron Microscopy studies on several higher nuclearity clusters have systematically shown that condensation processes of molecular clusters readily occur under the electron beam. For instance, in the case of $[HNi_{38}Pt_6(CO)_{48}]^{5-}$, it was observed that when the clusters were supported on amorphous carbon they initially appeared as homogeneous spherical spots of a size corresponding to the inner metal framework, yet, upon increasing the electron beam intensity, these coallesce into larger spots of sufficient homogeneity with an average diameter of *ca.* 40 Å. Interestingly, these much larger clusters display an elemental composition virtually identical, within experimental error, to that of the starting material and support the earlier suggestion that carbonyl clusters may have an unsuspectedly wider nuclearity range of existence. Further increasing the intensity of the electron beam caused the above clusters to lose CO and convert into larger particles with a Ni/Pt ratio still close to 6. Ultimately, formation of spots of the Ni_3Pt superordered alloy were ascertained by electron diffraction in a few particles. [401] In addition to the above stepwise transformation of a molecular compound into an alloy, the formation of metastable hollow spheres with diameters of *ca.* 10 000 Å but still displaying an elemental composition close to that of the starting material was constantly observed. Their occurrence is not a prerogative of $[HNi_{38}Pt_6(CO)_{48}]^{5-}$, as it was observed with several other clusters and depends mainly on the concentration of the solution used to prepare the microscope specimen. [402, 403] The appearance of these organometallic hollow spheres, made up of metals yet significantly coordinated by carbon monoxide, shows a formal resemblance to the hollow spherical powders of several inorganic oxides which can be obtained by aerosol or sol-gel techniques. [404, 405]

The results from a study of the platinum carbonyl clusters with ^{252}Cf Plasma Desorption Mass Spectrometry [406] might be related to the above observations. The presence in the spectra of species' with molecular weights corresponding to increasing multiples of that of the starting material was often observed. [406] It is unclear whether this reflects an oligomerization process of the cluster molecules through the formation of well defined M–M bonds, as in the intermolecular condensation processes examined in this and previous sections, or that it arises simply from the expulsion of increasingly larger chunks of cation-anion ensembles from the crystal network of the sample. In any case, this phenomenon seems to parallel the observations made by electron microscopy. As outlined throughout the chapter, several formal similarities between organometallic clusters and inorganic compounds can be noted as one follows the transition from the molecular to the solid state via the colloidal state. As a result of the close relationships which can be often recognized between the structures of discrete molecular clusters and solid inorganic materials, a growing interest in a molecular approach to inorganic materials using clusters as precursors is rapidly emerging. [407–411]

3.2.7 Conclusion

From the data and references presented herein, as well as the examination of the daily scientific literature, it is clear that the chemistry of low valent organometallic clusters has made tremendous progress throughout the years. Moreover, it continues to arouse a great deal of interest among experimentalists and theoreticians. As a result, the molecular structures of cluster compounds are becoming increasingly more sophisticated and one may envision their possible use as precursors for entirely new and interesting materials. The frontiers for the syntheses of known, as well as new, materials are beginning to be broadened since an increasing number of studies are entering the interface between molecular systems and the chemistry of colloids and extended solids. It is, therefore, likely that in the next few years this area of chemistry could provide not only compounds with novel structural features and interesting chemical, physical and spectroscopic properties, but they could find new developments and applications in neighboring fields.

Acknowledgments

We wish to thank Dr. Magda Monari and Emanuela Venturini for the art work with SCHAKAL88, [412] and Dr. Suzanne Mulley for correcting the manuscript.

References

[1] K. Wade, *Adv. Inorg. Chem. Radiochem.* **1976**, *18*, 1.
[2] J. Lahuer, *J. Am. Chem. Soc.* **1978**, *100*, 5305.
[3] R. B. King, D. H. Rouvray, *J. Am. Chem. Soc.* **1977**, *99*, 7834.
[4] B. K. Teo, *Inorg. Chem.* **1984**, *23*, 1251.
[5] D. M. P. Mingos, A. S. May in *The Chemistry of Metal Cluster Complexes* (Eds: D. F. Shriver, H. D. Kaesz, R. D. Adams) VCH, New York, **1990**, 11.
[6] D. M. P. Mingos, R. L. Johnston, *Struct. Bonding* **1987**, *68*, 29.
[7] J. Doedens, L. F. Dahl, *J. Am. Chem. Soc.* **1966**, *88*, 4847.
[8] B. F. G. Johnson, R. E. Benfield, *J. Chem. Soc. Dalton Trans.* **1978**, 1554.
[9] R. E. Benfield, B. F. G. Johnson, *J. Chem. Soc. Dalton Trans.* **1980**, 1743.
[10] B. F. G. Johnson, R. E. Benfield, *Top. Stereochem.* **1981**, *12*, 253.
[11] D. M. P. Mingos, *Inorg. Chem.* **1982**, *21*, 466.
[12] J. W. Lahuer, *J. Am. Chem. Soc.* **1986**, *108*, 1521.
[13] A. Sironi, *Inorg. Chem.* **1992**, *31*, 2467.
[14] C. P. Horwitz, E. M. Holt, D. F. Shriver, *Inorg. Chem.* **1984**, *23*, 2491.
[15] T. Beringhelli, G. D'Alfonso, H. Molinari, A. Sironi, *J. Chem. Soc. Dalton Trans.* **1992**, 689.
[16] B. K. Teo, N. J. A. Sloan, *Inorg. Chem.* **1985**, *24*, 4545.
[17] L. F. Dahl, M. S. Paquette, *J. Am. Chem. Soc.* **1980**, *102*, 6621.
[18] B. T. Heaton, *personal communication.*
[19] F. A. Cotton, G. Wilkinson, *Advanced Inorganic Chemistry,* 5th ed., Wiley, New York, **1988**.
[20] J. E. Huheey, *Inorganic Chemistry,* 2nd ed., Harper and Row, New York, **1978.**
[21] D. F. Shriver, P. W. Atkins, C. H. Langford, *Inorganic Chemistry,* Oxford University Press, Oxford, **1990**.
[22] K. F. Purcell, J. C. Kotz, *Inorganic Chemistry,* Saunders, Philadelphia, **1977.**
[23] P. Chini, *Inorg. Chim. Acta Rev.* **1968**, *2*, 31.
[24] D. Braga, F. Grepioni, *J. Organomet. Chem.* **1987**, *336*, C9.
[25] T. Chang, J. I. Zink, *J. Am. Chem. Soc.* **1987**, *109*, 692.
[26] W. L. Gladfelter, *Adv. Organomet. Chem.* **1985**, *24*, 41.
[27] E. Singleton, H. E. Oosthuizen, *Adv. Organomet. Chem.* **1983**, *22*, 209.
[28] A. J. Carty, *Pure Appl. Chem.* **1982**, *54*, 113.
[29] M. Green, J. A. K. Howard, M. Murray, J. L. Spencer, F. G. A. Stone, *J. Chem. Soc. Dalton Trans.* **1977**, 1509.
[30] Y. Yamamoto, K. Aoki, H. Yamazaki, *Organometallics* **1983**, *2*, 1377.
[31] *Inorganic and Organometallic Oligomers and Polymers,* (Eds: J. F. Harrod, R. M. Laine) Kluwer Academic Publishers, Dordrecht, The Netherlands, **1991**.
[32] P. W. R. Corfield, H. M. M. Shearer, *Acta Crystal.* **1966**, *21*, 957.
[33] P. W. R. Corfield, H. M. M. Shearer, *Acta Crystal.* **1966**, *20*, 502.
[34] M. S. Hussain, O. M. Abu-Salah, *J. Organomet. Chem.* **1993**, *445*, 295.
[35] F. A. Vollenbroek, J. J. Bour, J. W. A. Van der Velden, *Rec. Trav. Chim. Pays-Bas* **1980**, *99*, 137.
[36] C. P. Horwitz, D. F. Shriver, *Adv. Organomet. Chem.* **1984**, *23*, 219.
[37] R. H. Crabtree, M. Lavin, *Inorg. Chem.* **1986**, *25*, 805.
[38] P. Li, M. D. Curtis, *J. Am. Chem. Soc.* **1989**, *111*, 8279.
[39] F. A. Cotton, *Prog. Inorg. Chem.* **1976**, *21*, 1.
[40] R. J. Klinger, W. M. Butler, M. D. Curtis, *J. Chem. Soc.* **1978**, *100*, 5034.
[41] R. Colton, M. J. McCormick, *J. Coord. Chem. Rev.* **1980**, *31*, 1.
[42] C. Q. Simpson, M. B. Hall, *J. Am. Chem. Soc.* **1992**, *114*, 1641.
[43] A. Ceriotti, R. Della Pergola, G. Longoni, M. Manassero, M. Sansoni, *J. Chem. Soc. Dalton Trans.* **1984**, 1181.

[44] A. Ceriotti, G. Longoni, M. Manassero, M. Sansoni, R. Della Pergola, B. T. Heaton, D. O. Smith, *J. Chem. Soc. Chem. Comm.* **1982**, 886.
[45] E. Brivio, A. Ceriotti, F. Demartin, M. Manassero, *in press.*
[46] S. W. Ulmer, P. M. Skarstad, J. M. Burlich, R. E. Hughes, *J. Am. Chem. Soc.* **1973**, *95*, 4469.
[47] N. J. Nelson, N. Kime, D. F. Shriver, *J. Am. Chem. Soc.* **1969**, *91*, 5173.
[48] G. Schmid, V. Batzel, B. Stutte, *J. Organomet. Chem.* **1976**, *113*, 67.
[49] R. Colton, C. J. Commons, B. Hoskins, *J. Chem. Soc. Chem. Comm.* **1975**, 363.
[50] A. A. Pasynski, Yu. V. Skripkin, I. L. Eremenko, V. T. Kalinnikov, G. G. Aleksandrov, V. G. Andrianov, Yu. T. Struchkov, *J. Organomet. Chem.* **1979**, *165*, 49.
[51] W. A. Herrman, M. L. Ziegler, K. Weidenhammer, H. Biersack, *Angew. Chem. Int. Ed. Engl.* **1979**, *18*, 960.
[52] M. Manassero, M. Sansoni, G. Longoni, *J. Chem. Soc. Chem. Comm.* **1976**, 919.
[53] P. Brun, G. M. Dawkins, M. Green, A. D. Miles, A. G. Orpen, F. G. A. Stone, *J. Chem. Soc. Chem. Comm.* **1982**, 926.
[54] G. Longoni, A. Ceriotti, R. Della Pergola, M. Manassero, M. Perego, G. Piro, M. Sansoni, *Phil. Trans. R. Soc. Lond.* **1982**, *A308*, 47.
[55] J. W. Kolis, E. M. Holt, M. Drezdon, K. H. Whitmire, D. F. Shriver, *J. Am. Chem. Soc.* **1982**, *104*, 6134.
[56] J. W. Kolis, E. M. Holt, D. F. Shriver, *J. Am. Chem. Soc.* **1983**, *105*, 7307.
[57] P. Chini, G. Longoni, V. G. Albano, *Adv. Organomet. Chem.* **1976**, *14*, 285.
[58] P. Chini, B. T. Heaton, *Topics in Curr. Chem.* **1977**, *71*, 1.
[59] D. G. Evans, *J. Chem. Soc. Chem. Comm.* **1983**, 675.
[60] E. F. Paulus, *Acta Crystal.* **1969**, *B25*, 1205.
[61] M. McPartlin, W. J. H. Nelson, *J. Chem. Soc. Dalton Trans.* **1986**, 1557.
[62] L. Garlaschelli, S. Martinengo, P. L. Bellon, F. Demartin, M. Manassero, M. Y. Chiang, C. Wei, R. Bau, *J. Am. Chem. Soc.* **1984**, *106*, 6664.
[63] A. Fumagalli, S. Martinengo, G. Ciani, A. Sironi, B. T. Heaton, *J. Chem. Soc. Dalton Trans.* **1988**, 163.
[64] R. Hourihane, T. R. Spalding, G. Ferguson, T. Deeney, P. Zanello, *J. Chem. Soc. Dalton Trans.* **1993**, 43.
[65] V. G. Albano, D. Braga, F. Grepioni, *Acta Crystallogr.* **1989**, *B45*, 60.
[66] S. D'Addario, F. Demartin, L. Grossi, M. C. Iapalucci, F. Laschi, G. Longoni, P. Zanello, *Inorg. Chem.* **1993**, *32,* 1153.
[67] R. J. Lawson, J. R. Shapley, *J. Am. Chem. Soc.* **1976**, *98*, 7433.
[68] J. A. Connor in *Metal Clusters in Catalysis,* (Eds.: B. C. Gates, L. Guczi, H. Knozinger), Elsevier, Amsterdam, **1986**, 33.
[69] E. Band, E. L. Muetterties, *Chem. Rev.* **1978**, *78*, 641.
[70] B. T. Heaton, *Phil. Trans. R. Soc. Lond.* **1982**, *A308*, 95.
[71] S. Aime, L. Milone, *Prog. NMR Spectrosc.* **1977**, *11*, 183.
[72] D. Braga, *Chem. Rev.* **1992**, *92*, 633.
[73] S. Martinengo, B. T. Heaton, R. J. Goodfellow, P. Chini, *J. Chem. Soc. Chem. Comm.* **1977**, 39.
[74] S. J. Heyes, M. A. Gallop, B. F. G. Johnson, J. Lewis, C. M. Dobson, *Inorg. Chem.* **1991**, *30*, 3850.
[75] D. Braga, F. Grepioni, B. F. G. Johnson, H. Chen, J. Lewis, *J. Chem. Soc. Dalton Trans.* **1991**, 2559.
[76] M. P. Gomez-Sal, B. F. G. Johnson, J. Lewis, P. R. Raithby, A. H. Wright, *J. Chem. Soc. Chem. Comm.* **1985**, 1682.
[77] M. A. Gallop, B. F. G. Johnson, J. Lewis, P. R. Raithby, *J. Chem. Soc. Chem. Comm.* **1987**, 1809.
[78] L. H. Gade, B. F. G. Johnson, J. Lewis, *J. Chem. Soc. Dalton Trans.* **1992**, 933.
[79] S. R. Bunkhall, H. D. Holden, B. F. G. Johnson, J. Lewis, G. N. Pain, P. R. Raithby, M. J. Taylor, *J. Chem. Soc. Chem. Comm.* **1984**, 25.

[80] A. Fumagalli, S. Martinengo, V. G. Albano, D. Braga, F. Grepioni, *J. Chem. Soc. Dalton Trans.* **1989**, 2343.
[81] A. Moor, P. S. Pregosin, L. M. Venanzi, A. J. Welch, *Inorg. Chim. Acta* **1984**, *85*, 103.
[82] J. L. Vidal, W. E. Walker, R. L. Pruett, R. C. Schoening, *Inorg. Chem.* **1979**, *18*, 129.
[83] B. T. Heaton, L. Strona, R. Della Pergola, J. L. Vidal, R. C. Schoening, *J. Chem. Soc. Dalton Trans.* **1983**, 1941.
[84] C. Brown, B. T. Heaton, A. D. C. Towl, P. Chini, A. Fumagalli, G. Longoni, *J. Organomet. Chem.* **1979**, *181*, 233.
[85] D. A. Nagaki, L. D. Lower, G. Longoni, P. Chini, L. F. Dahl, *Organometallics* **1986**, *5*, 1764.
[86] F. A. Vollenbroek, J. P. Van der Berg, J. W. A. Van der Velden, J. J. Bour, *Inorg. Chem.* **1980**, *19*, 2685.
[87] J. W. A. Van der Velden, J. J. Bour, J. J. Steggerda, P. T. Beurskens, M. Roseboom, J. H. Nordik, *Inorg. Chem.* **1982**, *21*, 4321.
[88] K. P. Hall, D. M. P. Mingos, *Prog. Inorg. Chem.* **1984**, *32*, 237.
[89] L. J. Farrugia in *Organic Chemistry on Metal Clusters and Surfaces* S. Vittorio d'Alba (Italy), July 26–31, **1992**, Abstr. 67.
[90] P. F. Jackson, B. F. G. Johnson, J. Lewis, W. J. H. Nelson, M. McPartlin, *J. Chem. Soc. Dalton Trans.* **1982**, 2099.
[91] P. F. Jackson, B. F. G. Johnson, J. Lewis, M. McPartlin, W. J. H. Nelson, *J. Chem. Soc. Chem. Comm.* **1982**, 48.
[92] D. Braga, F. Grepioni, S. Righi, B. F. G. Johnson, P. Frediani, M. Bianchi, F. Piacenti, J. Lewis, *Organometallics* **1992**, *11*, 706.
[93] S. Drake, M. Basley, B. F. G. Johnson, J. Lewis, *Organometallics* **1988**, *7*, 806.
[94] V. G. Albano, F. Demartin, M. C. Iapalucci, G. Longoni, M. Monari, P. Zanello, *J. Chem. Soc. Dalton Trans.* **1992**, 497.
[95] M. D. Vargas, J. N. Nicholls, *Adv. Inorg. Chem. Radiochem.* **1986**, *30*, 123.
[96] H. Vahrenkamp, *Adv. Organomet. Chem.* **1983**, *22*, 169.
[97] W. L. Gladfelter, G. L. Geoffroy, *Adv. Organomet. Chem.* **1980**, *18*, 207.
[98] D. Braga, J. Lewis, B. F. G. Johnson, M. McPartlin, W. J. H. Nelson, M. D. Vargas, *J. Chem. Soc. Chem. Comm.* **1983**, 241.
[99] A. Bashall, L. H. Gade, J. Lewis, B. F. G. Johnson, G. J. McIntyre, M. McPartlin, *Angew. Chem. Int. Ed. Engl.* **1991**, *30*, 1164.
[100] E. Charalambous, L. H. Gade, B. F. G. Johnson, J. Lewis, M. McPartlin, H. R. Powell, *J. Chem. Soc. Chem. Comm.* **1990**, 688.
[101] A. J. Amoroso, L. H. Gade, B. F. G. Johnson, J. Lewis, P. R. Raithby, W. T. Wong, *Angew. Chem. Int. Ed. Engl.* **1991**, *30*, 107.
[102] S. Martinengo, P. Chini, *Inorg. Synth.* **1980**, *20*, 215.
[103] V. G. Albano, P. L. Bellon, *J. Organomet. Chem.* **1969**, *19*, 405.
[104] G. Ciani, A. Sironi, S. Martinengo, *J. Chem. Soc. Chem. Comm.* **1985**, 1757.
[105] G. Ciani, A. Sironi, S. Martinengo, *J. Chem. Soc. Dalton Trans.* **1981**, 519.
[106] V. G. Albano, G. Ciani, S. Martinengo, A. Sironi, *J. Chem. Soc. Dalton Trans.* **1979**, 978.
[107] V. G. Albano, A. Ceriotti, P. Chini, G. Ciani, S. Martinengo, W. M. Anker, *J. Chem. Soc. Chem. Comm.* **1975**, 859.
[108] J. L. Vidal, R. C. Schoening, *Inorg. Chem.* **1981**, *20*, 265.
[109] G. Ciani, A. Sironi, S. Martinengo, *J. Chem. Soc. Dalton Trans.* **1982**, 1099.
[110] G. Ciani, A. Moret, A. Sironi, S. Martinengo, *J. Organomet. Chem.* **1989**, *363*, 181.
[111] S. Martinengo, G. Ciani, A. Sironi, *J. Chem. Soc. Chem. Comm.* **1980**, 1140.
[112] J. L. Vidal, R. C. Schoening, *J. Organomet. Chem.* **1981**, *218*, 217.
[113] S. Martinengo, G. Ciani, A. Sironi, P. Chini, *J. Am. Chem. Soc.* **1978**, *100*, 7096.
[114] J. L. Vidal, L. A. Kapicak, J. M. Troup, *J. Organomet. Chem.* **1981**, *215*, C11.
[115] G. Ciani, A. Magni, A. Sironi, S. Martinengo, *J. Chem. Soc. Dalton Trans.* **1981**, 1280.

[116] S. Martinengo, G. Ciani, A. Sironi, *J. Am. Chem. Soc.* **1980**, *102*, 7564.
[117] J. L. Vidal, R. C. Schoening, J. M. Troup, *Inorg. Chem.* **1981**, *20*, 227.
[118] R. Della Pergola, F. Demartin, L. Garlaschelli, M. Manassero, S. Martinengo, N. Masciocchi, P. Zanello, *Inorg. Chem.* **1993**, *32*, 3670.
[119] R. Della Pergola, F. Demartin, L. Garlaschelli, M. Manassero, S. Martinengo, M. Sansoni, *Inorg. Chem.* **1987**, *26*, 3487.
[120] R. Della Pergola, L. Garlaschelli, M. Manassero, N. Masciocchi, P. Zanello, *Angew. Chem. Int. Ed. Engl.* **1993**, *32*, 1347.
[121] A. Ceriotti, P. Chini, R. Della Pergola, G. Longoni, *Inorg. Chem.* **1983**, *22*, 1595.
[122] P. Chini, G. Longoni, M. Manassero, M. Sansoni, *VIII Cong. Naz. Ass. Ital. Cristal.*, Ferrara (Italy), November 7–9, **1977**, Abstr. 85.
[123] R. W. Broach, L. F. Dahl, G. Longoni, P. Chini, A. J. Schultz, J. M. Williams, *Adv. Chem. Ser.* **1978**, *167*, 93.
[124] E. G. Mednikov, Y. L. Slovokhotov, Y. T. Sruchkov, *Metalloorg. Khim.* **1991**, *4*, 123.
[125] E. G. Mednikov, N. K. Eremenko, Y. L. Slovokhotov, Y. T. Struchkov, *J. Organomet. Chem.* **1986**, *301*, C35.
[126] E. G. Mednikov, *Metalloorg. Khim.* **1991**, *4*, 885.
[127] E. G. Mednikov, N. K. Eremenko, Y. L. Slovokhotov, Y. T. Struchkov, *J. Chem. Soc. Chem. Comm.* **1987**, 218.
[128] G. Longoni, P. Chini, *J. Am. Chem. Soc.* **1976**, *98*, 7225.
[129] L. F. Dahl, *unpublished results.*
[130] J. C. Calabrese, L. F. Dahl, P. Chini, G. Longoni, S. Martinengo, *J. Am. Chem. Soc.* **1974**, *96*, 2614.
[131] J. A. K. Howard, J. L. Spencer, D. G. Turner, *J. Chem. Soc. Dalton Trans.* **1987**, 259.
[132] S. S. Kurasov, N. K. Eremenko, Y. L. Slovokhotov, Y. L. Struchkov, *J. Organomet. Chem.* **1989**, *361*, 405.
[133] D. M. Washecheck, E. J. Wucherer, L. F. Dahl, A. Ceriotti, G. Longoni, M. Manassero, M. Sansoni, P. Chini, *J. Am. Chem. Soc.* **1979**, *101*, 6110.
[134] A. Ceriotti, P. Chini, G. Longoni, M. Marchionna, L. F. Dahl, R. Montag, D. M. Washecheck, *XV Cong. Naz. Chim. Inorg.*, Bari (Italy), Sept. 27–Oct. 1, **1982**, Abstr. A28.
[135] A. Ceriotti, P. Chini, G. Longoni, D. M. Washecheck, E. J. Wucherer, L. F. Dahl, *XIII Cong. Naz. Chim. Inorg.*, Camerino (Italy), Sept. 23–26, **1980**, Abstr. A2.
[136] J. Deutscher, S. Fadel, M. Ziegler, *Chem. Ber.* **1979**, *112*, 2413.
[137] J. Mielcke, J. Straehle, *Angew. Chem. Int. Ed. Engl.* **1992**, *31*, 464.
[138] T. Beringhelli, G. D'Alfonso, M. Freni, G. Ciani, A. Sironi, *J. Organomet. Chem.* **1985**, *295*, C7.
[139] G. Longoni, M. Manassero, M. Sansoni, *J. Am. Chem. Soc.* **1980**, *102*, 3242.
[140] G. Longoni, M. Manassero, M. Sansoni, *J. Am. Chem. Soc.* **1980**, *102*, 7973.
[141] G. Doyle, K. A. Eriksen, D. Van Engen, *J. Am. Chem. Soc.* **1986**, *108*, 445.
[142] G. Doyle, K. A. Eriksen, D. van Engen, *J. Am. Chem. Soc.* **1985**, *107*, 7914.
[143] V. G. Albano, L. Grossi, G. Longoni, M. Monari, S. Mulley, A. Sironi, *J. Am. Chem. Soc.* **1992**, *114*, 5708.
[144] E. Brivio, A. Ceriotti, R. Della Pergola, L. Garlaschelli, F. Demartin, M. Manassero, M. Sansoni, *CISCI91*, Chianciano (Italy), Oct. 6–11, **1991**, Abstr. 682.
[145] R. C. Adams, J. C. Lii, W. Wu, *Inorg. Chem.* **1992**, *31*, 2556.
[146] M. Fajardo, H. D. Holden, B. F. G. Johnson, J. Lewis, P. R. Raithby, *J. Chem. Soc. Chem. Comm.* **1984**, 24.
[147] J. M. Ragosta, J. M. Burlitch, *Organometallics* **1988**, *7*, 1469.
[148] D. Nagaki, J. V. Badding, A. M. Stacy, L. F. Dahl, *J. Am. Chem. Soc.* **1986**, *108*, 3825.
[149] A. Fumagalli, S. Martinengo, G. Ciani, *J. Chem. Soc. Chem. Comm.* **1983**, 1381.
[150] A. Fumagalli, S. Martinengo, G. Ciani, A. Sironi, *XVI Cong. Naz. Chim. Inorg.*, Ferrara (Italy), Sept. 12–16, **1983**, Abstr. 402.

[151] A. Fumagalli, S. Martinengo, G. Ciani, N. Masciocchi, A. Sironi, *Inorg. Chem.* **1992**, *31*, 336.
[152] A. Ceriotti, F. Demartin, G. Longoni, M. Manassero, G. Piva, G. Piro, M. Sansoni, B. T. Heaton, *J. Organomet. Chem.* **1986**, *301*, C5.
[153] A. J. Whoolery, L. F. Dahl, *J. Am. Chem. Soc.* **1991**, *113*, 6683.
[154] A. Ceriotti, F. Demartin, P. Ingallina, G. Longoni, M. Manassero, M. Marchionna, N. Masciocchi, M. Sansoni, *The Chemistry of the Platinum Group Metals*, Sheffield (UK), July 12–17, **1987**, Abstr. O30.
[155] A. Ceriotti, F. Demartin, G. Longoni, M. Manassero, M. Marchionna, G. Piva, M. Sansoni, *Angew. Chem. Int. Ed. Engl.* **1985**, *24*, 697.
[156] A. Fumagalli, S. Martinengo, V. G. Albano, D. Braga, F. Grepioni, *J. Chem. Soc. Dalton Trans.* **1993**, 2047.
[157] A. Fumagalli, S. Martinengo, P. Chini, D. Galli, B. T. Heaton, R. Della Pergola, *Inorg. Chem.* **1984**, *23*, 2947.
[158] V. G. Albano, D. Braga, S. Martinengo, *J. Chem. Soc. Dalton Trans.* **1981**, 717.
[159] B. F. G. Johnson, J. Lewis, G. L. Powell, P. R. Raithby, M. D. Vargas, J. Morris, M. McPartlin, *J. Chem. Soc. Chem. Comm.* **1986**, 429.
[160] L. H. Gade, B. F. G. Johnson, J. Lewis, M. McPartlin, T. Kotch, A. J. Lees, *J. Am. Chem. Soc.* **1991**, *113*, 8698.
[161] E. Charalambous, L. H. Gade, B. F. G. Johnson, T. Kotch, A. J. Lees, J. Lewis, *Angew. Chem. Int. Ed. Engl.* **1990**, *29*, 1137.
[162] S. Martinengo, G. Ciani, A. Sironi, *submitted.*
[163] S. Martinengo, G. Ciani, A. Sironi, *J. Organomet. Chem.* **1988**, *358*, C23.
[164] S. Martinengo, G. Ciani, A. Sironi, *J. Chem. Soc. Chem. Comm.* **1986**, 1742.
[165] B. F. G. Johnson, J. Lewis, M. A. Pearsall, L. G. Scott, *unpublished results* (cited in ref. 105).
[166] R. J. Goudsmit, J. G. Jeffrey, B. F. G. Johnson, J. Lewis, R. C. J. McQueen, A. J. Saunders, J. C. Liu, *J. Chem. Soc. Chem. Comm.* **1986**, 24.
[167] A. Ceriotti, F. Demartin, B. T. Heaton, P. Ingallina, G. Longoni, M. Manassero, M. Marchionna, N. Masciocchi, *J. Chem. Soc. Chem. Comm.* **1989**, 786.
[168] C. Allevi, B. T. Heaton, C. Seregni, L. Strona, R. J. Goodfellow, P. Chini, S. Martinengo, *J. Chem. Soc. Dalton Trans.* **1986**, 1375.
[169] A. Ceriotti, G. Piro, G. Longoni, M. Manassero, N. Masciocchi, M. Sansoni, *New J. Chem.* **1988**, *12*, 501.
[170] A. Fumagalli, S. Martinengo, D. Galli, C. Allevi, G. Ciani, A. Sironi, *Inorg. Chem.* **1990**, *29*, 1408.
[171] A. Ceriotti, R. Della Pergola, L. Garlaschelli, G. Longoni, M. Manassero, N. Masciocchi, M. Sansoni, P. Zanello, *Gazz. Chim. Ital.* **1992**, *122*, 365.
[172] A. Ceriotti, G. Longoni, M. Manassero, N. Masciocchi, G. Piro, L. Resconi, M. Sansoni, *J. Chem. Soc. Chem. Comm.* **1985**, 1402.
[173] R. C. Adams, G. Chen, J. G. Wang, *Polyhedron* **1989**, *8*, 2521.
[174] V. G. Albano, D. Braga, D. Strumolo, C. Seregni, S. Martinengo, *J. Chem. Soc. Dalton Trans.* **1985**, 1309.
[175] D. Strumolo, C. Seregni, S. Martinengo, V. G. Albano, D. Braga, *J. Organomet. Chem.* **1983**, *252*, C93.
[176] A. Ceriotti, R. Della Pergola, G. Longoni, M. Manassero, N. Masciocchi, M. Sansoni, *J. Organomet. Chem.* **1987**, *330*, 237.
[177] V. G. Albano, D. Braga, P. Chini, G. Ciani, S. Martinengo, *J. Chem. Soc. Dalton Trans.* **1982**, 645.
[178] V. G. Albano, D. Braga, A. Fumagalli, S. Martinengo, *J. Chem. Soc. Dalton Trans.* **1985**, 1137.
[179] V. G. Albano, F. Demartin, M. C. Iapalucci, F. Laschi, G. Longoni, A. Sironi, P. Zanello, *J. Chem. Soc. Dalton Trans.* **1991**, 739.
[180] G. Longoni, F. Morazzoni, *J. Chem. Soc. Dalton Trans.* **1981**, 1735.

[181] R. Della Pergola, L. Garlaschelli, C. Mealli, D. M. Proserpio, P. Zanello, *J. Cluster Science* **1990**, *1*, 93.
[182] P. Lemoine, *Coord. Chem. Rev.* **1988**, *83*, 169.
[183] P. Zanello, *Struct. Bond.* **1992**, *79*, 101.
[184] P. Zanello, *Coord. Chem. Rev.* **1988**, *83*, 199 and **1988**, *87*, 1.
[185] V. G. Albano, P. Chini, S. Martinengo, M. Sansoni, D. Strumolo, *J. Chem. Soc. Dalton Trans.* **1978**, 459.
[186] V. G. Albano, D. Braga, P. Chini, D. Strumolo, S. Martinengo, *J. Chem. Soc. Dalton Trans.* **1983**, 249.
[187] R. Mason, K. M. Thomas, D. M. P. Mingos, *J. Am. Chem. Soc.* **1973**, *95*, 3800.
[188] M. McPartlin, C. R. Eady, B. F. G. Johnson, J. Lewis, *J. Chem. Soc. Chem. Comm.* **1976**, 883.
[189] J. D. Roth, G. J. Lewis, L. K. Safford, X. Jiang, L. F. Dahl, M. J. Weaver, *J. Am. Chem. Soc.* **1992**, *114*, 6159.
[190] I. A. Salter, *Adv. Organomet. Chem.* **1989**, *29*, 249.
[191] R. Hoffmann, *Angew. Chem. Int. Ed. Engl.* **1982**, *21*, 711.
[192] V. G. Albano, D. Braga, S. Martinengo, P. Chini, M. Sansoni, D. Strumolo, *J. Chem. Soc. Dalton Trans.* **1980**, 52.
[193] D. Braga, K. Henrick, B. F. G. Johnson, J. Lewis, M. McPartlin, W. J. H. Nelson, M. D. Vargas, *J. Chem. Soc. Chem. Comm.* **1986**, 975.
[194] S. R. Drake, B. F. G. Johnson, J. Lewis, W. J. H. Nelson, M. D. Vargas, T. A. Adatia, D. Braga, K. Henrick, M. McPartlin, A. Sironi, *J. Chem. Soc. Dalton Trans.* **1989**, 1455.
[195] D. Braga, K. Henrick, B. F. G. Johnson, J. Lewis, M. McPartlin, W. J. H. Nelson, A. Sironi, M. D. Vargas, *J. Chem. Soc. Chem. Comm.* **1983**, 1131.
[196] M. J. Freeman, M. Green, A. G. Orpen, I. D. Salter, F. G. A. Stone, *J. Chem. Soc. Chem. Comm.* **1983**, 1332.
[197] V. Dearing, S. R. Drake, B. F. G. Johnson, J. Lewis, M. McPartlin, H. R. Powell, *J. Chem. Soc. Chem. Comm.* **1988**, 1331.
[198] A. Fumagalli, *personal communication*.
[199] S. R. Drake, K. Henrick, B. F. G. Johnson, J. Lewis, M. McPartlin, J. Morris, *J. Chem. Soc. Chem. Comm.* **1986**, 928.
[200] L. H. Gade, B. F. G. Johnson, J. Lewis, M. McPartlin, H. R. Powell, *J. Chem. Soc. Dalton Trans.* **1992**, 921.
[201] P. J. Bailey, B. F. G. Johnson, J. Lewis, M. McPartlin, H. R. Powell, *J. Chem. Soc. Chem. Comm.* **1989**, 1513.
[202] M. McPartlin, C. R. Eady, B. F. G. Johnson, J. Lewis, *J. Chem. Soc. Chem. Comm.* **1976**, 883.
[203] P. F. Jackson, B. F. G. Johnson, J. Lewis, P. R. Raithby, M. McPartlin, W. J. H. Nelson, K. D. Rouse, J. Allibon, S. A. Mason, *J. Chem. Soc. Chem. Comm.* **1980**, 295.
[204] M. R. Churchill, J. Wormald, *J. Am. Chem. Soc.* **1971**, *93*, 5607.
[205] A. J. Amoroso, B. F. G. Johnson, J. Lewis, P. R. Raithby, W. T. Wong, *J. Chem. Soc. Chem. Comm.* **1990**, 1208.
[206] P. Chini, S. Martinengo, G. Giordano, *Gazz. Chim. Ital.* **1972**, *102*, 330.
[207] G. Longoni, P. Chini, *Inorg. Chem.* **1976**, *15*, 3029.
[208] V. G. Albano, F. Azzaroni, M. C. Iapalucci, G. Longoni, M. Monari, S. Mulley in *Chemical Processes in Inorganic Materials: Metal and Semiconductor Clusters and Colloids*, MRS, **1992**, *272*, 115.
[209] S. Martinengo, A. Fumagalli, R. Bonfichi, G. Ciani, A. Sironi, *J. Chem. Soc. Chem. Comm.* **1982**, 825.
[210] A. Fumagalli, S. Martinengo, G. Ciani, *J. Organomet. Chem.* **1984**, *273*, C46.
[211] A. Albinati, A. Moor, P. S. Pregosin, L. M. Venanzi, *J. Am. Chem. Soc.* **1982**, *104*, 7672.
[212] T. Tanase, T. Horiuchi, K. Kobayashi, Y. Yamamoto, *J. Organomet. Chem.* **1992**, *440*, 1.

[213] B. T. Heaton, L. Strona, S. Martinengo, D. Strumolo, V. G. Albano, D. Braga, *J. Chem. Soc. Dalton Trans.* **1983**, 2175.
[214] S. Martinengo, D. Strumolo, P. Chini, V. G. Albano, D. Braga, *J. Chem. Soc. Dalton Trans.* **1984**, 1837.
[215] A. Ceriotti, G. Longoni, M. Manassero, M. Perego, M. Sansoni, *Inorg. Chem.* **1985**, *24*, 117.
[216] J. L. Vidal, R. A. Fiato, L. A. Cosby, R. L. Pruett, *Inorg. Chem.* **1978**, *17*, 2574.
[217] L. D. Lower, L. F. Dahl, *J. Am. Chem. Soc.* **1976**, *98*, 5046.
[218] J. P. Zebrowski, R. K. Hayashi, A. Bjarnason, L. F. Dahl, *J. Am. Chem. Soc.* **1992**, *114*, 3121.
[219] M. Schollenberger, B. Nuber, M. L. Ziegler, *Angew. Chem. Int. Ed. Engl.* **1992**, *31*, 350.
[220] P. T. Beurskens, J. J. Bour, W. P. Bosman, J. H. Noordik, M. Kolenbrander, J. A. Buskes, *Inorg. Chem.* **1984**, *23*, 146.
[221] V. G. Albano, M. Sansoni, P. Chini, S. Martinengo, D. Strumolo, *J. Chem. Soc. Dalton Trans.* **1976**, 970.
[222] B. K. Teo, X. Shi, H. Zhang, *J. Chem. Soc. Chem. Comm.* **1992**, 1195.
[223] B. K. Teo, X. Shi, H. Zhang, *J. Am. Chem. Soc.* **1991**, *113*, 4329.
[224] R. J. Goudsmith, B. F. G. Johnson, J. Lewis, P. R. Raithby, K. H. Whitmire, *J. Chem. Soc. Chem. Comm.* **1982**, 640.
[225] N. M. Boag, C. B. Knobler, H. D. Kaesz, *Angew. Chem. Int. Ed. Engl.* **1983**, *22*, 249.
[226] R. J. Goudsmith, B. F. G. Johnson, J. Lewis, P. R. Raithby, K. H. Whitmire, *J. Chem. Soc. Chem. Comm.* **1983**, 246.
[227] U. Bodensieck, H. Stoeckli-Evans, G. Suss-Fink, *Angew. Chem. Int. Ed. Engl.* **1991**, *30*, 1126.
[228] M. Fajardo, H. D. Holden, B. F. G. Johnson, J. Lewis, P. R. Raithby, *J. Chem. Soc. Chem. Comm.* **1984**, 24.
[229] P. Klufers, *Angew. Chem. Int. Ed. Engl.* **1984**, *23*, 307.
[230] P. Klufers, *Z. Kristal.* **1984**, *166*, 143.
[231] R. D. Ernst, T. J. Marks, J. A. Ibers, *J. Am. Chem. Soc.* **1977**, *99*, 2090.
[232] W. Gade, E. Weiss, *Angew. Chem. Int. Ed. Engl.* **1981**, *20*, 803.
[233] G. P. Elliot, J. A. K. Howard, C. M. Nunn, F. G. A. Stone, *J. Chem. Soc. Chem. Comm.* **1986**, 431.
[234] P. Betz, B. Krebs, G. Henkel, *Angew. Chem. Int. Ed. Engl.* **1984**, *23*, 311.
[235] S. P. Huang, M. G. Kanatzidis, *Angew. Chem. Int. Ed. Engl.* **1992**, *31*, 787.
[236] S. Martinengo, G. Siani, A. Sironi, *J. Chem. Soc. Chem. Comm.* **1991**, 26.
[237] C. E. Briant, B. R. C. Theobald, J. W. White, L. K. Bell, D. M. P. Mingos, *J. Chem. Soc. Chem. Comm.* **1981**, 201.
[238] K. P. Callahan, M. F. Hawthorne, *Pure Appl. Chem.* **1974**, *39*, 475.
[239] J. D. Corbett, *Chem. Rev.* **1985**, *85*, 383.
[240] A. J. Kahaian, J. B. Thoden, L. F. Dahl, *J. Chem. Soc. Chem. Comm.* **1992**, 353.
[241] D. F. Rieck, J. A. Gavney, R. L. Norman, R. K. Hayashi, L. F. Dahl, *J. Am. Chem. Soc.* **1992,** *114*, 10369.
[242] D. F. Rieck, R. A. Montag, T. S. McKechnie, L. F. Dahl, *J. Am. Chem. Soc.* **1986**, *108*, 1330.
[243] R. E. Des Enfants, J. A. Gavney, R. K. Hayashi, D. A. Rae, L. F. Dahl, *J. Organomet. Chem.* **1990**, *383*, 543.
[244] T. Weiske, J. Hrusak, D. Bohme, H. Schwarz, *Helv. Chim. Acta* **1992**, *75*, 79.
[245] T. G. Spiro, D. H. Templeton, A. Zalkin, *Inorg. Chem.* **1969**, *8*, 856.
[246] M. Tachikawa, E. L. Muetterties, *Prog. Inorg. Chem.* **1981**, *28*, 103.
[247] J. S. Bradley, *Adv. Organomet. Chem.* **1983**, *22*, 1.
[248] K. H. Whitmire, *J. Coord. Chem.* **1988**, *17*, 95.
[249] W. A. Herrmann, *Angew. Chem. Int. Ed. Engl.* **1986**, *25*, 56.
[250] K. C. C. Kharas, L. F. Dahl, *Adv. Chem. Phys.* **1988**, *70*, 1.

[251] A. K. Bandyopadhyay, R. Khatta, T. P. Fehlner, *Inorg. Chem.* **1989**, *28*, 4436.
[252] F. E. Hong, T. J. Coffy, D. A. McCarthy, S. G. Shore, *Inorg. Chem.* **1989**, *28*, 3284.
[253] C. E. Housecroft, D. M. Matthews, A. L. Rheingold, X. Song, *J. Chem. Soc. Chem. Comm.* **1992**, 842.
[254] S. M. Draper, A. D. Hattersley, C. E. Housecroft, A. L. Rheingold, *J. Chem. Soc. Chem. Comm.* **1992**, 1365.
[255] G. Ciani, G. D'Alfonso, P. Romiti, A. Sironi, M. Freni, *J. Organomet. Chem.* **1983**, *244*, C27.
[256] G. Ciani, G. D'Alfonso, M. Freni, P. Romiti, A. Sironi, *J. Chem. Soc. Chem. Comm.* **1982**, 705.
[257] T. Beringhelli, G. D'Alfonso, M. Freni, G. Ciani, A. Sironi, *J. Organomet. Chem.* **1985**, *295*, C7.
[258] R. Mason, W. R. Robinson, *J. Chem. Soc. Chem. Comm.* **1968**, 468.
[259] T. Chihara, R. Komoto, K. Kobayashi, H. Yamazaki, Y. Matsuma, *Inorg. Chem.* **1989**, *28*, 964.
[260] C. M. T. Hayward, J. R. Shapley, M. R. Churchill, C. Bueno, A. L. Rheingold, *J. Am. Chem. Soc.* **1982**, *104*, 7347.
[261] V. G. Albano, D. Braga, S. Martinengo, *J. Chem. Soc. Dalton Trans.* **1986**, 981.
[262] S. Martinengo, D. Strumolo, P. Chini, V. G. Albano, D. Braga, *J. Chem. Soc. Dalton Trans.* **1985**, 35.
[263] V. G. Albano, M. Sansoni, P. Chini, S. Martinengo, *J. Chem. Soc. Dalton Trans.* **1973**, 651.
[264] A. Ceriotti, G. Longoni, M. Manassero, N. Masciocchi, L. Resconi, M. Sansoni, *J. Chem. Soc. Chem. Comm.* **1985**, 181.
[265] A. Ceriotti, A. Fait, G. Longoni, G. Piro, L. Resconi, F. Demartin, M. Manassero, N. Masciocchi, M. Sansoni, *J. Am. Chem. Soc.* **1986**, *108*, 5370.
[266] A. Ceriotti, A. Fait, G. Longoni, G. Piro, F. Demartin, M. Manassero, M. Sansoni, *J. Am. Chem. Soc.* **1986**, *108*, 8091.
[267] K. M. MacKay, B. K. Nicholson, W. T. Robinson, A. W. Sims, *J. Chem. Soc. Chem. Comm.* **1984**, 1276.
[268] M. L. Blohm, D. E. Fjare, W. L. Gladfelter, *Inorg. Chem.* **1983**, *22*, 1004.
[269] F. Van Gastel, N. J. Taylor, A. J. Carty, *Inorg. Chem.* **1986**, *28*, 384.
[270] L. M. Bullock, J. S. Field, R. J. Haines, E. Minshall, D. N. Smit, *J. Organomet. Chem.* **1986**, *310*, C47.
[271] S. P. Colbran, F. J. Lahor, P. R. Raithby, J. Lewis, B. F. G. Johnson, C. J. Cardin, *J. Chem. Soc. Dalton Trans.* **1988**, 173.
[272] S. Martinengo, G. Ciani, A. Sironi, B. T. Heaton, J. Mason, *J. Am. Chem. Soc.* **1979**, *101*, 7095.
[273] R. Bonfichi, G. Ciani, A. Sironi, S. Martinengo, *J. Chem. Soc. Dalton Trans.* **1973**, 651.
[274] S. Martinengo, G. Ciani, A. Sironi, *J. Organomet. Chem.* **1988**, *358*, C23.
[275] S. Martinengo, G. Ciani, A. Sironi, *J. Chem. Soc. Chem. Comm.* **1986**, 1742.
[276] S. Martinengo, G. Ciani, A. Sironi, *J. Chem. Soc. Chem. Comm.* **1992**, 1405.
[277] S. Martinengo, G. Ciani, A. Sironi, *J. Am. Chem. Soc.* **1982**, *104*, 328.
[278] L. Garlaschelli, G. Ciani, S. Martinengo, *personal communication.*
[279] J. L. Vidal, W. E. Walker, R. C. Schoening, *Inorg. Chem.* **1981**, *20*, 238.
[280] J. L. Vidal, *Inorg. Chem.* **1981,** *20*, 243.
[281] J. L. Vidal, J. M. Troup, *J. Organomet. Chem.* **1981**, *213*, 351.
[282] L. Garlaschelli, A. Fumagalli, S. Martinengo, B. T. Heaton, D. O. Smith, L. Strona, *J. Chem. Soc. Dalton Trans.* **1982**, 2265.
[283] F. Demartin, M. C. Iapalucci, G. Longoni, *Inorg. Chem., in press.*
[284] T. Beringhelli, G. Ciani, G. D'Alfonso, A. Sironi, M. Freni, *J. Chem. Soc. Chem. Comm.* **1985**, 978.
[285] R. F. Boehme, P. Coppens, *Acta Crystal.* **1981**, *B37*, 1914.

[286] P. F. Jackson, B. F. G. Johnson, J. Lewis, P. R. Raithby, G. J. Will, M. McPartlin, W. J. H. Nelson, *J. Chem. Soc. Chem. Comm.* **1980**, 1190.
[287] E. H. Braye, L. F. Dahl, W. Hubel, D. L. Wampler, *J. Am. Chem. Soc.* **1962**, *84*, 4633.
[288] B. F. G. Johnson, J. Lewis, J. N. Nicholls, J. Puga, P. R. Raithby, M. J. Rosales, M. McPartlin, W. Clegg, *J. Chem. Soc. Dalton Trans.* **1983**, 277.
[289] P. F. Jackson, B. F. G. Johnson, J. Lewis, J. N. Nicholls, *J. Chem. Soc. Chem. Comm.* **1980**, 564.
[290] B. F. G. Johnson, J. Lewis, W. J. H. Nelson, J. N. Nicholls, J. Puga, P. R. Raithby, M. J. Rosales, M. Schroder, M. D. Vargas, *J. Chem. Soc. Dalton Trans.* **1983**, 2447.
[291] C. J. Adams, M. I. Bruce, B. W. Skelton, A. H. White, *J. Chem. Soc. Chem. Comm.* **1992**, 96.
[292] G. Gervasio, R. Rossetti, P. L. Stanghellini, G. Bor, *Inorg. Chem.* **1984**, *23*, 2073.
[293] S. G. Anema, K. M. MacKay, L. C. McLeod, B. K. Nicholson, J. M. Whittacker, *Angew. Chem. Int. Ed. Engl.* **1986**, *25*, 759.
[294] J. P. Zebrowski, R. K. Hayashi, L. F. Dahl, *submitted.*
[295] A. Gourdon, Y. Jeannin, *J. Organomet. Chem.* **1985**, *290*, 199.
[296] M. L. Blohm, W. L. Gladfelter, *Organometallics* **1985**, *4*, 45.
[297] G. Ciani, A. Sironi, *J. Organomet. Chem.* **1983**, *241*, 385.
[298] H. Lang, G. Huttner, L. Zsolnai, G. Mohr, B. Sigwarth, U. Weber, O. Orama, I. Jibril, *J. Organomet. Chem.* **1986**, *304*, 157.
[299] S. Luo, K. H. Whitmire, *J. Organomet. Chem.* **1989**, *376*, 297.
[300] V. G. Albano, F. Demartin, M. C. Iapalucci, F. Laschi, G. Longoni, A. Sironi, P. Zanello, *J. Chem. Soc. Dalton Trans.* **1991**, 739.
[301] C. K. Schauer, D. F. Shriver, *Angew. Chem. Int. Ed. Engl.* **1987**, *26*, 255.
[302] C. K. Schauer, E. J. Voss, M. Sabat, D. F. Shriver, *J. Am. Chem. Soc.* **1989**, *111*, 7662.
[303] R. D. Adams, I. T. Horvath, P. Mathur, *Organometallics* **1984**, *3*, 623.
[304] R. D. Adams, I. T. Horvath, L. W. Yang, *J. Am. Chem. Soc.* **1983**, *105*, 1533.
[305] R. D. Adams, I. T. Horvath, P. Mathur, B. E. Segmuller, L. W. Yang, *Organometallics* **1983**, *2*, 1078.
[306] B. Schiemenz, G. Huttner, *Angew. Chem. Int. Ed. Engl.* **1993**, *32*, 297.
[307] A. J. Wynd, S. E. Robins, D. A. Welch, A. J. Welch, *J. Chem. Soc. Chem. Comm.* **1985**, 819.
[308] Y. Do, C. B. Knobler, M. F. Hawthorne, *J. Am. Chem. Soc.* **1987**. *109*, 1853.
[309] K. H. Whitmire, R. R. Ryan, H. J. Wasserman, T. A. Albright, S. K. Kang, *J. Am. Chem. Soc.* **1986**, *108*, 6831.
[310] O. J. Scherer, H. Sitzman, G. Wolmershauser, *Angew. Chem. Int. Ed. Engl.* **1985**, *24*, 351.
[311] A. L. Rheingold, M. J. Foley, P. J. Sullivan, *J. Am. Chem. Soc.* **1982**, *104*, 4727.
[312] K. H. Whitmire, T. A. Albright, S. K. Kang, M. R. Churchill, J. C. Fettinger, *Inorg. Chem.* **1986**, *25*, 2799.
[313] A. L. Rheingold, P. J. Sullivan, *J. Chem. Soc. Chem. Comm.* **1983**, 39.
[314] J. Scott Leigh, K. H. Whitmire, K. A. Yee, T. A. Albright, *J. Am. Chem. Soc.* **1989**, *111*, 2726.
[315] S. Martinengo, G. Ciani, *J. Chem. Soc. Chem. Comm.* **1987**, 1589.
[316] E. Diana, G. Gervasio, R. Rossetti, F. Valdemarin, G. Bor, P. L. Stanghellini, *Inorg. Chem.* **1991**, *30*, 294.
[317] M. J. Colling, R. J. Gillespie, J. W. Kolis, J. F. Sawyer, *Inorg. Chem.* **1986**, *25*, 2057.
[318] D. Fenske, A. Hollnagel, *Angew. Chem. Int. Ed. Engl.* **1989**, *28*, 1390.
[319] L. C. Roof, W. T. Pennington, J. W. Kolis, *J. Am. Chem. Soc.* **1990**, *112*, 8172.
[320] L. C. Roof, W. T. Pennington, J. W. Kolis, *Angew. Chem. Int. Ed. Engl.* **1992**, *31*, 913.
[321] S. P. Huang, M. G. Kanatzidis. *J. Am. Chem. Soc.* **1992**, *114*, 5477.
[322] D. W. Hart, R. G. Teller, C. Y. Wei, R. Bau, G. Longoni, S. Campanella, P. Chini, T. F. Koetzle, *J. Am. Chem. Soc.* **1981**, *103*, 1458.

[323] C. R. Eady, B. F. G. Johnson, J. Lewis, M. C. Malatesta, P. Machin, M. McPartlin, *J. Chem. Soc. Chem. Comm.* **1976**, 945.

[324] C. R. Eady, P. F. Jackson, B. F. G. Johnson, J. Lewis, M. C. Malatesta, M. McPartlin, W. J. H. Nelson, *J. Chem. Soc. Dalton Trans.* **1980**, 353.

[325] P. J. Bailey, M. A. Beswick, B. F. G. Johnson, J. Lewis, P. R. Raithby, M. C. Ramirez de Arellano, *J. Chem. Soc. Dalton Trans.* **1992**, 3159.

[326] P. J. Bailey, B. F. G. Johnson, J. Lewis, M. McPartlin, H. R. Powell, *J. Organomet. Chem.* **1989**, *377*, C17.

[327] P. F. Jackson, B. F. G. Johnson, J. Lewis, M. McPartlin, W. J. H. Nelson, *J. Chem. Soc. Chem. Comm.* **1982**, 49.

[328] P. Chini, *J. Organomet. Chem.* **1980**, *200*, 37.

[329] P. Chini, G. Longoni, S. Martinengo, *Chim. Ind.* (Milano) **1978**, *60,* 989.

[330] B. T. Heaton, L. Strona, S. Martinengo, D. Strumolo, R. J. Goodfellow, I. H. Sadler, *J. Chem. Soc. Dalton Trans.* **1982**, 1499.

[331] P. L. Stanghellini, G. Longoni, *J. Chem. Soc. Dalton Trans.* **1987**, 685.

[332] D. Graham, J. Howard, T. C. Waddington, J. Tomkinson, *J. Chem. Soc. Faraday Trans.* **1983**, *2*, 1713.

[333] J. Eckert, A. Albinati, G. Longoni, *Inorg. Chem.* **1989**, *28*, 4055.

[334] P. Chini, G. Longoni, S. Martinengo, A. Ceriotti, *Adv. Chem. Ser.* **1978**, *167*, 1.

[335] S. Martinengo, B. T. Heaton, R. J. Goodfellow, P. Chini, *J. Chem. Soc. Chem. Comm.* **1977**, 39.

[336] J. S. Bradley, *Phyl. Trans. R. Soc.* London, **1982**, *A308*, 103.

[337] I. A. Oxton, S. F. A. Kettle, P. F. Jacson, B. F. G. Johnson, J. Lewis, *J. Chem. Soc. Chem. Comm.* **1979**, 687.

[338] I. A. Oxton, D. B. Powell, R. J. Goudsmit, B. F. G. Johnson, J. Lewis, W. J. H. Nelson, J. N. Nicholls, M. J. Rosales, M. D. Vargas, K. H. Whitmire, *Inorg. Chim. Acta Letters* **1982**, *64*, L259.

[339] C. M. Hay, B. F. G. Johnson, J. Lewis, R. C. S. McQueen, P. R. Raithby, R. M. Sorrell, M. J. Taylor, *Organometallics* **1985**, *4*, 202.

[340] V. G. Albano, P. Chini, S. Martinengo, D. J. McCaffrey, D. Strumolo, B. T. Heaton, *J. Am. Chem. Soc.* **1974**, *96*, 8106.

[341] H. Schmidbaur, B. Brachthauser, O. Steigelmann, *Angew. Chem. Int. Ed. Engl.* **1991**, *30*, 1488.

[342] J. A. Creighton, R. Della Pergola, B. T. Heaton, S. Martinengo, L. Strona, D. A. Willis, *J. Chem. Soc. Chem. Comm.* **1982**, 864.

[343] V. G. Albano, P. Chini, G. Ciani, M. Sansoni, D. Strumolo, B. T. Heaton, S. Martinengo, *J. Am. Chem. Soc.* **1976**, *98*, 5027.

[344] B. T. Heaton, L. Strona, S. Martinengo, *J. Organomet. Chem.* **1981**, *215*, 415.

[345] P. L. Stanghellini, G. Longoni, G. D'Alfonso, *Inorg. Chem.* **1987**, *26*, 2769.

[346] D. G. Bagley, *Nature* **1970**, *225*, 1040.

[347] B. F. G. Johnson, J. Lewis, S. W. Sankey, K. Wong, M. McPartlin, W. J. H. Nelson, *J. Organomet. Chem.* **1980**, *191*, C3.

[348] G. Ciani, G. D'Alfonso, M. Freni, P. Romiti, A. Sironi, *J. Chem. Soc. Chem. Comm.* **1982**, 339.

[349] D. Braga, K. Henrick, B. F. G. Johnson, J. Lewis, M. McPartlin, W. J. H. Nelson, A. Sironi, M. D. Vargas. *J. Chem. Soc. Chem. Comm.* **1983**, 1132.

[350] B. F. G. Johnson, R. D. Johnston, J. Lewis, *J. Chem. Soc. Chem. Comm.* **1967**, 1057.

[351] M. R. Churchill, J. Wormald, J. Knight, M. J. Mays, *J. Am. Chem. Soc.* **1971**, *93*, 3073.

[352] R. P. Stewart, U. Anders, W. A. G. Graham, *J. Organomet. Chem.* **1971**, *32*, C49.

[353] A. T. T. Hsieh, M. J. Mays, *J. Organomet. Chem.* **1972**, *37*, C53.

[354] J. K. Kolis, F. Basolo, D. F. Shriver, *J. Am. Chem. Soc.* **1982**, *104*, 5626.

[355] A. Ceriotti, L. Resconi, F. Demartin, G. Longoni, M. Manassero, M. Sansoni, *J. Organomet. Chem.* **1983**, *249*, C35.

[356] A. Ceriotti, P. Chini, G. Longoni, G. Piro, *Gazz. Chim. Ital.* **1982**, *112*, 353.
[357] J. W. Kolis, M. A. Drezdzon, D. F. Shriver, *Inorg. Synth.* **1989**, *26*, 246.
[358] B. F. G. Johnson, D. A. Kaner, J. Lewis, P. R. Raithby, M. J. Rosales, *J. Organomet. Chem.* **1982**, *231*, C59.
[359] M. J. Sailor, D. F. Shriver, *Organometallics* **1985**, *4*, 1476.
[360] M. J. Went, M. J. Sailor, P. L. Bogdan, C. P. Brock, D. F. Shriver, *J. Am. Chem. Soc.* **1987**, *109*, 6023.
[361] M. A. Drezdzon, D. F. Shriver, *J. Mol. Cat.* **1983**, *21*, 81.
[362] G. Bor, U. K. Dietler, P. L. Stanghellini, G. Gervasio, R. Rossetti, G. Sbrignadello, G. A. Battiston, *J. Organomet. Chem.* **1981**, *213*, 277.
[363] M. L. Blohm, W. L. Gladfelter, *Organometallics* **1985**, *4*, 45.
[364] A. S. Foust, M. S. Foster, L. F. Dahl, *J. Am. Chem. Soc.* **1969**, *91*, 5631.
[365] A. S. Foust, M. S. Foster, L. F. Dahl, *J. Am. Chem. Soc.* **1969**, *91*, 5633.
[366] J. D. Corbett, *Prog. Inorg. Chem.* **1976**, *21*, 129.
[367] L. E. Toth in *Transition Metal Carbides and Nitrides*, Academic Press, New York, **1971**.
[368] A. L. Bowman, G. P. Arnold, E. K. Storms, N. G. Nereson, *Acta Crystal.* **1972**, *B28*, 3102.
[369] C. N. R. Rao, J. Gapalakrishnan in *New Directions in Solid State Chemistry*, Cambridge University Press, Cambridge, UK, **1986**.
[370] K. Beechgaard in *Structure and Properties of Molecular Crystals* (Ed.: M. Pierrot), Elsevier, Amsterdam, **1990**, 235.
[371] G. J. Lewis, J. D. Roth, R. A. Montag, L. K. Safford, X. Gao, S-C. Chang, L. F. Dahl, M. J. Weaver, *J. Am. Chem. Soc.* **1990**, *112*, 2831.
[372] F. Demartin, M. Manassero, M. Sansoni, L. Garlaschelli, C. Raimondi, S. Martinengo, F. Canziani, *J. Chem. Soc. Chem. Comm.* **1981**, 528.
[373] J. C. Calabrese, L. F. Dahl, A. Cavalieri, P. Chini, G. Longoni, S. Martinengo, *J. Am. Chem. Soc.* **1974**, *96*, 2616.
[374] D. Braga, F. Grepioni, P. Milne, E. Parisini, *J. Am. Chem. Soc.* **1993**, *115*, 5115.
[375] R. Hesse, P. Jennische, *Acta Chim. Scand.* **1972**, *26*, 3865.
[376] P. Jennische, H. Anacker-Eickoff, A. Wolberg, *Acta Crystal* **1975**, *A31*, 5143.
[377] S. J. Lawton, W. J. Rohrbaugh, G. T. Kokotailo, *Inorg. Chem.* **1972**, *11*, 2227.
[378] A. P. Ginsberg, J. W. Koepke, J. J. Hauser, K. W. West, F. J. Di Salvo, C. R. Sprinkle, R. L. Cohen, *Inorg. Chem.* **1976**, *15*, 514.
[379] J. S. Miller, A. J. Epstein, *Prog. Inorg. Chem.* **1976**, *20*, 1.
[380] D. Braga, F. Grepioni, *Organometallics* **1992**, *11*, 1256.
[381] W. Deck, A. K. Powell, H. Vahrenkamp, *J. Organomet. Chem.* **1992**, *428*, 353.
[382] G. B. Ansell, M. A. Modrick, J. S. Bradley, *Acta Crystal* **1984**, *C40*, 365.
[383] L. H. Gade, B. F. G. Johnson, J. Lewis, M. McPartlin, H. R. Powell, *J. Chem. Soc. Chem. Comm.* **1989**, 110.
[384] P. J. Bailey, B. F. G. Johnson, J. Lewis, M. McPartlin, H. R. Powell, *J. Chem. Soc. Chem. Comm.* **1989**, 113.
[385] A. Sironi, *personal communication.*
[386] G. Allegra, E. Mostardini-Peronaci, R. Ercoli, *J. Chem. Soc. Chem. Comm.* **1966**, 549.
[387] G. Bor, G. Gervasio, R. Rossetti, P. L. Stanghellini, *J. Chem. Soc. Chem. Comm.* **1978**, 841.
[388] K. Deller, B. Eisenmann, *Z. Naturforsch.* **1976**, *31B*, 29.
[389] A. Gourdon, Y. Jeannin, *J. Organomet. Chem.* **1986**, *304*, C1.
[390] H. Lang, G. Huttner, L. Zsolnai, G. Mohr, B. Sigwarth, U. Weber, O. Orama, I. Jibril, *J. Organomet. Chem.* **1986**, *304*, 137.
[391] K. Blechsmith, H. Pfisterer, T. Zhan, M. L. Ziegler, *Angew. Chem. Int. Ed. Engl.* **1985**, *24*, 66.
[392] L. J. Arnold, K. M. MacKay, B. K. Nicholson, *J. Organomet. Chem.* **1990**, *387*, 197.

[393] A. L. Rheingold, K. H. Whitmire, M. D. Fabiano, J. S. Leigh, *unpublished results* (cited in ref. 263)

[394] K. H. Whitmire, K. S. Raghuveer, M. R. Churchill, J. C. Fettinger, R. F. See, *J. Am. Chem. Soc.* **1986**, *108*, 2778.

[395] F. Richter, H. Vahrenkamp, *Angew. Chem. Int. Ed. Engl.* **1978**, *17*, 444.

[396] R. D. Adams, D. Mannig, B. E. Segmuller, *Organometallics* **1983**, *2*, 149.

[397] W. Cen, K. J. Haller, T. P. Fehlner, *Inorg. Chem.* **1991**, *30*, 3120.

[398] R. L. Sturgeon, M. M. Olmstead, N. E. Schore, *Organometallics* **1991**, *10*, 1649.

[399] G. R. John, B. F. G. Johnson, J. Lewis, K. C. Wong, *J. Organomet. Chem.* **1979**, *169*, C23.

[400] W. Cen, P. Lindenfeld, T. P. Fehlner, *J. Am. Chem. Soc.* **1992**, *114*, 5451.

[401] B. T. Heaton, P. Ingallina, R. W. Devenish, C. J. Humphries, A. Ceriotti, G. Longoni, M. Marchionna, *J. Chem. Soc. Chem. Comm.* **1987**, 765.

[402] R. W. Devenish, P. J. Goodhew, B. T. Heaton, C. Jakob, S. Mulley, *Science Progress* **1990**, *74*, 513.

[403] R. W. Devenish, S. Mulley, B. T. Heaton, G. Longoni, *J. Mater. Res.* **1992**, *7*, 2810.

[404] A. M. Gadalla, H-F. Yu, *J. Mater, Res.* **1990**, *5*, 2923.

[405] J. Y. Ding, D. E. Day, *J. Mater. Res.* **1991**, *6*, 168.

[406] C. J. McNeal, J. M. Hughes, G. J. Lewis, L. F. Dahl, *J. Am. Chem. Soc.* **1991**, *113*, 372.

[407] M. L. Steigerwald, C. R. Sprinkle, *J. Am. Chem. Soc.* **1987**, *109*, 7200.

[408] M. L. Steigerwald, C. R. Sprinkle, *Organometallics* **1988**, *7*, 245.

[409] M. L. Steigerwald, T. Siegrist, S. M. Stuczynski, *Inorg. Chem.* **1991**, *30*, 2256.

[410] J. G. Brennan, T. Siegrist, P. J. Carroll, S. M. Stuczynski, L. E. Brus, M. L. Steigerwald, *J. Am. Chem. Soc.* **1989**, *111*, 4141.

[411] J. G. Brennan, T. Siegrist, Y. U. Kwon, S. M. Stuczynski, M. L. Steigerwald, *J. Am. Chem. Soc.* **1992**, *114*, 10334.

[412] E. Keller, *SCHAKAL88, Graphical Representation of Molecular Models*, University of Freiburg, FRG.

3.3 Metal Rich Large Clusters with P and N Ligands

Günter Schmid

3.3.1 Synthetic Aspects

As described in Chapter 3.2, the synthesis of small, medium sized, and also relatively large transition metal clusters in organic ligand shells usually proceeds through the step-wise build up of smaller precursors. Metal rich clusters with sulfur and selenium bridges, which can in principle be understood as ligand stabilized cutouts of the corresponding sulfide and selenide lattices, are spontaneously formed from simple molecular components (see Chapter 3.4). The cluster growth, probably kinetically controlled, is determined by the presence of appropriate ligands.

If metal ions in solution are reduced to atoms by an appropriate reducing agent, they normally will combine into colloids or into microcrystalline particles which finally precipitate out. Such processes can occur within seconds, as is known for instance from the formation of silver mirrors on glass by the reduction of silver ions. If the growth of the metal particles in solution can be retarded and if appropriate ligands are additionally offered, it may happen that metal rich ligand stabilized transition metal clusters of uniform size are formed.

Indeed, the synthesis of a series of ligand protected clusters is based on such events. Clusters which consist of only a few atoms up to those with a few thousand close packed atoms can be gained in this way. They represent excellent objects for studying the development of metallic states, because somewhere along the way from a few to some thousands of metal atoms metallic properties should become recognizable for the first time. In addition to their catalytic applications, these sorts of clusters have earned attention simply because of such fundamental questions. As will be shown in this chapter, clusters with sizes in the 1–2 nm range can be regarded as usable quantum dots in which only a few remaining metallic electrons are trapped.

In principle, the synthesis of such clusters could also succeed by the degradation of bulk particles, however, the experimental difficulties associated with this suggest the opposite way is better, namely to form clusters by the building up from atoms. It should be mentioned that up to now it doesn't seem possible to synthesize clusters with the aim of achieving a particular result. What matters is to find out the appropriate reaction conditions to generate, for example, a 55 atom or a 561 atom cluster.

The formation of clusters from atoms and their subsequent stabilization with ligands, in our case we consider mainly phosphorus and nitrogen containing molecules, is accompanied by a surprising phenomenon: most of the clusters obtained by this method have a so-called full shell structure. This means that the metal atoms tend to form cubic or hexagonal close packed structures like that found in the bulk with a complete and complimentary outer geometry. Thus, a central atom is surrounded by 12 others in a close packed manner such that 13 atoms form the first full shell cluster, a second shell of 42 atoms completes the two shell cluster with altogether 55 atoms, and so on. The *n*th shell of such a system consists of $10n^2 + 2$ atoms. It seems reasonable that nature tries to form such particles since the package corresponds to that found in the bulk state and the complete outer shell guarantees a minimum of energy. However, the formation of a stable full shell cluster cannot be assumed in any given case since icosahedral and polyicosahedral structures are known as well as close packed but imperfect structures.

Colloids and metallic precipitates are often formed despite the presence of ligand molecules. Nevertheless, this method enabled for the first time a route to a series of ligand protected cluster molecules consisting of such a large number of atoms that they were suited for the study of quantum properties and the formation of metallic states. The first such clusters were phosphine stabilized gold clusters and were described in 1965. [1] In the course of the following years the syntheses of numerous clusters having Au_9, [2–10] Au_{10}, [11] Au_{11}, [12–17] and Au_{13} [18, 19] cluster cores were reported. Their syntheses were normally carried out by the reduction of the corresponding gold phosphine complexes by $NaBH_4$. In some cases, the method of metal vapor synthesis was also applied. [20] In the Au_9, Au_{10}, and Au_{11} cluster cores, icosahedral structural elements are to be recognized, whereas the Au_{13} clusters consist of a perfect icosahedron with one gold atom lodged in the center. The molecular structure of $[Au_{13}(dppm)_6]^{4+}$, [19] formed by the reaction of $[Au_2(dppm)(NO_3)_2]$ (dppm = $Ph_2P\text{-}CH_2\text{-}PPh_2$) with $NaBH_4$, is shown in Figure **3-26**.

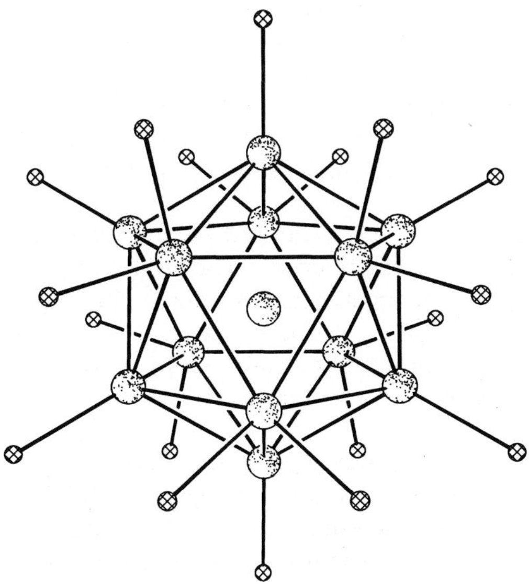

Figure 3-26. Molecular structure of the cation $[Au_{13}(dppm)_6]^{4+}$ (dppm = $Ph_2P\text{-}CH_2\text{-}PPh_2$). For clarity, only the P atoms of the dppm ligands are drawn.

A series of platinum containing phosphine stabilized gold clusters of the type $[Pt(AuL)_x]^{n+}$ has also been synthesized and, in most cases their structures investigated by single crystal X-ray analyses. [21–27] Copper, silver, and mercury containing species can be prepared by a subsequent addition of complexes of these elements, *e.g.* $[Pt(HgX)_2(AuPPh_3)_8]^{2+}$, $[Pt(Ag)(AuPPh_3)_8]^{3+}$, and $[Pt(CuCl)(AuPPh_3)_8]^{2+}$ can all be made from the reaction of $[Pt(AuPPh_3)_8]^{2+}$ with $[Hg_2X_2]$, [28] $AgNO_3$, [29] or CuCl. [30] The structures of all these mixed metal compounds are based on icosahedral frameworks.

Photolysis of Ph_3PAuN_3 leads to a reductive elimination of the azide group to yield Ph_3PAu^0 moieties which can then be used to combine with other complex fragments under formation of various bimetallic clusters: $[(Ph_3PAu)_5Mo(CO)_4]^+$, [31] $[(Ph_3PAu)_7Mo(CO)_3]^+$, [32] $[(Ph_3PAu)_6Mn(CO)_3]^+$, [33] $[(Ph_3PAu)_5Fe(CO)_3]^+$, [34] or $[(Ph_3PAu)_6AuCo_2(CO)_6]^+$. [35] The Au_5Mo skeleton forms a capped trigonal bipyramid with the Mo atom in the equatorial position, whereas the Au_7Mo cluster core consists of an icosahedral fragment as does the Au_6Mn combination. The cationic Au_5Fe cluster forms a bicapped tetrahedron. In the Au_6AuCo_2 system there are two Au_4Co trigonal bipyramids sharing one axial Au atom.

Clusters with more than one shell clearly prefer the more dense cubic (ccp) or hexagonal (hcp) close packed structures. This may not only be due to the increased density of ccp and hcp structures, but also to the fact that icosahedra cannot grow by a periodically ordered shell expansion, as the five fold symmetry does not allow it. In a M_{13} cluster (1+12), the symmetry generally does not yet play a decisive role. A hcp Rh_{13} cluster has already been mentioned in Chapter 3.2. The tendency to form dense packed structures as the number of metal atoms increases is confirmed by the larger phosphine stabilized gold cluster $[Au_{39}(PPh_3)_{14}Cl_6]Cl_2$, whose molecular structure has been determined by X-ray analysis (Fig. **3-27**). [36] The Au atoms are arranged in a slightly distorted hexagonal close packed structure. It is formed in a way similar to the Au_{13} cluster mentioned above, namely by $NaBH_4$ reduction of $HAuCl_4$ in the presence of PPh_3.

Some bimetallic phosphine stabilized silver/gold clusters are characterized by interpenetrating icosahedral structures. [37–42] The reason for this characteristic structural feature in these P(*p*-tolyl)$_3$ and PPh_3 protected $Au_{13}Ag_{12}$, $Au_{18}Ag_{20}$, and $Au_{18}Ag_{19}$ clusters, which have additional chlorine or bromine atoms coordinated is probably to be found in their having two different metaltypes. Since they cannot form ideal close packed structures, they therefore tend to make icosahedral building blocks, and as could be expected, multiple shell structures are avoided. Figure **3-28** illustrates the principle of interpenetrating icosahedra with the structure of $[\{(p\text{-tolyl})_3P\}_{12}Au_{18}Ag_{20}Cl_{14}]$. [41]

As representative for some other two shell M_{55} clusters, the synthesis of $[Au_{55}(PPh_3)_{12}Cl_6]$ shall be discussed in more detail. If $[Ph_3PAuCl]$, dissolved in benzene or toluene, is reacted with gaseous diborane B_2H_6 at ~60 °C, the brown/-black Au_{55} cluster is formed in about 30 % yield. [43] The unusual reducing agent B_2H_6 has a double function here. First, it reduces the Au(I) in Ph_3AuCl to Au with mainly zero valent character and second, its functions as a Lewis acid which leads to the formation of $H_3B\text{-}PPh_3$. Hereby, the concentration

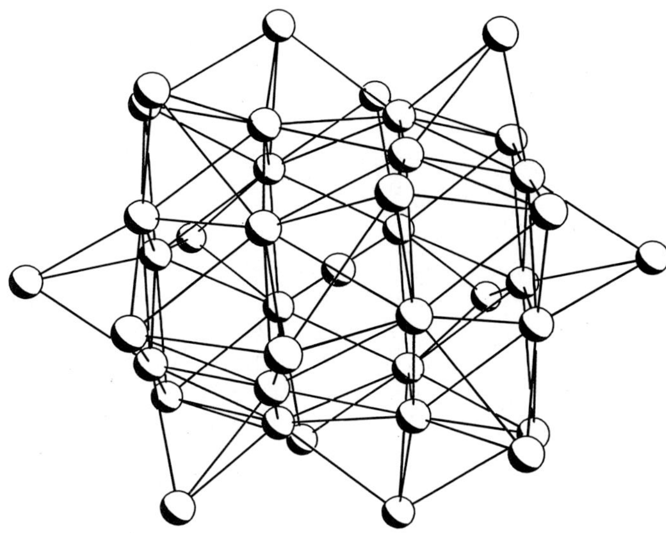

Figure 3-27. The slightly distorted hexagonal close packed Au_{39} core of the $[Au_{39}(PPh_3)_{14}Cl_6]^{2+}$ cation.

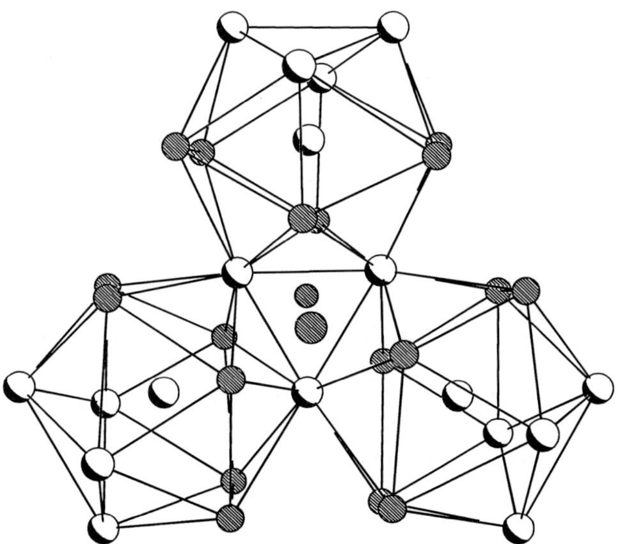

Figure 3-28. Metal skeleton of the cluster $[\{(p\text{-Tolyl})_3P\}_{12}Au_{18}Ag_{20}Cl_{14}]$, constructed from three interpenetrating icosahedra (silver atoms are striped).

of PPh_3 is decreased to such an extent that the development of Au_{55} clusters is enabled. Besides PPh_3, $P(p\text{-tolyl})_3$ can also be used as a ligand. The structural and physical properties of $Au_{55}(PPh_3)_3Cl_6$ will be discussed in detail in the following sections. Two shell clusters of platinum, rhodium and ruthenium can be synthesized by corresponding procedures. [44, 45]

Ligand exchange reactions have allowed the preparation of a novel cluster in just one case. If a dichloromethane solution of $Au_{55}(PPh_3)_{12}Cl_6$ is treated with an aqueous solution of $Ph_2P(m\text{-}C_6H_4SO_3Na)$, the PPh_3 ligands can be quantitatively substituted by the sulfonated phosphine molecules with the consequence that the Au_{55} cluster becomes water soluble. [46] While the organic phase becomes almost colorless, the aqueous phase changes color to dark brown due to the formation of $[Au_{55}\{Ph_2P(m\text{-}C_6H_4SO_3Na)\}_{12}Cl_6]$ (Eq. 3.34):

$$[Au_{55}(PPh_3)_{12}Cl_6] + 12\ Ph_2P(m\text{-}C_6H_4SO_3Na) \rightarrow [Au_{55}\{Ph_2P(m\text{-}C_6H_4SO_3Na)\}_{12}Cl_6] + 12\ PPh_3 \qquad (3.34)$$

Conductivity measurements on aqueous solutions show that the cluster is quantitatively dissociated into the twelvefold negatively charged $[Au_{55}\{Ph_2P(m\text{-}C_6H_4SO_3)\}_{12}Cl_6]^{12-}$ anion and 12 Na^+ cations. The Na^+ ions can be substituted by protons by means of ion exchange. Due to the high negative charge on the anion in aqueous solution, this cluster turns out to be considerably more stable than $[Au_{55}(PPh_3)_{12}Cl_6]$ in CH_2Cl_2 where it is degraded within the course of a few days into $[Ph_3PAuCl]$, smaller clusters like $[Au_{11}(PPh_3)_7Cl_3]$, and metallic gold.

The synthesis of the only four shell cluster which has been described up to now, $[Pt_{309}phen^*_{36}O_{30\pm10}]$, [47] also proceeds by the reduction of metal ions in solution, but here gaseous hydrogen is used. Phen* represents the so-called *batho*-phenanthroline, a 1,10-phenanthroline which is substituted by $p\text{-}C_6H_4SO_3Na$ groups at the 4,7 positions. Due to its hydrophilic sulfonate groups, the Pt_{309} cluster becomes readily soluble in water. The synthesis of the cluster starts with Pt(II)acetate, which is then reduced by gaseous hydrogen at room temperature in an acetic acid solution in the presence of equivalent amounts of phen*. The oxygen free $[Pt_{309}phen^*_{36}]$, which is probably charged by larger amounts of hydrogen, is extremely sensitive to air and very difficult to handle. Therefore, it is not isolated but instead carefully treated with oxygen to give the air stable $[Pt_{309}phen^*_{36}\text{-}O_{30\pm10}]$. The nature of the oxygen is not quite clear, but it presumably exists in a superoxidic or peroxidic state. Although the yields of the cluster are very low, X-ray powder diffraction studies on the black material clearly indicate a ccp packing of the Pt atoms as found in the bulk metal. Structural details are discussed in Section 3.3.2.

If an analogous procedure is carried out with palladium(II)acetate in the presence of phenanthroline (phen), the formation of three different full shell clusters can be observed. The five shell cluster $[Pd_{561}phen_{36}O_{200\pm20}]$ [46, 48] can be isolated in about 10 % yield with approximately 80–90 % of the palladium being used to form seven and eight shell clusters which, however, could not yet be separated from each other. From analytical and high resolution transmission electron microscopy (HRTEM) data (see Section 3.3.2) the idealized formula $[Pd_{1415}phen_{60}O_{\sim1100}]$ and $[Pd_{2057}phen_{84}O_{\sim1600}]$ can be concluded. [49] It is surprising that these clusters, which have core diameters of 3.15 and 3.60 nm respectively, can be observed by HRTEM as predominantly monocrystalline and well defined particles. Due to the covering with oxygen, these clusters are also air stable so that they can be easily handled, a tremendous advantage if these materials are to be used as catalysts (see Section 3.3.4.2). Modified reaction conditions and isolation procedures led to the first previously described Pd five shell cluster with the approximate formula $[Pd_{570\pm30}L_{60\pm3}OAc_{180\pm10}O_{180}]$ (L = 1,10-phenanthroline or 2,2'-bipyridine). [50, 51]

There exists today a series of full shell clusters whose members include one shell (M_{13}), two shell (M_{55}), four shell (M_{309}), five shell (M_{561}), seven shell (M_{1415}), and eight shell (M_{2057}) particles. With the seven and eight shell clusters we obviously reach the limit of clearly defined clusters. Although the formulas for the seven and eight shell clusters agree quite well with the experimental data, it must be emphasized that clusters of this size already show a certain particle size distribution and the ideal number of metal atoms cannot be reached in any particle even though the system tries to complete the full shell configuration. Without doubt, these particles represent the transition state to colloidal systems. The characteristics of metal colloids are typically polycrystalline structures and more or less pronounced particle size distributions.

3.3.2 The Characterization of Large Clusters by High Resolution Transmission Electron Microscopy (HRTEM) and by Scanning Tunneling Microscopy (STM)

Without doubt, the most attractive method to structurally characterize ligand stabilized transition metal clusters is X-ray analysis. Many of the smaller and medium sized carbonyl clusters (see Chapter 3.2) as well as those of the sulfide and selenide clusters (see Chapter 3.4) have been investigated by X-ray diffraction. The largest cluster of the sort discussed in this chapter which has been characterized by single crystal X-ray diffraction is $[Au_{39}(PPh_3)_{14}Cl_6]Cl_2$. [36] Suitable single crystals of any larger ligand protected clusters have not been grown for various reasons. On the one hand, there is a pronounced tendency for the very large clusters to decompose in solution. On the other hand, the spheric shape of these huge molecules works against long range order, especially due to the inclusion of solvent molecules in the spacious holes between the cluster molecules. For these reasons, high resolution transmission electron microscopy (HRTEM) has turned out to be the most successful method for the characterization of very large clusters. As a complementary method, scanning tunneling microscopy (STM) has begun to show promising developments.

At the present time, modern electron microscopy can give resolutions of less than 2 Å, which is sufficient in order to characterize cluster cores of the heavier transition metals at atomic resolution. Ligand molecules, of course, cannot be observed. The applications for HRTEM will increase because of the possibility for combining it with other analytical methods. Energy dispersive X-ray microanalysis (EDX) allows the identification of the elements which form the cluster and optical diffractograms give information about the symmetry of the metal core. However, HRTEM does have a disadvantage: since the energy of the electron beam can be as much as 400 keV the particles under investigation are heated up very rapidly and rearrangement and coalescence processes may result. This is especially true for smaller particles where the melting point is a function of the particle size. This fact is of great importance for particles in the 1–4 nm size range where the physical properties of the bulk material are partially lost (such quantum size effects will be discussed in detail in Section 3.3.3). HRTEM can also help in gaining information on the dynamics of crystal growth. Thus, rearrangements at faces, edges, and vertices as well as interactions of atomic clouds with distinct faces can be observed and videorecorded. [52, 53] Figure **3-29** shows a gold crystal with clouds of gold atoms above the (001) plane and Figure **3-30** shows a surface profile image of a growing gold crystal recorded along [110].

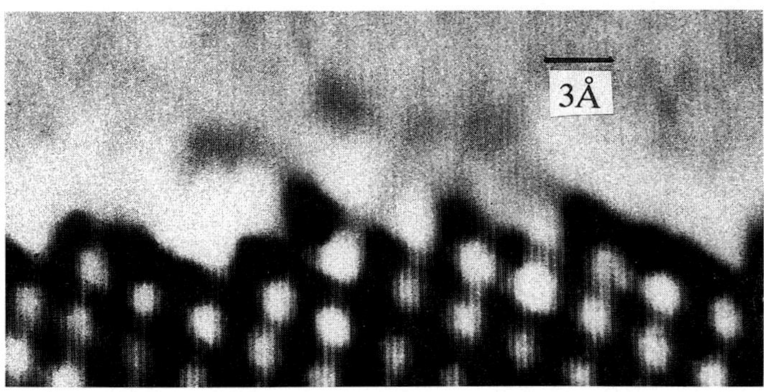

Figure 3-29. A gold surface recorded along [110] showing a cloud of atoms outside a (001) face. Time of registration 0.13s.

In addition to the crystal growth, a new "missing row" reconstruction can be observed in Figure **3-30a-d**. [53]

Besides these fascinating insights into the world of small metal particles, in most cases, HRTEM allows the characterization of metal clusters and metal colloids. The generation of clusters which are isolated from each other can normally be best accomplished from a very dilute solution, whereby a drop of which is put on the grid and the solvent evaporated off. Images which contain many isolated particles allow conclusions to be drawn about the particle size distribution. Highly magnified images of well oriented single clusters give information about the exact size and structure. The procedure will be illustrated by using an exemplary mixture of seven and eight shell palladium clusters. Figure **3-31a** shows a cutout of an amorphous graphite surface which is covered with clusters having sizes of about 3.2 and 3.6 nm. A detailed study of the particles shows that they have mainly monocrystalline character and that they exist in two different particle sizes: those with 15 atomic planes and those with 17 atomic planes (Figure **3-31b**). Thus, the existence of more or less perfect seven and eight shell clusters was verified.

The single cluster shown in Figure **3-31b** consists of 17 rows of cubic close packed palladium atoms and corresponds exactly to a cutout of the lattice of

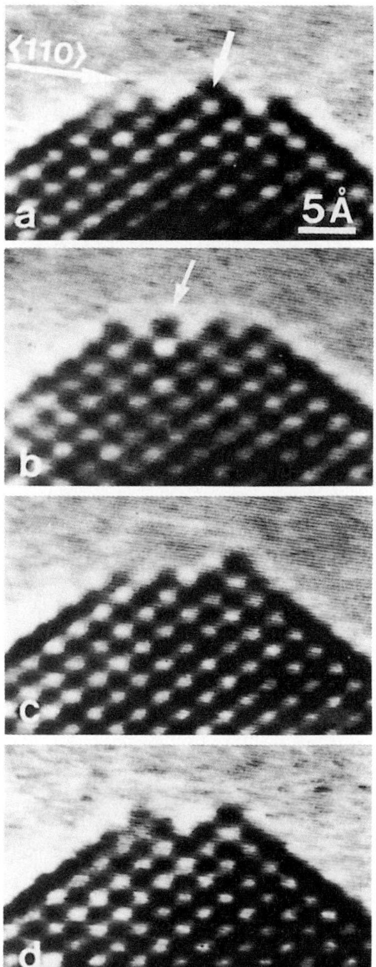

Figure 3-30. Surface profile image of a growing gold crystal recorded along [110]. The left arrow in a) shows a column of only 2–3 atoms. b) 3s later, two additional columns have been formed. Figures c) and d) were recorded within 0.5s and show how a new "missing row" reconstruction is accomplished. During the crystal growth columns are frequently removed.

bulk palladium. The smaller Pd_{561} five shell clusters can be imaged in the same way. An image of the four shell platinum cluster Pt_{309} is shown in Figure **3-32**, whereas in Figure **3-33** a single two shell Pt_{55} cluster at atomic resolution is shown. These images emphasize the value of HRTEM for the study of large clusters if X-ray structure analyses cannot be performed.

Scanning tunneling microscopy (STM) might develop into a useful method which compliments HRTEM. Although the first results are promising, atomic resolution will probably not be reached on more or less spheric cluster molecules. An assembly of ball-like cluster molecules of $[Pd_{561}phen_{36}O_{\sim 200}]$ can be seen in Figure **3-34** and represent the first successful STM attempts of this type. [54, 55] The exact determination of the cluster size is difficult and the particles are probably imaged together with their ligand shell. This assumption is supported by investigations on $[Au_{55}(PPh_3)_{12}Cl_6]$.

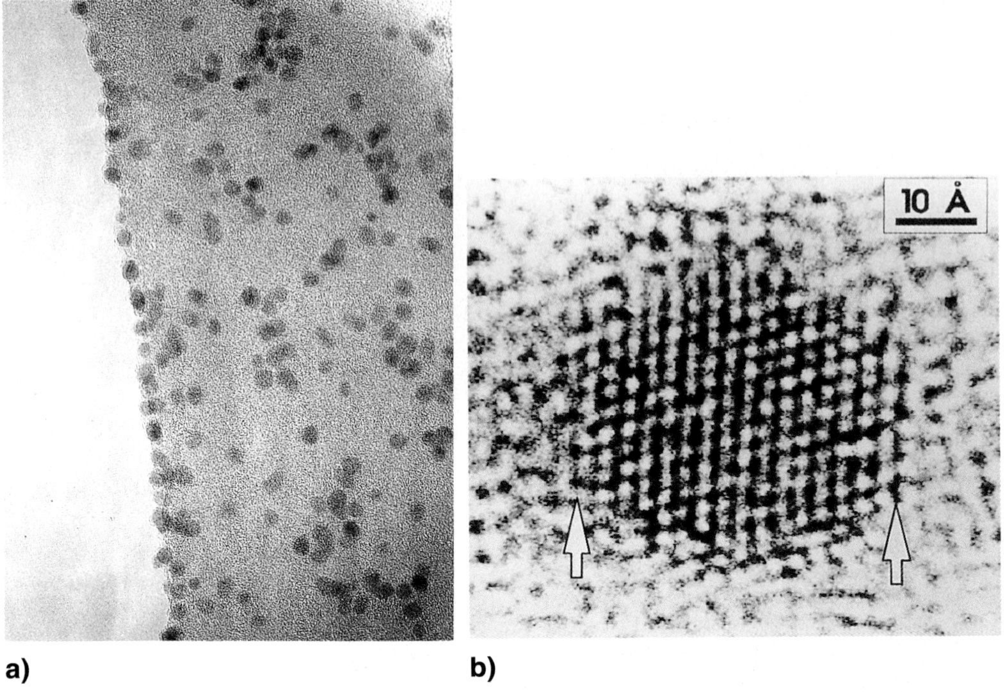

Figure 3-31. a) Image of a mixture of 3.2 and 3.6 nm Pd clusters (Pd7/8). b) A 3.6 nm Pd cluster particle consisting of 17 (111) layers of atoms. The arrows mark the cuboctahedral shape of a cubic close packed structure. The cluster is supported on amorphous carbon.

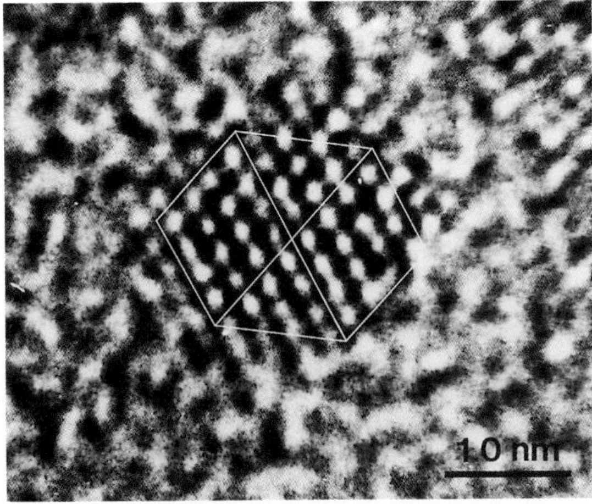

Figure 3-32. High resolution transmission electron micrograph of a four shell Pt_{309} cluster in the [110] direction. As in the bulk, the Pt atoms form a cubic close packed structure. Diameter 1.8 nm.

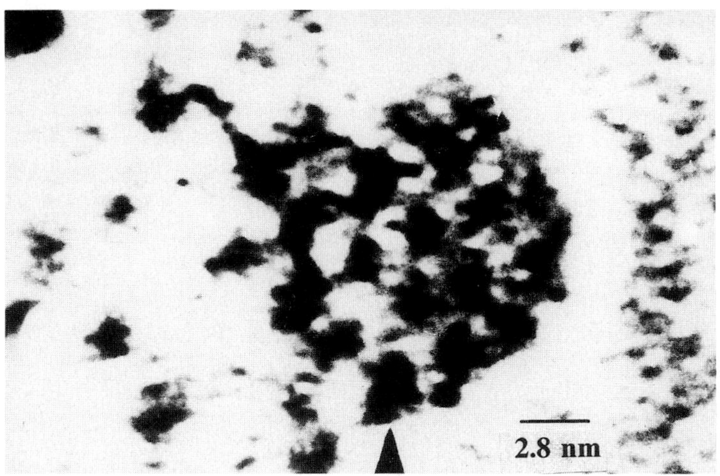

Figure 3-33. A single Pt_{55} cluster particle ([110] direction) showing five rows of cubic packed atoms.

Figure 3-34. Scanning tunnel microscopic image of five shell Pd cluster molecules $[Pd_{561}phen_{36}O_{\sim 200}]$. The ball like molecules are probably imaged together with their ligand shell (see also color plates).

What kind of image of ligand stabilized cluster molecules can be expected if the resolution increases? It can be assumed that the tip of the microscope cannot really come into contact with the gold atoms as they are about 4 Å below the outer atoms of the phenyl rings. Therefore, more or less randomly oriented phenyl rings will probably be imaged. Figure **3-35b** shows an image of a single $[Au_{55}(PPh_3)_{12}Cl_6]$ cluster molecule at "high resolution". Shown in Figure **3-35c** is a sort of twodimensional electron density profile of the cluster surface which was produced by rotating a dot electron model of the cluster molecule into various positions using XP-Software (SHELXTL-Plus). [56] The position which matched the STM image best was fixed and the hidden parts, which cannot be detected by STM, were distinguished. The images shown in Figures **3-35b** and **3-35c** are in reasonable agreement with the model. For a better understanding, a space filling model of $[Au_{55}(PPh_3)_{12}Cl_6]$ is shown in Figure **3-35a**.

It is clear that HRTEM and STM complement each other in a reasonable manner. Where HRTEM gives information on the size and structure of the metal core, STM helps to elucidate ligand orientations. However, it will be difficult to obtain confident results without having a detailed model for comparison.

3.3.3 Physical Properties

When discussing metal clusters which are protected by ligand shells so as to avoid coalescence of the metal particles, it must be remembered that the chemical interaction of the ligands with surface atoms causes electronic changes in the cluster cores. The influence of the ligands will be stronger the smaller the cluster core is. Whereas in a simple mononuclear metal complex it is extremely high, our present knowledge indicates that the influence of the ligand shell on the metal core in large clusters is mainly restricted to the surface atoms. This agrees well with recent results from surface chemistry studies. It has been shown that electron transfer processes between coordinated molecules and metal particles are mainly limited to the corresponding surface atoms. Sections 3.3.3.1–3.3.3.7 describe those physical properties of large transition metal clusters which afford electronic structure information which pertains to the dominating question on the evolution of the metallic state. The accuracy of the information will increase as the sections proceed from 3.3.3.1 to 3.3.3.7.

3.3.3.1 Magnetism

Magnetic susceptibility measurements on three different palladium clusters and colloids will be used as representative of many other results in the field of cluster magnetism and to illustrate that the study of the beginning (or ending) of the metallic state has many inherent difficulties. For example, although we know that bulk properties do not occur spontaneously as a certain number of atoms is reached in a cluster, the appearance of bulk behavior depends on the method of investigation. As we will see later, there exist methods capable of determining single delocalized electrons and to describe the electronic structure of different atoms in a cluster. Others, like those used to determine the magnetic properties discussed in this section, give information on the average collective behavior of the magnetically active electrons.

Palladium is well suited for the study of magnetic behavior. Bulk palladium is often described by a two-band model with an almost filled d-band and a partially filled s-band ($d^{10-x}s^x$). Whereas the s-band is considered to contain almost free electrons, the d-holes move in a narrow band. Palladium has a temperature dependent spin susceptibility and the free spin susceptibility, χ_o, is sensitive to

a)

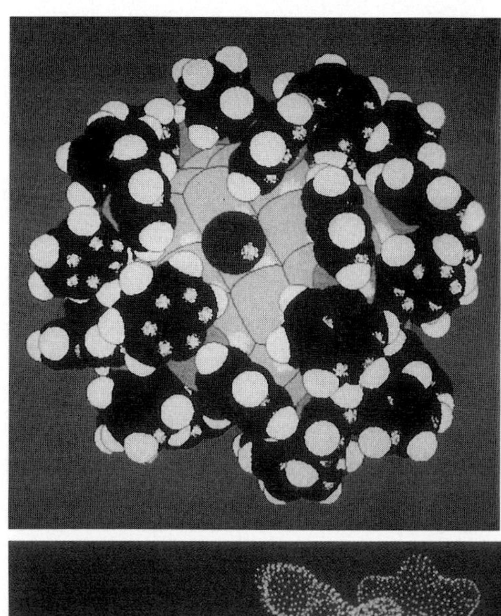

b)

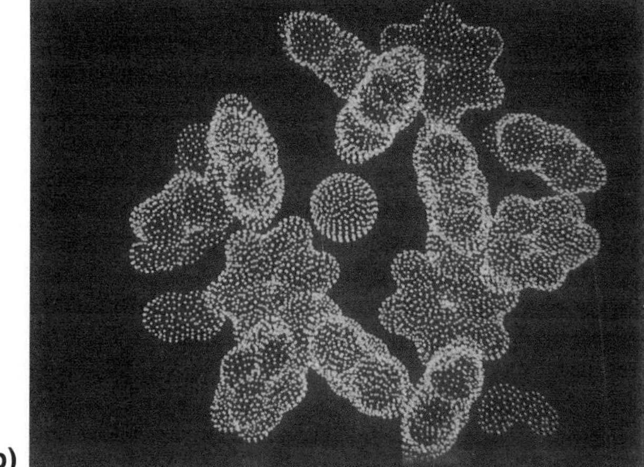

c)

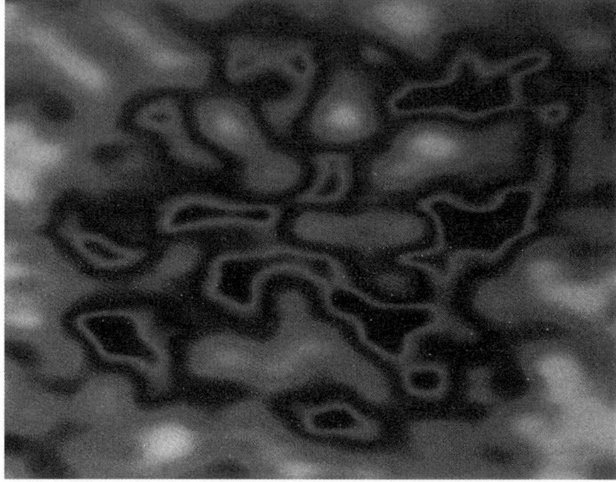

Figure 3-35. a) A computer simulated space filling model of $[Au_{55}(PPh_3)_{12}Cl_6]$. b) A computer simulated "two-dimensional" electron density image of $[Au_{55}(PPh_3)_{12}Cl_6]$ for comparison with c) the STM image of the same cluster molecule in probably the same direction. The similarities between both images are evident. A chlorine atom is positioned in the center of the images (see also color plates).

the density of states near the Fermi level E_F. Due to the complicated band structure near E_F (the density of states vary strongly with energy) as well as the position of E_F, which is close to both a major peak in the density of states and the top of the d-band, the Pauli susceptibility at low temperature is already dependent upon temperature. [57–59]

The density of states in small metallic particles is not a simple function of energy, in that the energies of the molecular orbitals themselves play a decisive role. For instance, the average electron level separation, ΔE, for the cluster $[Pd_{561}phen_{36}O_{\sim 200}]$ can be estimated to be on the order of 20K, according to the expression $\Delta E = 2E_F/N$ (N = number of s and d valence electrons). For thermal energies smaller than ΔE, quantum size effects (QSE) might be expected, however, spin-orbit couplings, which are appreciable in palladium, will drastically reduce QSE. Figure **3-36** shows the temperature dependent susceptibility of various ligand stabilized palladium particles compared to bulk Pd.

The cluster compounds of $[Pd_{561}phen_{36}O_{\sim 200}]$, a mixture of the seven and eight shell clusters $[Pd_{1415}phen_{60}O_{\sim 1100}]$ and $[Pd_{2057}phen_{84}O_{\sim 1600}]$ (see Section 3.3.1), and a ligand stabilized 15 nm Pd colloid have been used in the comparison with bulk palladium. [60] Diamagnetic corrections may incur relatively large errors, as one cannot start out from an exact stoichiometric composition of these large particles. Despite these insufficiencies, the development towards the typical bulk curve is apparent in Figure **3-36** and several observations can be noted. There is a substantial reduction in the susceptibility with decreasing cluster size and the temperature dependence of the susceptibility is reduced in the same direction. The small upturns at very low temperature, which can be observed for all particles except bulk Pd, must be attributed to paramagnetic impurities. A detailed analysis of these measurements would exceed the frame of this section. Very qualitatively, one can discuss the main results in terms of a model which is related to models that have been used to understand NMR results on small platinum particles.

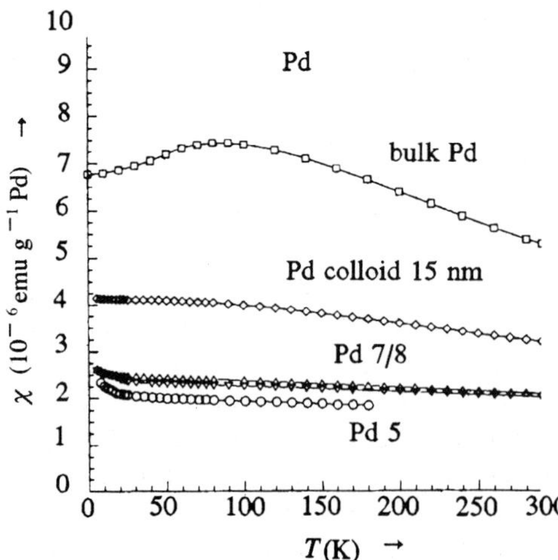

Figure 3-36. The temperature dependence of the susceptibility of various Pd particles compared to bulk palladium. The values are corrected for their diamagnetic contributions and normalized to the estimated weight of the Pd cores.

If carbon monoxide is chemisorbed on the surface of small platinum particles, the ^{195}Pt NMR spectra indicate that the local density of states, $D^d(r)$, decreases in an exponential manner from the bulk towards the surface. [61] The so-called Knight shift, K, as measured by NMR, consists of a K^s for the s-electrons and a K^d for the d-electrons. K^d is proportional to the local susceptibility, $\chi^d_P(r)$, on the nucleus. This local d-susceptibility is porportional to the local density of states, $D^d(r)$. In our model, atoms at different distances from the surface have different $D^d(r)$ and hence different Knight shifts. In the magnetic experiments described here, the susceptibilities of a macroscopic sample have been determined. This, of course, gives less detailed information than an NMR experiment, and so the NMR model must be somewhat simplified.

One can assume that at the surface of the cluster particles, the local density of states is reduced by a relative amount A compared to the bulk:

$$D^d(\text{surface}) = (1-A)D^d(\text{bulk}) \tag{3:1}$$

For spherical particles of radius R, the local density of states is

$$D(r) = D^s + D^d(\text{bulk})(1-Ae^{(r-R)/\lambda}) \tag{3:2}$$

where $R-r$ is the distance from the surface. The decrease in the density of states at the surface is $A = 0.32$ and the exponential decay constant of this decrease is $\lambda = 0.68$ nm.

Due to the low number of data points, the values for A and λ which give a best fit to χ are charged with large uncertainty. The relation between the susceptibility and the density of states is given by

$$\chi_o = \mu_o\mu_B{}^2(2/3\ D^s + \overline{D}^d) + \chi_{vv} + \chi_{dia} \tag{3:3}$$

The use of χ_o indicates that the so-called Stoner enhancement factor has not been included (the Stoner theory is a mean-field theory for itinerant electrons). In Equation (3:3), $\overline{D}^d$ is the averaged value of the varying D^d over the particles, χ_{vv} is the Van Vleck paramagnetism, and χ_{dia} is the Langevin (or Larmor) diamagnetism.

From Equation (3:2), it follows that the density of states, $D(r)$, depends exponentially on the distance from the surface, and Equation (3:3) describes the relation between the susceptibility and the density of states. Figure **3-37** illustrates the results impressively.

The curve follows the fit of the model describing the size dependence. The three experimental points match the curve quite well and we can now understand why even 15 nm palladium colloids show magnetic behavior still far from that of the bulk. Even for a 1 micron sample, one can expect a reduction in χ of about 1%! Values for A and λ obtained from different experiments for similarly sized particles agree quite well with the results discussed here. [61–64] Finally, we see that the magnetic properties of ligand stabilized particles, even if they are very huge particles like colloids, differ considerably from those found in the bulk. The

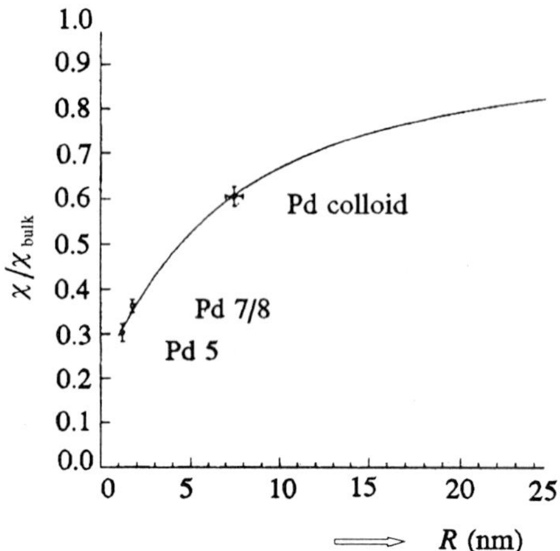

Figure 3-37. The susceptibility values of Pd5, Pd7/8 and 15 nm Pd colloids extrapolated to T = 0 and the best fit curve of the model describing the size dependence.

model used here to describe the experimental results helps to give insight into the physics involved. However, due to the very gradual change in χ as the radius of changes the particle this method is not very well suited to investigating the electronic details associated with the evolution of the metallic state.

3.3.3.2 Nuclear Magnetic Resonance (NMR)

Nuclear magnetic resonance spectroscopy is a powerful method to investigate ligand behavior on the cluster surfaces in solution. Although it can provide us with information about the ligand-metal bond, fluxional behavior, etc., in this section we will describe one example of a NMR investigation which was concerned with the metal core itself. [65, 66] The cluster $[Pt_{309}phen^*_{36}O_{\sim 30}]$ is well suited for an investigation by ^{195}Pt NMR spectroscopy in the solid state. As already mentioned, it consists of a four shell structure of ccp packed Pt atoms enveloped by the phen* ligands. Numerous ^{195}Pt experiments on small platinum particles have been carried out to date. [67–76] Platinum is best suited for such experiments because it exhibits one of the largest Knight shifts (K) of any metal and thus enables one to study the changing electronic properties when going from the bulk down to clusters or *vice versa*. As already described in the previous section, the local density of states at the Fermi level of the 5*d* electrons is an exponentially decreasing function of the distance from the center to the surface. This causes more or less large Knight shift gradients. A Knight shift in the NMR is considered to be one of the hallmarks of the metallic state and gives a very clear indication as to whether a metal cluster shows metallic behavior or not. Figure **3-38** shows the solid state ^{195}Pt NMR spectrum of $Pt_{309}phen^*_{36}O_{\sim 30}$ at 77 K. It clearly indicates the absence of a bulk signal which is to be expected at H_o/ν_o = 1.138 G kHz^{-1}.

On the other hand, two separate peaks can be distinguished. The low field peak at 1.096 G kHz^{-1} is typical for ^{195}Pt NMR signals in platinum complexes.

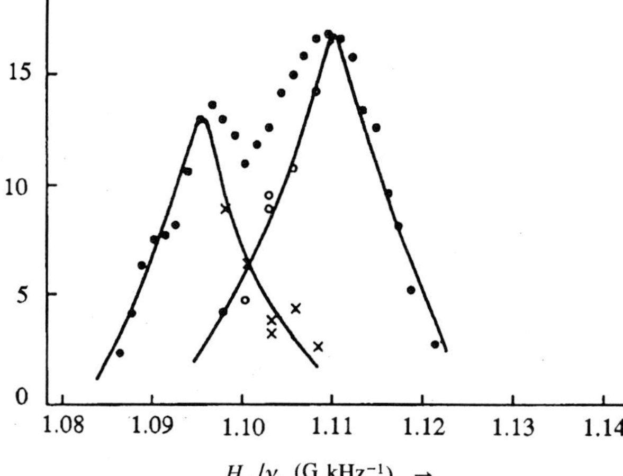

Figure 3-38. ^{195}Pt NMR lineshape of $[Pt_{309}phen^*_{36}O_{\sim 30}]$ at 77K. Black dots: spin echo intensities, corrected for T_1 and T_2; open circles: intensities of the short T_1 component; crosses: intensities of the long T_1 component. Note the absence of a bulk peak at H_o/ν_o = 1.138 G kHz.$^{-1}$

The high field peak at 1.110 G kHz^{-1} is caused by platinum atoms in a metallic environment. Simple calculations yield a metallic particle whose volume is represented by a sphere with a radius of about 8.7 Å. This agrees quite well with the radius of an inner cluster core consisting of 147 atoms. From the location of the metallic peak one can conclude that the density of states is remarkably reduced compared with bulk platinum, however, the Knight shift of the observed peak clearly indicates metal like character for the inner core of the cluster.

3.3.3.3 Extended X-Ray Absorption Fine Structure (EXAFS)

EXAFS studies have been carried out on mostly Au_{55} cluster nuclei. [77–79] They help to characterize the cluster cores by determining the coordinating environment of the metal atoms. AuL_3-edge extended X-ray absorption fine structure spectroscopy is well suited for elucidating important structural details in various gold cluster compounds where X-ray structure analyses fail. The nearest neighbor gold–gold distance in $[Au_{55}(PPh_3)_{12}Cl_6]$ at 80 K is 2.75 ± 0.01 Å. This represents a considerable contraction compared with the value of 2.875 Å in bulk gold at 80 K. [80] Previous calorimetric measurements on the same cluster compound, which showed that the two center gold–gold bonding energy is about 25% stronger than in bulk gold, [81] support these findings. The most important result is that there is only a single nearest neighbor Au–Au distance and shows that the Au_{55} cluster core has a cubic close packed structure. Icosahedral structures like those in the smaller gold clusters Au_9, Au_{10}, Au_{11}, and Au_{13} (see Section 3.3.1) or even polyicosahedral constructions like those in the mixed Au/Ag systems can be definitely ruled out. In contrast to these results, measurements on the icosahedral $[Au_{11}\{Ph_2P(p\text{-}ClC_6H_4)\}_7I_3]$ clearly show two different Au–Au distances of 2.66 and 2.87 Å, as must be expected for this structure type. Second and third nearest neighbor Au–Au distances also agree with the ideal cuboctahedral model of the

Au_{55} cluster. It is of interest that the crystallographically characterized gold cluster compound $[Au_{39}(PPh_3)_{14}Cl_6]Cl_2$ [36] shows a hexagonal close packed structure. It may be regarded as an intermediate stage on the way from the small icosahedral clusters to the bulk like ccp structure of $[Au_{55}(PPh_3)_{12}Cl_6]$.

The EXAFS studies also enabled the determination of the Au–P and the Au–Cl distances. The Au–P distance of 2.30 ± 0.02 Å agrees well with Au–P distances in structurally well characterized compounds. The same is true for the Au–Cl distance of 2.50 ± 0.01 Å. An EXAFS investigation on the related cluster compound $[Au_{55}\{Ph_2P(m\text{-}C_6H_4SO_3Na)\}_{12}Cl_6]$ gave very similar results. The nearest neighbor Au–Au distance is 2.79 ± 0.01 Å. The outer neighbor distances of 3.97 ± 0.03 and 4.90 ± 0.02 Å indicate the same cubic close packed arrangement of gold atoms as found in $[Au_{55}(PPh_3)_{12}Cl_6]$. The sulfonation of one of the phenyl rings of the phosphine ligands obviously does not cause a major effect on the geometry of the cluster core.

EXAFS measurements on the five shell cluster $[Pd_{561}phen_{36}O_{\sim 200}]$ revealed a nearest neighbor Pd–Pd distance of 2.73 ± 0.01 Å. In contrast to the smaller Au_{55} two shell cluster, the difference in the bond lengths between the cluster and the bulk material is small (2.744 Å at 80 K) as might be expected because of the more well developed metallic behavior associated with the larger number of atoms. Outer neighbor Pd–Pd distances are found to be equal to $\sqrt{2}$, $\sqrt{3}$, and $\sqrt{4}$ times the nearest neighbor distance and so prove the cubic close packed structure of the Pd_{561} cluster nucleus. The related cluster $[Pd_{570}phen_{60}OAc_{180}O_{180}]$ was originally described to have an icosahedral structure, [50] however, it now looks as though a ccp structure is also possible. [82] A ccp structure of the Pd_{561} core in $[Pd_{561}phen_{36}O_{\sim 200}]$ was already proposed from X-ray powder diffraction studies on a microcrystalline sample, as the characteristic but broadened reflections of the bulk palladium were observed. [46]

3.3.3.4 ^{197}Au Mössbauer Spectroscopy of Au_{55} Clusters

The Mössbauer effect spectroscopy provides an ideal microscopic tool to investigate the chemical environment of metal atoms in cluster molecules via their isomer shifts (I. S.) and their quadrupole splittings (Q. S.) Since the ccp structured Au_{55} core consists of chemically inequivalent gold atoms, these can be distinguished by means of their I. S. and Q. S. Figure **3-39a**, illustrates the different gold sites present and Figure **3-39b** shows the Mössbauer spectra in the 30–1.25 K temperature range.

According to the full shell model, the Au_{55} cluster core consists of a central Au atom surrounded by 12 further Au atoms which form the inner core of the Au_{55} nucleus. The 42 outer shell atoms are divided into 12 which bear phosphine ligands, 6 which link to chlorine atoms, and 24 which are uncoordinated surface gold atoms. In addition to the temperature dependence and the absorption intensity results, the Mössbauer spectra also give information on the Debye-Waller factors (*f*-factors or Mössbauer fractions) which are affected by intracluster vibrations whose values can also be calculated. [83] Since inequivalent gold atoms can

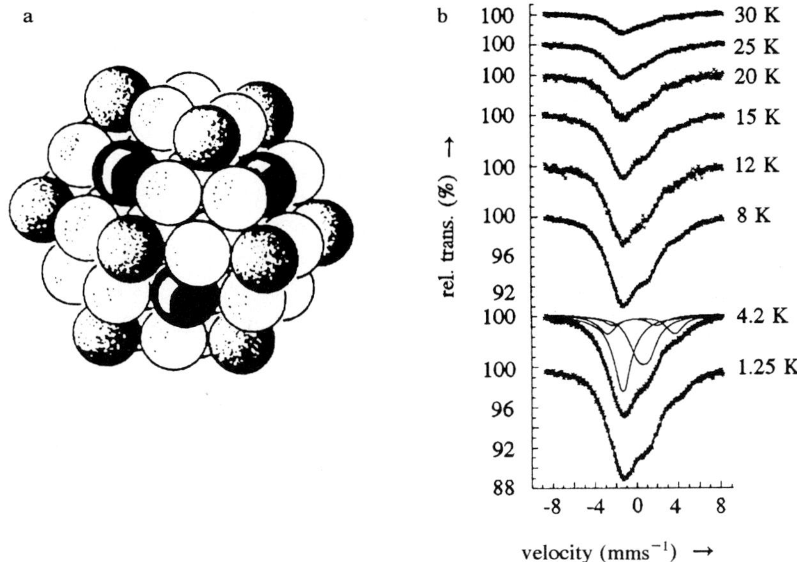

Figure 3-39. a) Schematic representation of the Au_{55} cluster core showing the various gold sites. b) ^{197}Au Mössbauer spectra of $[Au_{55}(PPh_3)_{12}Cl_6]$ in the 30 – 1.25 K temperature range. The spectrum at 4.2 K is fitted to the hyperfine parameters. [86]

be distinguished in the Mössbauer spectra of the Au_{55} clusters, the Debye-Waller factors of the different sites can be studied separately. The analysis of each Mössbauer spectrum at temperatures between 1.25 K and 30 K reveals one unsplit Mössbauer absorption line from the 13 core atoms and three quadrupole split (Q. S.) doublets from the three distinct surface sites. The isomer shift of the inner 13 gold atoms in the Au_{55} cluster nucleus is -1.4 ± 0.1 mm s^{-1}, very close to that observed for bulk gold (-1.224 mm s^{-1}) but clearly shifted from it. [84–87] In any case, it indicates an electronic environment of the inner cluster core similar to that in metallic gold. The inner Au atom in $[Au_{11}(PPh_3)_7(SCN)_3]$, for instance, has a completely different I. S. of +2.5 mm s^{-1} [83] which arises from an electronic charge transfer to the positively charged surface atoms. In the Au_{55} cluster, the core atoms are close to the oxidation state zero. The difference of about 0.2 ms^{-1} compared with bulk gold can be explained by a slightly lower 6*s* electron density. In Table **3-14** are the corresponding data listed.

The effective Einstein temperatures, Θ_E, are associated with the vibrations of the different types of gold atoms and have been calculated from the *f*-factors of the Mössbauer spectra. There is a distinct correlation between the coordination number of the different gold atoms and Θ_E. Figure **3-40** shows the temperature dependence of the *f*-factors together with the measured *f*-factors *vs.* the temperature of the different Au sites in $[Au_{55}(PPh_3)_{12}Cl_6]$.

It can be seen in Figure **3-39b** that the shapes of the spectra at different temperatures are identical over the temperature range and indicate that the intracluster vibrations remain constant. However, the total intensities of the spectra

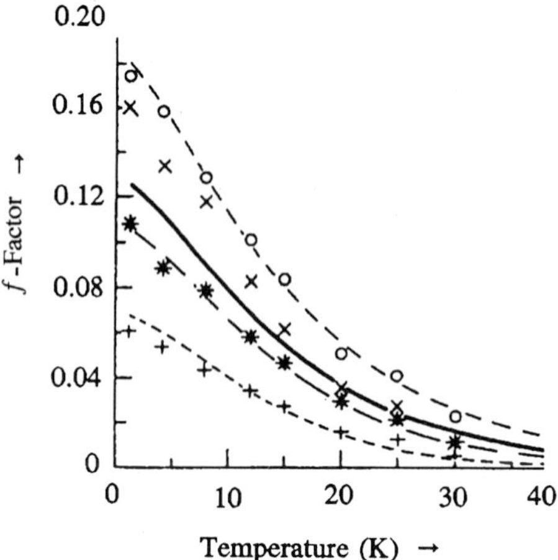

Figure 3-40. The temperature dependence of the Mössbauer fraction of $[Au_{55}(PPh_3)_{12}Cl_6]$. Circles: core atoms; crosses: Cl-bonded Au atoms; stars: unbounded surface atoms; plus signs: PPh_3 bonded Au atoms.

decrease drastically with increasing temperature and is a result of the varying intercluster vibrations.

As already mentioned, there are two other Au_{55} cluster molecules which have substituted phenyl rings on the phosphine ligands. The Mössbauer spectra of $[Au_{55}\{P(p\text{-tolyl})_3\}_{12}Cl_6]$ have also been measured in order to study the influence of these slightly chemically changed ligands on the electronic character of the Au_{55} nucleus. The spectra generally correspond to those of $[Au_{55}(PPh_3)_{12}Cl_6]$, but there are small differences in all the I. S. and the Q. S. values. There is an indication that the equivalent of about 1/2 a 6*s* electron partially transferred to 5*d* and/or 6*p* orbitals, is added to the Au_{55} core, compared with the PPh_3 ligated cluster. From a chemical point of view, these findings make sense, since $P(p\text{-tolyl})_3$ is a slightly stronger σ-donor ligand than PPh_3, due to the +I effect of the methyl substituents.

Table 3-14. The gold atom sites in $[Au_{55}(PPh_3)_{12}Cl_6]$, their multiplicities, coordination numbers, isomer shifts (I. S.), quadrupole splitting (Q. S.), and Einstein temperatures (Θ_E).

Au site	multiplicity	Coordination number	I. S. (mm s^{-1})	Q. S. (mm s^{-1})	Θ_E (K)
core	13	12	−1.4	0.0	136
bare surface	24	7	+0.3	1.4	99
PPh_3-bonded	12	5	+0.6	7.1	77
Cl-bonded	6	8	+0.1	4.4	119
bulk gold		12	−1.224	0.0	123

3.3.3.5 Photoelectron Spectroscopy (XPS)

Similar to the isomer shifts in Mössbauer spectroscopy, the results from photoelectron spectroscopy are sensitive to the local chemical environment of an atom. [88] However, whereas Mössbauer spectroscopy reflects the initial electronic structure of the system under investigation, XPS probes the state after elimination of an electron. [84] Furthermore, whereas the Mössbauer isomer shift gives informations on the *s* electron density within the nuclear volume, XPS results describe the overall electronic charge density of an atom. As such, Mössbauer spectroscopy and XPS compliment each other in an ideal manner.

Comparison of the data from XPS investigations on $[Au_{55}(PPh_3)_{12}Cl_6]$ with XPS data from some smaller gold cluster compounds and from bare gold on different substrates [89–90] gave three main results. Just as in the Mössbauer spectra, four different types of gold atoms could be identified in $[Au_{55}(PPh_3)_{12}Cl_6]$, each with its own 4f binding energy. The 5d spin-orbit splitting is greater than in the smaller gold clusters and it is close to the value for bulk gold. Figure **3-41a** shows the curve-fitting of the Au 4f peaks. [85] The four components are located at 84.0, 84.4, 85.3, and 86.3 eV and can be assigned to the bare surface sites, the core atoms, the PPh_3 coordinated gold atoms, and the Cl bonded sites respectively. Comparing these results with those from $[Au_{11}(PPh_3)_7Cl_3]$ and similar clusters, it can be seen that the central gold atoms in the Au_{55} core are less positively charged than in the smaller clusters, in agreement with the Mössbauer data. The Au 4f binding energy for the core gold sites in $[Au_{55}(PPh_3)_{12}Cl_6]$ is 0.4 eV higher than in bulk gold. EXAFS experiments described in Section 3.3.3.3 have already shown that the Au-Au distances in Au_{55} are significantly smaller than in bulk gold.

As repeatedly mentioned, one of the dominating questions concerned with cluster physics is that of the beginning (or ending) of the metallic state. The XPS valence band results clearly show that the Au 5*d* band in $[Au_{55}(PPh_3)_{12}Cl_6]$ closely resembles that of the 5d band in bulk gold. As can be seen in Figure **3-41b**, there is a strong indication of a finite density of states at the Fermi level (arrow), which suggests a metallic character in the Au_{55} cluster. Additionally, the 2.4 eV energy gap between the $5d_{5/2}$ and $5d_{3/2}$ components is close to that found in the bulk, and considerably larger than the 1.9 eV for the smaller Au_{11} cluster. Bare gold clusters of about the same size deposited on poorly conducting substrates show very similar results.

To summarize the XPS investigations on $[Au_{55}(PPh_3)_{12}Cl_6]$, one can say that all atoms of the Au_{55} cluster core are involved in the metallic binding, or in other words, the valence electrons are completely delocalized.

3.3.3.6 UV-Visible Spectroscopy

Freely mobile conduction electrons normally show a characteristic collective oscillation frequency. This plasma resonance can be observed as an absorption band in the UV-visible spectra of metal colloids. With decreasing particle size

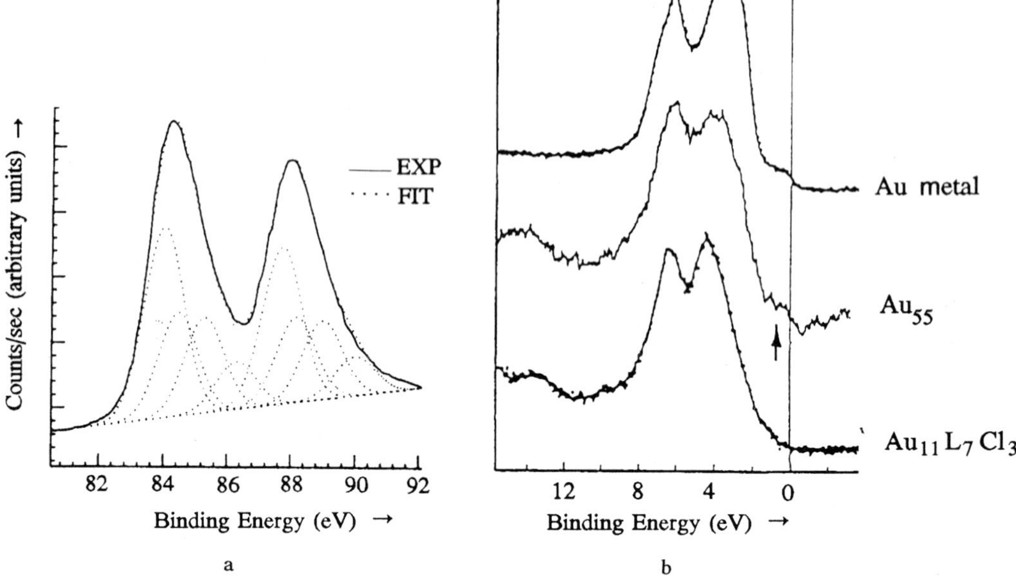

Figure 3-41. a) The Au $4f_{7/2}$ and Au $4f_{5/2}$ XPS peaks of $[Au_{55}(PPh_3)_{12}Cl_6]$ and the four site fit (dotted lines). With increasing binding energy the spectral lines correspond to the bare surface sites, the core sites, the PPh_3-bonded Au atoms, and the Cl-bonded Au atoms. b) XPS valence-band spectrum of $[Au_{55}(PPh_3)_{12}Cl_6]$, compared with the spectra of bulk gold and of $[Au_{11}(PPh_3)_7Cl_3]$. The arrow shows the position of the Fermi edge.

these absorptions weaken. The main feature of the UV-visible spectrum of $[Au_{55}(PPh_3)_{12}Cl_6]$ in solution is a broad absorption extending across the whole visible region from about 250 nm to 1350 nm and the spectrum of the water soluble $[Au_{55}\{Ph_2P(m\text{-}C_6H_4SO_3Na)\}_{12}Cl_6]$ is almost identical. [92] This findings differ completely from those of low nuclearity cluster molecules [93] and demonstrate that the Au_{55} clusters have electronic energy levels which are closely related to a developing band structure. Although similar to gold colloids, an expected plasma resonance which could be compared with colloidal behavior could not be established. [94]

The most probable explanation for the lack of a plasma resonance in Au_{55} is the damping and broadening of the resonance, as a result of size dependent, single electron 5d → 6p,6s interband transitions. Consequently, the Mie bandwidth is roughly double that of what would be predicted, that is the plasmon band is smeared out completely. [94]

Other interpretations of the UV-visible spectroscopy experiments are less plausible as they sometimes contradict other results. Furthermore, it should be mentioned that a plasma resonance absorption has a third power dependence on the particle diameter. Thus, even if the electrons perform a collective motion in the 1 nm particle, the absorption might be too weak to be recognized above the background.

Absorption maxima in the UV-visible region cannot be assigned to pure plasma resonance as the investigated metals have a smaller free electron behavior than gold, due to their considerable interband transition contributions. Using the Mie theory, the theoretical positions of the absorption maxima for various metals have been calculated. [95] Attempts to identify absorptions in the larger [$Pt_{309}phen^*_{36}O_{30}$] and [$Pd_{561}phen_{36}O_{\sim 200}$] have failed. [96] Only the $\pi \rightarrow \pi^*$ transition in the aromatic ligands at 292 nm for the platinum cluster and at about 230 nm for the palladium cluster are observed. Thus, at present, there is still no definite example of a clear plasma resonance absorption in a ligand stabilized cluster molecule.

3.3.3.7 Conductivity Measurements and Impedance Spectroscopy (IS)

In 1988, the so-called size induced metal insulator transition (SIMIT) was described. [97] It will be observed if the volume of a metal particle is reduced to such an extent that size dependent quantization effects occur. This is the beginning of the metal to semiconductor transition found at the very end of metallic behavior or, coming from the molecular state, the onset of metallic behavior. The SIMIT is already effective in 20 nm particles. Due to this quantization effect, standing electron waves with discrete energy levels are formed.

All the results from conductivity measurements on concentrated ligand stabilized transition metal clusters to date show an intercluster tunnel conductivity. The d. c. conductivity σ_{dc} *vs.* inverse temperature plots for the two different sized clusters [$Au_{55}(PPh_3)_{12}Cl_6$] and [$Pd_{561}phen_{36}O_{\sim 200}$] are shown in Figure **3-42**. [98] As can be seen, the conductivity increases with increasing temperature and, as could be expected, the temperature dependence of the smaller Au_{55} cluster is significantly more pronounced than for Pd_{561}. This frequency dependent intercluster conductivity is explained by a stochastic "multiple site" hopping mechanism, a percolation process in which electrons tunnel between cluster molecules [99–101]. It has been predicted that this multiple site process will be transformed into a "two-site" mechanism at higher frequencies. The conductivity measurements illustrated in Figure **3-42** do not give any indication for the separation of inter- and intracluster effects.

Low frequency impedance spectroscopy (IS) is a well suited method for the study of atomic processes by macroscopic measurements. It has not only been used to understand ionic relaxation processes in interface layers and in solid matrices but, at lower frequencies, it has allowed the observation of electronic relaxation processes. As a consequence of the SIMIT, metal particles in the microcluster size range should show novel electronic properties if their dimensions reach the de Broglie wave length, λ, of their valence electrons. Ligand stabilized metal clusters can be regarded as quantum dots in which freely mobile electrons are trapped. In the following, the "quantum box" [$Au_{55}(PPh_3)_{12}Cl_6$] is described as the very first example of a practical working quantum device building block. The primary difficulty in the application of metal clusters to microelectronic zero dimensional (0D) quantum devices is to isolate the neighboring

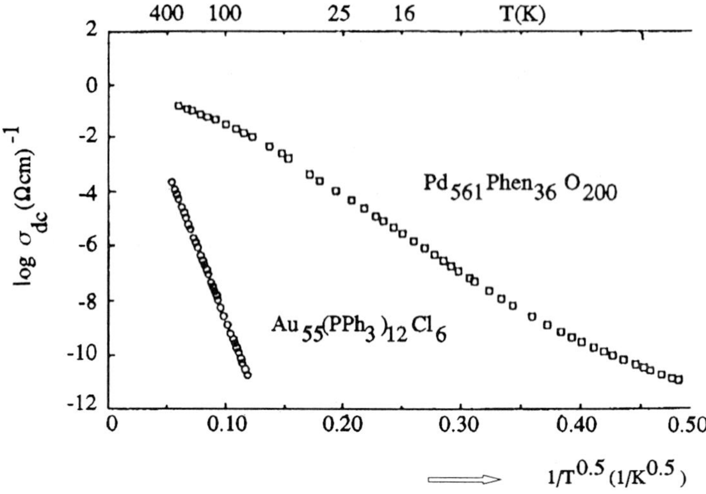

Figure 3-42. The DC conductivity σ as a function of temperature for $[Au_{55}(PPh_3)_{12}Cl_6]$ and $[Pd_{561}phen_{36}O_{\sim 200}]$.

quantum dots from one another and, at the same time, to contact them with one another. These conditions must be fulfilled if a system of quantum dots should act in a one, two or three dimensional working quantum device. Surprisingly, these conditions can be at least partially fulfilled in a relative simple manner: a powdery [56, 102, 103] material of $[Au_{55}(PPh_3)_{12}Cl_6]$ can be used to form disks having a 5 mm diameter and a 0.2–0.4 mm thickness by applying 2.5×10^8 Pa of pressure, whereby the gravimetrically determined density of the silicon like looking sample is 3.37 g cm^{-3}. Compared with the calculated density of 3.704 g cm^{-3} for a perfect close packed arrangement, this is an acceptable condition for the use of the disk as an approximately close packed assembly of the cluster molecules. As we will see later, some disorders in the sample do not disturb at all.

IS measurements have been performed in the 253–333 K temperature range and the results from the measurement at 293 K are shown in Figure **3-43**. The Argand diagram consists of two semicircles which can be understood and represented by a best fit of the data to an ideal Debye resistance (R_2)/capacitance (C_2) parallel link (lower frequency process **2**) in series with an additional Cole-Cole resistance (R_1)/CPE (constant phase element) parallel link (higher frequency process **1**). From the fit data, a Cole-Cole parameter $\alpha \approx 2/3$ results which is very typical for percolation mechanisms. The Ohmic part is the same for both processes: $R_1 \approx R_2$ and so $R_{total} = 2R_1$. The specific conductivity processes have $\sigma_1 = \sigma_2 = 3.23 \times 10^{-4}\ \Omega^{-1}m^{-1}$ at 293 K. Since both processes are thermally activated, they can be used to calculate the activation enthalpies EA_1 and EA_2. Thus, the Debye process activation enthalpy is $EA_2 = 0.16 \pm 0.03$ eV and the Cole-Cole process has $EA_1 = 0.15 \pm 0.03$ eV which indicates that $EA_2 \approx EA_1$ within experimental error. Since the relaxation times, τ_1 and τ_2, and the resistan-

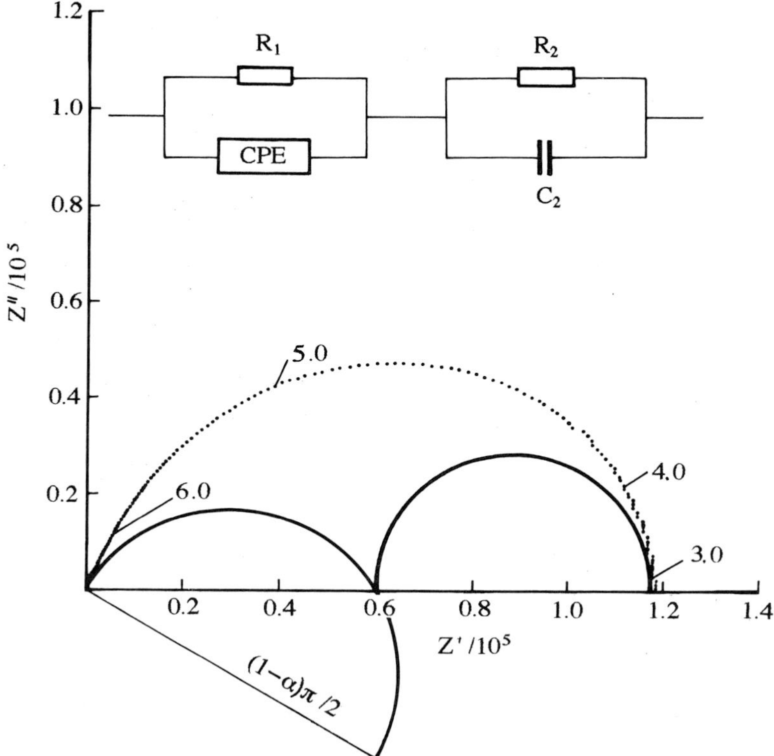

Figure 3-43. Result (dotted line) of an impedance measurement on $[Au_{55}(PPh_3)_{12}Cl_6]$ at 293 K. In the unscaled Argand diagram the real part of the impedance Z' is plotted versus the imaginary part of the impedance Z'' in the complex plane. The numbers along the plot indicate the logarithm of the measuring frequency ν. In the upper section, an equivalent circuit representing the electric behavior of the sample is shown. Below the measured plot are the Argand circles belonging to each separate link (solid lines).

ces, R_1 and R_2, show the same Arrhenius behavior, it follows from $\tau = R \times C$ that the capacitances C_2 and CPE must be temperature independent. As already mentioned, the frequency dependent intercluster conductivity has been explained by means of a stochastic "multiple site" hopping mechanism which should turn into a "two site" process at higher frequencies. Therefore we can now interpret the higher frequency Cole-Cole process as the relaxation of an intercluster process. This explanation is supported by the existence of a distributed CPE capacitance instead of a single capacitance. CPE is caused by a distribution in the jump times and in the distances between the clusters. The transfer of single electrons from one cluster nucleus to a neighboring one occurs by a tunneling through the fluctuating ligand shells.

The Debye process **2** must be interpreted as an intracluster step which is coupled with the intercluster process **1**. The immitance spectra show this Debye process as a sharp resonance, whereas the intercluster Cole-Cole process causes a broad signal. Some important conclusions can be drawn from these experimental data. The Au_{55} cluster acts as a quantum box at the very end of the SIMIT. The Pauli principle restricts the number of electronic states, N_{ex}, participating in the resonance process to be either 1 or 2 with energies E_{ex1} or E_{ex2}. These energies can be calculated from the capacitance C_2 = 48 pF and the experimental density of electronic states D_{ex} = 10.75 eV^{-1} respectively. Using the formula for 3D electron gas in a quantum box given in Equation (3:4):

$$x = \hbar\pi(2m\ E_F)^{-1/2} \tag{3:4}$$

where m = electron mass, E_F = Fermi energy, and x = lateral dimension of the quantum box, the two possible energies which result are E_{ex1} = 0.093 eV (E_{F1}) and E_{ex2} = 0.186 eV (E_{F2}). For E_{ex1} x = 2.01 nm and for E_{ex2} x = 1.42 nm. As we know, the diameter of the Au_{55} cluster nucleus is 1.4 nm. So, not only could the diameter of the Au_{55} core be determined independently, but it is shown that the 1.4 nm gold quantum box contains the last electrons which participate in the conduction processes. The density of states (DOS) at the Fermi level for a 3D electron gas $D(E_F)$ = $(m/\hbar^2)x^2$ with x = 1.42 nm can now be calculated as 4.40 eV^{-1} per cluster.

One of the important effects which is based on the quantum behavior of $[Au_{55}(PPh_3)_{12}Cl_6]$ is shown in Figure **3-44**. The total conductivity, σ_{total}, is doubled under resonance conditions such that σ_1 = σ_2 = $2\sigma_{total}$. At resonance, two (!) $[Au_{55}(PPh_3)_{12}Cl_6]$ cluster molecules can act as a molecular switch. The particle

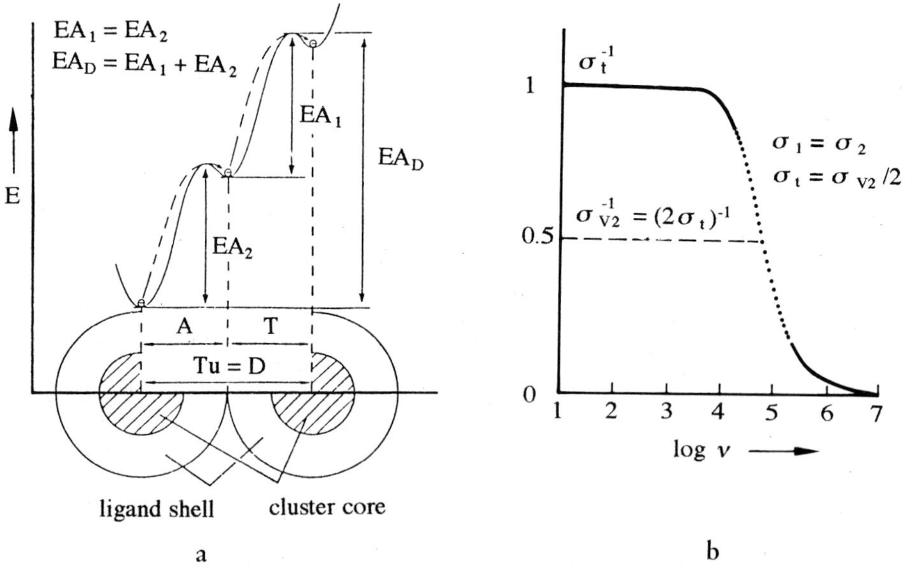

Figure 3-44. a) Scheme of the potential profile of a cluster pair. Activation step (A) and transport step (T) cause the tunneling step (T_u) which is equal to a disproportionation step (D) having an activation enthalpy EA_D. b) A sequence of both steps leading to a total conductivity σ_t which is doubled under resonance conditions to $\sigma_{1/2}$.

size implicates that two electrons occupy a kind of 'cluster valence orbital', whereas the intercluster conductivity is determined by single electron tunneling (SET) enabled by these 'valence electrons'. Even at room temperature, compact Au_{55} pellets are handy devices with quantum properties. The total disc can act as a tunnel resonance resistance (TRR), the resistance of which is divided in two under resonance conditions. The smallest units of a TRR would have the dimensions of two cluster molecules, that is about 4 nm. However, such minimized building blocks in quantum devices cannot yet be addressed. The resistance of a single cluster is coupled to that of a channel of cluster molecules in the dense packed pellet, or, in other words, the disc of cluster molecules consists of 1D quantum channels of correlated electrons. Figure **3-45** illustrates these conditions. Disorders which occur in individual clusters or in the disc package, or the occurrence of single holes, do not seriously disturb the function of the device. If single quantum dots or channels fail, they are either passed over or repaired through by-passes.

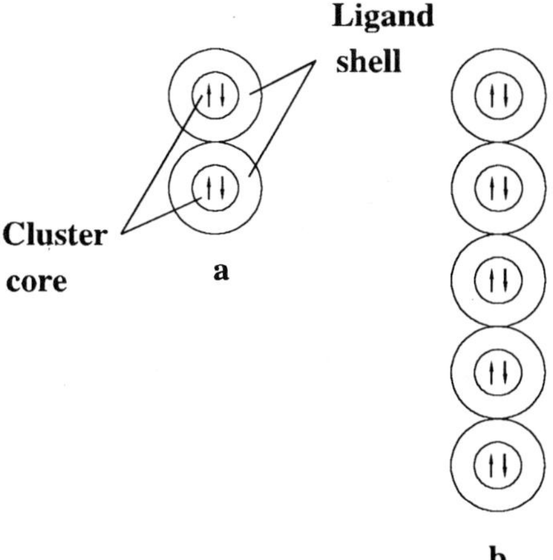

Figure 3-45. a) A smallest quantum device consisting of only two $[Au_{55}(PPh_3)_{12}Cl_6]$ clusters. This is also the smallest functional unit of a Tunnel Resonance Resistance (TRR). b) 1 D quantum channel, section of closest sphere packing in a compressed disk of $[Au_{55}(PPh_3)_{12}Cl_6]$ cluster molecules.

There seems to be a simple phase relation between the macroscopic frequency applied to the cluster sample and the microscopic frequency of the two electrons within the cluster cores: $\nu_{D\ macro} = 2\,|\,x\,|\,\nu_{D\ micro}$ (D = Debye). Using the macroscopic frequency $\nu_{D\ macro}$ for the Debye process a microscopic frequency of 10^{13}–10^{14} Hz results. This frequency does not depend on the geometry of the sample! If $\nu_{D\ macro}$ is increased by a factor 2 to ν_C (C = Cole-Cole), then for one of the two electrons, the microscopic frequency $\nu_{C\ micro}$ of the first excited state of the electron gas can be reached via a corresponding geometric phase coupling in which $\nu_{C\ micro} = 2\nu_{D\ micro}$ and $\nu_{C\ macro} = 2\nu_{D\ macro}$, as already mentioned above. Figure **3-46** illustrates these fundamental relations. If the energy of an excited single electron is higher than the Coulombic barrier, then it cannot be localized in the metal core further and will leave it through the ligand shells. This quantum mechanical tunneling causes oscillations in the charge of single electrons $\nu_{C\ micro}$. Over the

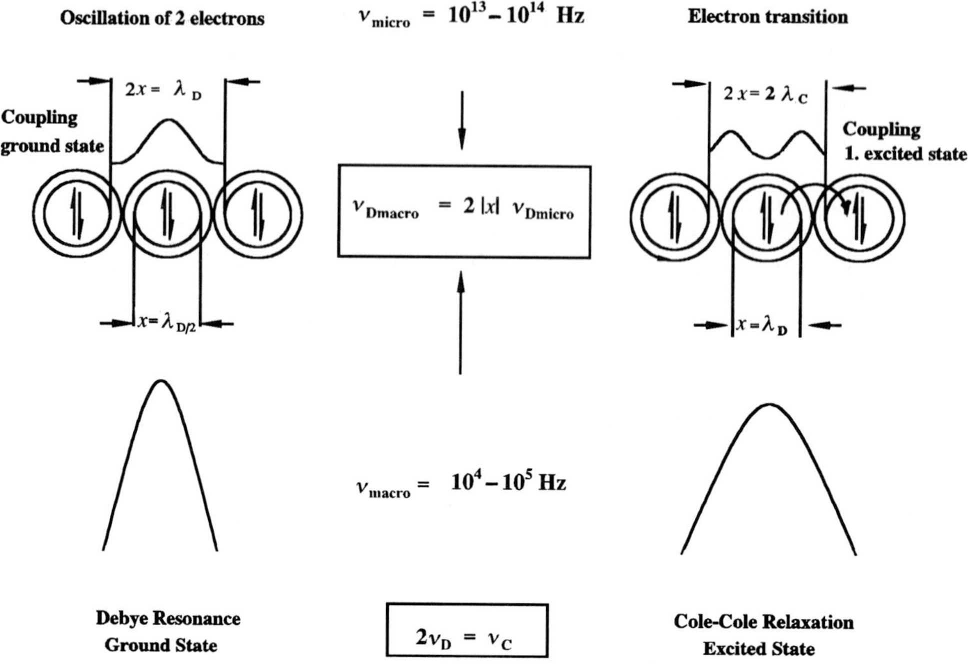

Figure 3-46. Explanation of the phase coupling problem between macro and micro frequencies. A doubling of the ground state frequency from $\nu_{\mathrm{D\ macro}}$ leads to an excited state (ν_{C}) where single electrons can tunnel through the ligand shells. The macroscopic frequency $\nu_{\mathrm{D\ macro}}$ is coupled with the intracluster microscopic frequency $\nu_{\mathrm{D\ micro}}$ simply by the diameter of the quantum box. The same is valid for $\nu_{\mathrm{C\ micro}}/\nu_{\mathrm{C\ macro}}$. The ν_{micro} is in the range of 10^{13}–10^{14} Hz, which falls in the visible and infrared region.

whole sample they are collectivized to $\nu_{\mathrm{C\ macro}} = 2\,|\,x\,|\,\nu_{\mathrm{C\ micro}}$, which is about 5×10^8 tunnel contacts per meter.

In the ground state, the sharp resonance of two localized electrons is observed which can also be collectivized over the whole sample by phase coupling. The difference between this energy quantized single electron tunneling (QSET) and the usual SET [104, 105] is the fact that QSET uses only single electrons with quantized states of energy, whereas SET starts out from an infinite number of electrons. The capacitance of a QSET unit is about 8×10^{-19} F. The application of the known SET materials has been limited up to now for a variety of reasons, for example:

- the difficulty in preparing miniaturized tunneling contacts
- the cooling to temperatures in the mK range
- the measurement of very small currents.

Such problems may not exist with QSET devices. Thus, $[Au_{55}(PPh_3)_{12}Cl_6]$ is a miniaturized tunneling contact, it functions at room temperature, and furthermore $I_{\mathrm{macro}} = I_{\mathrm{micro}} = 6 \times 10^{-5}$ A. The energy difference between the ground state and the first excited state lies between $\Delta E = 0.75 - 2$ eV at frequencies $\Delta E/h \approx 1.5 - 4 \times 10^{14}$ Hz, due to the wide band characteristics of CPE. Since this is in the visible and infrared region, optical effects can also be expected, although they have not yet been investigated in detail.

3.3.4 Chemical Properties

The chemistry of the large ligand stabilized clusters described in this section is, strictly speaking, rather limited. This is mainly due to their pronounced tendency to decompose in solution. As in many simple complexes, some of the ligands are dissociated and this aids coalescence processes between cluster particles which finally lead to the formation of precipitates. Higher temperatures promote these events. The extent of lability, of course, depends on both the metal and the ligand. Numerous ^{31}P-NMR investigations on M_{55} clusters, but especially on $[Au_{55}(PPh_3)_{12}Cl_6]$, in solution have shown that the phosphine ligands are highly fluxional. [106] Dissolved PPh_3 molecules exchange with coordinated phosphines. The average contact time with the cluster surface has been determined to be 3μs.

Ligand exchange reactions on $[Au_{55}(PPh_3)_{12}Cl_6]$ have also been studied. The water soluble phosphine $Ph_2P(m\text{-}C_6H_4SO_3Na)$ will substitute all PPh_3 ligands on the cluster surface to give $[Au_{55}\{Ph_2P(m\text{-}C_6H_4SO_3Na)\}_{12}Cl_6]$ which is, consequently, also water soluble. The stability of this cluster in aqueous solution is even higher and is probably due to the formation of a twelve fold negative anion by the dissociation of the Na^+ ions. Using ion exchange, the sodium cations may be substituted by protons.

Differential Scanning Calorimetry (DSC) measurements on $[Au_{55}(PPh_3)_{12}Cl_6]$ reveal a sharp exothermic decomposition signal at 156 °C. [81] Preparative thermolysis of the compound in solution results in the quantitative stoichiometric reaction (3.35):

$$[Au_{55}(PPh_3)_{12}Cl_6] \xrightarrow{\Delta T} 6[(Ph_3P)_2AuCl] + 49\ Au \qquad (3.35)$$

The formation of $[(Ph_3P)_2AuCl]$ is also observed in the DSC diagram by its endothermic melting point signal at 201 °C. The DSC investigations allowed the heat of decomposition at 156 °C to be evaluated as being 112 Jg^{-1} or 1590 $kJmol^{-1}$ of cluster. A nearest neighbor Au-Au interaction has been estimated being as 76.1 kJ. Compared with the bulk value, this is an increase of 14.8 kJ and agrees well with the EXAFS results, which indicated a shortening of the Au-Au distances compared with bulk gold (see Section 3.3.3.3).

Au_{55} cluster degradation and the formation of so-called superclusters (clusters of clusters) were observed during TOF-SIMS investigations (Time Of Flight Secondary Ion Mass Spectrometry). [107, 108] This method uses highly energetic Xe^+ ions in the kV range to bombard a solid cluster sample. The Xe^+ ions degrade the cluster molecules into various Au-ligand fragments which can be observed in the low mass range, and into more or less naked Au_{13} clusters arising from the inner cluster cores. Apparently, these Au_{13} particles coalesce into $[(Au_{13})_n]^+$ superclusters as they pass through the solid sample. The most important masses are summarized in Table **3-15**. There are three series of SI peaks observed in the positive spectra. The masses increase with the thickness of the sample. It is of interest that the masses 33.300 u and 139.000 u correspond to the superclusters $[(M_{13})_{13}]$ and $[(M_{13})_{55}]$ respectively, which represent full shell clusters of full shell clusters.

These results agree well with electrochemical experiments. If dichloromethane solutions of different M_{55} clusters are contacted to Pt electrodes to which 20 V dc is applied, the cluster molecules are degraded as a result of the contact with the electrodes. [109] Polarization effects may be the reason for the decomposition. Electrophoresis is observed without any indication of cluster decomposition if the platinum electrodes dip into water layers covering the organic phase in a U-tube. The black, thermodynamically unstable microcrystalline products formed on the Pt surfaces have been identified by X-ray powder diffraction to be novel $[(M_{13})_n]$ metal modifications. The results from the diffraction experiments indicate a structure consisting of cubic close packed M_{13} clusters which are linked via their triangular faces to form a kind of pseudo close packed structure with M_{13} clusters as building blocks.

The application of transition metal clusters in catalysis has been repeatedly prognosticated to become very important, however, these predictions have not been realized on the whole. Early attempts to use large transition metal clusters as homogeneous catalysts failed due to their instability, yet the use of small metal particles on supports has broad applications. [56] Numerous industrial processes are based on the catalytic activity of metal clusters and colloids. The simple separation of products from a heterogeneous catalyst and its repeated use are large advantages compared with homogeneous catalysts. The stereoselectivity in heterogeneous catalysis, however, is often less satisfactory than in homogeneous processes.

Table 3-15. Measured and Calculated Values[a] for the Peak Centers of $(Au_{13})_n^+$ in the Positive SIMS Spectra of $[Au_{55}(PPh_3)_{12}Cl_6]$.

first series			second series			third series		
measd	calcd	*n*	measd	calcd	*n*	measd	calcd	*n*
7700	7682	3	12800	12803	5	17900	17924	7
18000	17924	7	22500	23045	9	38400	38408	15
33300	33287	13	43500	43530	17	58900	58893	23
			84500	84499	33	79400	79378	31
						99500	99862	39
						119000	120347	47
						139000	140831	55

[a] In unified atomic mass units.

Ligand stabilized large clusters, as described in this chapter, have been under investigation for only a short time. Compared with the usual metal catalysts, which are generated on a support, ligand stabilized clusters can be prepared and characterized separately. In addition, they have a uniform size which may be of importance for it is known that metal particles in the 1–3 nm size range have special catalytic activity. [110] Since most of the clusters described in this chapter are within this range, they can be expected to show an increased activity over conventional catalysts with wide size distributions. Initial attempts with Rh_{55} clusters, chemisorbed on TiO_2 or zeolites, seem to fulfill these expectations in hydroformylation reactions. [111] The hydroformylation of propene to butanal using a 1% (by weight) catalyst of $[Rh_{55}\{P(t\text{-}Bu)_3\}_{12}Cl_{20}]$ on TiO_2 in an aqueous suspension at 100–120 °C and 300–100 bar proceeds at turnover rates between 400–600 $mol_{Propene}mol_{Rh}^{-1}min^{-1}$, clearly proving an extraordinary activity for the catalyst. However, the result must be viewed with caution for the ratio of *n*:*i*-butanal is 1:1, whereas the homogeneous rhodium catalyzed reaction yields about 90% of the preferred *n*-butanal. Another aspect is the eventual degradation of the Rh clusters by carbon monoxide to form soluble rhodium carbonyl compounds which can then take part in homogeneous catalysis. Such processes have been observed when the reaction takes places in an organic phase.

BET measurements and HRTEM investigations of cluster doted TiO_2 show some very interesting details. The chemisorption of only 1% weight of the cluster material on TiO_2 leads to a complete decrease of the BET (Brunauer, Emmet and Teller) activity of the support. Micrographs of such doted materials show that islands from small assemblies of 6–10 cluster molecules are formed on the TiO_2 surface. After heating this material at 200 °C for several hours, subsequent BET measurements show the original activity of adsorption. HRTEM investigations of the heated material reveal that the original cluster molecules have left their positions and have formed large aggregates on the TiO_2 surface. These observations lead to the conclusion that chemisorption of the cluster molecules from solution probably occurs around the entrances of the micropores. Their diameter of 8.5–10.5 Å is considerably smaller than the cluster diameter of about 20 Å, including the ligand shell, so that the cluster molecules quantitatively close the micropores. At temperatures of *ca.* 200 °C, the clusters begin to move and in doing so open the micropores again. DSC measurements confirm the desorption of the cluster molecules at about 200 °C.

The easy accessibility of ligand stabilized seven and eight shell palladium clusters makes them attractive candidates for hydrogenation reactions. Indeed, they can be chemisorbed on TiO_2, Al_2O_3, zeolites, or activated carbon from aqueous solutions and so form ideal heterogeneous hydrogenation catalysts. Various reactions prove that the activity again has increased compared with commercially available systems. The oxygen, which covers free surface atoms in addition to the phenanthroline ligands, is removed by the hydrogen in the starting period of the reaction. It can be assumed that the actual catalyst is a palladium-hydrogen system. The capability of a TiO_2 supported Pd catalyst (1% weight) shall be illustrated by means of the hydrogenation of 2-hexyne. The reaction occurs at room temperature and 1 atm of hydrogen pressure and is clearly heterogeneous. As can

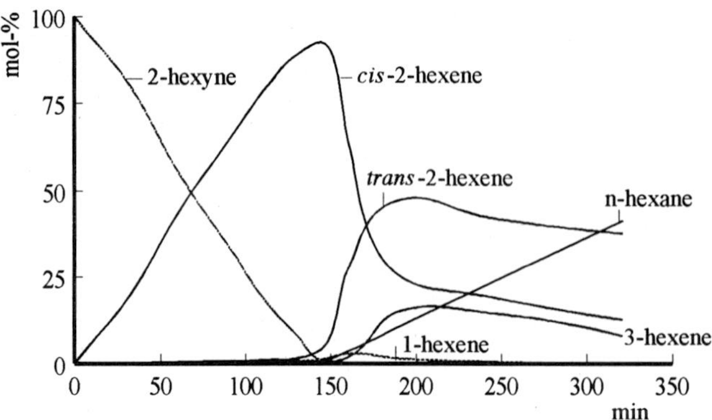

Figure 3-47. Hydrogenation of 2-hexyne with Pd7/8 on TiO_2 (1% weight) and H_2 at normal pressure and room temperature. During the first 2 hours, the 2-hexyne is almost quantitatively transformed into *cis*-2-hexene. Only then does the formation of *trans*-2-hexene and other isomers begin.

be seen from the linear ascent in Figure **3-47**, the hydrogenation of 2-hexyne follows a zero order rate law with respect to the educt concentration. This enables the calculation of the turnover frequency (TOF) for the formation of *cis*-2-hexene to be TOF = 75 mol $min^{-1}mol_{Pd}^{-1}$ with a selectivity of 95%. Compared with the best Pd catalyst used in practice, this corresponds to an increase in activity to about the threefold value with the same selectivity. Only after the semihydrogenation of 2-hexyne to *cis*-2-hexene has occured does the hydrogenation to *n*-hexene start. Whilst *trans*-2-hexene and the 3-hexenes appear as intermediates, 1-hexene is observed only in traces.

These very preliminary results of catalytic reactions with supported large ligand stabilized clusters are promising. The present experiments indicate a considerably increased activity compared with usual systems. For the present, one may interpret these results in terms of the uniformity of the cluster molecules used because the influence of the ligands is still not understood.

References

[1] L. Malatesta, L. Naldini, G. Simonetta, F. Cariat, *Chem. Commun.* **1965**, 212.
[2] P.C. Bellon, F. Cariati, M. Manassero, L. Naldini, M. Sansoni, *J. Chem. Soc., Chem. Commun.* **1971**, 1423.
[3] M. McPartlin, R. Mason, L. Malatesta, *J. Chem. Soc., Chem. Commun.* **1969**, 334.
[4] J.W.A. van der Velden, J.J. Bour, W.P. Bosman, J.H. Noordik, *J. Chem. Soc., Chem. Commun.* **1981**, 1218.
[5] M. Manassero, L. Naldini, M. Sansoni, *J. Chem. Soc., Chem. Commun.* **1979**, 385.
[6] F.A. Vollenbroek, W.P. Bosman, J.H. Noordik, J.J. Bour, P.T. Beurskens, *J. Chem. Soc., Chem. Commun.* **1979**, 387.
[7] M.K. Cooper, G.R. Dennis, K. Henrick, M. McPartlin, *Inorg. Chim. Acta* **1980**, *45*, L 151.
[8] K.P. Hall, B.R.C. Theobald, D.I. Gilmour, D.M.P. Mingos, A.J. Welch, *J. Chem. Soc., Chem. Commun.* **1982**, 528.
[9] C.E. Briant, K.P. Hall, D.M.P. Mingos, *J. Chem. Soc., Chem. Commun.* **1984**, 290.

[10] J. G. M. van der Linden, M. L. H. Paulissen, J. E. J. Schmitz, *J. Amer. Chem. Soc.*, **1983**, *105*, 1903.
[11] C. E. Briant, K. P. Hall, A. C. Wheeler, D. M. P. Mingos, *J. Chem. Soc., Chem. Commun.* **1984**, 248.
[12] M. McPartlin, R. Mason, *Chem. Commun.* **1969**, 334.
[13] V. G. Albano, P. L. Bellon, M. Manassero, M. Sansoni, *Chem. Commun.* **1970**, 1210.
[14] P. Bellon, M. Manassero, M. Sansoni, *J. Chem. Soc. Dalton Trans.* **1972**, 1481.
[15] F. Cariati, L. Naldini, *Inorg. Chim. Acta* **1971**, *5*, 172.
[16] J. M. M. Smits, P. T. Beurskens, J. W. A. van der Velden, J. J. Bour, *J. Cryst. Spectrosc.* **1983**, *13*, 373.
[17] W. Bos, R. P. F. Kanters, C. J. Van Halen, W. P. Bosman, H. Behm, J. M. M. Smits, P. T. Beurskens, J. J. Bour, L. H. Pignolet, *J. Organomet. Chem.* **1986**, *307*, 385.
[18] C. E. Briant, B. R. C. Theobald, J. W. White, C. K. Bell, D. M. P. Mingos, *J. Chem. Soc., Chem. Commun.* **1981**, 201.
[19] J. W. A. van der Velden, F. A. Vollenbroek, J. J. Bour, P. I. Beurskens, J. M. M. Smits, W. P. Bosman, *Rec. J. R. Neth. Chem. Soc.* **1981**, *100*, 148.
[20] F. A. Vollenbroek, P. C. P. Bouten, J. M. Trousten, J. P. van den Berg, J. J. Bour, *Inorg. Chem.* **1978**, *17*, 1345.
[21] R. P. F. Kanters, P. P. J. Schlebos, J. J. Bour, W. P. Bosman, H. J. Behm, J. J. Steggerda, *Inorg. Chem.* **1988**, *27*, 4034.
[22] R. P. F. Kanters, P. P. J. Schlebos, J. J. Bour, W. P. Bosman, J. M. M. Smits, P. T. Beurskens, J. J. Steggerda, *Inorg. Chem.* **1990**, *29*, 324.
[23] J. J. Bour, P. P. J. Schlebos, R. P. F. Kanters, W. P. Bosman, J. M. M. Smits, P. T. Beurskens, J. J. Steggerda, *Inorg. Chim. Acta* **1990**, *171*, 177.
[24] J. J. Steggerda, *Comments Inorg. Chem.* **1990**, *11*, 113.
[25] R. P. F. Kanters, P. P. J. Schlebos, J. J. Bour, J. Wijnhoven, E. van der Berg, J. J. Steggerda, *J. Organomet. Chem.* **1990**, *388*, 233.
[26] M. F. J. Schoondergang, J. J. Bour, G. P. F. van Strijdonck, P. P. J. Schlebos, W. P. Bosman, J. M. M. Smits, P. T. Beurskens, J. J. Steggerda, *Inorg. Chem.* **1991**, *30*, 2048.
[27] J. J. Bour, P. P. J. Schlebos, R. P. E. Kanters, M. F. J. Schoondergang, H. Addens, A. Overweg, J. J. Steggerda, *Inorg. Chim. Acta* **1991**, *181*, 195.
[28] J. J. Bour, W. v. d. Berg, P. P. J. Schlebos, R. P. F. Kanters, M. F. J. Schoondergang, W. P. Bosman, J. M. M. Smits, P. T. Beurkens, J. J. Steggerda, P. van der Sluis, *Inorg. Chem.* **1990**, *29*, 2971.
[29] R. P. F. Kanters, P. P. J. Schlebos, J. J. Bour, W. P. Bosman, J. M. M. Smits, P. T. Beurskens, J. J. Steggerda, *Inorg Chem.* **1990**, *29*, 324.
[30] M. F. J. Schoondergang, J. J. Bour, P. P. J. Schlebos, A. W. P. Vermeer, W. P. Bosman, J. M. M. Smits, P. T. Beurskens, J. J. Steggerda, *Inorg. Chem.* **1991**, *30*, 4704.
[31] G. Beuter, A. Brodbeck, M. Holzer, S. Maier, J. Strähle, *Z. anorg. allg. Chem.* **1992**, *616*, 27.
[32] G. Beuter, J. Strähle, *Angew. Chem. Int. Ed. Engl.* **1988**, *27*, 1094.
[33] J. Mielcke, J. Strähle, *Angew. Chem. Int. Ed. Engl.* **1992**, *31*, 464.
[34] G. Beuter, J. Strähle, *Z. Naturforsch.* **1989**, *44b*, 647.
[35] G. Beuter, J. Strähle, *J. Organomet. Chem.* **1989**, *372*, 67.
[36] B. K. Teo, X. Shi, H. Zhang, *J. Amer. Chem. Soc.* **1992**, *114*, 2743.
[37] B. K. Teo, K. Keating, *J. Amer. Chem. Soc.* **1984**, *106*, 2224.
[38] B. K. Teo, H. Zhang, X. Shi, *Inorg. Chem.* **1990**, *29*, 2083.
[39] B. K. Teo, X. Shi, H. Zhang, *J. Amer. Chem. Soc.*, **1991**, *113*, 4329.
[40] B. K. Teo, M. Hong, H. Zhang, D. Huang, *Angew. Chem. Int. Ed. Engl.* **1987**, *26*, 897.
[41] B. K. Teo, M. Hong, H. Zhang, D. Huang, X. Shi, *J. Chem. Soc., Chem. Commun.* **1988**, 204.
[42] B. K. Teo, H. Zhang, X. Shi, *J. Amer. Chem. Soc.* **1990** *112*, 8552.
[43] G. Schmid, *Inorg. Syntheses* **1990**, *7*, 214.
[44] G. Schmid, U. Giebel, W. Huster, A. Schwenk, *Inorg. Chim. Acta* **1984**, *85*, 97.

[45] G. Schmid, W. Huster, *Z. Naturforsch.* **1986**, *41b*, 1028.
[46] G. Schmid, B. Morun, J.-O. Malm, *Polyhedron* **1988**, *7*, 2321.
[47] G. Schmid, B. Morun, J.-O. Malm, *Angew. Chem. Int. Ed. Engl.* **1989**, *28*, 778.
[48] G. Schmid, *Mater. Chem. Phys.* **1991**, *29*, 133.
[49] G. Schmid, M. Harms, J.-O. Malm, J.-O. Bovin, J. van Ruitenbeck, H. W. Zandbergen, W. T. Fu, *J. Amer. Chem. Soc.* **1993**, *115*, 2046.
[50] M. N. Vargaftik, V. P. Zagarodnikov, I. P. Stolyarov, I. I. Moiseev, V. I. Likholobov, D. I. Kochubey, A. L. Chuvilin, V. I. Zaikowski, K. I. Zamaraev, G. I. Timofeeva, *J. Chem. Soc., Chem. Commun.* **1985**, 937.
[51] M. N. Vargaftik, I. I. Moiseev, D. I. Kochubey, K. I. Zamaraev, *Faraday Discuss* **1991**, *92*, 13.
[52] J.-O. Bovin, L. R. Wallenberg, D. J. Smith, *Nature* **1985**, *317*, 47.
[53] L. R. Wallenberg, J.-O. Bovin, *Naturwissenschaften* **1985**, *72*, 539.
[54] L. E. C. van de Leemput, J. W. Gerritsen, P. H. H. Rongen, R. T. M. Smokers, H. A. Wierenga, H. van Kempen, G. Schmid, *J. Vac. Sci. Technol.* **1991**, *B9*, 814.
[55] H. A. Wierenga, L. Soethout, J. W. Gerritsen, B. E. C. van de Leemput, H. van Kempen, G. Schmid, *Adv. Mater.* **1990**, *2*, 482.
[56] G. Schmid, *Chem. Rev.* **1992**, *92*, 1709.
[57] E. W. Elock, P. Rhodes, A. Teviodtdale, *Proc. Roy. Soc. (London)* **1954**, *A53,* 53.
[58] K. L. Liu, A. H. MacDonald, J. M. Daams, S. H. Vosko, *J. Magnetism and Magn. Mat.* **1979**, *12*, 43.
[59] W. Sänger, J. Voitländer, *Z. Phys.* **1978,** *B30*, 13.
[60] D. A. van Leeuwen, J. M. van Ruitenbeek, G. Schmid, L. J. de Jongh, *Physics Letters* **1992**, *A170*, 325.
[61] C. D. Makowa, C. P. Slichter, J. H. Sinfelt, *Phys. Rew.* **1985**, *B31*, 5663.
[62] N. Inoue, T. Sugawara, *J. Phys. Soc. Japan* **1978**, *45*, 450.
[63] G. K. Wertheim, S. B. DiCenzo, D. N. E. Buchanan, *Phys. Rev.* **1986**, *B33*, 5384.
[64] S. Ladas, R. A. Dalla Betta, M. Boudart, *J. Catalysis* **1978**, *53*, 356.
[65] D. van der Putten, H. B. Brom, L. J. de Jongh, G. Schmid *in Physics and Chemistry of Finite Systems: From Clusters to Crystals* (Eds.: P. Jena, S. N. Khanna, B. K. Rao), Kluwer Acad. Publ. **1992**, P. 1007.
[66] D. van der Putten, H. B. Brom, J. Wittereen, L. J. de Jongh, G. Schmid, *Z. Physik D, in press.*
[67] I. Yu, A. A. V. Gibson, E. R. Hunt, W. P. Halperin, *Phys. Rev. Lett.* **1980**, *44*, 348.
[68] I. Yu, W. P. Halperin, *J. Low Temp. Phys.* **1981**, *45*, 189.
[69] H. E. Rhodes, P.-K. Wang, H. T. Stokes, C. P. Slichter, J. H. Sinfelt, *Phys. Rev.* **1982**, *B26*, 3559.
[70] H. E. Rhodes, P.-K. Wang, C. D. Makowka, S. L. Rudaz, H. T. Stokes, C. P. Slichter, J. H. Sinfelt, *Phys. Rev.* **1982**, *B26*, 3569.
[71] T. Stokes, H. E. Rhodes, P.-K. Wang, C. P. Slichter, J. H. Sinfelt, *Phys. Rev.* **1982**, *B26*, 3575.
[72] J. J. van der Klink, J. Buttet, M. Graetzel, *Phys. Rev.* **1984**, **B29**, 6352.
[73] B. J. Pronk, H. B. Brom, A. Ceriotti, G. Longoni, *Solid State Comm.* **1987**, *64*, 7.
[74] C. D. Makowka, C. P. Slichter, J. H. Sinfelt, *Phys. Rev.* **1985**, *B31*, 5663.
[75] J. P. Bucher, J. J. van der Klink, *Phys. Rev.* **1988**, *B38*, 11038.
[76] J. P. Bucher, J. Buttet, J. J. van der Klink, *Surf. Sci.* **1989**, *214*, 347.
[77] M. C. Fairbanks, R. E. Benfield, R. J. Newport, G. Schmid, *Solid State Commun.* **1990**, *73*, 431.
[78] M. A. Marcus, M. P. Andrews, J. Zegenhagen, A. S. Bommannavar, P. Montano, *Phys. Rev.* **1990**, *B42*, 3312.
[79] P. D. Cluskey, R. J. Newport, R. E. Benfield, S. J. Gurman, G. Schmid, *Z. Phys. D, in press.*
[80] J. Donohoe in *The Structure of the Elements* Wiley, New York, **1974**, *224*, p. 216.
[81] R. E. Benfield, J. A. Creighton, D. G. Eadon, G. Schmid, *Z. Phys. D* **1989**, *12*, 533.

[82] M. N. Vargaftik, I. I. Moiseev, D. I. Kochubey, K. I. Zamaraev, *Farady Disc. Chem. Soc.* **1991**, *92*, 13.
[83] H. H. A. Smit, P. R. Nugteren, R. C. Thiel, L. J. de Jongh, *Physica* **1988**, *B153*, 33.
[84] R. C. Thiel, R. E. Benfield, R. Zanoni, H. H. A. Smit, M. W. Dirken, *Z. Phys. D,* **1993**, *81*, 1.
[85] R. C. Thiel, M. W. Dirken, R. E. Benfield, R. Zanoni, *Structure and Bonding, in press.*
[86] H. H. Smit, R. C. Thiel, L. J. de Jongh, *Z. Phys. D* **1989**, *12*, 193.
[87] G. Schmid, R. Pfeil, R. Boese, F. Bandermann, S. Meyer, G. H. M. Calis, J. W. A. van der Velden, *Chem. Ber.* **1981**, *114*, 5634.
[88] U. Gelius, *Phys. Scripta* **1974**, *9*, 133.
[89] G. K. Wertheim, J. Kwo, B. K. Teo, K. A. Keating, *Solid State Commun.* **1985**, *55*, 357.
[90] C. Battistoni, G. Mattogno, R. Zanoni, L. Naldini, *J. Electr. Spectr.* **1982**, *28*, 32.
[91] G. K. Wertheim, S. B. DiCenzo, S. E. Youngquist, *Phys. Rev. Lett.* **1983**, *51*, 2310.
[92] K. Fauth, U. Kreibig, G. Schmid, *Z. Phys. D* **1991**, *20*, 297.
[93] R. E. Benfield, *J. Organomet. Chem.* **1989**, *372*, 163.
[94] U. Kreibig, K. Fauth, C.-G. Granquist, G. Schmid, *Z. Physikal. Chemie* Neue Folge **1990**, *169*, 11.
[95] J. A. Creighton, D. G. Eadon, *J. Chem. Soc., Faraday Trans.* **1991**, *87*, 3881.
[96] R. E. Benfield, A. P. Maydwell, J. M. van Ruitenbeek, D. A. van Leeuwen, *Z. Phys. D, in press.*
[97] G. Nimtz, P. Marquard, H. Gleiter, *J. Cryst. Gr.* **1988**, *86*, 66.
[98] H. B. Brom, M. P. J. van Staveren, L. J. de Jongh, *Z. Phys. D* **1991**, *20*, 281.
[99] N. F. Mott, E. A. Davis, *Electronic Processes in Nanocrystalline Materials,* Clarendon Press, Oxford, **1971**.
[100] O. Madelung, *Festkörpertheorie III,* Springer, Berlin, Heidelberg, New York **1973**.
[101] S. Summerfield, *Philosoph. Mag.* **1985**, *52*, 9.
[102] U. Simon, G. Schön, G. Schmid, *Angew. Chem. Int. Ed. Engl.* **1993**, *32*, 250.
[103] U. Simon, G. Schmid, G. Schön in *P. D. Persans,* J. S. Bradley, R. R. Chianelli, G. Schmid, eds. *Materials Research Soc., Sympos. Proceed.* **1992**, *272*, 167.
[104] K. K. Licharew, T. Claeson, *Spektr. d. Wiss.* **1992**, issue 8, 62.
[105] A. Gladum, A. B. Zorin, *Phys. i. u. Zeit* **1992**, *23*, 159.
[106] G. Schmid, *Struc. Bonding* **1985**, *62*, 51.
[107] H. Feld, A. Leute, D. Rading, A. Benninghoven, G. Schmid, *J. Am. Chem. Soc.* **1990**, *112*, 8166.
[108] H. Feld, A. Leute, D. Rading, A. Benninghoven, G. Schmid, *Z. Phys. D* **1990**, *17*, 73.
[109] G. Schmid, N. Klein, *Angew. Chem. Int. Ed. Engl.* **1986**, *25*, 922.
[110] G. Schmid, *Aspects in Homogeneous Catal.* **1990**, *7*, 1 and references therein.
[111] G. Schmid, R. Küpper, H. Hess, J.-O. Malm, J.-O. Bovin, *Chem. Ber.* **1991**, *124*, 1889.

3.4 Transition Metal Clusters with Bridging Main Group Elements

Dieter Fenske

A very large number of organometallic transition metal clusters in which main group elements take on various functions were described in chapter 3.2. In the present chapter, predominantly clusters with main group elements acting in bridging function and with 'non organic' ligand shells will be discussed. As will be shown, many of these compounds can be regarded as nanoscaled cutouts of bulk materials, for instance, of transition metal sulfides and selenides. In some respects the discussions in this chapter lead up to those on condensed metal clusters, which will be described extensively in Chapter 5.

3.4.1 Transition Metal Clusters with E and ER Bridging Ligands (E = N, P, As, Sb; R = Organic Group)

3.4.1.1 Clusters with PR and P Bridging Ligands

It has been known for many years now that silylated phosphines react with transition metal halides to form polymeric species whose structures are still unknown. [1] This observation has received astonishingly little attention, despite the fact that a large number of reactions of silylated phosphines with main group halides could be clarified. [2] Of all the work performed on transition metal elements, that of H. Schäfer, from which a large number of complexes having phosphido, diphosphene, and diphosphorus ligands were isolated, deserves special mention. [3–11] These studies clearly show that the reaction of transition metal halides with, for example, $PPh(SiMe_3)_2$ proceeds by substitution at the metal-halide bond, with the driving force of the reaction being the formation of Me_3SiX. [12, 13] The synthesis of polynuclear metal complexes was not at first observed. It was the use of transition metal halide complexes with PR_3 ligands which led to the formation of characterizable cluster compounds.

If $[MCl_2(PR_3)_2]$ (M = Co, Ni) is reacted with $PPh(SiMe_3)_2$ in THF, the μ_3-PPh and μ_4-PPh bridged clusters **1** and **2** are obtained. [14]

In contrast, if $NiCl_2$, which is only sparingly soluble in THF, is allowed to react with PPh_3 and $PPh(SiMe_3)_2$, a different product, the oxygen sensitive complex **3**, is obtained; it is also formed from the reaction of **2** with an excess of $PPh(SiMe_3)_2$.

The high selectivity of these reactions is reflected in the up to 90 % yields of **1**, **2**, and **3**. Only clusters of cobalt and nickel have so far proved accessible via the

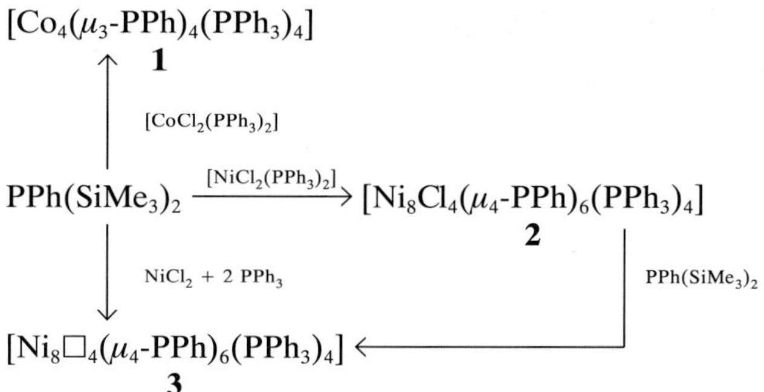

Scheme 3-13. Formation of PPh-bridged Co_4 and Ni_8 clusters.

reaction between $[MCl_n(PPh_3)_m]$ and $PPh(SiMe_3)_2$. The analogous reactions with $[FeCl_3(PPh_3)_2]$, $[HgCl_2(PPh_3)_2]$, or $[MCl(PPh_3)]$ (M = Hg, Ag, Au) lead to complexes of Fe^{2+} or metallic Hg, Ag and Au. The reactions with Rh, Pd, or Pt complexes have not yet been fully explored.

It can be seen in Figure **3-48** that **1** consists of a tetrahedral Co_4 cluster whose four faces are coordinated by μ_3-PPh ligands. Each Co atom has an additional PPh_3 ligand bound to it. The bonding parameters observed in **1** (T_d symmetry, Co–Co = 256(1) pm) are in good agreement with those found for a large number of known heterocubane clusters. [15–18] Compound **1** is diamagnetic and can be reversibly oxidized to $[\mathbf{1}]^+$ and $[\mathbf{1}]^{2+}$ in THF solution. [19] Consistent with this is the observation that **1** can also be oxidized by $CHCl_3$, CCl_4, or $C_2H_4Cl_2$ to give $[\mathbf{1}]^+Cl^-$. The cluster **1** will also react with CH_3COCl at temperatures above 330 K to form the paramagnetic species $[\mathbf{1}]^+[CoCl_3(PPh_3)]^-$ ($\mu_{eff} = 4.55\mu_B$ at 300 K). The molecular structure of **2** (Fig. **3-48**) consists of an almost regular Ni_8 cube (Ni–Ni = 259–262 pm, Ni-Ni-Ni angle = 89.6–90.3°) whose Ni atoms are coordinated alternately by PPh_3 and Cl ligands and each face is capped by a μ_4-PPh ligand. Its structure is thus comparable with $[Ni_8(CO)_8(\mu_4\text{-PPh})_6]$, which has already been described by Dahl et al. [20] The structure of **3** (Fig. **3-48**) is similar to that of **2**, whereby the Ni_8 cube in **3** is highly distorted with Ni-Ni-Ni bond angles ranging between 79.2 and 100.1°. This distortion is also reflected in the difference in the coordination geometry about Ni1, Ni3, Ni6, Ni7 (distorted tetrahedral coordination by the P atoms of one PPh_3 and three μ_4-PPh ligands) compared to Ni2, Ni4, Ni5, Ni8 (coordinated by three P atoms of the μ_4-PPh ligands and not coordinated to a terminal PPh_3 ligand). The usual electron counting rules indicate that the Ni2, Ni4, Ni5, and Ni8 atoms in **3** are associated with 16 valence electrons, whereas Ni1, Ni3, Ni6, and Ni7 are associated with 18. One consequence of this electron deficiency may be the shortening of the face diagonals between the $16e^-$ Ni atoms to 323–325 pm and the lengthening of the distances between the $18e^-$ Ni atoms to 384–398 pm.

The cluster **3** is unexpectedly stable. One reason for this may be the steric screening of the four coordinatively unsaturated Ni atoms (Ni2, Ni4, Ni5, and

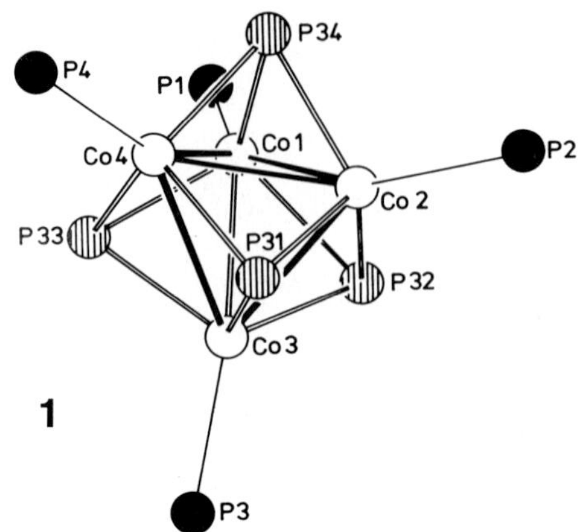

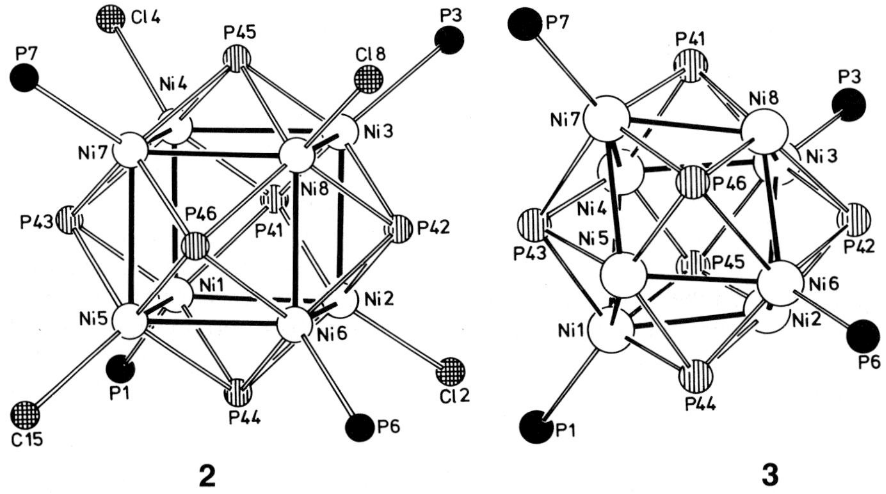

Figure 3-48. Molecular structures of the clusters **1**, **2**, and **3** (phenyl groups omitted). In **1**, P1–P4 and P31–P34 are the P atoms of the PPh_3 and μ_3-PPh ligands, respectively. In **2** and **3**, P1–P7 and P41–P46 are the P atoms of the PPh_3 and μ_4-PPh ligands respectively.

Ni8) by six phenyl groups of the PPh_3 and PPh ligands. This crown shaped arrangement leads to four identical free coordination sites in the molecule (indicated by □ in Scheme **3-13**), but with diameters restricted to only 400–500 pm. The space filling model illustrated in Figure **3-49** demonstrates this for the free coordination site at Ni5. Further coordination at these four Ni atoms by PPh_3 ligands is no longer possible because of the steric shielding around these atoms.

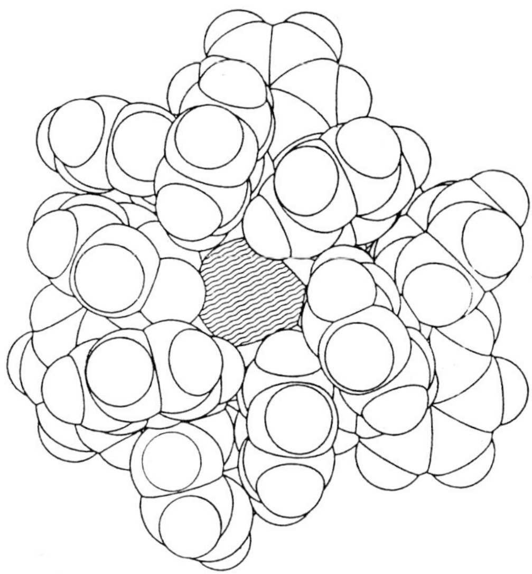

Figure 3-49. Space-filling model of **3**, showing the free coordination site (□) at atom Ni5 (shaded).

Reactions of [$Ni_8Cl_4(\mu_4$-PPh$)_6(PPh_3)_4$] and [$Ni_8\square_4(\mu_4$-PPh$)_6(PPh_3)_4$]

Although the number of structurally characterized clusters is now immense, the study of their reactivity has been mostly limited to smaller clusters. [21] Huttner et al., for example, were able to show that trinuclear, μ_3-ER bridged carbonyl metal clusters (E = N, P, As, Sb, Bi) are suitable for kinetic and mechanistic studies. [22] The peripheral ligands on such clusters are susceptible to exchange by addition and subsequent elimination. It is notable that the Ni_8 clusters **2** and **3** also lend themselves to studies of ligand substitution and addition reactions. The Cl ligands can be substituted in **2**, and neutral molecules can be added to the open coordination sites in **3**. Scheme **3-14** summarizes the reactions carried out so far.

The stability of the Ni_8 cluster framework is also demonstrated by the formation of $[\mathbf{2}]^+Br_3^-$ (**4**) from the reaction of **2** with concentrated methanolic solutions of bromine. With sodium amalgam, **2** forms the coordinatively unsaturated **3**. If the same reaction is carried out in the presence of solid $NiCl_2$, however, a "naked" nickel atom (presumably arising from the reduction of Ni^{2+}) is introduced at an open coordination site on **3** to form the Ni_9 cluster **7**. Whereas the dimensions of all the vacant coordination sites in **3** are identical, the unoccupied sites in **7** have smaller diameters and as a consequence, no further Ni atoms can be introduced.

The structure determination of **7** (Fig. **3-50**) revealed that the Ni-Ni distances in the Ni_8 cube are comparable to those in **3**. The Ni5–Ni9 bond length of 268.2 pm lies in the normal range of Ni–Ni single bonds. Since Ni9 is not bound to any other ligands, it can only be the steric screening by the six immediate phenyl groups that makes **7** is stabil. The nonbonding Ni $\cdots$ C(Ph) contacts of 326–380 pm make interactions between the π systems and Ni9 improbable. It is pos-

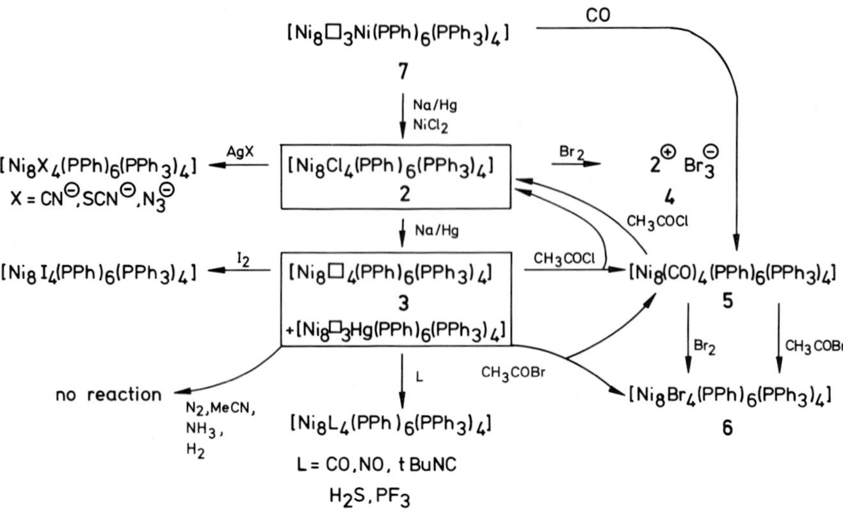

Scheme 3-14. Reactions leading to derivatives of the Ni_8 clusters.

sible, however, that weak bonding interactions to three of the phenyl groups protons are involved and would invoke a distorted tetrahedral coordination at Ni9 (Ni ··· H 247–294 pm). An NMR investigation to support this hypothesis is not feasible because of the paramagnetism of **7**.

Reaction of **2** with Na-amalgam has also led to the successful isolation of a complex which is isostructural with **7**, but in which the one free coordination site is now occupied by a Hg atom (see Scheme **3-14**). [23]

Attempts to incorporate Pd, Pt, or Au atoms into an open coordination site have led to, as yet, unidentified products. The unusually large unit cells (35000–55000 Å^3) might imply the presence of very large clusters. Other ligands (*e. g.*

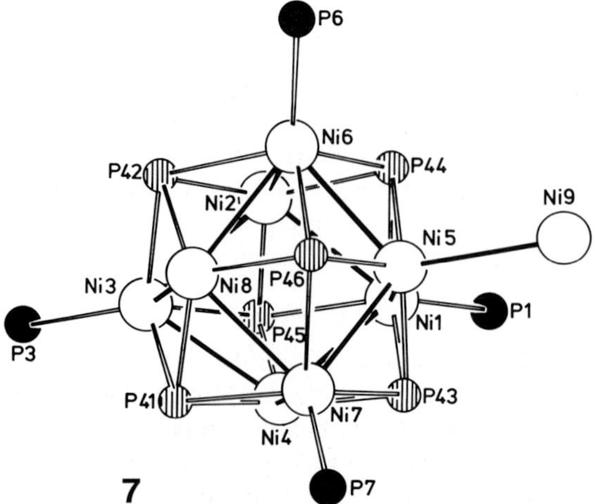

Figure 3-50. The molecular structure of **7** (phenyl groups omitted). P1–P7 and P41–P46 represent the P atoms of the PPh_3 and μ_4-PPh ligands respectively.

CN^-, SCN^-, N_3^-) can be attached to the cluster by the reaction of the corresponding silver salts with **2**. The alternation of PPh_3 and the newly introduced ligands is maintained and indicates a lack of mobility of the PPh_3 groups.

The coordinatively unsaturated **3** also serves as starting material for a variety of reactions (Scheme **3-14**). It reacts with CH_3COCl to form a mixture of **2** and **5** (Fig. **3-51**). A possible intermediate in this reaction is an acyl complex which would then decompose to form **5**. The cluster **5** can also be obtained from the reaction of **7** with CO. The bromo substituted cluster **6** is formed from the reaction of **5** with CH_3COBr.

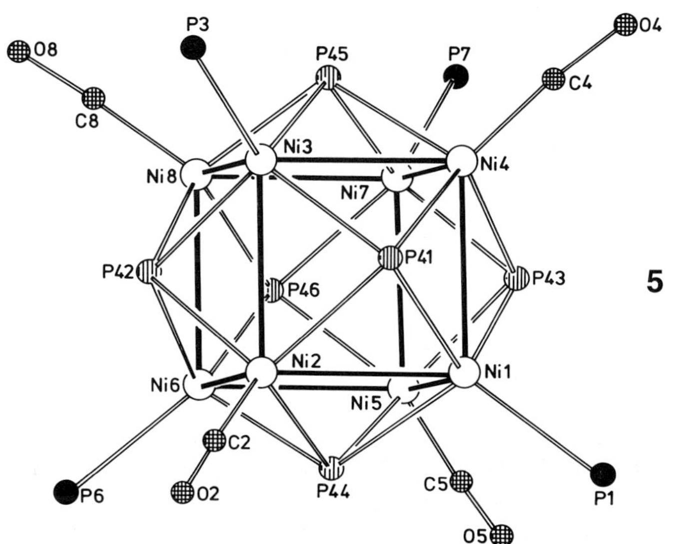

Figure 3-51. The structure of **5** (phenyl groups omitted). P1–P7 and P41–P46 represent the P atoms of PPh_3 and μ_4-PPh ligands respectively.

It can further be seen from Scheme **3-14** that the vacant coordination sites in **3** can be occupied by such ligands as CO, *t*BuNC, H_2S, and PF_3. No reaction is observed however, with MeCN, NH_3, or H_2 and no evidence for coordination of these molecules could be found in the solid state. Finally, N_2 does not coordinate to **3**, even under a pressure of 150 bar.

Structural, Spectroscopic, and Magnetic Properties of Ni_8 Clusters

A meaningful description of bonding in clusters requires an understanding of the relationships between the structures, the physical and chemical properties, and the number of valence electrons. The structural and chemical diversity of clusters has, however, inhibited the development of a general bonding theory. In spite of this, several bonding models have been conceived and applied with varying degrees of success; among these are the 18-electron rule, Wade's rules, semiempirical MO calculations, the topological electron counting rules of Teo and Mingos, and the isolobality principle of R. Hoffmann. [24–35].

According to the 18-electron rule, a valence electron count of 120 would be required for clusters with a cubic shape and this condition is fulfilled by e.g. $[Ni_8(CO)_8(PPh)_6]$. [20] The aforementioned compounds, however, often display appreciable deviations from this criterion; if PPh_3 and Cl^- are considered as $2e^-$ donors and μ_4-PPh as a $4e^-$ donor, then **2** and **3** contain 116 and 112 valence electrons respectively.

The information in Table **3-16** clearly shows that the Ni–Ni bond lengths are crucially influenced by the electron count. Surprisingly, the Ni–Ni bonds are shorter when fewer electrons are available for Ni–Ni bonding. The Ni–P bonds are less sensitive in this respect.

Table 3-16. Valence electron count (VE) and the Ni–Ni and Ni–P bond lengths in $[Ni_8X_4(PPh)_6(PPh_3)_4]$.

Compound	X	VE e^-	for Ni-Ni bonds	Ni-Ni [pm]	Ni-P [pm] PPh_3	PPh
3	□	112	16	251–254	221	212–216
						224–228
4	Cl	115	19	256–260	222–224	220–224
2	Cl	116	20	259–262	224–225	220–222
6	Br	116	20	259–262	224–225	220–222
5	CO	120	24	265–269	226–229	218–222

The magnetic properties of the Ni_8 clusters are likewise impossible to rationalize on the basis of simple assumptions. Magnetic susceptibility measurements can, however, provide information on the possibility of electron coupling within the cluster. Such measurements were carried out on the clusters $[Fe_4(\mu_3\text{-}S)_4(SPh)_4]^{n-}$ $(n = 2,3)$ some years ago by Holm et al. [36] As would be expected, **5** (120 valence electrons) is diamagnetic, but it is less easy to understand the paramagnetism of **3** (112 valence electrons), **2**, and **6** (each with 116 valence electrons). For example, a temperature dependent magnetic moment is observed for **2**: $\mu_{eff}(T) = 2.37\mu_B$ (300 K), 2.48 (200 K), 2.54 (100 K), 2.44 (25 K), 1.86 (10 K), and 0.94 (4 K). This can be rationalized if it is assumed that **2** contains Ni atoms in different oxidation states ($4\ Ni^0 + 4\ Ni^+$), whereby the magnetic properties would then corresponding to an antiferromagnetic coupling of the d^9 (Ni^+) centers.

Transition metal clusters can often undergo reversible redox processes without altering their metal framework. [37] Consistent with the observation that the clusters **2–6** tolerate large deviations from the $18e^-$ rule is the fact that they can also be oxidized and reduced reversibly.

The characterization of paramagnetic clusters by NMR methods is generally difficult because very broad signals are often observed. It is thus all the more surprising that the ^{1}H-NMR spectrum of **2** displays sharp and strongly temperature dependent resonances. The paramagnetism causes different isotropic shifts for the phenyl protons of the μ_4-PPh and PPh_3 ligands, allowing the *o*-, *m*-, and *p*-protons to be clearly differentiated. The following shifts are observed in $CDCl_3$ in the 323–213 K temperature range: PPh_3: *o*-H 3.15–0.93, *m*-H 10.60–12.75, *p*-H 3.37–1.27; PPh: *o*-H 6.82–6.01, *m*-H 6.93–7.20, *p*-H 5.59–5.22 ppm. A plot of the shifts against temperature shows that the dependence obeys Curie's law. [38] ^{31}P-NMR spectra show resonance signals only for diamagnetic derivatives. The observed chemical shifts of $\delta = 0–20$ ppm for PR_3 and 620–640 ppm for μ_4-PPh ligands (in THF) lie in their expected ranges. [39–41] The IR spectra of the compounds 2–7 are almost identical, except in the 1500–2000 cm^{-1} range, and have no diagnostic value.

Unexpected Reaction Products with -PR and -PR$_2$ Ligands

It is clearly apparent that the PR_3 ligands have an influence on the structure of the clusters formed. If, for example, $NiCl_2$ is reacted with $PhP(SiMe_3)_2$ in THF, a black oil is obtained that presumably consists of oligomers with the composition [NiPPh]. This observation was first reported by E. W. Abel, but he assumed that the mix was composed of hexameric units. [1] Depending on the reaction conditions, however, it is also possible to isolate $[Ni_{12}Cl_2(PPh)_2(P_2Ph_2)_4(PHPh)_8]$ (**8**) as a crystalline product. The terminal PHPh ligands are formed by reaction of $PhP(SiMe_3)_2$ with the solvent. The crystal structure analysis showed that the product is a Ni_{12} tetraasterane cluster (Fig. **3-52**), whose polyhedral faces are capped by μ_4-PPh and μ_6-P_2Ph_2 groups. In addition, the Ni atoms are coordinated to terminal $PHPh^-$ and Cl^- ligands. [42]

Only clusters of cobalt and nickel have so far proved accessible via the reaction between $[MCl_n(PPh_3)_m]$ and $PPh(SiMe_3)_2$. The analogous reactions employing $[FeCl_3(PPh_3)_2]$ or $[MCl(PPh_3)]$ (M = Hg, Ag, Au) lead only to complexes of Fe^{2+} and Cu^+ or metallic Hg, Ag or Au. In contrast, the reaction of CuCl with $PhP(SiMe_3)_2$ and PPh_3 gives very high yields of $[Cu_{12}(PPh)_6(PPh_3)_6]$ (**9**), whose framework derives from a cuboctahedron. An analogous reaction with PMe_3 led to the formation of the Cu_{14} cluster (**10**) whose structural relation to **9** can be seen in Figure **3-53**.

It is also possible to obtain Cu clusters from the reaction of CuCl with Ph_2PSiMe_3. In the presence of PR_3 or PR_2H, one can isolate clusters with 3, 4, 5, and 18 Cu atoms. Figure **3-54** shows the structure of the large complex $[Cu_{18}(PPh)_4(PPh_2)_{10}(PHPh_2)_3]$ (**11**), [43] prepared from CuCl, $PHPh_2$, Ph_2PSiMe_3 and $PhP(SiMe_3)_2$ (Eq. 3.36).

$$CuCl + PHPh_2 + Ph_2PSiMe_3 + PhP(SiMe_3)_2 \rightarrow \underset{\mathbf{11}}{[Cu_{18}(PPh)_4(PPh_2)_{10}(PHPh_2)_3]} \qquad (3.36)$$

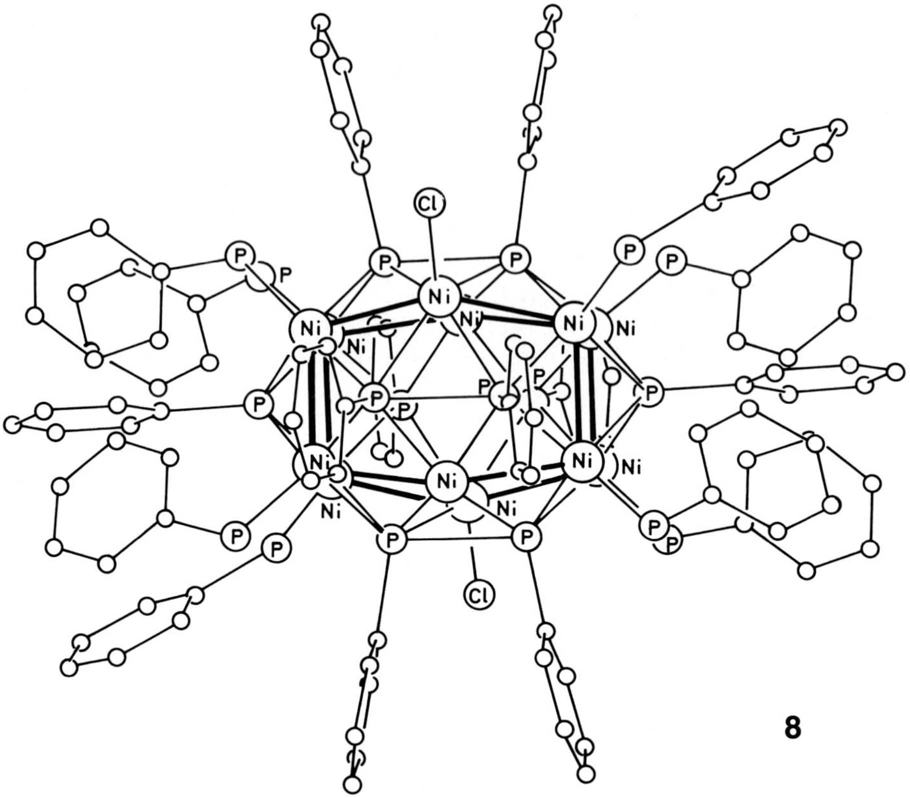

Figure 3-52. Molecular structure of $[Ni_{12}Cl_2(PPh)_2(P_2Ph_2)_4(PHPh)_8]$ (**8**).

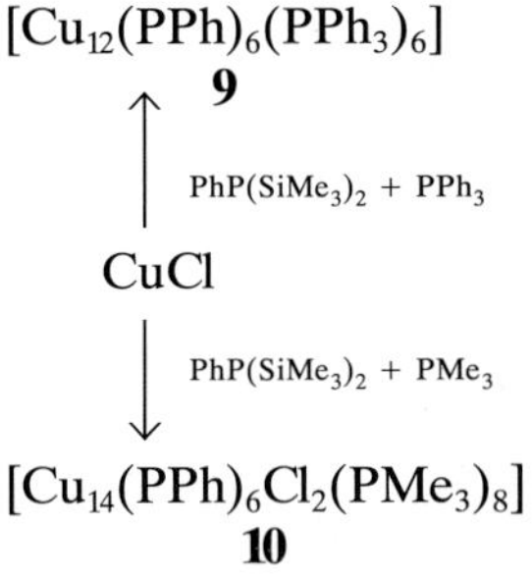

Scheme 3-15. Formation of Cu_{12} and Cu_{14} clusters from CuCl and $PhP(SiMe_3)_2$.

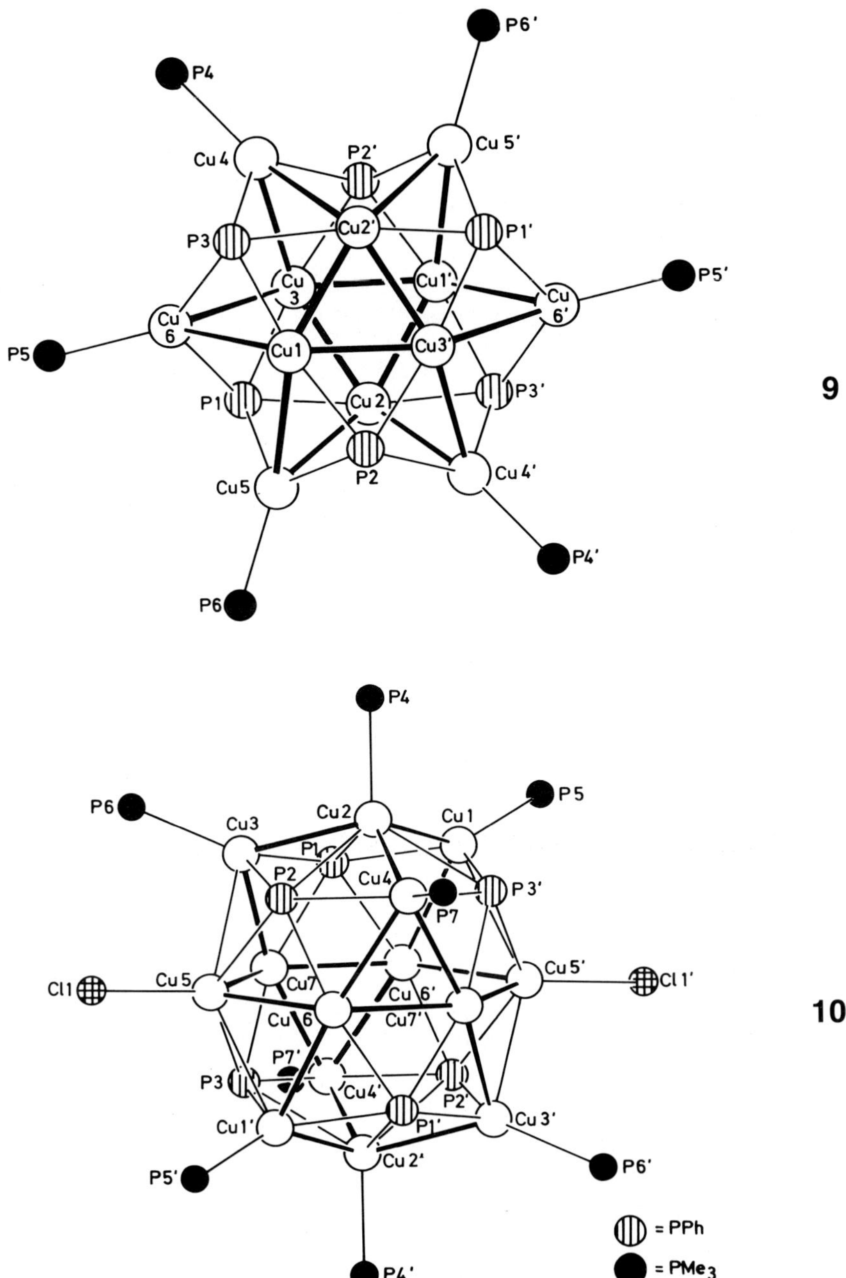

Figure 3-53. Molecular structures of **9** and **10** (C atoms are omitted). The terminal P atoms (black circles) are P atoms from the PPh_3 ligands in **9** or from the PMe_3 ligands in **10**. The other P atoms are from the bridging PPh ligands.

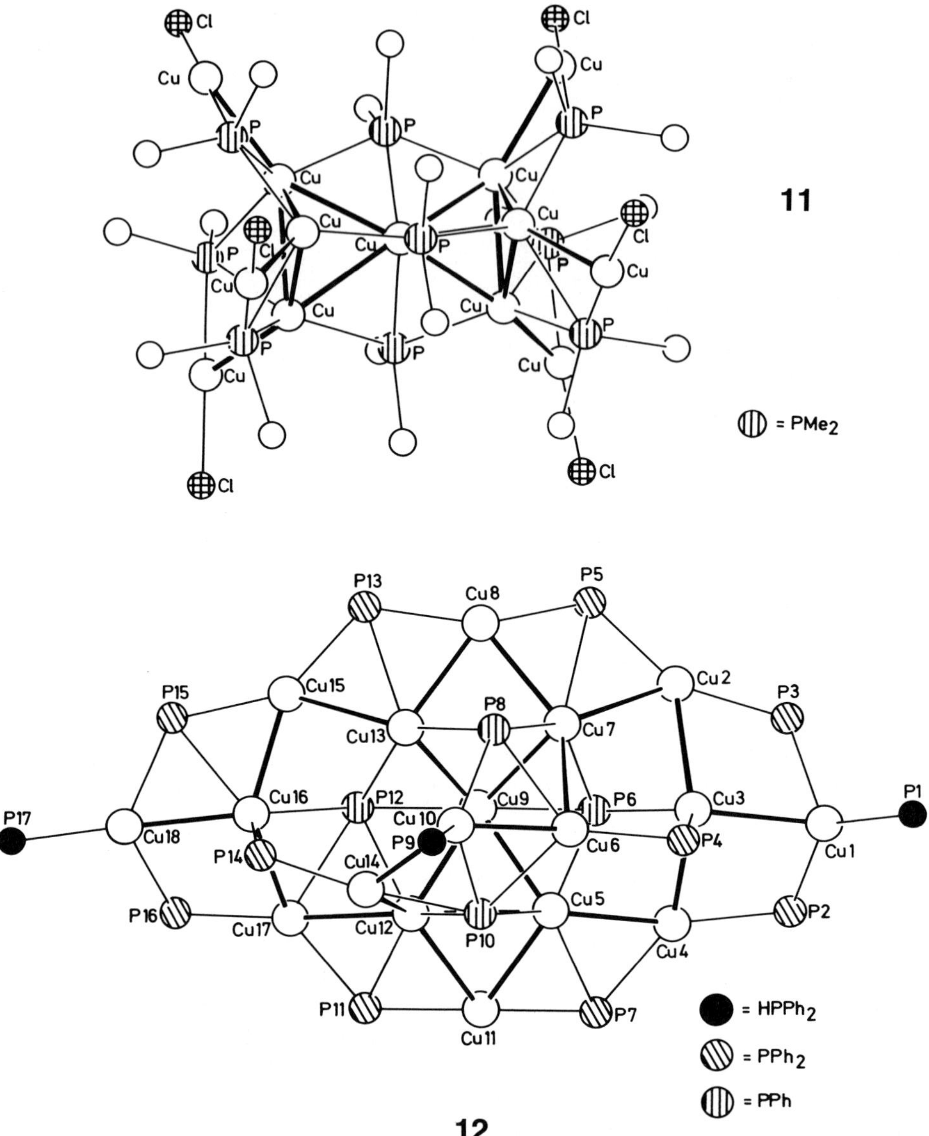

Figure 3-54. Molecular structure of the cluster $[Cu_{18}(PPh)_4(PPh_2)_{10}\text{-}(PHPh_2)_3]$ **11** and the cluster anion in **12**. The phenyl groups are omitted for clarity.

The 18 Cu atoms are bridged by PPh_2 ligands in various bonding modes. P2, P4, P14, and P16 are μ_2 ligands, whereas P3, P5, P7, P11, P13, and P15 can be described as μ_3 ligands. The bonding situations of the PPh bridges also differ. P6 and P8 are μ_4, P10 and P12 are μ_5 ligands. Knowing the conditions required for cluster synthesis, it is often a simple matter of replacing one or more of the reaction components by similar species in order to obtain new clusters. An example is shown in Equation (3.37).

$$CuCl + PMe_3 + Me_2PSiMe_3 \rightarrow \underset{\mathbf{12}}{[Cu(PMe_3)_4]_2[Cu_{13}Cl_6(PMe_2)_9]} \quad (3.37)$$

The deep red crystalline compound **12** is obtained in almost quantitative yield and consists of a Cu_{13} cluster anion. The six outer Cu atoms are bound to chloride ligands and the PMe_2 groups act as either μ_2 or μ_3 bridges between the Cu atoms (Fig. **3-54**). [43] It is possible that the cluster structures are in fact extracts of the insoluble oligomeric phosphido complexes formed from the reaction of $CuCl_2$ with $PhP(SiMe_3)_2$ or Ph_2PSiMe_3. [1] The fact that these compounds ultimately dissolve in the presence of tertiary phosphines to give multinuclear copper complexes [43] seems to back up this hypothesis. All the copper atoms in **9**, **10**, **11**, and **12** have a formal oxidation state of 1 + (d^{10} configuration). Although the contribution of d^{10}–d^{10} interactions to metal-metal bonding has been disputed, examples are known which show they could be of significance. [44–48] The Cu-Cu distances in **9–12** (**9**: 256.3–278.3 pm; **10**: 248–253 pm; **11**: 245–272 pm; **12**: 247–258 pm), however, show that only very weak Cu–Cu bonds can be present. Nevertheless, they do correspond to the values found in other multinuclear Cu complexes, specifically those with I^-, S^{2-}, Se^{2-}, and SR ligands. [49–56]

3.4.1.2 Clusters with As and Sb as Bridging Ligands

No evidence for the formation of complexes with bridging AsPh ligands has been found from attempts to extend the aforementioned synthetic principle to reactions of $PhAs(SiMe_3)_2$ with $[MCl_2(PPh_3)_2]$. A typical example is that of $[FeCl_2(PPh_3)_2]$ which upon reaction with $PhAs(SiMe_3)_2$ yields only metallic iron, $(AsPh)_n$, and Me_3SiCl. The reaction of $[CoCl_2(PPh_3)_2]$ with $PhAs(SiMe_3)_2$, however, leads to, together with small amount of metallic cobalt, the formation of the paramagnetic cluster $[Co_4As_6(PPh_3)_4]$ (**13**, 95 % yield). [57]

$$[CoCl_2(PPh_3)_2] + PhAs(SiMe_3)_2 \rightarrow \underset{\mathbf{13}}{[Co_4As_6(PPh_3)_4]} \quad (3.38)$$

The crystal structure determination (Fig. **3-55**) revealed a Co_4 tetrahedron whose faces are capped by three μ_3-As ligands and one μ_3-As_3 group. In addition, each Co atom is coordinated to a PPh_3 group. The cluster **13** has 56 valence electrons and may be thought of as being derived from a heterocubane $[Co_4(\mu_3As)_4(PPh_3)_4]$ cluster in which one μ_3-As ligand has been replaced by a μ_3-As_3 unit.

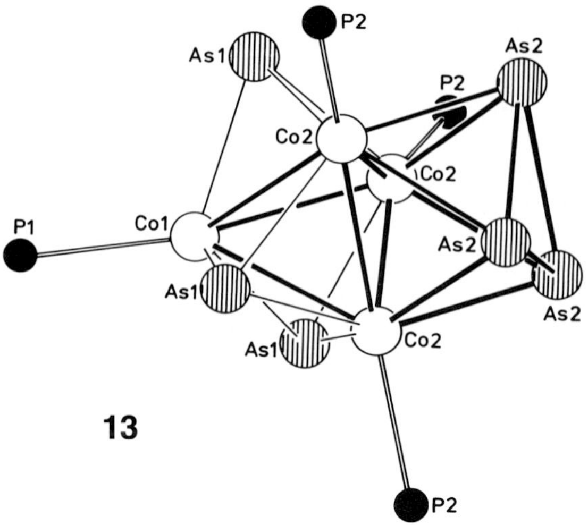

Figure 3-55. Molecular structure of **13**. Phenyl groups have been omitted for clarity. **13** has a three-fold crystallographic axis. The three crystallographically equivalent As atoms labelled As2 form the μ_3-As_3 unit, those labelled As1 are the μ_3-As ligands and P1 and P2 are the P atoms of the PPh_3 ligands.

Alternatively, it can be considered as a Co_3As_3 octahedron bonded over one Co_3 face to a $CoAs_3$ unit. Previously known complexes containing E'_3 units (E′ = P, As) have been generally considered as derivatives of a tetrahedral P_4 or As_4 unit, where one P or As atom has been replaced by an isolobal transition metal fragment. [58–63] The As–As bond lengths within the As_3 unit (246.3 pm) are comparable to those in As_4, $(AsPh)_6$, and $[(triphos)CoAs_3Co(triphos)]^{2+}$ and are only about 10 pm longer than the As–As bond lengths in the As_3 complexes $[As_3Co(CO)_3]$ and $[MoCpAs_3(CO)_2]$. [61–66] That the formation of **13** occurs at room temperature is surprising since it involves the cleavage of an As–C (phenyl) bond, however, examples which show the instability of As–C bonds have been known for some time. [67, 68]

Recent results confirm that $[NiCl_2(PPh_3)_2]$ will also react with $PhAs(SiMe_3)_2$ under cleavage of the As–C bonds and formation of As bridged complexes. The clusters **14** and **15** can be isolated from the reaction in THF (Eq. 3.39). [69] The structure of **14** is illustrated in Figure **3-56** and shows that the metal atom framework consists of a Ni_8 cube, in the center of which a further Ni atom is situated. The faces of the Ni_8 cube are capped by six μ_4-As ligands and the Ni atoms are coordinated either by Cl^- (Ni2, Ni4, and Ni8) or by PPh_3 (Ni1, Ni3, Ni5, Ni6, and Ni7) ligands. The Ni–Ni bond lengths within the Ni_8 cube (277–285 pm for Ni1–Ni8) are 40–50 pm longer than those to the central Ni9 atom (238–248 pm). In contrast, the bond lengths from the peripheral Ni atoms to the μ_4-As atoms (230–234 pm) are 30 pm shorter than the Ni9–As bonds (261–264 pm). Like the $[Au_9(PPh_3)_8]^+$ ion and the Rh_{14} core of the $[Rh_{14}(CO)_{25}]^{4-}$ ion, the centered cubic cluster **14** can be considered as a fragment of a larger centered cluster. [70–72] Preliminary results indicate that the reaction of $[NiCl_2(PPh_3)_2]$ with $PhAs(SiMe_3)_2$ can also lead to larger clusters by condensation of centered cubic clusters via common faces.

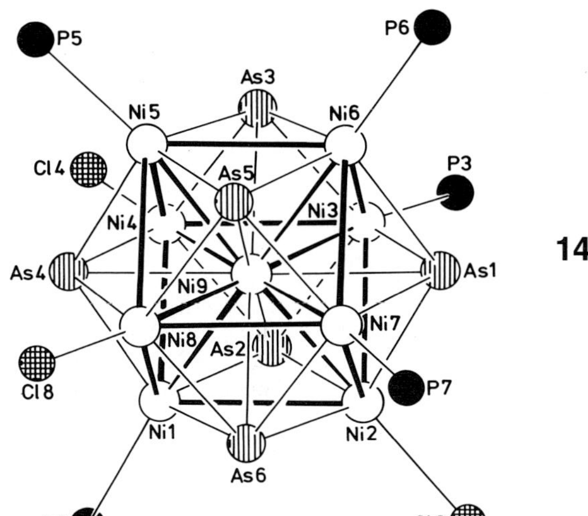

14

Figure 3-56. Molecular structure of **14** (phenyl groups omitted). P1–P7 represent the P atoms of the PPh_3 ligands.

$$[NiCl_2(PPh_3)_2] + PhAs(SiMe_3)_2 \rightarrow \underset{\mathbf{14}}{[Ni_9Cl_3(\mu_4\text{-}As)_6(PPh_3)_5]} + \underset{\mathbf{15}}{[Ni_9Cl_2(\mu_4\text{-}As)_6(PPh_3)_6]} \quad (3.39)$$

The cluster **14** contains 121 valence electrons, one electron less than predicted by the $18e^-$ rule. An electron counting rule for centered metal clusters, as proposed by Mingos, [30–33] states that the number of valence electrons in close-packed metal clusters should be $(12n_s + \Delta_i)$, where n_s is the number of peripheral metal atoms and Δ_i is 18 or 24 for centered cubic polyhedra. On this basis, **14** should possess either 114 or 120 valence electrons.

A further interesting feature of **14** is that it possesses analogous to **2**, three reactive Ni–Cl bonds. By breaking these bonds, it is possible to synthesize a coordinatively unsaturated complex similar to **3**. The vacant coordination sites thus formed are larger than those in **3** (because the μ_4-As ligands are smaller than μ_4-PPh ligands) and so can accomodate larger ligands. In the meantime, it has also been possible to characterize cluster analogues to **14** which have μ_4-P and μ_4-Sb ligands. [73]

Several cluster complexes of palladium can be synthesized by the reaction of $[PdCl_2(PPh_3)_2]$ with $E'(SiMe_3)_3$ (E' = As, Sb) as outlined in Scheme **3-16**.

In analogy to the Ni_9 cluster, both **16** and **16a** (Fig. **3-57**) consist of a body centered Pd_9 cube, but the Pd–Pd distances are very long. The 325.5 pm edge lengths of the Pd_9 cube in **16a** are about 15 pm longer than those in **16**. As a result, the distances to the central Pd atom are also longer (Pd1–Pd2: **16**: 268–270 pm; **16a**: 281–283 pm). Both **16** and **16a** have 124 valence electrons, which is two electrons more than the $18e^-$ rule predicts. Hückel calculations on this type of cluster with μ_4-Te ligands were performed by Wheeler [74] and

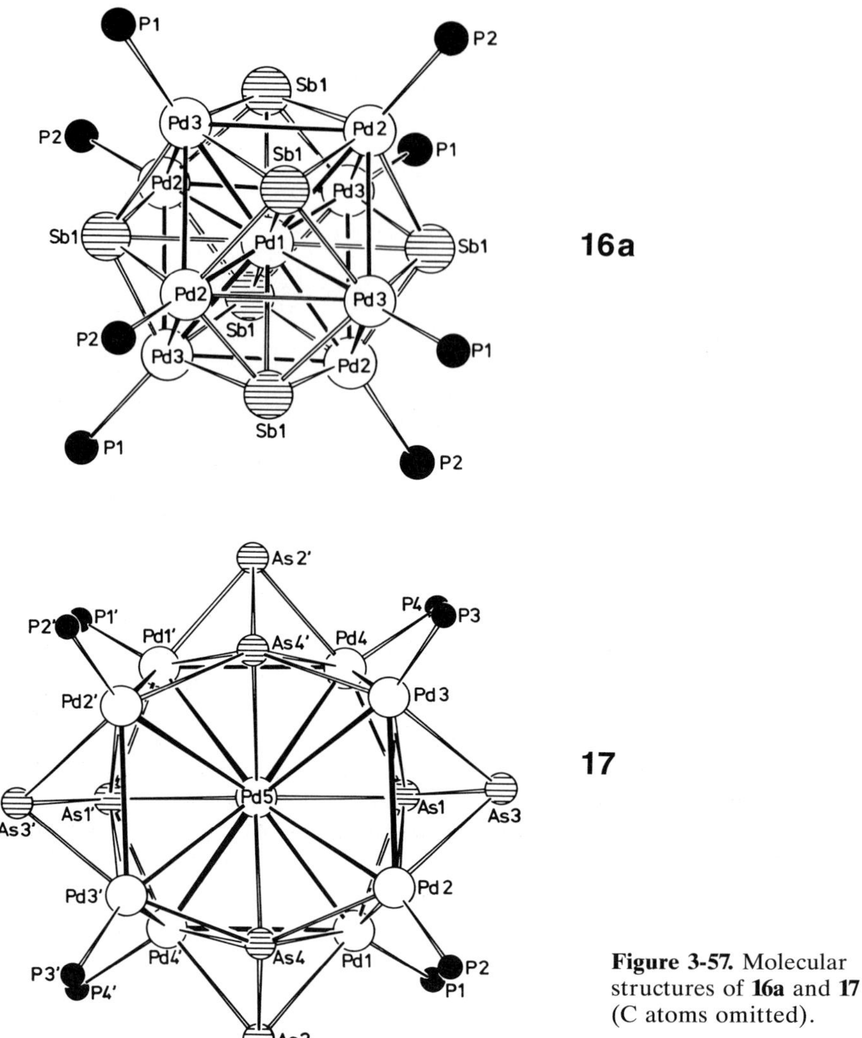

Figure 3-57. Molecular structures of **16a** and **17** (C atoms omitted).

showed that a valence electron count of either 114 or 130 electrons should be optimum. Electron counts of 124 or 120, however, are possible as a result of the influence which the ligands have on the ordering of the t_{1g}, e_g, and t_{2g} orbitals.

On the other hand, these are body centered cubic clusters which strictly obey the 18e$^-$ rule. An example for these can be found from the reaction of [$NiCl_2$-($PCyc_3)_2$] with $P(SiMe_3)_3$, from which a compound of composition [Ni_9P_6-$(PCyc_3)_6Cl_2$], containing 122 valence electrons, may be isolated.

The molecular structure of **17** (Scheme **3-16**) is shown in Figure **3-57**. Here, the nine palladium atoms build an approximate body centered cube however, because of the maximum twofold rotation axis, the structure may be considered as being derived from four [$Pd_2(PPh_3)_2$] units bridged via As_2 ligands with an additional

$$[PdCl_2(PPh_3)_2] \xrightarrow{As(SiMe_3)_3} [Pd_9As_6(PPh_3)_8]\ \mathbf{16} + [Pd_9As_8(PPh_3)_8]\ \mathbf{17} + [Pd_{20}As_{12}(PPh_3)_{12}]\ \mathbf{18}$$

$$[PdCl_2(PPh_3)_2] \xrightarrow{Sb(SiMe_3)_3} [Pd_9Sb_6(PPh_3)_8]\ \mathbf{16a}$$

Scheme 3-16. Formation of As and Sb containing Pd clusters from $[PdCl_2(PPh_3)_2]$.

Pd atom (Pd5) situated at the cube center. Such Pd_2As_2 units are also found in $[Pd_2As_2(PPh_3)_4]$. [75] The As–As distances in **17** fall within the range usually observed for these units. [76–80] From an alternative viewpoint, the structure of **17** may be regarded as an intermediate along the formation pathway of the body-centered Pd_8 cubes found in **16** and **16a**. As a result of the coordination of four of the cube faces by As_2 ligands, however, considerably longer Pd–Pd distances of 300–309 pm are found. The crystal structure of **18** (Scheme **3-16**) shows that it is formed from a palladium cube (Pd11–Pd16, Pd18, Pd20) with every edge being bridged by a μ_2-($PdPPh_3$) group (Fig. **3-58**). These groups lie approximately in the best plane defined by the four palladium atoms of a cube face, having deviations of only 28–43 pm. Alternatively, the cluster framework in **18** may be described as a polyhedron built up through the linkage of six Pd_6 rings whose faces are then capped by μ_5-As ligands.

The electron count for clusters with $n > 13$ are frequently not compatible with the schemes usually employed. If the As ligands are counted as $3e^-$ donors and the PPh_3 ligands as $2e^-$ donors, then **18** contains 260 electrons, that is 28 electrons less than predicted by the $18e^-$ rule and 12 electrons more than predicted by the $16e^-$ rule. According to the concept developed by Mingos ($n_s = 12$, $\Delta_i = 120$), **18** should contain 264 valence electrons, which is only 4 electrons more than that actually counted in **18**. Similarly large Pd clusters have been synthesized and structurally characterised by Mednikov et al. [81–83] and include $[Pd_{10}(CO)_{12}(PBu_3)_6]$, $[Pd_{23}(CO)_{22}(PEt_3)_{10}]$, and $[Pd_{38}(CO)_{28}(PEt_3)_{12}]$.

3.4.1.3 Nitrogen Bridged Clusters

Silylated derivatives of nitrogen have, until now, only been used to a limited extent in the synthesis of cluster complexes. Admittedly, a series of examples are known in which $N(SiMe_3)_3$ has been employed as a nitride transfer agent. [84] For instance, the reaction of WCl_6 with $N(SiMe_3)_3$ leads to the formation of the binuclear complex W_2NCl_9, from which $[W_3N_2Cl_{14}]^{2-}$ may be obtained. [85] A further example is the synthesis of $[\{Cp^*TaN(Cl)\}_3]$, obtained from the reaction of $[Cp^*TaCl_4]$ with $N(SiMe_3)_3$ [86] ($Cp^* = C_5Me_5$).

If $PhN(SnMe_3)_2$ is allowed to react with certain transition metal halide derivatives at higher temperatures (*e. g.* in refluxing toluene), the formation of NPh-bridged cluster compounds can be observed. [87] Some of these are outlined in Scheme **3-17**.

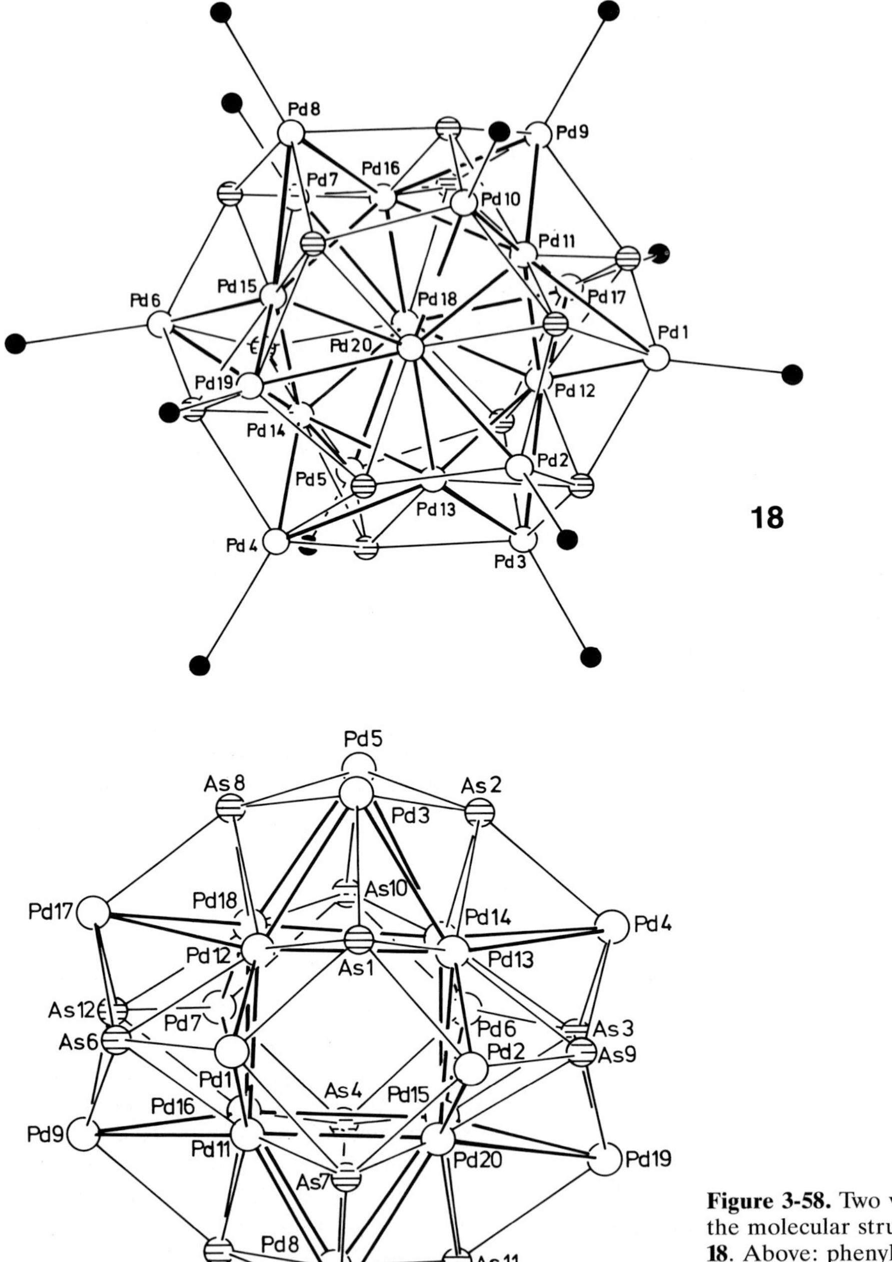

Figure 3-58. Two views of the molecular structure of **18**. Above: phenyl groups omitted; below: PPh_3 ligands omitted.

$$[Co_{11}(\mu_3\text{-}NPh)_6(\mu_2\text{-}NPh)_6(PPh_3)_3].$$
19

$\uparrow$ $[CoCl_2(PPh_3)_2]$

$$PhN(SiMe_3)_2 \xrightarrow{[NiCl_2(PMe_3)_2]} [Ni_3(\mu_2\text{-}NPh)_3(PMe_3)_3]^+ \ [SnCl_2Me_3]^-$$
20

$\downarrow$ $[CpCrCl_2]_x$

$$[Cp_3Cr_3(\mu_3\text{-}NPh)_2(\mu_2\text{-}NPh)]$$
21

Scheme 3-17. Formation of nitrogen bridged chromium, cobalt, and nickel clusters.

In **19** (Fig. **3-59**), the eleven Co atoms form part of a two-dimensional close-packed arrangement, where Co10 and Co11 lie above the plane formed by the other nine Co atoms (Co–Co: 246–288 pm). The twelve NPh ligands function either as μ_2 or μ_3 bridges, and only the Co atoms Co2, Co5, and Co8 are coordinated by PPh_3 ligands. Assuming the NPh ligands to have a charge of 2–, the Co atoms must have different formal oxidation states.

The ionic compound **20** contains a cluster cation, the structure of which is shown in Figure **3-59**. It consists of a Ni_3 ring, whose edges are bridged by μ_2-NPh ligands and each Ni atom is coordinated by an additional PMe_3 ligand. As a result, each Ni atom has an approximately trigonal planar coordination. Electron counting leads to a mixed valence description of the compound, with a Ni_3^{7+} cluster core. The low valence electron count of 41 is unusual and lies well below the prediction of the 18e$^-$ rule for M_3 clusters (48 electrons). The Ni–Ni bond lengths of 247–249 pm are considerably shorter than in $[Ni_3(PMe_3)_3Cl_3(\mu_3\text{-}NPMe_3)\text{-}(\mu_3\text{-}NH)]$ (263.9 pm), but longer than in $[Cp_3Ni_3(\mu_3\text{-}N\textit{t}Bu)]$ (221–234 pm). [88, 89] A Cr_3 ring, bridged by two μ_3-NPh ligands and one μ_2-NPh-ligand, is present in the structure of **21** (Fig. **3-59**). The Cr atoms have additional η^5-Cp groups coordinated and are accordingly in the +3 oxidation state (d^3 configuration). Any description of the bonding must take into consideration the large differences in the Cr–Cr distances (Cr1–Cr1′: 236 pm; Cr1–Cr2: 260 pm). From a very formal standpoint, it could be assumed that the Cr_3 ring contains two Cr–Cr single bonds and one Cr–Cr double bond. This would imply that **21** contains a single unpaired electron, which is not, however, in agreement with the measured magnetic moment of μ_{eff} = 1.45 BM (293K) per Cr atom. This behavior may be explained with ligand field theory, however, as Cr1 and Cr1′ have an approximate tetrahedral coordination, and Cr2 has a trigonal planar coordination. In a strong field, one unpaired electron per Cr atom would be expected. Extended Hückel calculations carried out on **21** have shown that only one unpaired electron per molecule should be expected. [90]

The observation that $[CpTiCl_2(PEt_3)_2]$ reacts with $N(SnMe_3)_3$ to form a Ti_4N_4 heterocubane (Fig. **3-60**) is interesting. [87]

$$[CpTiCl_2(PEt_3)_2] + N(SnMe_3)_3 \rightarrow \underset{\mathbf{22}}{[Cp_4Ti_4(\mu_3\text{-}NSnMe_3)_4]} \tag{3.40}$$

In **22**, whose structure is illustrated in Figure **3-60**, the Ti atoms are formally in the oxidation state +3 (d^1 configuration) and the cluster contains 52 valence electrons, 8 electrons less than found in many other heterocubane clusters. [91] Admittedly, other compounds are known whose electron counts fall short of this normal value and still others which exceed it. Extended Hückel calculations predict a Ti_4 tetrahedron having D_{2d} symmetry with four longer and two shorter Ti–Ti bonds for this type of structure with 52 valence electrons. [92, 93] Indeed, **22** contains two shorter Ti–Ti distances of 268.6(2) pm and four longer distances of 277.2–278.5(2) pm.

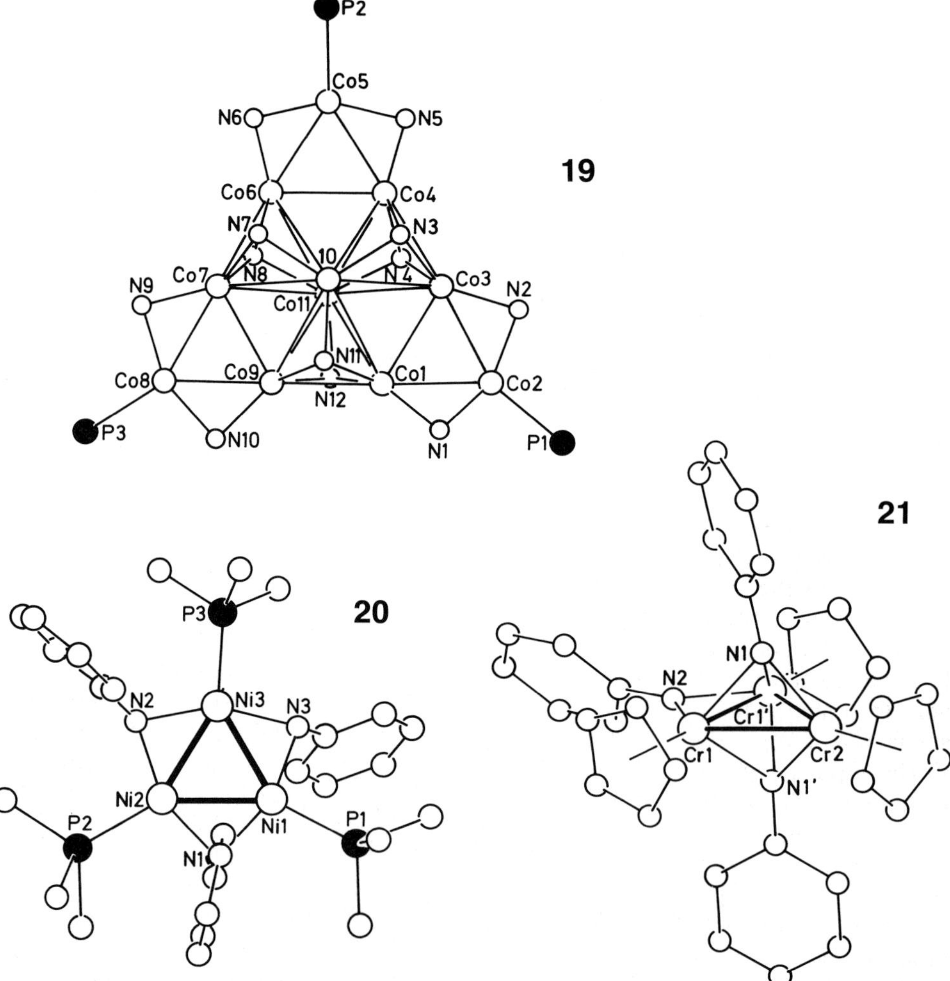

Figure 3-59. Molecular structures of **19** (phenyl groups omitted) and the cations in **20** and **21**.

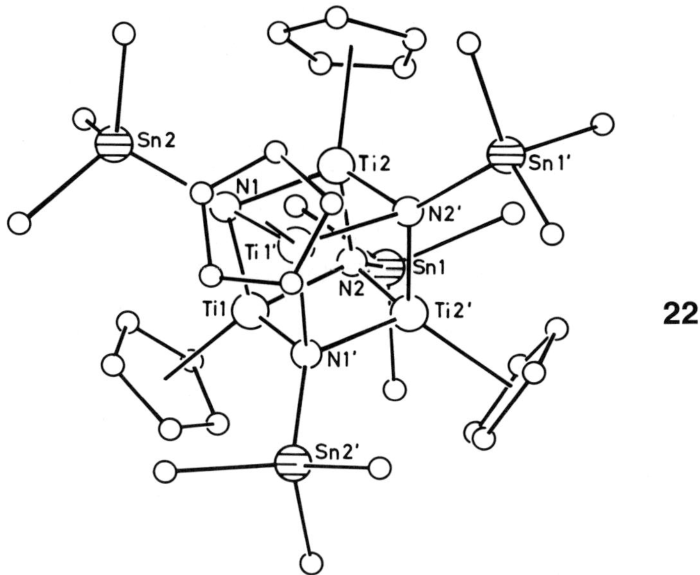

Figure 3-60. Molecular structure of **22**.

3.4.2 Clusters with ER Bridging Ligands (E = S, Se, Te; R = Organic Group).

Various different methods for the synthesis of clusters with ER (E = S, Se, Te; R = organic group) bridging ligands have been described in the literature. [94–104] A particularly large number of examples can be noted for complexes with SR ligands. Normally, the synthesis of these compounds is achieved by the use of sulphanes or metallated sulphanes from reactions with transition metal halides or carbonyls. Thus, the reaction of $NiCl_2$ with NaSR (R = Et, C_6H_{11}) leads to the complexes $[Ni_5(SEt)_{10}]$ and $[Ni_4(SC_6H_{11})_8]$. [105] In contrast, the reaction of $NiCl_2$ with NaS*t*Bu yields $[Ni_8S(S\textit{t}Bu)_9]^-$, a cluster anion that contains Ni atoms of mixed valence. [106] The structure of the starting material is certainly important for the analogous reaction of $(NEt_4)_2[NiBr_4]$ with NaS*i*Pr leads to a cubane type molecule with the composition $[Ni_4(S\textit{i}Pr)_8]$. [106] A number of other compounds such as $[M_4(SR)_{10}]^{2-}$ (M = Fe, Co : R = Et; M = Zn, Cd : R = Ph), [107, 108] $[Ag_{14}(S\textit{t}Bu)_{14}(PPh_3)_4]$, and $[Cd_{10}(SCH_2CH_2OH)_{16}]^{4+}$ [109, 110] have been synthesized using this reaction principle.

A series of clusters which contain other ligands in addition to the SR ligands has been reported. For example, the reaction of $Fe(SR)_3$ with Na_2S_2 led to the formation of $Na_4[Fe_6S_9(SR)_2]$ (R = Me, Et, $PhCH_2$) [111] and clusters having the composition $[M_{10}E_4(SPh)_{16}]$ (E = S, Se; M = Zn, Cd) resulted from the reaction of $[M_4(SPh)_{10}]^{2-}$ with elemental sulphur or selenium. [112] A further synthetic route from which a multinuclear metal complex having both E and ER ligands has been obtained is the partial oxidation of $[Fe(SEt)_4]^{2-}$ with sulphur. $[Fe_4S_6(SEt)_4]^{4-}$ was obtained in this way. [113] $[Fe_6Se_9(SMe)_2]^{4-}$ may be made in a similar manner by the reaction of $Fe(SMe)_3$ with Na_2Se_2. [114] In some cases, the use of NaSH has been reported, which reacts, for example, with $[Co_4(SPh)_{10}]^{2-}$ to give $[Co_8S_6(SPh)_8]^{n-}$ (n = 4,5). [115]

3.4.3 Clusters with E Bridging Ligands (E = S, Se, Te)

An extraordinarily large number of publications describing the synthesis of clusters with substituent free chalcogen ligands have appeared. [116–127, 94–103] These compounds are formed not only by solid state reactions, but also by other methods. Two examples are the reaction of Cs_2CO_3 with Re and S, which gives $Cs_4Re_6S_{13}$ [128] and salt melt reactions between K_2Te, Cu, and Te from which Cu-Te clusters such as $[Cu_5Te_5]^{2-}$ can be obtained. [129]

Some interesting synthetic methods are performed under hydrothermal conditions, which, in some cases, can lead to clusters with polychalcogen ligands. An example of this is the reaction of Mo with K_2Se_x from which $K_2Mo_3Se_{18}$ and the $[Mo_{12}Se_{56}]^{2-}$ cluster anion may be isolated. [130, 131]

These types of cluster compounds can also be obtained from reactions in solution. For example, $[Na(12\text{-}C\text{-}4)_2]_2[Cu_4(Se_4)_3]$ results from the reaction of CuCl, Na_2Se_x, and 12-C-4 in DMF. [132, 133]

A different way of synthesizing clusters which contain chalcogen bridging ligands is the desulphurization of mononuclear or binuclear complexes. For instance, from the reaction of $[(MeC_5H_4)_2V_2S_4]$ with PBu_3, a mixture of $[(MeC_5H_4)_4V_4S_4]$ and $[(MeC_5H_4)_5V_5S_6]$ is obtained. [93] The driving force of this reaction would appear to be the formation of $PSBu_3$. On the other hand, the direct use of phosphine chalcogenides in the synthesis of clusters has also been reported. For example, the reaction between $[Pd(PPh_3)_4]$, PEt_3, and $PTeEt_3$ has led to the formation of $[Pd_6Te_6(PEt_3)_8]$ [134] and from the reaction between $[Ni(COD)_2]$, PEt_3, and $PTeEt_3$ both $[Ni_9Te_6(PEt_3)_8]$ and $[Ni_{20}Te_{18}(PEt_3)_{12}]$ have been isolated. [135]

H_2E (E = S, Se) is often employed in reactions involving transition metal halide derivatives. Examples are the reactions of $Fe(BF_4)_2$ with PEt_3 and H_2S, or $Ni(BF_4)_2$ with $Na[BPh_4]$, PEt_3 and H_2S from which $[Fe_6S_8(PEt_3)_6][BPh_4]_2$, or $[Ni_9S_9(PEt_3)_6][BPh_4]_2$ and $[Ni_3S_2(PEt_3)_6][BF_4]_2$ can be isolated. [136, 137] The use of M_2E (E = S, Se, Te; M = Li, Na, K) has been described in a great number of articles. If Li_2S is reacted with $FeCl_3$ in the presence of Na[PhNCOMe], then $[Na_9Fe_{20}S_{38}]^{9-}$ and $[Na_2Fe_{18}S_{30}]^{8-}$ can be obtained. [138, 139] Polychalcogenides are also suitable materials for the synthesis of clusters and examples include the synthesis of $[Ni_4Se_4(Se_3)_5Se_4]^{4-}$ from Li_2Se_x, NEt_4Cl, and $Ni(S_2COEt)_2$, and the preparation of $[Cu_6S_{17}]^{2-}$ from the reaction between Cu(acac), PPh_4Br and S_x^{2-}. [140, 141] Polychalcogenide salts are likewise capable of reacting with transition metal carbonyls. Reaction of $[W(CO)_6]$ with K_2Te_3 in the presence of PPh_4Br yields $[W_4(CO)_{18}Te_2]^{2-}$, and with $(PPh_4)_2Te_4$ leads to the formation of

$[\{W(CO)_3\}_6(Te_2)_4]^{2-}$. [142, 143] The direct use of elemental chalcogen can also be profitably employed. For example, $[Co_2(CO)_8]$ and S react to give a mixture of $[Co_3(CO)_9S]^+$ and $[Co_3(CO)_7S_2]$, [144] $[Cp_2Ni_2(CO)_2]$ and sulphur lead to the formation of a mixture of $[Cp_3Ni_3S_2]$ and $[Cp_4Ni_5S_4]$, and the reaction of $Hg[Fe\text{-}(Co)_3NO]$ and sulphur leads to $[Fe_4(NO)_4S_4]$. [145, 146]

3.4.3.1 S and Se Bridged Clusters of Cobalt and Nickel having PR_3 Ligands (R = Organic Group)

The existence of the coordinatively unsaturated cluster **3** leads one to consider whether the synthesis of clusters with still larger vacant coordination sites might be possible. Since these vacant sites arise as a direct consequence of the steric screening of the metal atoms by peripheral ligands, several possible synthetic routes immediately suggest themselves:

a) *Substitution of PPh_3 by PR_3 ligands which have different steric requirements.* The reaction of $[NiCl_2(PR_3)_2]$ with $PPh(SiMe_3)_2$ leads to diphosphene complexes together with compounds of unknown structure when R = Me or Et. However, if R = C_4H_9 or cyclo-C_6H_{11}, then clusters having the composition $[Ni_8\square_4(PPh)_6(PR_3)_4]$ and showing properties analogous to those of **3** are obtained.
b) *Substitution of PPh_3 by $E'Ph_3$ (E' = As, Sb).* $[NiCl_2(E'Ph_3)_2]$ also reacts with $PhP(SiMe_3)_2$ to form clusters of the type $[Ni_8\square_4(PPh)_6(E'Ph_3)_4]$.
c) *Replacement of the μ_4-PPh ligand by a μ_4-PR ligand.* The reaction of $[NiCl_2(PPh_3)_2]$ with $RP(SiMe_3)_2$ leads to compounds of as yet unknown composition.
d) *Substitution of the μ_4-PPh ligand by less sterically demanding S, Se or Te ligands.* E $(SiMe_3)_2$ (E = S, Se, Te) react with derivatives of transition metal halides to form S, Se, and Te substituted clusters in a manner analogous to reactions involving the cleavage of the P-Si bond in $PhP(SiMe_3)_2$. Although many methods are known for the synthesis of clusters containing chalcogen ligands, the use of silylated chalcogenides offers ready access to a variety of transition metal clusters. [127, 104, 147–156]

To date, reactions of $E(SiMe_3)_2$ with PR_3 complexes of the following metal halides have been studied: $MnCl_2$, $FeCl_2$, $CoCl_2$, $RhCl_3$, $NiCl_2$, $PdCl_2$, $PtCl_2$, $CuCl_2$, AgCl, $ZnCl_2$, $CdCl_2$, and $HgCl_2$. Depending on the nature of both the metal and the coordinated ligands, these reactions lead either to the binary chalcogenides or to mixtures of various cluster compounds which may subsequently be separated by fractional crystallisation. Clearly, the solubility products of the chalcogenides and the stability constants of the clusters play important roles. The infrared spectra of these substances are dominated by the bands of the tertiary PR_3 ligands and thus the spectra of complexes with different metals but the same phosphine ligand are virtually identical. NMR spectra are sometimes difficult to obtain because of the very low solubility in most of the usual solvents and the ready oxidation in such chlorinated hydrocarbons as CH_2Cl_2 and $CHCl_3$. Therefore, identification of the products is generally dependent on crystal structure analyses.

The following factors exert a critical influence on the nature of the products:

a) *The M:E:PR_3 ratio.*
b) *The solvent.* The following have been used with success: toluene, MeCN, THF, and low melting (m.p. < 400K) salt mixtures.
c) *The conditions for crystal growth.* The solubility of the products is very dependent on the nature of the R substituents of the PR_3 ligands. Whereas species with PPh_3, $PtBu_3$, and $P(C_6H_{11})_3$ ligands are almost insoluble in all common solvents, the derivatives with PEt_3 and PBu_3 ligands are often soluble in such nonpolar solvents as heptane and diethylether. This initially surprising observation should enable a better control of the solubility of cluster products and ought to make possible the spectroscopic and cyclovoltammetric study of high molecularity clusters. It should be stressed, however, that mixtures of various clusters are often formed, the separation of which becomes more problematic as the cluster size increases.
d) *The nature of the transition metal halide derivative used as starting material is apparently of great significance.* Experiments have been performed on both neutral compounds such as $[CoCl_2(PR_3)_2]$, and ionic complexes such as $(NBu_4)[CoCl_3(PR_3)]$, and have found that they lead to different products.

The progress of the reaction may be followed, in certain cases, with the aid of time dependent ^{31}P NMR measurements. At the beginning of the reaction of $[NiCl_2(PBu_3)_2]$ with $Se(SiMe_3)_2$, for example, only the ^{31}P signal of the starting material can be observed. After only a few minutes, however, additional signals can be seen in the 0–15 ppm region which correspond to the formation of free PBu_3 and its phosphine selenide. During the course of the reaction, the intensity of the phosphine selenide signal increases. The formation of the phosphine selenide is an indication for the formation of larger clusters. It may well be that smaller cluster complexes react with the free phosphine molecules to give the phosphine chalcogenides. In this manner, cluster surfaces may be formed which then react further to form the more metal rich clusters.

The following discussion on the structures of various products is arranged according to the number of metal atoms.

M_3 and M_4 Clusters

No binuclear complexes have been isolated from the reactions between $[MCl_2(PR_3)_2]$ (M = Co, Ni; R = Et, Bu, C_2H_4Ph, Ph) and $E(SiMe_3)_2$ (E = S, Se, Te). The smallest clusters which can be prepared (albeit in low yield) are $[Ni_3Cl_2S_2(PPh_3)_4]$ (**23**) and $[Ni_3Cl_2Se_2(PEt_2Ph)_4]$ (**23a**). [157–159] The structures of **23** (see Fig. **3-61**) and **23a** consist of a Ni_3 triangle (Ni–Ni: **23**: 298 pm; **23a**: 317 pm) which is capped by two μ_3-E (E = S, Se) ligands and each Ni center is additionally bonded to two phosphines or to a phosphine and a Cl^-. The Ni atoms have different distorted planar coordinations (Ni1 and Ni3: S1, S2, PR_3 and Cl^-; Ni2: S1, S2, two PR_3). The $[Ni_3S_2(PEt_3)_6]^{2+}$ cluster, obtained by the reaction of $[NiCl_2(PEt_3)_2]$ with H_2S, has a similar structure. [136, 137, 160, 161]

M_4E_4 heterocubane clusters (Fig. **3-61**) may also be prepared from Co and Ni complexes. [162–164] Typical representatives of these are shown in Equations (3.41) and (3.42).

$$(NBu_4)[NiCl_3(PPh_3)] \xrightarrow{Se(SiMe_3)_2} \underset{\mathbf{24}}{[Ni_4Se_4(PPh_3)_4]} \quad (3.41)$$

$$[CoCl_2(PtBu_3)_2] \xrightarrow{Se(SiMe_3)_2} \underset{\mathbf{25}}{[Co_4Se_4(PtBu_3)_4]} \quad (3.42)$$

The crystal structure analyses on **24** and **25** showed that they are both constructed from tetrahedral M_4 clusters (M = Ni, Co), whose faces are capped by μ_3-Se ligands. The PR_3 ligands complete an irregular tetrahedral coordination of the metal atoms. [15–18, 147–154]

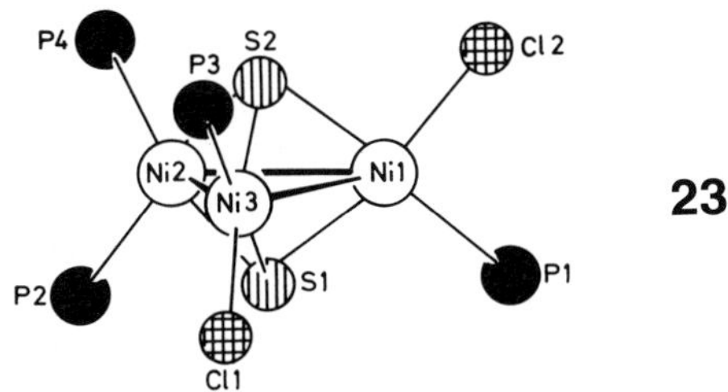

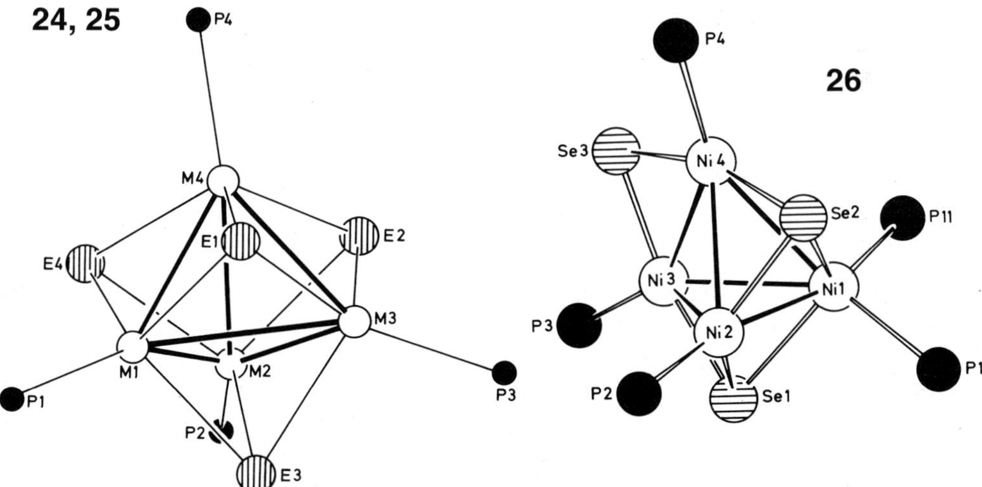

Figure 3-61. Molecular structure of **23** (P1–P4 are the P atoms of the PPh_3 ligands). Structure of the $[M_4E_4(PR_3)_4]$ framework (R groups omitted) in **24** (M = Ni) and **25** (M = Co). Molecular structure of **26**.

The metal–metal bond lengths within the Co_4Se_4 cluster (**25**: 264–270 pm) are, as would be expected, approximately equal, but they are 10–15 pm longer than in the PPh substituted cluster **1**. In contrast, in the $64e^-$ cluster **24**, the four electrons in excess of the 60 electrons required by the $18e^-$ rule give rise to four short (259, 263 pm) and two long (279, 293 pm) Ni–Ni bond lengths. As a consequence, the Ni_4 framework is distorted towards a structure with approximately D_{2d} symmetry, as was predicted for $64e^-$ heterocubanes by Dahl et al. [165] Otherwise, **24** has a structure similar to that of the Co_4 tetrahedron in **25**.

$[Ni_4Se_3(PPhe_3)_4]$ (**26**) can be prepared by the reaction of $[NiCl_2(PPhe_3)_2]$ (Phe = C_2H_4Ph) with $Se(SiMe_3)_2$. [166] The structure of **26** (Fig. **3-61**) displays a severely distorted Ni_4 tetrahedron (Ni–Ni: 239.3–288.9 pm), in which, as a result of the steric requirements of the $PPhe_3$ ligands, only two of the four tetrahedron faces are capped by μ_3-Se (Se1, Se2) ligands. The third selenium, Se3, acts as a μ_2 bridge between Ni3 and Ni4 which, in addition to the phosphine ligand distribution allows only Ni3 and Ni4 to have similar chemical environments; Ni2 has a trigonal planar coordination to P2, Se1, and Se2, and Ni1 has tetrahedral coordination. Although complex **26**, with its 60 valence electrons, satisfies the 18-electron rule for tetrahedral clusters, the Ni_4 cluster is, as mentioned previously, severely distorted.

M_5 Clusters

Cluster compounds composed of five Ni atoms (such as $[Ni_5Cl_2Se_4(PEt_2Me)_6]$ (**27**) [166] can be isolated from the reaction of $Se(SiMe_3)_2$ with $[NiCl_2(PEt_2Me)_2]$. This reaction leads to a mixture of products which also includes trinuclear and other multinuclear nickel complexes.

$$[NiCl_2(PEt_2Me)_2] \xrightarrow{Se(SiMe_3)_2} \underset{\mathbf{23b}}{[Ni_3Cl_2Se_2(PEt_2Me)_4]} + \underset{\mathbf{27}}{[Ni_5Cl_2Se_4(PEt_2Me)_6]} + \underset{\mathbf{50}}{[Ni_{21}Se_{14}(PEt_2Me)_{12}]} \quad (3.43)$$

The structure of **27** is shown in Figure **3-62** and consists of a planar spirocyclic Ni_5 cluster with $\bar{1}$ symmetry which is formed from two vertex-sharing Ni_3 units, both of which are capped by two μ_3-Se ligands. The structural similarity of **27** to the Ni_3 clusters **23** and **23b** is further illustrated by the fact that, in **27**, the Ni atoms Ni2 and Ni2′ are likewise bonded to two PEt_2Me ligands, with the Ni3 and Ni3′ centers to Cl^- and PEt_2Me ligands. This results in all the Ni atoms exhibiting a slightly distorted square-planar coordination geometry. The structure of **27** is similar to several Pd_5 clusters (see below) and a spirocyclic $HgNi_4$ cluster. [167] Based on the Ni–Ni distances in **27** (294.2–322.8 pm), it can not be assumed that there are bonding interactions between the Ni atoms.

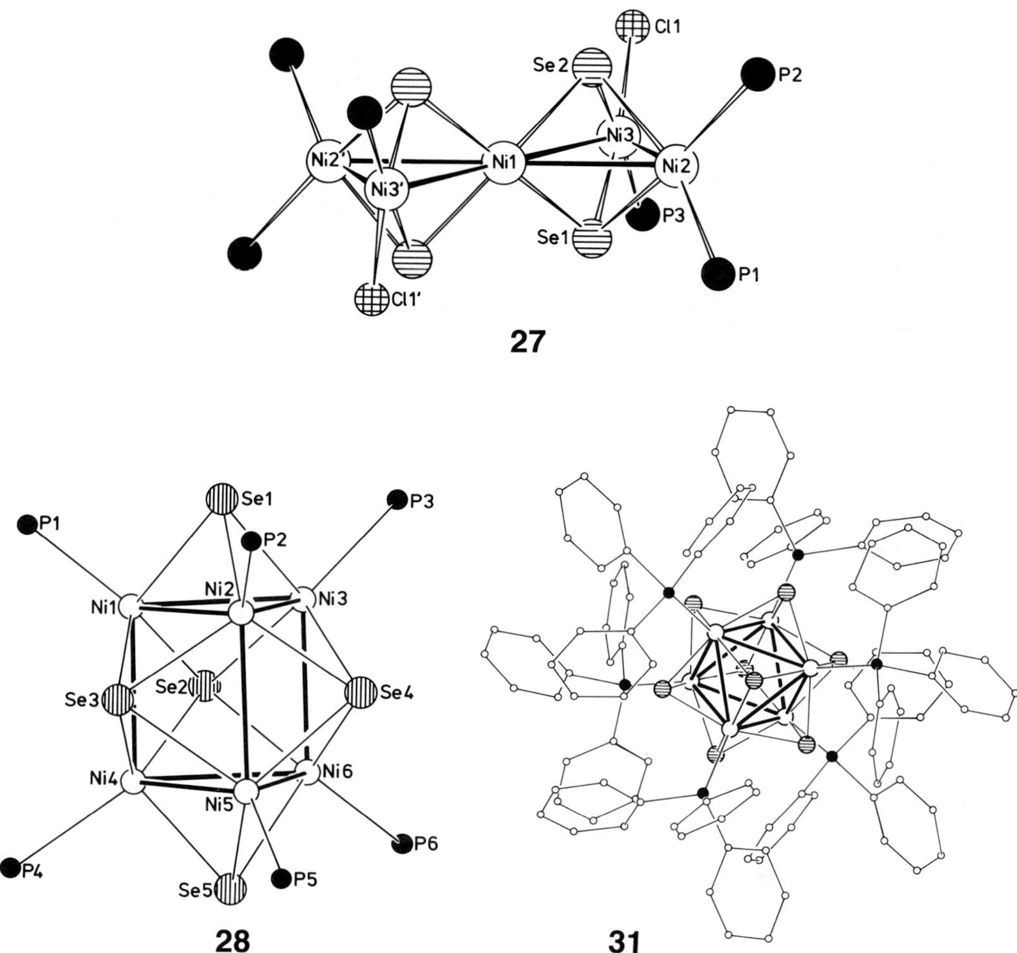

Figure 3-62. Molecular structure of **27** (C atoms omitted), **28** (without phenyl groups) and of $[Co_6E_8(PPh_3)_6]$ as exemplified by **31**.

M_6 Clusters

Two cluster compounds, **24** and **28**, can be isolated from the reaction of (nBu_4N)-$[NiCl_2(PPh_3)]$ with $Se(SiMe_3)_2$ in toluene. [168]

$$(n\mathrm{Bu_4N})[\mathrm{NiCl_2(PPh_3)}] \xrightarrow{\mathrm{Se(SiMe_3)_2}} \underset{\mathbf{24}}{[\mathrm{Ni_4Se_3(PPh_3)_4}]} + \underset{\mathbf{28}}{[\mathrm{Ni_6Se_5(PPh_3)_6}]} \quad (3.44)$$

As can be seen Figure **3-62**, **28** contains a Ni_6 trigonal prism in which all faces are capped using two μ_3-Se and three μ_3-Se atoms. Each Ni atom is also coordinated to the P atom of a PPh_3 group. The Ni–Ni bond lengths within the triangular faces (273.0–279.8 pm) are appreciably longer than those between these faces (2.60.2–266.3 pm). This distortion may be related to the fact that the cluster has 92 valence electrons, which is two more than predicted by either the $18e^-$ rule or topological considerations. [26–29] The ionic species $[Pt_3(\mu_2\text{-}CO)_3(CO)_3]_2^{2-}$ has a similar trigonal prismatic structure, [169] but in this case the metal–metal bonds within the triangular faces (266 pm) are shorter than those between them (304 pm).

The preparation of octahedral Ni_6 clusters seems to be impossible. In contrast, octahedral Co_6 clusters can be readily obtained from the reaction of either $[CoCl_2(PR_3)_2]$ or $(nBu_4N)[CoCl_3(PR_3)]$ with $E(SiMe_3)_2$ (E = S, Se, Te) in toluene or THF. Typical examples of these are: $[Co_6S_8(PnBu_3)_6]$ (**29**), $[Co_6S_8(PPh_3)_6]$ (**30**), $[Co_6Se_8(PPh_3)_6]$ (**31**), $[Co_6Se_8(PnBu_3)_6]$ (**32**), and $[Co_6Te_8(PEt_3)_6]$ (**33**). [173, 174] There are, in addition, numerous other ways of synthesizing this type of cluster. [171–177] Depending on the reaction conditions, the preparation of such ionic species as $[Co_6S_8(PPh_3)_6]^+[CoCl_3(THF)]^-$ (**34**) and $[Co_6Se_8(PPh_3)_6]^{2+}[CoCl_3(PPh_3)]_2^-$ is possible. [170] The basic structural principle in these compounds is illustrated in Figure **3-62** using **31** as an example. A central octahedron of Co atoms has each triangular face capped by a μ_3-E ligand. [170] The irregular tetrahedral coordination of each metal atom is completed by the P atom of a phosphine ligand. Table **3-17** compares selected structural parameters for the Co_6 clusters [170, 178] with those of the $[Co_6S_8(PEt_3)_6]^{n+}$ (n = 0,1) clusters prepared by Cecconi et al. [172, 175–177] Four significant points emerge:

a) The $97e^-$ and $98e^-$ clusters have respectively 13 and 14 electrons more than would be expected according to the $18e^-$ rule. This excess of electrons is also incompatible with other electron counting rules, such as the topological rules of Teo et al. [26–29] or Mingos et al. [30–33] The long Co–Co distances of

Table 3-17. Averaged structural parameters for the $[Co_6(\mu_3\text{-}E)_8(PR_3)_6]^{m+}$ (E = S, Se; R = Et, nBu, Ph; m = 0,1) series of isostructural clusters.

	Co–Co [pm]	E–E [pm]	Co–E [pm]	Co–P [pm]	Co–(μ_3-E)–Co [°]	Deviation of the Co atoms from the E_4 plane [pm]
$[Co_6S_8(PEt_3)_6]^+$	279.4	310.1	223.4	216.2	77.4	42–43
$[Co_6S_8(PEt_3)_6]$	281.7	309.5	223.3	213.8	78.2	45
$[Co_6S_8(PPh_3)_6]^+$ in **34**	281.9	308.6	222.8	218.0	78.5	45
$[Co_6S_8(PPh_3)_6]$ **30**	287.5	308.2	223.8	216.8	80.3	49
$[Co_6S_8(PnBu_3)_6]$ **29**	281,5	309.4	223.5	213.5	78.0	45
$[Co_6Se_8(PnBu_3)_6]$ **32**	294.6	327.2	235.4	214.7	77.6	45
$[Co_6S_8(PPh_3)_6]^+$	290.1	324.5	234.3	218.2	76.9	42–43
$[Co_6Se_8(PPh_3)_6]$ **31**	300.9	322.0	235.1	217.0	79.6	50
$[Co_6Te_8(PEt_3)_6]$ **33**	322	348.3	252.1	213.9	79.5	52

280–322 pm are quite possibly a consequence of the large number of valence electrons. Consistent with this view is the fact that the Co–Co contacts in the neutral complexes **29**, **30**, **31**, and **32** are up to 10 pm longer than those in their monocationic derivatives and implies that the one electron oxidation of **29**, **30**, and **31** involves the removal of an electron from an antibonding orbital. The symmetry of the clusters is unaffected by the change in valence electron number.

b) The Co–Co distances in the selenium bridged Co_6 octahedra of **31** and **32** are appreciably longer than those in the analogous sulphur derivatives **29** and **30**. The S–S and Se–Se contacts are about 50 pm less than the sum of the Van der Waals radii and could, therefore, be considered as weakly bonding. It is possible that the weakening of the Co–Co bonds is directly related to the increasing interactions between the S (or Se) ligands.
An alternative description of all the compounds in Table **3-17** is as cubic E_8 clusters (E = S, Se, Te), whose six square faces are each capped by $Co(PR_3)$ groups. Accordingly, the Co atoms lie 42–50 pm above the cube faces. This value is considerably larger than that of ± 14.5 pm found in the isostructural $[Mo_6Se_8]^{4-}$ cluster anion, in which weak interactions between the Se ligands are also presumed to occur. [179]
Recently, a tellurium analog to **29–34** has been synthesized. [173] The Co–Co distances of 322 pm and the Te–Te distances of 348.3 pm reinforce the hypothesis that the bonding situation in this cluster type is dominated by Co–E interactions and that any Co–Co interactions to be considered must be very weak. [180]

c) The PPh_3 substituted clusters have longer Co–Co bonds than the analogous derivatives with PEt_3 or PBu_3 ligands.

d) The Co–E bond lengths are influenced by neither the cluster charge nor the nature of the phosphine ligands. This implies that the metal-chalcogen bonding orbitals lie at lower energy than the Co–Co, Co–P, or E–E bonding orbitals. This in turn means that the Co–E bonds must make a considerable contribution to the stability of the cluster framework.

M_7 Clusters

The reactions of $[CoCl_2(PPh_3)_2]$ or $[CoCl_2(MeCN)_2]$ and PPh_3 with $S(SiMe_3)_2$ lead to the formation of compounds **35** and **36**, respectively, the metal frameworks of which are composed of Co_7 clusters as shown in Figure **3-63**. [170]

The central structural feature of **35**, **36**, and the cation in **37** is a metal cluster framework which may be considered as being derived from a cube from which one vertex has been removed. The resulting polyhedron is thus composed of three Co_4 and four Co_3 faces. With the exception of the triangular face described by Co2-Co2A-Co2B, all faces are capped with either μ_4-S or μ_3-S ligands. The clusters **35–37** differ in the number of chlorine atoms and phosphine ligands which are coordinated to their Co_7 frameworks, the sum of which, however, is always seven. Whereas **35** has two chlorine atoms and five phosphine ligands,

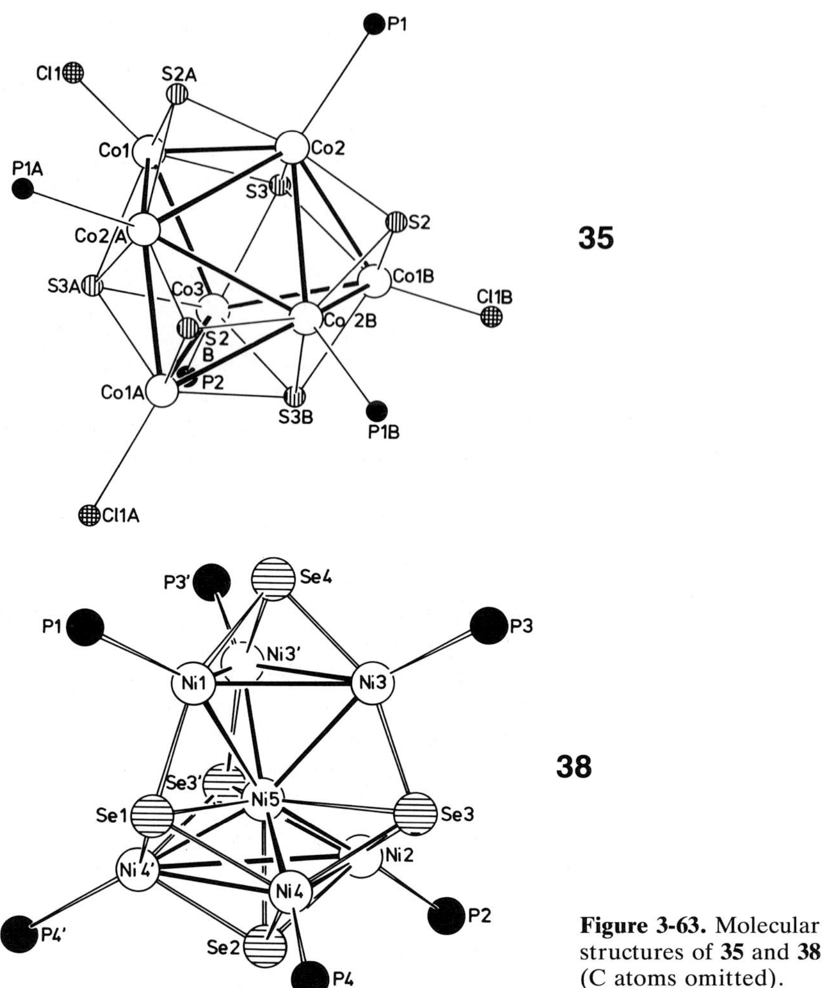

Figure 3-63. Molecular structures of **35** and **38** (C atoms omitted).

36 has three chlorine atoms and four phosphine ligands, and **37** has only one chlorine atom but six phosphine ligands.

There are nine short Co–Co bonds (**35**: 257.4–263.7 pm; **36**: 257.1–262.2 pm; **37**: 253.6–267.0 pm) and three longer contacts within the chalcogen free triangular face (**35**: 289.7 pm; **36**: 283.4 pm; **37**: 283.3–287.6 pm). Since **36** is isostructural with the $[Fe_7Cl_3S_6(PEt_3)_4]$ cluster prepared by Holm et al., it can be considered as having a geometry intermediate between prismatic and cuboctahedral. [181] Another isostructural compound is the $[Co_7I_3S_6(PEt_3)_4]$ cluster, which can be prepared quite simply by the reaction of CoI_2 and PEt_3 with $S(SiMe_3)_2$. [171] According to the $18e^-$ rule, the paramagnetic compounds **35** and **37** (each with 99 electrons) have a deficiency of two electrons and the diamagnetic **36** (with 98 electrons) a deficiency of three electrons. This is also reflected in the ready reactions of **35** and **36** with CO. The reaction with **35** leads to a compound with

$$[Co_7Cl_2S_6(PPh_3)_5] \xleftarrow{[CoCl_2(PPh_3)_2]} S(SiMe_3)_2 \xrightarrow[+\ PPh_3]{[CoCl_2(MeCN)_2]} [Co_7Cl_3S_6(PPh_3)_4]$$

35 **36**

$$\downarrow [CoCl_2(PEt_2Ph)_2]^-$$

$$[Co_7ClS_6(PEt_2Ph)_6]^+[CoCl_3(PEt_2Ph)]^-$$

37

Scheme 3-18. Routes to Co_7 clusters.

the composition $[Co_7(CO)_3Cl_2S_6(PPh_3)_5]$ whereby the addition of CO presumably occurs at the triangular face not occupied by a S ligand.

Figure **3-63** also shows the structure of the Ni_7 cluster $[Ni_7Se_5(P\mathit{i}Pr_3)_6]$ (**38**). An important characteristic of this cluster is the mirror plane which runs through the atoms Se1, Se2, Se4, Ni1, Ni2, and Ni5. The structure can be described as consisting of two vertex-sharing tetrahedra. In accordance with the principle of cluster condensation, this (previously unknown) cluster type contains 102 valence electrons. In the (Ni1, Ni3, Ni3′, Ni5) tetrahedron, the Ni–Ni distances range from 258.3–280 pm whereby the shorter bonds are to the central Ni5 atom of the molecule. The second tetrahedron (Ni2, Ni4, Ni4′, Ni5) has much shorter Ni–Ni5 distances of 233–234 pm. Similarly short Ni–Ni distances have only been found in binuclear complexes like $[Cp_2Ni_2X_2]$ (X = MeCN, Ph_2C_2,CO). [182–184] The 330 pm distances between Ni2, Ni4, and Ni4′ are considerably longer than those reported for normal Ni–Ni bonds. The polyhedral faces of the bitetrahedra are capped by one μ_3-Se ligand (Se4) and four μ_4-Se (Se1, Se2, Se3, and Se3′) ligands.

M_8 Clusters

Ni_8 clusters can be prepared by the direct reaction of $[NiCl_2(PR_3)_2]$ with $E(SiMe_3)_2$ (E = S, Se) and a variety of structure types can occur depending on the PR_3 ligand employed. The following Ni_8 clusters are derived from a cubic structure. [166, 170] $[Ni_8Cl_2S_6(PPh_3)_6]$ (**39**) $[Ni_8\square_2S_6(PPh_3)_6]$ (**40**) $[Ni_8\square_2Se_6(P\mathit{i}Pr_3)_4]$ (**41**) and $[Ni_8\square_2Se_6(PEt_2Ph)_6]$ (**42**). As an analogue to the Ni_8 clusters **2–7** which contain μ_4-PPh ligands (see section 3.1.), **39** (Fig. **3-64**) also consists of a slightly distorted Ni_8 cube. In contrast to **2–7**, however, all the cube faces are capped by μ_4-S ligands, six of the Ni atoms (Ni1, Ni2, Ni4, Ni1′, Ni2′, and Ni4′) are each bound to a PPh_3 group, and the other two Ni atoms (Ni3 and Ni3′) are each bound to a chlorine atom. [170] The Ni–Ni bond lengths lie within the range given in Table **3-16**. The cluster **39** can be reduced with Na/Hg in THF to form the related cluster **40**, which has lost two chlorides and the Ni3 and Ni3′ atoms are now only bound to three μ_4-S ligands. The vacant coordination sites (denoted by □) thus formed are topologically similar to those in **3**. Accordingly, **40** will react with CO to form $[Ni_8(CO)_2S_6(PPh_3)_6]$. The addition of larger molecules has not yet been attempted. Vacant coordination sites can also arise if **39** is treated with $AgPF_6$, which leads to the $[Ni_8\square_2S_6(PPh_3)_6]$ (PF_6) salt.

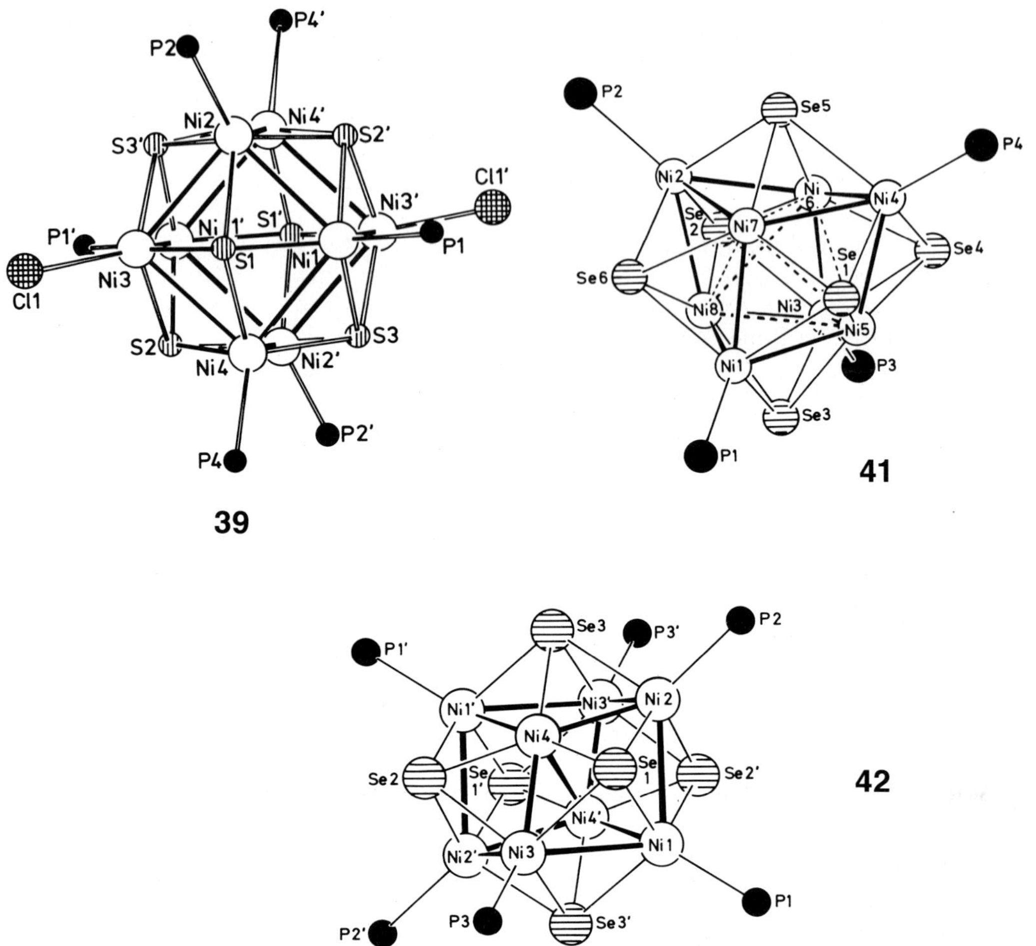

Figure 3-64. Molecular structures of **39**, **41**, and **42**. (C atoms are omitted).

Complex **41** (Fig. **3-64**) has 112 valence electrons and therefore fits into the series of known cubic Ni_8 clusters. A distortion of the Ni_8 cube is observed with increasing deviation from the expected valence electron count of 120. This distortion manifests itself by a shortening of certain Ni–Ni bonds and a lengthening of others. [23, 185] The ratio, *r*, of the length of the shorter face diagonal to the length of the edge of the cube can be used as a measure of this distortion. In clusters with 120 valence electrons, for example $[Ni_8Se_6(PnBu_3)_8]$, a regular Ni_8 cube (Ni–Ni: 270 pm) is found and $r = 1.41$. [23, 166, 185] In contrast, $r = 1.28$ for $[Ni_8\square_4(PPh)_6(PPh_3)_4]$ (Ni–Ni: 251.7–254.1 pm; Ni $\cdots$ Ni: 323–325 pm) with only 112 valence electrons, and for **41** this value is even further reduced to 1.14. The differing bonding situations of the Ni atoms also have an influence on the Ni–Se bond lengths. Ni atoms with the coordination number three (Ni5–Ni8) have Ni–Se bond lengths of 227.5–230.1 pm, which are about 15 pm shorter than those for the Ni atoms with the coordination number four (Ni1–Ni4: 242.2–245.6 pm).

The controlling influence of the ligands on the clusters structure can be clearly illustrated using **42** (Fig. **3-64**) as an example. As in **41**, the molecule consists of a distorted cube ($\bar{1}$ symmetry), whose square faces are all capped by μ_4-Se ligands. Six of the Ni atoms (Ni1, Ni2, Ni3, Ni1′, Ni2′, and Ni3′) have the coordination number four and are bonded to three Se atoms and one PEt_2Ph ligand. Ni4 and Ni4′ however are only coordinated to three Se atoms. This coordination deficiency leads to a distortion of the Ni_8 cube along the Ni4–Ni4′ diagonal (266.7 pm). The bonds from Ni4 to Ni1′, Ni2, and Ni3 (and from the symmetry related Ni4′ to Ni1, Ni2′, and Ni3′) measure 238.3 pm and are considerably shorter than the other Ni–Ni distances (266.7–277.9 pm). In **42**, as in **41**, the Ni–Se bonds to the Ni atoms with coordination number three are between 14 and 21 pm shorter than those to the Ni atoms with coordination number four. Complex **42** has, with 116 valence electrons, four electrons less than the usual electron counting rules would normally predict for a cubic cluster.

$$[NiCl_2(PPh_3)_2] \xrightarrow{S(SiMe_3)_2} \underset{\mathbf{39}}{[Ni_8Cl_2S_6(PPh_3)_6]} + [NiS(PPh_3)Cl]_n$$

$$[NiS(PPh_3)Cl]_n \xrightarrow{Zn} \underset{\mathbf{43}}{[Ni_8S_5(PPh_3)_7]}$$

Scheme 3-19. Formation of S bridged Ni_8 clusters.

The reaction of $[NiCl_2(PPh_3)_2]$ with $S(SiMe_3)_2$ in THF leads to the formation of **39** and a compound with the composition $[NiS(PPh_3)Cl]_n$ but whose structure is as yet unknown.

Although the latter can be recrystallized from THF or acetone, no single crystals have been obtained to date. If this compound is dehalogenated with Zn in THF, **43** is formed in 60% yield. This Ni_8 cluster is illustrated in Figure **3-65** and is constructed of two trigonal bipyramids which share a common edge. With the exception of Ni7, all the Ni atoms are bonded to a PPh_3 ligand. Not all the faces of the polyhedron can be capped by S atoms due to the large size of the PPh_3 ligands. S1 and S2 are μ_3-S ligands with Ni–S bond lengths of 211–217 pm. S5 can also be considered as a μ_3-ligand, albeit with unusual Ni–S–Ni bond angles in the 72.4–137.4° range. The S3 and S4 atoms are in the odd position of having two quite distinctly different bond lengths to their Ni atoms. In addition to the "normal" μ_3-S bond lengths of 213–219 pm to Ni, they show somewhat weaker contacts to the central Ni7 atom of 258.6–268.8 pm. If these weaker contacts are also to be considered as bonding interactions, S3 and S4 must then be described as μ_4-ligands. A realistic description of the bonding in this 114 valence electron cluster is difficult because of the wide range in the Ni–Ni bond lengths from 240–297 pm.

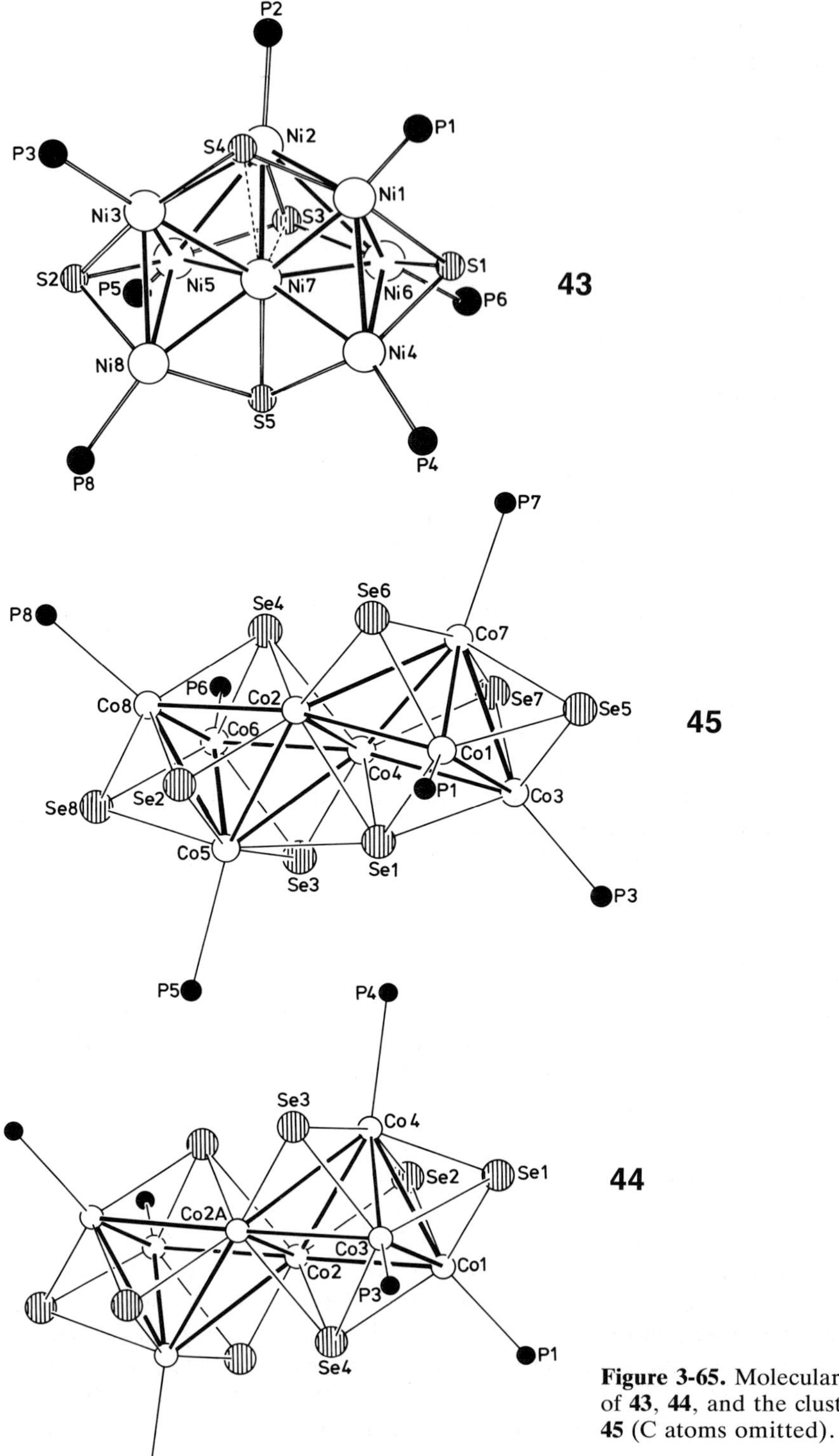

Figure 3-65. Molecular structures of **43**, **44**, and the cluster cation in **45** (C atoms omitted).

The important role played by the solvent can be demonstrated by the reaction of $(n\text{Bu}_4\text{N})[\text{CoCl}_3(\text{PPh}_3)]$ with $\text{Se(SiMe}_3)_2$ outlined in Equation (3.45).

$$(n\text{Bu}_4\text{N})[\text{CoCl}_3(\text{PPh}_3)] \xrightarrow{\text{Se(SiMe}_3)_2} \underset{\mathbf{31}}{[\text{Co}_6\text{Se}_8(\text{PPh}_3)_6]} + \tag{3.45}$$

$$\underset{\mathbf{44}}{[\text{Co}_8\text{Se}_8(\text{PPh}_3)_6][\text{Co}_6\text{Se}_8(\text{PPh}_3)_6]} + \underset{\mathbf{45}}{[\text{Co}_8\text{Se}_8(\text{PPh}_3)_6]^+[\text{CoCl}_3(\text{PPh}_3)]^-}$$

$$+ \underset{\mathbf{46}}{[\text{Co}_9\text{Se}_9(\text{PPh}_3)_6]}$$

If toluene is used as solvent, the products **31** (already discussed in the M_6 cluster subsection and the Co_9 cluster **46** (discussed in the next subsection) are obtained. If acetonitrile is used instead as solvent, the compounds **44** and **45** are obtained. Crystals of **44** contain both the uncharged cluster compounds $[\text{Co}_8\text{Se}_8(\text{PPh}_3)_6]$ (Fig. **3-65**) and $[\text{Co}_6\text{Se}_8(\text{PPh}_3)_6]$. In contrast, **45** consists of the charged cluster cation $[\text{Co}_8\text{Se}_8(\text{PPh}_3)_6]^+$ (Fig. **3-65**) and the complex anion $[\text{CoCl}_3(\text{PPh}_3)]^-$. The Co_8 polyhedra in both these compounds can be considered as being derived from two mutually trans square pyramids linked by a common edge. The faces are capped by either μ_3- or μ_4-Se ligands. The Co atoms making up the pyramid bases (**44**: Co1, Co2, Co3 and their symmetry related atoms; **45**: Co1, Co2, Co3, Co4, Co6, and Co8) are bonded to PPh_3 ligands. Unlike the neutral **44**, the $[\text{Co}_8\text{Se}_8(\text{PPh}_3)_6]^+$ cation of **45** does not possess an inversion centre. The bases of the two Co_5 square pyramids in **45** do not therefore constitute a single plane. This reduction in symmetry may have the following cause. The one electron oxidation of the neutral **44** promotes the formation of an additional weak bond between Co5 and Se1 (258 pm) in **45**. The consequence is a mutual tilting of the Co_5 pyramids, and the dihedral angle between the planes defined by (Co1, Co2, Co3, Co4)/(Co2, Co4, Co6, Co8) in the cation of **45** is 12.5°. The association of this structural change with the loss of one electron in the oxidation of **44** to **45** is quite remarkable. The Co–Co bond lengths however, remain largely unaltered (244.2–279.5 pm in **44**, 247.3–279.4 pm in **45**), and are appreciably shorter than those in **31**. This is consistent with the electron counts. Whereas **31** exceeds the requirements of the 18e^- rule by 14 electrons, **44** (116e^-) and **45** (115e^-) approach the optimum number of 114 valence electrons.

Clusters with Face-Sharing Co and Ni Octahedra

The condensation of octahedra via common vertices, edges or faces leads to cluster units of a type known since the end of the 1970's. Among these are the metal-rich subhalides of the early transition metals [186–188] and the Chevrel phases. [179, 189–192] The preparation of $[\text{Co}_9\text{Se}_{11}(\text{PPh}_3)_6]$ (**46**) (Fig. **3-66**) made it clear that this structural principle could be extended to the more electron-rich transition metals. [164, 168] All faces of the Co_9 cluster in **46** are capped by either μ_3- or μ_4-Se ligands. The Co atoms of the basal faces (Co1, Co4, Co9, Co6, Co7,

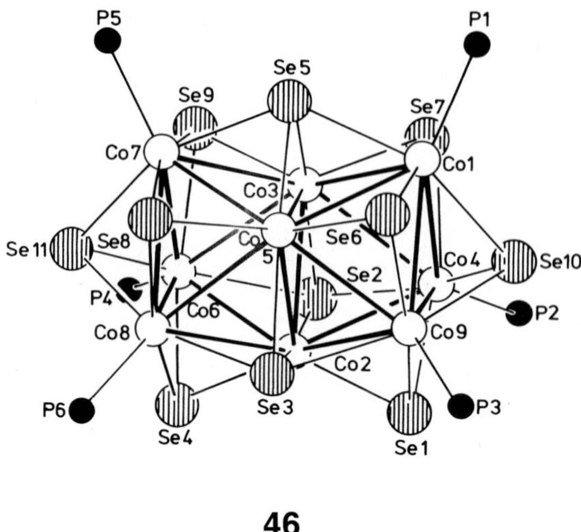

46

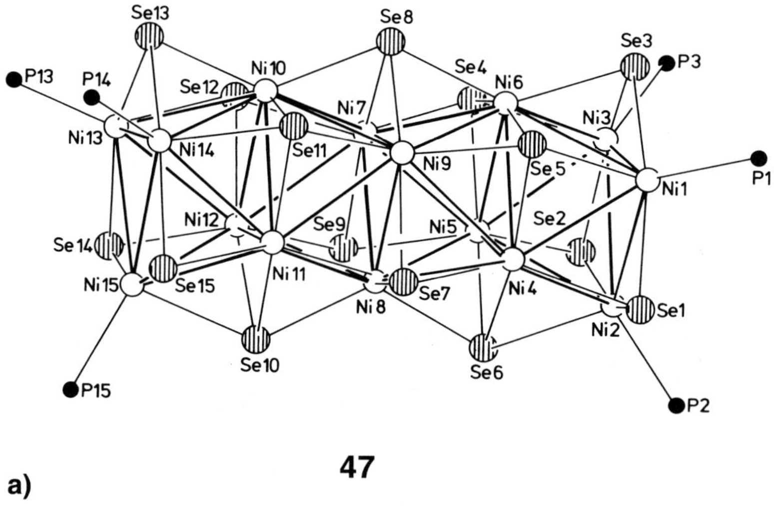

47

a)

Figure 3-66. Molecular structures of **46**, **47**, **48**, and the cluster cation in **49** (phenyl groups omitted in **46**, **48**, and **49**).

C2
C1
C4
C3
P1
C5
C16
Se4
Se2
Ni5
C15
C6
Ni1
C18
P3
Se5
Ni3
C17
Ni6
Se1
C14
C13
Ni2
Ni4
C12
Se3
Se6
P2
C11
C7
C8
C9
C10

48

P3
P10
Se1
Se6
Ni3
Se11
P11
Ni10
Se3
Se9
Ni4
Ni1
P1
Ni6
Ni2
P12
Ni11
Ni9
Se4
P2
Se8
Ni12
Se10
Ni5
Se2
Ni7
Se5
Ni8
P7
Cl
Se7
P8

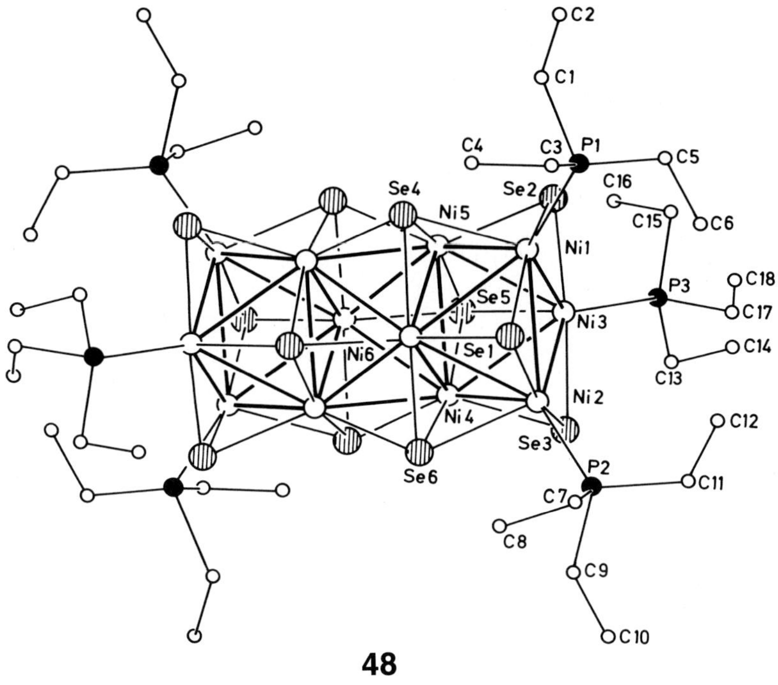

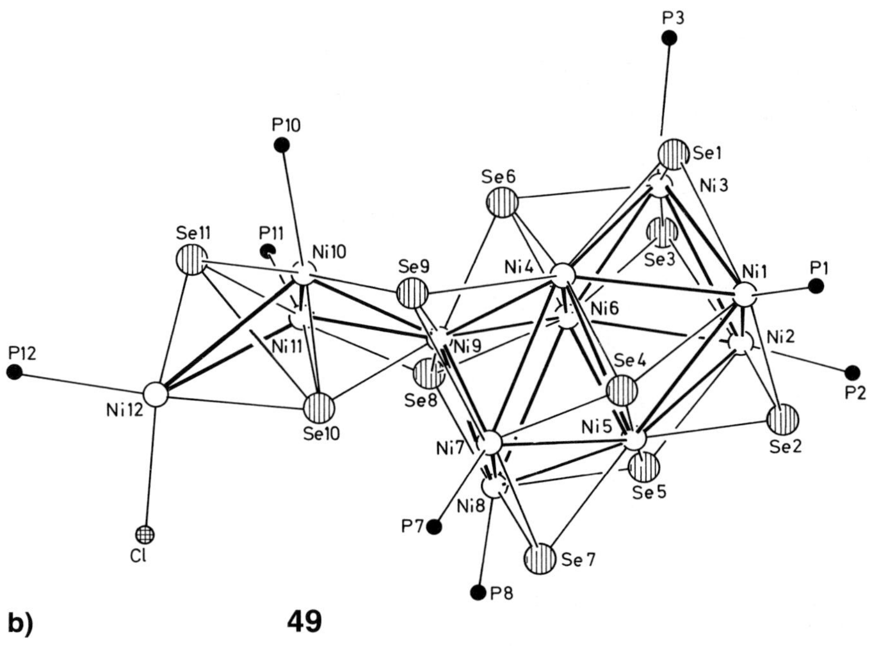

b) **49**

Co8) are each coordinated to an additional PPh_3 ligand. The Co–Co bond lengths lie between 271–298 pm; the shortest distances being found within the (Co2, Co3, Co5) triangular face common to both octahedra.

Clusters with similar structures are known in $[Ni_9S_9(PEt_3)_6]^{2+}$ and $[Rh_9(CO)_{19}]^{2-}$. [193–195] The condensation of two Co octahedra clearly permits greater tolerance with respect to the valence electron count, as can be seen by the fact that **46** contains 17 valence electrons more than predicted by the $18e^-$ rule.

The formation of **46** must proceed via smaller, more reactive complexes. The two Co_8 clusters in **44** and **45** are tenable candidates for being precursors or intermediates in the formation of the bioctahedral Co_9 cluster **46**. Additional redox processes could cause the mutual tilting of the Co_5 pyramid bases to become so pronounced that a suitable cavity would be formed for the incorporation of a further Co atom.

Face sharing metal atom octahedra are also observed in the Ni compounds **47** and **48** and in the cationic cluster in **49** (Fig. **3-66**).

$$\underset{\mathbf{47}}{[Ni_{15}Se_{15}(PPh_3)_6]} \xrightarrow[\text{MeCN}]{(nBu_4N)[NiCl_3(PPh_3)]} Se(SiMe_3)_2 \xrightarrow[\text{toluene}]{[NiCl_2(PEt_3)_2]} \underset{\mathbf{48}}{[Ni_{12}Se_{12}(PEt_3)_6]}$$

$$Se(SiMe_3)_2 \xrightarrow[\text{MeCN}]{[NiCl_2(PPh_3)_2]} \underset{\mathbf{49}}{[Ni_{12}Se_{11}(PPh_3)_8Cl]^{2+}[NiCl_3(PPh_3)]_2^-}$$

Scheme 3-20. Formation of face sharing octahedral nickel clusters.

On the other hand, it has not yet proved possible to sucessfully isolate a simple octahedral chalcogen-bridged Ni_6 cluster, although closely related compounds such as $[Ni_3(\mu_2\text{-}CO)_3(CO)_3]_2^{2-}$ are known. [169] Electronic factors are presumably responsible for this. A hypothetical "$[Ni_6Se_8(PPh_3)_6]$" cluster would have a total of 104 valence electrons. This is not only 20 electrons more than predicted by the $18e^-$ rule, but also six electrons more than in the analogous Co_6 complex **31**, whose long Co ··· Co contacts already imply the occupation of antibonding orbitals. The dicationic cluster in **49** consists of linked Ni_9Se_9 (Ni1–Ni9, Se1–Se9) and Ni_3Se_2 (Ni10–Ni12, Se10, Se11) units (Fig. **3-66**). These structural subunits are found in several other compounds such as $[Ni_9Se_9(PEt_3)_6]^{2+}$, $[Ni_3Se_2(PEt_3)_6]^{2+}$, $[Ni_3S_2Cp_3]$, and $[Ni_3Cl_2S_2(PPh_3)_4]$ (**23**). [193, 194, 196] The polyhedral faces of the bioctahedral subunit in **49** are capped by μ_3-Se (Se1, Se2, Se3, and Se7) and μ_4-Se (Se4, Se5, and Se6) ligands. Se8 and Se9 are also each bonded to a face of the bioctahedron but are also edge-bridged to the Ni_3 subunit (Ni11 and Ni10 respectively). These latter Ni–Ni bonds to Ni9 are relatively short (253–268 pm), whereas those within the Ni_3 cluster fragment itself (Ni10, Ni11, Ni12) are rather long (283–311 pm) and might indicate only weak Ni–Ni interactions between the Ni10, Ni11, and Ni12 atoms. The structural integrity of this Ni_3

triangle can therefore be ascribed primarily to the μ_4- and μ_3-ligands Se10 and Se11. The Se10–Se11 distance of 311 pm is very short and agrees well with the corresponding distance in the $[Ni_3Se_2(PEt_3)_6]^{2+}$ cation. With the exceptions of Ni4, Ni5, Ni6, and Ni9, all other Ni atoms are coordinated by a PPh_3 ligand. Ni12 is also bound to a Cl atom. The topology of the Ni and Se atoms means that the cation of **49** must be regarded as an intermediate en route to more highly condensed clusters of *trans* face-sharing metal atom octahedra. Obviously, the Ni–Cl bond is the reactive site of the molecule, where reaction with $Se(SiMe_3)_2$ and $[NiCl_2(PPh_3)_2]$ can promote further growth of the metal atom polyhedron.

In accord with the above, **47** and **48** are composed of four or three face-sharing, slightly distorted Ni_6 octahedra (Fig. **3-66**). Apart from the basal faces, all polyhedral faces in **47** and **48** are capped by μ_3- or μ_4-Se ligands (Ni–Se bond lengths: **47**: 231–249 pm; **48**: 232–244 pm). The Ni atoms of the basal faces are each bonded to a PPh_3 (**47**) or PEt_3 (**48**) ligand. Alternatively, **47** and **48** can be described as parallel stacks of layers each composed of three Ni and three Se atoms arranged in a hexagonal close-packing fashion. The interlayer distance is 216(3) pm with deviations from the plane restricted to only $\pm$10 pm. The Ni–Ni distances between the layers in **47–49** are shorter (**47**: 270–288 pm; **48**: 272–278 pm; **49**: 266–289 pm) than those within the individual layers (**47**: 288–301 pm; **48**: 288–297 pm; **49**: 268–313 pm). It is also observed that the 292–301 pm Ni–Ni bond lengths within the basal planes are appreciably longer than those within the inner layers. This may simply be a consequence of the fact that the Ni atoms in the basal planes are coordinated by phosphine ligands, whereas the Ni atoms in the inner planes are not.

The structures of the Chevrel phases, such as $[Mo_{12}Se_{14}]^{2-}$, $[Mo_{18}Se_{20}]^{4-}$, $[Mo_{24}Se_{26}]^{6-}$ and $[Mo_{30}Se_{32}]^{8-}$, are governed by the same structural principle as the Ni clusters **47–49** and the Co cluster **31**. [188, 197, 198] As regards their electron counts, however, there are considerable differences. For instance, the $222e^-$ cluster **47** and the $180e^-$ cluster **48** contain 30 and 24 electrons more, respectively, than predicted by the $18e^-$ rule, whereas in the Chevrel phases the $18e^-$ rule is largely obeyed. Clearly, antibonding orbitals are increasingly occupied in **47** and **48**. The differences can be further illustrated by a comparison of the valence electron concentrations. The Chevrel phases contain between 3.33–4.33 electrons per metal atom, whereas the Co clusters contain 6.17–6.78 electrons and the Ni clusters 8 valence electrons per metal atom. [192]

The composition of the Co and Ni compounds correspond to the general formula $[Co_{3n+3}E_{3n+5}(PR_3)_6]$ and $[Ni_{3n+3}E_{3n+3}(PR_3)_6]$, where E = S, Se and n is the number of condensed octahedra. The valence electrons of the nickel compounds are subject to a simple counting rule: $3 \times 10 + 3 \times \mathbf{4} = 42$ electrons for each Ni_3E_3 unit, plus $6 \times 2 = 12$ electrons for the terminal PR_3 ligands. It is entirely possible that still larger clusters which conform to this structural principle are formed in the reactions described above. Residues are often obtained, which, according to analytical results and general properties, seem to be of very high molecular weight. However, because of their increasing insolubility, their characterization is difficult.

$[Ni_{21}Se_{14}(PEt_2Ph)_{12}]$ and $[Ni_{34}Se_{22}(PPh_3)_{10}]$: Clusters with Unusual Structures

$[Ni_{21}Se_{14}(PEt_2Ph)_{12}]$ (**50**) and $[Ni_{34}Se_{22}(PPh_3)_{10}]$ (**51**) can be isolated from reaction product mixtures [164, 166] containing more than one cluster:

$$[Ni_8Se_6(PPh_3)_8] \xleftarrow{[NiCl_2(PPh_3)_2]} Se(SiMe_3)_2 \xrightarrow{[NiCl_2PEt_2Ph]} [Ni_8\square_2Se_6(PEt_2Ph)_6]\ \mathbf{42}$$

$$+ [Ni_{34}Se_{22}(PPh_3)_{10}]\ \mathbf{51} \qquad\qquad + [Ni_{21}Se_{14}(PEt_2Ph)_{12}]\ \mathbf{50}$$

Scheme 3-21. Formation of Ni_8, Ni_{21}, and Ni_{34} selenide clusters.

The structure of **50** (Fig. **3-67**, $\bar{1}$ symmetry) can be described as a central Ni_{13} cuboctahedron (Ni5–Ni11 and their symmetry related Ni atoms) which has two *trans* oriented Ni_4 faces (Ni5–Ni8, Ni5′–Ni8′) each bonded in a prismatic fashion to Ni_4 squares (Ni1–Ni4 and Ni1′–Ni4′) thereby forming two cube subunits. All Ni–Ni distances lie in the range 257.6–280.0 pm and correspond to the values observed in other selenium capped Ni clusters. This central, slightly distorted cuboctahedron represents a section of a face-centered cubic packed lattice. Previously reported compounds exhibiting this structure fragment include $[Pt_6Ni_{38}(CO)_{48}H_2]^{4-}$, $Pd_{38}(CO)_{28}(PEt_3)_{12}]$, and $[Pd_{23}(CO)_{22}(PEt_3)_{10}]$. [199, 200]

The structure of the Ni_{21} framework can be understood as an intermediate on the way to a series of full-shell clusters in which a central M_{13} cuboctahedron is surrounded by n shells each of which contain $(10n^2 + 2)$ metal atoms (for $n > 1$). This principle, as proposed by G. Schmid, is exemplified by clusters of the type $[M_{55}(PPh_3)_{12}Cl_6]$ (see Chapter 3.3). The atoms Ni1–Ni4 and Ni1′–Ni4′ can be regarded as being part of a second shell which can contain a maximum of 42 Ni atoms.

The metal framework is stabilized by 14 μ_4-Se ligands which cap the polyhedral Ni_4 faces (Ni–Se: 231.0–241.9 pm). The 12 peripheral Ni atoms (Ni1–Ni4, Ni9, Ni10 and the symmetry related atoms) are coordinated additionally by 12 PEt_2Ph ligands whose function is to provide a complete shielding of the $Ni_{21}Se_{14}$ core. The cluster **50** contains 290 valence electrons, which is 16 electrons more than predicted by the $18e^-$ rule if all Ni–Ni bonds shown in Figure **3-67** are considered as two-center, two-electron bonds. The principle for the condensation of clusters as formulated by D. M. P. Mingos leads to a valence electron number of 286, which is in relatively good agreement with the actual number counted for **50**.[30–33, 201, 202]

$[Ni_{34}Se_{22}(PPh_3)_{10}]$ (**51**) crystallizes from the filtrate of the reaction of $[NiCl_2(PPh_3)_2]$ with $Se(SiMe_3)_2$. The structure of **51** is governed by a previously unknown principle (Fig. **3-68**). The cluster has $\bar{1}$ symmetry and contains a central Ni_{14} unit consisting of two staggered planar Ni_5 rings (the Ni atoms of these rings are displayed as filled circles in Fig. **3-68** at a planar spacing of 448–456 pm with a parallel Ni_4 plane between them. Ni9 is disordered. The Ni_{34} polyhedron is completed by the addition of five Ni_4 "butterfly" fragments around thc equator

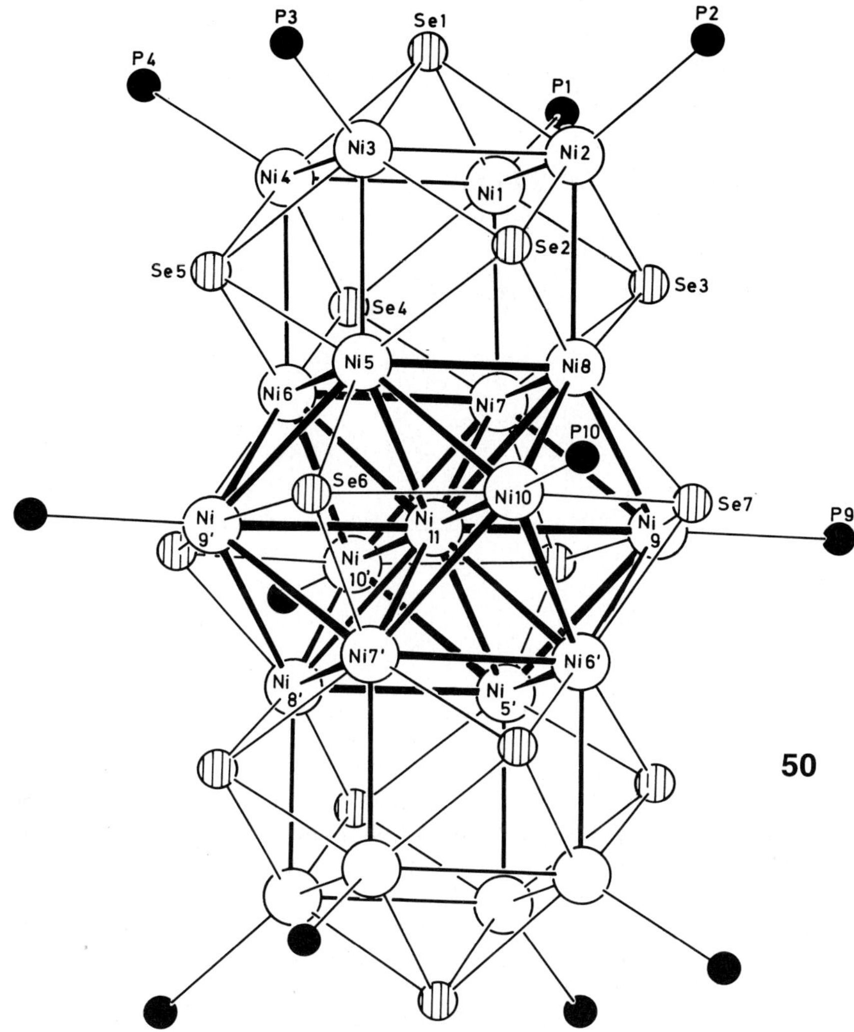

Figure 3-67. Molecular structure of **50** (C atoms are omitted).

of the pentagonal antiprismatic Ni_{14} unit. The surface of the Ni_{34} unit is then covered by twenty μ_4- and two μ_5-Se atoms (Ni–μ_4-Se: 226–246 pm, Ni–μ_5-Se: 236–246 pm). Ten peripheral Ni atoms (Ni5, Ni7, Ni8, Ni10, Ni13, and their symmetry equivalents) each have an additional coordination to one PPh_3 ligand. An alternative description, which stresses the highly symmetric nature of **51**, is based on the fivefold symmetry and marked layer structure of the Ni_{34} polyhedron. The fivefold symmetry of the Ni_5 and Se_5 planes is broken by the Ni_4 subunit in the center of the cluster, which is perhaps the reason for the disorder observed at Ni9.

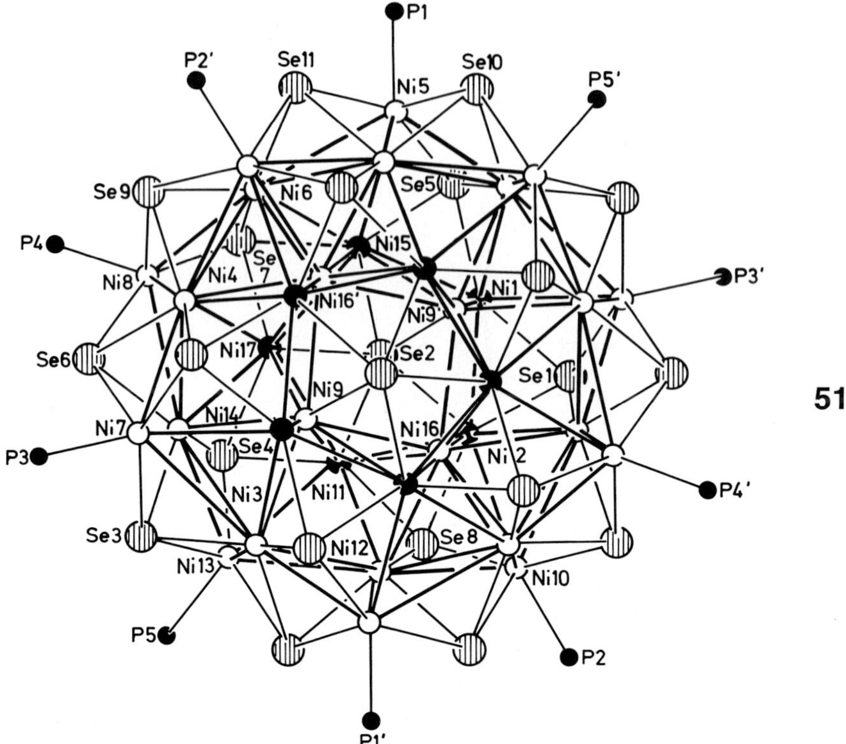

Figure 3-68. Molecular structure of **51** (phenyl groups omitted).

The cluster molecule has a diameter of about 3000 pm and the Ni–Ni distances in **51** lie in the range 236–312 pm. A description of the bonding in terms of the $18e^-$ rule is therefore not possible. **51** contains 448 valence electrons, which is in good agreement with the number predicted for M_n clusters ($n > 13$) on the basis of the Hume-Rothery rules for intermetallic phases ($440–444e^-$). [26–29] This indicates that the properties of cluster complexes will approach those of metals as the number of metal atoms increases. It should be emphasized, however, that the structure of the Ni_{34} unit cannot be described in terms of those principles derived for metals and intermetallic phases.

One can only speculate about the mechanism of formation of such complex molecules. It is a reasonable assumption, however, that aggregation of more highly condensed molecules occurs via smaller fragments. In this sense it is not unlikely that M_3 units (M = Ni, Co) play a decisive role in the syntheses. The most direct evidence for this is the existence of $[Ni_3Cl_2S_2(PPh_3)_4]$ (**23**) and $[Ni_3Cl_2Se_2(PEt_2Ph)_4]$ (**23 a**). It is notable that precisely such an Ni_3Se_2 subunit is found in $[Ni_{12}Se_{11}(PPh_3)_8Cl]^{2+}$ (**49**), being directly linked to a "precondensed" group: a Ni_9 cluster composed of three Ni_3 triangles. To illustrate the structure determining potential of the M_3 units, Scheme **3-22** presents schematically the metal atom frameworks of the sulphur and selenium bridged Ni clusters discussed here.

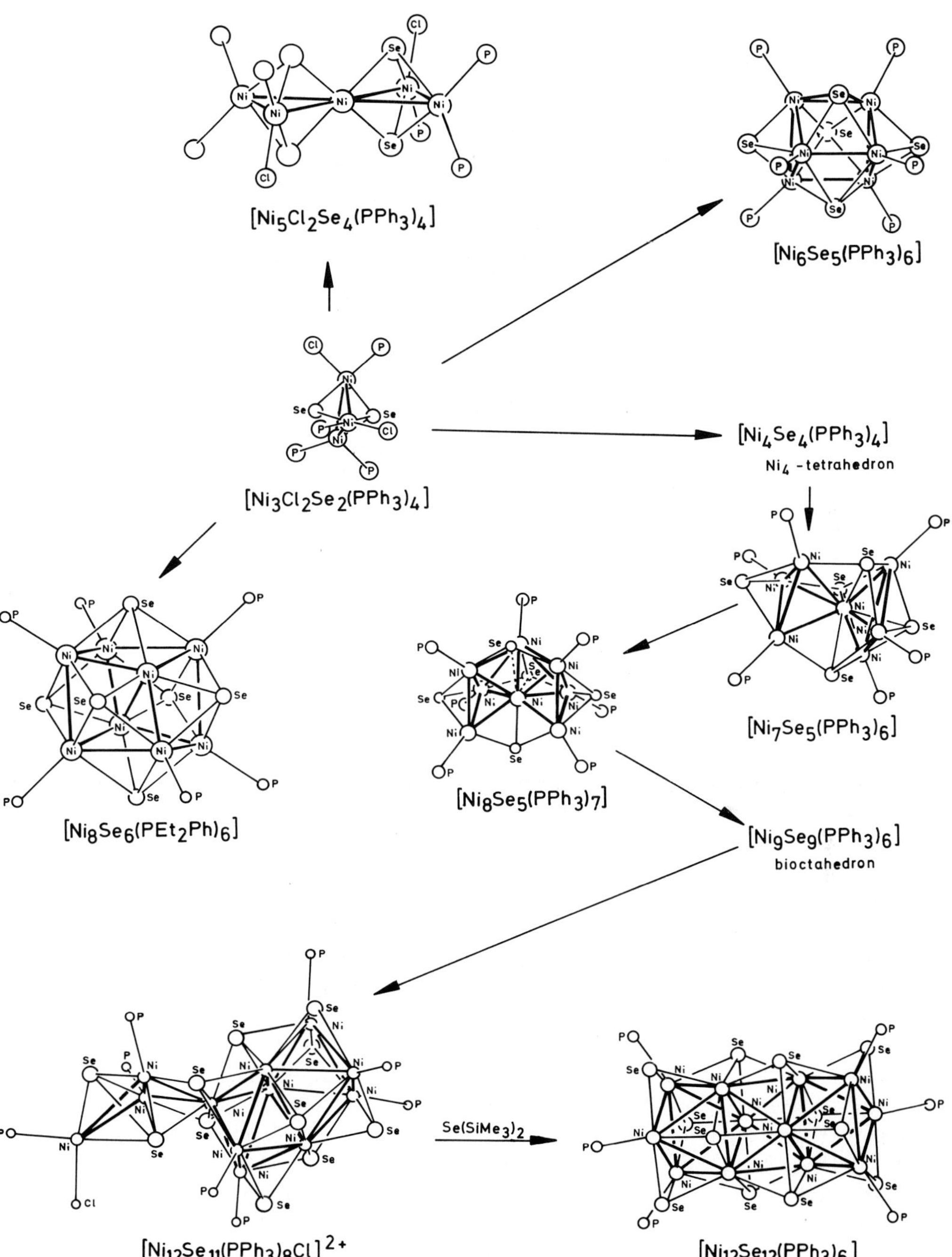

Scheme 3-22. Possible route for nickel cluster growth.

3.4.3.2 Palladium Clusters with S, Se, and Te Bridges

In the discussion on M_3 and M_4 clusters in Section 3.4.3.1, the possibility of the synthesis of trinuclear nickel clusters, such as $[Ni_3Cl_2S_2(PPh_3)_4]$ (**23**), was addressed. Performing the analogous reactions with $[PdCl_2(PPh_3)_2]$ and $E(SiMe_3)_2$ (E = S, Se, Te) leads to the formation of, among others, the trinuclear palladium complexes $[Pd_3Cl_2Se_2(PPh_3)_4]$ (**52**) and $[Pd_3(SeSiMe_3)_2Se_2(PPh_3)_4]$ (**53**). The relative amounts of **52** and **53** formed depend on the $Se(SiMe_3)_2$: $[PdCl_2(PPh_3)_2$ ratio of the starting materials (Scheme **3-23**). [158, 159]

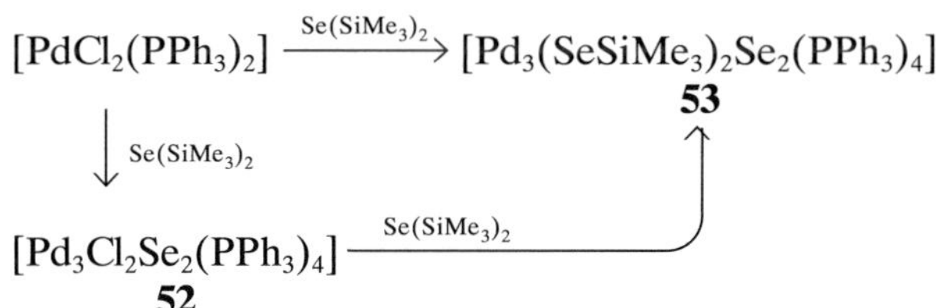

Scheme 3-23. Formation of Se bridged Pd_3 clusters.

It is also possible to synthesize the corresponding S and Te analogues in the same manner. The structures of **52a** (the sulfur analogue of **52**) and **53** are illustrated in Figure **3-69**. As can be seen, **52** consists of a structure analogous to that of the nickel compound **23**, with the difference that the chloride ligands in **52** are *cis*. Although the *trans* isomer of **52** has not been isolated, it seems to be present in the reaction mixture for only its presence can explain the formation of the Pd_3 complex **53**, which has two $SeSiMe_3$ groups in a *trans* orientation. In **52** and **53** the palladium triangles are capped on both sides by μ_3-Se ligands. The palladium atoms in these complexes are coordinated in an approximately square planar fashion, as is characteristic of the Pd^{2+} species (d^8 configuration). According to the usual electron counting procedures, each palladium atom has 16 valence electrons. It must be said, however, that the Pd–Pd distances of 308–318 pm and 323–337 pm in **52** and **53** respectively, are very long, and if bonding interactions do occur, they can only be of a very weak nature.

The large Pd–Pd distances lead to relatively short, nonbonding contacts between the μ_3-E (E = S, Se) bridging ligands (**52a**: S–S: 293 pm; **53**: Se–Se: 308.4 pm), which are significantly shorter than the sum of their van der Waals radii. [203, 204] Similar results and relationships have also been reported by other authors. [193, 194, 205]

Under different experimental conditions, the reaction of $[PdCl_2(PPh_3)_2]$ with $E(SiMe_3)_2$ (E = S, Se, Te) can also lead to other multinuclear palladium complexes. It is quite remarkable, however, that these complexes bear no relationship to the nickel complexes already mentioned. For example, there is no indication that a $[Pd_8E_6(PR_3)_8]$ cluster having a cubic Pd_8 framework, or an octahedral

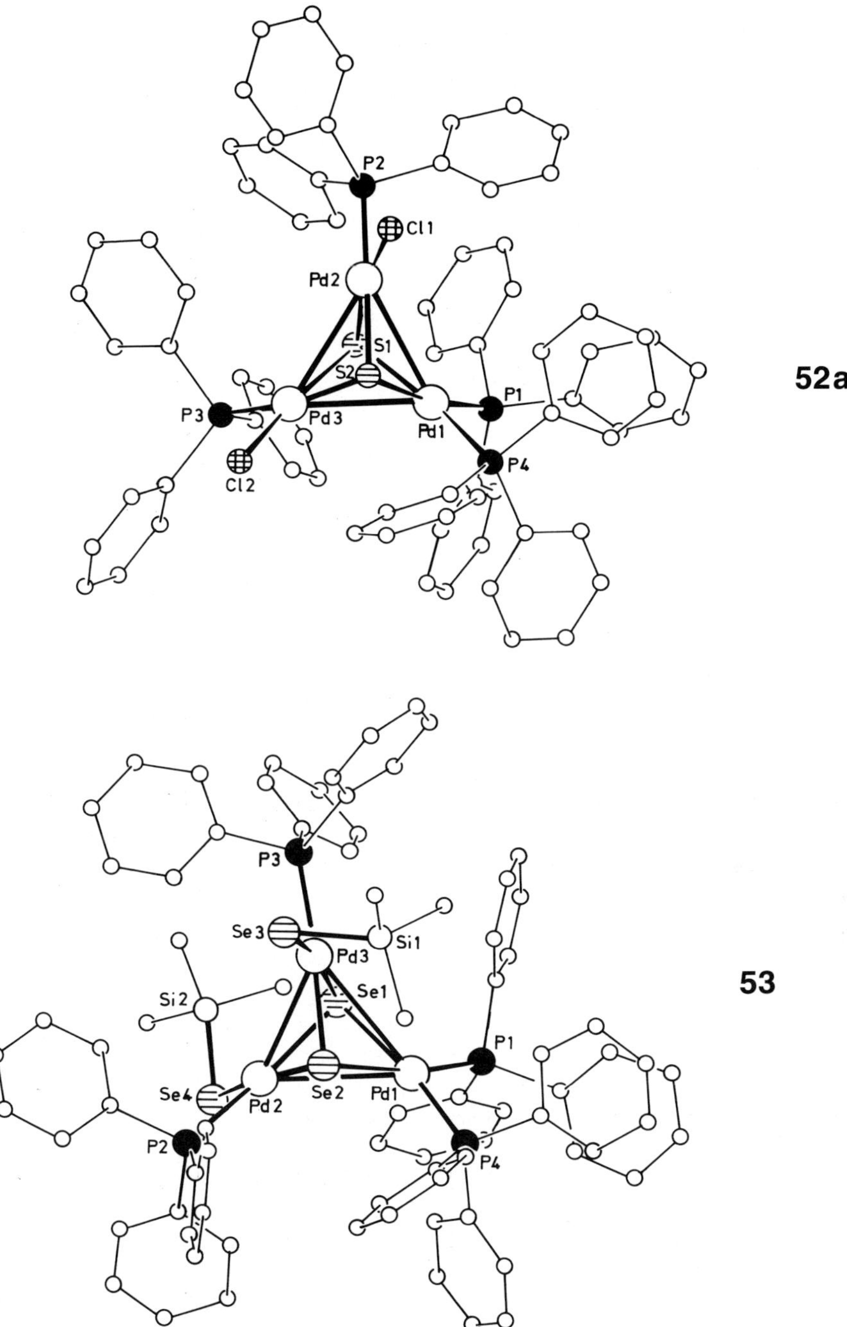

Figure 3-69. Molecular structures of **52a** and **53**.

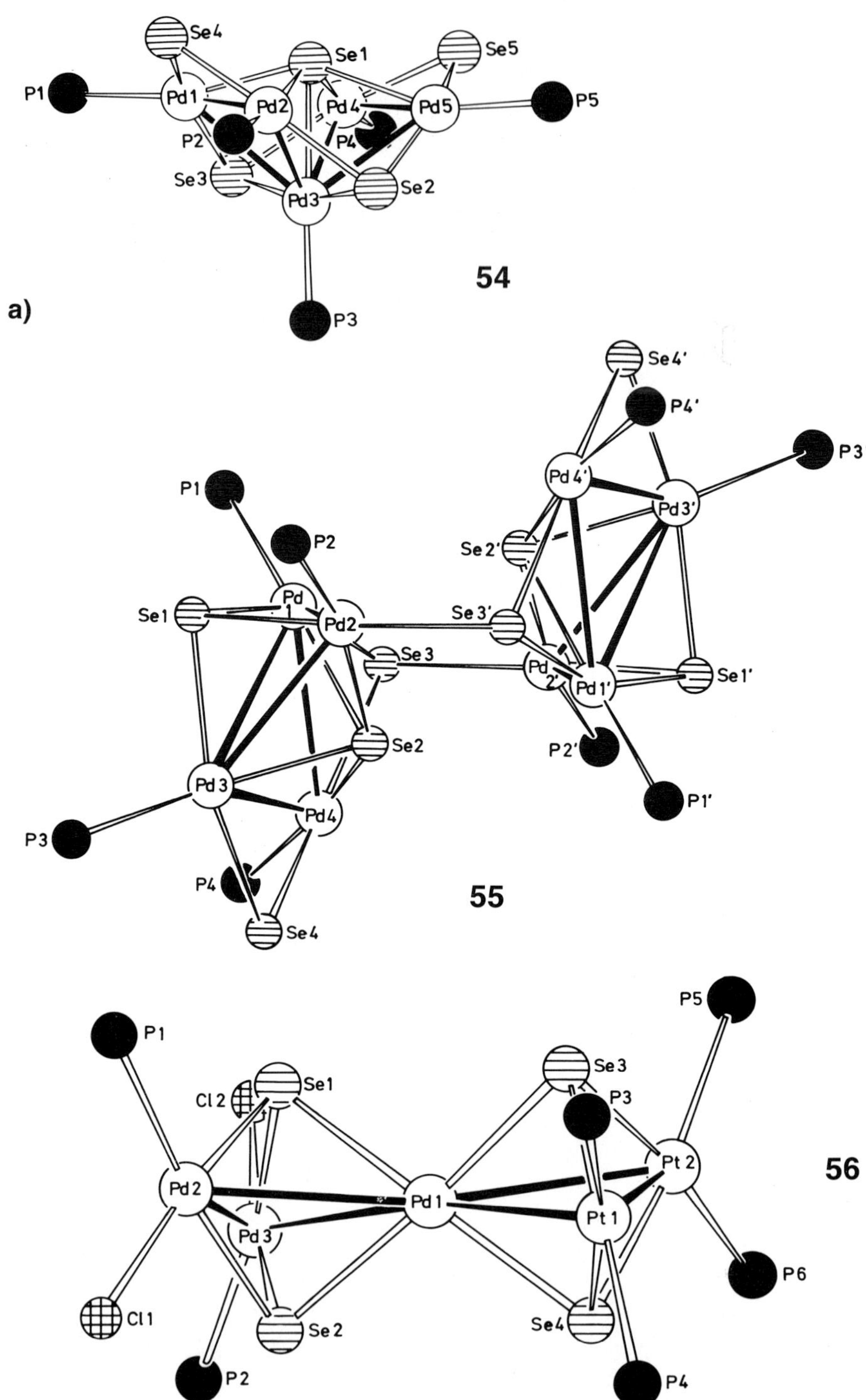
Se4
Se1
Se5
P1
Pd1
Pd2
Pd4
Pd5
P5
P2
P4
Se3
Se2
Pd3
54
a)
P3
Se4'
P4'
P3'
Pd4'
Pd3'
P1
P2
Se2'
Se1
Pd 1
Pd2
Se3'
Se3
Pd 2'
Pd1'
Se1'
Se2
P2'
Pd3
P1'
Pd4
P3
P4
55
Se4
P5
P1
Se3
Se1
Cl2
P3
Pt2
56
Pd2
Pd1
Pt1
Pd3
P6
Cl1
Se2
Se4
P2
P4

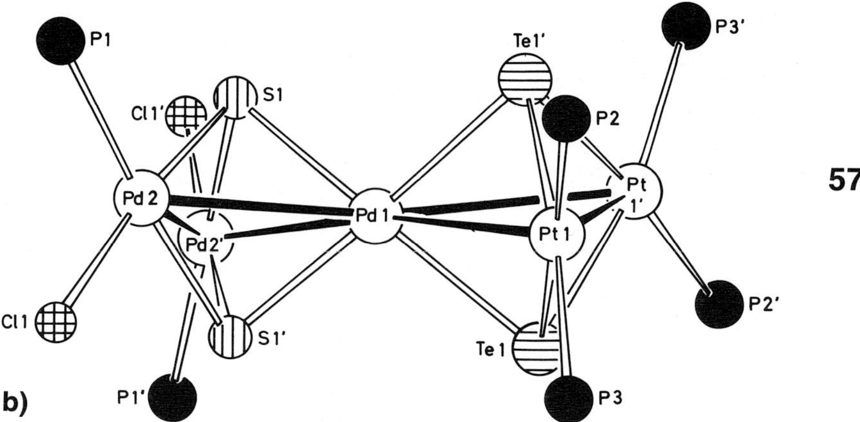

Figure 3-70. Molecular structures of **54**, **55**, **56**, and **57** (phenyl groups omitted).

$[Pd_6E_8(PR_3)_6]$ cluster can be synthesized. As already mentioned, the reaction of $[PdCl_2(PPh_3)_2]$ with $Se(SiMe_3)_2$ yields **52** and **53** but in addition to these, the palladium richer cluster complexes **54** and **55** may also be isolated (Eq. 3.46). [158, 159]

$$[PdCl_2(PPh_3)_2] \xrightarrow{Se(SiMe_3)_2} \mathbf{52} + \mathbf{53} + \underset{\mathbf{54}}{[Pd_5Se_5(PPh_3)_5]} + \underset{\mathbf{55}}{[Pd_8Se_8(PPh_3)_8]} \tag{3.46}$$

As can be seen in Figure **3-70**, **54** consists of a distorted square-pyramidal Pd_5 framework in which each palladium atom is coordinated by one PPh_3 ligand and three Se ligands. Each of these Se ligands has a different coordination function: Se1 acts as a μ_5-ligand, Se2 and Se3 as μ_3-ligands, and Se4 and Se5 as μ_2-ligands.

The clusters **52** and **53** have reactive PdCl and $PdSeSiMe_3$ groups which have been made to react with other transition metal complexes in a deliberate attempt to synthesize larger, heterometallic clusters. $[PtCl_2(PPh_3)_2]$ and **53** will react to form **56** under elimination of Me_3SiCl (Eq. 3.47).

$$\underset{\mathbf{53}}{[Pd_3(SeSiMe_3)_2Se_2(PPh_3)_4]} \xrightarrow{[PtCl_2(PPh_3)_2]} \underset{\mathbf{56}}{[Pd_3Pt_2Cl_2Se_4(PPh_3)_6]} + Me_3SiCl \tag{3.47}$$

The heterometallic cluster **56** is composed of three palladium atoms and two platinum atoms. It is approximately isostructural with the Ni_5 cluster **27** described in the discussion of M_5 clusters in Section 3.4.3.1 the only difference being that the chloride ligands in **56** are in *cis* positions (Fig **3-70**). Of note is the fact that the Pd_3 complex used as starting material remains intact. It is possible, however, to obtain an isomer of **56** in which the chloride ligands are *trans* to one another.[158, 159] The systematic build up of heterometallic clusters as presented in the aforementioned example can also be achieved by another method. The reaction of $[Pd_3Cl_2S_2(PPh_3)_4]$ (**52a**) with $[Pt_2(\mu_3\text{-}Te)_2(PPh_3)_4]$ leads to the formation of $[Pd_3Pt_2Cl_2S_2Te_2(PPh_3)_6]$ (**57**), a further spirocyclic Pd_3Pt_2 cluster. [158, 159, 206]

$$\underset{\mathbf{52a}}{[Pd_3Cl_2S_2(PPh_3)_4]} + [Pt_2(\mu_2\text{-}Te)_2(PPh_3)_4] \rightarrow \underset{\mathbf{57}}{[Pd_3Pt_2Cl_2S_2Te_2(PPh_3)_6]} \quad (3.48)$$

The structure of **57** (Fig. **3-70**) shows quite clearly that the structure in the **52a** starting material remains fully intact and an exchange between the S and Te ligands is not observed. These examples serve as evidence that spirocyclic compounds can be synthesized from trinuclear complexes by the replacement of PPh_3 ligands. The question remained, however, whether the pentanuclear complex **54** could also be formed in a similar manner, i.e. from a trinuclear precursor. Experiment showed that this is possible and indeed **54** can be synthesized from **52** by reaction with $Se(SiMe_3)_2$.

Remarkably, it has not proved possible to synthesize heterometallic clusters from **53** (the $SeSiMe_3$ derivative of **52**) or other transition metal halide derivatives (with the exception of the syntheses of **61** and **62**). [158, 159] It has been possible to synthesize larger palladium clusters, however. The reaction of **53** with $[CpCrCl_2(THF)]$ or $[PdCl_2(PPh_3)_2]$, for example, has led to the formation of Pd_6, Pd_7, and Pd_8 complexes (Scheme **3-24**). The isolation of these compounds is not an easy task.

$$\underset{\mathbf{53}}{[Pd_3(SeSiMe_3)_2Se_2(PPh_3)_4]} \xrightarrow{[CpCrCl_2(THF)]} \underset{\mathbf{52}}{[Pd_3Cl_2Se_2(PPh_3)_4]} + \underset{\mathbf{58}}{[Pd_6Cl_4Se_4(PPh_3)_6]} + \underset{\mathbf{60}}{[Pd_8ClSe_8(PPh_3)_8]^+[CpCrCl_3]^-}$$

$$\mathbf{53} \xrightarrow{[PdCl_2(PPh_3)_2]} \underset{\mathbf{59}}{[Pd_7(SeH)ClSe_6(PPh_3)_7]}$$

Scheme 3-24. Se bridged Pd clusters built up from Pd_3 species by $CpCrCl_2(THF)$.

Although these reactions indicate that very complicated processes must occur during the growth of the clusters, all the structures show an explicit topological relation to the starting material **53**, whose Pd_3 ring can be recognized in **58**, **59**, and **60** (Fig. **3-71**). The hexanuclear **58**, for example, is formed by reaction at the $SeSiMe_3$ sites under formation of Me_3SiCl and elimination of two PPh_3 ligands. The same structure type can be found in a cluster with bridging μ_2-EH (E = S, Se) ligands in place of the μ_2-Cl ligands. [158, 159] The Pd_3Se_2 unit in **53** (Pd1,

Pd2, Pd3, Se2, and Se3: Fig. **4-69**) is also to be observed in the structure of the heptanuclear **59** (Fig. **3-71**). It would appear that **59** is formed from an intermediate similar to the one formed in the synthesis of the hexanuclear **58**. In **59**, however, the second Pd_3Se_2 unit (Pd5, Pd6, Pd7, Se4, and Se5) has been further built upon, apparently via reaction at the site of the bridging Se4 ligand by the addition of a further Pd atom. The resulting compound represents a Pd_3 unit linked to a Pd_4 unit. The structure of the cation in **60** (Fig. **3-71**) can be understood as an extention of this structure. As in **59**, one Pd_3Se_2 unit (Pd1, Pd2, Pd3, Se1 and Se2) is linked to a second Pd_3Se_2 fragment (Pd4, Pd5, Pd6, Se5 and Se6), whereby this second unit has been enlarged by two further palladium atoms (Pd7 and Pd8). The result in this case has a Pd_3 unit linked to a Pd_5 unit through selenium bridges. The same growth principle is valid for **55** (Fig. **3-70**) in which two Pd_4 units (with the same structure as the Pd_4 unit in **59**) are linked via μ_3-Se ligands.

All the compounds **54–60** are built up of palladium atoms in the +2 oxidation state (d^8 configuration). Correspondingly, all the palladium atoms have a distorted square planar coordination. The Pd–Pd distances lie in the range 297–337 pm and are thus too long to justify a discussion of bonding interactions. With this in mind, the solid lines drawn between the palladium atoms in the structures illustrated in Figures **3-69–4-71** should not be regarded as bonds, but rather as geometrical lines to aid in the discussion of the structure. The word cluster is thus not strictly applicable to these compounds. It is better to describe them as multinuclear complexes in which each palladium atom has 16 valence electrons.

An interesting heterometallic cluster is formed from the reaction of **53** with $[AuCl(PPh_3)]$ (Eq. 3.49).

$$\underset{\textbf{53}}{[Pd_3(SeSiMe_3)_2Se_2(PPh_3)_4]} \xrightarrow{[AuCl(PPh_3)]} \underset{\textbf{61}}{[Au_3Pd_3Se_4(PPh_3)_6]^+Cl^-} \qquad (4\text{-}49)$$

The cluster complex **61** is formed as part of a mixture of products which also contains **52**, **56a**, and **58**. Figure **3-72** shows that the structure of the cation in **61** consists of a Pd_3 ring (Pd–Pd: 310–314 pm) which is bridged by a μ_3-Se ligand (Se1). The edges of the Pd_3 triangle are further bridged by μ_2-Se ligands (Se2, Se3, and Se4) who further act as bridges to the almost linearly coordinated gold atoms. All the palladium and gold atoms are each coordinated to one PPh_3 ligand. The Au–Pd distances are in the 293–294 pm range. If the assumption is made that the selenium atoms have an oxidation state of –2, then each gold atom has a charge of +1 and each palladium atom a charge of +2. The number of the valence electrons in the framework of **61** is 90. This sum can be divided in such a manner that each palladium atom has 16 valence electrons and each gold atom 14 valence electrons. The cluster **61** belongs to a class of Au–Pd substances which, until now, has been the subject of only limited study. [207–209]

The intriguing chemistry exhibited by the reactions of **53** with transition metal complexes is further demonstrated by its reaction with $[IrCl(COD)]_2$ (Eq. 3.50).

$$[Pd_3(SeSiMe_3)_2Se_2(PPh_3)_4] \xrightarrow{[IrCl(COD)]_2} [Pd_5Cl_2Se_4(PPh_3)_6] \quad (3.50)$$
53 **56a**

$$+ [Pd_6(SeH)_2Cl_2Se_4(PPh_3)_6] + [Ir_6PdSe_4(COD)_6]$$
58a **62**

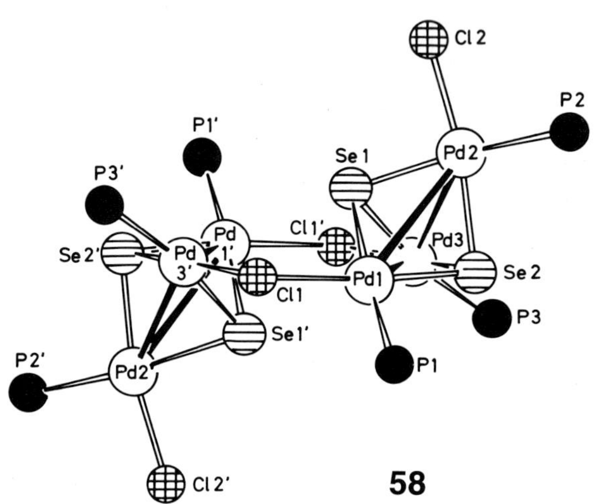

58

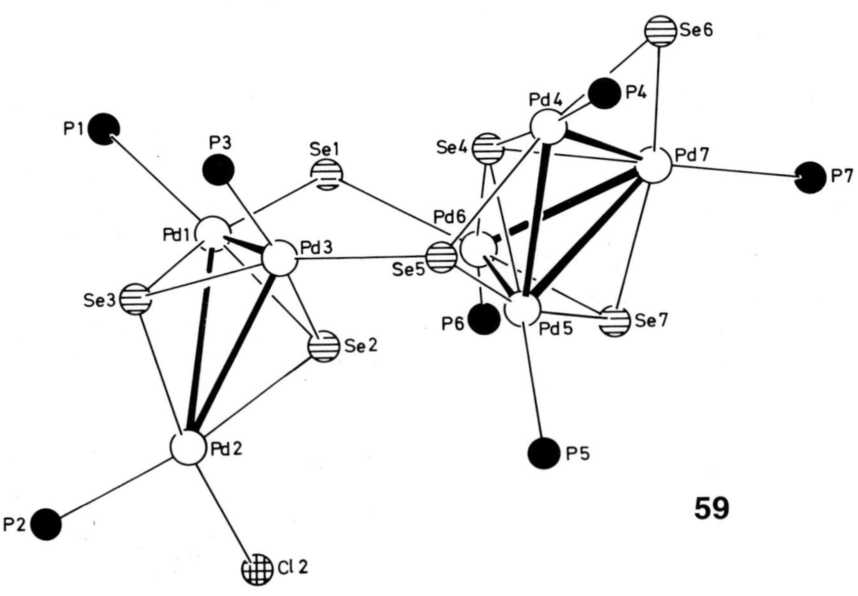

59

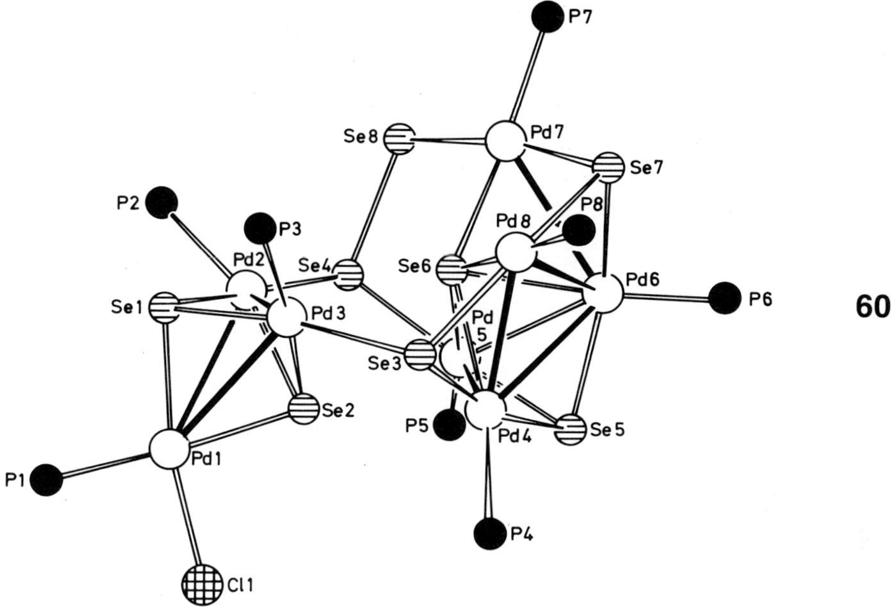

Figure 3-71. Molecular structures of **58**, **59**, and the cation in **60** (phenyl groups omitted).

In addition to the analogues of **56** and **58** formed from this reaction, the heptanuclear heterometallic cluster **62** can be obtained in very small yield. This compound is a spirocyclic complex consisting of two almost planar Ir_3Pd squares (Fig. **3-72**) which are twisted at approximately 90° to one another. Each of the squares is capped by two μ_4-Se ligands. Although **62** can be given a count of 104 valence electrons, a description of the bonding is difficult for the character of the metal-metal interactions is not at all clear (Ir–Ir: 284–285 pm; Ir–Pd: 278–281 pm).[158, 159]

Using the synthetic methods outlined above, it has only been possible to prepare palladium-chalcogen cluster complexes having a maximum of eight palladium atoms. However, a method for the synthesis of larger complexes was recently (re)discovered. Back in 1968, R. Ugo reported the synthesis of $[Pt_2S_2(PPh_3)_4]$ by the reaction of $[Pt(PPh_3)_4]$ with elemental sulphur. [210, 211] The reaction of $[Pd(PPh_3)_4]$ with sulphur or selenium in toluene leads to the formation of a number of cluster complexes, of which only a few have been structurally characterized to date. [212]

$$[Pd(PPh_3)_4] \xrightarrow{\text{S (or Se)}} \underset{\mathbf{54}}{[Pd_5Se_5(PPh_3)_5]} + \underset{\mathbf{64}}{[Pd_9Se_5(PPh_3)_8]} \quad (3.51)$$

$$+ \underset{\mathbf{63}}{[Pd_9S_9(PPh_3)_8]}$$

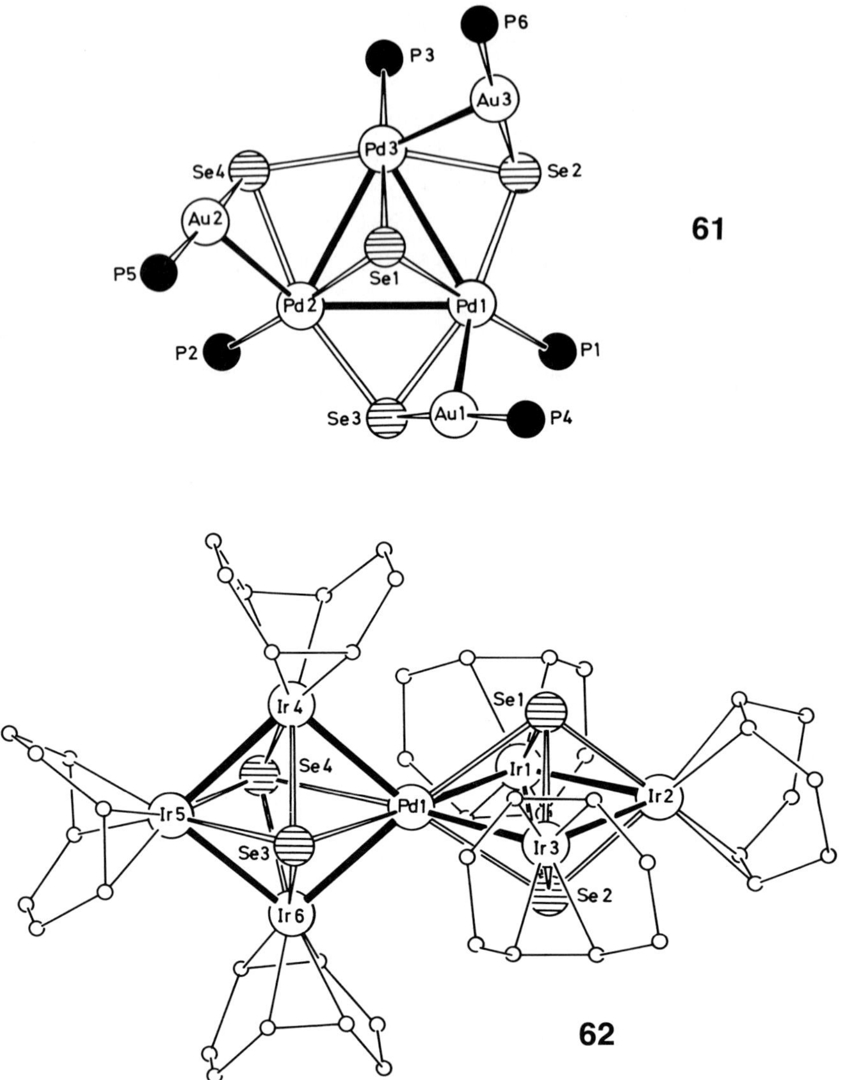

Figure 3-72. Molecular structures of the cation in **61** (phenyl groups omitted) and **62**.

This very simple synthetic route almost certainly holds more surprizes as the structures of the clusters formed are extremely dependent on the reaction conditions.

The structures of **63** and **64** are illustrated in Figure **3-73**. **64** can be derived from a strongly distorted Pd_9 cube in which four μ_3-Se ligands and one μ_4-Se ligand act as bridges between the peripheral palladium atoms. In **63**, two characteristic Pd_4S_3 units, as found in **55**, can be recognized. These Pd_4S_3 fragments are connected to each other through bridging μ_2-Se ligands and an additional Pd atom. It is worth mentioning that the analogous reactions involving $[Ni(PR_3)_4]$ lead to cluster compounds with quite different structures, but of the same type as those described in Section 3.4.3. Examples of these are $[Ni_4Se_4(PPh_3)_4]$ (**24**), $[Ni_6Se_5(PPh_3)_6]$ (**28**), $[Ni_8\square_4S_6(PPh_3)_4]$ (**41a**), $[Ni_{12}E_{12}(PEt_3)_6]$ (E = Se; S (**48**)).

3.4.4 Copper Clusters with Se and Te Ligands

Phosphine complexes of Cu, Ag, and Au halides have also been used in an attempt to synthesize clusters having chalcogenide ligands. [213, 214] If, for example, $[AgCl(PPh_3)_3]$ is allowed to react with $E(SiMe_3)_2$ (E = S, Se) in organic

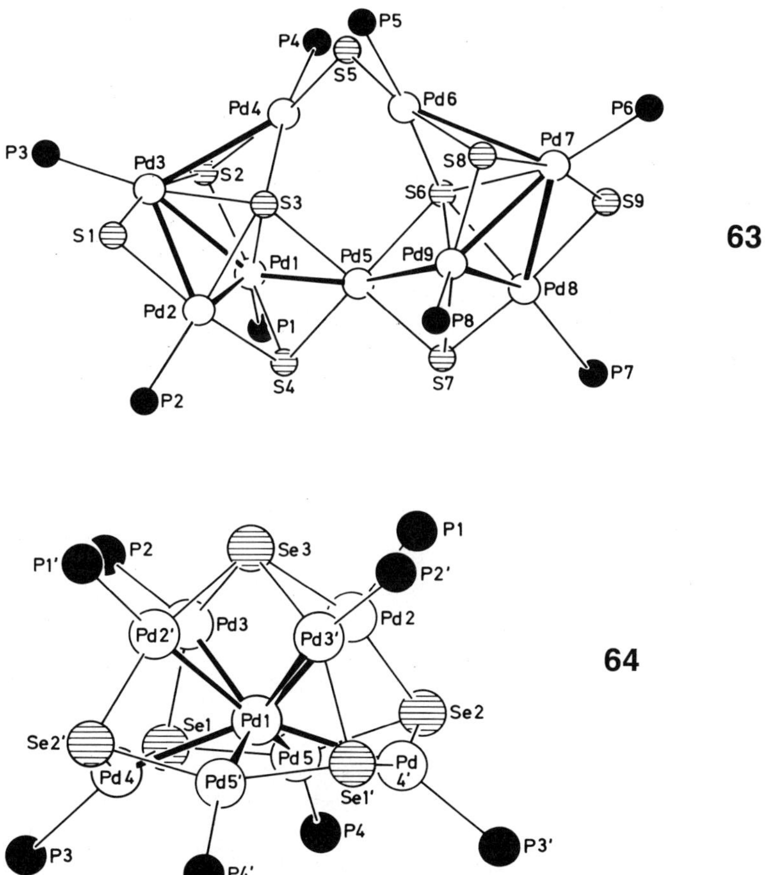

Figure 3-73. Molecular structures of **63** and **64** (phenyl groups omitted).

solvents, binary sulphides and selenides can be observed to result. In contrast, deeply colored solutions are formed from the analogous reactions with $[CuCl(PPh_3)]_4$, $[Cu_2Cl_2(PPh_3)_3]$, or $[CoCl(PPh_3)_3]$ in THF, from which black "crystals" precipitate. These "crystals" are amorphous, however, so no crystal structure analysis could be performed. According to elemental analyses, these "amorphous crystals" are metal-rich compounds. If diethylether is used as solvent for the reaction of CuCl with $PiPr_3$ and $Se(SiMe_3)_2$, then the rapid formation of a dark colored solution occurs, from which a mixture of **65** and **65a** crystallizes out within a few hours. Under the same reaction conditions, the use of $PtBu_3$ leads to the formation of **66** (Scheme **3-25**).

$$[Cu_{30}Se_{15}(PiPr_3)_{12}] + [Cu_{29}Se_{15}(PiPr_3)_{12}]$$
65 **65a**

$$\uparrow \; PiPr_3,\ Se(SiMe_3)_2$$

CuCl

$$\downarrow \; PtBu_3,\ Se(SiMe_3)_2$$

$$[Cu_{36}Se_{18}(PtBu_3)_{12}]$$
66

Scheme 3-25. Route to Se bridged high nuclearity Cu clusters.

Normally, microcrystalline products which are sparingly soluble in organic solvents are formed in these reactions. Crystals suitable for a structure analysis are obtained if the reaction is performed at low temperature and the temperature allowed to rise slowly from –30 °C to room temperature over the course of several days. Whereas **66** forms brown crystals, both **65** and **65a** form black ones (in thin layers they also appear brown). Their structures are almost identical, the most obvious difference being the additional Cu atom in the center of **65**.

The clusters **65** and **65a** (Fig. **3-74**) are built from five parallel layers of Cu and Se atoms. In both compounds the selenium atoms act as μ_4, μ_5, and μ_7 bridging ligands, connecting the layers to one another. Due to the different connectivities of the selenium ligands, the Cu–Se bonds are of varying length. Layers 1 and 5 have the composition $Cu_4Se_3(PiPr_3)_3$ and layers 2 and 4 the composition Cu_6Se_3. Layer 3 has the composition $Cu_9Se_3(PiPr_3)_6$ in **65a** and $Cu_{10}Se_3(PiPr_3)_6$ in **65**. As a result of the additional copper atom in the center of **65**, the Cu–Cu contacts in the middle layer are longer than those in **65a**.

The molecular structure of **66**, a large cluster with a two-fold rotation axis, is also shown in Figure **3-74**. It is composed of 36 copper atoms and bears no structural resemblance to either **65** or **65a**. As in **65** and **65a**, bridging selenium ligands bonding in μ_3 (Se7), μ_4 (Se1, Se9), μ_5 (Se2–Se6, Se8), and μ_6 (Se10) fashions are also found in **66**. The Cu–Se bond lengths lie in the same range as found for **65** and **65a**: Cu–(μ_3-Se): 237.6(6) pm; Cu–(μ_4-Se): 235.8–245.3(6) pm; Cu–(μ_5-Se):

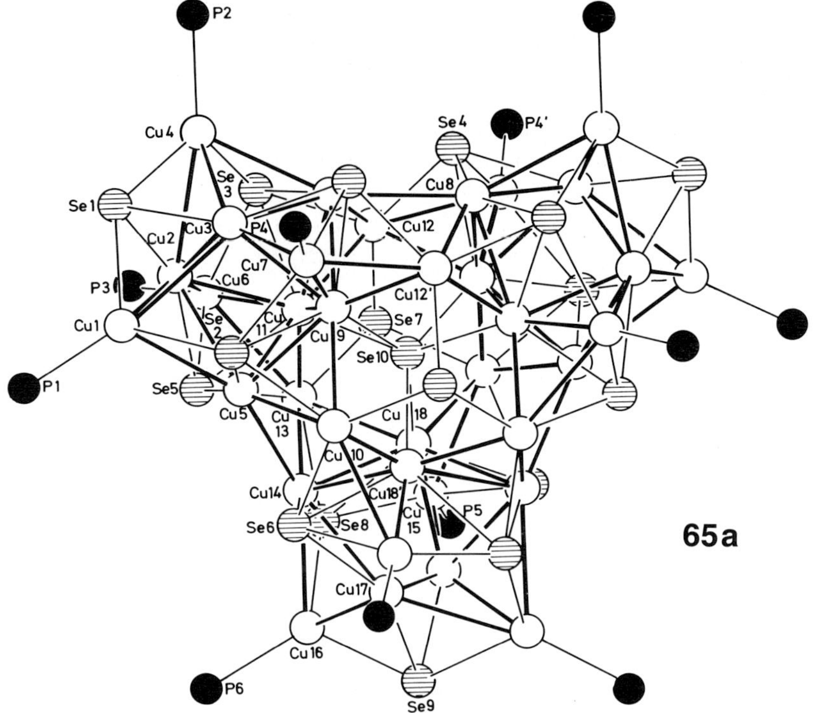

Figure 3-74. Molecular structures of **65** and **66** (C atoms omitted). (Cu–Cu distances of 290 pm or less are shown as solid lines.)

240.0–261.3(6) pm. Whereas the μ_3-, μ_4-, and μ_5-Se ligands cover the surface of the copper framework, the μ_6-Se10 resides in the center of the cluster and is surrounded by six copper atoms in a trigonal prismatic coordination (Cu–Se: 243.2(6) pm). Alternatively, the center of the cluster can be described as two Cu_6Se_2 cubes bridged by Se10. The twelve phosphine ligands are coordinated to the copper atoms on the cluster periphery and effect complete shielding of the central $Cu_{36}Se_{18}$ unit.

$[Cu_{36}Se_{18}(PtBu_3)_{12}]$ (**66**) has a molecular mass of 6136.8u. Attempts to vaporise the cluster have shown that it does not enter the gas phase without decomposition. With the aid of secondary ion mass spectroscopy (SIMS), it is possible to create molecular ions with mass distributions of around 3600u, 7200u and 11 000u. [215] The occurrence of multiples of a molecular building block was also observed in SIMS experiments with $[Au_{55}(PPh_3)_{12}Cl_6]$. [216] In the 2000–5000u range, molecular ions are found that correspond to particles of composition $[Cu_{36}Se_{18}–nCu_2Se]$, $[Cu_{35}Se_{17}–nCu_2Se]$, and $[Cu_{37}Se_{17}–nCu_2Se]$. Examples are $Cu_{37}Se_{18}$, Cu_{35} Se_{17}, $Cu_{33}Se_{16}$, $Cu_{31}Se_{15}$, $Cu_{29}Se_{14}$, $Cu_{27}Se_{13}$, $Cu_{25}Se_{12}$, $Cu_{23}Se_{11}$, $Cu_{21}Se_{10}$, $Cu_{19}Se_9$, $Cu_{17}Se_8$, and $Cu_{13}Se_6$. This means that only molecular ions of

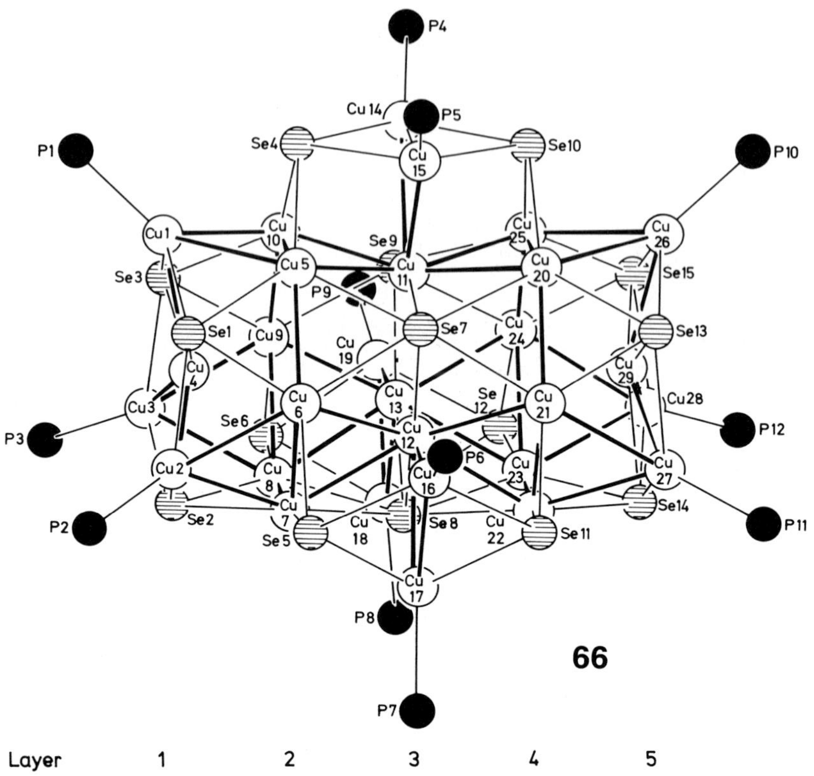

the general formula $Cu_{2n+1}Se_n$ appear. The PR_3 ligands are obviously removed from the clusters under the conditions of the experiment.

It is interesting to note that copper chalcogenide clusters are also found in the gas phase of Cu_2Se and Cu_2Te. I. Dance has reported, for example, that ions having the formulas $[Cu_{2n-1}S_n]^-$ and $[Cu_{2n-2}S_n]^-$ can be observed in the gas phase of Cu_2S after laser ablation. [217–219] Indeed, $[Cu_{29}E_{15}]^-$ (E = S, Se) is found to have a particularly high intensity in the mass spectrum. This is the same Cu:Se ratio as is found in the cluster **65a**. It is conspicuous that no molecular ions with the composition $[Cu_{2n}E_n]^-$ are observed. This Cu:Se ratio can only be observed when stabilized by a protecting shell of ligands as realised in the case of **65** and **66**. In the secondary ion mass spectrum of **66**, broad mass distributions are also found at around 7200u and 14400u. This indicates that the target cluster must have undergone some sort of transformation. If it is assumed that these peaks at higher masses may be taken as indications of the existence of PR_3 shielded copper chalcogenide clusters, it should be possible to synthesize complexes with compositions approximating to $Cu_{70}Se_{35}$, $Cu_{140}Se_{70}$, etc. That clusters of this size can indeed be realized is demonstrated by the syntheses of **67** and **69** (Scheme **3-26**).

$$\mathrm{CuCl} \xrightarrow{\mathrm{PEt_3\,+\,Se(SiMe_3)_2}} \underset{\mathbf{67}}{[\mathrm{Cu_{70}Se_{35}(PEt_3)_{22}}]} + \underset{\mathbf{68}}{[\mathrm{Cu_{20}Se_{13}(PEt_3)_{12}}]}$$

$$\mathrm{CuCl} \xrightarrow{\mathrm{PPh_3\,+\,Se(SiMe_3)_2}} \underset{\mathbf{69}}{[\mathrm{Cu_{146}Se_{73}(PPh_3)_{30}}]}$$

Scheme 3-26. Formation of high nuclearity copper selenide clusters from CuCl.

If the reaction between CuCl, PEt_3, and $Se(SiMe_3)_2$ is carried out in either THF or toluene, a deep brown solution is obtained from which no crystalline product can be isolated. If, on the other hand, the reaction is performed in diethylether, the reactants added at –80 °C, and the solution allowed to warm to room temperature over a period of several days, then single crystals suitable for structural analysis can be obtained. The yield of the smaller cluster **68** is generally very low.

The molecular structure of **67** is shown in Figure **3-75**, where it can be seen that the cluster contains a two-fold rotation axis. [214] A description of the arrangement of the copper atoms is problematic, but it is noticeable that the copper atoms are distributed in three distinct Cu_{13} units linked by layers of near planar Cu_6 rings. Fifteen peripheral copper atoms are bonded to the central unit.

The selenium atoms find themselves in varying bridging situations, with connectivities ranging from μ_4 to μ_6. With the exception of three atoms within the cluster framework, all selenium atoms are situated on the cluster surface. Only 22 peripheral copper atoms are coordinated by PEt_3 ligands which, in turn, effectively screen the cluster from the outside environment.

The best way to describe the structure of **67** is in terms of the packing of the selenium atoms. There are three distinct, parallel, and planar selenium layers; the two outer layers, 1 and 3, which contain 10 selenium atoms, and the inner layer, 2, which contains 15 selenium atoms. The layers are arranged in a hexagonal close-packed manner and are spaced at a distance of about 360 pm from each other. The distribution of the selenium atoms is reminiscent of the packing in β-Cu_2Se. In this solid state compound, however, the selenium atoms are arranged in a cubic close-packed fashion, with a layer separation of approximately 339 pm. Three copper atoms lie outside the layers 1 and 3, and 32 copper atoms each lie between the layers 1 and 2, and between the layers 2 and 3. The copper atoms occupy tetrahedral sites within the selenium atom matrix, but are displaced towards one Se_3 surface of the tetrahedron, giving rise to only a three-fold coordination.

The structure of **69** can be understood in the same way, but the selenium layers 1 and 3 are now composed of 21 atoms and the layer 2 of 31 atoms. 120 of the 146 copper atoms are situated between the selenium layers, and an additional 12 lie within the plane of layer 2. The remaining 14 copper atoms lie outside the layers 1 and 3. As in **67**, the copper atoms preferentially occupy tetrahedral sites, displaced in such a way as to be coordinated by only three selenium atoms,

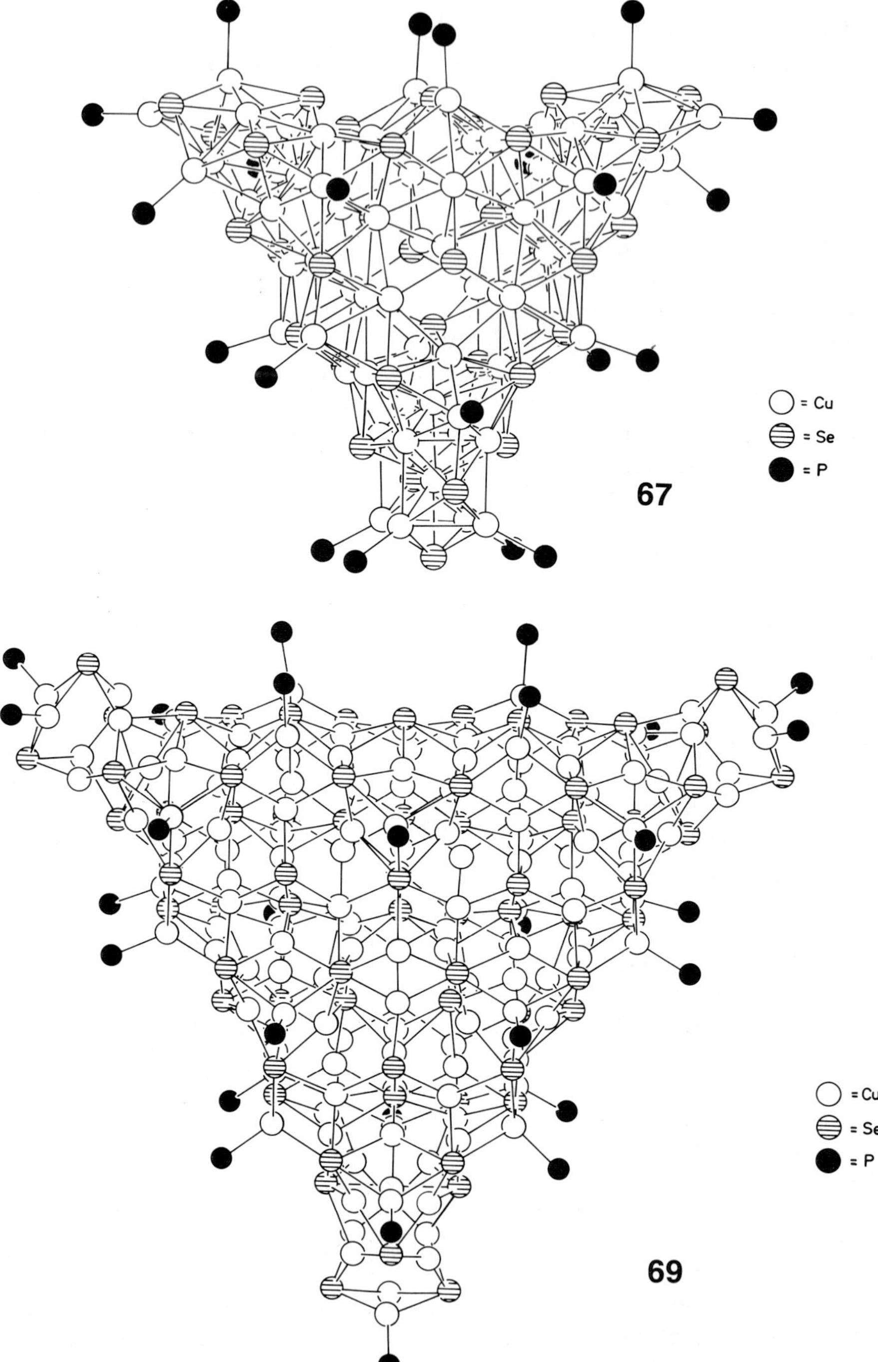

Figure 3-75. Molecular structures of **67** (Cu–Cu distances of 290 pm or less are shown as solid lines) and **69** (Cu–Cu bonds not shown for clarity). C atoms are omitted.

although some of the copper atoms have an approximately tetrahedral coordination by the selenium atoms. 30 copper atoms are tetrahedrally coordinated by three selenium ligands and a PPh_3 ligand. This situation is comparable to the coordination sphere found in $Cs_2Cu_5Se_4$. [220]

In the cubic β-Cu_2Se, one half of the copper atoms occupy tetrahedral sites and the remaining copper atoms are distributed either in octahedral sites or positions lying between Se_3 surface and tetrahedral sites. [221–224] It is obvious that **67** and **69** do not represent an exact cutout of the β-Cu_2Se structure, but there is a certain similarity regarding the atomic spacing and coordination relationships. This being the case, it is quite legitimate to ask the question of whether the ligandshielded $Cu_{70}Se_{35}$ and $Cu_{146}Se_{73}$ clusters have physical properties similar to those found in β-Cu_2Se. Since the cluster framework in **69** is completely enveloped by 30 PPh_3 ligands, the complete cluster has a diameter of about 40Å and a thickness of 15Å. Preliminary results have shown that the electrical conductivity of these compounds is dependent on their size. [220] $[Cu_{36}Se_{18}(PtBu_3)_{12}]$ for instance, is an insulator and $[Cu_{70}Se_{35}(PEt_3)_{22}]$ is a semiconductor with an electrical conductivity of 10^{-2} S cm^{-1}.

The d^{10} configuration of the copper atoms in **67** and **69** infers diamagnetic behavior and in fact the χ_{mol} measured at room temperature for **67** was -2.73×10^{-3} emu mol^{-1}. After subtracting the diamagnetic core contributions for Cu^+, Se^{2-}, and PEt_3, a weak temperature independent paramagnetism of $\chi_{mol} = +1.77 \times 10^{-3}$ emu mol^{-1} is obtained. This result corresponds well with the magnetic properties of Cu_2Se. [221–224] All Cu–Cu distances observed in the compounds **65**, **66**, **67** and **69** (all formally Cu^{1+}) lie in the 242–290 pm range. Similar bond lengths are found in other multinuclear copper complexes and indicate that only weak Cu–Cu interactions are present. [225]

The reaction between CuCl, PEt_3, and $Se(SiMe_3)_2$ also yields small amounts of $[Cu_{20}Se_{13}(PEt_3)_{12}]$ (**68**; Scheme **3-26** and Fig. **3-76**). The Cu:Se ratio in this complex is not 2:1. This 2:1 ratio is apparently not characteristic, for analogous reactions of CuCl with $Te(SiMe_3)_2$ in the presence of a tertiary phosphine also result in the formation of compounds with various Cu:Te ratios (Scheme **3-27**). [226]

In **70**, the four copper atoms form a butterfly structure bridged by two Te_2^{2-} ligands (Te–Te: 280.0–281.2(3) pm; Cu–Cu: 279.4–309.5(4) pm). Each copper atom has an additional bond to the P atom of a $PiPr_3$ ligand. All the copper atoms are thus coordinated in a distorted tetrahedral manner by three Te atoms and one P atom. The Cu–Te distances are of differing lengths such that each copper atom may be considered as having one shorter Cu–Te bond (257.4–258.5(2) pm) and two longer ones (265.8(3)–285.6(4) pm). Other authors have already reported on tetranuclear metal clusters with butterfly structures. [227–234]

The complexes **68, 71**, and **72** form spherical units. According to Figure **3-76**, **68** consists of a regular Cu_8 cube (Cu2–Cu2: 301 pm), each edge of which is bridged by a copper atom (Cu1–Cu1: 282 pm). The PEt_3 ligands are coordinated to these peripheral Cu1 atoms, and the shortest Cu–Cu distances of 268.0 pm in the resulting 20 atom polyhedron are those between Cu1 and Cu2. Alternatively, the copper framework can be described in terms of six trigonal prisms linked through common edges. This linking gives rise to planes made up of six copper

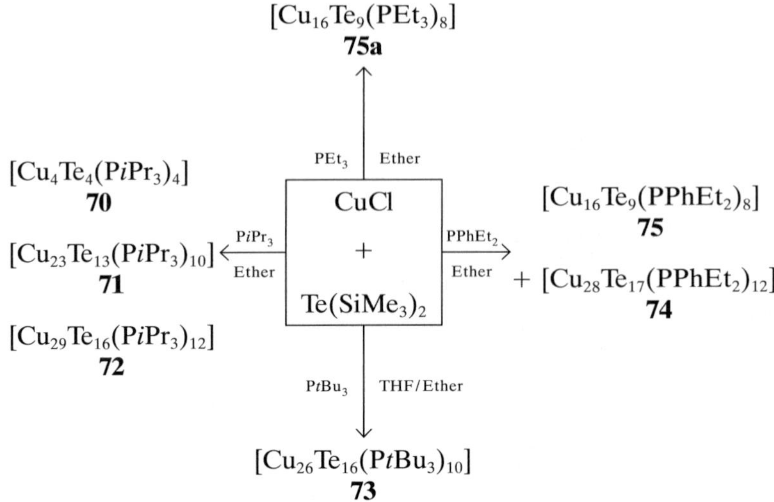

Scheme 3-27. Formation of copper telluride clusters from CuCl.

atoms. The metal framework is surrounded by twelve selenium ligands (Se2) which are arranged in such a fashion as to form an almost undistorted icosahedron (Se2 ··· Se2: 392–415 pm). The conceivable interpenetration of the two polyhedra, a Se_{12} icosahedron and a Cu_{20} pentagonal dodecahedron, is only partially realised, however. As a result of Cu–Cu interactions, the twelve Cu1 atoms lie above the faces of the Se_{12}, icosahedron, and the eight Cu2 atoms below the faces. The bonding situation in **68** is different than that in the structurally related $[Pd_{20}As_{12}(PPh_3)_{12}]$ (**18**), where the cluster center is unoccupied. As a consequence, the Pd–Pd distances in the central Pd_8 cube are significantly shorter (278.5–291.7 pm). The cluster **68** has 298 valence electrons. Assuming that all Cu–Cu distances under 290 pm are bonding, the $18e^-$ rule is exceeded by only two electrons.

The structure of **71** (Fig. **3-76**) can also be considered as being derived from a distorted icosahedron made up of the twelve peripheral tellurium atoms (Te ··· Te: 393.7–472.2 pm). Ten copper atoms coordinated by $PiPr_3$ ligands lie above ten of the twenty Te_3 faces. Ten further copper atoms lie below the remaining Te_3 faces. The cavity formed by this arrangement is occupied by three additional copper atoms (Cu2, Cu5, and Cu8) and a tellurium atom. It is possible that **71** is an intermediate in the formation of $[Cu_{29}Te_{16}(PiPr_3)_{12}]$ (**72**; Fig. **3-76**). The structure of **72** can be described in the same manner as $[Cu_{29}Se_{15}(PiPr_3)_{12}]$ (**65a**) mentioned earlier. The only difference between **72** and **65a** is that a chalcogen atom now occupies the center of the cluster. The fifteen peripheral tellurium atoms form a Frank-Kasper polyhedron (Te ··· Te: 392–480 pm) with 26 Te_3 faces. [235–236] Copper atoms, coordinated by $PiPr_3$ ligands, are found above twelve of the Te_3 faces. A further fourteen copper atoms are situated below the remaining Te_3 faces.

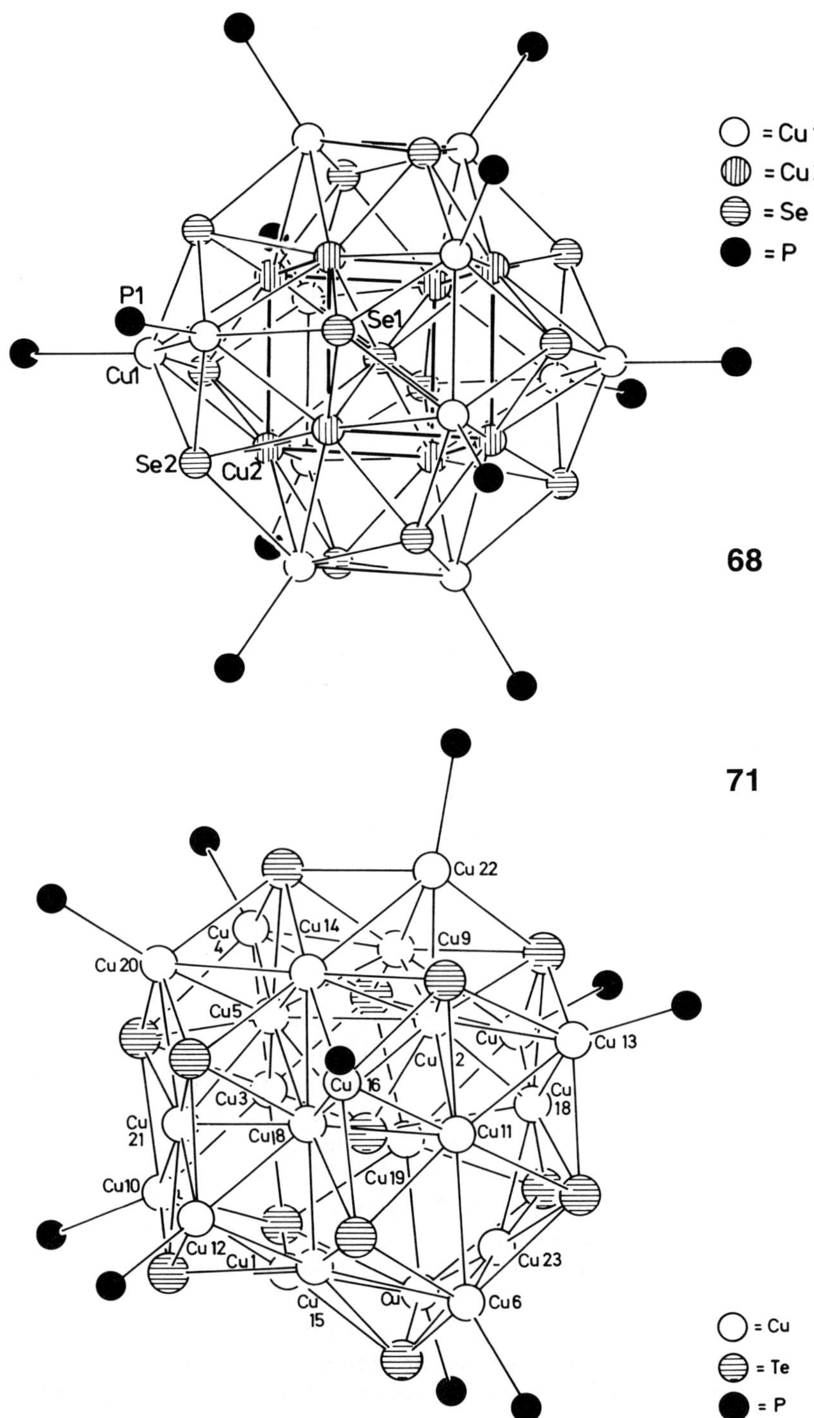
= Cu 1
= Cu 2
= Se
= P
P1
Cu1
Se1
Se2
Cu2
68
71
Cu 22
Cu 14
Cu 4
Cu9
Cu 20
Cu5
Cu 13
Cu 2
Cu 16
Cu3
Cu 18
Cu 21
Cu 8
Cu11
Cu10
Cu19
Cu 12
Cu1
Cu 23
Cu 15
Cu
Cu6
= Cu
= Te
= P

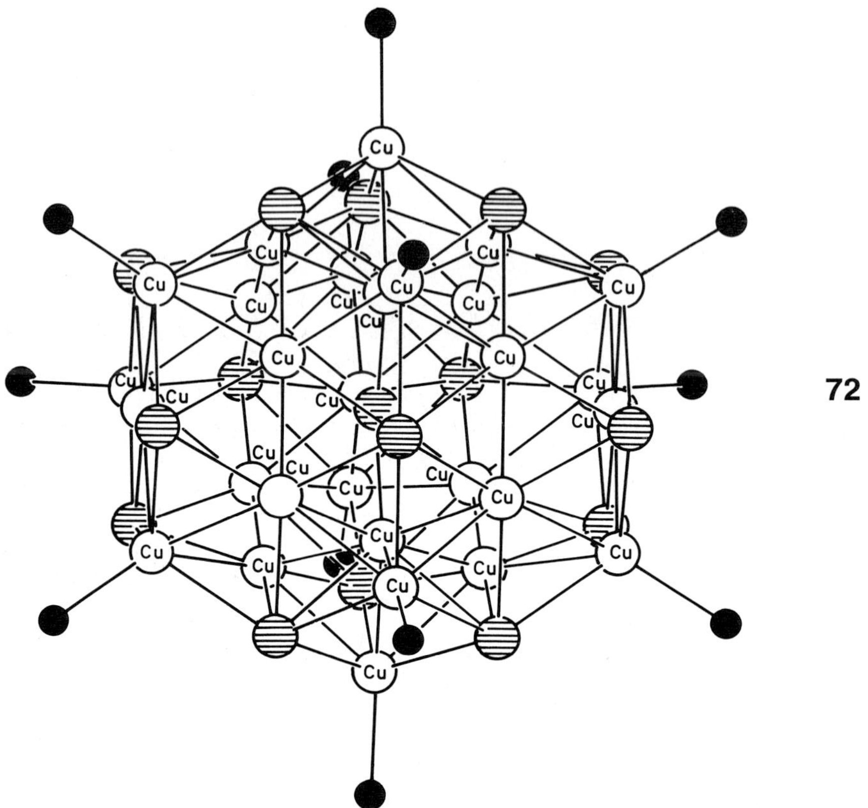

Figure 3-76. Molecular structures of **68**, **71**, and **72** (C atoms omitted).

Quite different products are obtained if a different phosphine is used instead of the sterically demanding P*i*Pr_3. $PPhEt_2$ leads to the formation of **74** and **75**, and PEt_3 to **75a**. The structural feature common to both clusters is shown in Figure **3-77** (**75** is used as the example). As can be seen, **75** has a mirror plane running through the atoms Te4, Te5, Te6, Cu8, and Cu9. The structure can be considered as being derived from a distorted Cu_6 trigonal prism (Cu6–Cu11) having Cu–Cu distances of 249.1(5)–306.0(5) pm and 329.2(4) pm in which six edges are bridged by copper atoms (Cu2, Cu4, Cu12, and Cu14) or by Cu_2 units (Cu3/Cu5, Cu13/Cu16). Finally, the metal framework is completed by Cu1 and Cu16. In this way, butterfly structure fragments analogous to those in **70** are formed by Cu1, Cu3, Cu5, and Cu6, and by Cu11, Cu13, Cu15, and Cu16. With the exception of the μ_6-Te ligand Te5, all the tellurium atoms function as μ_5-bridging ligands.

To describe the structure of **74** (Fig. **3-77**), it is appropriate to consider the sixteen phosphine free copper atoms first. These define a distorted Frank-Kasper polyhedron having a central tellurium atom. This polyhedron has 28 triangular faces, twelve of which are capped by copper atoms coordinated to $PPhEt_2$

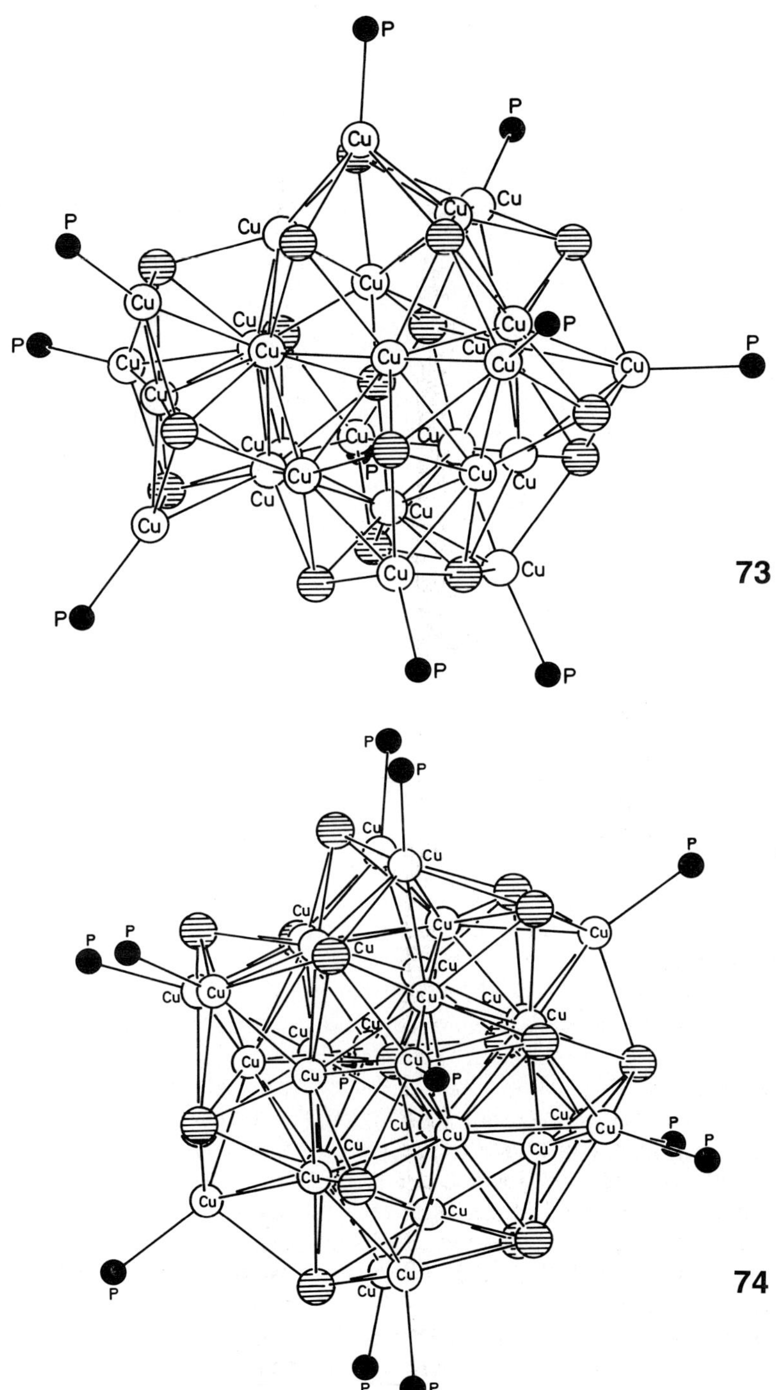
73
74

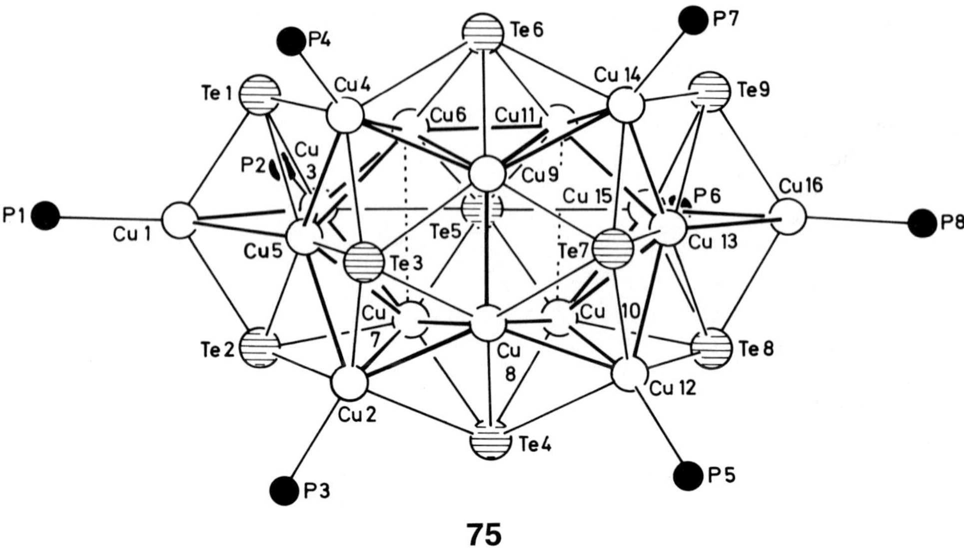

Figure 3-77. Molecular structures of **73**, **74**, and **75** (C atoms omitted).

ligands. The polyhedron is further coordinated by 16 peripheral tellurium ligands which themselves constitute a Frank-Kasper polyhedron, albeit severely distorted due to Te–Te interactions (Te ··· Te: 287.9(3)–469.8(7) pm).

The molecular structure of **73**, which contains the bulky $PtBu_3$ ligand, is also shown in Figure **3-77**. The fifteen peripheral tellurium atoms form a Frank-Kasper polyhedron with a central tellurium atom. This polyhedron is also very distorted due to Te–Te interactions. Copper atoms coordinated by $PtBu_3$ ligands are situated above ten of the Te_3 faces and sixteen Cu atoms are located below the remaining Te_3 faces.

None of the Cu/Te fragments in **71–75** bear a relationship to the structure of Cu_2Te. [237] Cu_2Te has a two dimensional network of layers of hexagonal prismatic Cu_{12} units which are connected via Te_2 bridges. In contrast, **71–75** have a more shell-like construction. The copper atoms which form the inner shell are surrounded by Te^{2-} ligands in the first sphere and by copper atoms coordinated to PR_3 in the second sphere. The structures show striking similarities to such intermetallic phases as the Laves phases, for example. [238] In both cases, Frank-Kasper polyhedra can be used to describe the structures.

The bonding situation in the Cu/Te clusters **70–75** cannot be described by simple electron counting rules. Based on the assumption that Te^{2-} and Te_2^{2-} ligands are always present, the copper atoms in **70** are in the formal oxidation state +1 (d^{10} configuration), and mixed valent states are to be assigned in **71–75**. Since the Cu–Cu distances in **70–75** (257–309 pm) are considerably longer than those for which Cu–Cu bonds are usually discussed, it must be assumed that bonding interactions between the copper atoms are of little significance. (44–56]

3.4.5 Chalcogen Bridged Transition Metal Clusters with η^5-Cyclopentadienyl, μ_3-Allyl, and CO Ligands

As was shown in the preceeding section, tertiary phosphines are capable of promoting the formation of clusters of the electron-rich transition metals. At the same time, they are able to suppress the formation of metal pnictides or chalcogenides that is always observed in the reaction of phosphine complexes of cobalt or nickel chlorides with $E(SiMe_3)_2$ (E = PPh, AsPh, S, Se).

It is possible to extend the synthetic principles which were presented above to the preparation of polynuclear complexes having Cp, allyl, and CO ligands. Preliminary results from the reactions of cyclopentadienyl, allyl, and CO complexes of the transition metal halides with $E(SiMe_3)_2$ (E = S, Se, Te) indicate that Me_3SiCl is eliminated and a series of novel compounds are formed. The reaction between $[CpFe(CO)_2Br]$ and $Se(SiMe_3)_2$ produces a diamagnetic compound composed of the heterocubane cluster anion $[Fe_4Se_4Br_4]^{2-}$ and two $[Se\{Fe(CO)_2Cp\}_3]^+$ cations (Eq. 3.52).

$$Se(SiMe_3)_2 \xrightarrow{[CpFe(CO)_2Br]} \underset{\mathbf{76}}{[Se\{Fe(CO)_2Cp\}_3]_2[Fe_4Se_4Br_4]} \qquad (3.52)$$

The structural parameters of the Fe_4Se_4 cage in **76** are essentially the same as those of other $[Fe_4E_4X_4]^{2-}$ clusters (E = S, Se; X = Cl, Br, I, SPh) reported by many authors. [239–245] As such, the Fe_4 tetrahedra display two short (276–279 pm) and four long (280.2–282.5 pm) Fe–Fe distances. The $[Se\{Fe(CO)_2Cp\}_3]^+$ cation is isolobal with the $(CH_3)_3Se^+$ cation. It has a Fe_3Se moiety in the shape of a compressed trigonal pyramid in which there are no bonding contacts between the Fe atoms (Fe $\cdots$ Fe: 400 pm). An identical cation was recently prepared by Herrmann et al. [246]

As was demonstrated by the synthesis of the salt **77**, a heterocubane cluster can also be formed from the reaction between $[CpMo(CO)_3Br]$ and $Se(SiMe_3)_2$ (Eq. 3.53). The cation in **77** (Fig. **3-78**; Mo–Mo: 294–301 pm) consists of three CpMo groups (Mo1, Mo2, Mo4), one $Mo(CO)_3$ group, and four capping Se ligands.[247–250]

$$[CpMo(CO)_3Br] \xrightarrow{Se(SiMe_3)_2} \underset{\mathbf{77}}{[(CpMo)_3Mo(CO)_3Se_4]^+[CpMo(CO)_3]^-} \qquad (3.53)$$

Organometallic derivatives of the less electron rich transition metals have been used by several investigators as precursors to polynuclear complexes. Cp-substituted transition metal halides, for example, have been allowed to react with M_2E_x (M = Li, Na; E = S, Se; x = 1,2) or elemental sulfur or selenium. [251–254]

$S(SiMe_3)_2$ and $Se(SiMe_3)_2$ have also been used successfully. Thus, the reaction of $[CpCrCl_2(P\mathit{n}Bu_3)]$ with $Se(SiMe_3)_2$ furnishes the heterocubane cluster $[Cp_4Cr_4Se_4]$. [255]

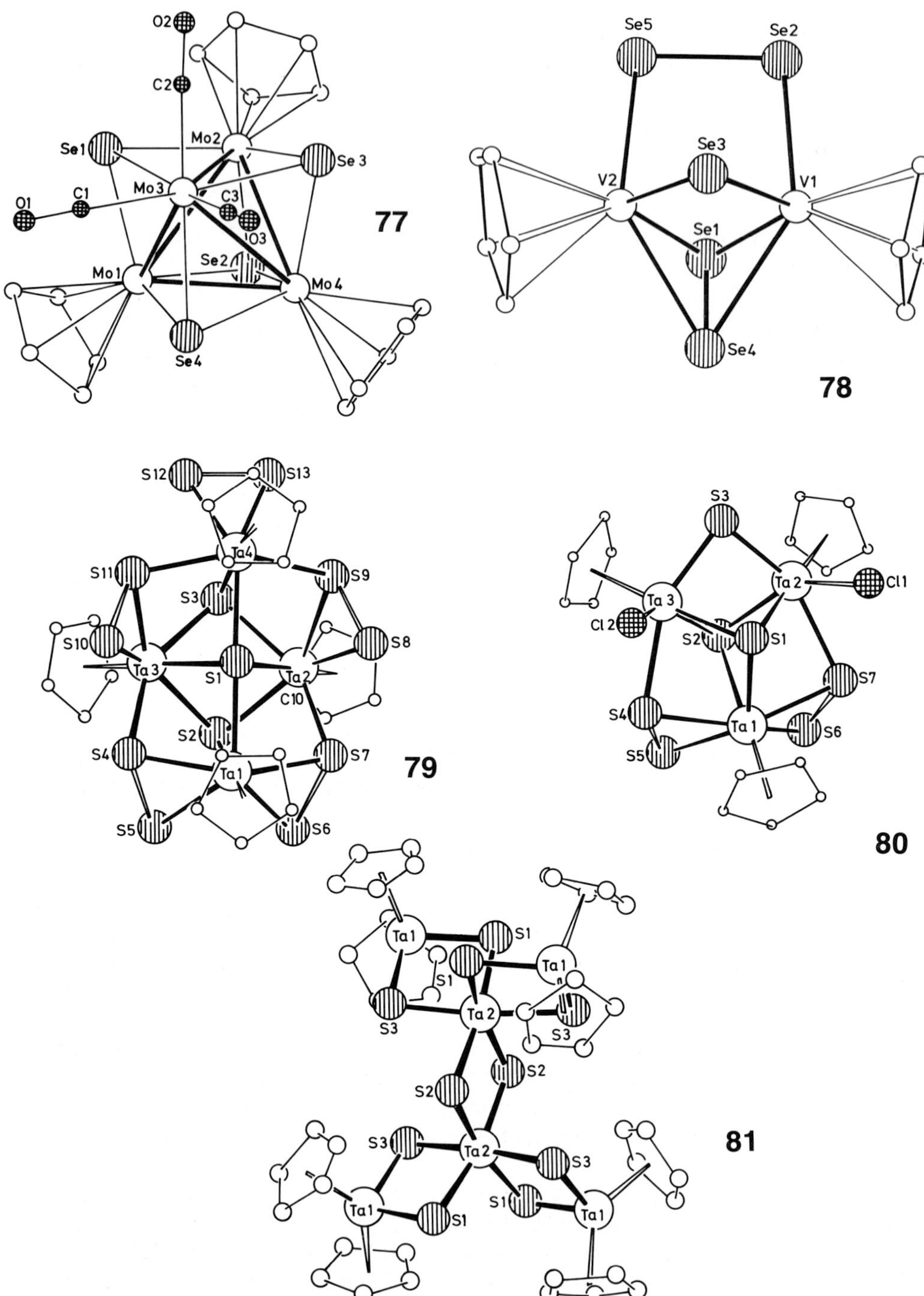

Figure 3-78. Molecular structures of **78**, **79**, **80**, and the cations in **77** and **81**.

Rauchfuss and Rheingold have reported the synthesis of the diamagnetic compound $[Cp_2V_2Se_5]$ (**78**) from the reaction of $[Cp_2VCl_2]$ with $Se(SiMe_3)_2$ (Eq. 3.54). [93, 256]

$$[CpVCl_2] \xrightarrow{Se(SiMe_3)_2} \underset{\mathbf{78}}{[Cp_2V_2Se_5]} \tag{3.54}$$

The structure of **78** is shown in Figure **3-78**. There are three different types of Se bridges: μ_2-Se (Se3) with V–Se: 237 pm; μ–η^1-Se_2 (Se2, Se5) with V–Se: 245–246 pm and Se–Se: 230.2 pm; and μ-η^2-Se_2 (Se1, Se4) with V–Se: 252–254 pm and Se–Se: 231 pm. Examples of these types of bridges are also known in other compounds. [257–260] Compared with $[Cp_2V_2S_5]$ (V–V: 266 pm), the V–V distance in **78** (277 pm) is unusually long. [93, 256] Surprisingly, the Ta analogue of **78** has not been prepared. Instead, the addition of a THF solution of $[CpTaCl_4]$ to $S(SiMe_3)_2$ leads to the formation of a crystalline mixture of **79** and **80** (Eq. 3.55). [261] If, however, a THF solution of $S(SiMe_3)_2$ is allowed to diffuse into a THF solution of $[CpTaCl_4]$, then the salt **81** can be isolated in high yield.

$$[CpTaCl_4] \xrightarrow{S(SiMe_3)_2} \underset{\mathbf{79}}{[Cp_4Ta_4S_{13}]} + \underset{\mathbf{80}}{[Cp_3Ta_3S_7Cl_2]} + \underset{\mathbf{81}}{[Cp_8Ta_6S_{10}][TaCl_5S]} \tag{3.55}$$

The structure of **80** (Fig. **3-78**) contains a Ta_3 triangle (Ta1 ··· Ta2, Ta1 ··· Ta3: 360 pm, Ta2 ··· Ta3: 324 pm) bonded by the two μ_3-S ligands S1 and S2 (S1–Ta: 245–260 pm, S2–Ta: 251 pm). Whereas the distances spanning Ta1 and Ta3, and Ta1 and Ta2, are each bridged by (μ-η^1, η^2)-S_2 groups (S–S: 205–207 pm, Ta1–S: 248–255 pm, Ta3–S4 and Ta2–S7: 268 pm), that between Ta2 and Ta3 is bridged simply by the μ_2-S ligand S3 (S3–Ta: 240–242 pm). Ta2 and Ta3 each display a distorted octahedral coordination and Ta1 has a seven fold coordination. The molecule **80** also contains two Ta–Cl bonds which act as reactive sites in reactions with $[CpTaCl_4]$ and $S(SiMe_3)_2$, from which the tetranuclear species $[Cp_4Ta_4S_{13}]$ (**79**) is formed. A comparison between the topologies of **79** and **80** (Fig. **3-78**) gives an indication the ligand reorientation necessary in this reaction. The μ_3-S (S1, S2) and μ_2-S (S3) ligands in **80** may become the μ_4-S (S1) and μ_3-S (S2, S3) ligands in **79**, whereby Ta4 becomes linked to Ta2 and Ta3 by two (μ_2-η^1, η^2)-S_2 fragments. The four Ta atoms in **79** thereby acquire a strongly distorted pentagonal bipyramidal coordination geometry. Assuming that there are both S_2^{2-} and S^{2-} bridges in **79** and **80**, then the Ta atoms in both compounds have a formal oxidation state of +5, which is consistent with the diamagnetism observed in these compounds.

The cation in **81** (Fig. **3-78**) contains octahedral TaS_6 (Ta2) and tetrahedral Cp_2TaS_2 groups (Ta1) linked together by μ_2-S ligands (Ta–S: 235–243 pm). There are no bonding contacts between the Ta atoms (Ta1 ··· Ta2: 338 pm, Ta2 ··· Ta2: 353 pm). In the meantime, evidence has been obtained which shows that linking tetrahedral Cp_2TaS_2 groups with octahedral TaS_6 groups can further lead to more highly condensed clusters.

The unique reactivity of compounds from the vanadium group elements can be seen with particular clarity in Equation (3.56) by the reaction of $[CpNbCl_4]$ with $Se(SiMe_3)_2$. Using toluene as solvent, no analogous complexes to **79–81** are obtained but instead, the brown compound **82** is formed exclusively. [261]

$$[CpNbCl_4] \xrightarrow{Se(SiMe_3)_2} \underset{\mathbf{82}}{[Cp_3Nb_3Se_5Cl_2]} \tag{3.56}$$

The structure analysis shows that **82** (Fig. **3-79**) contains three differently coordinated Nb atoms. Nb1 is coordinated by η^5-Cp, Se4, Se5, and Cl2 in a distorted tetrahedral geometry with Nb–C(Cp) = 236.6–244.2 pm, Nb1–Se = 239 pm, and Nb1–Cl2 = 236.6 pm. Nb3 has the coordination number 5, being coordinated by η^5-Cp, Se1, Se2, Se3, and Cl1 (Nb3–C(Cp): 240.1–244.0 pm, Nb3–Se1: 238.3 pm, Nb3–Se2(Se3): 261–263 pm, Nb3–Cl1: 242.9 pm). Nb2 has coordination number 6 and is coordinated by η^5-Cp and Se1–Se5 (Nb2–C(Cp): 235.0–

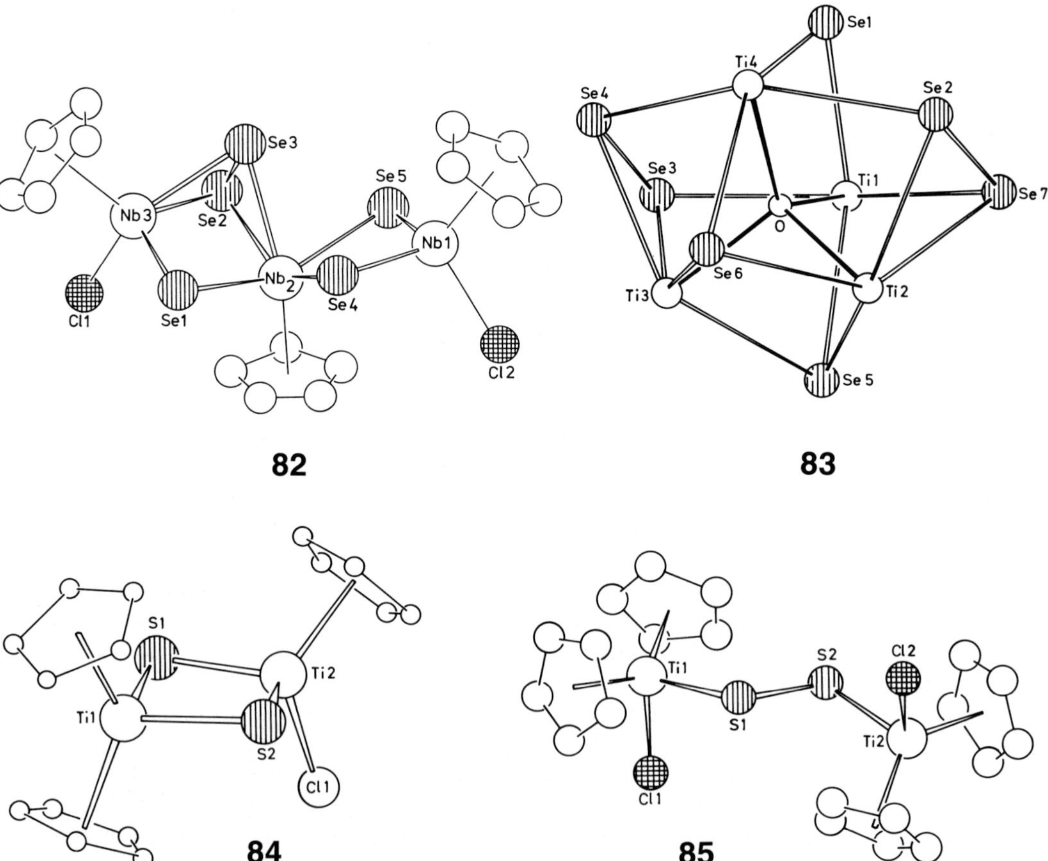

Figure 3-79. Molecular structures of **82**, **83** (without C atoms), **84**, and **85**.

241.8 pm, Nb2–Se: 266.5–271.7 pm). Assuming the charges on the ligands to be Cp^-, Se^{2-}, Se_2^{2-}, and Cl^-, then **82** can be regarded as a mixed valence compound with Nb1 and Nb3 having +4 and Nb2a +5 oxidation states. Correspondingly, the compound is paramagnetic, and its ESR spectrum in $C_2H_4Cl_2$ shows the expected 10 line resonance signal ($g = 2.0047$, $a_{Nb} = 11.6$ mT).

Cp complexes of Ti can also be used to form multinuclear compounds. Depending on the reaction conditions (e. g. solvent, temperature) employed, and the type of Cp ligands used, it is possible to prepare the compounds outlined in Scheme **3-28**.

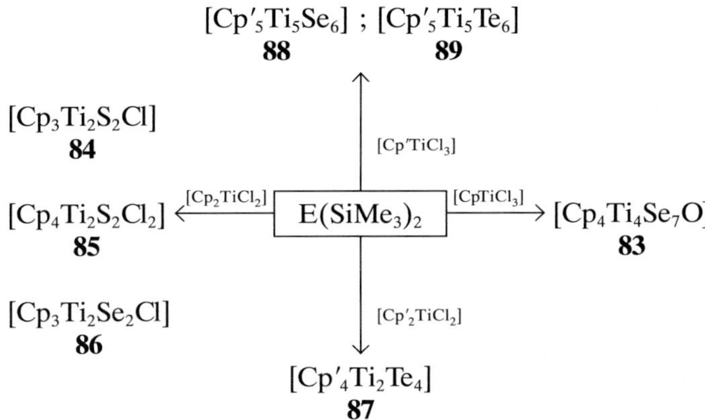

Scheme 3-28. Routes to S, Se, and Te bridged Ti clusters.

The diffusion of a THF solution of $E(SiMe_3)_2$ into a THF solution of $[Cp_2TiCl_2]$ leads to the formation of the binuclear complexes **84–86** in good yield. [261] A molecule of **84** (Fig. **3-79**) consists of a Cp_2Ti group bridged by two μ_2-S ligands to a CpTiCl group. Both Ti atoms thereby have a distorted tetrahedral coordination. Just as in the isostructural $[Cp_3Ti_2Se_2Cl]$ (**86**), the Ti–E bonds involving Ti2 (which is bonded to Cl1) are about 20 pm shorter than those involving Ti1 (**84**: Ti2–S: 223 pm, Ti1–S: 243 pm; **86**: Ti2–Se: 237 pm, Ti1–Se: 259 pm).

The structure of **85** (Fig. **3-79**) can be considered as being derived from two Cp_2TiCl fragments bridged by an S_2^{2-} group (S1–S2: 208.3 pm; Ti–S: 239 pm). Although both **84** and **86** have potentially reactive Ti–Cl bonds, larger complexes have not yet been synthesized.

If, on the other hand, $[CpTiCl_3]$ is allowed to react with $Se(SiMe_3)_2$ in THF, a crystalline residue is formed, from which the tetranuclear complex $[Cp_4Ti_4Se_7O]$ can be isolated in low yield by recrystallization from $C_2H_4Cl_2$. [261] The compound **83** has a structure which consists of a Ti_4 tetrahedron (Ti ··· Ti: 311.4–346.2 pm) enclosing a central O atom (Ti–O: 195–216 pm). The Ti atoms are six coordinate, and the Se ligands can be classified as μ_2-Se (Se1–Ti: 253 pm), μ_3-Se (Ti–Se5 / Se6: 260–274 pm), and μ_3-Se_2 bridges (Se3–Se4: 232 pm; Se2–Se7: 234

pm) with unequal Ti–Se distances (Ti3–Se3: 263 pm; Ti3–Se4: 266 pm, Ti1–Se3: 270 pm; Ti4–Se4: 268 pm). Rauchfuss, Rheingold et al. have recently presented the structure of $[(MeC_5H_4)_4Ti_4S_8O_2]$, which, in contrast to **83**, contains a Ti_4 cluster coordinated by four S_2 units and one μ_2-O ligand. [262] The same authors also obtained the heterocubane clusters $[(RC_5H_4)_4Ti_4S_4]$ (R = Me, *i*Pr) by the reduction of $[(RC_5H_4)TiCl_3]$ with Zn and subsequent reaction with $S(SiMe_3)_2$. [262]

The influence of the Cp ligands on the structure of the Ti complex formed becomes apparent when the C_5H_5 (Cp) group is replaced with the bulkier $C_5H_4CH_3$ (Cp′) group. The structural unit of **87** (Fig. **3-80**) is a six membered Ti_2Te_4 ring in a chair conformation. The Te–Te distances (272.6 pm) are shorter than the Te–Te distances found in the trigonal modification of elemental Te (283.5 pm). [263] Other examples having μ_2-Te_2 bridges include $[\{(triphos)M\}_2Te_2]$(M = Co, Ni; Te–Te: 280.2 pm) and $[Cp_2Mo_2Fe_xTe_2(CO)_7]$ (x = 1, 2; Te–Te: 313 pm). [264, 265]

The multinuclear compounds **88** and **89** (Scheme **3-28** and Fig. **3-80**) are isostructural and are based on a central pentanuclear Ti_5 unit in the form of a distorted trigonal bipyramid. All six triangular faces are capped by either μ_3-Se (**88**) or μ_3-Te ligands (**89**). In addition, all Ti atoms are coordinated by η^5-Cp′ ligands. This type of structure was recently described by Bottomley, who observed the formation of $[Cp_5Ti_5Se_6]$ as a result of the reaction of $[Cp_2Ti(CO)_2]$ with H_2S.[266] The Ti–Ti distances within the Ti_5E_6 polyhedron are dependent upon the nature of the bridging ligand. For example, in $[Cp_5Ti_5S_6]$, the Ti–Ti distances lie in the 307.6–321.4 pm range. As a result of the increase in the covalent radii on passing from sulphur to selenium and tellurium, the Ti–Ti contacts in **88** and **89** are also found to increase: 330.1–339.3 pm in **88** and 352.7–367.4 pm in **89**. All the Ti–Ti contacts lie outside the range for which bonding interactions may be considered. Theoretical calculations on $[Cp_5Ti_5S_6]$ show a possible Jahn-Teller distortion in Ti_5 polyhedra, [266–269] whereby the magnitude of this distortion is dependent upon the nature of the chalcogen ligand. [269] From a naive point of view, the three electrons available for Ti–Ti bonding could be divided in such a manner, so that the equatorial titanium atoms would have formal +3 charge (corresponding to a d^1 configuration), and the axial titanium atoms a +4 charge. The Ti–E distances are in accordance with this hypothesis.

A structural dependence on the chalcogen has also been found for clusters of vanadium. $[CpVCl_2(PMe_3)_2]$ reacts with $Te(SiMe_3)_2$ to form the heterocubane $[Cp_4V_4Te_4]$ cluster, which is isostructural to $[Cp'_4Ti_4S_4]$. [93, 270] Perhaps somewhat surprising are the reactions of $[CpZrCl_3]$ or $[CpVCl_2(PMe_3)_2]$ with $E(SiMe_3)_2$ to form the hexanuclear clusters **90** and **91** (Scheme **3-29**). [271]

$$\underset{\mathbf{90}}{[Cp_6Zr_6S_9]} \xleftarrow{[CpZrCl_3]} \underset{E\,=\,S,\,Se}{E(SiMe_3)_2} \xrightarrow{[CpVCl_2(PMe_3)_2]} \underset{\mathbf{91}}{[V_6Se_8O(PMe_3)_6]}$$

Scheme 3-29. Formation of Zr and V clusters with S and Se bridges.

The molecular structures of **90** and **91** are shown in Figure **3-80** and can be described as M_6 octahedra with face bridging μ_3-S and μ_3-Se ligands respectively. These M_6E_8 units are comparable with the M_6 clusters formed by the early transition metals. [272–280] All the chalcogen atoms in both **90** and **91** form regular cubes with nonbonding E–E contacts of 356.6–357.2 pm (S–S) and 351.4–359.2 pm (Se–Se). Although it is remarkable that these distances are approximately equal, it is probably a result of the different strengths of the metal-metal interactions. In **90**, the Zr–Zr distances of 364 pm are much longer than those found in $[Zr_6Cl_{12}(PMe_2Ph)_6$ [281] and therefore too long to allow for a discussion of bonding.

As a result of these long Zr–Zr contacts in **90**, the Zr atoms are found 79 pm above the faces of the S_8 cube. Clusters of the early transition metals typically have only a few valence electrons available for metal-metal bonding. The consequence of this is a tendency to fit hetero atoms into the cluster center. Some examples of this are $Rb_5[Zr_6Cl_{18}B]$, $[Zr_6Cl_{14}C]$, $[Zr_6Cl_{15}N]$, and $[Nb_6I_{11}H]$. [282–284] A similar situation can be found in **90**, where the center of the Zr_6 octahedron is occupied by an extra sulphur atom, S1. The distances from S1 to the surrounding zirconium atoms lie around 257 pm, which is slightly shorter than the 263.1–265.4 pm μ_3-S–Zr bond lengths.

In order to describe the bonding situation in **90**, it is helpful to regard the ligands as S^{2-} and Cp^-. In this "ionic model", which ignores the possibility of covalent bonding between Zr and S or Zr and Cp, **90** would be formulated as $[(Zr^{4+})_6(S^{2-})_9(Cp^-)_6]$. Accordingly, no valence electrons are available for Zr–Zr bonding in the Zr_6 octahedron. This finding is rather surprising because other octahedral Zr_6 molecules do have valence electrons available for Zr–Zr bonding. Examples of these include $[Zr_6Cl_{12}(PMe_2Ph)_6]$ and $(PPh_4)_2[Zr_6Cl_{18}Fe]$ with 12 and 8 available valence electrons respectively. [281–284] Extended Hückel calculations have been performed on a number of these types of clusters and verify the significance of the interstitial hetero atoms. Thus, the insertion of the central sulphur atom, S1, into the hypothetical $[Cp_6Zr_6(\mu_3\text{-}S)_8]$ cluster affords a stabilisation of about 4eV.

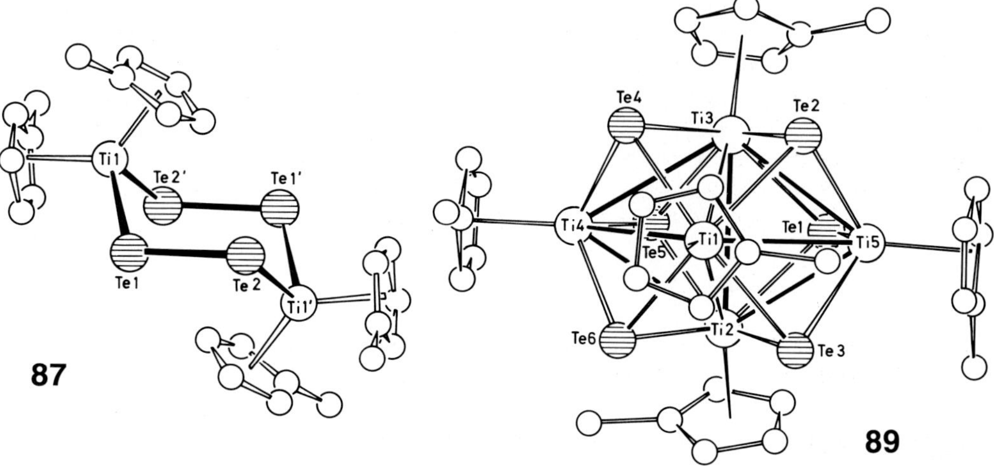

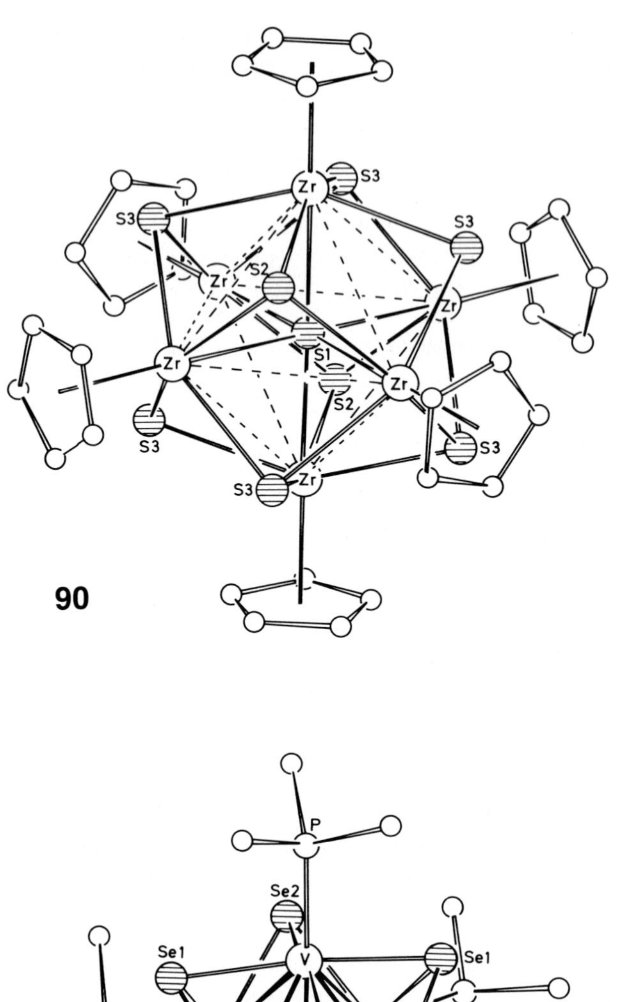

Figure 3-80. Molecular structures of **87**, **89**, **90**, and **91**.

The structure of **91** is quite comparable to that of **90**. Figure **3-80** shows that **91** consists of a V_6 octahedron, the faces of which are capped by μ_3-Se ligands at distances in the 251.4–252.8(3) pm range. Each vanadium atom is also coordinated to a PMe_3 ligand. An oxygen atom is located in the center of the V_6 octahedron. As a result of the significantly shorter V–V distances of 279.4–283.9(3) pm, the vanadium atoms lie only 21 pm above the faces of the Se_8 cube. The V–V distances found in **91** correspond fairly well with those in other vanadium clusters. Examples are $[Cp_5V_5O_6]$ (273.8–276.2 pm) and $[Cp_{11}V_{13}O_{18}(NMe_3)_2]$ (290.6(4) pm). [285, 286] The V–O bond lengths to the interstitial oxygen atom of 199.2(2) pm also show good agreement with those found in these compounds. Using the "ionic model" mentioned above, **91** can be considered as having the composition $[(V^{3+})_6(Se^{2-})_8(O^{2-})(PMe_3)_6]$ and should have 12 valence elelctrons available for V–V bonding. For comparison, $[Mo_6Cl_{12}]$ has 24 valence electrons which are available for Mo–Mo bonding. The difficulty in describing the bonding in **90** and **91** becomes apparent when one considers that the $18e^-$ rule predicts $84e^-$ for M_6 clusters. The cluster **90** contains 92 valence electrons and thus greatly exceeds this predicted value, whereas **91** contains only 80 valence electrons and has less than the predicted number. MO calculations have been made for clusters of the early transition metals having the composition $[Cp_mM_mE_x]$ (M = 3d or 4d element, E = O or S). [268] According to these, clusters with the composition $[Cp_6M_6E_8]$ (M = Ti, Zr) should have more valence electrons than predicted by the $18e^-$ rule, the excess number of electrons corresponding to the formula $6n-22$ ($n = 4$ for Ti and Zr). Therefore, this type of cluster should contain 86 valence electrons. The number of valence electrons in **90** exceeds this amount by exactly the number of electrons provided by the interstitial oxygen atom. No explanation has as yet been found for the electron deficitency in **91**, however.

Silylated derivatives of S, Se, and Te can also be used for the preparation of multinuclear complexes of the electron-rich transition metals containing allyl groups as ligands. $E(SiMe_3)_2$ reacts with $[(\eta^3\text{-}C_3H_5)MCl]_2$ (M = Ni, Pd) to form extremely unstable hexanuclear complexes. The sulphur derivatives of these had already been obtained by other methods. [287, 288] The analogous complexes with η^3-C_4H_7 ligands are appreciably more stable. According to the structure analysis on $[(\eta^3\text{-}C_4H_7)_6Pd_6Se_3]$ (**92**), it contains a slightly distorted M_6 trigonal prism (Fig. **3-81**). The Pd_4 faces are capped by μ_4-Se ligands and the Pd atoms are each bonded to an η^3-C_4H_7 group. The Pd–Pd bond lengths within the prism lie in the 288.4–313.2 pm range, but are considerably longer than those observed in other Pd clusters. [289–291]

The tetranuclear diamagnetic clusters **93** and **94** (Eq. 3.57) are formed in high yield from reaction of $[CpNiCl(PPh_3)]$ with $E(SiMe_3)_2$. Small amounts of **95** and **96** (which are also obtained from the reaction of PPh_3 with **93** and **94** respectively) are also formed. [292]

$$[CpNiCl(PPh_3)] \xrightarrow{E(SiMe_3)_2} \underset{\substack{E\,=\,Se:\ \mathbf{93}\\ E\,=\,Te:\ \mathbf{94}}}{[Cp_4Ni_4E_2]} + \underset{\substack{E\,=\,Se:\ \mathbf{95}\\ E\,=\,Te:\ \mathbf{96}}}{[Cp_4Ni_4E_3(PPh_3)_2]} \tag{3.57}$$

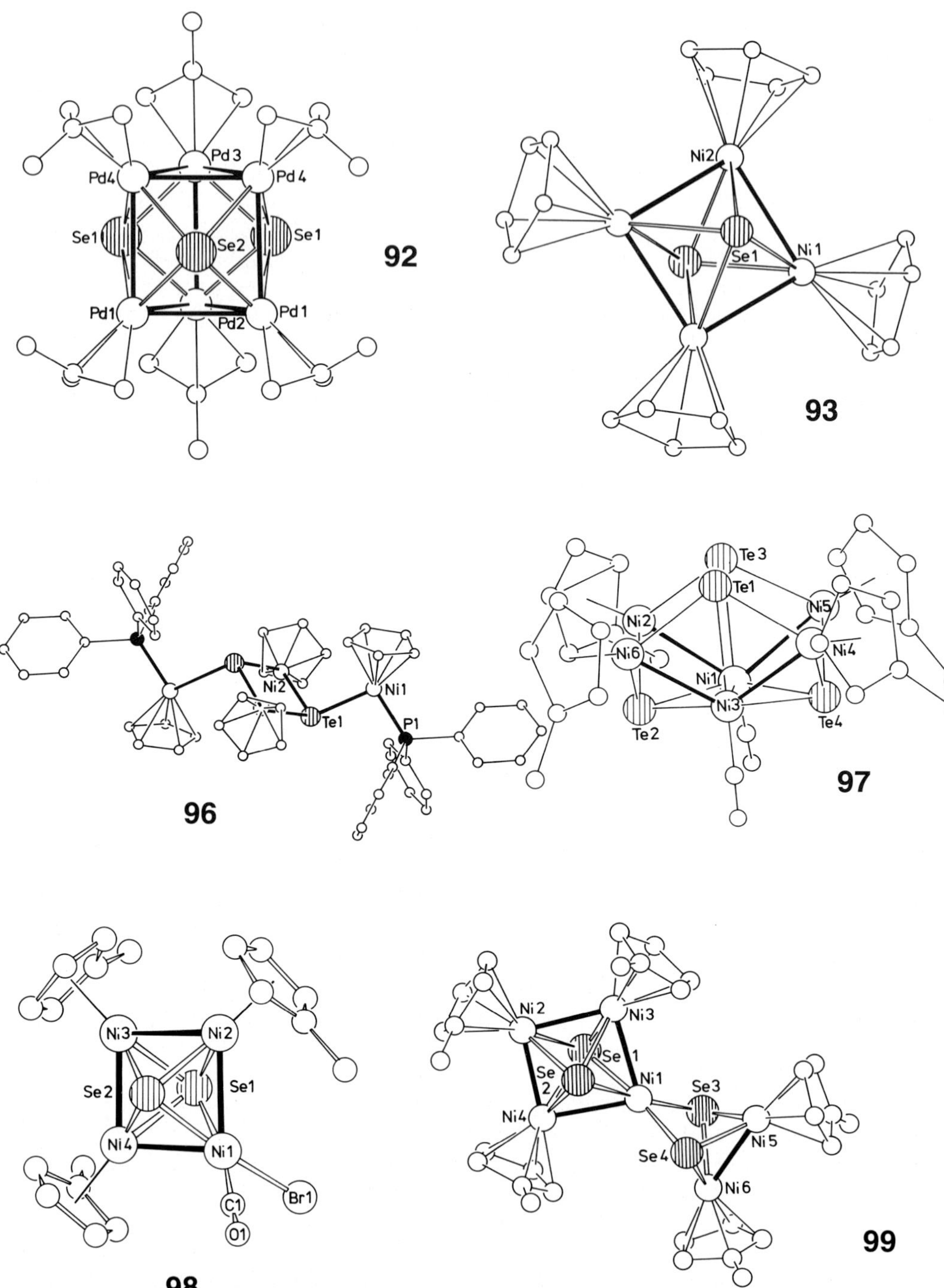

Figure 3-81. Molecular structures of **92**, **93**, **96**, **97**, **98**, and **99**.

The isostructural molecular structures of **93** (Fig. **4-81**) and **94** are built from an almost regular Ni_4 square (crystallographic $\bar{1}$ symmetry) in which each nickel is bonded to an η^5-Cp group and two μ_4-E ligands. Assuming that four Ni–Ni single bonds are present in **93** and **94**, these $68e^-$ clusters contain four electrons more than would be expected from the $18e^-$ rule. The Ni–Ni bond lengths of 257 pm lie in the expected range. [293–294] As in the Ni_3Se_2 fragment of **49**, the Se–Se contacts in **93** (303 pm) are shorter than the van der Waals contacts in either elemental selenium or Se_8^{2+}. [295, 296] It is possible that weak interactions between the Se ligands are responsible for this. [193, 194, 297] The structure of the isostructural **95** and **96** (Fig. **4-81**) can be viewed as arising from a breaking of all the Ni–Ni bonds in **93** or **94**. Further examples of the breaking of M–M bonds have been presented by other authors. [298–300] The compounds **95** and **96** contain four CpNi units linked by μ_3-E ligands, the outer Ni atoms also being coordinated by PPh_3 groups. Both compounds are diamagnetic and each Ni atom has an 18 electron configuration. Correspondingly, the Ni $\cdots$ Ni contacts (337–382 pm in **95**, 362–398 pm in **96**) give no indication of bonding interactions.

If the complex [Cp′Ni(CO)Br] (Cp′ = $C_5H_4CH_3$), presumably formed in the reaction of $[Cp'Ni(CO)]_2$ with Br_2 (Eq. 3.58), is allowed to react with $E(SiMe_3)_2$ (E = Se, Te), the structures of the various products clearly show the influence of the Cp′ ligands. [301, 302]

$$[Cp'\,Ni(CO)]_2 \xrightarrow[E(SiMe_3)_2]{Br_2} \underset{\mathbf{97}}{[Cp'_4Ni_6Te_4(CO)_2]^+}\ \underset{\mathbf{98}}{[Cp'_3Ni_4Se_2(CO)Br]} \tag{4.58}$$

$$+\ \underset{\mathbf{99}}{[Cp'_5Ni_6Se_4]} + \underset{\mathbf{100}}{[Cp'_8Ni_{15}Se_{10}(CO)_3]} + \underset{\mathbf{101}}{[Cp'_8Ni_{15}Se_{10}(CO)Cl_2]}$$

The molecular structure of **97** is depicted in Figure **3-81** and consists of two Ni_4 squares linked by a common edge. The Ni_4 squares are each capped by one μ_4-Te ligand (Te2 and Te4), and they are both coordinated to one Te_2^{2-}group (Te1–Te3: 288.5 pm). Ni1 and Ni3 are coordinated to CO ligands and the remaining nickel atoms by η_5-Cp′ ligands. The Ni–Ni distances vary greatly. The Ni1–Ni3 distance (280 pm) is considerably longer than the other Ni–Ni distances, all of which lie in the 250.8–254.3 pm range. The cluster **97** contains 98 valence electrons, whereby the nickel atoms coordinated to the Cp′ ligands (Ni2, Ni4, Ni5, and Ni6) can be considered as 17 electron fragments and the nickel atoms coordinated to the CO ligands (Ni1 and Ni3) as 15 electron fragments.

In contrast to the $68e^-$ cluster **93**, the $66e^-$ cluster **98** has only two electrons more than one would expect according to the $18e^-$ rule. The cluster **98** (Eq. 3.58 and Fig. **3-81**) can be viewed as a reactive intermediate in the synthesis of larger complexes when considering the successful isolation (Eq. 3.58) of the metal-rich compounds **99** (Fig. **3-81**), **100**, and **101** (Fig. **3-82**). As in **98**, **99** also contains a planar Ni_4 ring which is bridged by two μ_4-Se ligands. This Ni_4 ring in **99** is joined by two μ_3-Se ligands (Se3, Se4) to a Ni_2 dumbbell. With the exception of Ni1, which has square planar coordination to the four Se ligands, each Ni atom is bound to a η^5-Cp′ ligand. The Ni–Ni distances in the Ni_4 ring (250.2–261.3 pm)

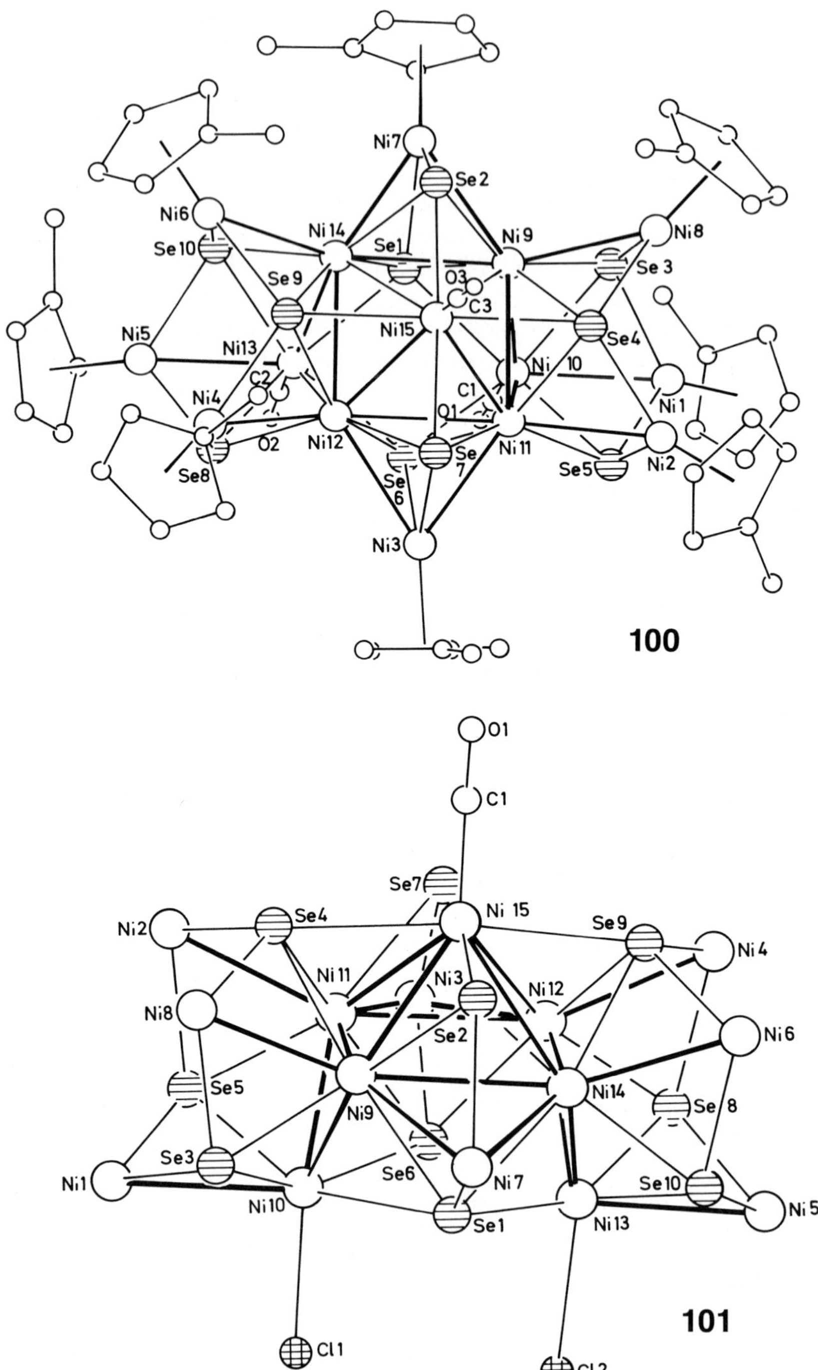

Figure 3-82. Molecular structures of **100** and **101** (without Cp′ ligands).

are comparable with those in **98** (252.8–263.1 pm). If one assumes that five Ni–Ni single bonds are present in **99**, then the paramagnetic cluster has three electrons more than would normally be expected.

The molecular structures of **100** and **101** (Fig. **3-82**) are almost identical. Whereas in **100**, three Ni atoms (Ni10, Ni13, Ni15) are bound to CO ligands, in **101**, two of these ligands are replaced by the chloride ligands Cl1 and Cl2. The structures of **100** and **101**, as depicted in Figure **3-82**, indicate that a previously unknown cluster condensation principle is operative. Accordingly, **100** consists of a central Ni_5 square pyramid (Ni9, Ni11, Ni12, Ni14, Ni15) in which the Ni–Ni distances within the Ni_4 base (**100**: 282–294.8 pm; **101**: 278.9–289.9 pm) are distinctly longer than the distances to the pyramid apex Ni15 (**100**: 271.8–273.5 pm; **101**: 270.3–273.9 pm). Eight further Ni atoms are bound to the Ni_5 cluster subunit (Ni2–Ni4, Ni6–Ni8, Ni10, Ni13), whose distances from each other are in the 252.3–268.3 pm range. The cluster is completed by the Ni1 and Ni5 atoms, which are bound to Ni10 and Ni13 respectively (Ni–Ni: 259.0–261.5 pm). Eight η^5-Cp′ groups are bound to the peripheral Ni atoms Ni1–Ni8 of **100** and **101**, and all faces of their respective Ni_{15} polyhedron are capped by Se ligands, which function as either μ_4-Se ligands (Se2, Se3, Se5, Se7, Se8, Se10) or μ_5-Se ligands (Se1, Se4, Se6, Se9). One could alternatively consider the structure of the Ni_{15} cluster as built from four Ni_3 triangles: (Ni1, Ni2, Ni8), (Ni9, Ni10, Ni11), (Ni12, Ni13, Ni14), and (Ni4, Ni5, Ni6). The Ni–Ni distances in the (Ni1, Ni2, Ni8) and (Ni4, Ni5, Ni6) outer rings are very long (368–378 pm). Those in the (Ni9, Ni10, Ni11), and (Ni12, Ni13, Ni14) inner rings are in the 258.5–283.0 pm range, *i. e.* in the range usually found for Ni–Ni bonds. The cluster core is completed by the atoms Ni3, Ni7, and Ni15.

If the ligands in **100** and **101** are regarded as Se^{2-}, Cp′, and Cl^-, then the Ni_{15} frameworks have the formal charges +28 and +30 respectively. Moreover, assuming that the Ni–Ni bonds shown in Figure **3-82** are single bonds, it is surprising to note that the valence electron counts of 226e$^-$ in **100**, and 236e$^-$ in **101** correspond reasonably well with the 234e$^-$ predicted by the 18-electron rule. The 18-electron rule is even fulfilled in the case of **101**, if the long Ni–Ni distances in the central Ni_4 ring (282–294 pm) are interpreted as nonbonding contacts.

$[Pd_{16}(P_2)_2(CO)_{12}(PPh_3)_8]$
102

↑ $[(C_4H_7)Pd(CO)Br]_n$

$P(SiMe_3)_3$

↓ CuCl PEt$_3$

$[Cu_{96}\{P(SiMe_3)_2\}_6P_{30}(PEt_3)_{22}]$
103

Scheme 3-30. The use of $P(SiMe_3)_3$ to form Pd_{16} and Cu_{96} clusters.

The potential of the synthetic principle described in this article is further illustrated by the observation that other silylated derivatives, such as $P(SiMe_3)_3$, can be employed to obtain new clusters. Most recent results show that $P(SiMe_3)_3$ will react with CuCl and PEt_3, and with $[(C_4H_7)Pd(CO)Br]_n$ and PPh_3 to form the unusual multinuclear complexes outlined in Scheme **3-30**.

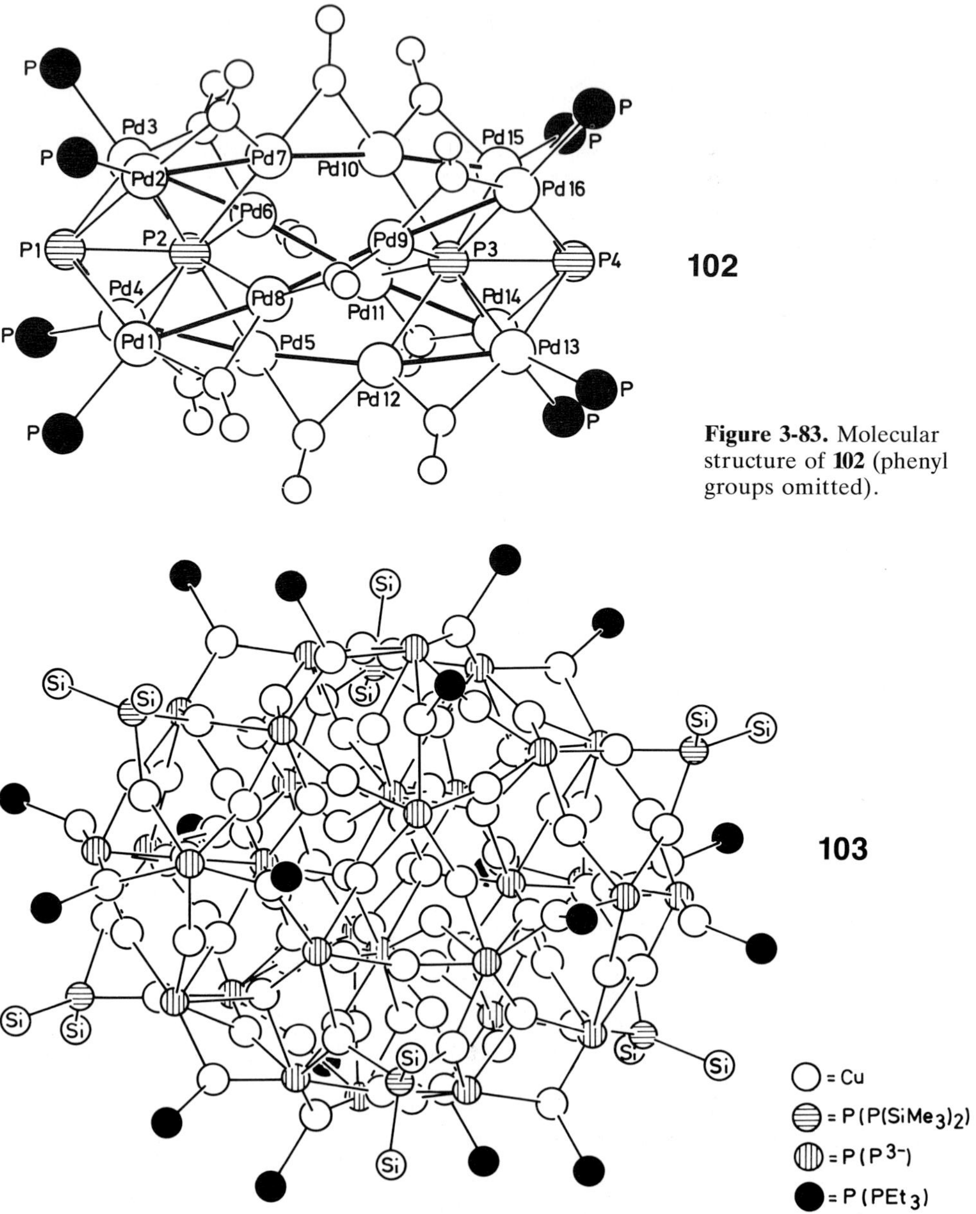

Figure 3-83. Molecular structure of **102** (phenyl groups omitted).

Figure 3-84. Molecular structure of **103** (C atoms omitted).

Figure **3-83** shows the molecular structure of **102**. Here, four parallel, square layers (Pd1–Pd4), (Pd5–Pd8), (Pd9–Pd12), and (Pd13–Pd16) are connected to each other through μ_2-CO bridging ligands. [303] Only those palladium atoms in the basal faces (Pd1–Pd4) and (Pd13–Pd16) are coordinated to PPh_3 ligands. Two P_2 units (P1–P2 and P3–P4) with short P–P bond lengths (221.5 and 226.3 pm) act as additional bridging ligands between the palladium layers. The reaction of $[(C_4H_7)Pd(CO)Br]_n$ and PPh_3 with $As(SiMe_3)_3$ leads to the formation of $[Pd_{16}(As_2)_2(CO)_{12}(PPh_3)_8]$ which is isostructural with 102.

The Cu_{96} cluster **103** shown in Figure **3-84** clearly represents an intermediate in the formation of still larger copper phosphido clusters. [304] The six μ_2-$P(SiMe_3)_2$ groups situated on the cluster periphery present reactive sites at which further cluster growth may occur. If it is assumed that 30 P^{3-} and 6 $P(SiMe_3)_2^-$ groups are present, then each copper atom has a formal oxidation state of +1. **103** can be regarded as a possible intermediate in the formation of Cu_3P.

References

[1] E. W. Abel, R. A. N. McLean, I. H. Sabherwal, *J. Chem. Soc. (A)* **1968**, 2371.
[2] G. Fritz, *Comments Inorg. Chem.* **1982**, *1*, 329.
[3] H. Schäfer, *Z. Naturforsch. B* **1978**, *33*, 351.
[4] H. Schäfer, *Z. Anorg. Allg. Chem.* **1980**, *467*, 105.
[5] H. Schäfer, *Z. Naturforsch. B* **1979**, *34*, 1358.
[6] H. Schäfer, J. Zipfel, B. Mifula, D. Binder, *Z. Anorg. Allg. Chem.* **1983**, *501*, 111.
[7] B. Deppisch, H. Schäfer, D. Binder, W. Leske, *ibid.* **1984**, *519*, 53.
[8] H. Schäfer, J. Zipfel, B. Gutekunst, U. Lemmert, *ibid.* **1985**, *529*, 157.
[9] B. Deppisch, H. Schäfer, *Acta Crystallogr. Sect. B* **1982**, *38*, 748.
[10] B. Deppisch, H. Schäfer, *Z. Anorg. Allg. Chem.* **1982**, *490*, 129.
[11] H. Schäfer, D. Binder, D. Fenske, *Angew. Chem.* **1985**, *97*, 523; *Angew. Chem. Int. Ed. Engl.* **1985**, *24*, 522.
[12] D. Fenske, K. Merzweiler, *Angew. Chem.* **1984**, *96*, 600; *Angew. Chem. Int. Ed. Engl.* **1984**, *23*, 635.
[13] D. Fenske, K. Merzweiler, *Angew. Chem.* **1986**, *98*, 357; *Angew. Chem. Int. Ed. Engl.* **1986**, *25*, 338.
[14] D. Fenske, R. Basoglu, J. Hachgenei, F. Rogel, *Angew. Chem.* **1984**, *96*, 160; *Angew. Chem. Int. Ed. Engl.* **1984**, *23*, 160.
[15] Trinh-Toan, W. P. Fehlhammer, L. F. Dahl, *J. Am. Chem. Soc.* **1972**, *94*, 3389.
[16] M. A. Neumann, Trinh-Toan, L. F. Dahl, *ibid.* **1972**, *94*, 3383.
[17] R. S. Gall, C. T.-W. Chu, L. F. Dahl, *ibid.* **1974**, *96*, 4019.
[18] G. Schmid, *Angew. Chem.* **1978**, *90*, 417; *Angew. Chem. Int. Ed. Engl.* **1978**, *17*, 392.
[19] 10^{-3} M solution in tetrahydrofuran ($[(C_4H_9)_4N][PF_6]$ (10^{-2} mol L^{-1}) as supporting electrolyte): $\mathbf{1} \xrightarrow{0.42\ V} \mathbf{1}^+ \xrightarrow{1.09\ V} \mathbf{1}^{2+}$.
[20] L. D. Lower, L. F. Dahl, *J. Am. Chem. Soc.* **1976**, *98*, 5046.
[21] H. Vahrenkamp, *Adv. Organomet. Chem.* **1983**, *22*, 169.
[22] G. Huttner, K. Knoll, *Angew. Chem.* **1987**, *99*, 765; *Angew. Chem. Int. Ed. Engl.* **1987**, *26*, 743.
[23] D. Fenske, J. Magull, *Z. Naturforsch. B* **1990**, *45*, 121.
[24] K. Wade, *Adv. Inorg. Chem. Radiochem.* **1976**, *18*, 1.

[25] F. A. Cotton, T. E. Haas, *Inorg. Chem.* **1964**, *3*, 10.
[26] B. K. Teo, *Inorg. Chem.* **1984**, *23*, 1251.
[27] B. K. Teo, G. Longoni, F. R. K. Chung, *ibid.* **1984**, *23*, 1257.
[28] B. K. Teo, *J. Chem. Soc. Chem. Commun.* **1983**, 1362.
[29] K. Wade, *J. Chem. Soc. Chem. Commun.* **1971**, 792.
[30] D. M. P. Mingos, *J. Chem. Soc. Chem. Commun.* **1985**, 1352.
[31] R. L. Johnston, D. M. P. Mingos, *Inorg. Chem.* **1986**, *25*, 1661.
[32] D. M. P. Mingos, *Chem. Soc. Rev.* **1986**, *15*, 31.
[33] D. M. P. Mingos, R. L. Johnston, *Struct. Bonding (Berlin)* **1987**, *68*, 29.
[34] F. G. A. Stone, *Angew. Chem.* **1984**, *96*, 85; *Angew. Chem. Int. Ed. Engl.* **1984**, *23*, 89.
[35] R. Hoffmann, *ibid.* **1982**, *94*, 725 and **1982**, *22*, 711.
[36] E. J. Laskowski, R. B. Frankel, W. O. Gillum, G. C. Papaefthymiou, J. Renaud, J. A. Ibers, R. H. Holm, *J. Am. Chem. Soc.* **1978**, *100*, 5322.
[37] Trinh-Toan, B. K. Teo, J. A. Ferguson, T. J. Meyer, L. F. Dahl, *J. Am. Chem. Soc.* **1977**, *99*, 408.
[38] F. H. Köhler, R. de Cao, K. Ackerman, J. Sedlmaier, *Z. Naturforsch. B* **1983**, *38*, 1406.
[39] P. S. Pregosin, R. W. Kunzin: ^{31}P and ^{13}C NMR of Transition Metal Phosphine Complexes, Springer, Berlin **1979**.
[40] J. Schneider, L. Zsolnai, G. Huttner, *Chem. Ber.* **1982**, *115*, 989.
[41] J. Schneider, G. Huttner, *ibid.* **1983**, *116*, 917.
[42] D. Fenske, J. Magull, *Z. Anorg. Allg. Chem.* **1991**, *594*, 29.
[43] A. Eichhöfer, D. Fenske, W. Holstein, *Angew. Chem.* **1993**, *105*, 257; *Angew. Chem. Int. Ed. Engl.* **1993**, *32*, 242.
[44] F. A. Cotton, X. Feng, M. Matusz, R. Poli, *J. Am. Chem. Soc.* **1988**, *110*, 7077.
[45] R. Mason, D. M. P. Mingos, *J. Organomet. Chem.* **1973**, *50*, 53.
[46] P. K. Mehrotra, R. Hoffmann, *Inorg. Chem.* **1978**, *17*, 2187.
[47] C. Kölmel, R. Ahlrichs, *J. Phys. Chem.* **1990**, *94*, 5536.
[48] K. M. Merz, R. Hoffmann, *Inorg. Chem.* **1988**, *27*, 2120.
[49] J. Beck, J. Strähle, *Angew. Chem.* **1985**, *97*, 419; *Angew. Chem. Int. Ed. Engl.* **1985**, *24*, 409.
[50] S. Gambarotta, C. Floriani, A. Chiesi-Villa, C. Guastini, *J. Chem. Soc. Chem. Commun.* **1983**, 1156.
[51] J. A. J. Jarvis, B. T. Kilbourn, R. Pearce, M. F. Lappert, *ibid.* **1973**, 475.
[52] H. Hartl, F. Mahdjour-Hassan-Abadi, *Angew. Chem.* **1984**, *96*, 359; *Angew. Chem. Int. Ed. Engl.* **1984**, *23*, 378.
[53] G. Henkel, P. Betz, B. Krebs, *J. Chem. Soc. Chem. Commun.* **1984**, 314.
[54] I. G. Dance, J. C. Calabrese, *Inorg. Chem. Acta*, **1976**, *L* 41.
[55] D. Coucouvanis, C. N. Murphy, S. K. Kanodia, *Inorg. Chem.* **1980**, 2993.
[56] M. Gernon, H. Kang, S. Liu, J. Zubieta, *J. Chem. Soc. Chem. Commun.* **1988**, 1031.
[57] D. Fenske, J. Hachgenei, *Angew. Chem.* **1986**, *98*, 165; *Angew. Chem. Int. Ed. Engl.* **1986**, *25*, 175.
[58] C. A. Ghilardi, S. Midollini, A. Orlandini, L. Sacconi, *Inorg. Chem.* **1980**, *19*, 301.
[59] A. Vizi-Orosz, V. Galamb, G. Palyi, L. Marko, G. Bor, G. Natile, *J. Organomet. Chem.* **1976**, *107*, 235.
[60] M. Di Vaira, L. Sacconi, *Angew. Chem.* **1982**, *94*, 338; *Angew. Chem. Int. Ed. Engl.* **1982**, *21*, 330.
[61] K. Hedberg, E. W. Hughes, J. Wiezer, *Acta Crystallogr.* **1961**, *14*, 369.
[62] M. Di Vaira, S. Midollini, L. Sacconi, F. Zanobini, *Angew. Chem.* **1978**, *90*, 720; *Angew. Chem. Int. Ed. Engl.* **1978**, *17*, 676.
[63] M. Di Vaira, S. Midollini, L. Sacconi, *J. Am. Chem. Soc.* **1979**, *101*, 1757.
[64] A. S. Foust, M. S. Foster, L. F. Dahl, *J. Am. Chem. Soc.* **1969**, *91*, 5631.
[65] I. Bernal, H. Brunner, W. Meier, H. Pfisterer, J. Wachter, M. L. Ziegler, *Angew. Chem.* **1984**, *96*, 428; *Angew. Chem. Int. Ed. Engl.* **1984**, *23*, 438.

[66] O. J. Scherer, H. Sitzmann, G. Wolmershäuser, *ibid.* **1984**, *96*, 979 and **1984**, *23*, 968.
[67] A. L. Rheingold, M. J. Foley, P. J. Sullivan, *Organometallics* **1982**, *1*, 1429.
[68] R. A. Jones, B. R. Whittlesey, *J. Am. Chem. Soc.* **1985**, *107*, 1078.
[69] D. Fenske, J. Ohmer, K. Merzweiler, *Angew. Chem.* **1988**, *100*, 1572; *Angew. Chem. Int. Ed. Engl.* **1988**, *27*, 1512.
[70] J. G. W. van der Linden, M. L. H. Paulissen, J. E. J. Schmitz, *J. Am. Chem. Soc.* **1983**, *105*, 1903.
[71] P. Chini, *J. Organomet. Chem.* **1980**, *200*, 37.
[72] G. Ciani, A. Sironi, S. Martinengo, *ibid.* **1980**, *192*, C42.
[73] D. Fenske, K. Vogt, unpublished results.
[74] R. A. Wheeler; Department of Chemistry, University of Oklahoma, Norman, OK 73019, unpublished results.
[75] D. Fenske, H. Fleischer, C. Persau, *Angew. Chem.* **1989**, *101*, 1740; *Angew. Chem. Int. Ed. Engl.* **1989**, *28*, 1665.
[76] M. Müller, H. Vahrenkamp, *J. Organomet. Chem.* **1983**, *252*, 95.
[77] A. S. Foust, M. S. Foster, L. F. Dahl, *J. Am. Chem. Soc.* **1969**, *91*, 5633.
[78] A. L. Rheingold, *Organometallics* **1982**, *1*, 1547.
[79] G. Huttner, B. Siegwarth, O. Scheidsteger, L. Zsolnai, O. Orama, Organometallics **1985**, *4*, 326.
[80] W. A. Herrmann, B. Koumbouris, T. Zahn, M. L. Ziegler, *Angew. Chem.* **1984**, *96*, 802; *Angew. Chem. Int. Ed. Engl.* **1984**, *23*, 812.
[81] E. G. Mednikov, N. K. Eremenko, V. A. Mikhailov, S. P. Gubin, Y. L. Slovokhotov, Y. T. Struchkov, *J. Chem. Soc. Chem. Commun.* **1981**, 989.
[82] E. G. Mednikov, N. K. Eremenko, Y. L. Slovokhotov, Y. T. Struchkov, *J. Organomet. Chem.* **1986**, *301*, C35.
[83] E. G. Mednikov, N. K. Eremenko, Y. L. Slovokhotov, Y. T. Struchkov, *J. Chem. Soc. Chem. Commun.* **1987**, 218.
[84] K. Dehnicke, J. Strähle, *Angew. Chem.* **1992**, *104*, 978; *Angew. Chem. Int. Ed. Engl.* **1992**, *31*, 955.
[85] T. Godemeyer, K. Dehnicke, D. Fenske, *Z. Naturforsch. B* **1985**, *40*, 1005.
[86] H. Plenio, H. W. Roesky, M. Noltemeyer, G. M. Sheldrick, *Angew. Chem.* **1988**, *100*, 1377; *Angew. Chem. Int. Ed. Engl.* **1988**, *27*, 1330.
[87] K. Maczek, D. Fenske, unpublished results.
[88] S. Otzuka, A. Nakamura, T. Yoshida, *Inorg. Chem.* **1968**, *7*, 261.
[89] H. F. Klein, S. Haller, H. König, M. Dartiguenave, Y. Dartiguenave, M. J. Menu, *J. Am. Chem. Soc.* **1991**, *113*. 4673.
[90] K. Maczek, J. Magull, D. Fenske, unpublished results.
[91] G. L. Simon, L. F. Dahl, *J. Am. Chem. Soc.* **1973**, *95*, 2164.
[92] F. Bottomley, F. Grein, *Inorg. Chem.* **1982**, *21*, 4170.
[93] J. D. Darkwa, J. R. Lockemeyer, P. D. W. Boyd, T. B. Rauchfuss, A. L. Rheingold, *J. Am. Chem. Soc.* **1988**, *110*, 141.
[94] O. J. Scherer, *Angew. Chem.* **1990**, *102*, 1137; *Angew. Chem. Int. Ed. Engl.* **1990**, *29*, 1104.
[95] G. Huttner, K. Knoll, *Angew. Chem.* **1987**, *99*, 765; *Angew. Chem. Int. Ed. Engl.* **1987**, *26*, 743.
[96] H. Vahrenkamp, *Angew. Chem.* **1975**, *87*, 363; *Angew. Chem. Int. Ed. Engl.* **1975**, *14*, 322.
[97] A. Müller, R. Jostes, F. A. Cotton, *Angew. Chem.* **1980**, *92*, 921; *Angew. Chem. Int. Ed. Engl.* **1980**, *19*, 875.
[98] A. Müller, E. Diemann, R. Jostes, H. Bögge, *Angew. Chem.* **1981**, *93*, 957; *Angew. Chem. Int. Ed. Engl.* **1981**, *20*, 934.
[99] W. A. Herrmann, *Angew. Chem.* **1986**, *98*, 57; *Angew. Chem. Int. Ed. Engl.* **1986**, *25*, 56.

[100] J. Wachter, *Angew. Chem.* **1989**, *101*, 1645; *Angew. Chem. Int. Ed. Engl.* **1989**, *28*, 1613.
[101] S. C. Lee, R. H. Holm, *Angew. Chem.* **1990**, *102*, 868; *Angew. Chem. Int. Ed. Engl.* **1990**, *29*, 840.
[102] I. A. Dance, *Polyhedron* **1986**, *5*, 1037.
[103] L. H. Gaede, *Angew. Chem.* **1993**, *105*, 25; *Angew. Chem. Int. Ed. Engl.* **1993**, *33*, 24.
[104] B. Krebs, G. Henkel, *Angew. Chem.* **1991**, *103*, 785; *Angew. Chem. Int. Ed. Engl.* **1991**, *30*, 769.
[105] M. Kriege, G. Henkel, *Z. Naturforsch. B* **1987**, *42*, 1121.
[106] G. Henkel, T. Krüger, B. Krebs, *Angew. Chem.* **1989**, *101,* 54; *Angew. Chem. Int. Ed. Engl.* **1989**, *28*, 61.
[107] K. S. Hagen, R. H. Holm, *Inorg. Chem.* **1984**, *23*, 418.
[108] K. S. Hagen, D. W. Stephan, R. H. Holm, *Inorg. Chem.* **1982**, *21*, 3928.
[109] I. G. Dance, L. J. Fitzpatrick, M. L. Scudder, D. J. Craig, *J. Chem. Soc. Chem. Commun.* **1984**, 17.
[110] S. Lacelle, W. C. Stevens, D. M. Kurtz, J. W. Richardson, R. A. Jacobson, *Inorg. Chem.* **1984**, *23,* 930.
[111] H. Strasdeit, B. Krebs, G. Henkel, *Inorg. Chem.* **1984**, *23*, 1816.
[112] I. G. Dance, A. Choy, M. L. Scudder, *J. Am. Chem. Soc.* **1984**, *106*, 6285.
[113] S. A. Al-Ahmad, J. W. Kampf, R. W. Dunham, D. Coucouvanis, *Inorg. Chem.* **1991**, *30*, 1164.
[114] H. Strasdeit, B. Krebs, G. Henkel, *Z. Naturforsch. B* **1987**, *42*, 565.
[115] G. Christou, K. S. Hagen, J. K. Bashkin, R. H. Holm, *Inorg. Chem.* **1985**, *24*, 1010.
[116] B. F. G. Johnson (Ed.): Transition Metal Clusters, Wiley, New York **1980**.
[117] R. B. King, *Prog. Inorg. Chem.* **1972**, *15*, 287.
[118] P. Chini, G. Longoni, V. G. Albano, *Adv. Organomet. Chem.* **1976**, *14*, 285.
[119] J. D. Corbett, *Prog. Inorg. Chem.* **1976**, *21,* 129.
[120] H. G. von Schnering, *Angew. Chem.* **1981**, *93*, 44; *Angew. Chem. Int. Ed. Engl.* **1981**, *20*, 33.
[121] A. Simon, *ibid.* **1981**, *93*, 23 and **1981**, *20*, 1.
[122] H. Schäfer, B. Eisenmann, W. Müller *ibid.* **1973**, *85*, 742 and **1973**, *12*, 649.
[123] H. Vahrenkamp, *Struct. Bonding (Berlin)* **1977**, *32*, 1.
[124] H. Vahrenkamp, *Angew. Chem.* **1978**, *90*, 403; *Angew. Chem. Int. Ed. Engl.* **1978**, *17*, 379.
[125] H. G. von Schnering, W. Beckmann, *Z. Anorg. Allg. Chem.* **1966**, *347*, 231.
[126] E. Röttinger, R. Müller, H. Vahrenkamp, *Angew. Chem.* **1977**, *89*, 341; *Angew. Chem. Int. Ed. Engl.* **1977**, *16*, 332.
[127] D. Fenske, J. Ohmer, J. Hachgenei, K. Merzweiler, *Angew. Chem.* **1988**, *100*, 1300; *Angew. Chem. Int. Ed. Engl.* **1988**, *27*, 1277.
[128] W. Bronger, A. Loevenisch, D. Schmitz, T. Schuster, *Z. Anorg. Allg. Chem.* **1990**, *587*, 91.
[129] Y. Park, M. G. Kanatzidis, *Chem. Mater,* **1991**, *3*, 781.
[130] J. Hsiou Liao, M. G. Kanatzidis, *Inorg. Chem.* **1992**, *31*, 431.
[131] J. Hsiou Liao, M. G. Kanatzidis, *J. Am. Chem. Soc.* **1990**, *112*, 7400.
[132] D. Fenske, A. Ahle, K. Dehnicke, *Chemiker-Zeitung* **1991**, *115*, 249.
[133] M. G. Kanatzidis, *Comments Inorg. Chem.* **1990**, *10*, 161.
[134] J. G. Brennan, T. Siegrist, S. M. Stuczynski, M. L. Steigerwald, *J. Am. Chem. Soc.* **1990**, *112*, 9233.
[135] J. G. Brennan, T. Siegrist, S. M. Stuczynski, M. L. Steigerwald, *J. Am. Chem. Soc.* **1989**, *111*, 9240.
[136] F. Cecconi, C. A. Ghilardi, S. Midollini, *J. Chem. Soc. Chem. Commun.* **1981**, 640.
[137] C. A. Ghilardi, S. Midollini, L. Sacconi, *J. Chem. Soc. Chem. Commun.* **1981**, 47.
[138] J.-Feng You, R. H. Holm, *Inorg. Chem.* **1991**, *30*, 1431.
[139] J.-Feng You, B. S. Snyder, R. H. Holm, *J. Am. Chem. Soc.* **1988**, *110*, 6589.

[140] J. M. McConnachie, M. A. Ansari, J. A. Ibers, *Inorg. Chim. Acta* **1992**, *198*, 85.
[141] A. Müller, M. Römer, H. Bögge, E. Krickemeyer, D. Bergmann, *Z. Anorg. Allg. Chem.* **1984**, *511*, 84.
[142] L. C. Roof, W. T. Pennington, J. W. Kolis, *Inorg. Chem.* **1992**, *31*, 2056.
[143] L. C. Roof, W. T. Pennington, J. W. Kolis, *J. Am. Chem. Soc.* **1990**, *112*, 8172.
[144] L. Marko, G. Bor, E. Klum, *Chem. Ber.* **1963**, *96*, 955.
[145] H. Vahrenkamp, L. F. Dahl, *Angew. Chem.* **1969**, *81*, 152; *Angew. Chem. Int. Ed. Engl.* **1969**, *8*, 144.
[146] R. S. Gail, C. Ting-Wa Chu, L. F. Dahl, *J. Am. Chem. Soc.* **1974**, *96*, 4019.
[147] F. Richter, H. Beurich, M. Müller, N. Gärtner, H. Vahrenkamp, *Chem. Ber.* **1983**, *116*, 3774.
[148] R. H. Holm, *Chem. Soc. Rev.* **1981**, *10*, 455.
[149] G. Henkel, W. Tremel, B. Krebs, *Angew. Chem.* **1983**, *95*, 314; *Angew. Chem. Int. Ed. Engl.* **1983**, *22*, 318.
[150] G. Christou, K. S. Hagen, R. H. Holm, *J. Am. Chem. Soc.* **1982**, *104*, 1744.
[151] A. Müller, W. Jaegermann, J. H. Enmark, *Coord. Chem. Rev.* **1982**, *46*, 245.
[152] F. Cecconi, C. A. Ghilardi, S. Midollini, *Cryst. Struct. Commun.* **1982**, *11*, 25.
[153] G. Christou, B. Ridge, H. N. Rydou, *J. Chem. Soc. Dalton Trans.* **1978**, 1423.
[154] R. D. Adams, I. T. Horvath, *Inorg. Chem.* **1984**, *23*, 4718.
[155] M. Laing, P. M. Kiernau, W. P. Griffith, *J. Chem. Soc. Chem. Commun.* **1977**, 221.
[156] C. E. Strouse, L. F. Dahl, *J. Am. Chem. Soc.* **1971**, *93*, 6032.
[157] D. Fenske, H. Fleischer, H. Krautscheid, J. Magull, *Z. Naturforsch. B* **1990**, *45*, 127.
[158] D. Fenske, H. Fleischer, H. Krautscheid, J. Magull, C. Oliver, S. Weisgerber, *Z. Naturforsch. B* **1991**, *46*, 1384.
[159] S. Fleischer, D. Fenske, H. Fleischer, unpublished results.
[160] F. Cecconi, C. A. Ghilardi, S. Midollini, A. Orlandini, A. Vacca, J. A. Ramirez, *J. Chem. Soc. Dalton Trans.* **1990**, 773.
[161] F. Cecconi, C. A. Ghilardi, S. Midollini, A. Orlandini, *Z. Naturforsch. B* **1990**, *46*, 1161.
[162] L. L. Nelson, F. Y.-K. Lo, A. D. Rae, L. F. Dahl, *J. Organomet. Chem.* **1982**, *225*, 309.
[163] M. A. Bobrik, E. J. Laskowski, R. W. Johnson, W. O. Gillum, J. M. Berg, K. O. Hodgson, R. H. Holm, *Inorg. Chem.* **1978**, *17*, 1402.
[164] D. Fenske, J. Ohmer, J. Hachgenei, *Angew. Chem.* **1985**, *97*, 993; *Angew. Chem. Int. Ed. Engl.* **1985**, *24*, 993.
[165] Trinh-Toan, B. K. Teo, J. A. Ferguson, T. J. Meyer, L. F. Dahl, *J. Am. Chem. Soc.* **1977**, *99*, 408.
[166] D. Fenske, H. Krautscheid, M. Müller, *Angew. Chem.* **1992**, *104*, 309; *Angew. Chem. Int. Ed. Engl.* **1992**, *31*, 321.
[167] Jinkang Gong, Jean Hüang, P. E. Fanwick, C. B. Kubiak, *Angew. Chem.* **1990**, *102*, 407; *Angew. Chem. Int. Ed. Engl.* **1990**, *29*, 396.
[168] D. Fenske, J. Ohmer, *Angew. Chem.* **1987**, *99*, 155; *Angew. Chem. Int. Ed. Engl.* **1987**, *26*, 148.
[169] J. C. Calabrese, L. F. Dahl, P. Chini, G. Longoni, S. Martinengo, *J. Am. Chem. Soc.* **1974**, *96*, 2614.
[170] D. Fenske, J. Hachgenei, J. Ohmer, *Angew. Chem.* **1985**, *97*, 684; *Angew. Chem. Int. Ed. Engl.* **1985**, *24*, 706.
[171] F. Cecconi, C. A. Ghilardi, S. Midollini, A. Orlandini, *Inorg. Chim. Acta* **1991**, *184*, 141.
[172] F. Cecconi, C. A. Ghilardi, S. Midollini, A. Orlandini, P. Zanello, *Polyhedron* **1986**, *5*, 2021.
[173] M. L. Steigerwald, T. Siegrist, S. M. Stuczynski, *Inorg. Chem.* **1991**, *30*, 2257.
[174] D. Fenske, S. Balter, umpublished results.
[175] F. Cecconi, C. A. Ghilardi, S. Midollini, *Inorg. Chim. Acta* **1981**, *64*, L47.

[176] F. Cecconi, C. A. Ghilardi, S. Midollini, A. Orlandini, *Inorg. Chim. Acta* **1983**, *76*, L183.

[177] M. Hong, Z. Huang, X. Lei, G. Wei, B. Kang, H. Liu, *Polyhedron* **1991**, *10*, 927.

[178] D. Fenske, J. Ohmer, K. Merzweiler, *Z. Naturforsch. B* **1987**, *42*, 803.

[179] O. Bars, J. Guillevic, D. Grandjean, *J. Solid State Chem.* **1973**, *6*, 6.

[180] P. Zanello, *Coord. Chem. Rev.* **1988**, *83*, 199.

[181] I. Noda, B. S. Snyder, R. H. Holm, *Inorg. Chem.* **1986**, *25*, 3851.

[182] R. D. Adams, F. A. Cotton, G. A. Rusholme, *Coord. Chem.* **1971**, *1*, 275.

[183] O. S. Mills, B. W. Shaw, *J. Organomet. Chem.* **1968**, *11*, 595.

[184] A. A. Hock, O. S. Mills, Advances in the Chemistry of Coordination Compounds, (Ed.: S. Kirshner), Macmillan, New York, **1961**, 640.

[185] J. K. Burdett, G. J. Miller, *J. Am. Chem. Soc.* **1987**, 109, 4081.

[186] A. Simon, *Angew. Chem.* **1981**, *93*, 23; *Angew. Chem. Int. Ed. Engl.* **1981**, *20*, 1.

[187] A. Simon, *J. Solid State Chem.* **1985**, *57*, 2.

[188] A. Simon, *Angew. Chem.* **1988**, *100*, 163; *Angew. Chem. Int. Ed. Engl.* **1988**, *27*, 159.

[189] R. Chevrel, M. Sergent, B. Seeber, O. Fischer, A. Grüttner, K. Yvon, *Mater. Res. Bull.* **1979**, *14*, 657.

[190] A. Grüttner, K. Yvon, R. Chevrel, M. Potel, M. Sergent, B. Seeber, *Acta Crystallogr. Sect.* **1979**, *B35*, 285.

[191] M. Potel, R. Chevrel, M. Sergent, M. Decourx, O. Fischer, C. R. *Hebd. Seances Acad. Sci. Ser.* **1979**, *C288*, 429.

[192] W. Hönle, H. G. von Schnering, A. Lipka, K. Yvon, *J. Less-Common Met.* **1980**, *71*, 135.

[193] C. A. Ghilardi, S. Midollini, L. Sacconi, *J. Chem. Soc. Chem. Commun.* **1981**, 47.

[194] F. Cecconi, C. A. Ghilardi, S. Midollini, *Inorg. Chem.* **1983**, *22*, 3802.

[195] S. Martinengo, A. Fumagalli, R. Bonifichi, G. Ciani, A. Sironi, *J. Chem. Soc. Chem. Commun.* **1982**, 825.

[196] H. Vahrenkamp, V. A. Uchtmann, L. F. Dahl, *J. Am. Chem. Soc.,* **1968**, *90*, 3272.

[197] R. Chevrel, M. Sergent, *Top. Curr. Phys.* **1982**, *32*, 25.

[198] R. Chevrel, P. Gougeon, M. Potel, M. Sergent, *J. Solid State Chem.* **1985**, *57*, 25.

[199] A. Ceriotti, F. Demartin, G. Longoni, M. Manassero, M. Marchionna, G. Piva, M. Sansoni, *Angew. Chem.* **1985**, *97*, 708; *Angew. Chem. Int. Ed. Engl.* **1985**, *24*, 697.

[200] E. G. Mednikov, N. K. Eremenko, Y. L. Slovokhotov, Y. T. Struchkov, *J. Organomet. Chem.* **1986**, *301*, C35.

[201] D. M. P. Mingos, *J. Chem. Soc. Chem. Commun.* **1983**, 706.

[202] D. M. P. Mingos, R. L. Johnston, *J. Organomet. Chem.* **1985**, *280*, 419.

[203] E. H. Henninger, R. C. Buschert, L. Heaton, *J. Chem. Phys.* **1967**, *46*, 586.

[204] R. K. McMullan, D. J. Prince, J. D. Corbett, *Inorg. Chem.* **1971**, *10*, 1749.

[205] H. Werner, W. Bertleff, U. Schubert, *Inorg. Chim. Acta* **1980**, *43*, 199.

[206] R. D. Adams, T. A. Wolfle, B. W. Eichhorn, R. C. Haushalter, *Polyhedron* **1989**, *8*, 701.

[207] R. Uson, A. Laguna, F. Fornies, I. Valenzuela, Q. Jones, G. Peter, G. M. Sheldrick, *J. Organomet. Chem.* **1984**, *273*, 129.

[208] L. N. Ito, B. F. Johnson, A. M. Muetting, L. H. Pignolet, *Inorg. Chem.* **1989**, *28*, 2028.

[209] L. N. Ito, A. M. P. Felicissimo, L. H. Pignolet, *Inorg. Chem.* **1991**, *30*, 988.

[210] R. Ugo, G. La Monica, S. Cenini, *J. Chem. Soc. A* **1971**, 522.

[211] R. Ugo, *Coord. Chem. Rev.* **1968**, *3*, 319.

[212] D. Fenske, C. Oliver, unpublished results.

[213] D. Fenske, H. Krautscheid, S. Balter, *Angew. Chem.* **1990**, *102*, 799; *Angew. Chem. Int. Ed. Engl.* **1990**, *29*, 796.

[214] D. Fenske, H. Krautscheid, *Angew. Chem.* **1990**, *102*, 1513; *Angew. Chem. Int. Ed. Engl.* **1990**, *29*, 1452.

[215] H. Feld, A. Benninghoven, D. Fenske, H. Krautscheid, unpublished results.

[216] H. Feld, A. Leute, D. Rading, A. Benninghoven, G. Schmid, *J. Am. Chem. Soc.* **1990**, *112*, 8166.
[217] J. H. El-Nakat, I. G. Dance. K. J. Fisher, G. D. Willet, *Inorg. Chem.* **1991**, *30*, 2958.
[218] J. H. El-Nakat, I. G. Dance, K-J. Fisher, G. D. Willet, *J. Chem. Soc. Chem. Commun.* **1991**, 746.
[219] J. H. El-Nakat, I. G. Dance, K. J. Fisher, D. Rice, G. D. Willet, *J. Am. Chem. Soc.* **1991**, *113*, 5141.
[220] W. Bronger, H. Schils, *J. Less-Common Met.* **1982**, *83*, 279.
[221] R. McLaren Murray, R. D. Heyding, *Can. J. Chem.* **1975**, *53*, 878.
[222] A. L. N. Stevels, F. Jellinek, *Recl. Trav. Chim.* **1971**, *90*, 273.
[223] P. Rahlfs, *Z. Phys. Chem. B* **1936**, *31*, 157.
[224] W. Borchert, *Z. Krist.* **1945**, *106*, 5.
[225] A. C. Kolbert, H. J. M. de Groot, D. van der Putten, H. B. Brom, L. J. de Jongh, G. Schmid, H. Krautscheid, D. Fenske, *Z. Phys. D* **1992**, in press.
[226] D. Fenske, J. Steck, *Angew. Chem.* **1993**, *105*, 254; *Angew. Chem. Int. Ed. Engl.* **1993**, *32*, 238.
[227] S. Stensvad, B. J. Helland, *J. Am. Chem. Soc.* **1978**, *100*, 6275.
[228] M. H. Chisholm, *J. Am. Chem. Soc.* **1982**, *104*, 2025.
[229] M. H. Chisholm, *J. Am. Chem. Soc.* **1979**, *101*, 7100.
[230] R. E. McCarley, in "Reactivity of Metal-Metal-bonds", *ACS Symp. Ser.* **1981**, *155*, 41.
[231] H. Chen, *J. Chem. Soc. Chem. Commun.* **1990**, *5*, 373.
[232] G. Granozzi, *J. Organomet. Chem.* **1983**, *244*, 383.
[233] P. L. Stanghellini, *Inorg. Chem.* **1987**, *26*, 2950.
[234] D. Fenske, J. Steck, unpublished results.
[235] C. F. Frank, J. S. Kasper, *Acta Cryst.* **1958**, *11*, 184.
[236] C. F. Frank, J. S. Kasper, *Acta Cryst.* **1959**, *12*, 483.
[237] H. Nowotny, *Metallforsch.* **1946**, *1*, 40, bzw. Structure Reports **1945–1946**, 50.
[238] W. B. Pearson, *Acta Cryst. B* **1968**, *24*, 1415.
[239] D. Fenske, P. Maué, K. Merzweiler, *Z. Naturforsch. B* **1987**, *42*, 928.
[240] M. A. Bobrik, K. O. Hodgson, R. H. Holm, *Inorg. Chem.* **1977**, *16*, 1851.
[241] M. G. Kanatzidis, D. Coucouvanis, A. Kostikas, A. Simopoulos, V. Papaefthymiou, *J. Am. Chem. Soc.* **1985**, *107*, 4925.
[242] W. Saak, S. Pohl, *Z. Naturforsch. B* **1985**, *40*, 1105.
[243] A. Müller, N. Schladerbeck, H. Bögge, *Chimia* **1985**, *39*, 24.
[244] B. A. Averill, T. Herskovitz, R. H. Holm, I. A. Ibers, *J. Am. Chem. Soc.* **1973**, *95*, 3523.
[245] K. S. Hagen, A. D. Watson, R. H. Holm, *Inorg. Chem.* **1984**, *23*, 2984.
[246] C. Hecht, E. Herdtweck, J. Rohrmann, W. A. Herrmann, W. Beck, P. M. Fritz, *J. Organomet. Chem.* **1987**, 330, 389.
[247] Y. Do, E. D. Simhon, R. H. Holm, *Inorg. Chem.* **1985**, *24*, 1831.
[248] Y. Do, E. D. Simhon, R. H. Holm, *ibid.* **1983**, *22*, 3809.
[249] J. R. Dorfman Girerd, E. D. Simhon, T. D. P. Stack, R. H. Holm, *ibid.* **1984**, *23*, 4407.
[250] D. Fenske, P. Maué, unpublished results.
[251] M. Draganjac, T. B. Rauchfuss, *Angew. Chem.* **1985**, *97*, 745; *Angew. Chem. Int. Ed. Engl.* **1985**, *24*, 742.
[252] C. E. Holloway, M. Melnik, *J. Organomet. Chem.* **1986**, *304*, 41.
[253] C. E. Holloway, M. Melnik, *ibid.* **1986**, *303*, 1, 39.
[254] C. E. Holloway, I. A. Walker, M. Melnik, *ibid.* **1987**, *321*, 143.
[255] W. A. Herrmann, J. Rohrmann, E. Herdtweck, H. Bock, A. Veltmann, *J. Am. Chem. Soc.* **1986**, *108*, 3135.
[256] A. L. Rheingold, C. M. Bolinger, T. B. Rauchfuss, *Acta Crystallogr. Sect.* **1986**, *C42*, 1878.

[257] H. Brunner, J. Wachter, E. Guggolz, M. L. Ziegler, *J. Am. Chem. Soc.* **1982**, *104*, 1765.
[258] M. R. Du Bois, D. L. Du Bois, M. C. van Derveer, R. C. Haltiwanger, *Inorg. Chem.* **1981**, *20*, 3064.
[259] G. J. Kubas, P. J. Vergamini, *ibid.* **1981**, *20*, 2667.
[260] M. Herberhold, D. Reiner, B. Zimmer-Gasser, U. Schubert, *Z. Naturforsch. B* **1980**, *35*, 1281.
[261] P. Maué, D. Fenske, *Z. Naturforsch. B* **1989**, *44*, 531.
[262] G. A. Zank, C. A. Jones, T. B. Rauchfuss, A. L. Rheingold, *Inorg. Chem.* **1986**, *25*, 1886.
[263] J. C. Jamieson, D. B. Mac Whan, *J. Chem. Phys.* **1965**, *43*, 1149.
[264] M. Di Vaira, M. Peruzzini, P. Stoppioni, *J. Chem. Soc. Chem. Commun.* **1986**, 374.
[265] L. E. Bogan, T. B. Rauchfuss, A. L. Rheingold, *J. Am. Chem. Soc.* **1985**, *107*, 3843.
[266] F. Bottomley, G. O. Egharevba, P. S. White, *J. Am. Chem. Soc.* **1985**, *107*, 4353.
[267] J. C. Hoffmann, J. G. Stone, W. C. Krusel, K. G. Caulton, *J. Am. Chem. Soc.* **1979**, *99*, 5829.
[268] F. Bottomley, F. Grein, *Inorg. Chem.* **1982**, *21*, 4170.
[269] D. Fenske, A. Grissinger, *Z. Naturforsch. B* **1990**, *45*, 1309.
[270] D. Fenske, M. Loos, unpublished results.
[271] D. Fenske, A. Grissinger, M. Loos, J. Magull, *Z. Anorg. Allg. Chem.* **1991**, *598*, 121.
[272] H. Schäfer, H. G. von Schnering, *Angew. Chem.* **1964**, *76*, 833.
[273] K. Brodersen, G. Thiele, H. G. von Schnering, *Z. Anorg. Chem.* **1965**, *337*, 120.
[274] R. Siepmann, H. G. von Schnering, H. Schäfer, *Angew. Chem.* **1967**, *79*, 650; *Angew. Chem. Int. Ed. Engl.* **1967**, *6*, 637.
[275] A. Simon, H. G. von Schnering, H. Schäfer, *Z. Anorg. Allg. Chem.* **1968**, *361*, 235.
[276] L. J. Guggenberger, A. W. Sleight, *Inorg. Chem.* **1969**, *8*, 2041.
[277] H. G. von Schnering, *Z. Anorg. Allg. Chem.* **1971**, *385*, 75.
[278] H. Okuyame, T. Taga, K. Osaki, K. Tsujikawa, *J. Bull. Chem. Soc. Jpn.* **1982**, *55*, 307.
[279] B. G. Hughes, L. J. Meyer, P. B. Fleming, R. E. McCarley, *Inorg. Chem.* **1970**, *9*, 1343.
[280] R. B. King, *Prog. Inorg. Chem.* **1972**, *15*, 287.
[281] F. A. Cotton, P. A. Kibala, W. Roth, *J. Am. Chem. Soc.* **1988**, *110*, 298.
[282] R. P. Ziebarth, J. D. Corbett, *J. Am. Chem. Soc.* **1989**, *111*, 3272.
[283] A. Simon, *Z. Anorg. Allg. Chem.* **1967**, *355*, 295.
[284] A. Simon, *Angew. Chem.* **1988**, *100*, 164; *Angew. Chem. Int. Ed. Engl.* **1988**, *27*, 159.
[285] F. Bottomley, D. F. Drummond, D. E. Paez, P. S. White, *J. Chem. Soc. Chem. Commun.* **1986**, 1752.
[286] F. Bottomley, D. E. Paez, P. S. White, *J. Am. Chem. Soc.* **1985**, *107*, 7226.
[287] B. Bogdanovíc, R. Goddard, P. Göttsch, C. Krüger, K. Schlichte, Y. Hung Tsay, *Z. Naturforsch. B* **1979**, *34*, 609.
[288] D. Fenske, A. Hollnagel, K. Merzweiler, *Z. Naturforsch. B* **1988**, *43*, 634.
[289] S. Otsuka, Y. Tatsuno, M. Miki, T. Aoki, M. Matsumoto, H. Yoshioka, K. Nakatsu, *J. Chem. Soc. Chem. Commun.* **1973**, 445.
[290] G. W. Bushnell, K. R. Dixon, P. M. Moroney, A. D. Rattray, Cheng Wan, *ibid.* **1977**, 709.
[291] P. M. Bailey, A. Keasey, P. M. Maitlis, *J. Chem. Soc. Dalton Trans.* **1978**, 1825.
[292] D. Fenske, A. Hollnagel, K. Merzweiler, *Angew. Chem.* **1988**, *100*, 978; *Angew. Chem. Int. Ed. Engl.* **1988**, *27*, 965.
[293] M. S. Paquette, L. F. Dahl. *J. Am. Chem. Soc.* **1980**, *102*, 6621.
[294] J. C. Calabrese, L. F. Dahl, A. Cavalieri, P. Chini, G. Longoni, S. Martinengo, *ibid.* **1974**, *95*, 2616.
[295] E. H. Henninger, R. C. Buschert, L. Heaton, *J. Chem. Phys.* **1967**, *46*, 586.
[296] R. K. McMullan, D. J. Prince, J. D. Corbett, *Inorg. Chem.* **1971**, *10*, 1749.

[297] R. C. Ryan, L. F. Dahl, *J. Am. Chem. Soc.* **1975**, *97*, 6904; C. H. Wei, L. F. Dahl, *Cryst. Struct. Commun.* **1975**, *4*, 583.
[298] G. Huttner, J. Schneider, H. Müller, G. Mohr, J. von Seyerl, L. Wohlfahrt, *Angew. Chem.* **1979**, *91*, 82; *Angew. Chem. Int. Ed. Engl.* **1979**, *18*, 76.
[299] M. Müller, H. Vahrenkamp, *Chem. Ber.* **1983**, *116*, 2311.
[300] J. S. Field, R. J. Haines, D. N. Smit, K. Natarajan, O. Scheidsteger, G. Huttner, *J. Organomet. Chem.* **1982**, *240*, C23.
[301] D. Fenske, A. Hollnagel, *Angew. Chem.* **1989**, *101,* 1412; *Angew. Chem. Int. Ed. Engl.* **1989**, *28*, 1390.
[302] J. Steck, D. Fenske, unpublished results.
[303] D. Fenske, C. Oliver, unpublished results.
[304] D. Fenske, W. Holstein, unpublished results.

4 Clusters in Cages

Sibudjing Kawi and Bruce C. Gates

4.1 Introduction

4.1.1 Clusters and Cages

The development of cluster science is causing a merging of molecular, colloid, and materials sciences. [1, 2] Contributing to this is the development of the science of clusters on surfaces and in the pores of solids. Among the best understood of these new materials are clusters which are confined within three dimensional cages that are about the same size as the clusters themselves. Such cages are part of the regular porous structures of crystalline molecular sieve zeolites, the pores of which are molecular scale cages connected to one another by windows.

4.1.2 Opportunities Offered by Cages: Cluster Confinement and Size Limitation

Clusters in cages are the subject of this chapter. Although the synthesis of clusters in cages is often analogous to that in solutions or on surfaces, the steric restrictions of the apertures limit what can enter the cages, and the steric restrictions of the cages limit what can be formed. [3–6] Encaged clusters are often structurally similar to those in solution. On the other hand, small cages can exert solvent-like effects on clusters and may cause their structures and properties to be different from those of clusters in solution or on nearly planar surfaces. [5] This confinement may hinder cluster interactions and increase cluster stability. Clusters in crystalline cages may also be easier to structurally characterize than clusters in other phases.

This chapter is a summary of the chemistry of encaged clusters, and its emphasis is on the unique properties associated with the confinement. This topic is part of the rapidly developing field of host-guest chemistry, which so far has been concerned primarily with small molecular or ionic guests in relatively small hosts such as multidentate chelating molecules like crown ethers. [7] There are also examples of cluster guests in micellar hosts [8–10] and in multilayer Langmuir-Blodgett films, [11, 12] but all these examples are beyond the scope of this chapter.

Zeolites and other molecular sieves having cage sizes on the order of 10 Å in diameter are hosts for clusters consisting of as many as tens of atoms. Molecular sieves with markedly larger cages have been discovered which might be used to enclose even larger clusters, although not those as large as colloids.

4.1.3 Materials with Cage Structures: Zeolites

Many crystalline materials have pores incorporated in their structures, generally with diameters of only a few Angstrom units. They are called molecular sieves because small molecules can intrude into their interior pore volumes, whereas molecules with critical dimensions larger than the aperture dimensions are sieved out. [13–15] Molecular sieving is exploited by industry for gas separation and selective catalysis of hydrocarbon conversion. Molecular sieving properties strongly influence the chemistry of clusters in crystalline porous solids, as will be described in detail in this chapter.

The most common and thoroughly investigated molecular sieves are the zeolites (crystalline aluminosilicates), as exemplified by the faujasites (zeolites X and Y), which have the typical unit cell formula $Na_j[(AlO_2)_j(SiO_2)_{192-j}] \cdot zH_2O$. The faujasites are typical of many molecular sieves in having structures comprised of cages which are connected by apertures.

To understand the structures of the zeolites, it is helpful to conceptually consider how they are constructed from SiO_4 and AlO_4 building block units (although this representation does not reflect the mechanisms of zeolite synthesis). These building blocks have the following tetrahedral structure:

Figure 4-1. Sodalite cage, shown as a ball-and-stick model (right) and in the more conventional representation of zeolite structures (left), whereby each vertex represents a Si or Al ion and the center of each line represents an O position. The sodalite cage is made up of 24 linked SiO_4 and AlO_4 tetrahedra. It is a building block in numerous zeolites, including zeolite A and the faujasites. The cage contains a void volume with a diameter of about 6.6 Å. Small molecules can penetrate into it through six membered oxygen rings, which have a diameter of 2.6 Å.

When these are linked together through shared oxygen atoms, they form polymers. The various geometries of the linkages lead to the various aluminosilicates. Since the Si is present as Si^{4+}, a SiO_4 tetrahedron linked by the sharing of O^{2-}

ions with four neighboring tetrahedra is electrically neutral. Since the Al is present as Al^{3+}, an AlO_4 tetrahedron similarly linked by the sharing of O^{2-} ions with four neighboring tetrahedra must bear a charge of −1.

When the tetrahedra are linked as shown in Figure **4-1**, they form a structure called a truncated octahedron or sodalite cage. This structure is shown in two ways in Figure **4-1**. The figure on the right indicates the individual ions as circles, and the tetrahedra are clearly evident (although some are incomplete as some oxygen ions projecting out of the page are omitted). The structure on the left in Figure **4-1** depicts the cage in a way which emphasizes the framework structure; this representation is conventional for zeolites. The center of a line represents an oxygen position, and the points connecting the lines represent Si or Al ions.

When the sodalite cages are connected as shown in Figure **4-2**, they form the mineral sodalite. If they are instead linked as shown in Figures **4-3** and **4-4**, they then form zeolite A and the faujasites, respectively. Each of these figures illustrates a porous structure consisting of cages and apertures. In some zeolites, as much as about 50 % of the volume is the void volume of the pores. In zeolite A and the faujasites, the pore structures comprise interconnected cages. The cage windows are the apertures through which molecules must pass if they are to enter the labyrinth.

The faujasite pore structure is three dimensional. Many zeolites have three dimensional structures which facilitate transport of molecules in and out of the interior. In contrast, some zeolites (*e.g.* mordenite) have virtual two dimensional

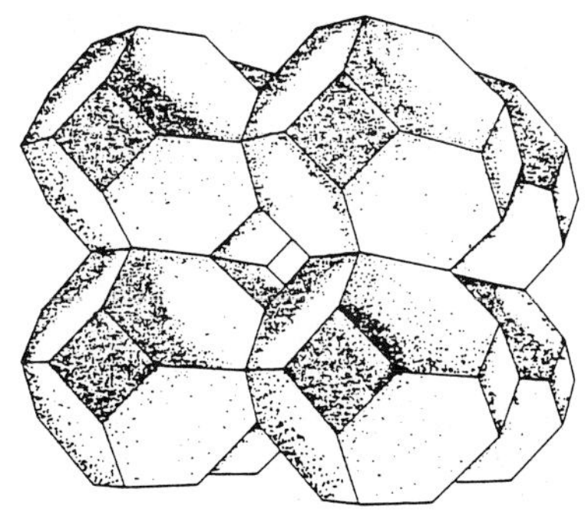

Figure 4-2. Structure of the mineral sodalite built from sodalite cages (shown in Fig. **4-1**) which share square faces. The representation is explained in the caption of Figure **4-1**.

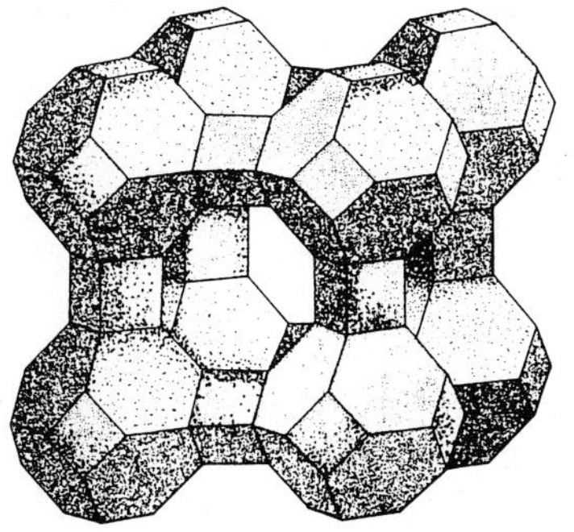

Figure 4-3. Framework structure of the molecular sieve zeolite A. The sodalite cages are linked through bridging oxygen ions connecting four membered oxygen rings. The void space in the center is called a supercage. Its 11.4 Å diameter is large enough to hold clusters such as $[Rh_6(CO)_{16}]$. The zeolite A structure is three dimensional. The windows connecting the supercages are formed from rings of 8 oxygen ions and have diameters of about 4.1 Å, with the exact value depending on the cations present within the structure.

pore structures, consisting of parallel nonintersecting channels. When the pore openings of such a zeolite are blocked, entry into the interior is prevented.

In faujasites, the largest cages (called supercages) are approximately spherical with diameters of 12.5 Å, and the apertures connecting them are approximately circular with diameters of 7.4 Å. The aperture rings contain 12 O^{2-} ions and 12 T ions, where T represents Si or Al. There are also smaller cages (sodalite cages) within the zeolite A and faujasite structures, whose diameters are 6.6 Å. Thus, the supercages are capable of holding molecules as large as $[Rh_6(CO)_{16}]$, which has a diameter of about 10 Å. Although it is small enough to fit in the supercage, it is too large to fit through the apertures and too large to fit in a sodalite cage.

Hundreds of molecular sieves are known, [15–17] many of which are naturally occurring zeolites, and a number have been efficiently synthesized on an industrial scale. The pore dimensions of some of the zeolites are summarized in Table **4-1**.

The aluminosilicate framework of a zeolite bears a negative charge equal to the number of Al^{3+} ions. This charge is compensated in the zeolites by additional cations, commonly Na^+, which can be exchanged for other cations by contacting of the solid with aqueous salt solutions. The exchangeable cations are identified in the name of the zeolite; for example, in the NaY zeolite, all the exchangeable cations are Na^+. The cations in the zeolite occupy various positions within the crystal structure. They influence not only the chemical properties but also, to a degree, the dimensions of the cages and apertures. They make the zeolites hydrophilic, and they also affect the chemistry of any clusters in the cages.

The negative charges of the aluminosilicate frameworks and the positive charges of exchangeable cations in the cages induce strong electric fields in zeolites. These fields affect the chemistry occurring in cages and, for example, facilitate the formation of occluded salts. [18]

Table 4-1. Zeolites and their aperture dimensions.

Zeolite	Number of Oxygens in the Ring	Aperture Dimensions, Å
Chabazite	8	3.6 × 3.7
Erionite	8	3.6 × 5.2
Zeolite A	8	4.1
ZSM-5	10	5.1 × 5.5; 5.4 × 5.6
ZSM-11	10	5.1 × 5.5
Ferrierite	10	4.3 × 5.5
Faujasite	12	7.4
Zeolite L	12 12	7.1 7.0
Mordenite	12	6.7 × 7.0
Offretite	12	6.4

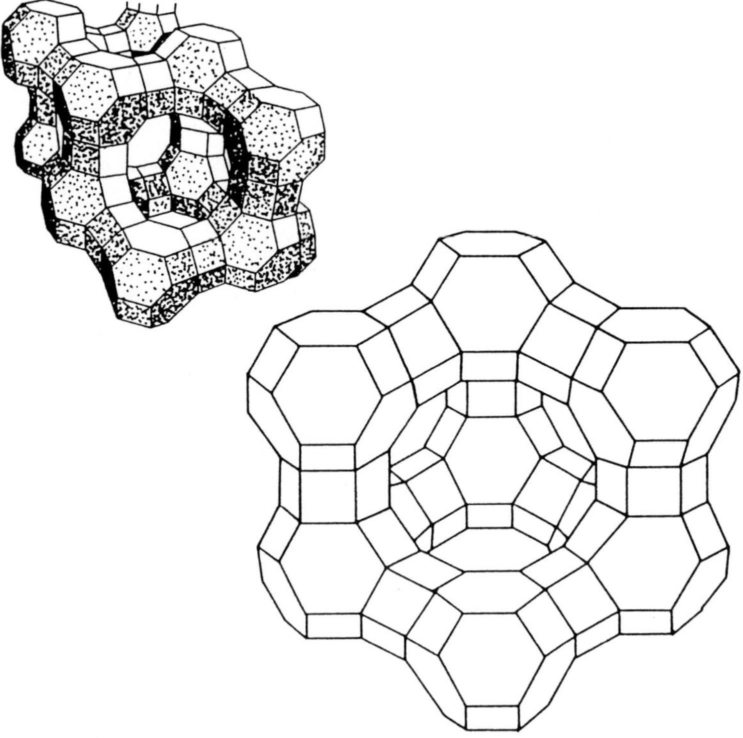

Figure 4-4. Framework structure of faujasite molecular sieves (zeolites X and Y). The sodalite cages are linked together through bridging oxygen ions connecting six membered oxygen rings. The void space in the center (supercage) has a diameter of 12 Å. Clusters such as $[Rh_6(CO)_{16}]$ and $[Ir_6(CO)_{16}]$ have been incorporated in these cages. The faujasite structure is three dimensional and the windows between the supercages are rings of 12 oxygen ions with diameters of about 7.4 Å.

When the exchangeable cations are H^+ (which are represented as being present in OH groups on the zeolite surface), the zeolite is a rather strong Brønsted acid. The acid-base properties can be varied by variation of the exchangeable cations. Whereas NaY zeolite is neutral or very weakly basic, CsY zeolite is basic. When excess CsOH (more than enough to exchange all the exchangeable cations with Cs^+) is present, the zeolite is a strong base. The zeolites also contain Lewis acid sites, namely, the Al^{3+} ions in the framework structure and some exchange ions. Brønsted acid sites can be converted into Lewis acid sites by a process called dehydroxylation, whereby water is removed from the surface by heating, as shown in the simplified two dimensional representation in Figure **4-5**.

Some zeolites have very low Al contents and thus have virtually nonpolar intrapore environments that are hydrophobic. Zeolites such as zeolite Y can be treated to remove Al from the framework, the product being called ultrastable Y zeolite.

There are many molecular sieves which are not zeolites. They have not been so well investigated as the zeolites because many have been discovered only recently

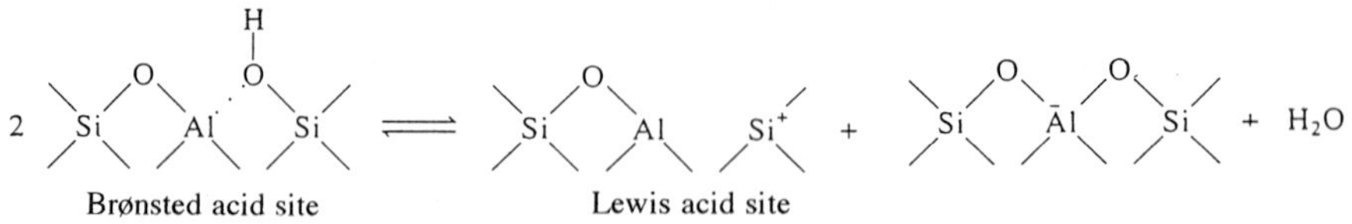

Figure 4-5. Schematic representation of the removal of water from the internal surface of a hydrogenform zeolite such as HY zeolite. The process is called dehydroxylation and converts two Brønsted acid sites (OH groups) into a Lewis acid site (exposed Al cation).

and because they have found far fewer applications than the zeolites. Among the best known is a family of aluminophosphates, $AlPO_4$, which are built up of AlO_4 and PO_4 tetrahedra (the latter containing P^{5+} and bearing a charge of +1). [19] Related materials are the silicoaluminophosphates, SAPOs (with Si as well as Al and P in the structure), and the MeAPOs, where Me is a metal such as Co. [20] Some of these nonzeolitic molecular sieves have crystal structures that are the equivalent of zeolite crystal structures, and some have structures that are not known among the zeolites. Some of these materials have neutral frameworks which thus lack any exchangeable cations or strong electric field gradients in the pores, and others are cation exchangers.

4.1.4 Properties of Clusters in Cages

Molecular sieve cages hold clusters having diameters which approach the cage diameters. Since many clusters are small enough to fit within the cages while being too large to fit through the apertures, they may be stably trapped in the cages and isolated from each other. Thus, the clusters may exist in a crystalline or nearly crystalline state without interacting with each other as pure phase clusters do in the solid state. In such a regular, dispersed state, the clusters may provide well defined models for characterizing the interactions between clusters and surfaces or clusters and isolated ions. Molecular sieves containing clusters may be expected to have properties that distinguish them from conventional materials. For example, some clusters in cages have unusual selectivities as catalysts. Clusters in neighboring cages may interact with each other in ways that impart novel properties to the materials, such as a semiconducting character. It has been suggested that clusters might have sharp optical absorption spectra that would allow them to function as quantum dot switches. [21–24] The hope for applications of the novel properties that these materials might exhibit has provided much of the motivation for the lively research in the field.

In the pages that follow, we summarize methods for the synthesis of clusters in cages, their structural characterization, reactivity, and catalytic and other properties. The literature of encaged clusters is limited to clusters in zeolites, and thus little is included here about other molecular sieves or potential cluster hosts. The literature citations are not comprehensive; rather, examples are cited to illustrate principles and to emphasize the unique properties of encaged clusters and the opportunities they offer for understanding cluster structures and properties that result from their confinement and from the uniformity of their structures in crystalline media. The literature has been selected to emphasize the samples that are relatively well characterized and relatively uniform in structure. Our assessments are intended to be critical, but we realize that not all of them will stand the test of time. The subject of clusters in cages is still new and rapidly developing, and the data available for determining structures and reactivities are fragmentary. For example, there are many reports of metal clusters supported in and on zeolites, including reports of industrial catalysts for processes such as hydrocracking, but since most of the materials referred to in these reports are structurally nonuniform and/or less than well characterized, they are barely considered here.

The encaged clusters considered in this chapter are almost exclusively metal carbonyls and metals (including bimetallics). The encaged metal carbonyls that have been most investigated include $[Rh_6(CO)_{16}]$, $[Ir_4(CO)_{12}]$, and the isomers of $[Ir_6(CO)_{16}]$; the crystal structures of the iridium clusters are shown in Figure **4-6**. Some of the most thoroughly characterized encaged metal clusters have been made from these metal carbonyls. Brief mention is made of metal oxide and also nonmetal clusters; ionic clusters are scarcely considered. Synthesis, characterization, reactivity, and catalytic and other properties are considered for these materials.

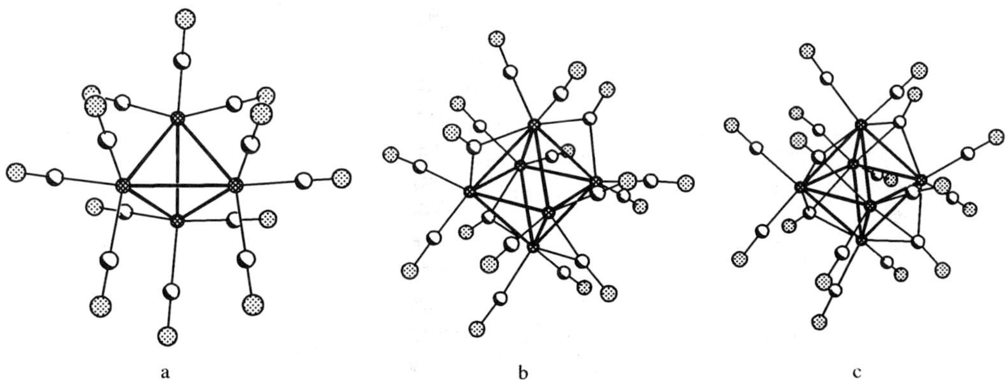

Figure 4-6. Structures of some of the most stable metal carbonyl clusters, each of which has been formed in the cages of NaY zeolite. $[Ir_4(CO)_{12}]$ has only terminal CO ligands (a). One of the two isomers of $[Ir_6(CO)_{16}]$ has edge bridging and terminal CO ligands (b), and the other has face bridging and terminal ligands (c). The CO ligands help to stabilize the metal frames of the metal carbonyl clusters. The iridium atoms are hatched, oxygen atoms are dotted, carbon atoms are shaded. Structure of $[Ir_4(CO)_{12}]$ reproduced from M. R. Churchill and J. P. Hutchinson, *Inorganic Chemistry,* **1978**, *17*, 3528, with permission of the American Chemical Society. Structures of $[Ir_6(CO)_{16}]$ reproduced from L. Garlaschelli et al., *Journal of the American Chemical Society,* **1984** *106*, 6664 with permission of the American Chemical Society.

4.2 Metal Carbonyl Clusters in Zeolites

The most thoroughly investigated and best understood clusters in cages are the metal carbonyls. A great advantage in using these types of clusters, from the standpoint of characterization, is that some of these very clusters can be formed in cages and are otherwise well known compounds which have already been characterized by numerous spectroscopic and X-ray methods. Thus, the structures of encaged metal carbonyl clusters are known with more confidence than those of other encaged clusters.

4.2.1 Synthesis

4.2.1.1 Sublimation of Polynuclear Metal Carbonyl Precursors

Due to the size limitation of the zeolite apertures (typically about 7 Å), it is usually difficult to directly introduce metal carbonyl clusters such as $[Ir_4(CO)_{12}]$ (about 8 Å in diameter) or $[Rh_6(CO)_{16}]$ (about 10 Å in diameter) into the cages. To overcome this difficulty, the metal carbonyl clusters can be synthesized within the cages themselves from ions or molecules small enough to enter the zeolite pores and serve as precursors.

Small metal carbonyl precursors can be introduced into zeolites by sublimation and sorption. Some metal carbonyl clusters can be introduced without fragmentation. $[Re_2(CO)_{10}]$ and $[Ru_3(CO)_{12}]$ have been introduced into the supercages of NaY zeolite without disruption of the clusters or loss of CO ligands; however, the structure of the $[Re_2(CO)_{10}]$ molecule becomes slightly distorted due to hydrogen bonding between the CO ligands and the hydroxyl groups of the zeolite. The original symmetry of the $[Ru_3(CO)_{12}]$ molecule is also perturbed. [25]

Often, metal carbonyls are mildly activated and react when they are introduced into the zeolites. For example, $[Co_2(CO)_8]$ sublimed into faujasites is rapidly converted, in vacuo, into $[Co_4(CO)_{12}]$, which is too large to diffuse out through the zeolite apertures. [26] However, there is a question about whether it is formed inside or outside the supercages or both. [27]

Bulky metal carbonyls deposited on the external surfaces of zeolite crystals may be fragmented and the fragments then transported into the zeolite cages, where the clusters may reform. Naccache et al. [28, 29] found that $[Rh_6(CO)_{16}]$, initially present on the external Y zeolite surface, will presumably fragment under vacuum as a result of thermal activation to give unidentified rhodium subcarbonyls which then react to reform $[Rh_6(CO)_{16}]$ in the cages. This cluster formation is an example of a reductive carbonylation reaction, discussed in the following section.

4.2.1.2 Ship-in-a-Bottle Syntheses

In ship-in-a-bottle syntheses, metal ions, metal complexes, or small metal clusters in zeolite cages can be converted in reactions with CO, or with CO + H_2 or CO + H_2O, to give (larger) metal carbonyl clusters that fit within the cages but are too large to pass through the pore windows. [3–5, 30] Similar reactions for metal cluster synthesis takes place in solution [31, 32] and on surfaces [33–37]. They are called reductive carbonylation reactions and are discussed below.

Reductive Carbonylation of Exchange Ions

Gentle treatment under a CO atmosphere is sufficient to convert some transition metal ions in zeolites into zeolite entrapped metal carbonyl clusters. Such clusters are usually formed under milder conditions in faujasite cages than in solution. This comparison illustrates the excellent solvating abilities of the cages.

Mantovani et al. [38] were the first to report on the formation of metal carbonyl clusters in zeolites by reductive carbonylation. They found that Rh^{3+} which had been exchanged into NaY zeolite by contacting with a $[Rh(NH_3)_6]^{3+}$ solution is converted under high pressures of CO + H_2 at 130 °C into $[Rh_6(CO)_{16}]$, as identified by infrared spectroscopy.

Gelin et al. [30] observed that Rh^{3+} ions formed from $[Rh(NH_3)_5Cl]^{2+}$ in NaY zeolite undergo the same transformations as in aqueous alkaline solutions in the presence of CO + H_2O, namely, successive formation of the carbonyl complexes $[Rh(CO)_2]$, $[Rh_4(CO)_{12}]$, and then $[Rh_6(CO)_{16}]$. Similarly, Gallezot et al. [39] reported that $[Ir_4(CO)_{12}]$ and subsequently $[Ir_6(CO)_{16}]$ were synthesized by reacting IrNaY zeolite with CO + H_2 at about 200 °C.

Platinum carbonyl clusters have also been formed in zeolite cages, but the literature does not yet present a consistent picture of the chemistry. [40–42] Several groups [40–42] have suggested the formation of $[Pt_9(CO)_{18}]^{2-}$ in zeolites occurs by reductive carbonylation. A discussion of the chemistry and a comparison of the results from several laboratories appears in a recent publication. [43]

Sachtler et al. [44–48] reported a new type of palladium carbonyl cluster in the NaY zeolite, and suggested it to be $[Pd_{13}(CO)_x]$, where x is unknown. The cluster was formed by carbonylating a PdNaY zeolite which had been gently reduced with H_2. Pd clusters with only carbonyl ligands are known to exist neither in solution nor in the solid state. Consequently, the compound is novel and its formation indicates the unique synthetic opportunities offered by the molecular scale cages. Since there are no good references to aid in the interpretation of the spectra, the suggestion that the cluster has 13 Pd atoms is speculative, as discussed below.

Under conditions similar to those used for the formation of this palladium carbonyl cluster, another palladium carbonyl cluster was formed in zeolite 5A. It has been suggested to be $[Pd_6(CO)_y]$, where y is unknown. Again, the suggestion [49] is speculative but the synthesis and the compound are novel.

In no case has the yield been determined for a metal carbonyl cluster which was formed in a zeolite by reductive carbonylation of exchange ions. Since the exchange ions can occupy sites that would likely be inaccessible to the CO reagent, it is likely that in many instances only a fraction of the exchange ions have been reductively carbonylated.

Synthesis from Organometallic Precursors

Small neutral metal carbonyls are convenient precursors for the synthesis of metal carbonyl clusters in cages. They offer the following advantages:

1) Because they are small, mononuclear metal carbonyls can fit into the cages where they can be converted into metal carbonyl clusters.
2) They may be restricted to the supercages of faujasites as a result of being too large to enter the sodalite cages. Thus, no unconverted metal may be left in the smaller cages, which usually happens when the precursors are exchange ions.
3) Any unconverted precursor that remains on the outside surface of the zeolite can be removed by washing. The precursor that remains is inside the zeolite and is the source of metal carbonyl clusters entrapped in the zeolite cages.

The chemistry of formation of metal carbonyl clusters from organometallic precursors in cages is similar to that observed in solution and on metal oxide surfaces. This is the most thoroughly investigated method of formation of well defined metal carbonyl clusters in cages. There are several approaches to the synthesis and these are discussed below.

4.2.1.3 Sorption of Metal Carbonyl Complexes

Mononuclear metal carbonyls or small metal carbonyl clusters introduced into zeolite supercages, either by sorption or chemical vapor deposition, may react to form metal carbonyl clusters having nuclearities (numbers of metal atoms) greater than those of the precursors. The resultant metal carbonyl clusters can be neutral or anionic, depending on the acid-base properties and the degree of hydration of the zeolite.

Iwamoto et al. [50] investigated the reactivity of iron carbonyl complexes in a hydrated NaY zeolite and found that $[Fe_2(CO)_9]$ and $[Fe_3(CO)_{12}]$ reacted in the cages to yield $[HFe_3(CO)_{11}]^-$, as characterized by infrared and ultraviolet spectroscopies. The infrared spectrum of $[HFe_3(CO)_{11}]^-$ within the zeolite shows a shift of the absorption bands of the bridging carbonyl ligand to lower energy relative to those in the spectrum of the anion in benzene, indicating ionic interactions between the oxygen of bridging carbonyl ligands and the Al^{3+} ions in the zeolite matrix. An analysis of the gas phase products formed during the sorption led to the stoichiometry postulated in Equation (4.1).

$$3Fe_2(CO)_9 + 2OH^- \rightarrow 2[HFe_3(CO)_{11}]^- + 3CO + 2CO_2 \quad (4.1)$$

Similarly, the formation of $[HFe_3(CO)_{11}]^-$ from $[Fe_3(CO)_{12}]$ sorbed in a NaY zeolite was postulated to result from a nucleophilic attack of OH^- on the coordinated CO ligand (Eq. 4.2).

$$Fe_3(CO)_{12} + OH^- \rightarrow [HFe_3(CO)_{11}]^- + CO_2 \qquad (4.2)$$

The formation of the anionic cluster was not observed in a dehydrated NaY zeolite, a result that indicates the importance of water or surface hydroxyl groups in the reaction.

In contrast, Ballivet-Tkatchenko and Coudurier [51] reported that when $[Fe_3(CO)_{12}]$ in dehydrated HY zeolite was heated to 60 °C, $[H_2Fe_3(CO)_{11}]$ was formed within the supercages and interacted with the zeolite through hydrogen bonding. This difference in the behavior of the iron carbonyls in zeolite Y indicates the importance of the acid-base properties of the zeolite.

In a strongly basic zeolite (prepared by treatment of NaY with NaN_3), adsorption of $[H_2Os(CO)_4]$ led to the formation of entrapped clusters which were suggested to be $[HOs_3(CO)_{11}]^-$. [52] This cluster was characterized by infrared and EXAFS spectroscopy [53] and suggested to have been formed in a reaction analogous to the known solution reaction. Thus, since $[H_2Os(CO)_4]$ is a weak Brønsted acid, it will be deprotonated to give $[HOs(CO)_4]^-$ in both basic solutions and on basic surfaces such as that of MgO. [39] $[HOs(CO)_4]^-$ will undergo condensation reactions to form trinuclear anionic species in protic media (Eqs. 4.3 and 4.4, B^- is a Brønsted base).

$$H_2Os(CO)_4 + B^- \rightarrow [HOs(CO)_4]^- + BH \qquad (4.3)$$

$$2H_2Os(CO)_4 + [HOs(CO)_4]^- \rightarrow [HOs_3(CO)_{11}]^- + CO + H_2 \qquad (4.4)$$

4.2.1.4 Reductive Carbonylation of Mononuclear Metal Carbonyl Complexes

The chemistry involved when mononuclear metal carbonyl complexes in zeolite cages are reductively carbonylated to give metal carbonyl clusters is analogous to that which occurs in solutions and on surfaces of amorphous metal oxides. The metal carbonyl clusters can be either neutral or anionic, depending on whether the zeolite is nearly neutral or basic, respectively.

This method was used [3–5] to prepare iridium carbonyl clusters from $[Ir(CO)_2$-(acac)] sorbed in faujasites. $[Ir(CO)_2(acac)]$ in the nearly neutral NaY zeolite was transformed in the presence of CO, first into the neutral iridium carbonyl cluster $[Ir_4(CO)_{12}]$, then into the isomer of $[Ir_6(CO)_{16}]$ with edge bridging and terminal CO ligands, and then into the isomer of $[Ir_6(CO)_{16}]$ with face bridging and terminal CO ligands (Fig. **4-6** and **4-7**). [3, 5] The metal carbonyl clusters were characterized by infrared and EXAFS spectroscopies and could not be extracted from the zeolite with a solution of tetrahydrofuran. Since they are slightly soluble in this solvent, the absence of extraction products indicates that the clusters were trapped in the cages, as expected considering the geometry of the cages, apertures, and clusters. The formation of neutral iridium carbonyl clusters in the nearly neutral NaY zeolite parallels the synthesis in neutral [54–56] and acidic [57, 58] solutions and on nearly neutral metal oxide surfaces, *e.g.*, γ-Al_2O_3 (Fig.

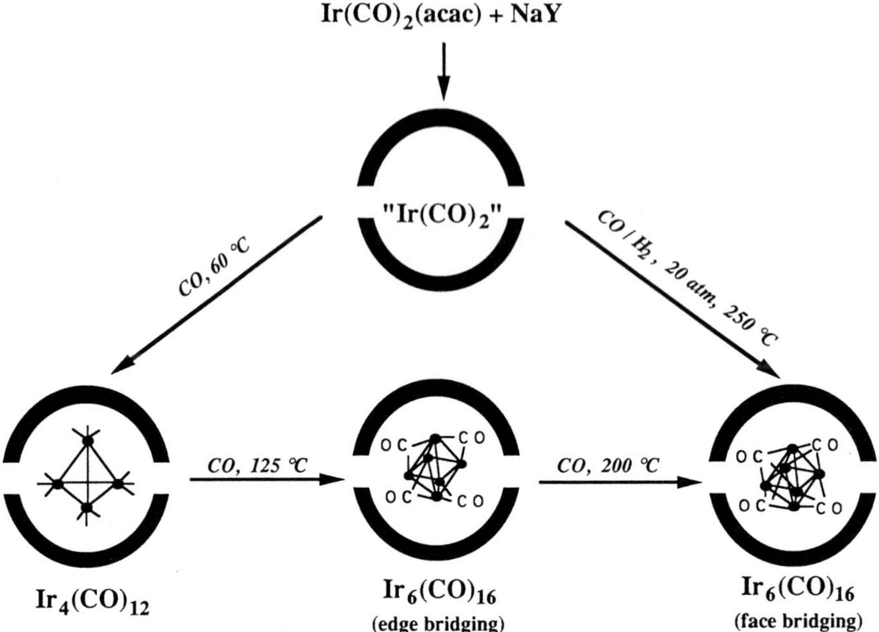

Figure 4-7. Ship-in-a-bottle synthesis of iridium carbonyl clusters. Schematic representation of the formation of $[Ir_4(CO)_{12}]$ and $[Ir_6(CO)_{16}]$ from $[Ir(CO)_2(acac)]$ in the cages of NaY zeolite. [3, 5] The precursor $[Ir(CO)_2(acac)]$ is small enough to diffuse into the zeolite supercages, where it reacts with CO to form the clusters, which are then trapped in the cages. For clarity, some of the CO ligands are not shown.

4-8). [59] $[Ir_4CO)_{12}]$ is synthesized in solution by the carbonylation of iridium salts [54–56] or by the carbonylation of $[Ir(CO)_2(acac)]$. [59] The two isomers of $[Ir_6(CO)_{16}]$ are stable in acidic solutions. [57, 58]

Kawi et al. [4, 60] reported on experiments comparing the iridium carbonyl chemistry in NaY zeolite with that in the more basic NaX zeolite (some of whose basicity should perhaps be attributed to the excess NaOH used in the preparation which was ultimately not washed out). The results showed that the supercages in the NaX zeolite were sufficiently basic to provide an efficient medium for the synthesis of anionic iridium carbonyl clusters. When $[Ir(CO)_2(acac)]$ in NaX zeolite was treated with CO, it was transformed into $[HIr_4(CO)_{11}]^-$ and then into $[Ir_6(CO)_{15}]^{2-}$. The anionic carbonyl clusters trapped in the cages were characterized by infrared and EXAFS spectroscopies and could not be extracted from the zeolite by ion exchange with bis(triphenylphosphine)iminium chloride, [PPN][Cl], in tetrahydrofuran solution.

The chemistry of anionic iridium carbonyl clusters in NaX zeolite parallels that in basic solutions and on the basic MgO surface (Fig. **4-9**). In basic solutions, the reductive carbonylation of $[Ir_4(CO)_{12}]$ with KOH in methanol under CO initially gives $[HIr_4(CO)_{11}]^-$, [61, 67] then $[Ir_8(CO)_{22}]^{2-}$, [62, 67] and finally $[Ir_6(CO)_{15}]^{2-}$.

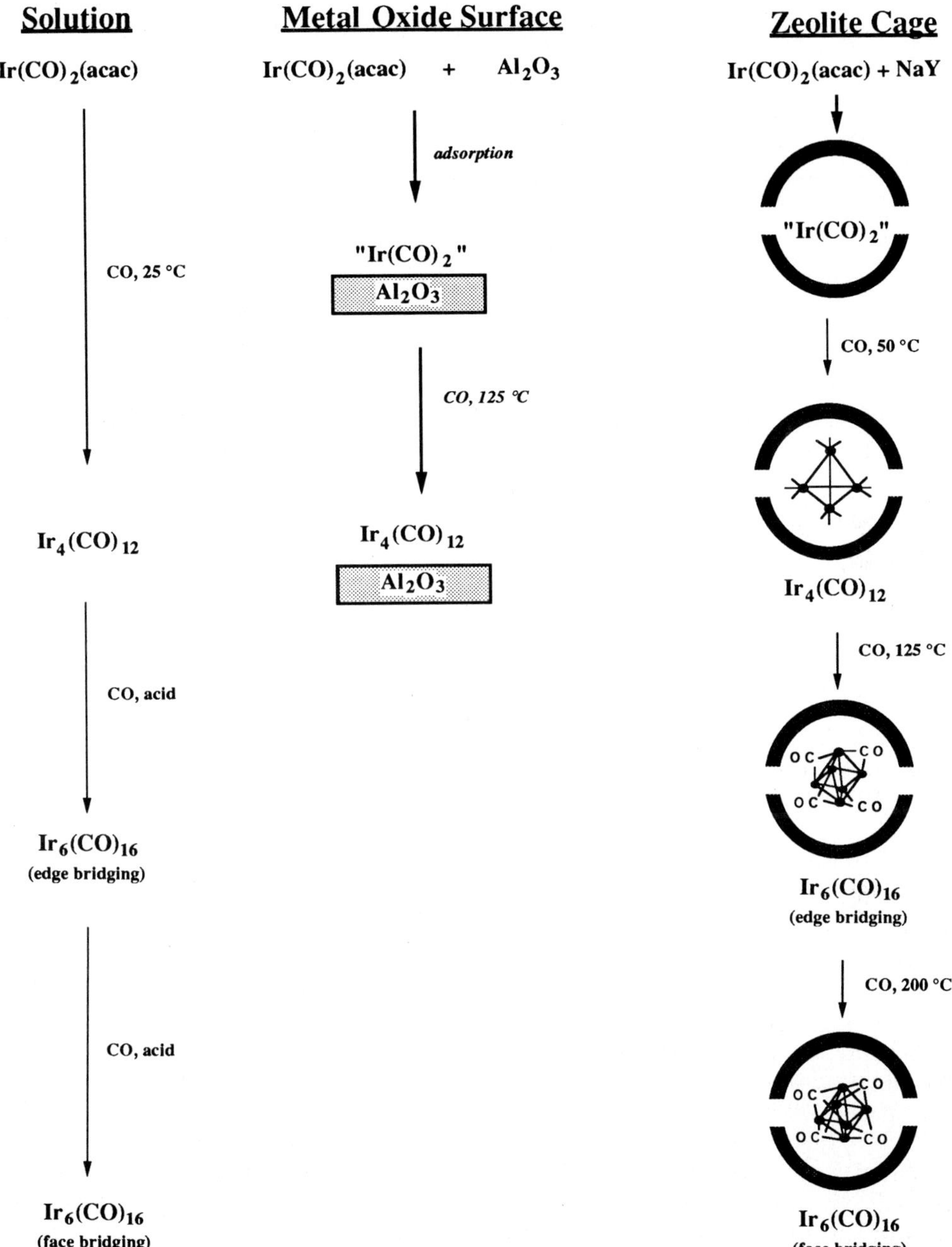

Figure 4-8. Synthesis of iridium carbonyl clusters in neutral solutions and on the nearly neutral surface of amorphous γ-Al_2O_3. The chemistry is very similar to that occurring in the cages of NaY zeolite (Fig. **4-7**). [3, 5] Whereas the clusters can be readily extracted from the surface of γ-Al_2O_3, under the same conditions they cannot be extracted from the zeolite because they are too large to fit through the cage windows and are thus trapped in the supercages.

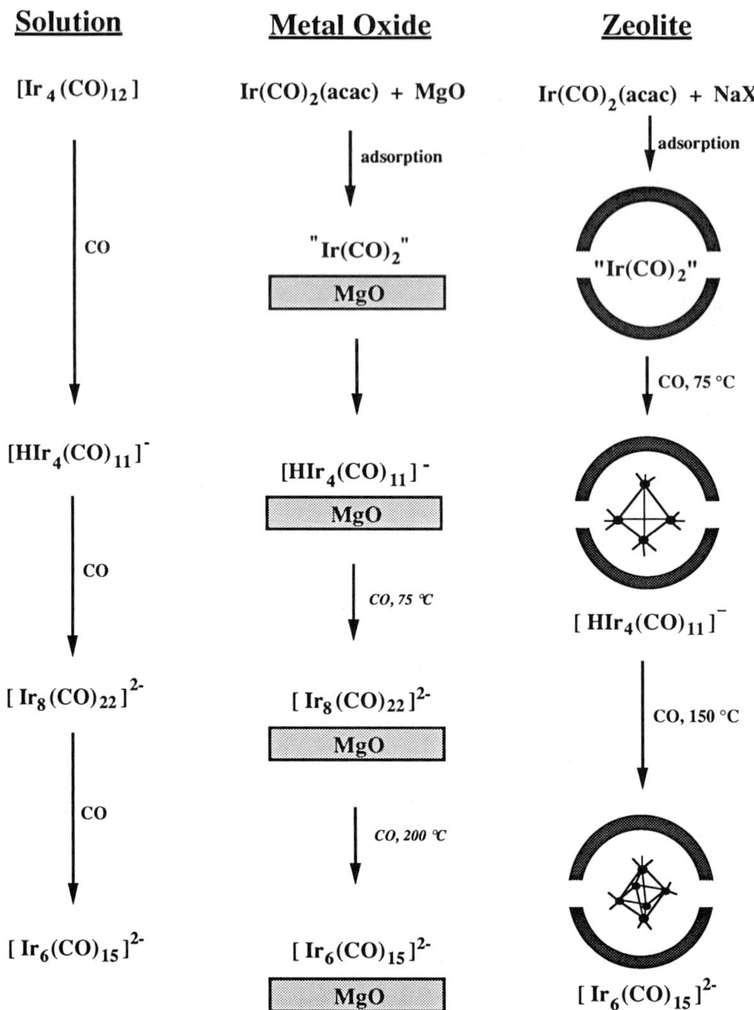

Figure 4-9. Synthesis of iridium carbonyl cluster anions in basic solutions, on the basic surface of MgO, and in the cages of basic NaX zeolite. [4, 37] The chemistry on the MgO surface is very similar to that occurring in the basic solution. The chemistry occurring in the zeolite cages is analogous except in one important respect: $[Ir_8(CO)_{22}]^{2-}$ is not observed to form in the cages although it is formed in high yields in solution and on MgO under the same conditions. The results indicate that the steric restriction within the cages limits what clusters can form. Evidently, $[Ir_8(CO)_{22}]^{2-}$, which is a dimer of tetrairidium carbonyls, is too large to fit in the cages. This is an illustration of the role which the cage plays in controlling the cluster chemistry. The clusters can be readily extracted from the surface of MgO by ion exchange, but under the same conditions they are not extracted from the zeolite because they are too large to fit through the cage windows and are thus trapped in the supercages.

[58, 63, 67] The chemistry of the successive formation of these anionic clusters on basic MgO surfaces is analogous to that in solution, [37] and each of these cluster anions could be extracted from MgO. Since the octairidium cluster (which is a dimer of the tetrairidium clusters) was evidently not formed in the NaX zeolite, it was inferred that the confinement of the iridium in the cages prevented its formation. This example illustrates the subtle control of the reactivity exerted by the cages.

The results are consistent with a ship-in-a-bottle synthesis taking place within the zeolite supercages. The $[Ir(CO)_2(acac)]$ precursor fits in the zeolite. $[Ir_4(CO)_{12}]$ (with a diameter of about 9 Å), the two isomers of $[Ir_6(CO)_{16}]$ (with diameters of about 11 Å), and $[Ir_6(CO)_{15}]^{2-}$ (with a diameter of about 11 Å) are small enough to fit in the supercages of zeolite Y (which have diameters of about 12 Å) but are too large to diffuse through the apertures (which have diameters of only about 7.4 Å). Thus, it can be inferred that the clusters, once formed, are trapped in the supercages and cannot be extracted.

A summary of the chemistry of metal carbonyl clusters in zeolite cages is presented in Table **4-2**. The table includes the characterization methods used for the various samples and give an indication of the degree of confidence that we place on the suggested structures.

4.2.2 Characterization Techniques

The chemistry of clusters in the cages of crystalline solids has developed rapidly in the last few years as powerful characterization techniques have emerged to allow their identification. Much of the early literature on these materials contains interpretations of spectra (usually infrared spectra) which are insufficient for a confident determination of the cluster structures. When metal carbonyl clusters are present in zeolite cages, their structures may be somewhat different from those in the crystalline state or in solution. The cages may appear to solvate the clusters for they may affect them through the polar interaction associated with the exchange cations or the negatively charged framework.

One of the principal issues complicating the synthesis and characterization of clusters in zeolites is the simultaneous formation of clusters or crystallites outside the zeolite crystals. In many preparations this is apparently unavoidable, yet most of the literature on zeolite supported clusters does not even address this issue. Much of the reported work on metal carbonyl clusters in zeolites has failed to include evidence which should establish whether all the clusters were actually confined within the cages. Characterization techniques are needed which can differentiate between the entrapped entities and those outside the pores by probing the ligand sphere and the metal species.

Most of the characterization methods discussed here are based on comparisons of spectra of entrapped species with those of well characterized reference compounds found in solution or in the solid state. The methods work best when the references have been structurally characterized by X-ray diffraction crystallography. Without good models for the entrapped species, structures inferred on the

Table 4-2. Metal carbonyl clusters in zeolites.

Metal	Precursor	Method	Cluster/Zeolite	Characterization	Refs.
Fe	$[Fe_2(CO)_9]$, $[Fe_3(CO)_{12}]$	sublimation	$[Fe_3(CO_{12}]$/HY	IR, gas phase analysis	a
Fe	$[Fe_2(CO)_9]$, $[Fe_3(CO)_{12}]$	sublimation	$[HFe_3(CO)_{11}]^-$/ NaY	IR, UV-vis, ESR, gas phase analysis	b
Ru	$[Ru_3(CO)_{12}]$	sublimation	$[Ru_3(CO)_{12}]$	IR, X-ray crystallography, volumetric measurement	c
Os	$[H_2Os(CO)_4]$	sublimation	$[Os_3(CO)_{11}]^{-\dagger}$/NaY	IR, EXAFS	d, e
Rh	$RhCl_3$, or $[Rh(NH_3)_6]Cl_3$	ion exchange followed by CO treatment	$[Rh_4(CO)_{12}]$/NaY $[Rh_6(CO)_{16}]$/NaY	IR, EXAFS	f, g, h
Ir	$[Ir(CO)_2(acac)]$	adsorption followed by CO treatment	$[Ir_4(CO)_{12}]$/NaY	IR, EXAFS	i, j
		ion exchange followed by CO treatment	$[Ir_4(CO)_{12}]$/NaY	IR	i, j
Ir	$[Ir(CO)_2(acac)]$	adsorption followed by CO treatment	$[HIr_4(CO)_{11}]^-$/NaX	IR	k
Ir	$[Ir(CO)_2(acac)]$	adsorption followed by CO treatment	$[Ir_6(CO)_{16}]$ (edge)*/ NaY	IR	l, m
Ir	$[Ir(CO)_2(acac)]$	adsorption followed by CO treatment	$[Ir_6(CO)_{16}]$ (face)*/ NaY	IR, EXAFS	l, m
Ir	$[Ir(NH_3)_5Cl]Cl_2$	ion exchange, calcination, then treatment in $CO + H_2O$ or $CO + H_2$	$[Ir_6(CO)_{16}]$ (face)*/ NaY	RED	n
Ir	$[Ir(CO)_2(acac)]$	adsorption followed by CO treatment	$[Ir_6(CO)_{15}]^{2-}$/NaX	IR, EXAFS	k

Table 4-2. (Continued).

Metal	Precursor	Method	Cluster/Zeolite	Characterization	Refs.
Pd	$[Pd(NH_3)_4](NO_3)_2$	ion exchange calcination, then CO treatment	$[Pd_{13}(CO)_x]^{\dagger}$/NaY	IR, EXAFS	o, p, q
Pd	$[Pd(NH_3)_4](NO_3)_2$	ion exchange, calcination, reduction then CO treatment	$[Pd_6(CO)_x]^{\dagger}$/ NaA	IR, EXAFS	r, s
Pt	$Pt(NH_3)_4Cl_2$	ion exchange followed by CO treatment	$[Pt_9(CO)_{18}]^{2-\dagger}$/NaY	IR, UV-vis, EXAFS	t, u
Pt	$Pt(NH_3)_4Cl_2$	ion exchange followed by CO treatment	$[Pt_{12}(CO)_{24}]^{2-\dagger}$/NaY	IR, UV-vis, EXAFS	u
Pt	$Pt(NH_3)_4Cl_2$	ion exchange followed by CO treatment	$[Pt_{15}(CO)_{30}]^{2-\dagger}$/NaX	IR	u

† Structures are regarded as speculative; see discussion in text.
* Edge refers to edge bridging CO ligands; face to face bridging CO ligands.

(a) D. Ballivet-Tkatchenko, G. Coudurier, *Inorg. Chem.* **1979**, *18*, 558.
(b) M. Iwamoto, S. Nakamura, H. Kusano, S. Kagawa, *J. Phys. Chem.* **1986**, *90*, 5244.
(c) P. Gallezot, G. Coudurier, M. Primet, B. Imelik, in *Molecular Sieves-II* (Ed.: J. R. Katzer), American Chemical Society, Washington, DC, USA, **1977**, p. 144.
(d) P.-L. Zhou, S. D. Maloney, B. C. Gates, *J. Catal.* **1991**, *129*, 315.
(e) S. Maloney, P.-L. Zhou, M. J. Kelley, B. C. Gates, *J. Phys. Chem.* **1991**, *95*, 5409.
(f) E. Mantovani, N. Palladino, A. Zanobi, **1977–78**, *3*, 385.
(g) L.-F. Rao, A. Fukuoka, N. Kosugi, H. Kuroda, M. Ichikawa, *J. Phys. Chem.* **1990**, *94*, 5317.
(h) E. J. Rode, M. E. Davis, B. E. Hanson, *J. Catal.* **1985**, *96*, 574.
(i) S. Kawi, B. C. Gates, *Catal. Lett.* **1991**, *10*, 263.
(j) S. Kawi, J.-R. Chang, B. C. Gates, *J. Phys. Chem.*, in press.
(k) S. Kawi, B. C. Gates, *J. Chem. Soc. Chem. Commun.* **1992**, 702.
(l) S. Kawi, B. C. Gates, *J. Chem. Soc. Chem. Commun.* **1991**, 994.
(m) S. Kawi, J.-R. Chang, B. C. Gates, *J. Am. Chem. Soc.* **1993**, *115*, 4830.
(n) G. Bergeret, P. Gallezot, F. Lefebvre, *Stud. Surf. Sci. Catal.* **1986**, *28*, 401
(o) L.-L. Sheu, H. Knözinger, W. M. H. Sachtler, *J. Am. Chem. Soc.* **1989**, *111*, 8125.
(p) Z. Zhang, H. Chen, L. L. Sheu, W. M. H. Sachtler, *J. Catal.* **1991**, *127*, 213.
(q) Z. Zhang, H. Chen, W. M. H. Sachtler, *J. Chem. Soc., Faraday Trans.* **1991**, *87*, 1413.
(r) Z. Zhang, W. M. H. Sachtler, *J. Mol. Catal.* **1991**, *67*, 349.
(s) Z. Zhang, A. P. Cavalcanti, W. M. H. Sachtler, *Catal. Lett.* **1992**, *12*, 157.
(t) A. De Mallmann, D. Barthomeuf, *Catal. Lett.* **1990**, *5*, 293.
(u) G.-J. Li, T. Fujimoto, A. Fukuoka, M. Ichikawa, *Catal. Lett.* **1992**, *12*, 171.

basis of only one kind of spectra are often not accurate. In general, reliable characterizations are expected only when sufficient complementary characterization methods are used in combination. A selection of techniques that have been used to characterize zeolite supported metal carbonyls is summarized in Table **4-3**, the details of which will be addressed in the following paragraphs.

4.2.2.1 Chemical Methods

Most of the techniques described here cannot establish whether a metal carbonyl cluster is present on the surface of a zeolite particle or within the intracrystalline cages. Chemical treatments have been used to determine whether the clusters are inside or outside, for example by the use of probe molecules that are too large to fit in the pores and those that do fit.

Extraction

Clusters adsorbed on the outside surfaces of zeolites can often be easily extracted with neutral solvents or with salt solutions which remove the cluster ions by cation metathesis (ion exchange). Comparison of the infrared spectra of the extracted species and those of known species helps identify the encaged species. When treatment of a sample with such solutions fails to remove sorbed clusters, they are inferred to be trapped within the cages. The inference is supported when the same clusters adsorbed on the surface of a large pored material such as an amorphous metal oxide are removed by extraction.

For example, $[Ir_6(CO)_{15}]^{2-}$ in the NaX zeolite could not be extracted with [PPN][Cl] in tetrahydrofuran solution, [4, 60] whereas this cluster on the surface of MgO was completely extracted. [37]

Intrusion of Gas Phase Probes

Clusters on the outside surface of a zeolite can sometimes be distinguished from those within the cages by reactions with probe molecules of various sizes, whereby some are small enough to enter the pores and some are too large to enter. The probes most commonly used are phosphines, as they react with many metal carbonyls. The products are typically characterized by infrared spectroscopy.

For example, Rode et al. [64] found that the infrared bands characteristic of $[Rh_6(CO)_{16}]$ in a NaY zeolite were not affected by exposure of the sample to *n*-hexyldiphenylphosphine, which has a critical diameter larger than that of the NaY apertures (about 7.4 Å). The authors concluded that the clusters were located in the supercages of the zeolite, consistent with the inferences of earlier researchers. [30, 38]

4.2.2.2 Temperature-Programmed Desorption

Temperature-programmed desorption and temperature-programmed decomposition are valuable because they give quantitative measures of the composition and reactivity of supported organometallic clusters. In these experiments, the temperature of a sample in an inert carrier gas stream is ramped at a known rate, and the gaseous effluent is characterized, often continuously, as the effluent stream flows through a thermal conductivity detector, and sometimes to a gas chromatograph and/or mass spectrometer. The data provide a quantitative profile of the products formed and a measure of the energy required to desorb ligands such as CO from the cluster. One must be alert to the possibility that zeolitic OH groups may oxidize the metal carbonyl clusters and cause the evolution of CO, H_2, CO_2, or CH_4, leaving metal oxide fragments in the zeolite. [69] Unfortunately, there is a lack of quantitative results to illustrate the method for metal carbonyl clusters in zeolites.

4.2.2.3 Infrared spectroscopy

Infrared spectroscopy is the most easily and widely applied method for the identification of metal carbonyl clusters. Spectra taken in the carbonyl stretching region (2200–1600 cm^{-1}) provide information about CO ligand coordination and, hence, a basis for structural characterization. CO stretching vibrations arising from terminal carbonyl ligands (2140–1800 cm^{-1}) are usually well separated from those from bridging carbonyl ligands (1850–1600 cm^{-1}). The infrared absorptions of terminal carbonyl ligands normally have high extinction coefficients and provide easily measured fingerprints of molecular complexes. CO stretching frequencies are also sensitive to the metal oxidation state. The frequency of a vibrational mode for isoelectronic binary metal carbonyls decreases monotonically as the formal oxidation state of the metal decreases. These trends are also observed for CO ligands in metal clusters and CO adsorbed on metal surfaces. Shifts in the CO stretching frequencies provide indirect evidence of changes in the metal oxidation states.

The identification of metal carbonyl clusters in cages has commonly been inferred from comparisons of their spectra with those of analogous molecular clusters in solution or the solid state. The identification is most reliable when the clusters are small, since the spectra of small metal carbonyls are more nearly definitive fingerprints than those of large metal carbonyl clusters. For example, the infrared spectra in the carbonyl stretching region can be used as fairly good indicators for the presence of small clusters such as $[Ir_4(CO)_{12}]$ and the isomers of $[Ir_6(CO)_{16}]$ (Fig. **4-10**). These spectra are less useful for identifying larger clusters such as $[Os_{10}C(CO)_{24}]^{2-}$ and $[Rh_{12}(CO)_{30}]^{2-}$ because the spectra are similar to the spectra of numerous related metal carbonyl clusters. When close analogues are not available, infrared spectra are not sufficient for identification of metal carbonyl clusters. Even when data for authentic compounds are available, these spectra may be insufficient.

Table 4-3. Techniques that have been used to characterize encaged clusters.

Method	Types of Cluster	Information	Comment
Extraction	metal carbonyl clusters	internal or external location of metal carbonyl cluster in zeolite	Carbonyl cluster encaged in the zeolite cages cannot diffuse through the zeolite aperture and cannot be extracted out; effective for anionic clusters but not effective for some neutral carbonyl clusters that are difficult to dissolve in solvent.
Intrusion of gas phase probes	metal carbonyl clusters	internal or external location of metal carbonyl cluster in zeolite	Large phosphine molecule cannot diffuse through zeolite aperture into the zeolite cages to react with encaged carbonyl clusters; effective for highly reactive carbonyl clusters.
Metal-catalyzed reduction of methylviologen	metal clusters	internal or external location of metal cluster in zeolite	Characteristic blue color of the reduced viologen radicals makes it easy to locate clusters.
Metal-catalyzed reduction of iron cyanide	metal clusters	internal or external location of metal cluster in zeolite	$[Fe(CN)_6]^{3-}$ does not enter the anionic framework of zeolites L and Y and can be reduced only by metal outside the zeolite cages.
IR	metal carbonyl clusters	structures of metal carbonyl clusters; interaction of metal carbonyl clusters with the zeolite framework	Identifications of metal carbonyl clusters in cages are inferred from comparisons of their spectra with those of analogous molecular clusters in solution or in solid state; most reliable for very small cluster; shifts of carbonyl bands identify the interaction of carbonyl clusters with the zeolite framework.
NMR (^{13}C, Xe, metal)	metal carbonyl clusters; metal clusters	structure, size, and location of cluster in zeolite cages	Chemical shifts of Xe NMR are sensitive to xenon pressure, temperature, type of cations, and size of zeolite crystals.
EXAFS	metal carbonyl clusters; metal clusters; bimetallic clusters; semiconductor clusters	structures of metal, metal carbonyl, and semiconductor clusters; interaction of these clusters with zeolite framework	Structures are derived from coordination number and distance of metal-metal, metal-support, and metal-adsorbate contributions; useful for characterization of metal-support interface; interpretation of EXAFS data should be based on the spectra of well-characterized standards of molecular analogues of the supported species that have been characterized by X-ray crystallography.

Table 4-3. (Continued).

Method	Types of Cluster	Information	Comment
XANES	metal clusters	change of oxidation state of metal	Qualitative information only; influence of metal cluster size on the white line intensity needs to be accounted for.
XPS	metal carbonyl clusters; metal clusters; bimetallic clusters	changes of metal oxidation state; internal or external location of cluster on zeolite	Technique requires ultrahigh vacuum and hence lack of stability of samples complicates the analysis; air sensitive materials are not suitable with this technique due to air exposure during sample loading.
UV-visible	metal carbonyl clusters; semiconductor clusters	structure and size of cluster	Complementary structural information; suitable for highly colored materials; does not provide a detailed fingerprint of surface structures; band shifts related to cluster size.
TEM	metal clusters	size and location of cluster on zeolite	Metal cluster size and location can be modified by the electron beam, due to high local temperature rises; lack of resolution prevents detection of smallest metal clusters.
Raman	metal carbonyl clusters; metal clusters	metal-metal bonds	Complementary structural information; samples are subject to destruction by laser beams, and fluorescence often prevents measurement of useful spectra.
EPR	metal carbonyl clusters; ionic clusters	structure and environment of cluster in zeolite	Useful for characterization of species having unpaired electrons; due to its high sensitivity, very small percentages of total metal loading are measurable.
X-ray diffraction	metal carbonyl clusters; ionic clusters; metal clusters	structure of cluster	Structural information (bond lengths and coordination numbers around a given metal atom) based on radial electron distribution from X-ray data; RED gives only metal-metal, not metal-support or metal-adsorbate contributions.

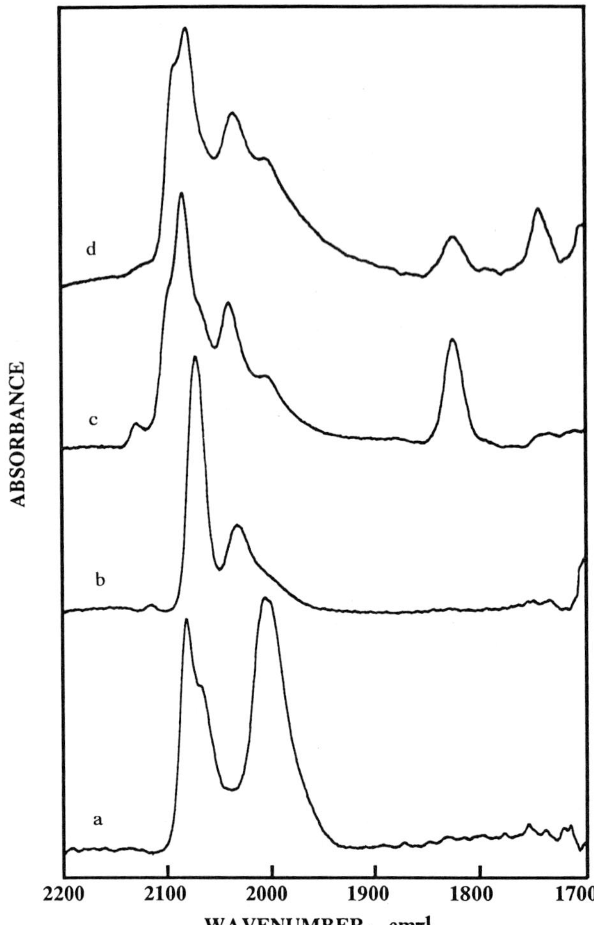

Figure 4-10. Infrared spectra of iridium carbonyls in the cages of the NaY zeolite. [5] a) The precursor $[Ir(CO)_2(acac)]$, b) $[Ir_4(CO)_{12}]$, c) $[Ir_6(CO)_{16}]$ isomer with edge bridging ligands, and d) a mixture of the isomers of $[Ir_6(CO)_{16}]$ with edge bridging and face bridging ligand. Reproduced from the *Journal of the American Chemical Society* with permission of the American Chemical Society.

The spectrum attributed to $[Rh_6(CO)_{16}]$ in the NaY zeolite [38, 64, 65] resembles that of crystalline $[Rh_6(CO)_{16}]$, but the bridging CO band is shifted by about 40 cm^{-1} to lower frequency. This may be an indication of the formation of an adduct (explained below) between the oxygen of the triply bridged CO ligand and the Lewis acid sites on the internal zeolite wall (*e.g.* Al^{3+} or Na^+).

An example in which the infrared results gave evidence for ion pairing of a cluster anion, $[Ir_6(CO)_{15}]^{2-}$, with the zeolite NaX is shown in Table **4-4** where the frequencies for $[Ir_6(CO)_{15}]^{2-}$ in tetrahydrofuran solution are listed for comparison.

4.2.2.4 Nuclear Magnetic Resonance (NMR) Spectroscopy

Nuclear magnetic resonance (NMR) spectroscopy has found only little application in the study of zeolite entrapped metal carbonyl clusters. With modern high-field instrumentation, however, increased attention to this method is

Table 4-4. Infrared Spectral Data in the Carbonyl Stretching Region for NaX Supported Iridium Carbonyl Species and Related Molecular Iridium Clusters. [78]

Sample	ν_{CO}, cm^{-1}	Refs.
$[HIr_4(CO)_{11}]^-$/NaX	2072 w, 2044 sh, 2035 s, 2011 m, 2000 sh, 1765 mw	see text
$[NEt_4][HIr_4(CO)_{11}]$ in THF	2067 w, 2030 sh, 2017 vs, 1986 m, 1978 m, 1832 m	a
$Na[HIr_4(CO)_{11}]$ in diethyl ether	2072 w, 2039 s, 2020 vs, 1990 m, 1984 m, 1830 w, 1730 m	a
$[P(CH_2C_6H_5)(C_6H_5)_3][HIr_4(CO)_{11}]$ in THF	2015 vs, 2005 vs, 1985 s, 1970 s, 1800 m	b
$[Ir_6(CO)_{15}]^{2-}$/NaX	2001 s, 1993 s, 1710 m	see text
$[NMe_4]_2[Ir_6(CO)_{15}]$ in THF	2030 sh, 1970 s, 1910 sh, 1775 s, 1735 s	c
$Na_2[Ir_6(CO)_{15}]$ in THF	1993 s, 1984 s, 1928 w, 1788 m, 1735 m	d

(a) D. M. Vandenberg, T. C. Choy, P. C. Ford, *J. Organomet. Chem.* **1989**, *366*, 257.
(b) R. Bau, M. Y. Chiang, C.-Y. Wei, L. Garlashelli, S. Martinengo, T. F. Koetzle *Inorg. Chem.* **1984**, *23*, 4758.
(c) M. Angoletta, L. Malatesta, G. L. Caglio, *J. Organomet. Chem.* **1975**, *94*, 99.
(d) S. D. Maloney, M. J. Kelley, D. C. Koningsberger, B. C. Gates, *J. Phys. Chem.* **1991**, *95*, 9406.

expected. ^{13}C NMR spectroscopy employing magic angle spinning (MAS) and cross polarization (CP) techniques has recently been used to characterize organometallics in zeolites. Since the chemical shift tensors of terminal carbonyl ligands are axially symmetric and highly anisotropic, and the chemical shift tensors of bridging carbonyl ligands are much less anisotropic and significantly asymmetric, [70] terminal and bridging carbonyl ligands can be easily distinguished from one another and from CO that has reacted with hydrogen or other organic groups. For example, ^{13}C CP-MAS (cross polarization-magic angle spinning) NMR spectra of $[Rh_4(CO)_{12}]$ in faujasites were reported by Gelin et al., [71] who observed chemical shifts almost identical to those of the molecular cluster in a Nujol mull.

4.2.2.5 Extended X-ray Absorption Fine Structure (EXAFS) Spectroscopy

EXAFS spectroscopy is a powerful method for determining the atomic structure of metal and metal carbonyl clusters. [72] Analysis of the results provides coordination numbers as well as distances of metal-metal, [72] metal-support, [73] and metal-adsorbate contributions. [74, 75] In an EXAFS experiment, X-ray absorption measurements are made in the region of the absorption edges of the metal. The fine structure observed at energies slightly greater than that of the absorption edge, at which core electrons are ejected, is a consequence of the interference between the photoelectrons ejected from the absorbing atom and the photoelectrons back-scattered from the neighboring atoms. The Fourier transform of the EXAFS oscillations gives a radial distribution, from which the interatomic distances can be determined by a fitting of the data. The exact determination of the absorber-backscatterer distances requires corrections for phase shifts and backscattering amplitudes determined from EXAFS data from well characterized standards. The most appropriate standards are molecular analogues of the supported species which have been further characterized by X-ray crystallography. Alternatively, compounds composed of near neighbor elements of the absorbing atom can be used without significant loss of accuracy.

A limitation of EXAFS spectroscopy is that although it may provide relatively accurate information about the nearest neighbors of the absorbing atoms, only little quantitative information can be obtained about the arrangement of atoms at greater distances (*i.e.* in the higher coordination spheres). EXAFS spectroscopy gives accurate determinations of distances between absorber and backscatterer atoms (distances often determined within 0.1 Å), but the uncertainties in the coordination numbers are often ± 20 % or more. In particular, EXAFS spectroscopy can give evidence for metal–C–O interactions because the near linearity of this moiety implies a significant multiple scattering effect which helps in the identification of the terminal CO ligands. [76] However, since multiple scattering effects are much less for bridging CO ligands, when a mixture of the two kinds is present, it is difficult to determine precise structural information. [77]

EXAFS spectroscopy is one of the most powerful methods for determining the structures of zeolite entrapped metal carbonyl clusters because it gives quantitative structural information. EXAFS spectroscopy can, in principle, be used to characterize samples in reactive atmospheres. However, most of the reported results have been obtained for samples in unreactive atmospheres, and usually at liquid nitrogen temperature where the signal-to-noise ratio is markedly greater than that at room temperature.

The method has been applied to NaY encaged $[Ir_4(CO)_{12}]$ (Table **4-5**). The results confirm the structure of the zeolite-entrapped species that had been inferred from the infrared spectra (Fig. **4-10**). However, the EXAFS data indicate that $[Ir_4(CO)_{12}]$ is not the only iridium species present in the zeolite. The data suggest that some (unidentified) mononuclear iridium species (perhaps iridium carbonyls) are also present and comprise about 20 % of the iridium. [78] EXAFS spectroscopy is perhaps the best suited technique for the detection of such minority species because infrared spectroscopy alone is often not sufficient to detect them.

The literature includes only a few examples of zeolite encaged clusters which have been characterized with EXAFS spectroscopy with a thorough analysis of the data. Examples of relatively thorough data analysis are those for clusters inferred to be triosmium carbonyls [53] in NaN_3-treated NaY zeolite, $[Ir_4(CO)_{12}]$ [67] and $[Ir_6(CO)_{16}]$ [5] in NaY zeolite, and $[Ir_6(CO)_{15}]^{2-}$ [60] in NaX zeolite.

With good data, the EXAFS method may be almost sufficient to determine cluster nuclearities of 3 or 4, but it becomes difficult with values of about 6 and

Table 4-5. EXAFS results characterizing $[Ir_4(CO)_{12}]$ in NaY zeolite.[a, b] [78]

Shell	N	R, Å	$\Delta\sigma^2$, Å^2	ΔE_0, eV	EXAFS reference
Ir–Ir	2.6	2.68	0.0004	4.88	Pt–Pt
Ir–CO:					
Ir–C	1.8	1.87	0.0017	6.03	Ir–C
Ir–O*	2.0	3.02	0.0028	–4.63	Ir–O*
Ir–$O_{support}$	0.76	2.13	0.0075	–3.78	Pt–O

[a] Notation: N, coordination number for absorber-backscatterer pair; R, distance; Δk, limits used for forward Fourier transformation (k is the wave vector); Δr, limits used for shell isolation (r is distance); n, power of k used for Fourier transformation.
[b] Estimated precision: N, ± 20 % (Ir–$O_{support}$, ± 30 %); R, ± 1 % (Ir–O*, Ir–C, Ir–$O_{support}$, ± 2 %); $\Delta\sigma^2$, ± 30 %; ΔE_0, ± 10 %.

beyond the capability of currently available methods with nuclearities of about 10. For example, the suggested identification of palladium carbonyl clusters in Y zeolite as $[Pd_{13}(CO)_x]$ referred to above [44–48] was based primarily on infrared and EXAFS spectra. The infrared spectra gave clear evidence for terminal and bridging CO ligands and suggested unique species; [44–46] however, they were not sufficient to identify which cluster(s) because there is no known palladium cluster having only carbonyl ligands to serve as a comparison. The EXAFS spectra indicated a Pd–Pd first shell coordination number of 5.8, but higher shell data were not reported. [47, 48] The analysis is therefore incomplete. EXAFS evidence for Pd–CO interactions was lacking in the analysis because it was restricted to a narrow region emphasizing only the metal-metal interactions. Thus, the suggestion that the clusters contained 13 Pd atoms, though not inconsistent with the available data, is not fully demonstrated by the data, and the reported results are apparently not sufficient to determine the cluster nuclearity.

Similar statements are true for the EXAFS data and their analysis which characterized palladium carbonyl clusters in zeolite A as being $[Pd_6(CO)_y]$. [49] There was no EXAFS evidence given for Pd–CO interactions, since the analysis was restricted to a narrow range of the data that emphasized only the first shell Pd-Pd interactions. A thorough analysis of EXAFS data for a metal carbonyl cluster requires an accounting of multiple shells, including the higher shell metal-metal contributions and the metal-carbonyl contributions. Since the technique determines average structural data, the presence of small amounts of compound, such as larger metal clusters, outside the zeolite crystallites could lead to misinterpretation. Thus, the results were not sufficient to determine the cluster nuclearity of this sample either.

Similarly, EXAFS data used to characterize platinum carbonyl clusters as being $[Pt_9(CO)_{18}]^{2-}$ are open to question because the Pt–CO interactions are not evident above the noise in the spectra, and it is not clear that the data quality was sufficient for the reported multiple shell analysis. [41, 42]

4.2.2.6 X-ray Diffraction and Scattering

These methods have been seldom used but are potentially quite valuable as they provide quantitative structural data. The diffraction method involves determination of a radial electron distribution (RED) function which gives a distribution of distances and coordination numbers of atoms surrounding a metal atom. This method has been used to characterize the formation of rhodium [79] and iridium [39] carbonyl clusters in NaY zeolite.

The predominant feature of the radial electron distribution pattern of a sample inferred to contain $[Ir_6(CO)_{16}]$ in NaY zeolite [39] is a strong peak at 2.78 Å, which is typical of the first neighbor Ir–Ir distance of the hexanuclear iridium carbonyl in the crystalline state. Peaks observed at atomic distances exceeding 8 Å in the RED pattern of the crystalline form of the cluster were not observed with the zeolite sample and suggest that $[Ir_6(CO)_{16}]$ is isolated in separate cages.

4.2.2.7 Ultraviolet-Visible Spectroscopy

Many metal complexes and clusters are colored and have distinctive ultraviolet-visible spectra. [80] The method offers the advantage of ease of application, but it has been used only seldom in the characterization of zeolite entrapped organometallics. The spectra may provide evidence of metal–metal bonds, as has been shown for carbonyl clusters of Fe, Ru, and Os, [81, 82] but there are hardly any data for zeolite entrapped clusters. The absorption bands of clusters are shifted to lower energy as the cluster nuclearity increases. [83] Ultraviolet-visible spectroscopy has been used to detect the formation of $[HFe_3(CO)_{11}]^-$ in NaY zeolite [50] and of clusters suggested to be $[Pt_9(CO)_{18}]^{2-}$ in NaY zeolite. [40–42] Since the spectra do not provide highly specific structural information, the method is of secondary importance.

On the other hand, the fact that many metal carbonyl clusters are colored [80] often allows their formation in zeolite cages to be qualitatively monitored visually. The color of a zeolite which contains a cluster may not be quite the same as that of the cluster in the pure solid state. For example, $[Ir_4(CO)_{12}]$ in NaY zeolite is yellow, a color almost the same as that of the cluster in solution. [3, 5, 66] The color of the $[Ir_6(CO)_{16}]$ isomer with edge bridging CO ligands in a NaY zeolite is yellow, [3, 5] but the color of this cluster in tetrahydrofuran solution is red. [57, 58]

4.2.2.8 Electron Paramagnetic Resonance (EPR) Spectroscopy

Electron paramagnetic resonance (EPR) [or electron spin resonance (ESR)] spectroscopy is useful for the characterization of species having unpaired electrons. Since most molecular clusters are diamagnetic, the application of this technique is limited. However, paramagnetic species may be formed under certain conditions when metal carbonyl clusters are formed in zeolite cages. Due to the high sensitivity of the EPR technique (as few as 10^{11} spins can be detected in favorable cases), very small percentages of the total metal loading become measurable, and information about the structure, environment, and electronic configuration can be gained. [70]

Abdo and Howe [84] observed the EPR signal of a radical containing two equivalent molybdenum nuclei formed from activated samples prepared from $[Mo(CO)_6]$ in HY zeolite and suggested that dinuclear anionic carbonyl species analogous to the known complex $[Mo_2(CO)_{10}]^{2-}$ may have formed. Similarly, EPR spectroscopy has been used to investigate the conversion of $[Fe_2(CO)_9]$ to $[HFe_3(CO)_{11}]^-$ in HNaY zeolite. [50]

4.2.2.9 X-ray Photoelectron Spectroscopy (XPS)

X-ray photoelectron spectroscopy (XPS) provides an indication of the oxidation states of the metals in clusters by comparisons of their binding energies with those of standards. Typically, the determinations are not exact and need further confirmation by other methods. XPS is especially useful for detection of changes in oxidation states. Since the technique requires ultrahigh vacuum, instability and volatility of the samples are often complications. This technique has been used to characterize the formation of $Rh(CO)_2$ in NaY zeolite. [85]

4.2.2.10 Raman Spectroscopy

The other vibrational spectroscopies, although less easily applied, may provide complementary structural information. Raman spectroscopy has been used to detect metal–metal bonds in metal oxide supported osmium [86] and iridium [87] clusters. This method might be expected to find application in the study of zeolite supported metal carbonyl clusters, but it is still far from routine since samples are subject to destruction by laser beams, and fluorescence often prevents measurement of useful spectra.

4.2.3 Reactivity of Metal Carbonyl Clusters in Zeolites

This section is a summary of the reactivities of metal carbonyls supported in cages (Table **4-6**).

4.2.3.1 Formation of Lewis Acid-Base Adducts

Metal carbonyls in zeolite cages are often present as Lewis acid-base adducts, with the basic oxygen of the carbonyl groups interacting with the Lewis acid sites of the zeolite. The basicity of the oxygen in carbonyl ligands of metal clusters depends on the CO coordination geometry. [90] Triply bridging carbonyl ligands (*i.e.* those bonded to three metal atoms) are more basic than doubly bridging carbonyl ligands, which are more basic than terminal carbonyl ligands.

A general pattern appears to be the following: interaction of a metal carbonyl with a Lewis acid via the oxygen of a carbonyl group typically results in a large decrease in the infrared absorption frequency of the carbonyl group, whereby the stretching frequencies of the noninteracting carbonyl ligands all shift to slightly

Table 4-6. Reactions of metal carbonyl clusters in zeolite cages.

Reaction class	Cluster	Zeolite	Resultant structure in zeolite	Characterization methods	Refs.
Nucleophilic attack on CO	$[Fe_3(CO)_{12}]$	NaY	$[HFe_3(CO)_{11}]^-$	IR, UV-vis, EPR, gas phase analysis	a
Lewis acid-base adduct formation	$[Ir_6(CO)_{16}]$ (edge)*	NaY	$[Ir_6(CO)_{16}]$ (edge)* $\cdots Al^{3+}$	IR	b
	$[Ir_6(CO)_{16}]$ (face)*	NaY	$[Ir_6(CO)_{16}]$ (face)* $\cdots Al^{3+}$	IR, EXAFS	b
	$[Rh_6(CO)_{16}]$	NaY	$[Rh_6(CO)_{16}] \cdots Al^{3+}$	IR, EXAFS	c
Brønsted acid-base interaction	$[Co_4(CO)_{12}]$	NaY	$[Co_4(CO)_{12}] \cdots HO-$	IR, gas phase analysis	d
	$[Fe_3(CO)_{12}]$	HY	$[Fe_3(CO)_{12}] \cdots HO-$	IR, volumetric measurements	e
Ion pair formation	$[HIr_4(CO)_{11}]^-$	NaX	$[HIr_4(CO)_{11}]^- \cdot\cdot Na^+$	IR	f, g
	$[Ir_6(CO)_{15}]^{2-}$	NaX	$[Ir_6(CO)_{15}]^{2-} \cdots Na^+$	IR, EXFAS	f, g
	$[HFe_3(CO)_{11}]^-$	NaY	$[HFe_3(CO)_{11}]^- \cdots Al^{3+}$	IR, UV-vis, EPR, gas phase analysis	a
Redox reactions	$[Ir_6(CO)_{16}]$	NaY	$Ir(CO)_2\{OZ\}_x$†	IR	b
Disproportionation	$[Co_2(CO)_8]$	NaY	Co^{2+} and $[Co(CO)_4]^-$	IR, gas phase analysis	d

* Edge and face refer respectively to the clusters with edge bridging and face bridging ligands.

† Z refers to the zeolite matrix.

(a) M. Iwamoto, S. Nakamura, H. Kusano, S. Kagawa, *J. Phys. Chem.* **1986**, *90*, 5244.
(b) S. Kawi, J.-R. Chang, B. C. Gates, *J. Am. Chem. Soc.* **1993**, *115*, 4830.
(c) L.-F. Rao, A. Fukuoka, N. Kosugi, H. Kuroda, M. Ichikawa, *J. Phys. Chem.* **1990**, *94*, 5317.
(d) R. L. Schneider, R. F. Howe, K. L. Watters, *Inorg. Chem.* **1984**, *23*, 4600.
(e) D. Ballivet-Tkatchenko, G. Coudurier, *Inorg. Chem.* **1979**, *18*, 558.
(f) S. Kawi, B. C. Gates, *J. Chem. Soc. Chem. Commun.* **1992**, 702.
(g) S. Kawi, J.-R. Chang, B. C. Gates, *J. Catal.* **1993**, *142*, 585.

higher frequencies. Shriver et al. [91, 92] reported shifts in the terminal and bridging carbonyl bands as a result of adduct formation between compounds such as $[(C_5H_5)Fe(CO)_2]_2$ and Lewis acids such as $Al(C_2H_5)_3$, which led to shifts of 100–300 cm^{-1} to lower energy for the bridging carbonyl bands and shifts of 30–70 cm^{-1} to higher energy for the terminal carbonyl bands. Tessier-Youngs et al. [93] reported shifts in the bridging carbonyl bands as a result of adduct formation between the triply bridging CO ligands of $[CpFe(CO)]_4$ and the Al^{3+} ions on the surface of alumina.

Similar shifts have been observed for $[Rh_6(CO)_{16}]$ in a NaY zeolite and were attributed to the interactions of the oxygen atoms of the face bridging carbonyl ligands with the Al^{3+} ions. [65] This pattern is also borne out by shifts in the stretching frequencies of the face bridging and edge bridging carbonyl ligands in the two isomers of $[Ir_6(CO)_{16}]$, which have been formed separately in NaY zeolites. [3, 5] The shifts indicate a stronger interaction between the Lewis acid sites and the oxygen of the face bridging CO ligands than with the oxygen of edge bridging CO ligands. The shifts in the frequencies relative to those of the respective clusters in tetrahydrofuran solution are the following: [5] the shifts in the frequencies for the terminal carbonyl ligands of the face bridging and edge bridging isomers are 20 cm^{-1} and 5 cm^{-1} to higher wavenumbers, respectively, and those for the bridging ligands are 35 cm^{-1} and 10 cm^{-1} to lower wavenumbers. In contrast, $[Ir_4(CO)_{12}]$ in the zeolite, which has only terminal carbonyl ligands, is characterized by insignificant shifts in the C–O stretching frequencies, consistent with the weak basicity of the ligands. [3, 5]

In summary, the interactions of the carbonyl ligands of rhodium and iridium carbonyl clusters with the Lewis acid sites in zeolites are indicated by shifts in the stretching frequencies of the CO ligands, and the pattern parallels that observed for metal carbonyl clusters in solutions containing Lewis acids and on surfaces of metal oxides containing Lewis acid sites.

An important consequence of the effects of the adduct formation on the ν_{CO} infrared spectra of encaged metal carbonyl clusters is that the spectra of clusters with bridging carbonyl groups, which are rather strongly basic, are significantly shifted from those of the clusters in neutral solvents, and identification by comparison with the spectra of the same clusters in solution is usually not straightforward. This point has not always been appreciated in the literature.

4.2.3.2 Formation of Brønsted Acid-Base Adducts

The chemistry of adduct formation may, alternatively, involve hydrogen bonding of the clusters' carbonyl groups with proton donor OH groups in the zeolite. Schneider et al. [26] investigated the reactions of $[Co_2(CO)_8]$ with Y zeolites. When $[Co_2(CO)_8]$ is sublimed into the Y zeolite, $[Co_4(CO)_{12}]$ quickly forms as the predominant organometallic species, as shown by the infrared spectrum. This chemistry resembles that of $[Co_2(CO)_8]$ on the surface of weakly acidic SiO_2 and leads to the following suggestion for the reaction stoichiometry:

$$2Co_2(CO)_8 \rightarrow Co_4(CO)_{12} + 4CO \tag{4.5}$$

The reaction is inhibited by CO, an observation which is consistent with this stoichiometry. When $[Co_2(CO)_8]$ is sublimed in the presence of CO onto Y zeolite, the resultant adsorbed species is the molecularly adsorbed $[Co_2(CO)_8]$. Upon sorption of $[Co_2(CO)_8]$ into the zeolite, the OH band of the zeolite is shifted by about 300–400 cm^{-1} to lower energy, the terminal carbonyl bands of the cluster are shifted by about 10–20 cm^{-1} to higher energy, and the bridging carbonyl bands of the cluster are shifted by about 40–60 cm^{-1} to lower energy. These shifts are evidence for hydrogen bonding between the OH groups and the bridging carbonyl oxygens.

Ballivet-Tkatchenko et al. [51] observed similar infrared evidence for hydrogen bonding interactions between zeolite OH groups and clusters believed to be $[H_2Fe_3(CO)_{11}]$, formed from $[Fe_3(CO)_{12}]$ in the HY zeolite by heating to 60 °C.

4.2.3.3 Ion Pairing

The nature of the interactions between anionic metal carbonyls and alkali metal ions in solution [94, 95] and on surfaces [37, 96] has been the subject of a number of investigations over the years. When the carbonyl oxygens of anionic metal clusters interact as ion pairs with alkali metal ions, there is a large decrease in the frequency of the infrared absorption for the carbonyl group interacting with the cation and slight shifts to higher frequency for the noninteracting carbonyl ligands.

The major terminal carbonyl band in the infrared spectrum of $[HIr_4(CO)_{11}]^-$ in the NaX zeolite is shifted by about 15 cm^{-1} to higher frequency and the bridging carbonyl band is shifted by about 55 cm^{-1} to lower frequency with respect to those for $[PPN][HIr_4(CO)_{11}]$ in tetrahydrofuran solution. [4, 60] These shifts are consistent with an ion pairing of $[HIr_4(CO)_{11}]^-$ with the Na^+ ions in the "solvating" zeolite cage, similar to the ion pairing of $[HIr_4(CO)_{11}]^-$ with the Mg^{2+} ions on a MgO surface [37] and that of $[HIr_4(CO)_{11}]^-$ with Na^+ in diethyl ether solution. [97] The relatively large shift in the characteristic bridging carbonyl band of the encaged $[HIr_4(CO)_{11}]^-$ suggests a strong interaction between the anion's bridging carbonyl oxygen and the Na^+ ions in the cage. Such an interaction is expected to result in a net electron withdrawal from the cluster and lead to a decrease in the backbonding to the terminal carbonyl ligands, a strengthening of the carbon-oxygen bonds, and a shift in the terminal carbonyl bands to higher frequencies, all consistent with the observations.

Similar trends in the infrared band shifts have been observed for $[Ir_6(CO)_{15}]^{2-}$ in the NaX zeolite. [4, 60] A comparison of the spectrum of this encaged cluster with the spectrum of $[PPN]_2[Ir_6(CO)_{15}]$ in tetrahydrofuran solution and with that of $[NEt_4]_2[Ir_6(CO)_{15}]$ in tetrahydrofuran solution shows a shift of the major terminal carbonyl band of the encaged species of about 12 cm^{-1} to higher frequency and a shift of the bridging carbonyl band of about 60 cm^{-1} to lower frequency. Again, these results suggest ion pairing of $[Ir_6(CO)_{15}]^{2-}$ with the Na^+ in the zeolite cage, similar to the ion pairing of $[Ir_6(CO)_{15}]^{2-}$ with the Mg^{2+} ions on the MgO surface. [37]

Consistent with this pattern, Iwamoto et al. [50] observed shifts in the carbonyl bands of $[HFe_3(CO)_{11}]^-$ in a hydrated NaY zeolite, which was formed by sublimation of $[Fe_2(CO)_9]$ or $[Fe_3(CO)_{12}]$ into the zeolite. The shift to lower energy of the absorption band of the bridging carbonyl ligand is attributed to a strong interaction of this ligand with an Al^{3+} ion in the zeolite.

4.2.3.4 Nucleophilic Attack at CO Ligands

Binary metal carbonyls can also react with the hydroxyl groups of zeolites to form zeolite encaged anions, the reactivity being analogous to that of metal carbonyls with surface hydroxyl groups on metal oxide surfaces or with hydroxide ions in solution (Eq. 4.6). [98, 99]

$$M_n(CO)_x + \{Z\text{–}OH\} \rightarrow [HM_n(CO)_{x-1}]^- \{Z\} + CO_2 \qquad (4.6)$$

Here the braces denote surface groups and Z represents the zeolite.

Iwamoto et al. [50] investigated the sorption of $[Fe_3(CO)_{12}]$ and $[Fe_2(CO)_9]$ in hydrated NaY zeolite. They concluded from infrared and ultraviolet-visible spectra that the sorption of $[Fe_2(CO)_9]$ led to the formation of $[HFe_3(CO)_{11}]^-$, which interacts via the bridging carbonyl ligand with an OH group, as shown in Equation (4.1).

Volumetric measurements of the products evolved when $[Fe_3(CO)_{12}]$ was sorbed in the zeolite, combined with infrared and ultraviolet spectra of the solid, also indicated the formation of $[HFe_3(CO)_{11}]^-$, as shown in Equation (4.2). The observation that the sorption of $[Fe_3(CO)_{12}]$ was much slower than that of $[Fe_2(CO)_9]$ was inferred to be a consequence of their size differences. The critical molecular dimension of $[Fe_3(CO)_{12}]$ (*ca.* 10.5 × 7.5 Å) is close to the diameter of the zeolite window (about 7.4 Å). The anion was inferred to have been generated inside the zeolite supercages.

4.2.3.5 Disproportionation Reactions

Disproportionation has been observed to occur in cages as it does in solution. For example, $[Co_2(CO)_8]$, which has zero valent cobalt, reacts readily with NaY zeolite [26] to give $[Co(CO)_4]^-$ and Co^{2+}. A similar result has also been observed on the basic metal oxide MgO. [100]

4.2.3.6 Oxidative Fragmentation and Reductive Condensation

When metal carbonyl clusters in zeolite cages are subjected to severe thermal conditions under, for example, vacuum or an O_2 atmosphere, complex transformations typically occur. The most commonly observed reactions involve changes in the nuclearity of the organometallic species (*i.e.*, fragmentation and/or aggregation of the clusters), which are often accompanied by changes in the metal's oxidation state.

Typically, the thermal activation (with decarbonylation) of a zeolite encaged metal carbonyl cluster leads to a fragmentation of the metal framework and an oxidation of the metal centers, possibly via a redox process involving the oxidative addition of surface hydroxyl groups. Oxidative fragmentation thus produces mononuclear metal complexes (metal subcarbonyls) in which the metals are in low oxidation states. For example, $[Rh_6(CO)_{16}]$ in a NaY zeolite was converted into rhodium subcarbonyls upon treatment in O_2 at 157 °C, [65] and the $[Ir_6(CO)_{16}]$ isomer with edge bridging ligands in the same zeolite was converted into iridium subcarbonyls in the presence of O_2 at 125 °C. [5]

Oxidative fragmentation in cages can sometimes be reversed by treatment of the sample with CO. In NaY zeolite, the iridium subcarbonyls formed by the fragmentation of the $[Ir_6(CO)_{16}]$ isomer having edge bridging ligands can be reductively carbonylated to regenerate the cluster. [5] There are clear parallels between reductive carbonylation in basic solution and in zeolite cages. In both cases, redox condensation reactions are expected to play a central role. Some of the molecular clusters formed have the nuclearity of the original cluster precursor, a result that indicates a low mobility of the reaction intermediates and suggests that both the clusters and the fragmentation products may be entrapped in the zeolite cages and hindered from migrating to the external zeolite surface.

These results also illustrate the advantages of using zeolite encaged clusters as precursors for well defined metal catalysts. Some encaged clusters appear to be stable to cycling through oxidation and reduction without forming crystallites on the external zeolite surface.

The available data suggest several hypotheses for how to maximize the stability of metal carbonyl clusters on supports.

1) Choosing metals which give stable metal frameworks (e.g. Os and Ir).
2) Using optimum zeolite/metal combinations (e.g. basic NaX zeolite to stabilize cluster anions such as $[Ir_6(CO)_{15}]^{2-}$).
3) Using high CO partial pressures to stabilize the clusters.

4.2.4 Catalysis

There has been a heightened interest in catalysis by encaged clusters in recent years that is probably motivated by the following.

1) Since molecular metal clusters have structures different from those of mononuclear complexes of the same metal and offer unique opportunities for the bonding of reactant ligands, they have been suggested to be likely to have novel catalytic properties (especially selectivities), an idea put forth by Muetterties. [101–103] However, most molecular metal clusters are fragile and only few examples of catalysis by them have emerged. Zeolites are attractive in prospect as supports for clusters, since they may stabilize them by entrapment.
2) Zeolites have found wide industrial application in shape selective catalysis, [104] whereby the molecular sieving properties of the zeolites are exploited.

Almost all the examples involve acid catalysis, and researchers have been enticed by the possibilities of shape selective catalysis by metals in zeolites. [104–106]

3) Metal clusters in zeolite cages are small and structurally well defined relative to the metal crystallites (often called clusters) present in typical metal oxide supported metal catalysts used in industry. Thus, researchers have investigated zeolite supported metals in attempts to better understand the structures of supported metals, the interactions of metals with supports, and the dependence of catalytic properties on cluster size and the nature of the interactions with the support. [105, 107, 108]

Most molecular metal clusters are metal carbonyls, and the CO ligands help to stabilize the metal frames. Consequently, it is logical to suppose that reactions involving CO might be good candidates for cluster catalysis. This supposition is, at least partially, borne out by the observations. Metal carbonyl clusters in solution are catalysts or catalyst precursors for alkene hydroformylation (Eq. 4.7) and the water-gas shift reaction (Eq. 4.8). [109, 110]

$$\text{R–CH=CH}_2 + \text{CO} + \text{H}_2 \rightarrow \text{R–CH}_2\text{–CH}_2\text{–CHO} \tag{4.7}$$

$$\text{CO} + \text{H}_2\text{O} \rightarrow \text{CO}_2 + \text{H}_2 \tag{4.8}$$

Below we summarize the results obtained by using metal carbonyl clusters in zeolites as catalysts or catalyst precursors for reactions involving CO (Table **4-7**).

4.2.4.1 CO Hydrogenation

The catalytic hydrogenation of carbon monoxide is catalyzed by supported crystallites of a number of metals (*e.g.* Ru). The reaction is also catalyzed by metal complexes in solution, although requiring very high pressures. [111] The products are usually methane and higher hydrocarbons (alkanes and alkenes), and the reaction is known as the Fischer-Tropsch reaction. Some supported metal catalysts, such as Cu and some others, like Rh, supported on basic metal oxides, yield oxygenated products, including methanol and higher alcohols. [112, 113]

The practical value of the Fischer-Tropsch reaction is limited by the unfavorable Schulz-Flory distribution of hydrocarbon products that is indicative of a chain growth polymerization mechanism. In attempts to increase the yields of lower hydrocarbons such as ethylene and propylene (potentially valuable as feedstocks to replace petrochemicals), researchers have used zeolites as supports for the metals in attempts to impose a shape selectivity on the catalysis [114] or to control the performance through particle size effects. [115] These attempts have been partially successful, giving unusual distributions of products, such as high yields of C_3 [114] or C_4 hydrocarbons. [116] However, the catalysts are often unstable because the metal is oxidized or because it migrates out of the zeolite cages to form crystallites, which then give the Schulz-Flory product distribution.

Zeolite entrapped metal carbonyl clusters prepared by the methods described above are potentially more stable and selective CO hydrogenation catalysts

Table 4-7. Reactions involving CO which are catalyzed by zeolites containing metal carbonyl clusters. The clusters are regarded as catalyst precursors; the catalytically active species are generally not known.

Cluster/Zeolite	Reaction	Conditions	Comment	Refs.
$[HFe_3(CO)_{11}]^-$/NaY	water gas shift	60–180 °C, 1 bar, 2.4 kPa H_2O	Activity high and comparable to that in the homogeneous phase at high pressure (40 bar).	a
$[Os_3(CO)_{11}]^{-*}$/NaY	CO hydrogenation	300 °C, 19 bar, $CO/H_2 = 1$ (molar)	Catalyst stable on stream more than 20 days and gave a non Schulz-Flory distribution of C_1–C_5 hydrocarbons with high alkene to alkane ratios.	b
$[Rh_6(CO)_{16}]$/NaY	alkene hydroformylation	80 °C, 80 bar, $CO/H_2 = 1$ (molar)	Catalyst active for hydroformylation, giving a selectivity for normal aldehydes similar to that for homogeneous rhodium carbonyl catalysts.	c
	CO hydrogenation	180–250 °C, 30 bar, $CO/H_2 = 0.5$ (molar)	Methanol observed in products.	
$[Ir_6(CO)_{16}]$/NaY	CO hydrogenation	250 °C, 20 bar, $CO/H_2 = 1$ (molar)	Catalyst stable on stream more than 8 days and gave a non Schulz-Flory distribution of C_1–C_5 hydrocarbons, deviating at C_3.	d, e
$[Ir_6(CO)_{15}]^{2-}$/NaX	CO hydrogenation	250 °C, 20 bar, $CO/H_2 = 1$ (molar)	Catalyst stable on stream at least 3 days and gave a non Schulz-Flory distribution of C_1–C_5 hydrocarbons, deviating at C_4.	f, g
$[Pt_{12}(CO)_{24}]^{2-*}$/NaY	water gas shift	27–150 °C, < 1 bar	Catalyst active, but stability not reported.	i
	NO reduction by CO	200 °C	NO breaks Pt–Pt bonds to give Pt subcarbonyls.	j, h
$[Pd_6(CO)_x]^*$/NaA	CO hydrogenation	290 °C, 11 bar, $CO/H_2 = 1$ (molar)	Methanol and dimethyl ether the sole products besides methane.	k

* Precursor structure not determined with confidence.

because 1) the metal clusters may be too large to migrate through the zeolite apertures, and 2) the small clusters, which are limited in size by the cages, might have selectivities different from those of conventional catalysts.

Zhou et al. [52] showed that strongly basic NaY zeolites containing entrapped osmium carbonyl clusters (described above) catalyzed CO hydrogenation and gave a non Schulz-Flory distribution of C_1–C_5 hydrocarbons with high alkene to alkane ratios. The catalyst was stable, operating at 300 °C and 19 bar with a CO/H_2 molar ratio of 1, for more than 20 days with only a small loss in activity and only a moderate loss in selectivity. However, the activity was markedly lower than that of conventional supported metal catalysts.

Infrared and EXAFS studies [52, 53] of the used catalyst, which was yellow, indicated the presence of osmium carbonyl clusters, possibly $[HOs_3\text{-}(CO)_{11}]^-$. This cluster might have been a catalyst precursor, and catalysis by small (undetected) amounts of Os metal could not be ruled out.

$[Ir_6(CO)_{16}]$ in NaY zeolite [5] and $[Ir_6(CO)_{15}]^{2-}$ in NaX zeolite, [4, 60] prepared by carbonylation of sorbed $[Ir(CO)_2(acac)]$ as described above, have been investigated as catalysts for CO hydrogenation at 250 °C and 20 bar of a reactant mixture of equimolar CO + H_2. Again, the catalysts had low activities, but they were stable for at least several days. Infrared and EXAFS data characterizing the used catalysts indicated the presence of predominantly $[Ir_6(CO)_{16}]$ and $[Ir_6(CO)_{15}]^{2-}$ in the NaY and NaX zeolites, respectively. However, other iridium species, thought to be mononuclear but not structurally identified, were indicated by the EXAFS data. Both catalysts gave non Schulz-Flory distributions of C_1–C_5 hydrocarbons, with both giving methane as the primary product. The NaY zeolite catalyst gave a local maximum at C_3 hydrocarbons and the NaX zeolite catalyst a local maximum at C_4 hydrocarbons. The catalysts were relatively stable, with only little loss in activity and selectivity during several days of operation in a flow reactor. It is possible that the iridium carbonyl clusters were the catalysts in the CO hydrogenation reaction, but the possibility that other species (such as mononuclear iridium complexes or undetected amounts of iridium metal) were the catalytically active species could not be ruled out.

Table 4-7. (Continued).

(a) P.-L. Zhou, S. D. Maloney, B. C. Gates, *J. Catal.* **1991**, *129*, 315.
(b) M. Iwamoto, S. Nakamura, H. Kusano, S. Kagawa, *J. Phys. Chem.* **1986**, *90*, 5244.
(c) E. Mantovani, N. Palladino, A. Zanobi, *J. Mol. Catal.* **1977–78**, *3*, 385.
(d) S. Kawi, B. C. Gates, *J. Chem. Soc. Chem. Commun.* **1991**, *994.*
(e) S. Kawi, J. R. Chang, B. C. Gates, *J. Am. Chem. Soc.* **1993**, *115*, 4830.
(f) S. Kawi, B. C. Gates, *J. Chem. Soc. Chem. Commun.* **1992**, 702.
(g) S. Kawi, J. R. Chang, B. C. Gates, *J. Catal.* **1993**, *142*, 585.
(h) R.-J. Wang, T. Fujimoto, T. Shido, M. Ichikawa, *J. Chem. Soc. Chem. Commun.* **1992**, 962.
(i) G.-J. Li, T. Fujimoto, A. Fukuoka, M. Ichikawa, *J. Chem. Soc. Chem. Commun.* **1991**, 1337.
(j) G.-J. Li, T. Fujimoto, A. Fukuoka, M. Ichikawa, *Catal. Lett.* **1992**, *12*, 171.
(k) Z. Zhang, F. A. P. Cavalcanti, W. M. H. Sachtler, *Catal. Lett.* **1992**, *12*, 157.

The used NaY zeolite catalyst containing iridium carbonyl clusters was yellow when removed from the flow reactor after three days of continuous operation at 250 °C and 20 bar, consistent with the presence of the $[Ir_6(CO)_{16}]$ isomer with face bridging ligands. [5] The catalyst used under the same conditions but supported in the NaX zeolite was reddish brown after a day of operation, consistent with the presence of $[Ir_6(CO)_{15}]^{2-}$. [4, 60]

Zeolites containing carbonyl clusters of Rh [71, 117, 118] and of Pd [68] show similar behavior, but they are structurally less well characterized, and the catalytically active species are unknown.

4.2.4.2 Alkene Hydroformylation

NaY zeolite containing $[Rh_6(CO)_{16}]$ has been shown to be a catalyst for the hydroformylation of alkenes in the liquid phase at 80 °C and 80 bar. [38] The catalysts were selective for the formation of butyraldehydes, with a *n*- to *iso*-butyraldehyde ratio similar to that shown by mononuclear rhodium carbonyls in solutions. [119] The catalyst was also tested for the hydroformylation of nonconjugated dialkenes and produced mainly dialdehydes. [38] The catalyst was recovered after use by decantation and reused with only a small loss in activity.

Rode et al. [64] investigated propylene hydroformylation by such a catalyst and measured the infrared spectra of the functioning catalyst. They concluded that the rhodium carbonyl cluster was not the catalytically active species, but, instead, a mononuclear rhodium species was suggested to be the active species.

4.2.4.3 Water Gas Shift Reaction

Several metal carbonyl clusters (*e.g.* $[Ru_3(CO)_{12}]$ and $[Ir_4(CO)_{12}]$) have been investigated in solution as possible catalysts for the water gas shift reaction. [120] A plausible common mechanism has been proposed: Nucleophilic attack by H_2O or OH^- on an electrophilic metal center of the cluster to form an unstable carbohydroxy metal complex, which is then decarboxylated to give a metal hydride from which H_2 is eliminated. This reaction is one of the relatively few for which there is good evidence for catalysis by clusters themselves, rather than fragments or metal aggregates formed from them.

Some zeolites containing metal carbonyl clusters also are catalytically active for the shift reaction. Iwamoto et al. [50] investigated the catalytic activity of zeolite NaY incorporating $[HFe_3(CO)_{11}]^-$ at 60–180 °C and 1 bar. The catalytic activity was comparable to that reported for the solution phase reaction which uses $[Fe(CO)_5]$ as the catalyst precursor at 180 °C and 40 bar. Infrared and ultraviolet spectra indicated that $[HFe_3(CO)_{11}]^-$ was stable under the reaction conditions. A plausible mechanism was proposed to explain the observations (Eqs. 4.9–4.12).

$$[HFe_3(CO)_{11}]^- + H_2O \rightarrow H_2Fe_3(CO)_{11} + OH^- \quad (4.9)$$

$$H_2Fe_3(CO)_{11} \rightarrow H_2 + Fe_3(CO)_{11} \quad (4.10)$$

$$Fe_3(CO)_{11} + CO \rightarrow Fe_3(CO)_{12} \quad (4.11)$$

$$Fe_3(CO)_{12} + OH^- \rightarrow [HFe_3(CO)_{11}]^- + CO_2 \quad (4.12)$$

Kinetic and spectroscopic results indicated that the reaction between $[HFe_3(CO)_{11}]^-$ and H_2O was rate determining.

Wang et al. [121] reported that the NaY zeolite containing Pt carbonyl clusters of the family $[Pt_3(CO)_6]_n^{2-}$ (n = 3, 4) is catalytically active for the shift reaction at 27–150 °C and that the photocatalytic reaction was about 38 times faster than the catalytic reaction at 25 °C. Since the platinum carbonyls present in the used catalyst were identified only by infrared spectroscopy, there remains doubt about what they were as well as what the catalytically and photocatalytically active species might have been.

4.2.4.4 Summary

In summary, it is difficult to determine the catalytically active species in any supported catalyst, and there are still no well documented examples of catalysis by metal carbonyl clusters themselves in zeolites. There is, however, substantial indirect evidence that metal carbonyl clusters in zeolite cages may be either the catalysts or the catalyst precursors for a number of reactions involving CO. In some cases, these clusters are the only detectable organometallic or metallic species, and they are stable under the conditions of the catalytic reactions. Some of the catalysts retain the colors and the infrared spectra of the metal carbonyl clusters even after weeks of catalytic operation. In the few instances when EXAFS data were available, the presence of metal carbonyl clusters within the zeolites was indicated; however, evidence for other species that are plausible catalyst precursors was also obtained.

Most of the catalytic reactions are restricted to low temperatures. The instability of the clusters at higher temperatures is a major limitation to their application as catalysts. However, some metal carbonyl clusters are stable at relatively high temperatures (300 °C). They appear to be stabilized by a CO atmosphere which can provide stabilizing ligands, and they may be stabilized by the zeolite cages as well. It is hypothesized that the cages hinder migration of the clusters and thus their coalescence to form larger clusters or metal crystallites.

4.2.5 Uniqueness of Zeolite Cages as Media for Cluster Synthesis and Stabilization

For the most part, good analogies can be drawn between the chemistry of metal carbonyl clusters in solution and that on surfaces as well as that in zeolite cages. The differences are related primarily to the steric restrictions imposed by the cages. For example, in basic solutions and on the basic surface of MgO, $[Ir(CO)_2(acac)]$ is converted sequentially into $[HIr_4(CO)_{11}]^-$, $[Ir_8(CO)_{22}]^{2-}$, and $[Ir_6(CO)_{15}]^{2-}$. [37, 67] In contrast, in the NaX zeolite, the sequential formation proceeds under similar conditions, but without the formation of $[Ir_8(CO)_{22}]^{2-}$, which is believed not to form because it is too large to fit into the zeolite supercage. [4, 60] It is also suggested that entrapment in the zeolites helps to stabilize the metal carbonyl clusters more than they might be in solution or on the nearly planar surfaces. Thus, whereas $[Ir_6(CO)_{15}]^{2-}$ in the NaX zeolite and the $[Ir_6(CO)_{16}]$ isomer with face bridging ligands in the NaY zeolite, for example, are stable in the presence of $CO + H_2$ (equimolar) at 250 °C, the same clusters are not stable under the same conditions on the surface of MgO or γ-Al_2O_3, respectively.

4.3 Metal Clusters in Zeolites

Metal containing species in zeolite cages can be prepared in sizes ranging from that of an isolated ion to that of a cluster with ten or more metal atoms. For this size range of metal clusters, a number of properties are different from those of the bulk metal and are sensitive to the cluster size. For example, a transition from Pauli to Curie type magnetic behavior is predicted to occur on going from bulk to cluster as the continuum of electronic energy levels is replaced by discrete levels. [122] Hence, for studies geared towards a fundamental understanding of structure-property relationships in nanoclusters, clusters in uniform cages are especially appealing. Furthermore, the entrapment in cages may affect the stability and accessibility of the clusters and be of advantage in applications such as catalysis.

Characterization is relatively simple since the clusters contain only metal atoms, and usually of only a single element. However, there is one problem associated with the characterization of metal clusters in cages. In contrast to the situation for metal carbonyl clusters, there is no base data set for these metal clusters themselves in a pure state to be used for comparison. This means that spectra of encaged metal clusters cannot be compared with those of their analogues in the liquid or solid state because they simply are not known. Thus the basis for structure determination is in a sense weaker than that for metal carbonyl clusters.

4.3.1 Synthesis

There are numerous reports of attempts to prepare metal clusters (as distinguished from metal carbonyl clusters) in zeolite cages, [105, 108, 123–125] most often by reduction of exchange ions in the cages or by decarbonylation of metal carbonyl clusters. One of the challenges has been to confine the resultant clusters within the cages, and often the literature reports have failed to provide sufficient information to determine the actual fraction of metal remaining within the cages. The synthetic conditions, including the nature and degree of hydration of the zeolite, the types and amounts of metal ions or other species to be reduced, the types of cocations, the pretreatment conditions, the nature of the reducing agents, the temperature, etc. all influence the metal dispersion, migration, and growth of metal clusters or aggregates.

4.3.1.1 Decomposition of Metal Carbonyl Clusters

Researchers have attempted to drive off the carbonyl ligands of molecular metal carbonyl clusters in the hope of preparing naked clusters of the same nuclearity. It is now evident, although this assertion contradicts some of the claims in the literature, that most of the attempts have failed and have instead led to increases in cluster nuclearity and loss of structural simplicity. Usually, the lack of sufficient characterization has prohibited the determination of the nuclearities of all the resultant species, and often only the larger species (clusters) have been detected. With the increasing availability of techniques such as high resolution transmission electron microscopy and EXAFS spectroscopy this field is expected to develop rapidly.

A relatively thorough investigation has been carried out with iridium clusters in the NaY zeolite. [5] The results indicate that the molecular metal carbonyl clusters can be decarbonylated to give clusters that may be iridium hydrides. The authors reported that $[Ir_6(CO)_{16}]$ in the NaY zeolite could be decarbonylated by treatment in He followed by H_2 at 325 °C. The decarbonylated clusters were characterized by EXAFS spectroscopy. The characteristic first shell Ir–Ir coordination number of 3.6 for the decarbonylated sample is the same (within the $\pm 20\%$ experimental error) as the crystallographically determined value of 4.0 for $[Ir_6(CO)_{16}]$ (Fig. **4-6**). [57] The result suggests that the structure of the Ir cluster frame after decarbonylation resembles that of the octahedral frame of $[Ir_6(CO)_{16}]$. The lack of any significant peaks in the Fourier transforms which would correspond to higher shell Ir neighbors suggests that there was no significant formation of larger Ir clusters or crystallites in the cages or on the outer zeolite surface.

$[Ir_4(CO)_{12}]$ in NaY zeolite was similarly decarbonylated. [3, 78] The EXAFS data used to characterize the decarbonylated cluster (again, perhaps an iridium hydride) showed that the first shell Ir–Ir coordination number (3.4) is, within the experimental error of about $\pm 20\%$, equal to the value of 3.0 determined crystallographically for $[Ir_4(CO)_{12}]$ (Fig. **4-6**). The result is consistent with the suggestion that the nuclearity and the tetrahedral frame of the precursor were retained in the decarbonylated clusters. Furthermore, no Ir–Ir contributions were detected at distances greater than the first Ir–Ir coordination shell, implying that there was no significant formation of larger iridium clusters or crystallites.

Similarly, $[Rh_6(CO)_{16}]$ in NaY zeolite has been decarbonylated. [65, 126] It was observed that the decarbonylation was reversible, which suggests that the decarbonylated species were Rh_6. However, this suggestion is speculative as will be discussed below. Rao et al. [65] showed that $[Rh_6(CO)_{16}]$ in NaY zeolite could be decarbonylated in H_2 at 200 °C. Their EXAFS results showed that the first shell Rh–Rh coordination number increased from 3.1 to 4.6 as the clusters were decarbonylated, indicating a growth in the cluster nuclearity. Thus, even though there is an analogy between the chemistry of the iridium carbonyls and the rhodium carbonyls in NaY zeolite, the analogy has not been shown to extend to the simple decarbonylation of the clusters.

Nagy et al. [127] used thermal decomposition of $[Fe(CO)_5]$ sorbed in HY zeolite to form highly dispersed iron particles, which were observed by transmission electron microscopy to be 10–50 Å in size. Photochemical decomposition gave smaller particles (too small to be observed with transmission electron microscopy), presumably because of the strong interaction between the carbonyl clusters and the surface resulting from the electronic excitation of the metal carbonyl during decomposition.

4.3.1.2 Reduction of Exchange Ions

Zeolite supported metal clusters have most commonly been prepared by ion exchange followed by reduction. [104, 108, 123–125] An advantage of this method is that the amount of a transition metal in a zeolite can be controlled by varying the exchange temperature, the concentration of the ion-exchange solution, or the number of ion-exchange steps. Unfortunately, however, samples prepared by this method are not easily reproducible. The ease of reduction of the transition metal ions in the zeolites depends on the nature and loading (and thus location) of the cations, the presence of oxidizing agents such as hydroxyl groups, the presence of residual water, and the structure and composition of the zeolite matrix. Appropriate treatment of the ion-exchanged zeolites is needed in order to prevent migration and sintering of the metal during the course of the reduction. Since samples which have been prepared by the methods described in this section are for the most part less than uniform in structure and less than well characterized, the section is brief. The reduction often leads to metal clusters or particles that are too large to fit within the cages of the zeolite, and it may be accompanied by the breakup of the zeolite framework, apparently caused by the growing clusters. [128]

Reduction by H_2

Cations of noble metals in zeolites are easily reduced by H_2, and this is the method most commonly applied. Proper activation and reduction treatments are essential to obtain the highest metal dispersions. Empirically determined treatment conditions are usually employed to obtain well dispersed metals, and generalizations about them are difficult. For example, prior to the final reduction in H_2, it is usually necessary to eliminate any NH_3 produced during the thermal decomposition of an amine complex or a NH_4^+ ion in the zeolite. The reduction of metal ions in the presence of evolving NH_3 can easily lead to the formation of agglomerated metal. Furthermore, Dalla Betta and Boudart [129] pointed out that the direct reduction of noble metal cations by H_2 at high temperatures might lead to the formation of mobile neutral metal hydrides, which can cause metal agglomeration and low metal dispersions (Eqs. 4.13 and 4.14).

$$Pt(NH_3)_4^{2+} + 2H_2 \rightarrow Pt(NH_3)_2H_2 + 2NH_3 + 2H^+ \quad (4.13)$$

$$Pt(NH_3)_2H_2 \rightarrow Pt^0 + 2NH_3 + H_2 \quad (4.14)$$

Thermal activation of the complexes in flowing O_2 prior to the reduction by H_2 is essential if highly dispersed Pt clusters in zeolites are to be obtained.

If, however, more oxophilic metals than Pt are used (e.g. Ru), then pretreatment in O_2 before reduction should be avoided because this leads to the formation of metal oxides such as RuO_2. In these cases, pretreatment under vacuum or in an inert gas is recommended, with the reduction to be carried out in the presence of H_2.

Complications may ensue because of the distribution of the exchange ions over sites in the sodalite cages or the supercages of the faujasites. Calcination may lead to migration of the cations into the smaller cages. The migration of the cations into the smaller cages can be minimized by blocking the windows to these cages with cations having a high charge density prior to the introduction of the transition metal ions. This technique is particularly useful for difficult to reduce metals such as Ni and Co, which require high temperature for reduction. For example, Ca^{2+}, Sr^{2+}, and Mn^{2+} have been used to force Ni^{2+} and Co^{2+} to remain within the supercages of the faujasites. [130–132]

Reduction by Metal Vapors

Methods employing H_2 as a reductant are not applicable to the metal cations of the first transition row because the redox potentials are inappropriate; therefore, stronger reducing agents are needed. In addition, reduction by H_2 leads to the formation of H^+ in the zeolite (present in OH groups), and since some zeolites in the H^+ form are unstable (e.g. zeolite A), they are destroyed. In such cases, metal vapors have been used as reducing agents instead of H_2. The metals vapors (e.g. Na, Cd) which are oxidized remain as cations within the zeolite and often complicate matters by, for example, combining with the reduced metals in unknown ways. Furthermore, sintering of the metal from the supercages to the external surface of the zeolite seems to be a general complication. [115, 133, 134]

Reduction with Ethylene Glycol

A method for reducing first row transition metal cations (e.g. Cu^{2+}, Ni^{2+}) in NaY zeolite with ethylene glycol has been reported (the polyol process). [135] The metal ions are reduced to the zero valent state, with the degree of reduction being dependent on the amount and location of the cations. The greater reducing efficiency of ethylene glycol over H_2 is believed to be a result of its ability to coordinate to the metal ions, thereby shielding them from electrostatic interactions with the framework oxygens. [135] The high efficiency of the relatively large ethylene glycol molecule suggests that the cations migrate to accessible sites before reduction. The process might be useful for the preparation of structurally uniform metal clusters in zeolites, but further investigations are needed.

4.3.1.3 Solution Phase Metal Atom Technique

Metal atoms have been introduced into zeolites as "solvated metal atoms" and then converted into metal clusters. [136] Introduction of a metal vapor into a slurry of dehydrated zeolite in toluene gives *bis*(toluene) metal(0) compounds. These labile organometallics can be sorbed into zeolites at low temperature, and as the temperature is increased, the volatile ligands are removed by vaporization and metal clusters are formed. This method has been used to prepare iron and cobalt clusters in zeolites. [136] The technique is expected to give smaller metal clusters with more uniform structures than the techniques requiring high temperature reduction. However, only large pore zeolites can be loaded in this way because of the large size of most of these organometallic complexes. Questions still remain about the uniformity of the clusters formed in this way.

4.3.1.4 Impregnation with Salt Solutions

When zeolites have little capacity for ion exchange, the introduction of metal ions by ion exchange gives only low loadings. Instead, metal salts may be introduced into the pores by contacting the solid with a salt solution followed by drying and reduction. The method is usually not appropriate for the incorporation of metals in particular cages, because the initial distribution, in contrast to what might be attainable by ion exchange, is unpredictable.

4.3.2 Characterization Techniques

Many of the techniques used to characterize metal clusters in zeolites are the same as those already discussed for metal carbonyl clusters in zeolites. In the following section we will emphasize additional techniques and comment on the value of some of the techniques used for metal clusters in particular in addressing issues such as the oxidation state of the metal and the uniformity of the clusters.

4.3.2.1 EXAFS Spectroscopy

EXAFS spectroscopy was already discussed for metal carbonyl clusters in Section 4.2.2.5. Since metal clusters are simpler in composition than metal carbonyl clusters, the EXAFS method [72, 73, 137] can provide more information about the former than about the latter. In particular, information about the structure of the cluster-zeolite interface has been obtained, for example, for samples that were suggested [5, 78] to contain predominantly Ir_4 or Ir_6 units in NaY zeolite prepared by the decarbonylation of $[Ir_4(CO)_{12}]$ and $[Ir_6(CO)_{16}]$, respectively. The formation of these samples is discussed below.

The sample prepared by the decarbonylation of $[Ir_4(CO)_{12}]$ is characterized by a first shell Ir–Ir coordination number of 3.3 (±20%, Table **4-8**). Although the data are consistent with the suggestion that there are Ir_4 clusters present in the zeolite, they are not sufficient to demonstrate whether these are the only encaged Ir species. The Ir–O contributions are evidence of the interaction of the clusters with the cage walls. There are no significant peaks in the Fourier transforms corresponding to any higher shell Ir neighbors, which suggests that there was no significant sintering of the Ir to form larger clusters or crystallites. These observations indicate that the structure of the Ir cluster frame after decarbonylation under mild conditions is tetrahedral Ir_4.

Similar data are listed in Table **4-9** for clusters formed from $[Ir_6(CO)_{16}]$ in the NaY zeolite. The first shell data indicate an Ir–Ir coordination number of 3.6 (±20%), which is consistent with the presence of Ir_6 octahedra, but does not rule out other structures in addition to these. It was suggested that the Ir_6 clusters are stabilized by the rigid environment of the zeolite cages so that migration and sintering occur less rapidly than on a support with larger pores.

4.3.2.2 X-ray Diffraction and Scattering

X-ray diffraction and scattering are valuable characterization methods. The accuracy of the results, however, depends strongly on the loading of metal in the zeolite.

Table 4-8. EXAFS results characterizing the iridium clusters formed by decarbonylation of $[Ir_4(CO)_{12}]$ in NaY zeolite at 325 °C in He followed by H_2.[a, b] [78]

Shell	N	R, Å	$\Delta\sigma^2$, Å^{2c}	ΔE_0, eV^d	EXAFS reference
Ir–Ir	3.2	2.70	0.0030	0.06	Pt–Pt
Ir–$O_{support}$:					
Ir–O_l	1.5	2.68	0.0036	–6.11	Pt–O
Ir–O_s	1.7	2.19	0.0012	–3.56	Pt–O
Ir–C	0.7	1.94	0.0028	–9.50	Ir–C

[a] Notation: N, coordination number for absorber-backscatterer pair; R, distance; Δk, limits used for forward Fourier transformation (k is the wave vector); Δr, limits used for shell isolation (r is distance); n, power of k used for Fourier transformation; the subscripts s and l refer to short and long, respectively.
[b] Estimated precision: N, ± 20% (Ir–$O_{support}$, ± 30%); R, ± 1% (Ir–C, Ir–O_l, and Ir–O_s, ± 2%); $\Delta\sigma^2$, ± 30%; ΔE_0, ± 10%.
[c] Debye-Waller factor.
[d] Inner potential correction.

Table 4-9. EXAFS results characterizing the iridium clusters formed by decarbonylation of $[Ir_6(CO)_{16}]$ with face bridging CO ligands in NaY zeolite at 300 °C in H_2.[a, b] [5]

Shell	N	R, Å	$\Delta\sigma^2$, Å^2	ΔE_0, eV	EXAFS reference
Ir–Ir	3.60	2.71	0.0035	–2.00	Pt–Pt
Ir–$O_{support}$:					
Ir–O_s	1.01	2.17	0.0070	–6.20	Pt–O
Ir–O_l	1.85	2.71	0.0036	–6.11	Pt–O

[a] Notation: N, coordination number for absorber-backscatterer pair; R, radial absorber-backscatterer distance; $\Delta\sigma^2$, Debye-Waller factor (difference with respect to reference compounds); ΔE_0, inner potential correction (correction of the edge position); the subscripts s and l refer to short and long, respectively.
[b] Estimated precision: N, ± 20 % (Ir–O_l + Ir–O_s, ± 30 %); R, ± 2 % (Ir–Ir, ± 1 %); $\Delta\sigma^2$, ± 30 %; ΔE_0, ± 10 %.

X-ray diffraction crystallography can be used to determine not only the framework structures but also the locations of the isolated metal ions in the zeolites. The method is applicable when the metals are periodically isolated within the crystalline zeolite structure and not agglomerated outside of it. Although the method is, in principle, appropriate for the characterization of all types of metal clusters in zeolites, its use has been largely restricted to ionic clusters, [138, 139] which are beyond the scope of the present review.

Small angle X-ray scattering (SAXS) [140] and wide angle X-ray scattering (WAXS) [141] provide good measures of the size distributions of zeolite supported metal clusters since the zeolites themselves produce only weak small angle scattering because of the regularity of their porous structures. X-ray scattering is also useful for the determination of the specific surface areas of the clusters. The range of cluster sizes which can be measured extends down to the atomic size, and the sensitivity is sufficient to allow measurements with metal concentrations as low as 0.5–1 wt%. [142, 143]

The radial electron distribution (RED) determined from X-ray diffraction data has been frequently used to characterize the structures of encaged metal clusters. In contrast to EXAFS spectroscopy, the RED gives only metal–metal distances, not metal–support and metal–adsorbate distances.

4.3.2.3 Transmission Electron Microscopy (TEM)

Transmission electron microscopy (TEM) has been frequently used to identify the size and location of metal clusters in zeolites. Clusters as small as about 5 Å have been observed, for example, for Pt clusters supported in and on NaY zeolites. [145] The cluster size distribution determined from the electron micrographs (Fig. **4-11**) indicates the presence of clusters ranging in diameter from

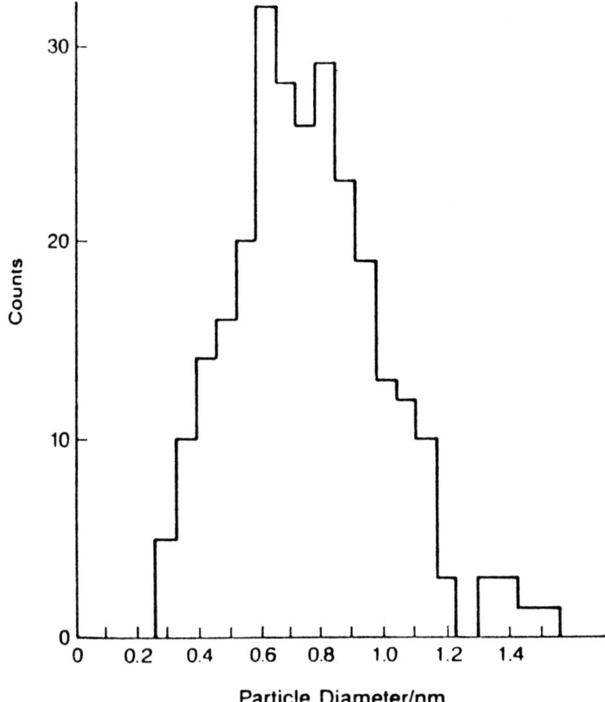

Figure 4-11. Size distribution of Pt clusters in and on zeolite Y as determined by transmission electron microscopy. [145] The distribution of sizes indicates the nonuniformity of the clusters, which is typical of zeolite supported metals. Reproduced from *Ultramicroscopy* with permission of North Holland Publishing Co.

about 3 to 15 Å. These results illustrate the typical nonuniformity of cluster sizes in samples prepared by the reduction of the metal with H_2.

There are some important limitations to TEM. 1) Electron beams are liable to produce high local temperature rises, and hence the metal cluster size and the location can be modified by the powerful electron beam, [146] 2) hundreds of metal clusters from many different micrographs should be measured in order to obtain a statistical representation of the size distribution, and 3) electron beam absorption by the zeolite crystal can hamper the detection of the particles. The best images seem to be those obtained from thin slabs of the zeolite crystals which have been cut at different heights and at different places with an ultramicrotome equipped with a diamond knife. The best images are from end-on views looking down into the pores with stacks of clusters in the cages.

4.3.2.4 Chemical Probes

A chemical means of identifying the location of metal clusters has been demonstrated for Pt clusters in Y and L zeolites. [147] One of the probe reactions was the Pt catalyzed reduction of methylviologen cations (abbreviated as MV^{2+}). In the presence of H_2 and Pt^0, MV^{2+} is reduced to the cation radical ($MV^{\cdot +}$) at pH > 7 (Eq. 4.15).

$$MV^{2+} + 1/2\ H_2 \rightarrow MV^{\cdot +} + H^+ \qquad (4.15)$$

MV^{2+} will exchange up to a maximum of 1 and 8 molecules per unit cell within the zeolites L and Y respectively. The excess remains in the external solution. Since both the rate of exchange between zeolite bound and solution phase methylviologen is relatively low and because the potential of the $MV^{2+}/MV^{\cdot +}$ couple is slightly more positive inside the zeolite than in the solution, zeolites which hold Pt only within their pores will not catalyze the rapid reduction of solution phase MV^{2+}. [147] Thus, a zeolite having all its Pt inside the pores took on the characteristic blue color of the reduced viologen radicals, and the solution remained colorless. A zeolite containing Pt both internally and externally also turns blue, but so does the external solution. [147]

A shape selective chemical probe involves the reduction of $[Fe(CN)_6]^{3-}$, which does not enter the anionic framework of zeolites L and Y. [147] In the presence of H_2 and Pt, $[Fe(CN)_6]^{3-}$ is reduced to $[Fe(CN)_6]^{4-}$. In the presence of zeolites containing Pt only within the pores, ultraviolet-visible spectra of the solution taken before and after 6 h of contact with H_2 were found to be identical. Under the same conditions but in the presence of zeolites containing Pt both within the pores and outside them, there was a complete disappearance of the absorbance maxima at 302 and 418 nm attributed to $[Fe(CN)_6]^{3-}$. In the former case, only internal Pt was present, whereas in the latter, external Pt particles catalyzed the reduction of $[Fe(CN)_6]^{3-}$.

4.3.2.5 NMR Spectroscopy

Zeolite supported metal clusters have been characterized by several kinds of NMR spectroscopy. ^{129}Xe NMR spectroscopy provides a sensitive probe of the contents of the cages and NMR of spin active metals provide evidence for the various metals in zeolites. Spectra of the sorbed xenon give information about the chemical shifts associated with the collisions of the xenon atoms with the cage walls and with the encaged species. Hence, the method can be used to estimate the average number of metal atoms per cluster. [148] The line shape and the chemical shift are sensitive to the metal species present within the cages or channels. For example, zeolite supported metal clusters show much larger chemical shifts than the zeolite alone under the same experimental conditions.

The results based on chemical shifts should be interpreted with caution because the difference between the chemical shifts of the supported sample and those of the zeolite support depends on the xenon pressure, [149] on the type of cations exchanged into zeolite (e.g. divalent vs. monovalent, [150–153]) and on the temperature and size of the zeolite crystals. [154, 155] A recent publication by Ryoo et al. [149] illustrates the application of the technique in characterizing clusters of several different transition metals in zeolite Y.

Metal NMR spectroscopy is also beginning to gain favor as a technique for the characterization of encaged clusters. [156] For example, Zhang et al. [157] used ^{59}Co spin echo NMR spectroscopy to characterize the size and location of the Co clusters in the cages of NaY zeolite.

4.3.2.6 X-ray Absorption Near Edge Spectroscopy (XANES)

X-ray absorption near edge spectroscopy (XANES) can be used to qualitatively estimate the oxidation state of the metal atoms of the carbonyl clusters by determining the intensities of the X-ray absorption edges or white lines. The white line (the large absorption observed in the near edge region) results from the transitions of the electrons from the 2p to the d states and has been suggested to provide information about the charge on the absorbing metal atoms in metal catalysts. [72] The influence of the metal cluster size on the white line intensities must be accounted for since the white line intensity decreases as the cluster size increases, as has been shown by Hartree-Fock-Slater LCAO calculations [158] as well as by XANES experiments. [159] XANES has been used in attempts to understand the transfer of electrons between various cations and Pt clusters in Y zeolites. [160]

4.3.2.7 Far Infrared Spectroscopy

Vibrations in zeolite frameworks are observed in the range 250–400 cm^{-1}, [161] whereas those of metal atoms, ions, and clusters within the framework are expected in the range 30–250 cm^{-1}. [161, 162] Far infrared spectroscopy is therefore an informative method for determining the locations of metal ions in zeolites. The method has been used to characterize various cations (e.g. Na^+, Cs^+, Co^{2+}) in zeolite Y, and has provided information about the cations' location, population, and distribution. [163, 164] The method is sometimes more effective than X-ray diffraction to determine cation locations.

4.3.2.8 Sorption of Gases

Sorption and desorption of gases, usually H_2, have been widely used to characterize metal clusters in zeolites. [165] The methods are standard for supported metal catalysts and provide reliable measures of surface area and metal dispersion. However, when the clusters are as small as those in zeolite cages, there are important limitations to the method. For example, the stoichiometries of the reactions of H_2 with metal clusters are not well known and vary from metal to metal. The method will be mentioned again in Section 4.3.3.5 for Pt clusters in the zeolite L, where the results are in good agreement with the EXAFS results.

The measurement of the sorption of CO and of NO in combination with infrared spectroscopy gives valuable structural information about encaged clusters. Primet et al. [166] showed that the smaller the Pt cluster, the higher the vibrational frequency ν_{NO} of NO sorbed on the metal. A similar trend is expected for CO. When using this method one must be aware of the possibility that the CO or NO may cause oxidative fragmentation of the clusters or lead to cluster agglomeration.

4.3.2.9 X-ray Photoelectron Spectroscopy (XPS)

XPS provides indications of the oxidation states of the metals, but alone they are far from definitive. The technique is most valuable when characterizing changes in metal oxidation states. XPS can also be useful for determining whether the metal is inside the cages or present on the external surface. If it is outside the cages, then surface enrichment in metal is observed and the ratio of the intensity of an XPS line of a metal to that of a framework atom (e.g. Si) is larger than the corresponding ratio when the clusters are in the cages. [167]

4.3.2.10 Other Physical Techniques

Other techniques expected to be of value with the appropriate metals (such as iron containing samples) are Mössbauer effect [168] and ESR [169] spectroscopies and magnetic susceptibility [170] measurements.

4.3.2.11 Summary

Zeolite encaged metal clusters are among the simplest and best characterized supported metal clusters. They provide opportunities for demonstrating the power of certain characterization techniques including EXAFS spectroscopy, high resolution transmission electron microscopy, and metal NMR spectroscopy. What sets the zeolite supported clusters apart from others is their isolation in cages and their near uniformity of structure; however, even the best characterized of these materials are not quite uniform in structure, and it is difficult to determine what other species are present. It appears that the most nearly uniform encaged clusters are those prepared by the decarbonylation of encaged metal carbonyl clusters prepared by ship-in-a-bottle syntheses.

4.3.3 Reactivities

Although the structures of supported metal clusters cannot be determined as precisely as the structures of metal clusters in the crystalline state, EXAFS and RED structural data show that the metal skeletons of encaged metal clusters are not rigid pieces of bulk metal. Adsorbates can modify the interatomic distances and the lattice symmetry in the same way as ligand shells control the geometry of molecular clusters, and they may induce reactions that lead to massive structural changes. The following sections summarize the important reactions of metal clusters in zeolite cages.

4.3.3.1 Redispersion of Metal Clusters by Oxidative Fragmentation

CO has often been used as a convenient probe for investigating the electronic structures of metal oxide and zeolite supported metals. However, CO is not simply an innocent probe. For example, oxidative fragmentation occurs when CO reacts with Rh clusters on metal oxides or in zeolites. [171, 172] The breakup by CO of small clusters of metals that have some oxophilic character is consistent with the tendency of CO to fragment carbonyl clusters of these metals in solution. [31]

Bergeret et al. [79] observed that upon reaction of CO with 10 Å Rh clusters encaged in Y zeolite at room temperature, total destruction of the metal cluster occurred, as evidenced by the disappearance of Rh–Rh distances in the radial electron distribution and by the appearance of absorption bands for $Rh(I)(CO)_2$ species in the infrared spectra. Gelin et al. [173] observed the same phenomenon upon the adsorption of CO on 10 Å iridium clusters encaged in NaY zeolite. They found that at low surface coverage, CO was merely adsorbed onto the Ir surface but as the coverage increased, $Ir(I)(CO)_2$ species formed. The authors ruled out the possibility of zeolitic protons as the oxidizing agent, and, in fact, CO was identified as the oxidizing agent. The authors postulated that owing to the heat evolved during CO adsorption, CO was subsequently dissociated, which caused partial oxidation of the metallic iridium to the monovalent state. Further coordination of CO molecules resulted in the formation of monovalent iridium dicarbonyl species. The details of the chemistry remain to be elucidated.

Cluster destruction was also observed to result from NO adsorption on Pd aggregates in Y zeolite at room temperature. [174] Other examples are summarized in Table **4-10**.

4.3.3.2 Agglomeration of Metal Clusters in the Presence of CO

Adsorption of CO on supported metal clusters does not always lead to cluster disruption. Instead, adsorption of CO on noble metal clusters (e.g. Pd, Pt) may lead to cluster agglomeration both in zeolites and on metal oxides. [47, 180] Zhang et al. [47] reported that small Pd clusters could be formed within the supercages of zeolite Y after calcination at 500 °C and subsequent reduction at 350 °C. EXAFS data used to characterize the clusters indicated an average coordination number of 4, which was taken as evidence for the presence of Pd clusters having a nuclearity of 6. When CO was adsorbed on these clusters, the average coordination number increased from 4 to 6 and led the authors to suggest that larger Pd clusters (assumed to have a nuclearity of 13, as discussed in Section 4.2.1.2) formed in the supercages. It was further postulated that CO adsorption weakens not only the Pd–Pd bonds in the clusters but also the interactions of the noble metal with the zeolite.

Table 4-10. Redispersion of encaged metal clusters.

Reactant	Metal/ zeolite	Cluster size	Temperature, °C	Resultant species	Characterization methods	Refs.
O_2	Pd/Y	10 Å	200	Pd^{2+}	X-ray techniques	[176]
	Ag/Y	200 Å	357	Ag^+	XRD	[177]
CO	Rh/Y	10 Å	27	$Rh^I(CO)_2$	IR, XRD	[79]
	Ir/Y	10 Å	23	$Ir^I(CO)_2$	IR, calorimetric and volumetric study	[174b]
NO	Pd/Y	20 Å	27	Pd^{2+}	XRD, ESR, IR, mass spectroscopy	[174a]
	Ni/mordenite	80 Å	200	Ni^{2+}	UV, magnetic measurements	[175]
Cl_2	Pd/Y	–	–	–	–	[178]
	Pt/L, ZSM-5	10–100 Å	357	Pt^{4+}	TEM, XRD, TPR	[179]
NH_3	Pd/Y	–	–	–	–	[178]

4.3.3.3 Reversible Recarbonylation of Molecular Metal Clusters

NaY zeolite encaged clusters, which had been formed by decarbonylation of zeolite encaged $[Ir_6(CO)_{16}]$ and inferred from EXAFS spectra to be predominantly Ir_6, could be reversibly recarbonylated. [3, 5, 181] The recarbonylation at liquid nitrogen temperature initially gave CO adsorbed on the clusters, but as the temperature was raised to about –10 °C, mononuclear iridium carbonyl species were formed, followed by the formation of $[Ir_4(CO)_{12}]$ at about 40 °C, and finally the isomer of $[Ir_6(CO)_{16}]$ with edge bridging ligands at about 125 °C (Fig. **4-12**).

As stated above (Section 4.3.3.1), the more usual effect of adsorbing CO on supported group 8 metal clusters is to cause significant morphological changes, including cluster fragmentation or aggregation. These processes are usually irreversible, with the products generally being multinuclear in character and nonuniform. The details of the surface chemistry depend upon the metal cluster size and the hydroxyl group content of the support surface.

Thus, it appears that the iridium/zeolite combination in our initial example may be nearly optimal for allowing the reversible decarbonylation and recarbonylation reactions which retain the cluster nuclearity. It may also be that iridium is not so noble as to undergo agglomeration and not so oxophilic as to

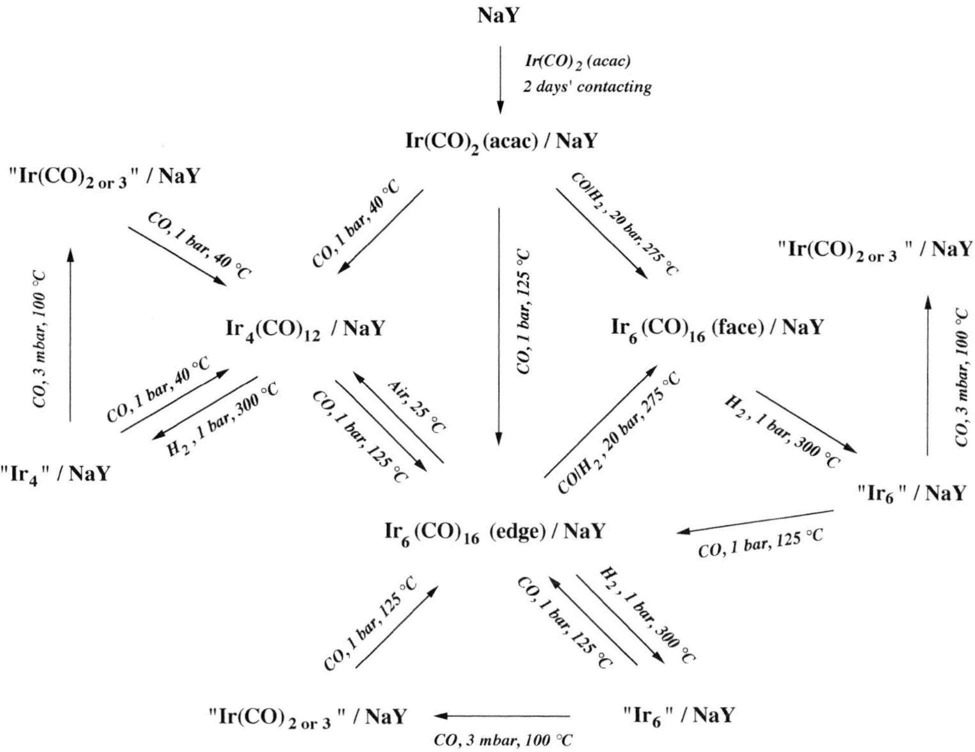

Figure 4-12. Schematic representation of the decarbonylation and recarbonylation of iridium clusters in the cages of NaY zeolite. [181] The decarbonylation-recarbonylation process is not simply reversible. The recarbonylation proceeds through mononuclear intermediates, and $[Ir_4(CO)_{12}]$ is an intermediate in the formation of $[Ir_6(CO)_{16}]$.

undergo fragmentation. Furthermore, the zeolite cages may help to stabilize the hexairidium clusters.

Similar chemistry appears to occur with rhodium clusters in Y zeolite, but the products of the recarbonylation are not simply the initial $[Rh_6(CO)_{16}]$, but include a not well characterized cluster that appears to have edge bridging CO ligands. [65]

4.3.3.4 Adduct Formation of Metal Particles and Protons

The reduction of metal ions by H_2 can form protons which bind to oxygen and become part of OH groups in the zeolite. Sheu et al. [46] postulated that a fraction of the protons created during palladium ion reduction remain attached to Pd clusters and form positively charged metal-proton adducts. Evidence supporting this suggestion comes from the observation that when small Pd clusters chemisorbed CO and coalesced to give larger particles, the intensity of the OH band increased.

4.3.3.5 Effects of Sulfur on Encaged Pt Clusters

Extremely small and nearly uniform Pt clusters have been formed in zeolite L by reduction of Pt salts. Zeolite L has a structure with straight channels. The pores are unidimensional and consist of ellipsoidal cages with internal dimensions of 4.8 × 12.4 × 10.7 Å connected by windows with diameters of 7.4 Å. Pt clusters in the BaKL zeolite have been characterized by infrared spectroscopy (with adsorbed CO), [182] TEM (Fig. **4-13**), [182–184] chemisorption of H_2, [184, 185] and EXAFS spectroscopy. [184, 185] The clusters are characterized by a first shell Pt–Pt coordination number of 3.7, which indicates clusters of 5 or 6 Pt atoms in the cages. This result is consistent with the electron micrograph illustrated in Figure **4-13**. These materials are highly selective catalysts for the dehydrocyclization of *n*-hexane to give benzene. [186] The catalyst is highly sensitive to poisoning by sulfur-containing compounds. The effect of sulfur has been elucidated with EXAFS spectroscopy. [185] Sulfur interacts strongly with the Pt, thereby weakening its interactions with the zeolite cages by forming Pt–S bonds, and causing the first shell Pt–Pt coordination number, as determined by EXAFS

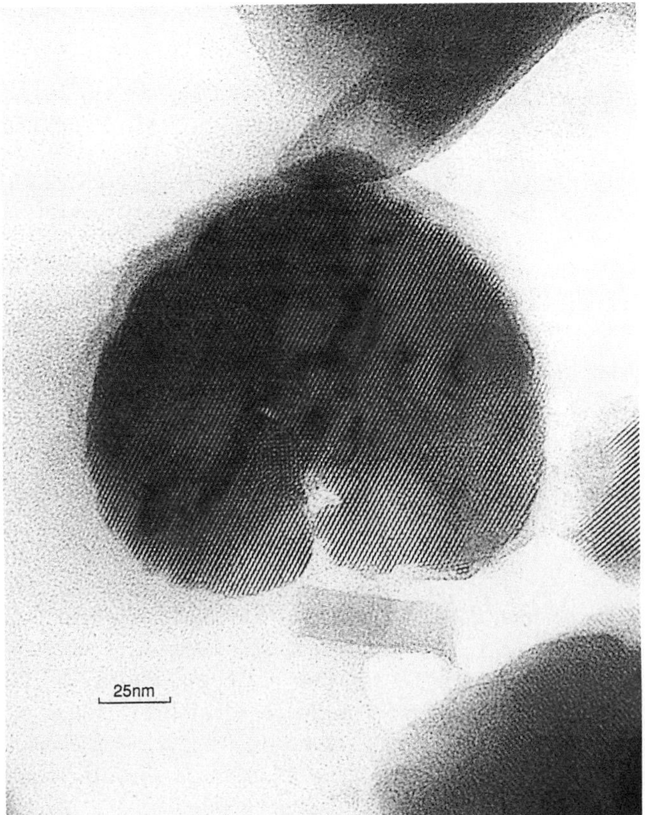

Figure 4-13. Transmission electron micrograph of Pt clusters in zeolite L. [184] Virtually all the clusters are in the zeolite pores. They have an average of 5 or 6 atoms each, as determined by EXAFS spectroscopy. The material shown here is a highly selective catalyst for the conversion of *n*-hexane into benzene and H_2. Catalysts of this type will soon be applied industrially for this reaction. Reproduced from *Catalysis Letters* with permission of J. C. Baltzer AG.

spectroscopy, to increase from 3.7 to 5.5. [185] Evidently, clusters of this size can block the entrances of the straight pores and thus make the Pt in the interior inaccessible to reactants, causing a marked loss in catalytic activity. [185, 187]

4.3.4 Modification of Electronic Structure

Zeolite supported metal clusters are important catalysts. The zeolite matrix not only imposes steric constraints on reacting molecules (shape selective catalysis) and provides acidic sites, but it also apparently affects the electronic properties of the encaged metal clusters.

The modification of the electronic and catalytic properties of metal clusters in zeolites has been reviewed by Gallezot. [108] He showed that in addition to changes in the electronic structure of metal clusters as a result of intrinsic size effects, the electronic structure of a metal cluster can also be modified by the cluster environment, for example, by electron transfer from the metal clusters to electron acceptor sites in the zeolite lattice. He considered Lewis acid sites, Brønsted acid sites, and multivalent cations to be potential electron acceptors. Electron deficient clusters appear to be resistant to poisoning by sulfur, which is advantageous in catalytic applications. The issues of cluster size effects and electronic effects are complex and continue to be vigorously debated.

4.3.5 Catalysis

Metals catalyze many reactions of practical importance, and the most common form of an industrial metal catalyst is a supported metal consisting of clusters or particles on a high surface area support such as a metal oxide. The support may also be a zeolite as in the Pd supported in faujasite catalyst used for hydrocracking. For the most part, zeolite supported metals are structurally complex and are beyond the scope of this chapter. They have been reviewed elsewhere. [105]

Zeolites containing proton donor groups are immensely important industrial catalysts, primarily for hydrocarbon conversions. The most notable characteristic of zeolite catalysts is their shape selectivity. There are several classes of shape selectivity: 1) *reactant shape selectivity* – some reactants will fit into the zeolite pores where they can react, but other reactants which are too large to fit cannot react to any significant degree, 2) *transition state shape selectivity* – reaction by which the products are formed from transition states of relatively small size can take place in cages where other potential reactions are excluded because the transition states are too large to be formed in the cages, and 3) *product shape selectivity* – some products are small enough to diffuse out through the apertures whereas others are too large to fit and are further converted into smaller products in the cages.

Examples of shape selective catalysis for each class are well illustrated with the acidic zeolites, but there are far fewer examples of such catalysis with metal containing zeolites. Reactant shape selectivity is illustrated by the early work of

Weisz et al., [188] who used Pt clusters in zeolite A for selective catalysis of the hydrogenation (or the oxidation) of straight chain hydrocarbons in mixtures with branched hydrocarbons.

Evidence for transition state shape selectivity in metal containing zeolites is less persuasive. One example involves Pd clusters in the zeolite ZSM-5, which is a medium pore zeolite (Table **4-1**) with a set of straight parallel pores intersected by a set of zig-zag pores. The acidic form of the Pd containing zeolite is selective for the conversion of acetone to methylisobutyl ketone. This reaction occurs by an Aldol condensation which is catalyzed by the acidic groups followed by a hydrogenation which is catalyzed by the metal clusters. [189] Pt in zeolite Y, however, is much less selective, presumably because the larger cages do not impose a shape selectivity.

Among the most remarkable zeolite supported metal catalysts is the Pt in KL zeolite (or BaL or BaKL) mentioned above (Section 4.3.3.5). It is now being applied industrially for the production of benzene from *n*-hexane. There is much debate over the origin of the selectivity. Some authors have attributed it to the geometry of the narrow pores [190, 191] and others have attributed it to the small size of the Pt clusters. [192]

Boudart and others have presented a case for electronic effects in catalysis by metal clusters in zeolite Y. [193] They found that the catalytic activity of approximately 10 Å Pt clusters for neopentane isomerization and hydrogenolysis is 40 times higher than that of Pt clusters supported on Al_2O_3 or on SiO_2. This increase in activity was attributed to an electron transfer from the Pt clusters to the zeolite support. It was suggested that, as a result of this transfer, Pt behaves more like Ir, its near neighbor in the periodic table and a metal exceeding Pt in its catalytic activity for the neopentane conversion by more than two orders of magnitude.

4.4 Synthesis and Characterization of Bimetallic Clusters in Zeolites

Supported bimetallic clusters have found important applications as catalysts, such as for naphtha reforming. [137] For example, Re-Pt clusters supported on η-Al_2O_3 have properties similar to those of Pt clusters but exhibit superior operational stability.

One of the major challenges to understanding bimetallic catalysts is to obtain supported clusters with relatively well understood and uniform compositions. Supported bimetallic catalysts are usually prepared by deposition of metal salts onto a support followed by calcination and reduction by H_2 at elevated temperature. This approach leads to nonuniformities in cluster composition and structure and even physical separation of the two metals. These difficulties have prompted research on catalysts prepared from molecular metal clusters since they provide precursors of well defined stoichiometry and structure.

4.4.1 Decomposition of Metal Carbonyls on Reduced Metals

Zeolite supported bimetallic clusters can be prepared by adsorption and decomposition of an organometallic complex on metal clusters in a zeolite. The metal carbonyl should be small enough to pass through the zeolite pores and undergo thermal decarbonylation at relatively low temperatures. This technique has been used to prepare Pt-Mo clusters supported in zeolites. [194] Here, Pt clusters of about 10 Å in diameter are first obtained by the usual ion exchange and reduction process, and then $Mo(CO)_6$ vapors are adsorbed and decomposed under reducing atmospheres. The catalysts have been characterized by techniques including H_2 and CO chemisorption, TEM, and TPD [194, 195] as well as EXAFS and RED measurements. [196] The Mo atoms were concluded to be deposited on the Pt atoms in an fcc array and inhibited chemisorption on the Pt. Similarly, Sachtler et al. [197, 198] prepared Pt-Re clusters in NaY zeolite cages by sorption and decomposition of $[Re_2(CO)_{10}]$ on Pt clusters.

These bimetallic clusters may be selective catalysts. For example, Gallezot et al. [195, 199, 200] investigated the catalytic properties of Pt-Mo clusters in NaY zeolites for alkane hydrogenolysis and CO hydrogenation. It was suggested that H_2 is preferentially dissociated by the Pt atoms, whereas hydrocarbon or CO are preferentially adsorbed on the Mo atoms. Since the two elements are associated in the same cluster, the Pt atom provides the dissociated hydrogen required to hydrogenate the fragments adsorbed on the Mo atom.

4.4.2 Decomposition of Bimetallic Carbonyl Clusters

The difficulties in preparing supported bimetallic clusters have prompted research on clusters prepared from molecular bimetallic clusters since they are precursors with well defined compositions and structures. Although the cluster framework may break up during thermal treatment or catalysis, the two metal components may still remain in intimate contact with each other in the cages. It is necessary that the precursors be small enough to enter the zeolite pores, and it has been suggested to be advantageous if one of the metals forms bonds with the OH groups of the zeolite in order to anchor and stabilize the resultant intrazeolitic species. The latter requirement can be achieved by choosing one of the metal constituents to be an oxophilic metal (e.g. Mo, Re). The anchoring is intended to minimize agglomeration of the metals under reducing conditions.

Borvornwattananont et al. [201–203] introduced heterobimetallic compounds into zeolite Y in attempts to prepare zeolite supported catalysts. The complexes were anchored to the zeolite framework by the oxophilic metal, and reactions were suggested to be catalyzed by the second metal center. The chemistry of the anchoring process, and the thermal stability and reactivity of $[Me_3SnMn(CO)_5]$ and two germylene complexes $[Cl_2(THF)GeM(CO)_5]$ (M = Mo, W) in zeolite Y have been investigated. The heterometallic bonds were maintained up to about 100 °C, but broken at higher temperatures. The complexes interacted with the zeolite framework through the Ge and Sn containing moieties respectively.

4.4.3 Reduction of Ion Exchanged Metal Complexes

The idea behind this method is the ion exchange of two different metal ions in the faujasite cages which is then followed by reduction. Random collisions of the metal atoms in the supercages will then presumably lead to the formation of bimetallic clusters. It has been postulated that bimetallic clusters are favored if the precursor of the less reducible metal is mobile in the zeolite cages. For instance, if Pt is the noble (easily reduced) component of a bimetallic cluster, it will be reduced first, and the subsequent reduction of the other mobile precursor in contact with a Pt would then be catalyzed by the Pt. Hence, the temperature required to reduce the less reducible element is markedly lower than that required for its reduction in the absence of Pt.

If both metal ions are uniformly distributed in the supercages, bimetallic clusters of homogeneous composition could presumably be generated in the supercages after reduction. A typical example is the formation of PtRh clusters in NaY zeolite. [204] On the other hand, if one kind of metal ion (Pt^{2+}) is located in the supercages and the other (Cu^{2+}) in the sodalite cages, then PtCu clusters can be formed by the transfer of Cu atoms from the sodalite cages to Pt clusters formed in the supercages. [205] The results obtained with the Pt-Cu/NaY catalyst are in agreement with the catalyzed reduction model. [206] The resultant clusters consist of a Pt core and a Cu mantle.

4.4.4 Reductive Carbonylation of Mixed Metal Complexes

Ichikawa et al. [207, 208] prepared mixtures of rhodium and iridium carbonyls in zeolite Y by reductive carbonylation of their salts. The carbonyls were suggested to be $[Rh_xIr_{6-x}(CO)_{16}]$ on the basis of infrared, EXAFS, and other results. The complexity of the mixture made it difficult to determine which clusters were present and in what proportions. The cluster mixture was decarbonylated, giving clusters that were suggested to be RhIr clusters of various compositions.

Similar experiments have been done to prepare FeRh clusters from such precursors as $[Rh_4(CO)_{12}]$, which were stated to have been formed in the cages of zeolite Y. [209] There is a question about the location and composition of the resultant clusters, in part because $[Rh_4(CO)_{12}]$ is simply too large to enter the cages.

4.5 Metal Ion Clusters in Zeolites

Ionic clusters in zeolite cages are stabilized by their ionic interactions with the zeolite framework. They have metal–metal distances intermediate between those in metals and those in metal oxides. The subject of ionic clusters in zeolites is dealt with only briefly here, with sodium clusters chosen as an example.

Rabo and Kasai [210, 211] were the first to show that zeolites can stabilize neutral as well as charged sodium clusters. The ESR spectrum of the material resulting from the interaction of Na vapor with the zeolite NaX was attributed to an Na_6^{5+} paramagnetic unit. Harrison et al. [212] showed that the charged clusters are located in the sodalite cages of zeolites A, X, and Y and can only be obtained when an excess of sodium is introduced into the zeolite.

An alternate method for the preparation of such clusters [213] is based on the impregnation of a dehydrated zeolite with alcoholic solutions of sodium azide, followed by the controlled decomposition of the azide. The ionic sodium clusters were shown by ESR spectroscopy to be Na_4^{3+} and are formed in the zeolite pores. The clusters exhibit catalytic properties for both isomerization and hydrogenation reactions of alkenes and alkynes.

These examples are included only to give a brief introduction to this subject, and many others are not cited here.

4.6 Semiconductor Clusters in Zeolite Cages

Nanometer sized semiconductor clusters, expected to have properties different from those of molecular and bulk semiconductors of the same composition, represent a new class of materials. Interest in their preparation and potential applications as photocatalysts and device components used in quantum electronics and nonlinear optics is growing rapidly. Isolated, so-called nanophase semiconductors can be considered as zero and one dimensional quantum dots and quantum wires. Their electronic, optical, and photochemical properties change with cluster size. Wider electronic band gaps and new absorption maxima in the electronic spectra have been observed as the size of these materials decreases and have been interpreted as quantum size effects. For example, as the dimensions of the semiconductor particle are reduced, a shift to higher energy in the absorption spectrum relative to that of the bulk is generally observed.

The synthesis of well defined semiconducting clusters having a homogeneous morphology and size distribution would be a big step forward towards understanding the physical origins of quantum size effects. It would be advantageous if the quantum dot clusters could be fixed, isolated, and regulated in such a way that every quantum dot were embedded in a material with a band gap larger than that of the cluster so as to enable quantum inclusions. The use of lithographic and epitactic procedures to construct quantum dots seems to be a promising method. [214, 215] Alternatively, the use of chemically synthesized clusters as quantum dots is also a promising direction in research.

In principle, many chemical approaches exist for the preparation of quantum dot clusters. Classical preparative methods include wet colloidal techniques, growth in dielectric glass matrices, and growth in polymers. In these materials, however, cluster or particle sizes are usually nonuniform, and agglomeration of individual particles often occurs. Thus, the cluster size effects on the optical and other properties are obscured.

Chemical synthetic techniques are attractive alternatives, and the zeolites have an advantage over the above mentioned matrices because they offer the opportunity for the inclusion of small clusters with nearly uniform size distributions. By using zeolites with various architectures, the spatial arrangement of the clusters can, in principle, be controlled and their growth constrained to the cage dimensions. The zeolite structures may also provide stabilizing media. Furthermore, by filling all the internal pores of a zeolite crystal, one has the potential of forming novel three dimensional so-called supercluster structures that are built from individual clusters interconnected through the cage windows. When different zeolites are used as the templates, superclusters exhibiting a variety of structures and electronic properties could be prepared.

Several reviews [216–219] summarize the current status of the materials synthesis and understanding of the size dependent electronic and photochemical properties of zeolite supported semiconductors.

The characterization of encaged semiconductor clusters presents a challenge because, at least so far, there is a lack of encaged semiconductor structures which have molecular cluster analogues. Thus, the advantages found for the characterization of metal carbonyl clusters (Section 4.2.2) are lacking, and the structures and uniformity of the encaged species are not as well understood as for encaged metal carbonyl (and metal) clusters.

4.6.1 Synthesis

4.6.1.1 Ion exchange

Precursor cations can be introduced into the host zeolite framework and subsequently converted into the semiconductor phase by chemical or thermal treatment. The ion exchange process can lead to changes in the cation locations, as mentioned above. This process must be carefully controlled in order to give materials that can be consistently reproduced.

4.6.1.2 Organometallic (Metallorganic) Chemical Vapor Deposition (MOCVD)

MOCVD is traditionally a "planar" surface technique used for the deposition of very thin layers of materials from the vapor phase. This technique offers a wide versatility in the synthesis of clusters in zeolites. The advantage of this technique, in comparison with the ion exchange technique, is that the location of the clusters in the zeolites can be easily controlled by the choice of organometallic precursors (*e.g.* a metal carbonyl or metal alkyl), such that they can be sterically restricted to large channels or cages. For example, with zeolite Y, one can select precursors that are too large to enter into the sodalite cages but small enough to pass into the supercages, that is, the precursors may be introduced specifically into the supercages of the host. Once confined in the supercages, the precursors can be transformed into clusters by thermal or photochemical dissociation or by other chemical reactions.

Intrazeolitic MOCVD chemistry is relatively new. A synthetic chemistry for the preparation of encaged clusters from metal alkyl precursors (alkyl is often methyl) has been suggested by a few researchers. [220, 221] In the first step, a vapor phase precursor is introduced into the zeolite supercages. One of the methyl groups can then react with a Brønsted acid site, thereby producing CH_4 and a concurrent anchoring of the organometallic precursor to the zeolite internal surface (Eq. 4.16, Z represents the zeolite framework).

$$\text{Z–OH} + (\text{CH}_3)_n\text{M} \rightarrow \text{Z–O–M}(\text{CH}_3)_{n-1} + \text{CH}_4 \qquad (4.16)$$

It has been suggested [221] that the reactivity of the precursor towards the Brønsted acid sites is determined by the nucleophilicity of the methyl groups, whereby the initial replacement of an electron releasing methyl group with an electron withdrawing oxide ligand of the zeolite inhibits further reaction by decreasing the nucleophilicity of a second methyl group. Thus, the maximum loading of the organometallic precursor into the zeolite is determined by the number of Brønsted acid sites, which is ultimately controlled by the Si/Al ratio of the zeolite framework and by the ion exchange.

The semiconductor clusters can then be formed in the zeolite by a ship-in-a-bottle synthesis (Eqs. 4.17 and 4.18). Here, the precursor anchored in the supercages of zeolite is allowed to react with an appropriate reagent (*e.g.* H_2S, H_2Se, or PH_3) at a slightly elevated temperature whereby the remaining methyl groups of the anchored precursor are released as CH_4 and a labile species is formed (Eq. 4.17, X = S, Se, or P). Condensation reactions then follow (Eq. 4.18).

$$\text{Z–O–M–}(\text{CH}_3)_n + n\text{H}_2\text{X} \rightarrow \text{Z–O–M–}(\text{XH})_n + n\text{CH}_4 \qquad (4.17)$$

$$\text{Z–O–M–XH} + \text{X–H–M–O–Z} \rightarrow \text{Z–O–M–X–M–O–Z} \qquad (4.18)$$

The reaction is driven to completion when excess H_2S or H_2Se is removed, leaving the semiconductor clusters in the cages.

4.6.2 Characterization

4.6.2.1 Metal sulfides

Metal sulfides have been incorporated into zeolites X, Y, and A. [222, 223] It was concluded that superclusters made up of quantized semiconductors can be formed by completely filling the internal zeolite cavities to the percolation threshold. The supercluster is created by the three dimensional interconnection of intrazeolite semiconductors which yield materials with properties intermediate between those of discrete molecular semiconductor clusters and those of bulk semiconductors.

For example, CdS clusters were synthesized in the cages of zeolites X, Y, and A by allowing the calcined Cd^{2+} form of each zeolite to react with flowing H_2S at 100 °C. [222, 223] EXAFS spectra and Reitveld refinement of the powder X-ray diffraction results provided evidence that led the authors to suggest that discrete, cubic $(CdS)_4$ clusters (interpenetrating Cd and S tetrahedra) having Cd–S bond lengths of 2.47 Å were formed strictly inside the smaller sodalite cages (Fig. **4-14**). At low Cd^{2+} loadings these clusters were isolated, and showed an optical absorption maximum around 280 nm. As the loading within the zeolite increased, the wavelength of the optical absorption shifted to about 350 nm. It was concluded that the discrete $(CdS)_4$ clusters began to electronically interconnect by quantum tunnelling to create a supercluster structure. The supercluster was suggested to be formed by a percolative process with a threshold of 4 ± 1 wt% CdS.

These superclusters can be viewed as three dimensional arrays of mutually interacting clusters having structures constrained to the zeolite pore structure. Different spatial arrangements of these clusters might be prepared by using different zeolites as templates. The effort has been extended to other semiconductors as guests and sodalite as the host. [224] The cluster size dependence of the shift of the optical absorption spectra or of the increase in the band gap of the semiconductor materials (*e.g.* ZnS, CdS, and PbS) is now well documented and is reasonably well described by the electron-hole in a sphere model reported by Brus. [225]

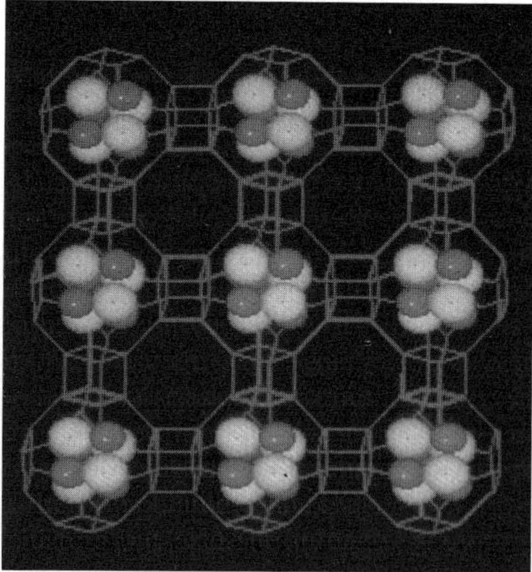

Figure 4-14. Structures proposed for $(CdS)_4$ clusters in the sodalite cages of zeolite A. [217] Reproduced from *Science* with permission of the American Association for the Advancement of Science (see also color plates).

4.6.2.2 Metal oxides

Ozin, Özkar, et al. [226–229] reported a new approach to synthesizing transition metal oxide structures in molecular sieve hosts. $[W(CO)_6]$ located within the supercages of NaY zeolite was photochemically dissociated in the presence of O_2. This method provides a mild and quantitative synthetic pathway to tungsten(VI) oxide moieties which are encapsulated within the supercages of the zeolite Y (Eq. 4.19).

$$n[W(CO)_6] \;/\; NaY + 9/2\; n\; O_2 \rightarrow n[WO_3]/NaY + 6\; nCO_2 \quad (4.19)$$

After the photooxidation reaction, one half of the supercage volume in NaY is freed so that subsequent impregnation/photooxidation of the precursor can be carried out in a stepwise fashion. These materials were thermally treated under vacuum at 300 and 400 °C to give encaged structures inferred to have the composition WO_{3-x}, with $x = 0.5$ and 1 respectively. This reduction process can be reversed by treatment of the sample at 300 °C in an O_2 atmosphere.

EXAFS, TEM, X-ray diffraction, XPS, EPR, ^{29}Si MAS-NMR, ^{23}Na NMR, ultraviolet-visible, and gravimetry data, among others, indicated that at a loading of one W atom per supercage, only WO_3 and $WO_{2.5}$ moieties were present in the form of the μ-dioxo bridged W_2O_6 and μ-oxo bridged W_2O_5 dimers, respectively. Each of these is anchored to the zeolite cages through terminal oxo-tungsten bonds to two of the four site II Na^+ ions located in a supercage (Fig. **4-15**). The optical absorption peak of the clusters (3.5 eV) was shifted to higher energy relative to that of bulk WO_3.

As the WO_3 loading was increased to the maximum value of about four units per supercage, the absorption peak abruptly shifted to a limiting value of 3.3 eV. Infrared, MAS-NMR, XPS, and XRD results suggested the formation of cubane-like $(WO_3)_4$ tetramers within the supercages. These clusters were inferred to be anchored via terminal oxo-tungsten bonds to all four site II Na^+ ions in the supercage.

Since the oxygen content of the clusters can be quantitatively adjusted by a thermal vacuum induced reversible reductive-elimination oxidative-addition of dioxygen, the electronic properties of these oxide clusters can be easily manipulated as a result of their facile redox interconvertibility. Ozin et al. [230] discovered how to alter local electrostatic fields experienced by the zeolite encaged tungsten oxide by varying the ionic potential of the constituent supercage M^+ ions across the alkali metal series. This method provides the first opportunity to fine tune the band gap of tungsten oxide clusters.

The characterization of these materials is challenging, and it is especially difficult to determine the uniformity of the encaged species. The reported characterizations are based on data obtained with almost all the available physical methods, and the proposed structures seem to be consistent with the observations. In view of the potential value of these materials, it would seem to be appropriate to pursue the characterization even more deeply, such as by synthesis and full structural characterization of molecular analogues of the proposed

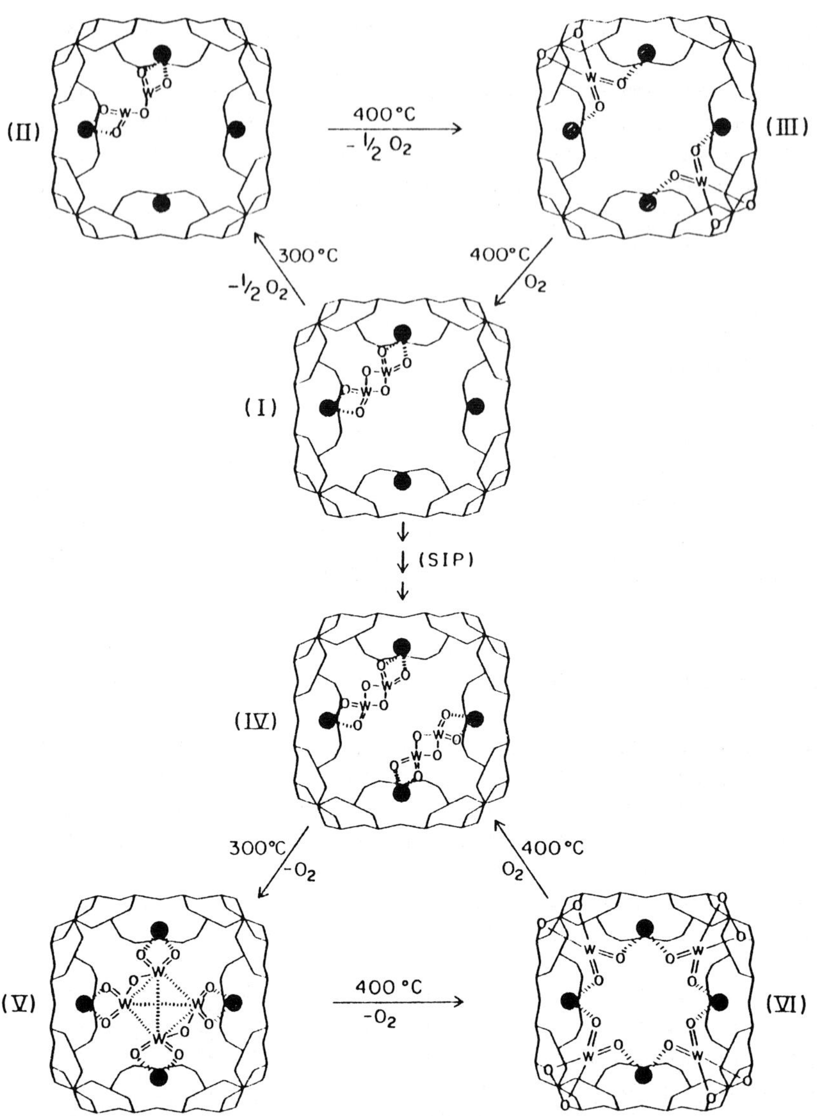

Figure 4-15. Structures suggested on the basis of EXAFS spectroscopy and other physical methods for tungsten oxide clusters in the sodalite cages of zeolite NaY. [229] The structures correspond to half loaded and fully loaded cages. Reproduced from *Journal of the American Chemical Society* with permission of the American Chemical Society.

encaged structures so as to allow comparison of their spectra with those of the encaged species.

There is a rapidly developing literature describing other metal oxides in zeolites, with TiO_2 being well represented. [231–234] The characterization of these materials is largely based on ultraviolet-visible spectroscopy, and the structures are less than well understood. Whether the encaged species are clusters and whether they are uniform in structure remains to be determined. Similar statements can also be made about the materials described in the following sections (4.6.2.3 and 4.6.2.4).

4.6.2.3 Metal selenides

Cadmium selenide is of interest for photosensitized electron transfer reactions used for solar energy conversion and photocatalysis. Small ensembles of CdSe have been synthesized within the cages of zeolite Y by ion exchange with Cd(II) and subsequent treatment with H_2Se. [235] The presence of CdSe clusters with Cd–Se bond lengths of 2.60 Å was inferred from EXAFS data measured at both the Cd and Se absorption edges.

4.6.2.4 Selenide Chains and Rings

Elemental semiconductor clusters encaged in zeolites provide a valuable opportunity for gaining a fundamental understanding of semiconductor clusters because stoichiometry is not a concern in the synthesis. Selenium is of interest because it has an intermediate electrical conductivity and a negative coefficient of resistivity in the dark; hence it is markedly photoconductive. It has uses in, for example, photoelectric devices and xerography. When Se is sorbed into a molecular sieve, it gives markedly different optical absorption spectra from those of the bulk material.

Solid Se exists in several forms: a trigonal form containing helical chains, three monoclinic forms containing Se_8 rings, a rhombohedral form containing Se_6 rings, and possibly an orthorhombic form containing Se_7 rings. [236] The thermodynamically favored form of Se under ambient conditions is trigonal Se, which consists of helical chains having three Se atoms per repeat unit. Upon sorption into a molecular sieve matrix, any of these forms (or a combination) are possible. In zeolite A, the optical spectra suggest that rings are formed, similar to those of the monoclinic Se_8 bulk form. [237] Solid state ^{77}Se MAS-NMR spectra showed that several different Se allotropes exist in the larger cages of zeolite Y. [238] ^{77}Se MAS-NMR and EXAFS results showed that selenium exists only in a helical chain conformation similar to that in the trigonal allotrope in $AlPO_4$-5 and mordenite. [237–239]

4.7 Prospects: Clusters in Fullerenes

Since the discovery of its stable soccerball structure [240] and of the arc-discharge preparation method for synthesizing large quantities of C_{60} (buckminsterfullerene, Fig. **4-16**) and other fullerenes, [241] research on these materials has been progressing at a furious pace. The chemistry of the fullerenes has been recently summarized, [242, 243] and new reviews are rapidly making even rather recent ones out of date.

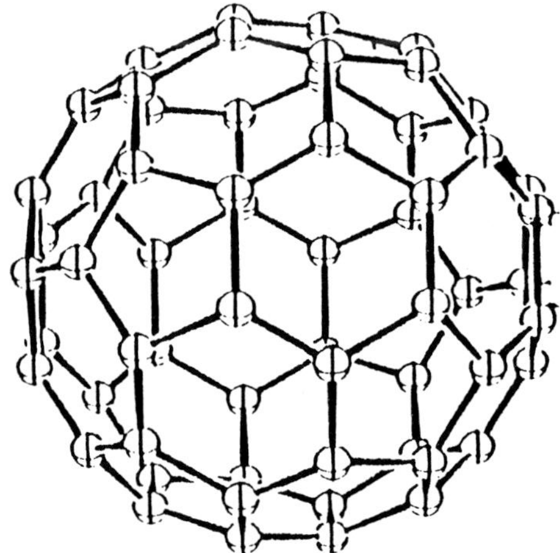

Figure 4-16. Schematic representation of the structure of fullerene-60. [242] Reproduced from *Chemical Reviews* with permission of the American Chemical Society.

Of great interest was the discovery of high temperature superconductor properties in C_{60} doped with combinations of alkali and other metals. [244, 245] These can be viewed as insertion compounds with the metal center occupying octahedral and tetrahedral interstitial sites in the fcc lattice of C_{60}. Charge transfer occurs from the metal center to a vacant conduction band formed from interaction between the C_{60} and metal valence orbitals in the close packed lattice. Since the fullerenes have stable hollow cage structures, there is a possibility that clusters might be synthesized and stabilized within them. These hypothetical materials might be expected to have novel properties, including superconductivity.

4.8 Summary and Evaluation

The molecular scale cages of crystalline aluminosilicate zeolites (molecular sieves) are uniform, microscopic, and solvent-like media which can be used for synthesis of new materials, often called nanoclusters. Clusters synthesized in the cages include metal carbonyl, metal, and metal oxides. The best characterized of these are those clusters that are also well known in molecular chemistry, that is, such metal carbonyls as $[Ir_4(CO)_{12}]$, $[Rh_6(CO)_{16}]$, and $[Ir_6(CO)_{16}]$. These are formed in the supercages of NaY zeolite. Anions such as $[HIr_4(CO)_{11}]^-$ and $[Ir_6(CO)_{15}]^{2-}$ have been formed in the cages of NaX zeolite.

The presence of such clusters in zeolites has been well documented by a variety of spectroscopic methods, but there is still no evidence that clusters have been made in pure form in any zeolite. The highest reported yields are roughly 80 %. Most of characterizations of the species in the cages were not quantitative.

For the most part, the chemistry involved in synthesis of metal carbonyl clusters in cages is similar to that occurring in solutions and on the surfaces of amorphous metal oxides. Syntheses that take place in neutral solvents and on nearly neutral surfaces also take place within the cages of nearly neutral zeolites such as NaY. The synthesis of cluster anions like $[HIr_4(CO)_{11}]^-$ takes place in basic zeolites such as NaX, in much the same way as it does in basic solutions and on basic surfaces.

Cluster precursors must be small enough to enter the cages, and the most common are cations which can be introduced at exchange sites in the cages or such organometallic precursors as $[Ir(CO)_2(acac)]$. The cluster formation is often described as a ship-in-a-bottle synthesis because the clusters formed (typically by reaction of the precursor with CO) may just fit within the supercage (about 12 Å in diameter in zeolites X and Y) yet be too small to leave by passing through the windows connecting the cages (about 7.4 Å in diameter in these zeolites).

The confinement affects what can be formed in the cages. For example, when iridium carbonyl clusters are synthesized either on the surface of MgO or in solution, $[HIr_4(CO)_{11}]^-$ is formed first, followed by $[Ir_8(CO)_{22}]^{2-}$ (which is a dimer of the tetrairidium cluster), and finally $[Ir_6(CO)_{15}]^{2-}$. In contrast, within the supercages of NaX zeolite, $[HIr_4(CO)_{11}]^-$ and then $[Ir_6(CO)_{15}]^{2-}$ are formed, but $[Ir_8(CO)_{22}]^{2-}$ is not observed, presumably because it is too large to fit in the cage and therefore cannot exist there.

Clusters of many metals have been formed in zeolites, but most of the materials are structurally nonuniform and less than well characterized. The most nearly uniform and thoroughly characterized of the encaged metal clusters are those formed by decarbonylation of metal carbonyl clusters such as $[Ir_4(CO)_{12}]$ and $[Ir_6(CO)_{16}]$. These can be decarbonylated reversibly and apparently without significant changes in the metal frame, giving encaged Ir_4 and Ir_6 in high yields. These materials are among the simplest supported metals and are attractive candidates for theoretical investigations.

The chemistry associated with the recarbonylation of these iridium clusters in zeolite NaY is partially understood. When the decarbonylated clusters are brought in contact with CO at liquid nitrogen temperature, iridium carbonyl clusters of poorly defined structure are formed. As the sample is then heated in the presence of CO, the clusters fragment to give mononuclear complexes (iridium subcarbonyls) which are converted into $[Ir_4(CO)_{12}]$ and then into $[Ir_6(CO)_{16}]$.

These are but a few of the reactions of metal carbonyl and metal clusters in zeolite cages. The patterns of reactivity are broadly consistent with those observed for analogous species in solution and on metal oxide surfaces, except for the constraints imposed by the narrow pores and small cages.

Metal carbonyl clusters (*e.g.* of Rh) can be used as precursors to form catalysts for reactions involving CO, including the water gas shift reaction, alkene hydroformylation, and CO hydrogenation. Although the catalysts exhibit some unusual selectivities, such as in the hydrogenation of CO to give non Schulz-Flory distributions in their hydrocarbon products, they are not highly active relative to some of the more conventional catalysts. The species in the zeolites that are formed from the cluster precursors and which are the actual catalytically active species have not yet been elucidated.

Metal clusters in zeolites are catalysts for a number of reactions, including alkene hydrogenation and alkane hydrocracking. The former is an example of shape selective catalysis, whereby straight chain alkenes can enter the zeolite pores and react but branched alkenes cannot enter and so do not substantially react. The latter have been applied commercially. Pt clusters in the zeolites KL and BaKL are remarkably selective catalysts for the dehydrocyclization of *n*-hexane to give benzene, and they are now applied commercially. The origin of the selectivity is still not fully understood, but it may be primarily a consequence of the smallness of the Pt clusters, which consist of only about 5 or 6 atoms on average, as determined by EXAFS spectroscopy, H_2 chemisorption, and electron microscopy.

Semiconductor materials such as CdS and tungsten oxides have been formed in the cages of zeolites. The confined semiconductor clusters are evidently stabilized through encapsulation and coordination to the host framework. The structures of the zeolite cages offer the opportunity to control the cluster geometry and size distribution, making it possible, in principle, to synthesize quantum semiconductor superclusters that is, those connectal through the pores of the zeolite structure. Cluster electronic properties can be altered by varying the dielectric and charge properties of the nanoporous host.

This subject is still in an early stage of development, and there is much to be learned about the synthetic chemistry and the precise structures of the intrazeolitic materials. It would be helpful to prepare clusters with known molecular analogues for comparison.

The subject of clusters in cages is restricted to clusters in zeolites. There are many opportunities for expansion of this class of materials to include new clusters and new cages. The materials offer good opportunities for answering fundamental scientific questions, such as how the size and composition of clusters affect their properties (e.g. optical and catalytic properties). Comparisons between the clusters in zeolites and those in cages of other materials, such as aluminophosphate molecular sieves, might be useful in determining the influence of electric field gradients and framework composition on the properties of confined clusters. Some newly discovered molecular sieves have cavities larger than the supercages of the faujasites and might offer opportunities for the preparation of encaged clusters larger than those already known. Examples of these include 1) VPI-5, [246] an aluminophosphate molecular sieve with an essentially circular 18-T-atom ring and a one dimensional channel structure with a diameter of about 12–13 Å, 2) cloverite, [247] a gallophosphate molecular sieve with a 20-T-atom cloverleaf shaped entrance window to a 29–30 Å diameter supercage, and 3) JDF-20, [248] an aluminophosphate molecular sieve with an elliptical 20-T-atom ring and a one-dimensional channel structure. These molecular sieves, however, lack the stability and the large cage, small window structures of some of the zeolites.

It is likely that the greatest motivation for continuing work on encaged nanoclusters is the prospects they offer for applications. Catalytic applications are already a reality, as mentioned above, and these seem to be only a beginning, since the catalytic research that has been done with metal containing molecular sieves pales in comparison with that which has been done with acidic molecular sieves. We believe that there are good prospects for shape selective catalysts in reactions catalyzed by metal complexes and metal clusters in zeolites (e.g. in the pharmaceutical industry). There are good indications that confined clusters of semiconductor materials in cages will allow the tuning of their optical and electronic properties. Such materials could facilitate progress in the application of much smaller devices than are now available, for example, as optical switches in computers.

Acknowledgments

We thank Professor M. E. Davis of the California Institute of Technology for helpful comments. The work was supported by the National Science Foundation (grant CTS-9300754).

References

[1] D. F. Shriver, H. D. Kaesz, R. D. Adams, *The Chemistry of Metal Cluster Complexes,* VCH Publishers, New York, **1990**.
[2] G. Ertl, in *Metal Clusters in Catalysis* (Eds.: B. C. Gates, L. Guczi, H. Knözinger), Elsevier, Amsterdam, **1986**, p. 577.
[3] S. Kawi, B. C. Gates, *J. Chem. Soc., Chem. Commun.* **1991**, 904.
[4] S. Kawi, B. C. Gates, *J. Chem. Soc., Chem. Commun.* **1992**, 702.
[5] S. Kawi, J. R. Chang, B. C. Gates, *J. Am. Chem. Soc.* **1993**, *115*, 4830.
[6] B. C. Gates, *Angew. Chem. Int. Ed. Engl.* **1993**, *32*, 228.
[7] G. A. Ozin, C. Gil, *Chem. Rev.* **1989**, *89*, 1749.
[8] M. L. Steigerwald, L. E. Brus, *Acc. Chem. Res.* **1990**, *23*, 183, and references cited therein.
[9] N. Herron, Y. Wang, H. Eckert, *J. Am. Chem. Soc.* **1990**, *112*, 1322.
[10] Y. Wang, N. Herron, *J. Phys. Chem.* **1991**, *95*, 525, and references cited therein.
[11] I. Moriguchi, I. Tanaka, Y. Teraoka, S. Kagawa, *J. Chem. Soc. Chem. Commun.* **1991**, 1401.
[12] E. S. Smotkin, C. Lee, A. J. Bard, A. Campion, M. A. Foxe, T. E. Mallouk, S. E. Weber, *Chem. Phys. Lett.* **1988**, *152*, 265.
[13] D. W. Breck, *Zeolite Molecular Sieves,* Wiley, New York, **1974**.
[14] R. M. Barrer, *Zeolites and Clay Minerals as Sorbents and Molecular Sieves,* Academic Press, London, **1978**.

[15] R. Szostak, *Molecular Sieves,* Van Nostrand Reinhold, New York, **1988**.
[16] (a) P. J. Grobet, W. J. Mortier, E. F. Vansant, G. Schulz-Ekloff, *Innovation in Zeolite Materials Science,* Elsevier, Amsterdam, **1988**.
(b) H. van Bekkum, E. M. Flanigen, J. C. Jansen, *Introduction to Zeolite Science and Practice,* Elsevier, Amsterdam, **1991**.
(c) A. Dyer, *An Introduction to Zeolite Molecular Sieves,* John Wiley & Sons, New York, **1988**.
[17] (a) E. G. Derouane, F. Lemos, C. Naccache, F. R. Ribeiro, *Zeolite Microporous Solids: Synthesis, Structure, and Reactivity,* Kluwer Academic, Dordrecht, **1992**.
(b) M. L. Occelli, H. Robson, *Molecular Sieves,* Van Nostrand Reinhold, New York, **1992**.
[18] E. G. Derouane, in *Surface Organometallic Chemistry: Molecular Approaches to Surface Catalysis* (Ed.: J. M. Basset et al.), Kluwer Academic Publishers, Dordrecht, **1988**, p. 299.
[19] S. T. Wilson, in *Introduction to Zeolite Science and Practice* (Ed.: H. van Bekkum, E. M. Flanigen, J. C. Jansen), Elsevier, Amsterdam, **1991**, p. 137.
[20] J. Chen, G. Sankar, J. M. Thomas, R. Xu, G. N. Greaves, D. Waller, *Chem. Mater.* **1992**, *4*, 1373, and references cited therein.
[21] G. A. Ozin, A. Kuperman, A. Stein, *Angew. Chem. Int. Ed. Engl.* **1989**, *28*, 359, and references cited therein.
[22] G. D. Stucky, J. MacDougall, *Science* **1990**, *247*, 669, and references cited therein.
[23] Y. Wang, N. Herron, *Res. Chem. Interm.* **1991**, *15*, 17.
[24] J. E. MacDougall, H. Eckert, G. D. Stucky, N. Herron, Y. Wang, K. Moller, T. Bein, D. Cox, *J. Am. Chem. Soc.* **1989**, *111*, 8006.
[25] P. Gallezot, G. Coudurier, M. Primet, B. Imelik, in *Molecular Sieves-II* (Ed.: J. R. Katzer), American Chemical Society, Washington, DC, USA, **1977**, p. 144.
[26] R. L. Schneider, R. F. Howe, K. L. Watters, *Inorg. Chem.* **1984**, *23*, 4600.
[27] M. C. Connaway, B. E. Hanson, *Inorg. Chem.* **1986**, *25*, 1445.
[28] P. Gelen, Y. Ben Tarrit, C. Naccache, *J. Catal.* **1979**, *59*, 357.
[29] R. Shannon, J. C. Vedrine, C. Naccache, F. Lefebvre, *J. Catal.* **1985**, *88*, 43.
[30] P. Gelin, F. Lefebvre, B. Elleuch, C. Naccache, Y. Ben Taarit in *Intrazeolite Chemistry* (Eds.: G. D. Stucky, F. G. Dwyer) *ACS Symp. Ser.* **1983**, *218*, p. 455.
[31] P. Chini, G. Longoni, V. G. Albano, *Adv. Organomet. Chem.* **1976**, *14*, 285, and references cited therein.
[32] G. L. Geoffroy, in *Metal Clusters in Catalysis* (Eds.: B. C. Gates, L. Guczi, H. Knözinger), Elsevier, Amsterdam, **1986**, p. 1.
[33] A. K. Smith, F. Hugues, A. Theolier, J. M. Basset, R. Ugo, G. M. Zanderighi, J. L. Bilhou, V. Bilhou-Bougnol, W. F. Graydon, *Inorg. Chem.* **1979**, *18*, 3104.
[34] J. M. Basset, A. Theolier, D. Commereuc, Y. Chauvin, *J. Organomet. Chem.* **1985**, *279*, 147.
[35] H. H. Lamb, B. C. Gates, *J. Am. Chem. Soc.* **1986**, *108*, 81.
[36] H. H. Lamb, A. S. Fung, P. A. Tooley, J. Puga, T. R. Krause, M. J. Kelley, B. C. Gates, *J. Am. Chem. Soc.* **1989**, *111*, 8367.
[37] S. Kawi, B. C. Gates, *Inorg. Chem.* **1992**, *31*, 2939.
[38] E. Mantovani, N. Palladino, A. Zanobi, *J. Mol. Catal.* **1977–1978**, *3*, 385.
[39] G. Bergeret, P. Gallezot, F. Lefebvre, *Stud. Surf. Sci. Catal.* **1986**, *28*, 401.
[40] A. DeMallmann, D. Barthomeuf, *Catal. Lett.* **1990**, *5*, 293.
[41] G.-J. Li, T. Fujimoto, A. Fukuoka, M. Ichikawa, *J. Chem. Soc. Chem. Commun.* **1991**, 1337.
[42] G.-J. Li, T. Fujimoto, A. Fukuoka, M. Ichikawa, *Catal. Lett.* **1992**, *12*, 171.
[43] J. R. Chang, Z. Xu, S. K. Purnell, B. C. Gates, *J. Mol. Catal.* **1993**, *80*, 49.
[44] L.-L. Sheu, H. Knözinger, W. M. H. Sachtler, *Catal. Lett.* **1989**, *2*, 129.
[45] L.-L. Sheu, H. Knözinger, W. M. H. Sachtler, *J. Mol. Catal.* **1989**, *57*, 61.
[46] L.-L. Sheu, H. Knözinger, W. M. H. Sachtler, *J. Am. Chem. Soc.* **1989**, *111*, 8125.

[47] Z. Zhang, H. Chen, L. L. Sheu, W. M. H. Sachtler, *J. Catal.* **1991**, *127*, 213.
[48] Z. Zhang, H. Chen, W. M. H. Sachtler, *J. Chem. Soc. Faraday Trans.* **1991**, *87*, 1413.
[49] Z. Zhang, W. M. H. Sachtler, *J. Mol. Catal.* **1991**, *67*, 349.
[50] M. Iwamoto, S. Nakamura, H. Kusano, S. Kagawa, *J. Phys. Chem.* **1986**, *90*, 5244.
[51] D. Ballivet-Tkatchenko, G. Coudurier, *Inorg. Chem.* **1979**, *18*, 558.
[52] P.-L. Zhou, S. D. Maloney, B. C. Gates, *J. Catal.* **1991**, *129*, 315.
[53] S. Maloney, P.-L. Zhou, M. J. Kelley, B. C. Gates, *J. Phys. Chem.* **1991**, *95*, 5409.
[54] L. Malatesta, G. Caglio, M. Angoletta, *Inorg. Synth.* **1972**, *13*, 95.
[55] R. Della Pergola, L. Garlaschelli, S. Martinengo, *J. Organomet. Chem.* **1987**, *331*, 271.
[56] F. P. Pruchnik, K. Wajda-Hermanowicz, M. Koralewicz, *J. Organomet. Chem.* **1990**, *384*, 381.
[57] L. Garlaschelli, S. Martinengo, P. L. Bellon, F. Demartin, M. Manassero, M. Y. Chiang, C.-Y. Wei, R. Bau, *J. Am. Chem. Soc.* **1984**, *106*, 6664.
[58] R. E. Stevens, P. C. C. Lin, W. L. Gladfelter, *J. Organomet. Chem.* **1985**, *287*, 133.
[59] S. Kawi, J.-R. Chang, B. C. Gates, *J. Phys. Chem.* **1993**, *97*, 5375.
[60] S. Kawi, J.-R. Chang, B. C. Gates, *J. Catal.* **1993**, *142*, 585.
[61] R. Bau, M. Y. Chiang, K. Wei, K. Garlaschelli, S. Martinengo, T. F. Koetzle, *Inorg. Chem.* **1984**, *23*, 4758.
[62] F. Demartin, M. Manassero, L. Garlaschelli, C. Raimondi, S. Martinengo, F. Canziani, *J. Chem. Soc. Chem. Commun.* **1981**, 528.
[63] F. Demartin, M. Manassero, M. Sansoni, L. Garlaschelli, S. Martinengo, *J. Chem. Soc. Chem. Commun.* **1980**, 903.
[64] E. J. Rode, M. E. Davis, B. E. Hanson, *J. Catal.* **1985**, *96*, 574.
[65] L.-F. Rao, A. Fukuoka, N. Kosugi, H. Kuroda, M. Ichikawa, *J. Phys. Chem.* **1990**, *94*, 5317.
[66] S. Kawi, B. C. Gates, *Catal. Lett.* **1991**, *10*, 263.
[67] M. Angoletta, L. Malatesta, G. L. Caglio, *J. Organomet. Chem.* **1975**, *94*, 99.
[68] Z. Zhang, A. P. Cavalcanti, W. M. H. Sachtler, *Catal. Lett.* **1992**, *12*, 157.
[69] A. Brenner, in *Metal Clusters* (Ed.: M. Moskovits), Wiley, New York, **1986**, p. 249.
[70] H. Knözinger, in *Metal Clusters in Catalysis* (Eds.: B. C. Gates, L. Guczi, H. Knözinger), Elsevier, Amsterdam, **1986**, p. 187, and references cited therein.
[71] F. Lefebvre, P. Gelin, C. Naccache, Y. Ben Taarit, in *Proceedings of The Sixth International Zeolite Conference* (Eds.: D. Olson, A. Bisio), Butterworth, Guilford, **1984**, p. 435.
[72] D. C. Koningsberger, R. Prins, *X-ray Absorption: Principles, Applications, Techniques of EXAFS, SEXAFS, and XANES* Wiley, New York, **1988**.
[73] B. C. Gates, D. C. Koningsberger, *CHEMTECH* **1992**, 301, and references cited therein.
[74] F. B. M. Duivenvoorden, D. C. Koningsberger, Y. S. Uh, B. C. Gates, *J. Am. Chem. Soc.* **1986**, *108*, 6254.
[75] J. R. Chang, L. U. Gron, A. Honji, K. M. Sanchez, B. C. Gates, *J. Phys. Chem.* **1991**, *95*, 9944.
[76] B. K. Teo, *EXAFS: Basic Principles and Data Analysis,* Springer-Verlag, New York, **1986**, p. 183.
[77] S. D. Maloney, M. J. Kelley, D. C. Koningsberger, B. C. Gates, *J. Phys. Chem.* **1991**, *95*, 9406.
[78] S. Kawi, J. R. Chang, B. C. Gates, *J. Phys. Chem.,* in press.
[79] G. Bergeret, P. Gallezot, P. Gelin, Y. Ben Taarit, F. Lefebvre, C. Naccache, R. D. Shannon, *J. Catal.* **1987**, *104*, 279.
[80] P. Chini, *Pure Appl. Chem.* **1970**, *23*, 489.
[81] D. B. W. Yawney, F. G. A. Stone, *J. Chem. Soc. (A)* **1969**, 502.
[82] D. R. Tyler, R. A. Levenson, H. B. Gray, *J. Am. Chem. Soc.* **1978**, *100*, 7888.
[83] P. Chini, S. Martinengo, *J. Chem. Soc. Chem. Commun.* **1969**, 1092.

[84] S. Abdo, R. F. Howe, *J. Phys. Chem.* **1983**, *87*, 1722.
[85] M. Primet, J. C. Vedrine, C. Naccache, *J. Mol. Catal.* **1978**, *4*, 411.
[86] M. Deeba, B. J. Streusand, G. L. Schrader, B. C. Gates, *J. Catal.* **1981**, *69*, 218.
[87] G. Mestl, N. D. Triantafillou, H. Knözinger, B. C. Gates, *J. Phys. Chem.* **1993**, *97*, 666.
[88] M. D. Ward, J. Schwartz, *Organometallics* **1982**, *1*, 1030.
[89] T. N. Huang, J. Schwartz, *J. Am. Chem. Soc.* **1982**, *104*, 5244.
[90] H. H. Lamb, Ph. D. Thesis, University of Delaware, **1986**, chapter 1.
[91] D. F. Shriver, *J. Organomet. Chem.* **1975**, *94*, 259.
[92] C. P. Horwitz, D. F. Shriver, *Advan. Organomet. Chem.* **1984**, *23*, 219.
[93] C. Tessier-Young, F. Correa, D. Ploch, R. L. Burwell, Jr., D. F. Shriver, *Organometallics* **1983**, *2*, 898.
[94] J. P. Collman, R. G. Finke, P. L. Matlock, R. Wahren, R. G. Komoto, J. I. Brauman, *J. Am. Chem. Soc.* **1978**, *100*, 1119.
[95] M. Y. Darensbourg, *Prog. Inorg. Chem.* **1985**, *33*, 221.
[96] J. P. Scott, J. R. Budge, A. L. Rheingold, B. C. Gates, *J. Am. Chem. Soc.* **1987**, *109*, 7736.
[97] D. M. Vandenberg, T. C. Choy, P. C. Ford, *J. Organomet. Chem.* **1989**, *366*, 257.
[98] H. H. Lamb, B. C. Gates, H. Knözinger, *Angew. Chem., Int. Ed. Engl.* **1988**, *27*, 1127.
[99] F. Hugues, J. M. Basset, Y. Ben Taarit, A. Choplin, M. Primet, D. Rojas, A. K. Smith, *J. Am. Chem. Soc.* **1982**, *104*, 7020.
[100] N. Homs, A. Choplin, P. Ramirez de la Piscina, L. Huang, E. Garbowski, R. Sanchez-Delgado, J.-M. Basset, *Inorg. Chem.* **1988**, *27*, 4030.
[101] E. L. Muetterties, T. N. Rhodin, E. Band, C. F. Brucker, W. R. Pretzer, *Chem. Rev.* **1979**, *79*, 91.
[102] E. L. Muetterties, *J. Organomet. Chem.* **1980**, *200*, 170.
[103] E. L. Muetterties, J. Stein, *Chem. Rev.* **1979**, *79*, 479.
[104] W. O. Haag, N. Y. Chen, in *Catalyst Design* (Ed.: L. Louis Hegedus), John Wiley & Sons, New York, **1987**, p. 163.
[105] P. A. Jacobs, in *Metal Clusters in Catalysis* (Eds.: B. C. Gates, L. Guczi, H. Knözinger), Elsevier, Amsterdam, **1986**, p. 357, and references cited therein.
[106] (a) P. B. Weisz, V. J. Frilette, R. W. Maatman, E. B. Mower, *J. Catal.* **1962**, *1*, 307.
(b) M. E. Davis, C. Saldariaga, J. A. Rossin, *J. Catal.* **1987**, *103*, 520.
[107] M. Ichikawa, *Adv. Catal.* **1992**, *38*, 283, and references cited therein.
[108] P. Gallezot, in *Metal Clusters* (Ed.: M. Moskovits), Wiley-Interscience, New York, USA, **1986**, p. 219.
[109] L. Markó, A. Vizi-Orosz in *Metal Clusters in Catalysis* (Eds.: B. C. Gates, L. Guczi, H. Knözinger), Elsevier, Amsterdam, **1986**, p. 89, and references cited therein.
[110] W. L. Gladfelter, K. J. Roesselet in *The Chemistry of Metal Cluster Complexes* (Eds.: D. F. Shriver, H. D. Kaesz, R. D. Adams), VCH, New York, **1990**, p. 329, and references cited therein.
[111] (a) J. L. Vidal, W. E. Walker, *Inorg. Chem.* **1987**, *109*, 4518.
(b) B. T. Heaton, J. Jonas, T. Eguchi, G. A. Hoffman, *J. Chem. Soc. Chem. Commun.* **1981**, 331.
(c) B. T. Heaton, L. Strona, J. Jonas, T. Eguchi, G. A. Hoffman, *J. Chem. Soc. Dalton Trans.* **1982**, 1159.
[112] H. Knözinger, B. C. Gates, in *Metal Clusters in Catalysis* (Eds.: B. C. Gates, L. Guczi, H. Knözinger), Elsevier, Amsterdam, **1986**, p. 531, and references cited therein.
[113] (a) M. Ichikawa, *Bull. Soc. Chem. Jpn.* **1978**, *51*, 2268.
(b) ibid, **1978**, *51*, 2273.
[114] P. A. Jacobs, H. H. Nijs, *J. Catal.* **1980**, *65*, 328.
[115] D. Fraenkel, B. C. Gates, *J. Am. Chem. Soc.* **1980**, *102*, 2478.
[116] L. F. Nazar, G. A. Ozin, F. Hugues, J. Godber, D. Rancourt, *J. Mol. Catal.* **1983**, *21*, 313.

[117] T. J. Lee, B. C. Gates, *Catal. Lett.* **1991**, *8*, 15.
[118] T. J. Lee, B. C. Gates, *J. Mol. Catal.* **1992**, *71*, 335.
[119] L. Markó, in *Aspects of Homogeneous Catalysis, vol 2* (Ed.: R. Ugo), Riedel Publishing, Dordrecht, **1974**.
[120] D. M. Vandenberg, Ph. D. Thesis, University of California at Santa Barbara, **1986**.
[121] R.-J. Wang, T. Fujimoto, T. Shido, M. Ichikawa, *J. Chem. Soc. Chem. Commun.* **1992**, 962.
[122] R. Kubo, A. Kanabata, S. Kobayaski, *Am. Rev. Mater. Sci.* **1984**, *14*, 49.
[123] P. Gallezot, G. Bergeret, in *Metal Microstructures in Zeolites* (Eds.: P. A. Jacobs, N. I. Jaeger, P. Jiru, G. Schultz-Ekloff), Elsevier, Amsterdam, **1982**, p. 167.
[124] P. A. Jacobs, in *Metal Microstructures in Zeolites* (Eds.: P. A. Jacobs, N. I. Jaeger, P. Jiru, G. Schultz-Ekloff), Elsevier, Amsterdam, **1982**, p. 71.
[125] P. Gallezot, *Catal. Rev.-Sci. Eng.* **1979**, *20*, 121.
[126] T. Joh, M. Taakachi, *Mem. Inst. Ind. Res. Osaka Univ.* **1989**, *46*, 37.
[127] J. B. Nagy, M. van Eenoo, E. G. Derouane, *J. Catal.* **1979**, *58*, 230.
[128] D. Exner, N. I. Jaeger, K. Möller, R. Nowak, H. Schrubbers, G. Schulz-Ekloff, P. Ryder, in *Metal Microstructures in Zeolites* (Ed.: P. A. Jacobs, N. I. Jaeger, P. Jiru, G. Schulz-Ekloff) Elsevier Scientific, Amsterdam, **1982**, p. 205.
[129] R. A. Dalla Betta, M. Boudart, in *Proceedings of the Fifth International Congress on Catalysis* (Ed.: J. W. Hightower), North Holland, Amsterdam, **1973**, p. 1329.
[130] M. Suzuki, K. Tsutsumi, H. Takahashi, Y. Saito, *Zeolites* **1989**, *9*, 98.
[131] D. Oliver, M. Richard, L. Bonneviot, M. Che, in *Growth and Properties of Metal Clusters* (Ed.: J. Bourdon), Elsevier, Amsterdam, **1980**, p. 165.
[132] H. J. Jiang, M. S. Tzou, W. M. H. Sachtler, *Catal. Lett.* **1988**, *1*, 99.
[133] J. A. Rabo, C. L. Angell, P. H. Kasai, V. Schomaker, *Disc. Faraday Soc.* **1966**, 329.
[134] F. Schmidt, W. Gunsser, J. Adolph, *A. C. S. Symp. Ser.* **1977**, *40*, 291.
[135] P. B. Malla, P. Ravindranathan, S. Komarneni, E. Breval, R. Roy, *Mat. Res. Soc. Symp. Proc.* **1991**, *233*, 207.
[136] L. F. Nazar, G. A. Ozin, F. Hugues, J. Godber, D. Rancourt, *J. Mol. Catal.* **1983**, *21*, 313.
[137] J. H. Sinfelt, *Bimetallic Catalysts: Discoveries, Concepts, and Applications,* Exxon Monograph, Wiley, New York, **1983**.
[138] S. H. Song, Y. Kim, K. Seff, *J. Phys. Chem.* **1991**, *95*, 9919.
[139] S. H. Song, U. S. Kim, Y. Kim, K. Seff, *J. Phys. Chem.* **1992**, *96*, 10937.
[140] A. Renouprez, C. Hoang Van, P. Compagnon, *J. Catal.* **1974**, *34*, 411.
[141] P. Gallezot, A. I. Bienenstock, M. Boudart, *Nouv. J. Chim.* **1978**, *2*, 263.
[142] P. Gallezot, A. Alarcon-Diaz, J. A. Dalmon, A. J. Renouprez, B. Imelik, *J. Catal.* **1975**, *39*, 334.
[143] A. Renouprez, C. Hoang-Van, P. A. Compagnon, *J. Catal.* **1974**, *34*, 411.
[144] P. Ratnasamy, A. J. Leonard, *Catal. Rev.-Sci. Eng.* **1972**, *6*, 293.
[145] M. Boudart, M. G. Samant, R. Ryoo, *Ultramicroscopy* **1986**, *20*, 125.
[146] M. Pan, J. M. Cowley, I. Y. Chan, *Catal. Lett.* **1990**, *5*, 1.
[147] L. Persaud, A. J. Bard, A. Campion, M. A. Foxe, T. E. Mallouk, S. E. Webber, J. M. White, *Inorg. Chem.* **1987**, *26*, 3825.
[148] L.-C. de Menorval, J. P. Fraissard, T. Ito, *J. Chem. Soc. Faraday Trans. 1* **1982**, *78*, 403.
[149] R. Ryoo, S. J. Cho, C. Pak, J.-G. Kim, S.-K. Ihm, J. Y. Lee, *J. Am. Chem. Soc.* **1992**, *114*, 76.
[150] T. Ito, J. Fraissard, *J. Chem. Phys.* **1982**, *76*, 5225.
[151] T. Ito, J. Fraissard, *J. Chem. Soc., Faraday Trans. 1* **1987**, *83*, 451.
[152] A. Gedeon, J. L. Bonardet, T. Ito, J. Fraissard, *J. Phys. Chem.* **1989**, *93*, 2563.
[153] N. Bansal, C. Dybowski, *J. Phys. Chem.* **1988**, *92*, 2333.
[154] O. B. Yang, S. I. Woo, and R. Ryoo, *J. Catal.* **1990**, *123*, 375.
[155] R. Ryoo, C. Pak, B. F. Chmelka, *Zeolites* **1990**, *10*, 790.

[156] A. N. Murty, M. Seamster, A. N. Thorpe, R. T. Obermyer, V. U. S. Rao, *J. Appl. Phys.* **1990**, *67*, 5847.

[157] Z. Zhang, Y. D. Zhang, W. A. Hines, J. I. Budnick, W. M. H. Sachtler, *J. Am. Chem. Soc.* **1992**, *114*, 4843.

[158] W. Ravenek, A. P. J. Jansen, R. A. van Santen, *J. Phys. Chem.* **1989**, *93*, 6445.

[159] F. B. M. van Zon, S. D. Maloney, B. C. Gates, D. C. Koningsberger, *J. Am. Chem. Soc.*, in press.

[160] M. G. Samant, M. Boudart, *J. Phys. Chem.* **1991**, *95*, 4070.

[161] E. M. Flanigen, H. A. Szymanski, H. Khatami, *Advan. Chem. Ser.* **1971**, *101*, 201.

[162] T. Stock, D. Dombrowski, J. Fruwert, H. Ratajczak, *J. Chem. Soc. Faraday I* **1983**, *79*, 2773.

[163] C. A. Ozin, M. D. Baker, J. M. Parnis, *Angew. Chem.* **1983**, *95*, 813.

[164] (a) M. D. Baker, J. Godber, G. A. Ozin, *J. Phys. Chem.* **1985**, *89*, 2299.
(b) G. A. Ozin, M. D. Baker, J. Godber, C. J. Gil, *J. Phys. Chem.* **1989**, *93*, 2899.

[165] P. Gallezot, in *Proceedings of the Sixth International Zeolite Conference* (Ed.: D. Olson, A. Bisio), Butterworths, Guildford, **1984**, p. 352.

[166] M. Primet, J. M. Basset, E. Garbowski, M. V. Mathieu, *J. Am. Chem. Soc.* **1975**, *97*, 3655.

[167] K. H. Minachev, G. V. Antoshin, E. S. Shpiro, Y. Usifov, in *Proceedings of the Sixth International Congress on Catalysis,* **1977**, p. 621.

[168] E. Schmidt, W. Gunsser, J. Adolph, in *Molecular Sieves-II* (Ed.: J. R. Katzer), American Chemical Society, Washington, DC, USA, **1977**, p. 291.

[169] J. C. Vedrine, M. Dufaux, C. Naccache, B. Imelik, *J. Chem. Soc. Faraday Trans. 1* **1978**, *74*, 440.

[170] M. F. Guilleux, D. Delafosse, G. A. Martin, J. A. Dalmon, *J. Chem. Soc. Faraday Trans. 1* **1979**, *75*, 165.

[171] H. F. J. Van't Blik, J. B. A. D. Van Zon, T. Huizinga, J. C. Vis, D. C. Koningsberger, R. Prins, *J. Amer. Chem. Soc.* **1985**, *107*, 3139.

[172] F. Solymosi, M. Pasztor, *J. Phys. Chem.* **1985**, *89*, 47.

[173] P. Gelin, A. Auroux, Y. Ben Taarit, P. C. Gravelle, *Appl. Catal.* **1989**, *46*, 227.

[174] (a) M. Che, J. F. Dutel, P. Gallezot, M. Primet, *J. Phys. Chem.* **1976**, *80*, 2371.
(b) P. Gelin, A. Auroux, Y. Ben Taarit, P. C. Gravelle, *Appl. Catal.* **1989**, *46*, 227.

[175] E. D. Garbowski, C. Mirodatos, M. Primet, M. V. Mathieu, *J. Chem. Phys.* **1983**, *87*, 303.

[176] G. Bergeret, P. Gallezot, B. Imelik, *J. Chem. Phys.* **1981**, *85*, 411.

[177] H. K. Beyer, P. A. Jacobs, J. B. Uytterhoeven, *J. Chem. Soc. Faraday 1* **1976**, *72*, 674.

[178] O. Feeley, W. M. H. Sachtler, *Appl. Catal.* **1990**, *67*, 141.

[179] O. Feeley, W. M. H. Sachtler, *Appl. Catal.* **1991**, *75*, 93.

[180] J. R. Chang, D. C. Koningsberger, B. C. Gates, *J. Am. Chem. Soc.* **1992**, *114*, 6460.

[181] T. Beutel, S. Kawi, S. K. Purnell, H. Knözinger, B. C. Gates, *J. Phys. Chem.* **1993**, *97*, 7284.

[182] C. Besoukhanova, J. Guidot, D. Barthomeuf, *J. Chem. Soc., Faraday Trans. I* **1981**, *77*, 1595.

[183] S. B. Rice, J. Y. Koo, M. M. Disko, M. M. J. Treacy, *Ultramicroscopy* **1990**, *34*, 108.

[184] M. Vaarkamp, J. V. Grondelle, J. T. Miller, D. J. Sajkowski, F. S. Modica, G. S. Lane, B. C. Gates, D. C. Koningsberger, *Catal. Lett.* **1990**, *6*, 369.

[185] M. Vaarkamp, J. T. Miller, F. S. Modica, G. S. Lane, D. C. Koningsberger, *J. Catal.* **1992**, *138*, 675.

[186] J. R. Bernard, in *Proceedings Fifth International Conference Zeolites* (Ed.: L. V. C. Rees), Heyden, London, **1980**, p. 686.

[187] G. B. McVicker, J. L. Kao, J. J. Ziemiak, W. E. Gates, J. L. Robbins, M. M. J. Treacy, S. B. Rice, T. H. Vanderspurt, V. R. Cross, A. K. Ghosh, *J. Catal.* **1993**, *139*, 48.

[188] (a) P. B. Weisz, V. J. Frilette, *J. Phys. Chem.* **1960**, *64*, 382.
(b) P. B. Weisz, V. J. Frilette, R. W. Maatman, E. B. Mower, *J. Catal.* **1962**, *1*, 307.

[189] T. J. Huang, W. O. Haag, U. S. Patent 4 339 606 1982, assigned to Mobil Oil Corp.
[190] E. G. Derouane, D. J. Vanderveken, *Appl. Catal.* **1988**, *45*, L15.
[191] S. J. Tauster, J. J. Steger, *J. Catal.* **1990**, *125*, 387.
[192] S. B. Hong, E. Mielczarski, M. E. Davis, *J. Catal.* **1992**, *134*, 349.
[193] M. Boudart, G. Djéga-Mariadassou, *Kinetics of Heterogeneous Catalytic Reactions,* Princeton University Press, Princeton, **1984**.
[194] T. M. Tri, J. P. Candy, P. Gallezot, J. Massardier, M. Primet, J. C. Vedrine, B. Imelik, *J. Catal.* **1983**, *79*, 396.
[195] T. M. Tri, J. Massardier, P. Gallezot, B. Imelik, *J. Mol. Catal.* **1984**, *25*, 151.
[196] M. Samant, G. Bergeret, G. Meitzner, P. Gallezot, M. Boudart, *J. Phys. Chem.* **1988**, *92*, 3542.
[197] C. M. Tsang, S. M. Augustine, J. B. Butt, W. M. H. Sachtler, *Appl. Catal.* **1989**, *46*, 45.
[198] C. Dossi, J. Schaefer, W. M. H. Sachtler, *J. Mol. Catal.* **1989**, *52*, 193.
[199] T. M. Tri, J. Massardier, P. Gallezot, B. Imelik, *J. Catal.* **1984**, *85*, 244.
[200] F. Borg, P. Gallezot, J. Massadier, V. Perrichon, *C_1 Mol. Chem.* **1986**, *1*, 397.
[201] A. Borvornwattananont, K. Moller, T. Bein, *J. Chem. Soc. Chem. Commun.* **1990**, 28.
[202] A. Borvornwattananont, K. Moller, T. Bein, *J. Phys. Chem.* **1992**, *96*, 6713.
[203] A. Borvornwattananont, T. Bein, *J. Phys. Chem.* **1992**, *96*, 9447.
[204] M. S. Tzou, K. Asakura, Y. Yamazaki, H. Kuroda, *Catal. Lett.* **1991**, *11*, 33.
[205] M. S. Tzou, M. Kusunoki, K. Asakura, H. Kuroda, G. Moretti, W. M. H. Sachtler, *J. Phys. Chem.* **1991**, *95*, 5210.
[206] G. Moretti, W. M. H. Sachtler, *J. Catal.* **1989**, *115*, 205.
[207] M. Ichikawa, L.-F. Rao, T. Ito, A. Fukuoka, *J. Chem. Soc. Faraday Disc.* **1989**, *87*, 232.
[208] M. Ichikawa, L.-F. Rao, T. Kimura, A. Fukuoka, *J. Mol. Catal.* **1990**, *62*, 15.
[209] A. Fukuoka, L.-F. Rao, N. Kosugi, H. Kuroda, M. Ichikawa, *Appl. Catal.* **1988**, *50*, 295.
[210] J. A. Rabo, C. L. Angell, P. H. Kasai, V. Schomacker, *Disc. Faraday Soc.* **1986**, *41*, 328.
[211] J. A. Rabo, P. H. Kasai, *Progr. Sol. St. Chem.* **1975**, *9*, 1.
[212] M. R. Harrison, P. P. Edwards, J. Klinowski, J. M. Thomas, *J. Sol. St. Chem.* **1984**, *54*, 330.
[213] L. R. M. Martens, P. J. Grobet, P. A. Jacobs, *Nature* **1985**, *315*, 568.
[214] K. Ploog, *Angew. Chem. Int. Ed. Engl.* **1988**, *27*, 593.
[215] I. P. Herman, *Chem. Rev.* **1989**, *89*, 1323.
[216] G. A. Ozin, A. Kuperman, A. Stein, *Angew. Chem. Int. Ed. Engl.* **1989**, *28*, 359.
[217] G. D. Stucky, J. E. MacDougall, *Science* **1990**, *247*, 669.
[218] Y. Wang, N. Herron, *Res. Chem. Interm.* **1991**, *15*, 17.
[219] Y. Wang, N. Herron, *J. Phys. Chem.* **1991**, *95*, 525.
[220] G. A. Ozin, *Adv. Mater.,* in press.
[221] G. A. Ozin, C. L. Bowes, M. R. Steele, *Mat. Res. Soc. Symp. Proc.* **1992**, *277*, 105.
[222] (a) Y. Wang, N. Herron, *J. Phys. Chem.* **1987**, *91*, 257.
(b) Y. Wang, N. Herron, *J. Phys. Chem.* **1988**, *92*, 4988.
[223] N. Herron, Y. Wang, M. M. Eddy, G. D. Stucky, D. E. Cox, K. Moller, T. Bein, *J. Am. Chem. Soc.* **1989**, *111*, 530.
[224] G. D. Stucky, V. I. Srdanov, W. T. A. Harrison, T. E. Gier, N. L. Keder, K. L. Moran, K. Haug, H. I. Metiu, in *Supramolecular Architecture* (Ed.: T. Bein), *ACS Symp. Ser.* **1992**, *499*, 294.
[225] (a) L. E. Brus, *J. Phys. Chem.* **1986**, *90*, 2555.
(b) M. L. Steigerwald, L. E. Brus, *Annu. Rev. Mater. Sci.* **1989**, *19*, 471.
[226] G. A. Ozin, S. Özkar, *J. Phys. Chem.* **1990**, *94*, 7556.
[227] G. A. Ozin, S. Özkar, K. Moller, T. Bein, *J. Am. Chem. Soc.* **1991**, *112*, 9575.
[228] K. Moller, T. Bein, S. Özkar, G. A. Ozin, *J. Phys. Chem.* **1991**, *95*, 5276.
[229] G. A. Ozin, R. A. Prokopowicz, S. Özkar, *J. Am. Chem. Soc.* **1992**, *114*, 8953.

[230] G. A. Ozin, A. Malek, R. A. Prokopowicz, P. M. Macdonald, S. Özkar, K. Moller, T. Bein, *Mat. Res. Soc. Proc.* **1991**, *233*, 109.
[231] J. S. Krueger, C. Lai, Z. Li, J. E. Mayer, T. E. Mallouk, in *Inclusion Phenomena and Molecular Recognition* (Ed.: J. Atwood), Plenum Press, New York, **1990**, p. 365.
[232] Y. I. Kim, R. L. Riley, M. J. Huq, S. Salim, A. N. Le, T. E. Mallouk, *Mater. Res. Soc. Symp. Proc.* **1991**, *233*, 145.
[233] X. Liu, K.-K. Iu, J. K. Thomas, *Chem. Phys. Lett.* **1992**, *195*, 163.
[234] R. J. Davis, *Chem. Mater.* **1992**, *4*, 1410.
[235] K. Moller, M. M. Eddy, G. D. Stucky, N. Herron, T. Bein, *J. Am. Chem. Soc.* **1989**, *111*, 2564.
[236] P. Cherin, P. Unger, *Inorg. Chem.* **1967**, *6*, 1589.
[237] J. B. Parise, J. E. MacDougall, N. Herron, R. Farlee, A. W. Sleight, Y. Wang, T. Bein, K. Moller, L. M. Moroney, *Inorg. Chem.* **1988**, *27*, 221.
[238] K. Takmura, S. Hosokawa, H. Endo, S. Yamasaki, H. Oyangi, *J. Phys. Soc. Jpn.* **1986**, *55*, 528.
[239] Y. Katayama, M. Yao, Y. Ajiro, M. Inui, H. Endo, *J. Phys. Soc. Jpn.* **1989**, *58*, 1811.
[240] H. W. Kroto, J. R. Heath, S. C. O'Brien, R. F. Curl, R. E. Smalley, *Nature* **1985**, *318*, 162.
[241] W. Kratschmer, L. D. Lamb, K. Fostiropoulos, D. R. Huffman, *Nature* **1990**, *347*, 354.
[242] H. W. Kroto, A. W. Allaf, S. P. Balm, *Chem. Rev.* **1991**, *91*, 1213.
[243] S. W. McElvany, M. M. Ross, J. H. Callahan, *Acc. Chem. Res.* **1992**, *25*, 162.
[244] R. C. Haddon, *Acc. Chem. Res.* **1992**, *25*, 127, and references cited therein.
[245] P. J. Fagan, J. C. Calabrese, B. Malone, *Acc. Chem. Res.* **1992**, *25*, 134, and references cited therein.
[246] M. E. Davis, C. Saldarriaga, C. Montes, J. Garces, C. Crowder, *Nature* **1988**, *331*, 698.
[247] M. Estermann, L. B. McCusker, C. Baerlocher, A. Merrouche, H. Kessler, *Nature* **1991**, *352*, 320.
[248] Q. Huo, R. Xu, S. Li, Z. Ma, J. M. Thomas, R. H. Jones, A. M. Chippindale, *J. Chem. Soc. Chem. Commun.* **1992**, 875.

5 Discrete and Condensed Transition Metal Clusters in Solids

Arndt Simon

5.1 Introduction

The assembly of a quasi infinite number of closely spaced atoms in a crystal will have significant consequences for their properties. Nonstoichiometry is an immanent problem and cooperative phenomena are much more obvious than in molecules. Besides these real peculiarities, there are a number of fictitious ones which originate from the various descriptive viewpoints of the different fields of chemistry. To give an example, compounds with the brutto compositions CH_2 and CeH_2 do not seem to have much in common. Indeed, the first formula is a representative of those compounds having strong covalent and localized bonds, be it in molecules like C_2H_4 and C_6H_{12} or in a polymer chain like $(CH_2)_n$. In contrast, CeH_2 forms a three dimensional extended solid with considerable ionic and metallic bonding. The very general feature of their bonding, however, is identical, namely that only a part of the valence electrons of C (Ce) is used for heteronuclear bonds to the H atoms, while the remainder of the electrons are involved in homonuclear C–C or Ce–Ce bonding.

Whereas certain bonding features are considered most unusual in a molecule, similar features are quite normal in solid state compounds. To give an example, the unusual coordination of the carbon atom in the discrete molecule $[Li_6C]$ is suggestive of hypervalency, [1, 2] however the occurrence of octahedrally coordinated carbon atoms is often observed in transition metal carbides having the rocksalt type structure, thus a quite normal situation. From a simple ionic representation as $[(Li^+)_6C^{4-}(e^-)_2]$, the only surprise is the surplus of electrons. Yet, even this feature is familiar with such interstitial carbides as V_2C in which, the valence electrons of V (Li) which are not used for heteronuclear V–C (Li–C) bonds are used to form homonuclear V–V (Li–Li) bonds, as suggested by the (8–N) rule. [3]

The structures and properties of metal rich compounds are well suited to illustrate the sometimes different yet complimentary views in molecular and solid state chemistry. The examples already mentioned illuminate important facets of the discussion to come. Metal rich compounds contain a larger number of metal atoms than necessary to saturate the maximum valencies of the nonmetal atoms. The excess of metal centered valence electrons can be involved in bonding, nonbonding, or even antibonding interactions between adjacent metal atoms. Of primary interest here are the electron poor transition metals whereby the bonding interactions dominate and result in the formation of discrete metal clusters or extended M–M bonded arrays.

At this point it seems appropriate to briefly mention and discuss the formalism which will be used throughout the text in counting the electrons assigned to M–M bonding states. This is done according to the Zintl-Klemm concept which utilizes a purely ionic description of bonding between metal and nonmetal atoms. There are cases where the results from this formal procedure nearly represent the actual chemical bonding situation. Thus the notation $[(Cs^+)_{11}(O^{2-})_3(e^-)_5]$ for the cesium suboxide $Cs_{11}O_3$ not only yields a precise number for the amount of electrons in purely M–M bonding states which is in quantitative agreement with experiment, [4] but also matches the chemical bonding in the compound. For the compound Gd_2Cl_3 which is formed from the highly electropositive metal gadolinium, however, the formula $[(Gd^{3+})_2(Cl^-)_3(e^-)_3]$ oversimplifies the nature of the chemical bonding but still leads to a precise count for the number of electrons in M–M bonding states. Last but not least, the description of $MoCl_2$ as $Mo^{6+}(Cl^-)_2\ (e^-)_4$ is entirely unrealistic with regards to its real bonding features, yet the formula indicates that four electrons per metal atom could be available for M–M interactions, no matter whether the other valence electrons of the metal atom are involved in predominantly ionic or covalent bonding with the nonmetal atoms. Although the surplus valence electrons can be used for M–M bonding as found for Mo_6Cl_{12}, the case of $CrCl_2$ illustrates that they need not always be used as such.

Metals in their elemental state mostly crystallize in close packed structures or in the body centered cubic lattice. For a face centered cubic metal the unit cell shown in Figure **5-1** is normally chosen as the conventional description of the structure. As an alternative, the close packed arrangement can be decomposed into a network of condensed regular octahedra which share corners and edges as indicated in Figure **5-1** for the case of cubic close packing, or of deformed octa-

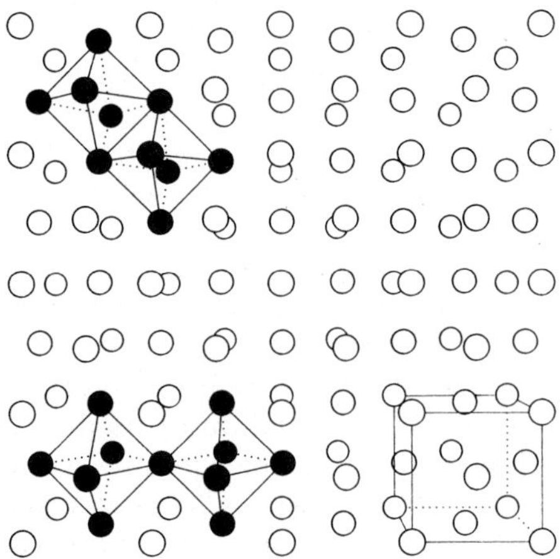

Figure 5-1. Part of a cubic close packed lattice. The conventional fcc unit cell (right side) and edge and corner condensed octahedra (left side) are drawn.

hedra in the body centered cubic lattice. In a hexagonal close packed lattice the octahedra are condensed via edges and faces.

Much of the structural chemistry of metal rich transition metal compounds containing p elements can be discussed in terms of fragments of the metal structures which are cut from the fcc, hcp, or bcc element structures and surrounded in a suitable way by nonmetal atoms. The octahedral fragment of the element structures, whether regular (fcc, hcp) or deformed (bcc), is the most abundant in these compounds. The following discussion will therefore focus on structures which are built from octahedral M_6 units and will extend somewhat to other kinds of M_n units.

Two premises govern these structures. To a first approximation, the relative amount of nonmetal atoms in the compound determines whether the M_6 unit is discrete and well surrounded by nonmetal atoms. If not, oligomeric units or polymeric structures of varying dimensionality result from a condensation of such M_6 units. With a sufficient number of valence electrons, strong M–M bonding exists and the M_6 units are empty. If there is an electron deficiency, then the M_6 units must be stabilized by interstitial atoms which can be either nonmetal or metal atoms.

In the past thirty years, a tremendous amount of work has been concentrated in this field of research. Since the early summary of the research by Schäfer and Schnering [5] numerous reviews have appeared on the chemistry and physics of these cluster compounds. Access to them is provided by just citing some more recent ones. [6–16]

5.2 Empty Octahedral Metal Clusters

Clusters with octahedral M_6 units are most frequently found with those metals (Nb, Ta, Mo, W, Re, and Tc) which have a maximum number of electrons available for M–M bonding and hence are distinguished by high melting points and large values for the heat of atomization due to their strong M–M bonding. Chemical bonding in these clusters and the patterns of interconnection are well known and greatly understood. There exist two kinds of clusters, one with eight nonmetal atoms above the faces, $[M_6X_8]$ and one with twelve X atoms above the edges of the M_6 octahedron $[M_6X_{12}]$, (Fig. **5-2a**).

Early interest in octahedral clusters was due to the work of both Brosset and Pauling. [17, 18] In these clusters every M atom has a tetragonal coordination to the X atoms. The chemical bonding in these clusters has been repeatedly analysed. [19–24] The number of M–M bonding states can be determined in a simple way. [25] Each M atom bonds four X atoms through a mixture of its d_{xy}, s, and p orbitals. In the case of the $[M_6X_8]$ cluster the remaining four d atom orbitals mix to yield a fourfold degenerate valence state that forms (slightly bent) bonds along the octahedral edges and leads to twelve 2 electron-2 center bonds.

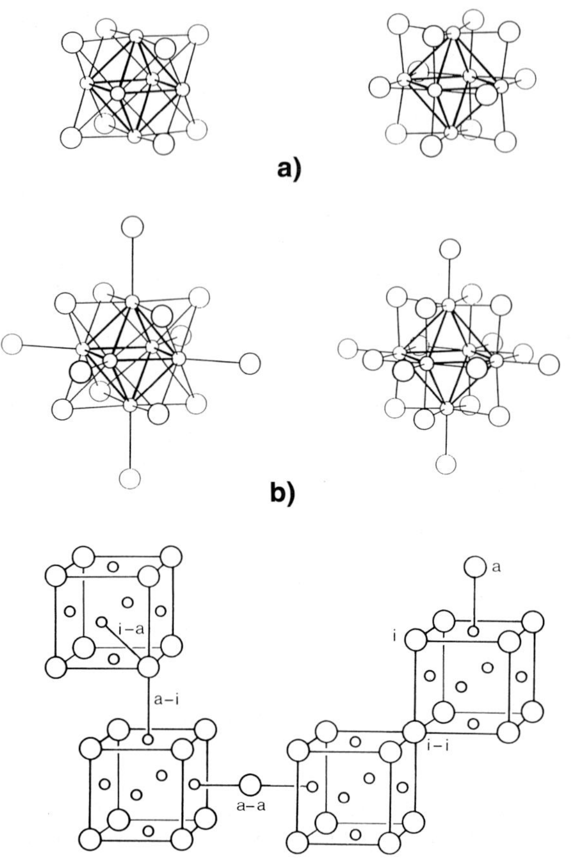

Figure 5-2. a) $[M_6X_8]$ (left) and $[M_6X_{12}]$ clusters, the X^i atoms capping the faces und the edges respectively, of the M_6 octahedron. b) $[M_6X_{14}]$ (left) and $[M_6X_{18}]$ clusters, the X^a atoms bonding to the apices of the M_6 octahedron. c) Interconnection patterns and notations for octahedral metal clusters via X^i and X^a type atoms.

In the case of the $[M_6X_{12}]$ cluster, the MX_4 planes are rotated by 45° and the corresponding d valence hybrids overlap slightly above the octahedral faces to yield eight $2e^-$-3c bonds. Therefore, ions like $[Mo_6Cl_8]^{4+}$ and $[Nb_6Cl_{12}]^{2+}$ are effectively isoelectronic for both have 20 electron pairs occupying M–X and M–M bonding states.

Exceptions to these closed shell configurations appear rather frequently and an explanation for their systematic occurrence is provided by MO calculations. The results of these indicate 12 and 8 orbitals for the $[M_6X_8]$ and $[M_6X_{12}]$ clusters respectively, which have M–M bonding but also M–X antibonding character. They group with increasing energies into a_{1g}, t_{1u}, t_{2g}, t_{2u}, and e_g for the $[M_6X_8]$ and a_{1g}, t_{2g}, t_{1u}, and a_{2u} for the $[M_6X_{12}]$ clusters. The stable electron concentrations very much depend on the balance between M–M bonding and M–X antibonding interactions. Thus, the $[Nb_6X_{12}]^{2+}$ cluster with 16 electrons available for M–M bonding is stable for the halides X = Cl and Br but can be easily oxidised to the 3+ and 4+ species, whereas for X = O the Nb–O antibonding interactions dominate over the Nb–Nb bonding interactions of the a_{2u} orbital. The occupation of this orbital is therefore avoided and the preference for a 14 electron cluster species is given. [16]

Before getting more deeply involved in the structural aspects of these clusters, the preparative situation should be briefly addressed. All the cluster compounds having octahedral M_6 cores discussed in the following are prepared from high temperature reactions. The clusters are not formed in a rational stepwise synthesis as is all too well known in organic synthesis. Of course, once the clusters have formed many of them can be brought into solution and modified by methods typical for solution chemistry. Numerous approaches proved unsuccessful in their attempts to form the M_6 core from smaller fragments in solution via a controlled synthetic route. A first such synthesis from M_3 fragments [26] is therefore a remarkable step in the chemistry of these cluster compounds

To come back to structural aspects of these compounds, in addition to the atoms represented by X^i which sit above the edges or faces of the M_6 octahedra in the early systematic notation of Schäfer and Schnering, [5]* there are X^a atoms which bond to the octahedral apices (Fig. **5-2b**). In compounds such as $K_4[Nb_6Cl_{18}]$(= $K_4[Nb_6Cl^i_{12}Cl^a_6]$) [27] or $Hg[Mo_6Cl_{14}]$ (= $Hg[Mo_6Cl^i_8Cl^a_6]$) [28] all coordination sites of the cluster are occupied by their own X atoms and so there is no linkage between the clusters. There is a variety of different ways to interconnect adjacent clusters and these are indicated schematically for the $[M_6X_8]$ cluster in Figure **5-2c**. For sake of clarity, only bridging between two clusters is shown. Bridges between three discrete clusters are the exception, although they exist, for example, in $Nb_6Cl^i_{12}I^{a-a}_{6/3}$. [29] Later, in the discussion on condensed cluster systems it will be seen that bridges between 3 or 4 adjacent M_6 units are frequently found.

The electronic balance and the degree of coverage of the M_6 units by X atoms critically determine the physical properties of the compounds. A total coverage always means insulating or semiconducting behavior, even when the important condition for metallic conductivity, namely the partial occupation of the M–M bonding cluster states is met. An increasing degree of cluster linkage leads to a stepwise convergence of the M_6 units, especially if the X^i atoms are involved, and following the general idea of the Herzfeld-Mott criterion, [30–32] the metallic state will eventually be reached. The stepwise transition from insulators to metals can be seen with the compounds listed in Table **5-1**.

The $[M_6X_8]$ clusters found in the iodides Nb_6I_{11}, [33] $CsNb_6I_{11}$, [34] $Nb_6I_8(NR_3)_6$, [35] and the unusual chain compound Nb_6I_9S [36] are rather an exception for the valence electron poor metal Nb. This type of cluster however, is common with the (chalcogenide) halides of the more electron rich metals Mo [7, 9, 37, 38] and W as well as the (halide) chalcogenides of Re [7, 9, 39] and Tc. [40] There is a sufficient number of d electrons in the dihalides of Mo and W to fill

* Throughout the book brackets are used to denote the composition of a discrete molecular cluster. In the present chapter this notation is kept for discrete clusters only. For interconnected clusters, the structure precise notation of Schäfer and Schnering is used, and in the case of condensed cluster systems the connectivity is indicated by the indices only.

Table 5-1. Molybdenum and rhenium M_6X_8 type halide, halide chalcogenide and chalcogenide clusters. [7]

General formula of compounds	Known compounds	Cluster structures	Physical properties
$Re_6\hat{X}_4X_{10}$	$Re_6Se_4Cl_{10}$, $Re_6S_4Br_{10}$, $Re_6Se_4Br_{10}$	$[Re_6\hat{X}^{i}_{4}X^{i}_{4}X^{a}_{6}]$	Insulator, dielectric relaxation transition
$ARe_6\hat{X}_5X_9$	A(+) = Li, Na, K, Rb, Cu, Ag	$[Re_6\hat{X}^{i}_{5}X^{i}_{3}X^{a}_{6}]$	Insulator, dielectric relaxation
AMo_6X_{14}	A(2+) = V, Cr, Mn, Fe, Co, Ni, Zn, Cd, Hg, Sn, Pb, Mg, Ca, Sr, Ba, Eu, Yb X = Cl	$[Mo_6X^{i}_{8}X^{a}_{6}]$	Insulator
$ARe_6\hat{X}_6X_8$	A(2+) = Cr, Mn, Re, Co, Cu, Pb, Cd	$[R_6\hat{X}^{i}_{6}X^{i}_{2}X^{a}_{6}]$	Insulator
$A_2Mo_6X_{14}$	$A_2Mo_6Cl_{14}$; A(+) = Li, K, Rb, Cs, Cu, $Cs_2Mo_6Cl_8Br_6$	$[Mo_6X^{i}_{8}X^{a}_{6}]$	Insulator
AMo_6X_{13}	AMo_6Cl_{13}; A(+) = Na, Ag	$Mo_6X^{i}_{8}X^{a}_{4}X^{a\text{-}a}_{2/2}$	Insulator
$Re_6\hat{X}_5X_8$	$Re_6Se_5Cl_8$	$Re_6\hat{X}^{i}_{5}X^{i}_{3}X^{a}_{4}X^{a\text{-}a}_{2/2}$	Insulator, dielectric relaxation
Mo_6X_{12}	X = Cl, Br, I	$Mo_6X^{i}_{8}X^{a}_{2}X^{a\text{-}a}_{4/2}$	Insulator
$Re_6\hat{X}_6X_6$	$Re_6Se_6Cl_6$	$Re_6\hat{X}^{i}_{6}X^{i}_{2}X^{a}_{2}X^{a\text{-}a}_{4/2}$	Insulator
$Mo_6X_{10}\hat{X}$	X = Cl, Br; $\hat{X}$ = S, Se, Te X = I; $\hat{X}$ = Se, Te	$Mo_6X^{i}_{7}\hat{X}^{i}X^{a\text{-}a}_{6/2}$	Insulator, dielectric relaxation
$Re_6\hat{X}_7X_4$	X = Br; $\hat{X}$ = S, Se	$Re_6\hat{X}^{i}_{7}X^{i}X^{a\text{-}a}_{6/2}$	Insulator
$ARe_6\hat{X}_{11}$	A(2+) = Sr, Ba, Eu; $\hat{X}$ = S	$Re_6\hat{X}^{i}_{8}\hat{X}^{a\text{-}a}_{6/2}$	
$A_4Re_6\hat{X}_{11+n}$	A(+) = alkali metal; $\hat{X}$ = S, Se, (S-S)	$Re_6\hat{X}^{i}_{8}\hat{X}^{a\text{-}a}_{(6-n)/2}(\hat{X}\text{–}\hat{X})^{a\text{-}a}_{n/2}$	
$Re_2\hat{X}_5$	Re_2Te_5	$Re_6\hat{X}^{i}_{8}\hat{X}^{a\text{-}a}_{7\,6/6}$	
$Mo_6X_8\hat{X}_2$	X = Br; $\hat{X}$ = S X = I; $\hat{X}$ = S, Se	$Mo_6X^{i}_{5}\hat{X}^{i}\hat{X}^{i\text{-}i}_{2/2}X^{a\text{-}a}_{6/2}$	Insulator, dielectric relaxation anisotropy: $\varepsilon\|/\varepsilon\perp \approx 10^2$
$Re_6\hat{X}_8X_2$	$Re_6Se_8Cl_2$, $Re_6\hat{X}_8Br_2$($\hat{X}$ = S, Se)	$Re_6\hat{X}^{i\text{-}a}_{4/2}X^{a}_{2}\hat{X}^{a\text{-}i}_{4/2}$	Anisotropic semiconductor $\varrho\|/\varrho\perp \approx 10^{-2}$
$Mo_6X_6\hat{X}_3$	$Mo_6Br_6S_3$	$Mo_6X^{i}_{4}\hat{X}^{i\text{-}i}_{2/2}\hat{X}^{i\text{-}a}_{2/2}X^{a\text{-}a}_{4/2}X^{a\text{-}i}_{2/2}$	Semiconductor, 0.037 eV for 100 K $< T <$ 293 K; 0.020 eV for $T <$ 10 K

Table 5-1. (Continued).

General formula of compounds	Known compounds	Cluster structures	Physical properties
$Mo_6X_2\hat{X}_6$	$Mo_6X_2S_6$ (X = Br, I) $Mo_6X_x\hat{X}_{8-x}$ $\hat{X}$ = Se, Te; X = Cl, Br, I $0 < x < 3$	$Mo_6X^i_2\hat{X}^{i-a}_{6/2}\hat{X}^{a-i}_{6/2}$	Superconductor $Mo_6Br_2S_6$, $T_c = 13.8$ K $Mo_6I_2S_6$, $T_c = 14$ K $Mo_6X_xSe_{8-x}$ X = Cl, $T_c \leq 6.5$ K X = Br, $T_c \leq 7.1$ K X = I, $T_c \leq 7.6$ K $Mo_6I_xTe_{8-x}$, $T_c \leq 2.6$ K
$Mo_6\hat{X}_8$	$\hat{X}$ = S, Se, Te	$Mo_6\hat{X}^i_2\hat{X}^{i-a}_{6/2}\hat{X}^{a-i}_{6/2}$	Superconductor
$Cs_{0.6}Mo_6\hat{X}_7$	$\hat{X}$ = S	$Mo_6\hat{X}^{i-i}_{2/2}\hat{X}^{i-a}_{6/2}\hat{X}^{a-i}_{6/2}$	Superconductor $T_c = 7.7$ K

all M–M bonding states in the $[M_6X_8]$ cluster. When the halogen atoms X are successively replaced by divalent chalcogen atoms $\hat{X}$, the d electron concentration is simultaneously kept high by a stepwise reduction in the (X, $\hat{X}$): Mo ratio. In the case of the more valence electron rich Re, a larger number of $\hat{X}$ atoms is needed in order to avoid the occupation of M–M antibonding states. The decrease in the nonmetal content of the cluster compounds of Re occurs to a lesser extent than in those of Mo.

The stepwise increase in the dimensionality of the cluster linkages as a function of the nonmetal to metal ratio is particularly well documented with the compounds of Mo and Re formed from the halogens and chalcogens. Table **5-1**, which is taken from reference, [7] summarizes the large amount of knowledge in this field. The different patterns of interconnection and the increasing degree of cluster interconnection with decreasing (X, $\hat{X}$): M ratio are shown schematically for selected compounds in Figure **5-3**. Discrete clusters are present in the structures of the chloromolybdates and the ternary molecular compound $[Re_6S_4Cl_{10}]$ $(=[Re_6S^i_4Cl^i_4Cl^a_6])$. [41] The increasing degree of interconnection via X^a atoms gives rise to one, two, and three dimensional frameworks. The structures of AMo_6Cl_{13} and $Re_6Se_5Cl_8$ [42] contain cluster chains. The structures of the Mo dihalides [43] and of the isotypic compound $Re_6Se_6Cl_6$ [42] are made up of layers. The complete connection of clusters via X^{a-a} bridges in all directions according to $M_6X^i_8X^{a-a}_{6/2}$ is achieved in the structures of Nb_6I_{11} [33] and the isotypic compounds $Mo_6X_{10}\hat{X}$. [44, 45] $Li_4Re_6S_{11}$ [46] is isostructural with respect to the cluster substructure and isoelectronic with the latter compounds, if a full charge transfer from the Li atoms to the clusters is assumed.

It is interesting to note that M–M and X–X bonding may occur simultaneously. Besides simple bridges of the X^{a-a} type, polyanionic bridges have been known for a long time from the structure of W_6Br_{16} $(=W_6Br^i_8Br^a_4\ (Br_4)^{a-a}_{2/2})$, [47] and more recently $K_4Re_6S_{12}$ $(=K_4Re_6S^i_8S^{a-a}_{4/2}(S_2)^{a-a}_{2/2})$. [48] For Re_2Te_5 $(=Re_6Te^i_8(Te_7)^{a-a}_{6/6})$ a particularly interesting structure has been found. [49, 50] Here, the bridges consist

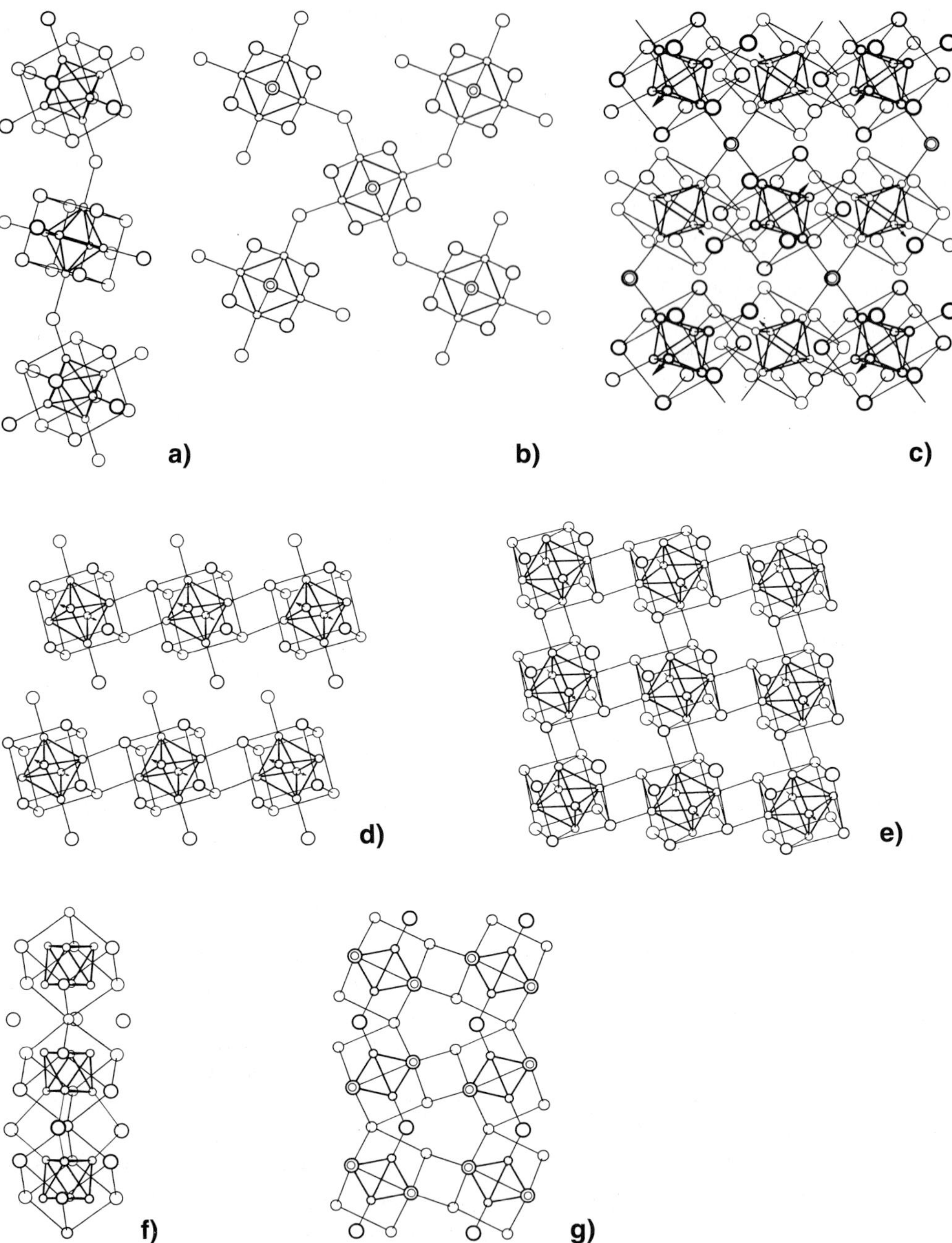

Figure 5-3. Increasing connectivity of $[M_6(X, \hat{X})^i_8]$ clusters in halides, halide chalcogenides, and chalcogenides of the metals Nb, Mo, and Re. $X^{a\text{-}a}$ links in a) $Re_6Se_5Cl_8$ (chain), b) Mo_6Cl_{12} (layer), and c) Nb_6I_{11} (network); $X^{i\text{-}a}$ links in d) $Re_6Se_8Cl_2$ (layer) and e) $Mo_6S_6Br_2$ (network); $X^{i\text{-}i}$ ($+X^{i\text{-}a}$) links in f) $Mo_6Br_8S_2$ (chain) and g) $Mo_6Br_6S_3$ (layer). M, $\hat{X}$, and X atoms are drawn with increasing size taken from ref. [6] (cf. Table **6-1** and text).

of Te_7 units containing a central Te atom having a square planar environment of Te atoms, which is a quite typical atomic arrangement in polytellurides. [51]

At low nonmetal contents, $(X, \hat{X}) \leq 10$, the $X^{i\text{-}a}$ and $X^{i\text{-}i}$ linkages become important. In the structure of $Re_6Se_8Cl_2$ ($= Re_6Se^{i}_{4}Se^{i\text{-}a}_{4/2}Se^{a\text{-}i}_{4/2}Cl^{a}_{2}$) [52] the clusters are connected via $X^{i\text{-}a}$ bonds to form layers that are surrounded by Cl atoms and are held together by van der Waals forces. When these X^a type Cl atoms are omitted, the $M_6\hat{X}_8$ type structure results and consists of $[M_6\hat{X}_8]$ clusters which are connected in all directions via $X^{i\text{-}a}$ bonds according to $M_6\hat{X}^{i}_{2}\hat{X}^{i\text{-}a}_{6/2}\hat{X}^{a\text{-}i}_{6/2}$ (Chevrel phases). If $X^{i\text{-}i}$ type interconnections are formed, they always occur via opposite atoms of the cluster and lead to linear chains. Such chains, which are van der Waals bonded to each other, are found in the structure of the compound $Mo_6I_8Se_2$. [53] In contrast to the Mo compound which shows disorder of the Se atoms, the previously mentioned Nb compound Nb_6I_9S [36] exhibits a nice differentiation in the functionalities of the I and S atoms according to $Nb_6I^{i}_{6}S^{i\text{-}i}_{2/2}I^{a\text{-}a}_{6/2}$.

The connection of $M_6(X, \hat{X})_{10}$ type chains via $X^{i\text{-}a}$ bridges results in layers as they occur in the structure of $Mo_6Br_6S_3$ [54] according to $Mo_6Br^{i}_{4}S^{i\text{-}i}_{2/2}S^{i\text{-}a}_{2/2}S^{a\text{-}i}_{2/2}Br^{a\text{-}a}_{4/2}$. When these layers are linked via $X^{i\text{-}a}$ contacts the Mo-S framework of $Cs_{0.6}Mo_6S_7$ is formed [55] which corresponds to the formula $Mo_6S^{i\text{-}i}_{2/2}S^{i\text{-}a}_{6/2}S^{a\text{-}i}_{6/2}$. At this point, a critical limit is reached in the structural chemistry of these cluster compounds. A further reduction in the X:M ratio leads to a condensation of the M_6 octahedra themselves.

All $[M_6X_8]$ compounds having 24 electrons per M_6 unit in M–M bonding states, *i. e.* with a fully occupied valence band, are either insulators or semiconductors, even when close contacts occur be interconnections via X^i atoms. In the case of $Mo_6Br_6S_3$, the 0.037 eV energy gap at room temperature is nearly closed. It was mentioned earlier that a partial occupation of M–M bonding states does not result in metallic conduction when the electrons of the neighboring clusters are kept well separated from one another due to the $X^{a\text{-}a}$ bridges and thus remain localized on the clusters. The brownish-black Nb_6I_{11} ($= Nb_6I^{i}_{8}I^{a\text{-}a}_{6/2}$) is a semiconductor with a band gap of approximately 0.4 eV, [56] although only 19 electrons instead of 24 fill M–M bonding states in the cluster.

The discussion so far has concentrated on the structural chemistry of the electron rich clusters of the $[M_6X_8]$ type. To prepare for the next paragraph, interest is now focussed on metals with smaller numbers of valence electrons which then prefer to form $[M_6X_{12}]$ type clusters. A tremendous amount of work has been performed on $[M_6X_{12}]$ type halide clusters of the metals Nb and Ta both in the solid state and in solution. Clusters with 14, 15, and 16 electrons in M–M bonding states are well known. Examples of compounds with this cluster type but formed with divalent anions had been very scarce, being represented by only two which contained the $[Nb_6O_{12}]$ cluster. [57, 58] Recently, a rather broad chemistry of reduced oxoniobates based on this cluster has emerged. [16]

The compounds characterized so far by single crystal X-ray analysis are summarized in Table **5-2**. With only a few exceptions, these are complicated phases of mixed valency compounds which contain niobium in simultaneously two, three, or even more different bonding situations. To take an example, the structure of $Na(Si_xNb_{1-x})Nb_{10}O_{19}$ contains not only $[Nb_6O_{12}]$ clusters but, in addition, M–M

Table 5-2. Unit cell parameters [pm] and properties of compounds containing discrete Nb_6O_{12} clusters. [16]

compound	space group	*a*	*b*	*c*	*Z*	colour	electrical[a] property
$Mg_3Nb_6O_{11}$	$P\bar{3}m1$	604.2	–	746.6	1	dark green	s
$Mn_3Nb_6O_{11}$	$P\bar{3}m1$	608.0	–	762.7	1	dark green	s
$Na(V_{3-x}, Nb_x)Nb_6O_{14}$ ($x \approx 1.0$)	$P6_3/m$	1603.4	–	1807.9	3	golden[b]	s
$Na_3Al_2Nb_{34}O_{64}$	$R\bar{3}$	784.4	–	7065.0	3	golden[b]	s
$Na(Si, Nb)Nb_{10}O_{19}$	$R\bar{3}$	784.1	–	4221.8	6	red	s
$K(Si, Nb)Nb_{10}O_{19}$	$R\bar{3}$	783.7	–	4236.0	6	golden[b]	–
$K_3Al_2Nb_{34}O_{64}$	$R\bar{3}$	783.0	–	7051.0	3	golden[b]	–
$K_4Al_2Nb_{35}O_{70}$	$P\bar{3}$	782.2	–	2632.0	1	brown	–
$KSiNb_{10}O_{19}$	$P6_3$	779.2	–	1443.0	2	golden[b]	–
$Rb_4(Si,Nb)_2Nb_{35}O_{70}$	$P\bar{3}$	782.1	–	2651.3	1	golden[b]	–
$Ba_3Si_2Nb_{20.8}O_{44}$	$P\bar{3}$	777.4	–	1676.5	1	black	–
$BaSiNb_{10}O_{19}$	$P6_3$	777.7	–	1451.0	2	black	s
$Ba_2Nb_{15}O_{32}$	$R\bar{3}$	777.7	–	3551.8	3	golden[b]	s
KNb_8O_{14}	*Pbam*	935.2	1026.8	592.9	2	golden[b]	s
$KNb_8O_{14-x}F_x$ ($x \approx 1.0$)	*Pbam*	932.7	1028.0	593.7	2	golden[b]	s
$SrNb_8O_{14}$	*Pbam*	925.7	1030.1	595.4	2	golden[b]	s
$BaNb_8O_{14}$ (I)	*Pbam*	931.1	1033.1	595.4	2	golden[b]	s
$BaNb_8O_{14}$ (II)	*Pcab*	2371.0	1036.0	942.2	8	golden[b]	s
$LaNb_8O_{14}$	*Pbam*	919.4	1033.7	596.2	2	golden[b]	s
$EuNb_8O_{14}$	*Pbam*	927.7	1031.1	596.4	2	golden[b]	s
$NaNb_{10}O_{18}$	$P2_1/c$ ($\beta = 110.48°$)	784.9	931.6	1028.1	2	golden[b]	m

[a] s = semiconducting, m = metallic. [b] extrinsic property

bonded Nb_2 units characteristic for Nb^{4+} and also single Nb atoms in octahedral voids which indicate they are in the pentavalent state.

The comparison of the structures of two of these compounds, $Na(Si_xNb_{1-x})$-$Nb_{10}O_{19}$ and $Na_3Al_2Nb_{34}O_{64}$, [59] in Figure **5-4** gives an impression of their very complex atomic arrangements, but also indicates that large portions of these structures are identical. In fact, an analysis of all the structures shows that all the known cluster oxoniobates are built from common close packed layers of Nb_6

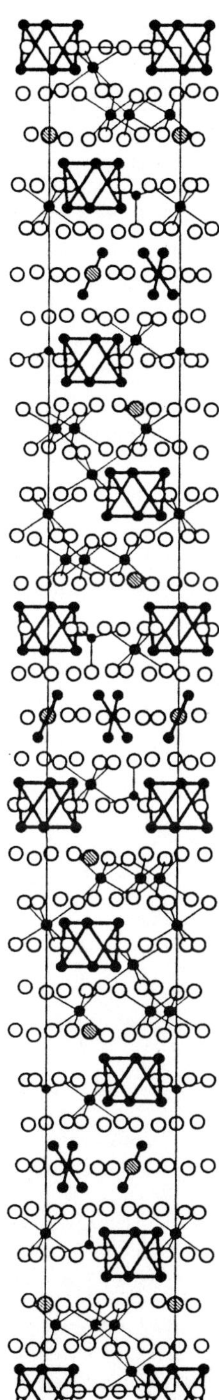

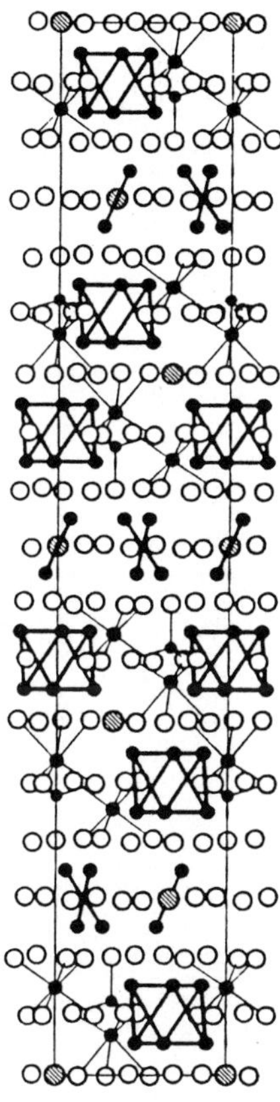

Figure 5-4. Projections of the hexagonal unit cells of (left) $Na_3Al_2Nb_{34}O_{64}$ and (right) $Na(Si_xNb_{1-x})Nb_{10}O_{19}$ along [100]. Atoms are drawn with increasing size for Al, Si (filled), Nb (filled), O (open), and Na (shaded).

octahedra and O atoms, layers formed by large cations and O atoms as well as pure O atom layers. Voids in these close packings are occupied by smaller cations including those of Nb which in turn may form small Nb_2 or Nb_3 cluster units. The complicated mixed valent character of these cluster oxoniobates and the unknown extent of charge transfer between the different Nb_n entities makes any definite assigment of the oxidation states for the Nb atoms in the $[Nb_6O_{12}]$ clusters a hazardous task. The only compounds which allow for an unambiguous prediction of the number of electrons in M–M bonding states, 14, are the semiconductors $Mg_3Nb_6O_{11}$ and $Mn_3Nb_6O_{11}$. [58, 60] From band structure calculations, however, there is some evidence that 14 is indeed the preferred number of electron to occupy M–M bonding states in the $[Nb_6O_{12}]$ clusters.

Figure **5-5** shows the structure of $SrNb_8O_{14}$ together with the result from Extended Hückel (EH) calculations. The structure contains one $[Nb_6O_{12}]$ cluster per formula unit besides two single Nb atoms which fill octahedral voids. The total density of states compared with the projected density of states for those Nb atoms belonging to the $[Nb_6O_{12}]$ cluster shows that the lowest band near –16 eV has mainly oxygen character, whereas the narrow bands around the Fermi level are essentially metal atom centered. They exhibit only minor dispersion and nicely represent the sequence of molecular orbitals a_{1g}, t_{1u}, t_{2g}, and a_{2u} of the discrete $[M_6X_{12}]$ type cluster. As the crystal orbital overlap population (COOP)

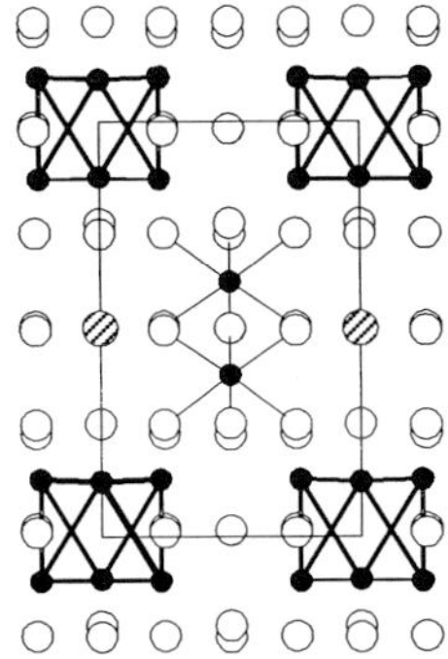

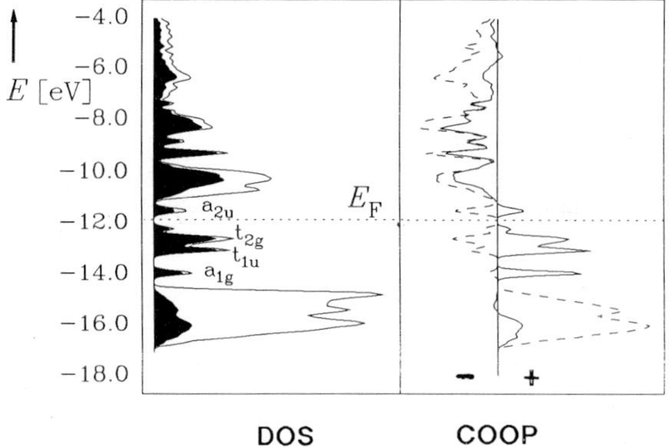

Figure 5-5. (top) Projection of the orthorhombic unit cell of $SrNb_8O_{14}$ along [010] in the range $-0.25 \leqq y \leqq +0.25$ Nb (filled), O (open), Sr (shaded). (bottom) Total DOS for $SrNb_8O_{14}$, projected DOS for the Nb atoms of the cluster (dark shaded), and COOP curves for the Nb–Nb and Nb–O interactions in the cluster (solid and dashed lines respectively).

analysis [61–63] demonstrates, all these narrow bands have strong Nb–Nb bonding character. Only the highest lying band, a_{2u}, is dominated by a Nb–O antibonding contribution which causes this band to have an overall antibonding character. The electron occupation of the cluster states is such so as to avoid the filling of the a_{2u} band. As discussed earlier, although the occupation of 14 electrons in the M–M bonding states of the $[M_6X_{12}]$ cluster seems to be the preferred situation for the $[Nb_6O_{12}]$ cluster, from an analysis of the Nb–Nb distances in the various compounds there is some doubt as to whether this condition strictly holds. [64]

5.3 Clusters Containing Interstitial Atoms

Clusters with only a few valence electrons available for M–M bonding can be stabilized by interstitial atoms. The consequence of this is a partial substitution of weak M–M bonds by strong bonds between the metal atoms and the interstitials. Thus, the stabilization gained by the insertion of interstitial heteroatoms into the cluster cavities occurs at the expense of a further weakening of the (anyhow weak) M–M bonding. It is clear from these considerations that this "stabilization" may finally lead to a "destruction" of the cluster in the sense that all electrons are removed from M–M bonding states to provide bonding with the interstitial. As will be shown, there are cases where nothing is left but a topology which is reminescent of a metal cluster but lacks all M–M bonding.

The occupation of the cavity within the M_6 unit of a discrete cluster was first demonstrated with the hydrogenation reaction of Nb_6I_{11} to yield $Nb_6I_{11}H$. [65] The central position of the H atom within the octahedron was originally concluded from neutron powder diffraction data of the hydrogenated and deuterated compound, and later verified by the use of large single crystals of $Nb_6I_{11}D$, see Figure **5-6**. [66] Interestingly, the first example of a molecular cluster with an interstitial, $[Ru_6(CO)_{17}C]$, was discovered at the same time [67] as the solid state example. In the meantime, a multitude of both molecular and solid state examples of clusters with interstitials has been reported. Our discussion will be confined to solid state examples and the list in Table **5-3** indicates the broad spectrum of interstitial atoms and structure types known to date. The rare earth metals and the elements of the Ti group form a series of cluster halides that are in many cases isostructural with the corresponding compounds of the more electron rich neighboring metals, but differ by the additional atoms inside the clusters, frequently in the centers of the M_6 octahedra. Clearly, the relative amount of halogen atoms determines the pattern of interconnection between the clusters.

Interstitial atoms within $[M_6X_8]$ type clusters have only been observed with $Nb_6I_{11}H$, [33] $CsNb_6I_{11}H$, [34] and more recently, Nb_6I_9SH. [36] There is no indication for interstitial atoms in the clusters of the Chevrel phases, despite the

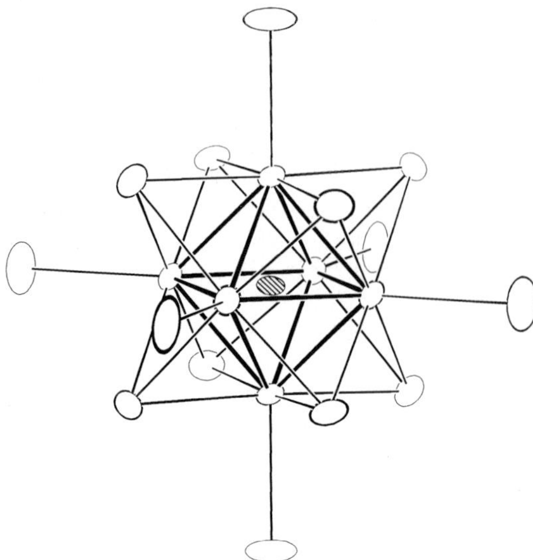

Figure 5-6. $[Nb_6I^i_8I^a_6D]$ cluster from a single crystal neutron investigation of $Nb_6I_{11}D$ at 350 K.

Table 5-3. $[M_6M_8]$ and $[M_6X_{12}]$ halide clusters having interstitial atoms, their interconnection patterns and their corresponding empty cluster structure types. Taken from [6, 10, 13] and additional references in the Table.

	"Empty" Clusters	"Filled" Clusters	Interconnection
$[M_6X_8]$	$Mo_6I_8Se_2$ [53] Nb_6I_{11} $CsNb_6I_{11}$	Nb_6I_9SH [36] $Nb_6I_{11}H$ $CsNb_6I_{11}H$	$M_6X^i_6X^{i-i}_{2/2}X^{a-a}_{6/2}$ $M_6X^i_8X^{a-a}_{6/2}$ $M_6X^i_8X^{a-a}_{6/2}$
$[M_6X_{12}]$	–	$Y_6I_{10}Z$ (Z = Co, Ni, Ru, Rh, Os, Ir, Pt) [80] $Y_6I_{11}C_2$	$M_6X^i_2X^{i-i}_{4/2}I^{i-a}_{6/2}I^{a-i}_{6/2}$ $M_6X^i_4X^{i-i}_{2/2}I^{i-a}_{6/2}I^{a-i}_{6/2}$
	–	$Sc_7Cl_{12}Z$ (Z = B, N) $Sc_7I_{12}Z$ (Z = Co, Ni) $Y_7I_{12}Z$ (Z = Fe, Co) $Pr_7I_{12}Z$ (Z = Mn, Fe, Co, Ni, Cu, Ru, Rh, Pd, Re, Os, Ir, Pt) [80] $Gd_7I_{12}Z$ (Z = Ni, Cu, Ru, Rh, Pd, Os, Ir, Pt, Au) [80]	$M_6X^i_6X^{i-a}_{6/2}X^{a-i}_{6/2}$
	–	$Zr_6I_{12}Z$ (Z = B, C) $Zr_6Cl_{12}Z$ (Z = H, Be, B) $Zr_6Br_{12}B$ [81] $Zr_6I_{12}Mn$	
	–	$CsEr_6I_{12}C$ [82] $CsPr_6I_{12}C_2$ [83]	$M_6X^i_6X^{i-a}_{6/2}X^{a-i}_{6/2}$ $M_6X^i_6X^{i-a}_{6/2}X^{a-i}_{6/2}$

Table 5-3. (Continued).

"Empty" Clusters	"Filled" Clusters	Interconnection
–	$Zr_6X_{13}B$ (X = Cl, Br) (K, Rb) $Zr_6Cl_{13}Be$	$M_6X^{i}_{10}X^{i-i}_{2/2}X^{a-a}_{6/3}$
Nb_6Cl_{14}	$Zr_6Cl_{14}Z$ (Z = B, C) $Zr_6X_{14}Fe$ (X = Cl, Br, I) $Th_6Br_{14}C$ [84] $Zr_6Br_{14}C$ [80]	$M_6X^{i}_{10}X^{i-a}_{2/2}X^{a-i}_{2/2}X^{a-a}_{4/2}$
–	$AZr_6I_{14}C$ (A = K_X-Cs) $AZr_6Cl_{14}B$ (A = Li-Cs, Tl) [81] $Cs_xZr_6I_{14}Z$ (Z = Al, Si, Mn, Fe, Co)	$M_6X^{i}_{10}X^{i-a}_{2/2}X^{a-i}_{2/2}X^{a-a}_{4/2}$
Nb_6F_{15}	$Zr_6Cl_{15}Z$ (Z = N, Co, Ni) [85] $Li_2Zr_6Cl_{15}Mn$, $LiZr_6Cl_{15}Fe$ [85] $Th_6Br_{15}Z$ (Z = Mn, Fe, Co, Ni) [81] Th_6Br_{15} (H, D)$_7$ [84]	$M_6X^{i}_{12}X^{a-a}_{6/2}$
Ta_6Cl_{15}	$Zr_6Cl_{15}N$	$M_6X^{i}_{12}X^{a-a}_{6/2}$
$NaNb_6Cl_{15}$	$Na_2Zr_6Cl_{15}B$	$M_6X^{i}_{12}X^{a-a}_{6/2}$
$CsNb_6Cl_{15}$	$AZr_6Cl_{15}C$ (A = K-Cs) $KZr_6Cl_{15}N$ $AZr_6Br_{15}Fe$ (A = Rb, Cs) $AZr_6Cl_{15}B$ (A = KCs, Rb_2)	$M_6X^{i}_{12}X^{a-a}_{6/2}$
–	$K_2Zr_6Cl_{15}Z$ (Z = Be, B)	$M_6X^{i}_{12}X^{a-a}_{6/2}$
–	$A_3Zr_6Cl_{15}Be$ (A = K, Rb)	$M_6X^{i}_{12}X^{a-a}_{6/2}$
–	$A_3Zr_6Cl_{15}Be$ (A = K, Rb, Cs)	$M_6X^{i}_{12}X^{a-a}_{6/2}$
–	$A_3Zr_6Cl_{16}Z$ (A = Cs; Z = B, C) [85]	$M_6X^{i}_{12}X^{a-a}_{4/2}X^{a}_{2}$
–	$A_4Zr_6Cl_{16}Be$ (A = Na, Cs) [86]	$M_6X^{i}_{12}X^{a-a}_{4/2}X^{a}_{2}$
–	$Ba_2Zr_6Cl_{17}B$ [87]	$M_6X^{i}_{12}X^{a-a}_{2/2}X^{a}_{4}$
–	$Cs_2Zr_7Cl_{18}Be$, $KLaZr_6Cl_{18}C$ $CsLaZr_6Cl_{18}Fe$ $Cs_2Ln_7Cl_{18}C$ [82]	$M_6X^{i}_{12}X^{a}_{6}$
–	$Ba_3Zr_6Cl_{18}Be$ [88]	$M_6X^{i}_{12}X^{a}_{6}$
–	$Rb_5Zr_6Cl_{18}B$ [89]	$M_6X^{i}_{12}X^{a}_{6}$
–	$Li_6Zr_6Cl_{18}H$	$M_6X^{i}_{12}X^{a}_{6}$

electron deficiency they would have as discrete clusters $[M_6\hat{X}_8]$: 20 electrons in M–M bonding states. Their electron deficiency is turned into an energetically favorable situation only through the packing of clusters in the crystal.

$Nb_6I_{11}H$ and $CsNb_6I_{11}H$ are formed by heating the binary and ternary compounds respectively, in the presence of H_2. Hence, the $[Nb_6I_8]$ cluster is known both with and without an interstitial atom. The hydrogen is only lost under vacuum at such a high temperature that further decomposition of the residue occurs through the formation of gaseous NbI_4. If this decomposition reaction is suppressed by the inclusion of the sample in an Nb ampoule, which is permeable to H_2, the hydrogenation reaction can be performed reversibly.

The magnetic properties of Nb_6I_{11} change drastically upon the absorption of H_2. $Nb_6I_{11}H$ is nonmagnetic at low temperature, in contrast to the Curie-Weiss paramagnet Nb_6I_{11}. Apparently, the H atom adds its electron to the electron deficient cluster system. Yet, the H atom does not play the role of a simple electron donor as the externally located Cs atom does in $CsNb_6I_{11}$. The large difference in the electronegativities of H and Nb indicates that H is rather hydridic in nature. In fact, according to calculations, electron density cumulates at the H atom in the cluster. [68]

The solution to this apparent contradiction [69–71] leads to simple general conclusions on the bonding of interstitial atoms in clusters and the meaning of "counting rules" for this case. The most simple case of a hydrogen atom bound in the cluster cavity is well suited to exemplify the general bonding situation when p and d elements are the interstitials as well as when the interstitials are groups of atoms which might be bonding or nonbonding with themselves.

The interaction of the H 1s orbital function with the fully occupied a_{1g} state of the cluster results in a drastic lowering of this level. The antibonding combination lies well above the HOMO-LUMO gap. Therefore, the number of bonding cluster states is unchanged when the interstitial is inserted into the cluster, and only the number of electrons in these states is increased by the one electron of the H atom. The stabilization of the cluster is not primarily due to the added electron since it enters the nearly nonbonding HOMO, but is instead due to the lower energy of the originally M–M bonding a_{1g} that becomes essentially localized at the H atom. This description of the chemical bonding of the H atom in a cluster is identical with the description of the bonding in metal hydrides except for the different terms used. In NiH_x or PdH_x an "impurity band" having strong H 1s character is formed from low lying d band states and occurs below the conduction band. The additional electrons added by the interstitial H atoms fill holes in the band at the Fermi level. [72–74] This general view will be discussed once again when we analyse the bonding in hydride halides of the rare earth metals in section 6.5.

Moving from Nb to its more electron deficient neighbor Zr creates a qualitatively new situation. The d elelctron concentration is too low to allow for the formation of the $[M_6X_8]$ cluster. The number of electrons is even too low for the $[M_6X_{12}]$ type cluster. Unlike Nb_6I_{11}, where the cluster centers can be reversibly occupied by impurity atoms, interstitial atoms are a necessary requirement for the formation of the $[Zr_6X_{12}]$ unit. All the $[Zr_6Cl_{12}]$ cluster compounds which have only 9 or 10 electrons in M–M bonding states and were claimed to be free of any interstitials later turned out to be indeed stabilized by interstitials. The presence of interstitials is also a necessary condition for the occurrence of the M_6X_{12} cluster with the even more electron deficient rare earth metals.

The $[M_6X_{12}]$ cluster is able to incorporate a variety of atoms as clearly seen in Table **5-3**. Those examples having single B, C, or N atoms are not the most surprising ones but rather those which contain atoms of other less electropositive transition metals in the cluster centers. The previously discussed bonding scheme for an interstitial hydrogen atom needs only a minor modification in order to understand the bonding of these other interstitial atoms. [75, 76] When nonmetal atoms are the interstitials, the 2s (a_{1g}) and 2p (t_{1u}) states of the central atom strongly mix with the cluster states that have the same symmetries. The antibonding combinations lie well above the HOMO-LUMO gap. Therefore, the relevant number of orbitals is the same as for the empty cluster. Only their energies are lower and the number of electrons in these states is increased by the number of valence electrons of the central atom. The highest lying effectively nonbonding a_{1u} level can be either occupied or not as seen in the series of compounds $Zr_6I_{12}C$, $CsZr_6I_{14}C$, and $Zr_6I_{14}C$ in much the same way as in the case of empty clusters like $[Ta_6Cl_{12}]^{n+}$, $n = 2, 3$, or 4. [77, 78] The occupation of the M–M bonding states by 14 electrons is somewhat preferred for the interstitial stabilized $[Zr_6X_{12}]$ clusters and is similar to the situation with the empty $[Nb_6O_{12}]$ clusters. In spite of this similarity, there is a marked difference between both cluster species. Whereas the $[Zr_6X_{12}]$ cluster incorporates Si as an interstitial in $Cs_xZr_6I_{14}Si$ [79] so as to raise the number of electrons by four, Si is located outside the $[Nb_6O_{12}]$ cluster as part of a $[SiO_4]$ tetrahedron from where it simply donates its four electrons to the cluster in, for example, $Na(Si_xNb_{1-x})Nb_{10}O_{19}$. [59] This clearly indicates that the competition in forming different types of bonds (Si–Zr *vs.* Si–I or Si–Nb *vs.* Si–O) finally decides the position of the additional atom, that is, whether it resides in the cluster center or between the clusters.

The cluster in $Zr_6I_{12}C$ has the same number of available MO's and electrons as the empty cluster in Nb_6Cl_{14}. The essential difference, however, is that all 16 electrons in the $[Nb_6Cl_{12}]$ cluster occupy M–M bonding states, whereas in the $[Zr_6I_{12}C]$ cluster the strong bonds between the anionic C and the surrounding cationic Zr atoms only allow a few direct M–M bonds. The population analysis leads to a charge of –1.8 on the C atom. [24] It is therefore clear that although the formulation $[(Zr^{4+})_6(I^-)_{12}C^{4-}\ (e^-)_8]$ ignores the covalency between Zr and C (and I), it still yields a realistic estimate of the number of electrons involved in M–M bonding.

Of particular interest is the introduction of transition metal atoms into the $[M_6X_{12}]$ clusters of the rare earth metals Zr and Th. The chemical bonding of the 3d metal atom is illustrated by the MO diagram for the $[Zr_6I^i_{12}FeI^a_6]^{4-}$ cluster in Figure **5-7**. [90] A more recent treatise of the bonding of interstitial transition metal atoms in zirconium and rare earth metal clusters has been given. [91, 91a] There are two main differences between the bonding of interstitial nonmetal atoms and that of interstitial d metal atoms. The polarity of the interstitial bonds is smaller in the case of a d metal atom. The a_{1g} state is primarily Zr–Zr bond-

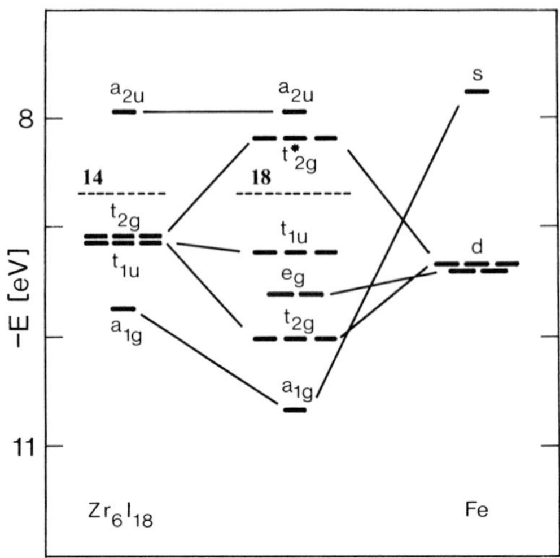

Figure 5-7. MO diagram for a regular $[Zr_6I^i_{12}I^a_6]$ cluster with a central Fe atom.

ing, and moreover, four additional electrons can be localized at the interstitial d metal atom (e_g), which have no bonding interaction with the Zr_6 frame. This calculation suggests 18 electrons in metal centered orbitals as the magic number for this type of cluster. Since, however, the energy of the t_{2g}^* level depends on the interstitial, even more electrons can be accommodated in this cluster type as seen with $Th_6Br_{15}Ni$. [84]

What criterion decides whether an element is able to contribute as an interstitial atom to the stabilization of an electron deficient cluster? Obviously, the M_6 octahedron bears some relation to a microscopically small piece of metal. The vibrational frequency of the H atom in the $[Nb_6H]$ unit of $Nb_6I_{11}H$ is nearly identical to the frequency of H in the metallic hydride NbH_x. [92] Since the bonding in the cluster and that in the extended metal lattice are so similar, the obvious question to ask is which elements form stable compounds with the bulk metal that represents the cluster atoms. The answer to this question yields a qualitative explanation for the fact that the electron deficient Nb_6 unit incorporates H, while the Mo_6 octahedron does not. Zr forms numerous intermetallic compounds with Be, Al, and other d metals and, obviously, does not loose this ability when only six Zr atoms are joined. Inspection of the experimental data for the relevant binary systems or the use of Miedema's concept for the stability of intermetallic systems [93, 94] proves helpful in the search for possible interstitial atoms, and naturally limiting the search to atoms of appropriate size to fit into the octahedral site. Of course, if the intermetallic compounds are very stable they could also compete with the cluster compound formation.

The interstitials discussed so far have been single atoms which reside in the center of the M_6 octahedron. In the large octahedral voids presented by the rare earth metals, also molecular groups like C_2 can be accommodated. They play an important role in numerous phases containing condensed clusters which are discussed later. Figure **5-8a** shows the $[Sc_6I_{18}C]$ cluster as it occurs in the compound

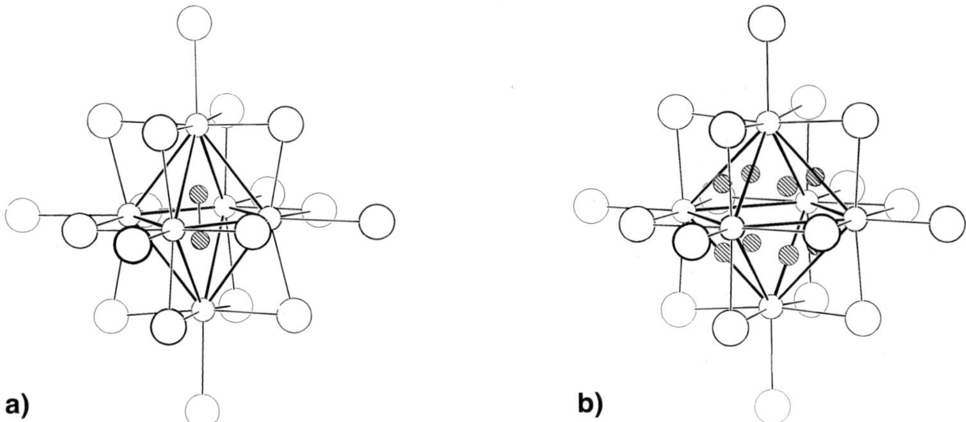

Figure 5-8. a) $[Sc_6I^i_{12}I^a_6C_2]$ cluster in the compound $Sc_6I_{11}C_2$. b) $[Th_6Br^i_{12}Br^a_6D_7]$ cluster in the compound $Th_6Br_{15}D_7$ (7 D atoms are statistically distributed over 8 equivalent positions above the faces the octahedron).

$Sc_6I_{11}C_2$. [95] As with another example of a C_2 unit in a discrete $[M_6X_{12}]$ type cluster, $Cs_2Pr_6I_{12}C_2$, [83] the distance d(C–C) = 139 pm is significantly shorter than that of a conventional single bond. The question arises for the first time as to how the excess electrons of the cluster framework (7 for $Sc_6I_{11}C_2$ and 8 for $Cs_2Pr_6I_{12}C_2$) are to be distributed in the competition between filling molecular orbitals of the C_2 units and filling M–M bonding orbitals of the cluster. Molecular orbital calculations on C_2 centered single [95] and double octahedra clusters [96] will be discussed later in detail and give a clear answer to this question. The MO's of the C_2 unit is filled up to the π^* level. Hence, the C_2 unit accepts 6 electrons. The surplus of electrons then enters a purely M–M bonding level which, in the case of the $[M_6X_{12}C_2]$ cluster, belongs to the b_{2g} representation derived from the t_{2g} of the regular cluster upon tetragonal distortion. The short C–C bond is evidence for a backbonding from the filled C π^* orbitals into empty metal d states. Such covalency effects lead to very short distances between the C atoms and the apical M atoms in the cluster.

A unique cluster which is stabilized by a large number of “interstitial” hydrogen atoms was found recently. The compound $Th_6Br_{15}D_7$ [84, 97] has the same arrangement of the heavy atoms as found in the structure of Nb_6F_{15} (Fig. **5-8b**). The seven D atoms are statistically distributed over the eight equivalent positions above the octahedral faces. They lie slightly out of the octahedron. This displacement is easily explained in terms of the electrostatic repulsion between the deuteride ions which exhibit very short D–D distances of 205 pm. Recalling the valence bond model for the closed configuration of the $[M_6X_{12}]$ type cluster – eight $2e^-$-3c bonds with maximum overlap slightly above the octahedral faces – the positions of the D atoms are also optimal for covalent bonding. The structure provides a nice example for the coloration of chemical bonds. There is a $2e^-$-3c bond in one face, and the other multicenter bonds are decorated by a hydrogen atom in the center. Thus, the cluster reaches its magic number of 16 electrons in the skeleton. The hydrogen atoms formally add their seven electrons to the electron deficient empty cluster which itself has 24–15 = 9 electrons available for M–M bonding. Of course, this counting of electrons does no justice to the chemical bonding. As in the simple case of $Nb_6I_{11}H$, the hydrogen atoms are rather hydridic. The a_{1g}, t_{2g}, and t_{1u} derived orbitals are no longer M–M bonding and it is only the a_{2u} orbital which is left as a weakly M–M bonding (occupied) level in $Th_6Br_{15}H_7$. [84] It will be seen later that very similar arguments can be used to understand the band structures of condensed clusters with hydride halides.

The example of $Th_6Br_{15}H_7$ shows that sites other than the central positions can be occupied by the "interstitials" which stabilize an electron deficient cluster. The result should be transferable to other metal clusters, too. The empty 10 electron cluster $[Zr_6X_{12}]^{2+}$ was prepared from solution [98] and will easily take up hydrogen. As calculations show, [99] an energy minimum is reached with a tetrahedral arrangement of 4 H atoms above the faces of the Zr_6 octahedron.

A particularly interesting development concerns the solution chemistry of $[Zr_6Cl_{12}Z]$ clusters. [100, 101] Alkali metal salts of the kind $A_xZr_6Cl_{12+x}Z$ can be dissolved (e. g. in CH_3CN in the presence of a cryptand) and the clusters are then isolated as R_4N^+ or $(C_6H_5)_4P^+$ salts. This solution chemistry of interstitially stabilized clusters entered via solid state preparations promises great versatility and adds new facets to the traditional cluster chemistry of the neighboring elements Nb and Ta.

5.4 Condensed Empty Clusters

As described in section 5.3 the stepwise lowering of the nonmetal to metal ratio in compounds having $[M_6X_8]$ and $[M_6X_{12}]$ clusters leads to an increased interconnection and an associated change from semiconducting to metallic behavior. Still, in the most metal rich systems, *i. e.* the $[Mo_6\hat{X}_8]$ Chevrel phases, the relative number of $\hat{X}$ atoms is sufficiently high so as to completely occupy all the inner positions of the cluster and even the X^a positions are all occupied by $\hat{X}$ atoms of adjacent clusters. Intercluster M–M bonding has played only a minor role up to this point.

Further reduction of the nonmetal to metal ratio creates a qualitatively new situation as now not only nonmetal atoms but also metal atoms have to be shared between clusters. Such condensation can occur via corners, edges, or faces of the M_6 octahedron. In fact, numerous structures of metal rich compounds composed of d metals and p elements can be interpreted in terms of condensed clusters which form characteristic partial structures. A more comprehensive treatise of the concept has been given earlier. [102] At that time, emphasis was laid on a scheme designed to order a large body of related structures. The concept was looked upon as a rather didactic tool. [103] In the years since, the concept proved to serve as a valuable guideline to unravel the chemistry of new systems that have the character of "oligomers" and "polymers" formed from monomer clusters.

The condensation of $[M_6X_8]$ or $[M_6X_{12}]$ type clusters via corners of the M_6 octahedron is the most simple case since the monomeric unit is kept complete in this process. Whereas many such structures based on the $[M_6X_8]$ type arrangement had been known, the structural chemistry of corner condensed $[M_6X_{12}]$ clusters gained importance only recently. It will be discussed later.

The condensation of $[M_6X_8]$ clusters via opposite corners results in a straight chain which has the composition $M_{2/2}M_4X_{8/2} = M_5X_4$. The M as well as the X atoms are shared between adjacent clusters. This type of cluster chain was first found in the structure of Ti_5Te_4. [104] There are quite a number of isotypic compounds for M = V, Nb,Ta,Mo and X = S,Se,Te,As,Sb. Ti_5Te_4 (12 electrons) and Mo_5As_4 (18 electrons) [105] indicate the wide range in the electronic balance which is possible. Calculations for the discrete chains which took into account the distortions observed in the real compounds led to a result showing that electronic gaps or regions of low densities of states existed for the ranges 12 to 13 and 17 to 18 electrons, making the experimental observation of a certain concentration of compounds with these specific electron counts understandable. [106]

The Ti_5Te_4 structure is shown as a projection along the tetragonal axis in Figure **5-9**. The complete $[M_6X_8]$ clusters in the chains are easily recognized. The interatomic contacts ($M–X^{a-i}$) are the same as found in the Chevrel phases. For a single chain, the bonding is easily derived from the bonding in the discrete cluster. [6, 25] The ordering of the d states remains principly the same as in the dis-

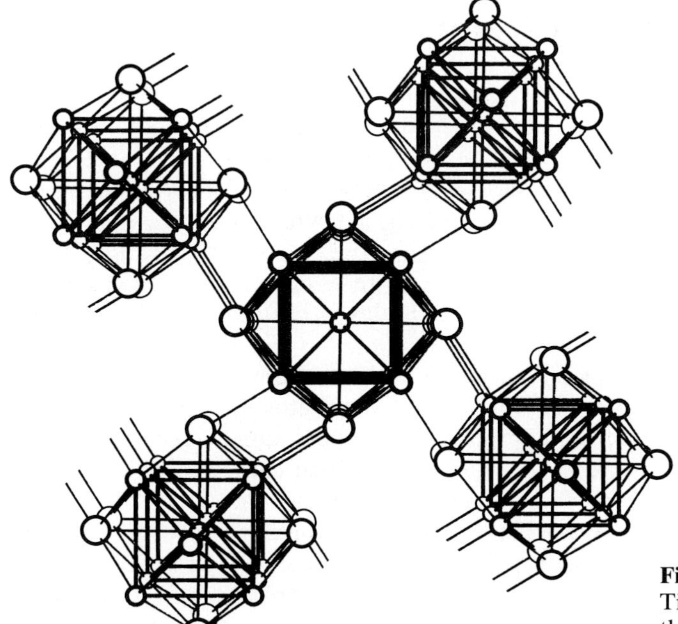

Figure 5-9. Structure of Ti_5Te_4 projected along [001] of the tetragonal unit cell.

crete cluster. The four atoms in the octahedron base are not involved in the condensation and their valence orbitals combine ($2e^-$-2c bonding) to form four bonding and four antibonding MO's. In contrast, the axial M atoms each bond to the atoms of two adjacent octahedron bases via four degenerate $2e^-$-3c bonds and give rise to four bonding, nonbonding, and antibonding combinations. Despite some rather large dispersions in the bands, their origin from the discrete MO's of the cluster is still recognizable, at least for low energies.

The full band structure calculation for the three dimensional M_5X_4 structure, however, leads to a greatly modified view of the bonding. [107] The deviations are due to large distortions of the M_6 octahedra in conjunction with the close approach of adjacent cluster chains. The M_6 octahedra in the structures of all the M_5X_4 phases are compressed in such a way that the M–M distances in the octahedron bases are only slightly shorter than the intercluster distances (322 pm *vs.* 343 pm in Ti_5Te_4). The M–X distances between chains (d(Ti–Te) = 277 pm) are of the same length or even shorter than those in the chains (277, 282, 295 pm). [104] The population analysis reveals no bonding along the edges of the octahedron bases. The atomic as well as the electronic structure both indicate that the metal substructure in the M_5X_4 phases is closely related to the bcc arrangement as previously proposed by Franzen. [108]

The M_5X_4 chain is a characteristic structural component of a variety of compounds with quite different compositions. The beautiful structure of $Nb_{21}S_8$ [109] is show as one example in Figure **5-10**. The structure contains M_5X_4 chains as well as parallel linear units formed by the condensation of four such chains. In this process of condensation, the X atoms of the single chains are largely substituted by M atoms of an adjacent chain. All Nb_6 octahedra are compressed to such a degree that the inner atoms of the unit have a cubic environment very similar to that found in the bcc structure of Nb metal. The structure gives the impression of an eutectic separation which stops within atomic dimensions and leads to a suspension of metallic Nb in a sulfide matrix. In fact, as mentioned in the Introduction, it is only a matter of the viewpoint taken whether the structures of metal rich compounds are discussed in terms of condensation of monomers or as pieces cut out from the structures of the elemental metals. In this sence, the polycyclic aromates could also be considered as derivatives of either benzene or graphite.

$Nb_{21}S_8$ represents a particularly impressive intergrowth structure of M_5X_4 type chains and chains formed from four of these single chains according to the formula $Nb_{21}S_8 = Nb_5S_4 \cdot Nb_{12}S_4 \cdot Nb_4$ whereby the additional Nb atoms are found between the chains. Figure **5-11** shows that a family of metal rich compounds composed of d metals with p elements exists which exhibits structures strictly bases on the same kind of units, namely nonmetal atoms generally in trigonal prismatic environments together with one dimensional fragments of the bcc structure. Considering the many possible ways of varying both the number of chains which are to be condensed to form such fragments, and the way in which the different types of units can be put together, the relatively few known examples of structures containing condensed M_5X_4 chains seem to represent a tip of an iceberg. The structure of Ta_2Se is just a recent result [110] which shows the extension into a two dimensional M–M bonded framework based on condensed M_5X_4 type chains.

Figure 5-10. Projection of the tetragonal structure of $Nb_{21}S_8$ along [001].

A few structures formed from $[M_6X_{12}]$ type clusters condensed via corners of the M_6 octahedra have been known for a long time. Thus, the structure of the low temperature modification of TiO is an assembly of parallel chains of *trans* corner sharing $[Ti_6O_{12}]$ clusters. [102] The single chain has the composition $Ti_{2/2}Ti_4O_{8/2}O_4 = Ti_5O_8$, and it is obvious from the composition that in the structure of TiO the O atoms must be shared between adjacent chains. In the structure of NbO, the clusters are condensed via all apex atoms leading to a rigid three dimensional framework with extended M–M bonding [5] according to $Nb_3O_3 = Nb_{6/2}O_{12/4}$. The close connection between the chemical bonding in an isolated $[M_6X_{12}]$ cluster and in the extended arrays of TiO and NbO has been impressively illustrated by band structure calculations. [111, 112]

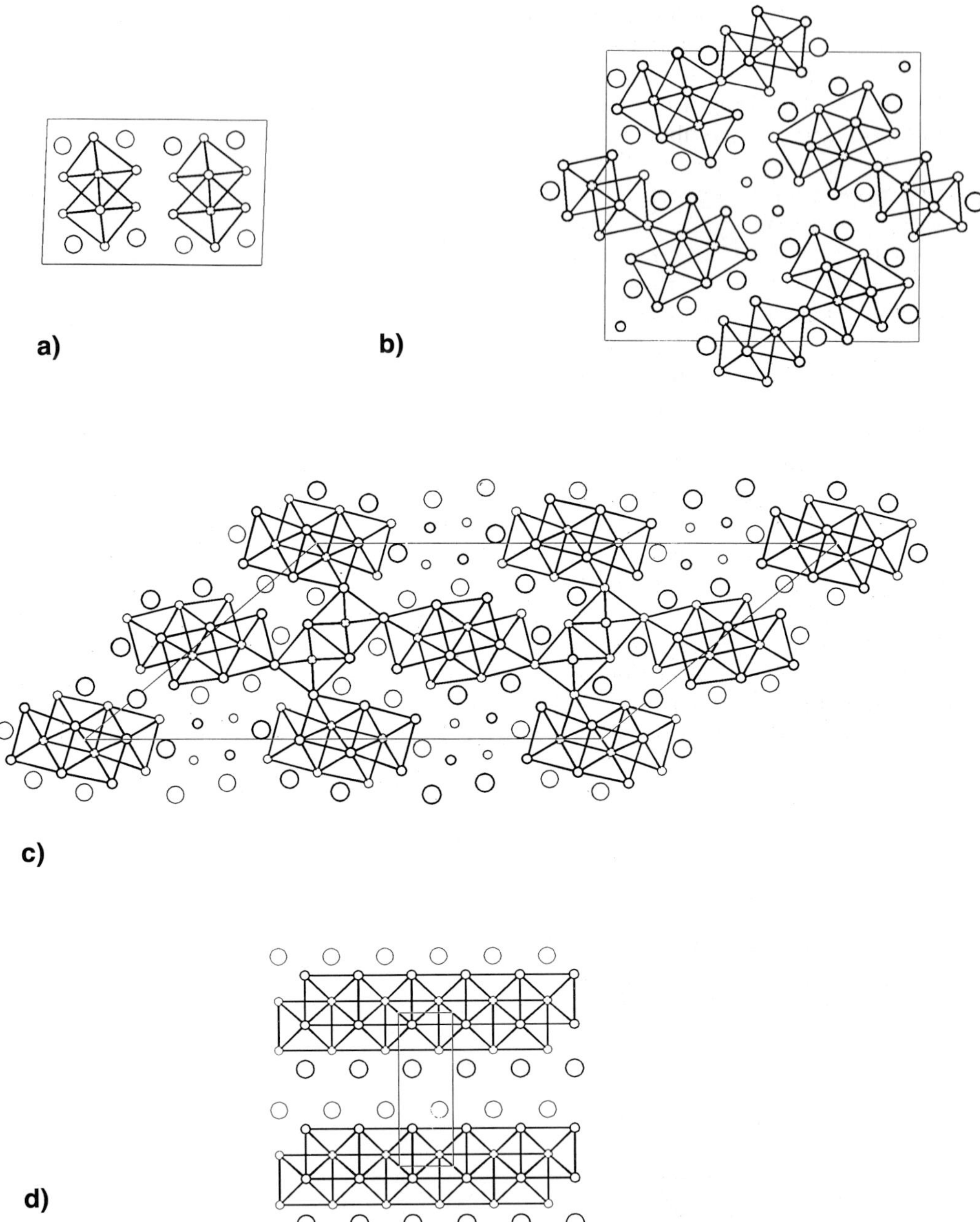

Figure 5-11. Structure based on condensed M_5X_4 type chains (references cited in [102], for Ta_2Se see [110]). a) Nb_2Se, b) $Nb_{14}S_5$, c) Ti_8S_3, and d) Ta_2Se. (Projections along the short axes; thin and bold lines are used for atoms in height 0 and 1/2 respectively; unit cells outlined).

The structures of ordered TiO and of NbO are closely related to the rocksalt structure but with 1/6 and 1/4 of the respective positions of both M and O atoms being unoccupied. The reduced lattice energy due to the vacancies has to be compensated for by M–M bonding. The proportion of vacancies therefore increases with the number of d electrons available for M–M bonding. The rocksalt structure is stable for d^1 systems (*e. g.* TiN, NbC, etc), the structure of the TiO d^2 system lies between the rocksalt structure and the structure of the NbO d^3 system.

There is an interesting relation between TiO and compounds composed of discrete electron deficient clusters which are stabilized by interstitial atoms. TiO is an obvious borderline case. Having only a few electrons which can be used for M–M bonding, the arrangement with only empty clusters is easily transformed into a (statistically disordered) rocksalt structure at elevated temperature. Now, a portion of the ligand atoms plays the role of the interstitials.

The discrete $[Nb_6O_{12}]$ cluster had been known for many years to occur in reduced oxoniobates [58, 113] and the final product of the condensation of this cluster via all corners of the octahedron (NbO) is all too familiar also. One could expect a rich variety of condensed clusters which fall between these two extremes, and this is indeed the case. [16] The first step in the condensation is realized with the $[Nb_{11}O^i_{20}]$ cluster shown in Figure **5-12** and it is formed from two apex sharing $[Nb_6O^i_{12}]$ type clusters. It occurs in the compound $K_4Al_2Nb_{11}O_{21}$. [114]. All outer terminal positions on the 10 Nb atoms which are not involved in the condensation are occupied by X^a type atoms. In contrast to the M_5X_4 chain, the octahedra are rather regular in the $Nb_{11}O_{20}$ cluster, and in fact the Nb–Nb distances involving the central Nb atom are the longest ones. Although the inter-

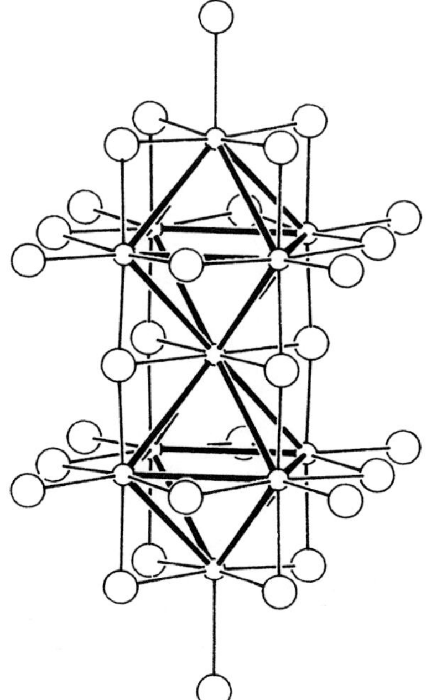

Figure 5-12. $[Nb_{11}O^i_{20}O^a_{10}]$ cluster in $K_4Al_2Nb_{11}O_{21}$ (small circles Nb, large O).

cluster Nb–Nb distances of 308 pm are also rather short compared to the average intracluster distance of 284 pm, the bonding analysis via COOP calculations indicated that any M–M bonding between the clusters is negligibly small.

Figure **5-13** shows the development of the M–M bonding orbitals for oxoniobate clusters with an increasing degree of condensation and includes the hypothetical $[Nb_{16}O^{i}_{28}O^{a}_{14}]$ cluster. The condensation of two Nb_6O_{12} clusters raises the energy of an e′ orbital originating from the t_{2g} state of the single octahedral cluster. A full band structure calculation for $K_4Al_2Nb_{11}O_{21}$ together with a COOP analysis indicates that this e′ orbital is dominated by Nb–O antibonding interaction in a similar way as the a_{2u} orbital is in the single octahedral cluster. This is due to the increased Nb–O π^* interaction for those O atoms which are "squeezed" between the two octahedra, and due to the weakening of the M–M bonding character of this orbital because of the symmetry determined node at the central Nb atom. The $[Nb_6O_{20}]$ cluster should therefore be most stable with 24 electrons in M–M bonding states. In fact, this optimal electron count could be experimentally realized with the synthesis of $K_4Al_2Nb_{11}O_{20}F$.

Whereas our knowledge on "oligomeric" clusters built from corner sharing $[Nb_6O_{12}]$ "monomers" is restricted to this one example, there exists a rich chemistry of "polymers" (Table **5-4**).

The topological similarity of the $[Nb_6O_{12}]$ cluster fragment in the NbO structure and a corresponding fragment of the perovskite type $BaNbO_3$ structure, illustrated in Figure **5-14**, is one important prerequisite for an intergrowth of both structures.

Figure **5-15** shows the fascinating High Resolution Transmission Electron Microscopy (HRTEM) image of a so-called phasoid [115] formed in the BaO/

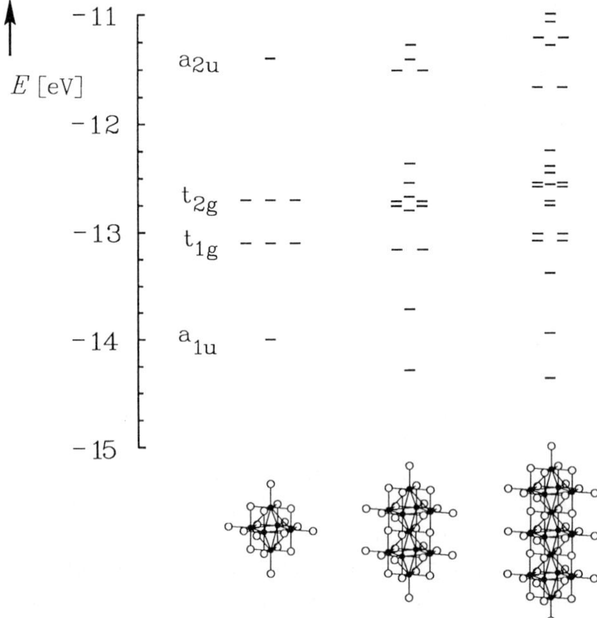

Figure 5-13. Molecular orbital diagrams for the $[Nb_6O^{i}_{12}O^{a}_{6}]$ and $[Nb_{11}O^{i}_{20}O^{a}_{10}]$ as well as the hypothetical $[Nb_{16}O^{i}_{28}O^{a}_{14}]$ cluster.

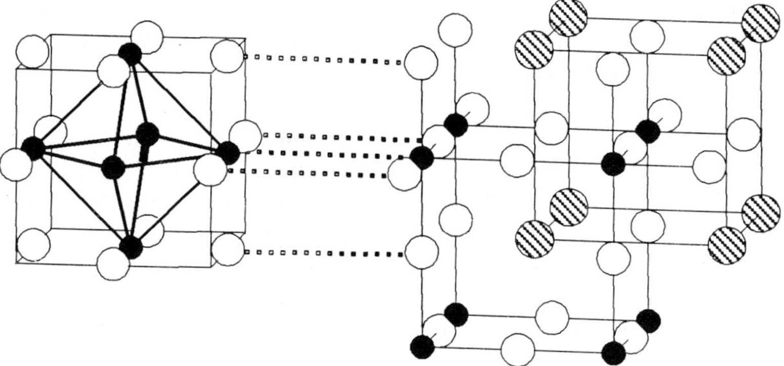

Figure 5-14. Topological similarity between the structures of NbO (left; a_o = 421 pm) and $BaNbO_3$ (right; a_o = 408.5 pm for $Ba_{0.95}NbO_3$).

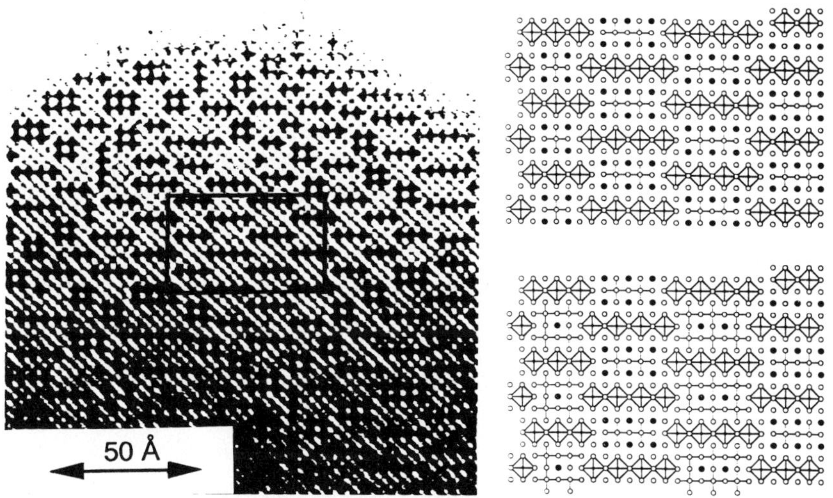

Figure 5-15. (left) HRTEM image of the α-phasoid in the Ba-Nb-O system. Chains of corner sharing Nb_6 octahedra are seen as dark crosses, the Ba and Nb atoms of the perovskite blocks as smaller spots in between. (right) Possible structural interpretations of the marked area.

NbO/NbO_2 system. [116] The metastable crystallite of such a phasoid exhibits considerable local order. Each dark cross corresponds to a chain of *trans* corner sharing Nb_6 octahedra. Several of these chains are condensed into larger one dimensional units which are intergrown with additional fragments of the perovskite structure represented by the light parts in the image. This phasoid clearly offers itself as a source of many different compounds having condensed cluster structures. It even contains well ordered microdomains which could possibly be "extracted" as single phases by an appropriate preparation. The compound $Ba_4Nb_{14}O_{23}$ might serve as an example. Its structure consists of triple chains of corner sharing [Nb_6O_{12}] clusters embedded in a matrix of the perovskite type structure. Figure **5-16** presents a HRTEM image [117] of the phase which was later also characterized by X-ray structure analysis. [117a]

The phasoid is composed of an intergrowth of one dimensional structural units. Any large excess of either NbO or $BaNbO_3$ seems to result in a layer-like intergrowth as shown in Figure **5-17** where single NbO type layers are suspended in a perovskite type matrix and vice versa. [118] An example of an ordered intergrowth as seen with $Ba_2Nb_5O_9$ is shown in Fig. **5-17c**.

The possible intergrowth structures between NbO and $BaNbO_3$ can be systematized [58, 115] by using the notations [p·q·r] and {s·t·u} respectively, where p, q, and r indicate the number of condensed Nb_6 octahedra in the three directions of space and s, t, and u characterise the perovskite type fragment. Hence, the structure of NbO corresponds to [∞·∞·∞] and the layered oxoniobates $Ba_2Nb_5O_9$, $BaNb_4O_6$, $BaNb_7O_9$, and $Sr_2Nb_8O_{12}$ – Figure **5-18** shows the unit cells – are described as [1·∞·∞] {2·∞·∞}, [1·∞·∞] {1·∞·∞} [2·∞·∞] {1·∞·∞} and [2∞·∞] {2·∞·∞}, respectively.

Those compounds containing corner condensed [Nb_6O_{12}] clusters which have been characterized by X-ray investigations are summarized in Table **5-4**. Some of them are depicted in Figure **5-19**.

Table 5-4. Unit cell parameters [pm] and properties of compounds containing condensed Nb_6O_{12} clusters. [16]

compound	space group	*a*	*b*	*c*	*Z*	colour	electrical[a] property
$K_4Al_2Nb_{11}O_{21}$	*I4/mmm*	879.9	–	1252.7	2	black	s
$K_4Al_2Nb_{11}O_{21-x}F_x$ ($x \approx 0.2$)	*I4/mmm*	878.6	–	1263.7	2	black	s
$Ba_{1-x}Nb_5O_8$ ($x = 0.2$)	*P4/m*	660.8	–	410.7	1	black	m
$Ba_4Nb_{14}O_{23}$	*Cmmm*	2080.7	1245.2	415.1	2	black	–
$Ba_{4-x}Nb_{17}O_{26}$ ($x = 1$)	*P4/m*	1210.0	–	414.8	1	black	–
KNb_4O_6	*P4/mmm*	413.9	–	825.4	1	black	m
$BaNb_4O_6$	*P4/mmm*	418.2	–	821.3	1	black	m
$K_2Nb_5O_9$	*P4/mmm*	413.8	–	1210.1	1	black	m
$Sr_2Nb_5O_9$	*P4/mmm*	414.1	–	1204.4	1	black	m
$Ba_2Nb_5O_9$	*P4/mmm*	417.2	–	1224.0	1	black	m
$Eu_2Nb_5O_9$	*P4/mmm*	413.8	–	1203.7	1	black	m
$BaNb_7O_9$	*P4/mmm*	419.5	–	1242.6	1	black	m
NbO	*Pm3m*	421.1	–	–	1	golden	m

[a] s = semiconducting, m = metallic

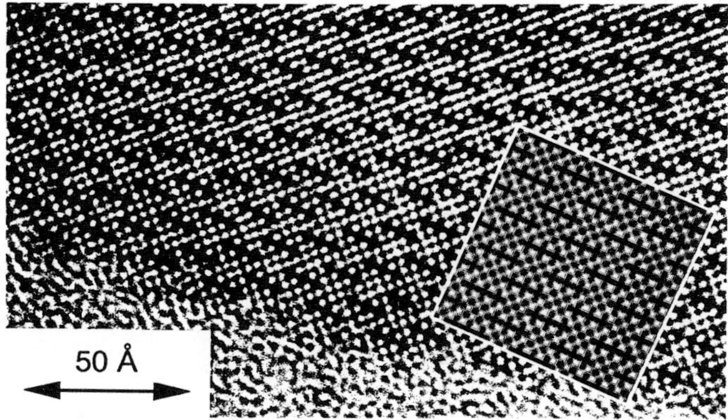

Figure 5-16. HRTEM image of $Ba_4Nb_{14}O_{23}$ (interpretation as in Figure **6-15**). The calculated image is shown as inset.

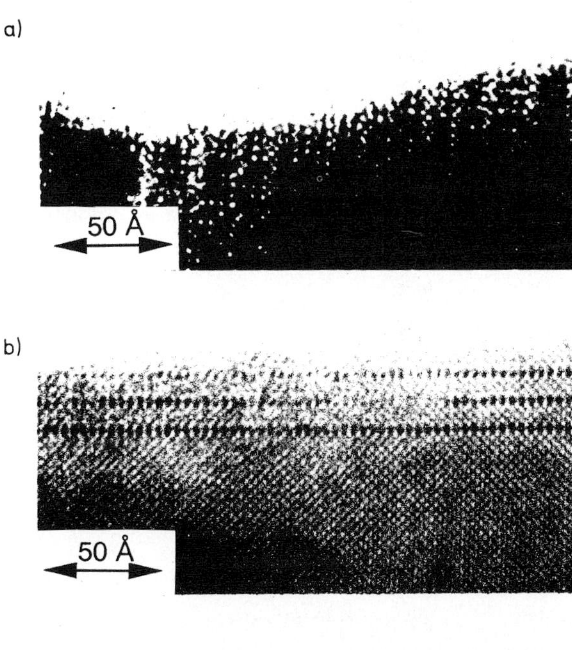

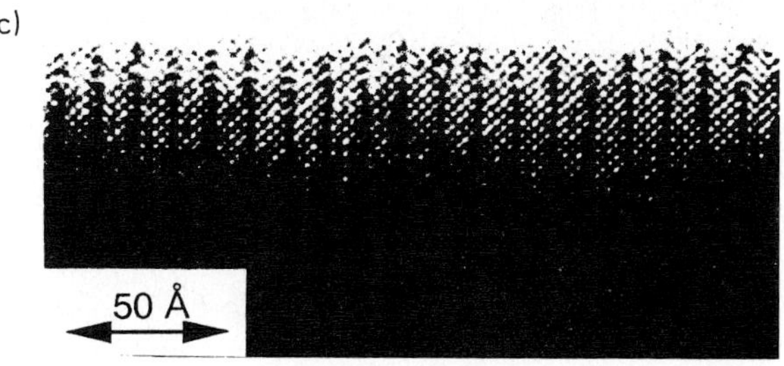

Figure 5-17. HRTEM images of layered intergrowth in the Ba-Nb-O system. a) perovskite layers in NbO, b) NbO layers in perovskite, and c) ordered intergrowth, in the compound $Ba_2Nb_5O_9$.

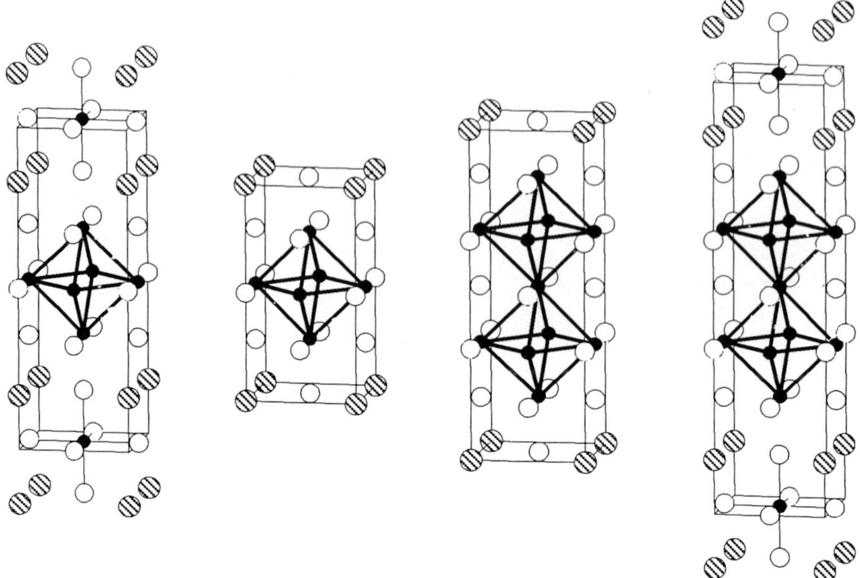

Figure 5-18. Structural relationships between (from left to right) $Ba_2Nb_5O_9$, $BaNb_4O_6$, $BaNb_7O_9$, and $Sr_2Nb_8O_{12}$ [Svensson, unpublished].

In order to arrive at a preparative strategy for the search of compounds with new kinds of fragments of the NbO structure, "magic" electron counts for known and yet unknown fragments need to be derived. Band structure calculations prove to be very helpful to this end. [58]

Figure **5-20** compares the band structures of compounds which contain chains ($BaNb_5O_8$), layers ($BaNb_4O_6$) and the three dimensional network (NbO). The results have to be compared with the band structure of a compound containing discrete Nb_6O_{12} clusters shown in Figure **5-5**. The DOS curves are in agreement with the metallic nature of the samples for a dimensionality of one and above. The COOP curves nicely illustrate the optimized band filling, that is the bands with predominately M–M bonding contributions are occupied, but as soon as Nb–O antibonding character becomes dominant they stay empty. Obviously the correct electron count is essential for the stability of these phases. Based on this conclusion, a simple rule of stability with high predictive power can be derived, for example, by starting from the optimized electron counts of 14 for the $[Nb_6O_{12}]$ cluster, 24 for the $[Nb_{11}O_{20}]$ cluster, and 3 for the NbO fragment (*i. e.* 3 valence electrons per formula unit left for M–M bonding).

As indicated in Figure **5-21**, the two terminal atoms in the $[Nb_{11}O_{20}]$ cluster are bonded as in the $[Nb_6O_{12}]$ cluster and therefore call for 2.33 electrons each. The central atom has an environment as in NbO and contributes 3 electrons. This leaves approximately 2.0 electrons from each of the other eight atoms in order to reach a number near the 24 electrons found in the M–M bonding states of the $[Nb_{11}O_{20}]$ cluster. By assigning the appropriate number of electrons to the atoms

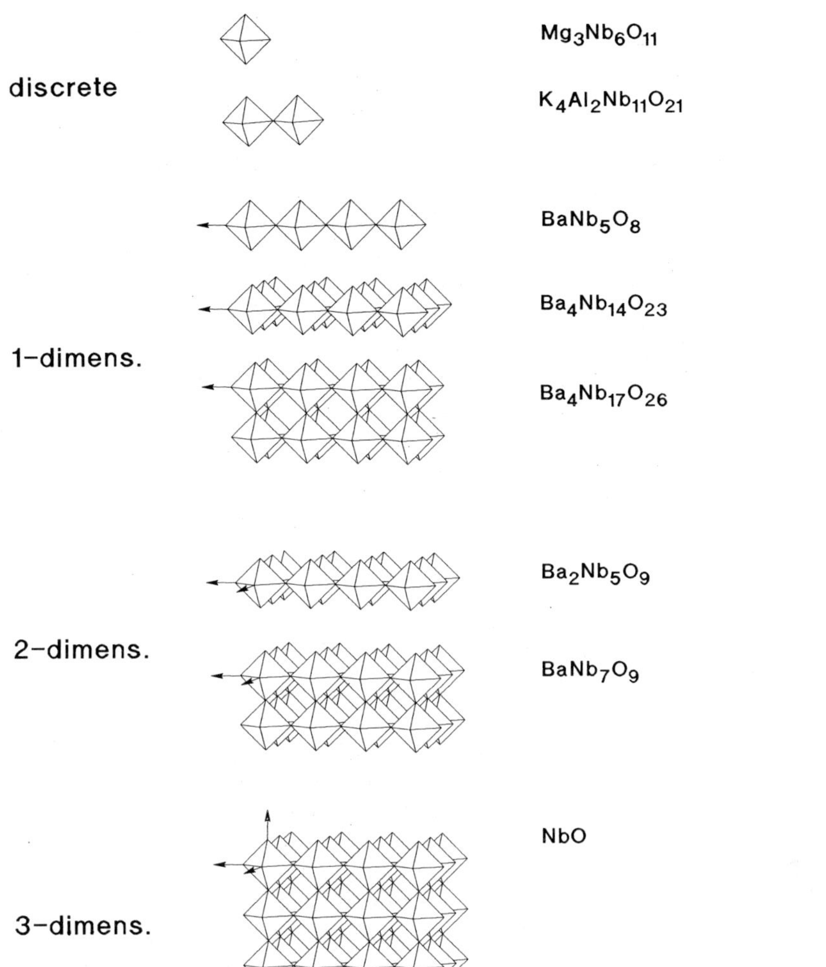

Figure 5-19. Condensed $[Nb_6O_{12}]$ clusters in reduced oxoniobates (schematic with only Nb_6 octahedra drawn).

according to their functionality in an arbitrary framework of corner condensed $[M_6O_{12}]$ clusters a prediction of the optimal electron concentrations for all kinds of oligomeric or polymeric structures exhibiting this mode of condensation is made in a simple way. As far as these predictions can be tested so far, the comparison with known compounds (Table **5-5**) indicates an excellent agreement with experiment.

When M_6 octahedra are condensed via apices the entire $[M_6X_8]$ and $[M_6X_{12}]$ clusters stay complete and the directions of the valence orbitals are not significantly changed as long as there is no severe distortion in the clusters due to condensation. The situation must change in the case of a condensation via edges or faces of the octahedra. In this mode of condensation, one X atom is replaced by

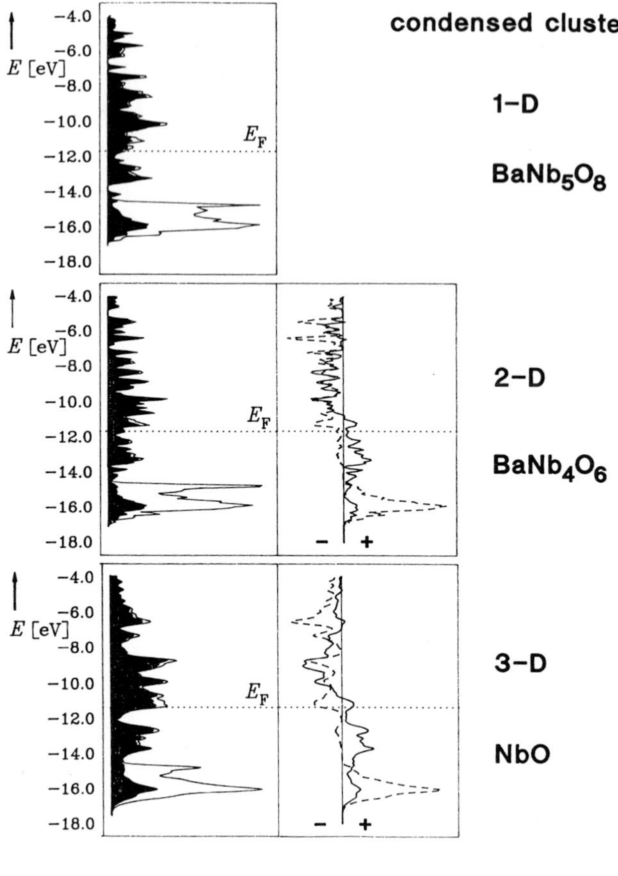

Figure 5-20. Band structures for (from top to bottom) $BaNb_5O_8$ (cluster chains), $BaNb_4O_6$, (layers), and NbO (three dimensional framework). On left side are the total DOS's (projected DOS's for the Nb atoms of the clusters in black) and on the right side the COOP curves for nearest neighbor Nb–Nb and Nb–O interactions (full and broken lines, respectively).

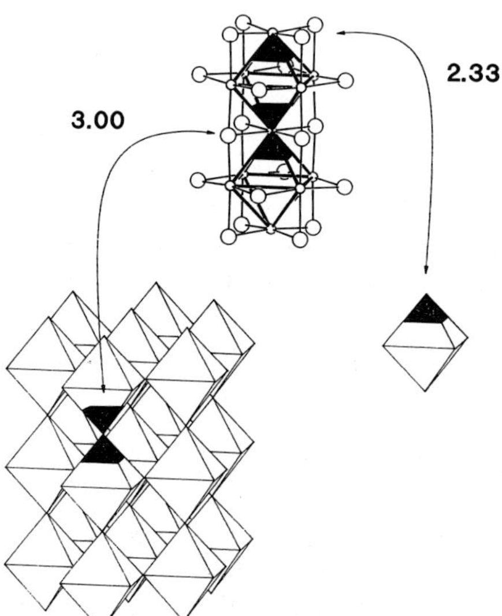

Figure 5-21. Derivation of the appropriate number of electrons for M atoms in M_6O_{12} type clusters condensed via corners of the M_6 octahedra. The M atoms are distinguished according to three different functionalities (see text).

Table 5-5. Optimal electron numbers for condensed Nb_6O_{12} clusters. [16]

composition	cluster framework	predicted NVE*/cluster	observed NVE*/cluster
discrete $[Nb_6O_{12}]$ clusters	$[1 \cdot 1 \cdot 1]$	14	13–15
$K_4Al_2Nb_{11}O_{21}$	$[1 \cdot 1 \cdot 2]$	23.67	24
$BaNb_5O_8$	$[1 \cdot 1 \cdot \infty]$	11	11
$Ba_4Nb_{14}O_{23}$	$[1 \cdot 3 \cdot \infty]$	31	31–32
$Ba_4Nb_{17}O_{26}$	$[2 \cdot 2 \cdot \infty]$	40	40–41
KNb_4O_6	$[1 \cdot \infty \cdot \infty]$	10	9
$BaNb_4O_6$	$[1 \cdot \infty \cdot \infty]$	10	10
$K_2Nb_5O_9$	$[1 \cdot \infty \cdot \infty]$	10	9
$Sr_2Nb_5O_9$	$[1 \cdot \infty \cdot \infty]$	10	10–11
$Ba_2Nb_5O_9$	$[1 \cdot \infty \cdot \infty]$	10	10–11
$Eu_2Nb_5O_9$	$[1 \cdot \infty \cdot \infty]$	10	10–11
$BaNb_7O_9$	$[2 \cdot \infty \cdot \infty]$	19	19
NbO	$[\infty \cdot \infty \cdot \infty]$	9	9

* Number of valence electrons

two or three M atoms respectively, which can only be bonded via drastically reoriented orbitals (rehybridization) compared to the discrete cluster. So a correlation of the electronic band structures of extended systems and discrete clusters is difficult. One might therefore ask whether a description of, for example, a chain of octahedra in terms of condensed clusters is appropriate. However, numerous compounds have been found in recent years which contain both "polymeric" and "oligomeric" clusters based on edge and face sharing M_6 octahedra. A kind of "macromolecular chemistry" with metal clusters has developed. Table **5-6** summarizes the different crystal structures of the known reduced oxomolybdates which contain chains of *trans* edge condensed Mo_6 octahedra.

The simple and particularly aestetic structure of $NaMo_4O_6$ [119] is shown as a projection along [001] in Figure **5-22a**.It contains parallel chains of edge sharing Mo_6 octahedra in which O atoms lie above all free edges. Thus, the fragments are related to the $[M_6X_{12}]$ type arrangement as illustrated with the side-on view of the chain in Figure **5-22b**.

The chain has the composition $Mo_{4/2}Mo_2O_2O_{8/2} = Mo_4O_6$. The Mo and O atoms which are not involved in the condensation process interconnect the chains according to $Mo_4O^{i}_4O^{i\text{-}a}_2$ which is similar to the connectivity in rutile. The cations

Table 5-6. Oxomolybdates whose structures contain chains of edge sharing Mo_6 octahedra. [11, 120–127]. The number, z, of metal centered valence electrons was derived from (a) the ionic formula and (b) from the bond-order sums Σs_i (Mo–O) for all Mo atoms. [128] Since compound *2* contains an excess of Mo in octahedral voids, the value of z derived according to (a) is minimal.

	Compounds		Interconnections of the chains	z/Mo_4 (a)	(b)
1	$A_xMo_4O_6$	A = Li–Rb	$Mo_4O^i_4O_2^{i\text{-}a}$	13.0	12.8
		In		13.0	13.1
		$Pb_{0.8}$		13.6	13.3
		$Sn_{0.9}$		13.8	13.4
	$Ba_5(Mo_4O_6)_8$			13.25	
2	$A_{2-x}A'_xMo_4O_7$	$A_{2-x}A'_x$ = $Sc_{0.75}Zn_{1.25}$	$Mo_4O^i_4O^{i\text{-}i}_{2/2}O^a_2$	14.7	14.5
		$Ti_{0.5}Zn_{1.5}$		14.5	14.4
		$Sc_{0.5}Fe_{1.5}$		14.5	14.3
3	$Mn_{1.5}Mo_8O_{11}$		$Mo_4O^i_3O^{i\text{-}a}O^{i\text{-}i}_{1/2}O^{i\text{-}a}$	14.5	14.0
4	$A_xMo_8O_{10}$	A_x = Li	$Mo_4O_4^{i\text{-}a}O^{i\text{-}i}_{2/2}$	14.5	14.3
		Zn		15.0	14.7
5	$Ca_{5.45}Mo_{18}O_{32}$		$Mo_4O^i_4O^a_2$	14.6	14.6
	$Gd_4Mo_{18}O_{32}$			14.4	14.4
6	$Ho_4Mo_4O_{11}$		$Mo_4O^i_4O^a_2$	14.4	14.1

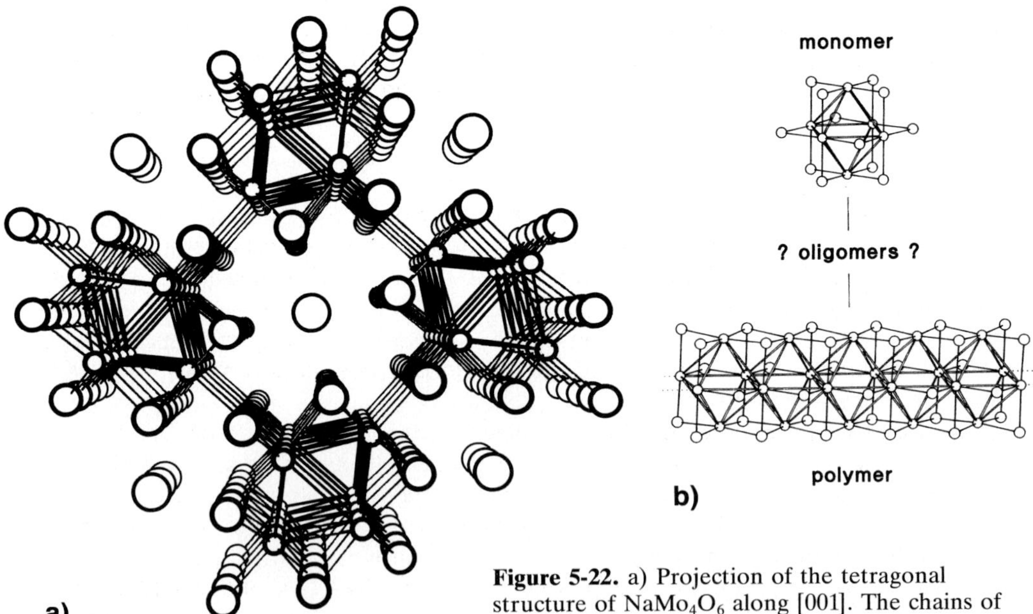

Figure 5-22. a) Projection of the tetragonal structure of $NaMo_4O_6$ along [001]. The chains of Mo_6 octahedra are outlined. b) Comparison of a discrete $[Mo_6O_{12}]$ cluster ("monomer") to the infinite Mo_4O_6 chain ("polymer").

occupy the cubic voids in the channels between the chains and the same structure is formed regardless of the cation employed. Pb occupies an acentric position in the cube and is coordinated in the typical pyramidal configuration of the "lone pair" Pb^{2+} ion. When the cations are In or Sn, the position occupied is approximately in the cube faces and the coordination is square planar. This special coordination, in conjunction with the short interatomic distances (*e. g.* *d*(In–In) = 286.3 pm), is evidence that M–M bonds occur in both the cationic and anionic parts of the structure.

The structures of the reduced oxomolybdates having cluster chains summarized in Table **5-6** have different patterns of interconnection between the chains. Whereas in **1**, **2**, **3**, **5**, and **6** the chains are parallel, in **4** they form layers which are stacked crosswise. In addition, **5** and **6** contain chains of single Mo atoms and ribbons of edge sharing M_4 rhomboids. Such ribbons are the only structural element in $NaMo_2O_4$. The bridging O atoms in these oxomolybdates are coordinated by up to four Mo atoms. Oxygen atoms coordinated in three- or fourfold planar as well as SF_4 like fashion are quite common.

The number of electrons in M–M bonding states can be estimated in several ways. The formal counting procedure via oxidation states used so far (*e. g.* $[Na^+(Mo^{6+})_4(O^{2-})_6(e^-)_{13})]$ gives results which are very similar to those obtained from a detailed analysis of all the Mo–O distances based on an empirical bond length-bond strength relationship. [128] According to this analysis, 13 to 15 d electrons populate M–M bonding orbitals in the Mo_4 fragment. At high electron counts, the sum of the M–M bonds as determined by Pauling's formula from the Mo–Mo distances is somewhat low. The logarithmic equation of Pauling might be too simple for an analysis of M–M bonds between heavy transition metal atoms [129] even more as the contact distances between the Mo atoms are strongly influenced by the bridging O atoms. On the other hand, the apparently low bond order sums could indicate that a part of the d electrons have entered nonbonding or antibonding states since chain distortions are typically found at high electron counts. For example, in **2** the Mo_6 octahedra are tilted around the common edges leading to alternating short and long distances between the apical atoms. In **3**, **4**, **5**, and **6** the apical atoms are pairwise associated and the corresponding common edge in the octahedron base is considerably lengthened. Of course, the cation distribution within the structures as well as the interchain connections will influence these distortions. In any event, the distortions are electronically favored, as indicated from Extended Hückel calculations on an undistorted chain compared to a chain deformed as in **2**. [61]

The essential features of the band structures of compounds containing Mo_4O_6 chains can be worked out from a one dimensional calculation if the dispersion of the bands is assumed to be small in directions which are orthogonal to the chain. The analysis of the bonding and antibonding contributions of the different kinds of Mo–Mo contacts in the projected density of states is particularly instructive. The corresponding diagrams are plotted in Figure **5-23**. As one would expect considering the crystal structure, the bonds along the joined edges are strongest and the bonds between apical atoms weakest. Thirteen electrons per Mo_4 fragment enter the bands without occupying antibonding states. Yet, even with 15

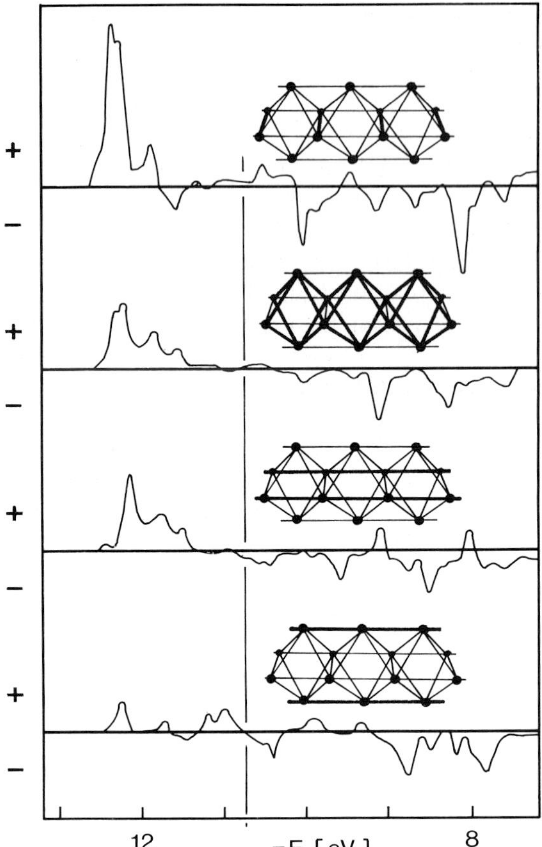

Figure 5-23. COOP diagrams for the different kinds of nearest neighbor Mo–Mo interactions in the chains of Mo_6 octahedra of $NaMo_4O_6$. The analysed types of interactions are indicated by bold lines. The vertical line refers to the Fermi level at E_F.

electrons the energy balance is still good because the weakening of the bonds along the chain direction is compensated by a strengthening of the bonds making up the joined edges. At high electron counts, the formation of one strong bond from two weak bonds between the apical atoms of adjacent octahedra leads to an lower energy.

The structures of the previously discussed oxomolybdates contain quasi infinite chains. This holds, for example, for the In molybdate $InMo_4O_6$, although it should be mentioned that superstructure reflections are always observed and have not been satisfactorily explained so far. Under weakly oxidizing conditions, phases of the general composition $In_{1+x}Mo_4O_{6+2x}$ can be prepared and are built up from finite parts of the infinite chain structure. X-ray and HRTEM investigations indicate that many structural variations are possibly present.

$In_{11}Mo_{40}O_{62}$ (*i. e.* $x = 0.1$) was the first compound from this series of phases to be discovered [130] and despite its complex structure it is readily accessible. The structure of $In_{11}Mo_{40}O_{62}$ is formed from alternating layers which contain clusters built from four and five octahedra respectively. The intercluster Mo–Mo distances are long enough to allow for a discussion of the clusters as discrete oligomeric units. As in the monomeric M_6X_{12} cluster, the oligomeric ones with edge sharing Mo_6 octahedra are surrounded by O atoms above all free edges. The structure of $In_{11}Mo_{40}O_{62}$ therefore contains alternating layers of parallelly oriented $[Mo_{18}O_{30}]$ and $[Mo_{22}O_{36}]$ clusters which are interconnected at the top and bottom according to $(Mo_{18}O_{26}O^{i\text{-}i}_{4/2})(Mo_{22}O_{32}O^{i\text{-}i}_{4/2})$. Bridging $X^{i\text{-}a}$ atoms link the clusters within the layers. In the channels between these clusters, one finds chain fragments of 5 and 6 In atoms respectively.

The $[Mo_{18}O_{30}]$ cluster and the In_5 chain with their nearest neighbor O atoms are shown in Figure **5-24**. Here, we have the interesting situation that both M–M bonded polycations and polyanions occur simultaneously in $In_{11}Mo_{40}O_{62}$. The In_5 chain can be formally decomposed into $(2 \times In^{2+}) + (3 \times In^{+})$ and thus described as In_5^{7+}. The summation of all In–O bond orders as calculated via the appropriate bond length-bond strength relationship affords the same value for the ionic charge as that for the anionic charge derived from all Mo–O distances. The formal charge on the In_6 chain fragment is calculated to be In_6^{8+} in the same way. The same assignment of charges for the polycations follows from an Extended Hückel calculation, [131] however, according to this calculation the charge distribution within the oxomolybdate framework rather corresponds to the presence of $[Mo_{18}O_{28}]^{6-}$ and $[Mo_{22}O_{34}{}^{9-}]$ anions.

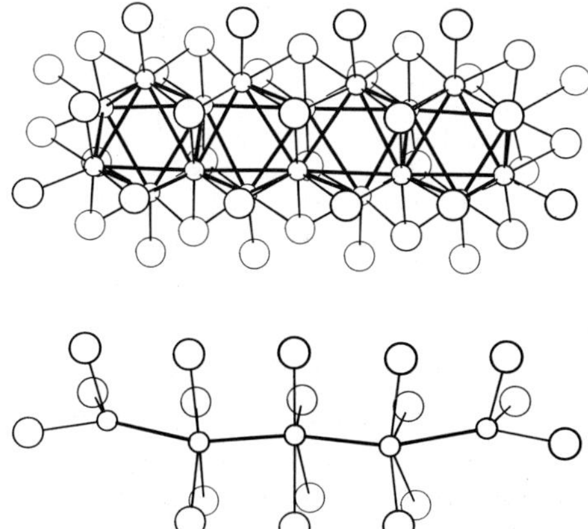

Figure 5-24. $[Mo_{18}O_{30}]$ cluster together with the polycationic In_5 chain in $In_{11}Mo_{40}O_{62}$.

X-ray investigations on single crystals always indicate some disorder, and HRTEM images allow a detailed analysis of the defects. [132, 133] Figure **5-25a** shows that in addition to a regular alternation of layers of clusters formed from four and five octahedra respectively, it is possible to have adjacent layers of short clusters followed by the same number (here two) of layers formed by long clusters without changing the overall composition. It has been found that fragments of crystals may exhibit a quite irregular pattern of such layers. Other regions contain well ordered domains which contain only the small or the large clusters (Fig. **5-25b**) and can be formulated as the discrete phases $In_5Mo_{18}O_{28}$ ($\simeq In_{1.11}Mo_4O_{6.22}$) or $In_3Mo_{11}O_{17}$ ($\simeq In_{1.09}Mo_4O_{6.18}$) respectively. It is clear from a comparison with the stoichiometric phase $In_{11}Mo_{40}O_{62}$ ($\simeq In_{1.10}Mo_4O_{6.20}$) that the

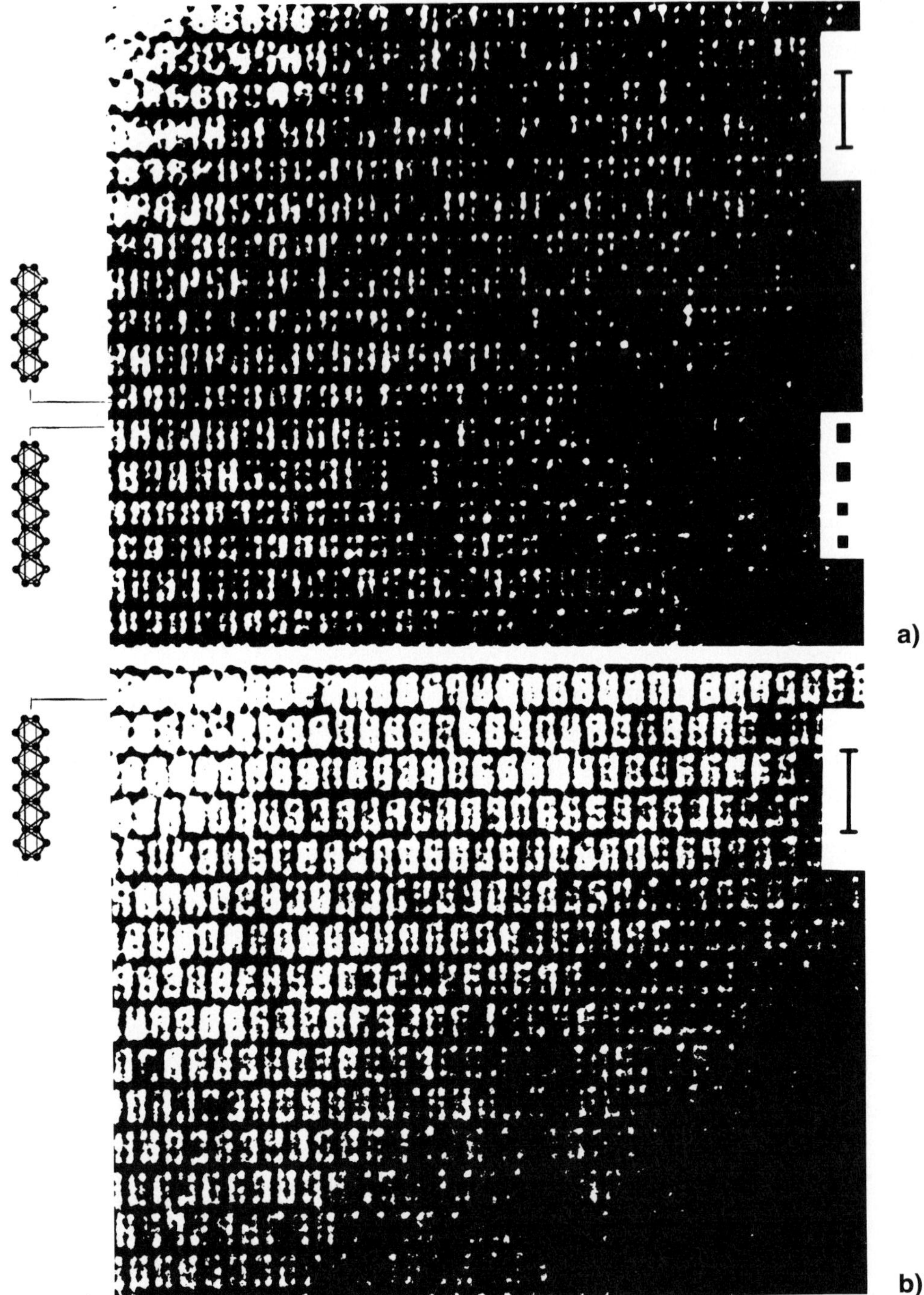

Figure 5-25. HRTEM images of a) $In_{11}Mo_{40}O_{62}$ exhibiting a regular sequence of layers of different cluster lengths which is interrupted by a layered fault as indicated, and b) $In_3Mo_{11}O_{17}$ showing layers of clusters formed from 5 condensed Mo_6 octahedra.

preparation of microscopically homogeneous products carries some severe intrinsic problems since neighboring phases differ only marginally in composition as illustrated by the given examples.

With the Ba oxoniobates we again meet the same "phasoid" problem discussed earlier. In spite of these inherent problems, discrete phases built from oligomeric oxomolybdate clusters could be isolated and characterized by single crystal X-ray analysis. As one expects, none of them is perfectly ordered. In addition to the local inhomogeneities discussed above, twinning frequently occurs and the cluster layers can be stacked in different ways to yield, in the simplest cases, either an orthorhombic or a monoclinic structure. In the structure of $In_{11}Mo_{40}O_{62}$, [130] the orthorhombic arrangement is the same as it is in the structure of $In_3Mo_{11}O_{17}$. [134] Domains of both stacking variants have been observed in the HRTEM images [133] of $In_5Mo_{18}O_{28}$. When even smaller clusters are present, the monoclinic structure always seems to be preferred [135–137] and this has interesting consequences for the intercluster contacts.

Besides the anionic cluster framework, the counter cations in the channels between the clusters also exhibit very interesting features in that they frequently form hitherto unknown polycations. For instance, indium in a +1 oxidation state normally has a lone pair configuration which can be either stereochemically active or inactive according to the known salt chemistry of this element. In the structures of these oxomolybdates, however, the rigid matrix of the cluster anions forces the In^+ cations into such close proximity that the lone pair configuration is transformed into two In–In bonds for each In atom. The bonds manifest themselves by an even closer approach of the atoms within the In chain (d(In–In) $\geq$ 262 pm) than required by the matrix. Hence, these oxomolybdates offer the possibility to stabilize hitherto unkown linear polycations of In with the help of suitable anionic clusters.

The compound $Tl_{0.8}Sn_{0.6}Mo_7O_{11}$ has presented a similar surprise. [136] It contains layers formed from oligomeric clusters of three edge sharing Mo_6 octahedra as shown in Figure **5-26**. The voids within the layers are occupied by Sn^{2+} and

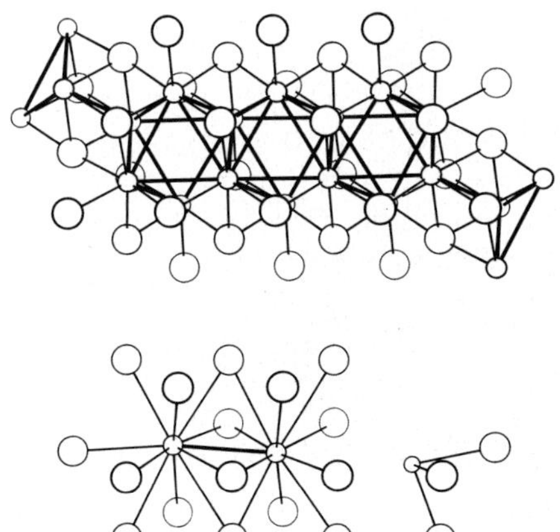

Figure 5-26. $[Mo_{14}O_{24}]$ cluster in $Tl_{0.8}Sn_{0.6}Mo_7O_{11}$, intercluster bonds indicated together with Tl_2^{4+} and Sn^{2+} ions which lie between the clusters.

Tl_2^{4+} ions in such a way that each channel contains only one of these ions, the ratio of them being closely linked with the correct electron balance for the oxomolybdate structure (*vide infra*). The Tl–Tl distance of 284 pm corresponds to a single bond.

Mentioning electron balances leads to the question of whether similar rules can be derived for determining an optimal electron count for condensed cluster oxomolybdates as was done for the oxoniobates. Extended Hückel calculations on oligomeric oxomolybdate clusters do not reveal a well defined HOMO-LUMO gap. Rather the envelope to the molecular orbital distribution quickly approaches the characteristic DOS for the infinite M_4O_6 chain. [131] A rough estimate of an optimized electron concentration is derived by an alternative route. Compounds from the $In_{1+x}Mo_4O_{6+x}$ phase which are based on the structural principle of the indium oxomolybdates in which the clusters are arranged in layers follow the general formula In_{n+1} $(Mo_{4n+2}O_{6n+4})$ if only one polyanionic cluster is present. The number of Mo_6 octahedra in the cluster is designated as n. The oligomeric clusters can be fragmented into $[M_4O_7]$ (top) and $[M_2O_5]$ (bottom) which together constitute a divided $[M_6O_{12}]$ cluster together with n–1 inserted $[M_4O_6]$ fragments taken from the infinite chain. The fragment orbitals correspond to those in the monomeric cluster and in the polymeric chain. Therefore, 14 electrons per $[M_6O_{12}]$ fragment in analogy to the $[Nb_6O_{12}]$ cluster and 13 to 15 per $[Mo_4O_6]$ fragment should be the appropriate numbers of electrons required by the metal centered d states in the oligomeric clusters. Within these limits, the allowed range of electron concentrations per Mo atom has been calculated for different values of n and is plotted in Figure **5-27**. The drawing shows a limited region where oligomeric clusters of specific lengths should be stable. A too low electron concentration does not allow sufficiently strong M–M bonding, and a too high electron concentration should also destabilize the cluster due to the occupation of bands with M–M antibonding character. Compared to an earlier presentation, [6] the region of stability is more narrow here because the electron count for the discrete $[M_6O_{12}]$ cluster is now restricted to 14. The occurrence of a 16 electron $[Mo_6O_{12}]$ cluster in $Ca_{16.5}Mo_{13.5}O_{40}$ [137a] is rather an exception in oxides, however raises the question, whether the restriction to 14 electrons for the $[Mo_6O_{12}]$ fragment in oligomeric clusters is allowed.

The plot was used as a guideline for the preparation of reduced oxomolybdates. [6] Working hypotheses are always subject to the possibility of surprises. According to the graph, the optimal electron concentration drastically decreases with decreasing cluster size. It was suggested, for example, that clusters with $n = 3$ should not be stabile with electron concentrations that are larger than 3.1 e^-/Mo. However, oxomolybdates composed of even smaller clusters, including the $[M_6O_{12}]$ entity itself found in the compound $BaMo_6O_{10}$, [138] can be prepared at significantly higher electron counts. In these cases, the excess of metal centered electrons is used for intercluster M–M bonding and is quite significant for $1 \leq n \leq 3$ [135–138c] due to a much closer approach of the clusters in the monoclinic structures than in $In_{11}Mo_{40}O_{62}$. The analysis for one of the compounds, $Tl_{0.8}Sn_{0.6}Mo_7O_{11}$, shows that the Mo–Mo bonds between adjacent clusters sum up to nearly a single bond per cluster. Taking this into account the 3.1 e^-/Mo

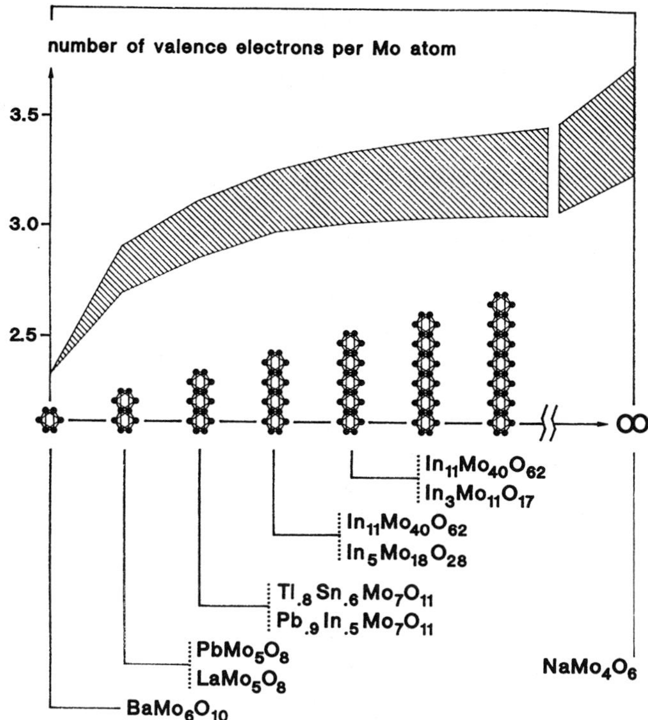

Figure 5-27. Stability range for discrete oligomeric clusters derived from a fragmentation procedure (see text). Representative compounds which contain these clusters are indicated.

value for intracluster bonding then closely matches the stability range presented in Figure **5-27**.

Last but not least, a particularly interesting result should be mentioned. The cluster $[Mo_8O_{24}]$ as it is found in the structure of $NdMo_8O_{14}$ represents a slightly different fragment of the Mo_4O_6 chain. [139, 140] The cluster consists of the complete $[Mo_6O_{12}]$ entity but has two additional Mo atoms attached above adjacent trigonal faces of the Mo_6 octahedron as seen in Figure **5-28**. The arrangement can also be viewed as a condensate of a complete $[Mo_6X_{12}]$ type cluster and one of its fragments, the butterfly shaped $[Mo_4O_9]$ cluster with an additional O atom in the center. The structure reminds us that the concept of cluster condensation has to make use of fragments of the octahedral clusters as well as totally different kinds of clusters for completeness. This aspect will be briefly addressed again in Section 5.6.

Of course, one may argue that cutting a structure down into "clusters" becomes an arbitrary procedure if intra- and intercluster distances are of comparable length. This arbitrariness may come into play even one step earlier, namely when the specific condensation of the $[M_6O_{12}]$ clusters via edges is accompanied by a loss of ligand positions and therefore the bonding pattern has to change with respect to the monomeric unit. These considerations also hold for the chemistry of a last variant of cluster condensation, that is, the condensation via faces of the M_6 octahedron which is particularly impressive and well developed.

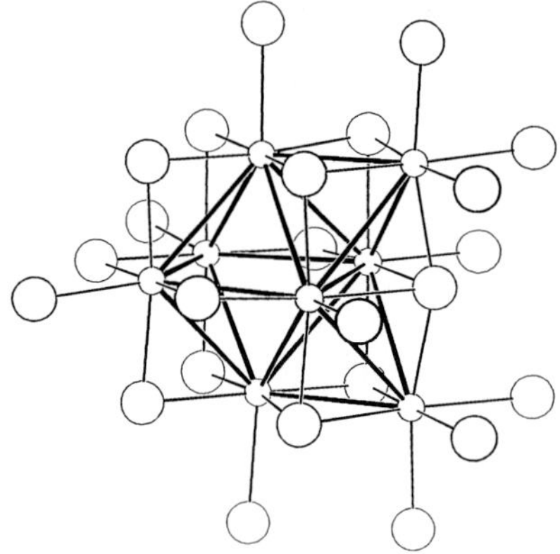

Figure 5-28. $[Mo_8O_{24}]$ cluster in the structure of $NdMo_8O_{14}$.

Table 5-7. Ternary molybdenum chalcogenides $A_xMo_{3n+3}\hat{X}_{3n+5}$ characterized by single crystal investigations containing (**A**) one individual kind of cluster with n face condensed Mo_6 octahedra and (**B**) different kinds of clusters in one compound. The number of electrons per Mo atom, z/Mo, in M–M bonding states as derived from the composition is compared to the optimum range estimated through a fragmentation of the clusters. [6]

n	compound	cluster,	charge (–)	z/Mo experimental	z/Mo estimated
A					
1	$A_XMo_6\hat{X}_8$ ($\hat{X}$ = S, Se, Te)	$Mo_6\hat{X}_8$	0–4	4.00–3.33	
2	$Ag_{2.3}CsMo_9Se_{11}$	Mo_9Se_{11}	3.3	3.92	4.11–3.67
2	$Ag_{4.4}ClMo_9Se_{11}$	Mo_9Se_{11}	3.4	3.93	4.11–3.67
2	$Ag_{3.6}Mo_9Se_{11}$	Mo_9Se_{11}	3.6	3.96	4.11–3.67
3	$Cs_2Mo_{12}Se_{14}$	$Mo_{12}Se_{14}$	2	3.83	4.17–3.83
5	$Rb_4Mo_{18}Se_{20}$	$Mo_{18}Se_{20}$	4	4.00	4.22–4.00
7	$Cs_6Mo_{24}Se_{26}$	$Mo_{24}Se_{26}$	6	4.08	4.25–4.08
9	$Cs_8Mo_{30}Se_{32}$	$Mo_{30}Se_{32}$	8	4.13	4.27–4.13
∞	$TlMo_3Se_3$	$Mo_{6/2}Se_{6/2}$	1	4.33	
B					
1	$In_{\sim 3}Mo_{15}Se_{19}$	Mo_6Se_8			
2		Mo_9Se_{11}			
1	$In_2Mo_{15}Se_{19}$	Mo_6Se_8			
2		Mo_9Se_{11}			
1	$Tl_4Mo_{18}S_{22}$	Mo_6S_8			
3		$Mo_{12}S_{14}$			

The final product from the condensation of $[Mo_6\hat{X}_8]$ clusters via opposite faces of the octahedra is the $Mo_3\hat{X}_3$ chain. It occurs in the compounds $AMo_3\hat{X}_3$ for the univalent elements A = K, Tl, etc. Numerous compounds have been synthesized and structurally characterized [37, 38, 141] which contain such intermediate oligomeric clusters of stepwise increasing sizes as are shown in Figure **5-29**. Those phases which have been characterized by single crystal structure investigations are summarized in Table **5-7**.

The compounds have the general composition $A_xMo_{3n+3}\hat{X}_{3n+5}$ and can contain either one kind of cluster exclusively or a combination of clusters in an ordered array. With the exception of the Ag compounds, the structures which contain only one kind of cluster exhibit a particularly close relationship with the structure of the starting member of the series. All compounds crystallize in the space group $R\overline{3}$ as do the Chevrel phases $A_x[Mo_6\hat{X}_8]$.

In order to derive the oligomeric units in a formal sense, the $[Mo_6\hat{X}_8]$ cluster is stretched along its three fold axis, and $n-1$ planar $[Mo_3\hat{X}_3]$ fragments are inserted into the (now elongated) clusters. The corresponding number of A atoms are found in the channels between clusters.

This genesis of the oligomeric clusters allow once again the use of a suitable fragmentation procedure in order to estimate the optimal electron counts. All intra- and intercluster contacts of the end $[Mo_3\hat{X}_4]$ fragments (top and bottom) are the same as those in the Chevrel phases. Two of these end fragments together can have 20 to 24 d electrons available for M–M bonding. The number of electrons in M–M bonding states per $[Mo_3\hat{X}_3]$ fragment is 13 in the infinite chain. Therefore, the number of electrons, z, in M–M bonding states of an oligomeric cluster with n condensed Mo_6 octahedra should lie within the limits $13n+7 \leq z \leq 13n+11$, where the upper limit corresponds to the closed configuration.

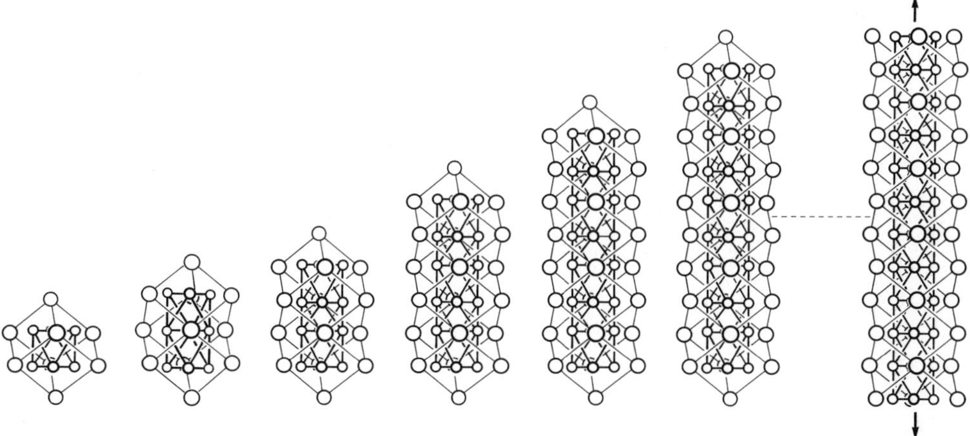

Figure 5-29. Monomeric and oligomeric $[Mo_3X_3]$ clusters formed from Mo_6 octahedra condensed via opposite faces. The experimentally verified clusters have the compositions $[Mo_6X_8]$, $[Mo_9X_{11}]$, $[Mo_{12}X_{14}]$, $[Mo_{18}X_{20}]$, $[Mo_{24}X_{26}]$, and $[Mo_{30}X_{32}]$. The final member is represented by the linear chain $[Mo_3X_3]$.

This estimate is in agreement with the results from MO calculations for the ions $[Mo_9S_{11}]^{4-}$ and $[Mo_{12}S_{14}]^{6-}$. [23] Taking $z = 13n + 11$ where $n = 2$ and 3, one calculates $z = 37$ and $z = 50$, respectively. The MO calculation finds 50 electrons in M–M bonding states for $n = 3$. In the case of $n = 2$ (and all clusters with even n) there is some ambiguity in the electron count due to a nonbonding state at the HOMO-LUMO gap. The cluster contains 38 electrons in the case that this level is occupied and 36 if not occupied.

All the examples in Table **5-7** have fewer d electrons than expected for the closed configuration. The electron counts are safe for the large cluster compounds having alkali metal cations provided there is no homogeneity range around the stoichiometric phases. In each case, four electrons are missing for the closed configuration.

Band structure calculations for the quasi infinite $Mo_3\hat{X}_3$ chain [23, 142] yield a band with very large dispersion for the 6 equivalent bonds to the $[Mo_3\hat{X}_3]$ fragment along the chain direction. This band is half filled with an electron count of 13. The compounds $AMo_3\hat{X}_3$ are metallic but with distinctly one dimensional character. The Peierls distortion expected for such a system obviously occurs with those alkali metal derivatives which show a gradual transition to semiconducting behavior with decreasing temperature. [143] In contrast, the electronic coupling of the chains through the Tl ions in $TlMo_3Se_3$ is apparently strong enough to suppress the metal to semiconductor transition. The compound remains metallic at low temperatures and becomes superconducting at $T < 6K$. [144]

Chemically, the one dimensional character of these phases expresses itself in a fascinating reaction. $LiMo_3Se_3$ forms colloidal suspensions in polar solvents such as DMSO. Electron microscopy reveals single $Mo_3\hat{X}_3$ chains in such systems. [145] Numerous topochemical reactions with these cluster compounds proceed without a disassembly of the entire structural framework ("soft chemistry"). [146] As an example, the indium can be extracted from $InMo_3Se_3$ as InCl by the reaction with HCl at 429 °C. Electrochemically, the residual Mo_3Se_3 can be reversibly intercalated by up to 4.5 Li, [147] making Mo_3Se_3 an interesting candidate for a solid state battery. The theoretical energy density of a Li/Mo_3Se_3 cell (referred to the volume) is nearly the same as for the widely used Li/TiS_2 cell.

The topochemical reaction described for $InMo_3Se_3$ can be used to prepare new binary and ternary molybdenum chalcogenides which are not accessible directly. Starting from either $In_2Mo_{15}Se_{19}$ or $In_{\sim 3}Mo_{15}Se_{19}$, which are prepared from the elements, pure $Mo_{15}Se_{19}$ can be produced by the removal of In. [148, 149] As the arrangement of the clusters is different in the two starting componds, $Mo_{15}Se_{19}$ is obtained in two modifications. The empty cluster frameworks can be reloaded by a direct reaction with metals (*e. g.* alkali metals, Zn, Cd, Tl, Sn, Pd) at elevated temperatures or by electrochemical reactions, and new ternary compounds are formed.

The last mentioned reactions leave the cluster framework unchanged. Attempts to disassemble the framework in a chemically controlled way have not been successful as yet. Are such reactions possible and will they open a solution chemistry on oligomeric clusters comparable to that offered by the well defined bulk solids?

5.5 Condensed Clusters with Interstitials

Interstitial atoms frequently occur in discrete clusters but even more so in structures that can be described in terms of condensed clusters. As discussed in Section 5.3 for discrete clusters, there are compounds that allow the reversible insertion of interstitial atoms under equilibrium conditions as well as many compounds that are only stable with the inclusion of interstitial atoms.

The monohalides of zirconium, ZrCl and ZrBr, are examples of condensed cluster systems which can exist both with and without interstitials. Their structures [150] contain characteristic X–Zr–Zr–X layer units composed of single closepacked X and Zr layers. The units can be discussed in terms of edge condensed Zr_6 octahedra which are surrounded by X atoms above all free faces as in the $[M_6X_8]$ cluster.

The compounds reversibly absorb H_2. [151] At a composition $ZrXH_{0.5}$, tetrahedral voids within the twin Zr atom layers are occupied. [152] It is remarkable that for a composition ZrX = H, two H atoms are located in each trigonally elongated Zr_6 octahedron. [153] As in the case of $Th_6Br_{15}H_7$, the hydrogen atoms occupy the centers of opposing trigonal faces. For reasons of electrostatic repulsion, the X atoms move above the edges of the octahedra during the hydrogenation reaction. Hence, the Zr/X arrangement is now related to the $[M_6X_{12}]$ cluster.

The absorption of H_2 by the halides ZrX provides chemical evidence for an electron deficiency in the M–M bonding states. The hypothetical chalcogenides $Ta\hat{X}$, which would be isoelectronic with ZrX, do not exist at all. However, the compound Ta_2S_2C, which has C atoms in all interstices of the Ta_6 octahedron (again the $\hat{X}$ atoms are above the edges of the octahedron), has been prepared. [154] The "MX" compounds formed from the valence electron poor metals Sc, Y, and Ln exhibit the ZrX type structure and were originally mistaken as being binary compounds. [155–158] Later, it was found that they were stabilized by interstitial hydrogen. [159–162] Gd_2Br_2C [163] can also be prepared and it is isostructural with Ta_2S_2C.

An important borderline is reached with the last example. Whatever the actual distribution of valence electrons in Ta_2S_2C may be, the formal ionic limit of $[(Ta^{5+})_2(S^{2-})_2C^{4-}(e^-)_2]$ leaves electrons for (extended) M–M bonding. In contrast, all electrons are used for strong heteroatomic bonds in Gd_2Br_2C ($\triangleq [(Gd^{3+})_2$-$(Br^-)_2C^{4-}]$). Here, the "stabilization of the clusters" by the interstitial atoms has finally led to a destruction of the cluster, at least if M–M bonding is a requirement for using the term "cluster". Gd_2Br_2C is a normal and simple ionic compound just as the isostructural and isoelectronic La_2O_2S. [164]

Gd_2Br_2C is just one example from the broad condensed cluster chemistry of the rare earth metals which has come to light in recent years. From the above, it is obvious that this chemistry lies on the borderline of M–M bonded cluster compounds and salts. The clusters can exist as any combination of empty or filled and discrete or condensed. Bonding within these clusters ranges from M–M bonded species that may be stabilized by additional strong heteropolar bonding to interstitial atoms on one hand to species that have only ionic bonding and are simple salts on the other. The latter still have atomic arrangements in common with cluster compounds, yet have no electrons for M–M bonding.

The new chemistry of metal rich halides of the rare earth metals has recently been described in detail. [13] Here, it might be sufficient to work out certain features of this chemistry at the expense of a complete coverage of compounds that exist.

The compound Gd_2Cl_3 and its isotypic phases (*e. g.* Gd_2Br_3, Y_2Cl_3, Tb_2Cl_3, Tb_2Br_3) seem to be the only binary metal rich rare earth halides which have been discussed in terms of condensed cluster structures. The structure of Gd_2Cl_3 [165, 166] is shown in Figure **5-30** and contains parallel chains of *trans* edge sharing Gd_6 octahedra which are coordinated by Cl atoms above all free faces as in the $[M_6X_8]$ cluster. The Gd–Gd interatomic distances along the joined edges are shorter than in elemental Gd (337 pm as compared to 357 pm). The octahedron is extremely elongated in chain direction (390 pm), and the shortest Gd–Gd distances between the basal and apex atoms are 373 and 378 pm. It is clear from these data that the Gd–Gd distances are much longer than in the case of the previously discussed clusters of the 4d and 5d metals where they generally range below 300 pm. The larger cages formed with lanthanide clusters have interesting consequences with respect to the incorporation of interstitial atoms as will be shown later.

The photoelectron spectrum of Gd_2Cl_3 [167] reveals the essentials of its bonding when compared with the spectrum of Gd. As Figure **5-31** shows, the Gd 4f band of Gd_2Cl_3 lies 10 eV below the Fermi level and exhibits no trace of multiplet

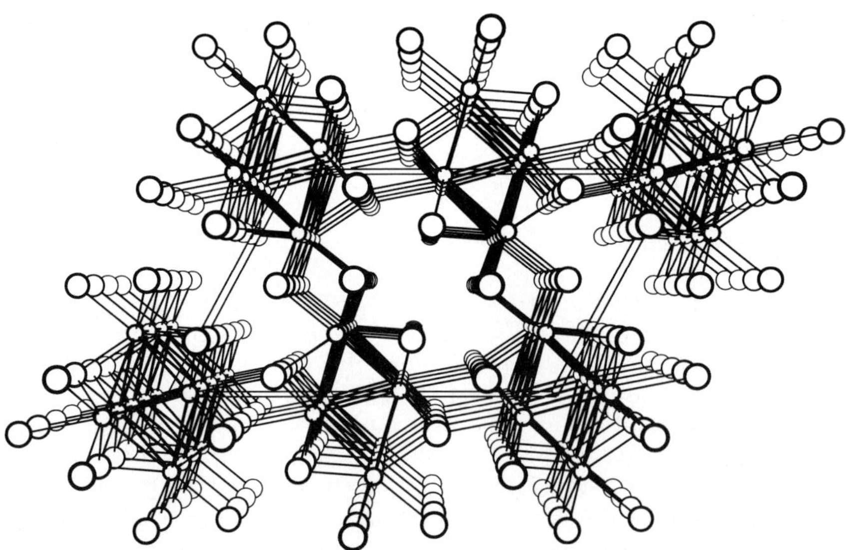

Figure 5-30. Projection of the structure of Gd_2Cl_3 along [010]. The chains of edge sharing Gd_6 octahedra (small circles) are enhanced.

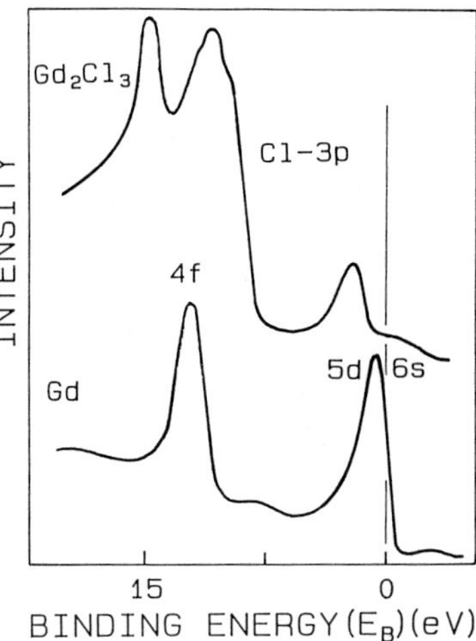

Figure 5-31. He II photoelectron spectra of Gd and Gd_2Cl_3.

splitting. It therefore corresponds to a $4f^7$ configuration which is typical for the ionic core of Gd^{3+}. The low lying Cl 3p band gives evidence for strongly polar Gd–Cl bonds. Residual metal valence electrons are present according to the formulation $[(Gd^{3+})_2(Cl^-)_3(e^-)_3]$. These electrons have d character and fill M–M bonding states. Surprisingly, one finds electronic localization in spite of the non integral number of 1.5 electrons per metal atom. The lack of any density of states at the Fermi level and the band gap of 0.85 eV, as determined from electrical measurements, [168] are clear evidence for electronic localization in the black compound. This feature can be understood with the use of band structure calculations [169] which indicate that the semiconducting properties of Gd_2Cl_3 result from an avoided crossing of bands as a consequence of the symmetry of the extended solid. From a molecular point of view the electron localization might be explained in terms of the formation of multiple bonds within the pairs of Gd atoms exhibiting short Gd–Gd distances rather than a distribution of electrons over the Gd_6 octahedra in the chains.

Few valence electrons are available for M–M bonding in Gd_2Cl_3. The subchloride is only weakly stable against decomposition into Gd and $GdCl_3$. The enthalpy of this reaction is 30 ± 15 kJmol^{-1} as determined from calorimetric studies on the enthalpy of formation of Gd_2Cl_3. [170] Cocrystallization experiments with Gd_2Cl_3 indicate that Tb is the only lanthonoid that can be incorporated to a significant amount. [171] Obviously, the island of stability for the subhalides of the lanthanides is small, however, it may extend to the actinides. [172]

The "stabilization" of Gd_2Cl_3 by two interstitial nitrogen atoms per Gd_6 octahedron leads to the colorless compound Gd_2Cl_3N. [173] According to the formulation $[(Gd^{3+}{}_2(Cl^-)_3N^{3-}$ all electrons are removed from M–M bonding states by N and are involved in strong Gd–N bonds. Despite this, the average length of the distances between neighboring Gd atoms is nearly unchanged when compared with the structures of Gd_2Cl_3. If one compares the topology of the structure of Gd_2Cl_3 to that of its nitride, Gd_2Cl_3N, it is clear that the mutual repulsion between the N^{3-} ions deforms the Gd_6 octahedron into two edge sharing Gd_4 tetrahedra with each being centered by an N atom. The Cl^- ions are rearranged around the chains of tetrahedra. The Cl and N atoms together form the typical coordination polyhedra of a tricapped trigonal prism around the Gd atoms.

The band structure calculation and population analysis for Gd_2Cl_3N [169] yield a more quantitative interpretation of the bonding compared to the extreme ionic model used above. The result of these calculations is summarized by the approximate formula $[(Gd^{2+})_2(Cl^{0.7-})_3N^{1.8-}]$ which clearly indicates strong covalent bonding to be present in the compound by the significant reduction in the ionic charges. This formulation does not, however, change the essential point, namely that there are no M–M bonds between the Gd atoms.

No nitrogen deficiency could be detected in the compound Gd_2Cl_3N. Another nitride chloride, $Gd_2Cl_3N_{0.8}$ (the isotypic yttrium analogue $Y_2Cl_3N_{0.8}$ also exists), has been reported and described by a Gd-Cl partial structure that is identical with the Gd_2Cl_3 structure in every detail, including the volume of the unit cell. [174] The N atoms reside in the tetrahedral voids between the octahedra of the undistorted chain and formed by the Gd atoms making up the joined edges and the adjacent apical Gd atoms. One would expect this phase to be a metal due to

Table 5-8. Gd carbide halides. Crystallographically different C–C distances deviate by less than ± 5 pm from the given values. The number of electrons z (per formula unit) in M–M bonding states is derived via the ionic limit on the basis of the experimental C–C distances ($(C_2)^{4-}$ and $(C_2)^{6-}$ corresponding to 130 and 145 pm, respectively).

Compound	"Linkage" of the Gd_6 Octahedra	C_n	d(C–C) [pm]	z
$Gd_7I_{12}C$	one octahedron	C	–	5
$Gd_{10}Cl_{18}C_4$	two octahedra	C_2	145	0
$Gd_{10}Cl_{17}C_4$	two octahedra	C_2	145	1
$Gd_{10}I_{16}C_4$	two octahedra	C_2	145	2
Gd_4I_5C	straight chain	C	–	3
$Gd_{12}X_{17}C_6$ (X = Br, I)	folded chain	C_2	145	1
$Gd_7I_{10}C_2$	twin chain	C	–	3
$Gd_6X_7C_2$ (X = Br, I)	twin chain	C	–	3
Gd_3I_3C	twin chain	C	–	2
Gd_2Br_2C	planar layer	C	–	0
$Gd_2X_2C_2$ (X = Cl, Br)	planar layer	C_2	130	0
Gd_2XC (X = Cl, Br, I)	planar layer	C	–	1
$Gd_6Cl_5C_{3.0-3.5}$	undulated layer	C	–	1–0
Gd_3Cl_3C	network	C	–	2

the partial filling of M–M bonding bands, and one would also expect a range of homogeneity to exist with respect to the nitrogen content.

The transition from Gd_2Cl_3 to Gd_2Cl_3N sketched above is an inconvincingly formal one since a topochemical reaction of this kind is not possible. The difference between the black Gd_2Cl_3 and the colorless Gd_2Cl_3N is too large to help decipher the details of what happens when the borderline between cluster systems and simple salts is crossed. On the other hand, the multitude of known carbide halides of the lanthanides allows smaller steps to be observed.

There exists a large number of rare earth carbide halides. Although many have already been described, [13] more examples have recently been added. [175] Table **5-8** summarizes just those carbide halides of Gd that have been structurally characterized. [6, 13, 176] All the structures contain octahedral Gd_6 units which are either discrete or condensed via edges and are surrounded above all free edges by halogen atoms as in the M_6X_{12} cluster. As a peculiarity, C_2 units as well as single C atoms may occupy the centers of the metal atom octahedra. The occurrence of C and C_2 species and the observation of a variety of C–C distances in the C_2 moiety can be used to test the electron counting scheme which has been applied throughout this text on a finely graded chemical scale.

Two octahedral $[Gd_6X_{12}]$ units are condensed together via edge sharing in the compounds $Gd_{10}Cl_{18}C_4$, $Gd_{10}Cl_{17}C_4$ (= $Gd_{10}Cl^{i}_{16}Cl^{i\text{-}i}_{2/2}$), and $Gd_{10}I_{16}C_4$ (= $Gd_{10}I^{i}_{14}I^{i\text{-}i}_{4/2}$. [177, 178] The same type of cluster is also observed in $Cs[Er_{10}I_{18}I_{18}C_4]$. [179] The C_2 units are oriented parallel to one another and fill both octahedral voids (Figure **5-32**). The $Gd_{10}Cl_{18}C_4$ arrangement corresponds to the first step in the condensation of filled M_6 octahedra via edges and closely follows the pattern found for the (empty) oligomeric clusters and chains in the reduced oxomolybdates discussed in Section 6.4. Even the $X^{i\text{-}i}$ type interconnection in $Gd_{10}I_{16}C_4$ is identical to that in $La_2Mo_{10}O_{16}$.

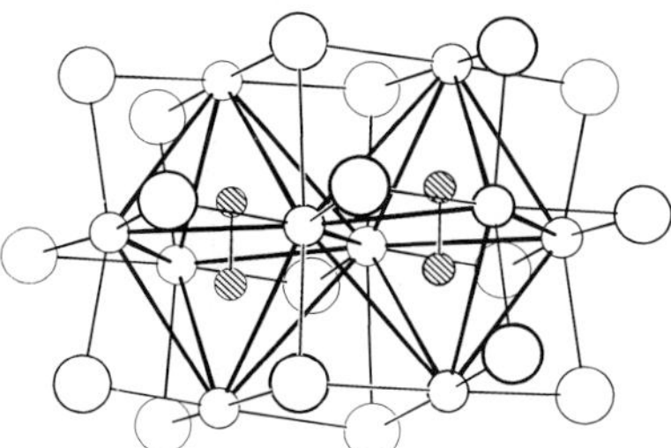

Figure 5-32. $Gd_{10}Cl_{18}C_4$ unit in the compound of identical composition (C, Gd, and Cl atoms drawn with increasing sizes).

The increasing degree of condensation of the C interstitial Gd_6X_{12} units via edges of the octahedra is easy to visualize for the examples chosen in Figure **5-33**. The straight chains of *trans* edge linked Gd_6C octahedra in the structure of Gd_4I_5C (Fig. **5-33a**) together with their surrounding anions closely correspond to the chains in $NaMo_4O_6$, however, the interconnections of the chains are of a different type. The next step in cluster condensation would involve the coupling of two parallel M_4X_6 chains under loss of the X atoms above the joined edge. This has not yet been realized with empty clusters but it is known with Gd_6C octahe-

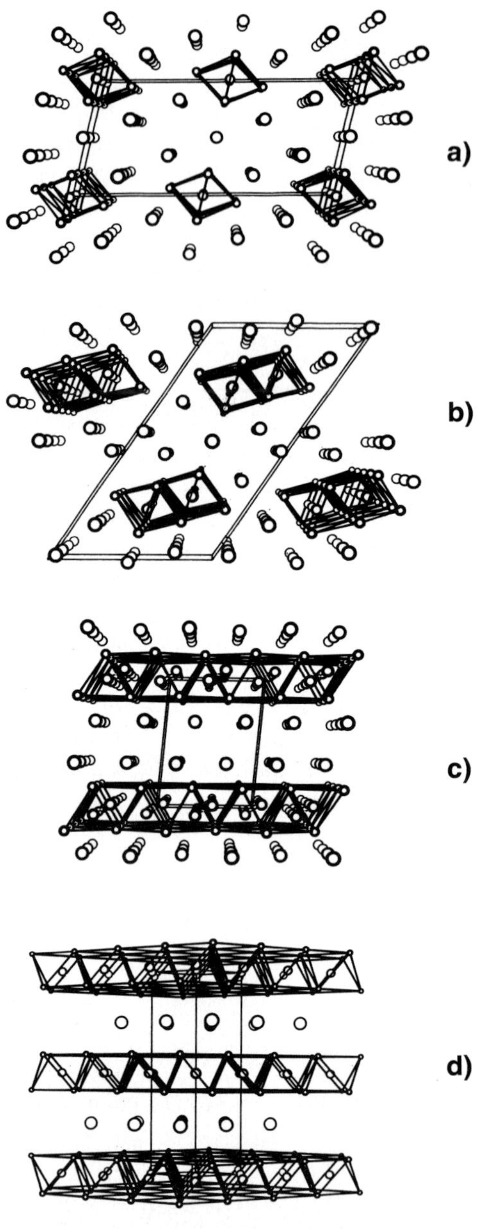

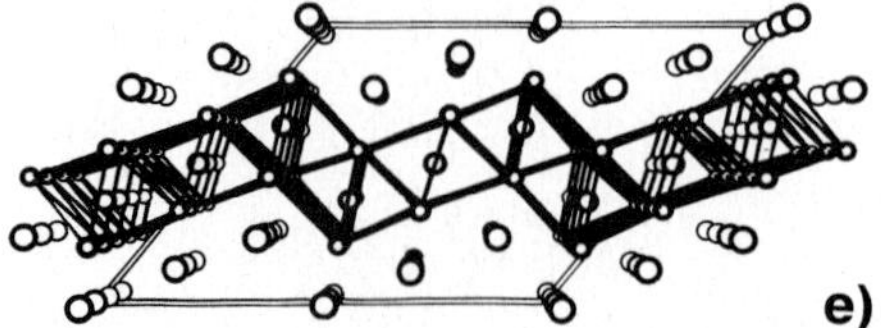

Figure 5-33. Central projections along the shortest axes of the structures of a) Gd_4I_5C (chains), b) $Gd_6Br_7C_2$ (twin chains), c) $Gd_2Cl_2C_2$ (discrete layers), d) Gd_2IC (layers bridged via I atoms), and e) $Gd_6Cl_5C_3$ (tilted layers). The metal octahedra are outlined.

dra for example, in the structures of $Gd_6X_7C_2$ (X = Br, I; Fig. **5-33b**) and Gd_3I_3C. The structure is built from twin parallely oriented chains formed from edge sharing Gd_6C octahedra. The extension of this construction principle leads to a hypothetical series of compounds with the general formula $Gd_{2a+2}X_{2a+2}$ for the M/X framework, where *a* is the number of condensed octahedra chains. So far, except for the initial members of the series with $a = 1$ and 2, only the final members with $a = \infty$ have been realized in the compounds $Gd_2X_2C_2$ (X = Cl, Br; Fig. **5-33c**). These have typical layer structures with van der Waals bonds between adjacent X atom layers. In the structures of the compounds of general composition Gd_2XC (X = Cl, Br, I; Fig. **5-33d**), the layered slabs are linked via X^{i-i} bridges. It is interesting to note dramatic influence which the geometry about these bridges has on the magnetic properties of these compounds. In the case of a trigonal prismatic environment around the X atoms the compounds order as ferromagnets, and for a trigonal antiprismatic environment they are antiferromagnets. [180] On increasing the degree of condensation still further, the structure of Gd_3Cl_3C is finally reached which contains a three dimensional network of edge condensed Gd_6C octahedra. Numerous other structural variations are possible as with the folded chains in $Gd_{12}I_{17}C_6$ or the undulated layers in $Gd_6Cl_5C_{3.0-3.5}$ (Fig. **5-33e**). More variations have been identified by their X-ray patterns, but have not yet been characterized in detail.

The way in which the structures of the lanthanide carbide halides have been drawn in Figure **5-33** emphasizes the relation of the structures to those of the reduced d metal compounds discussed earlier. One should be careful not to be misled by the drawings, for the lines between the metal atoms might have entirely different meanings. In $NaMo_4O_6$, these lines indicate M–M bonding (although the single line does not necessarily mean a $2e^-$-2c bond!), while in the case of $Gd_2Cl_2C_2$, where there are no electrons available for M–M bond formation, the lines are only used to highlight the topology of the structure.

The borderline which separates cluster compounds from simple salts is well illustrated with the example of Gd_2ClC. The structure is built from close packed layers of Gd (abc), Cl (ABC) and C atoms (α,β,γ) in a cubic sequence AcβaC-bαcBaγb ... This arrangement bears a close similarity to the structure of the monofluoride Ag_2F [181] which contains layers of condensed but empty Ag_6 octahedra. In the structure of Gd_2ClC, all octahedral voids are occupied by interstitial C atoms which stabilize the clusters. If the distinction between the different kinds of anions is omitted (*i. e.* ABC $\simeq \alpha\beta\gamma$), then Gd_2ClC has the rocksalt structure.

The interstitial C_2 units will commonly have C–C distances of around 145 pm, which corresponds to a shortened single bond, but the C–C distance of approximately 130 pm found in $Gd_2X_2C_2$ (X = Cl, Br) is indicative of a double bond. The C–C stretching vibration in the Raman spectra is observed at 1158 cm^{-1} for $Gd_{10}Cl_{18}C_4$ and 1175 cm^{-1} for both $Gd_{10}Cl_{17}C_4$ and $Gd_{12}Br_{17}C_6$. In $Gd_2Cl_2C_2$ the streching vibration is found at 1578 cm^{-1}. According to Siebert [182] the bond orders derived from the corresponding force constants are 1.2 and 1.9.

As expected, the hydrolysis of $Gd_{10}Cl_{18}C_4$ with H_2O at room temperature yields a hydrocarbon mixture containing 95 % C_2H_6. It has to be admitted, however, that the unambiguous assignment of a specific C_n species in a crystal from the products of its solvolysis reactions, as can be done with CaC_2, is rather an exception. It has been known for a long time that particularly when delocalized electrons are present in the solid, the observed gas phase species need not be related to the actual carbon units in the solid. The gases liberated during the hydrolysis of Gd_2ClC under comparable conditions contain 55 % CH_4, 30 % H_2, and 15 % C_2 hydrocarbons. Although these might still be within the range of expectation for the decomposition of a "methanide", side reactions have obviously occurred. The latter dominate in the case of $Gd_2Cl_2C_2$ where one finds only 15 % C_2H_4 in addition to 45 % C_2H_2, 30 % C_2H_6 and, in particular, some amount of C_4 hydrocarbons. [183]

An important question with respect to the electron counting procedure is whether the different C species within the structures of the lanthanide carbide halides can be rationalized in terms of a simple stabilization of the electron deficient clusters by interstitial atoms. It has been demonstrated with the most simple example of $Nb_6I_{11}H$ that the interstitial atom actually depletes the M–M bondig states of an electron deficient cluster. This effect is further enhanced in the case of the electropositive lanthanides, because the Fermi level of the metals is high compared to the s and p states of carbon. In order to give a quantitative statement about M–M bonding one only needs to determine how many electrons are left with the metal atoms after bonding the X atoms since, in the case of the carbide halides, these electrons will be transferred to empty σ_p, π,π^*, and σ^*_p states of the neutral C_2 unit. For $Gd_2Cl_2C_2$ = $[(Gd^{3+})_2(Cl^-)_2(C_2^{4-})]$ the framework donates four electrons to the C_2 unit and so two antibonding states are effectively not occupied. The observed short C–C distance therefore corresponds to a double bond. In $[Gd_{10}Cl_{18}C_4]$($\triangleq$ $[(Gd^{3+})_5(Cl^-)_9(C_2^{6-})_2$, only the highest antibonding state remains unoccupied and so a C–C single bond results. Finally, since in Gd_2Br_2C (= $[(Gd^{3+})_2(Br^-)_2C^{4-}]$) a total of eight electrons would have been donated from the framework to the two carbon atoms and all antibonding states would therefore be occupied in a C_2 unit, single C atoms occur. To summarize, all compounds listed in Table **5-8** conform to the naive electron counting scheme based on an ionic limit according to the generalized 8–N rule [3] in every detail.

The examples chosen to demonstrate the electron counting are special in the sense that the available number of electrons from the framework is just equal to the number which the C_2 unit accepts. In the more general case, these numbers are different and any excess of electrons over this balance enters M–M bonding states. Admittedly, the presented view is simplified. In a more detailed analysis it is obvious that the relative position of the C centered bands to the Fermi level determines whether M–M bonding might persist even in the presence of a double bond entity C_2^{4-}, as it was recently found for $Gd_4I_5C_2$. [184]

In spite of such deviations, the electronic balance as derived above leads to several obvious conclusions. The occurrence of single interstitial C atoms or the formation of double or single bonds in the interstitial C_2 units clearly indicate that the number of electrons donated by the metal framework and, hence, the number of electrons which are still available for M–M bonding conform to the ionic limit. There is evidence for an influence of the cage size on the amount of electron transfer which can occur between the M atoms and the C_2 units. Last but not least, different carbide species, for example $C^{4-}$and C_2^{6-} or C_2^{4-} and C_2^{6-} can occur simultaneously in a compound if the structure allows it. The range of homogeneity of $Gd_6Cl_5C_{3.0-3.5}$ is rationalized in terms of a partial substitution of C^{4-} by C_2^{6-}, which proceeds until all M–M bonding states are depleted. [185]

In the previous discussion, the electron concentration in M–M bonding states was derived by assuming that the ions carry their conventional formal charges (*e. g.* Gd^{3+} or C_2^{6-}). So as to avoid any misunderstanding about the relevance of such a procedure, Figure **5-34** presents the result from a band structure calculation for $[Gd_{10}Cl_{18}C_4]$. [96] The MO's of the weakly interacting (free) C_2 units shown on the right hand side of the drawing are lowered to such an extent when the C_2 units are introduced into the $[Gd_{10}Cl_{18}C_4]$ cage that the π^* level falls below the Fermi level and becomes occupied. The Gd 5d bands are empty. Of course, there is a significant Gd 5d contribution to the valence bands, especially due to back bonding from π^* into the empty d states. The participation of d states in the bonding corresponds to a covalency in the Gd–C and Gd–Cl bonds according to

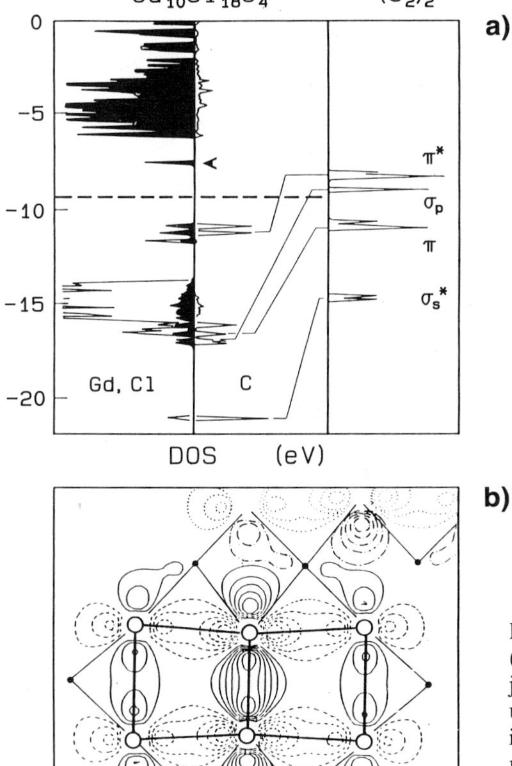

Figure 5-34. a) Left: total and Gd projected (shaded) DOS for $Gd_{10}Cl_{18}C_4$, center: C projected DOS, right: MO's for interacting C_2 units (σ_s and σ_p* omitted). The dashed line indicates the Fermi level. b) Real space representation of the split-off d band shown by an arrow in a). Gd and Cl atoms in the base of the bioctahedron are connected by lines.

the more realistic bonding description $[(Gd^{1.3+})_{10}(Cl^{0.45-})_{18}(C_2^{2.4-})_2]$, however, without any Gd–Gd bonds being formed. Only the lowering of the halogen content in the compounds $Gd_{10}Cl_{17}C_4$ and $Gd_{10}I_{16}C_4$ leaves electrons with Gd and leads to the occupation of the narrow band which is split off from the Gd 5d block. This band has 90% d character and, as can be seen in Figure **5-34b**, it clearly has M–M bonding character in the equatorial plane of the bioctahedron and, in particular, along the joined edge between the two octahedra. The experimentally observed shortening of this edge by 9 pm in $Gd_{10}Cl_{17}C_4$ compared to $Gd_{10}Cl_{18}C_4$ is in agreement with this calculation. The description of this bonding is closely related to that in the single octahedron Sc_6C_2 species. [95]

The structures give hardly any indication for the fact that a borderline between the simple salt $Gd_{10}Cl_{18}C_4$ and the cluster compound $Gd_{10}Cl_{17}C_4$ has been crossed. The "cluster" in $[Gd_{10}Cl_{18}C_4]$ also results from a fcc type packing of the Cl^- and C_2^{6-} anions in the ratio 9:1. Only the favorable octahedral voids around the highly charged anions are filled by the cations. $Gd_{10}Cl_{18}C_4$ is simply a defect variant of rocksalt.

The discussion shows that a simple ionic model is generally suitable for counting the electrons in M–M bonding states. Physical properties, particularly the electrical conductivity, are in agreement with these electronic balances but need to be critically analysed. Thus, metallic behavior is only expected when an excess of metal valence electrons is present in systems having extended M–M bonding through their chain, twin chain, layer, or 3 D networks. One might argue that this is a necessary condition, but it is not sufficient as the semiconducting properties of Gd_2Cl_3 and other chain systems indicate. [13] Extended M–M bonding is, however, not even a necessary prerequisite for metallic conductivity as demonstrated by the phases $R_2X_2C_2$. They form gold or bronze colored crystals which exhibit two dimensional metallic conductivity, even though all bands with M–M-bonding character should be empty.

The explanation for this fictitious disagreement is trivial. The statement about a filled valence and an empty conduction band does not give any clue to the energy separation of these bands. Metallic properties might occur "accidentally" if filled valence and empty conduction bands are not separated by an energy gap. Electrical measurements show $Gd_2Cl_2C_2$ to be a metal and the photoelectron spectrum of $Gd_2Cl_2C_2$ (Fig. **5-35**) exhibits electron density at the Fermi level. The energy of the Gd 4f band might give an indication for the cause of the nonvanishing density at the Fermi level, namely, backbonding from occupied C_2 π^* states into empty Gd d states. The analysis of the bonding in $Gd_2Cl_2C_2$ from Extended Hückel calculations [186] supports this interpretation. The calculated density of states (4f omitted) and the projected densities for Gd 5d (shaded) are plotted in Figure **5-36** (top). The character of the band states at the Fermi level is to be seen in the corresponding COOP diagrams for the C–C, Gd–C, and Gd–Gd interactions. As expected, they are C–C antibonding (π^*), but they are bonding with respect to the Gd–C interactions and, in particular, the bonding interactions involve those Gd atoms which are collinear with the C_2 unit. The occupation of these bands represents a compromise between the weakening of the C–C bonding and the formation of bonds between the C and Gd atoms. It is evident from

Figure **5-36** that the comparatively high Gd 5d contribution to the bands at E_F is almost entirely due to the covalent Gd–C bonds and does not lead to M–M bonding. According to the population analysis, the bonding is best described as $[(Gd^{1.36+})_2(Cl^{0.46-})_2C_2^{0.90-}]$.

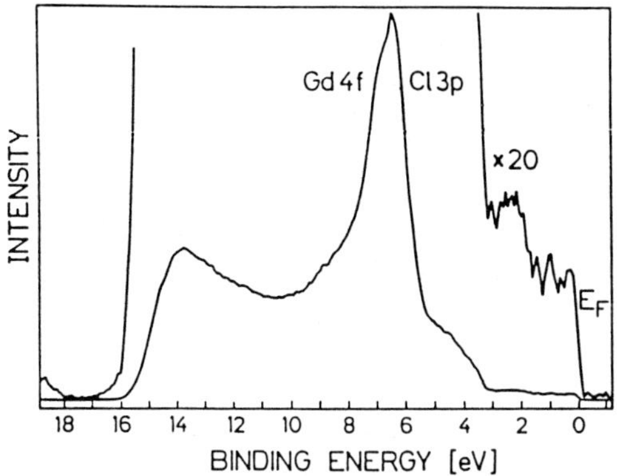

Figure 5-35. He I photoelectron spectrum of $Gd_2Cl_2C_2$ showing the nonvanishing electron density at the Fermi level.

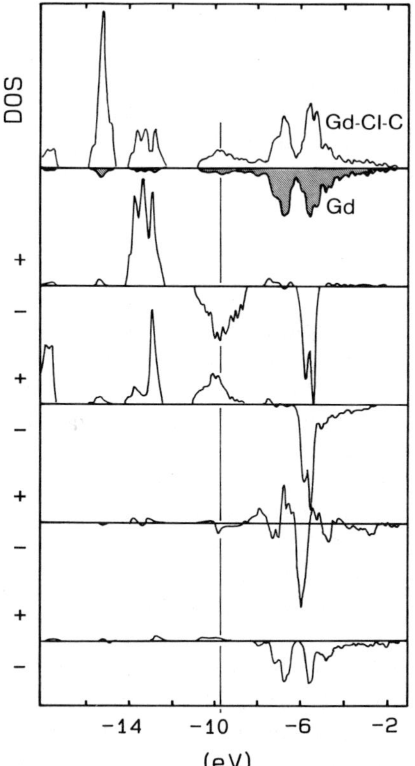

Figure 5-36. (top to bottom) Total and Gd projected (shaded) DOS for $Gd_2Cl_2C_2$. The bands are analysed via COOP curves for nearest neighbor C–C, C–Gd_{apical}, Gd_{basal}–Gd_{basal} and Gd_{apical}–Gd_{basal} interactions. The Fermi level is marked.

The formation and disruption of M–M bonds through chemical reactions can be performed reversibly with the hydride halides of the trivalent lanthanides whose structures are closely related to that of $Gd_2Cl_2C_2$.

TbClH mimics the structure of ZrBr when one considers only the TbCl framework. Tb_6 octahedra are condensed via edges and form layers which are surrounded by Cl atoms above the octahedral faces as in the M_6X_8 cluster. The Tb atoms are close packed, and neutron diffraction investigations on the deuterated compound $TbClH_{0.8}$ show that the tetrahedral voids within the Tb atom twin layers are occupied by D atoms (Fig. **5-37a**). [187] The compound has a homogeneity range of $TbClD_{0.67-1.00}$ which arises from the D atoms' partial occupancy of the void.

$TbClH_{1-x}$ looks like graphite and is a good two dimensional metallic conductor, as expected from the formulation $[Tb^{3+}Cl^{-}(H^{-})_{1+x}(e^{-})_{1+x}]$. If the phase is heated in a H_2 atmosphere, it topochemically reacts to form $TbClH_2$. A neutron diffraction study on the isotypic compound $TbBrD_2$ [188] has shown that the Tb_6 "octahedra" are considerably elongated and are each filled by two H(D) atoms which reside near the basal planes of the trigonal antiprisms (Fig. **5-37b**) in a structure similar to ZrClH. In addition, all Cl atoms are moved from their positions above faces to positions above edges. The geometrical conditions are closely related to those found in $Th_6Br_{15}D_7$.

The reaction of $TbClH_{1-x}$ with H_2 is accompanied by a dramatic change in the physical properties. $TbClH_2$ forms transparent crystals and is an insulator. In $TbClH_{1-x}$, the excess valence electrons of the metal are delocalized in a band that has M–M bonding character. The reaction with H_2 results in a localization of all

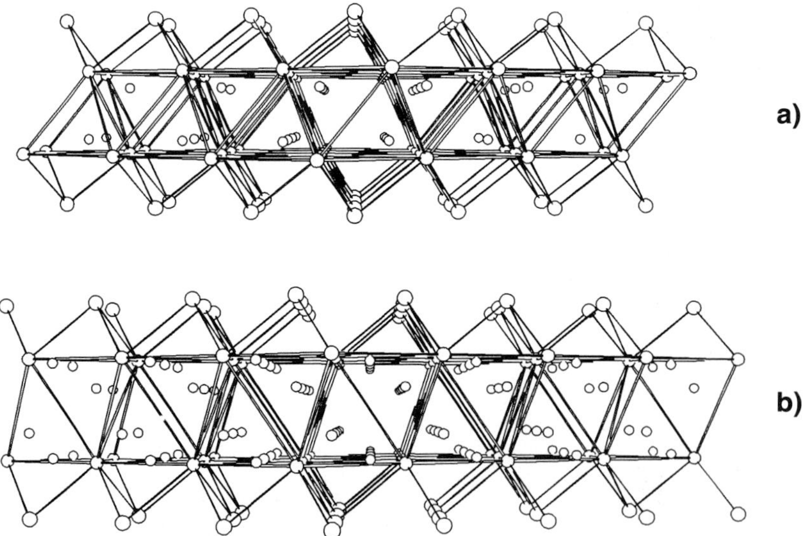

Figure 5-37. Single slabs in the layered structures of a) $TbClD_{0.8}$ and b) $TbBrD_2$. The halogen, terbium, and hydrogen (D) atoms are drawn with decreasing size.

electrons in a narrow band that has essentially H 1s character. When $TbClH_2$ is heated under vacuum, hydrogen is lost and the M–M bonds are "switched on" as a consequence. As shown in Figure **5-38** the resistivity of such a sample drops by several orders of magnitude upon formation of $TbClH_{1-x}$.

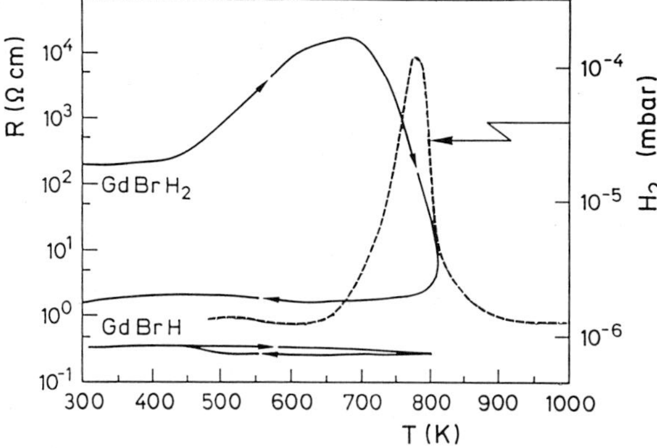

Figure 5-38. Change in the electrical resistivity of a polycrystalline $GdClH_2$ pellet upon heating in vacuo together with the detection of H_2 evolution.

The phase relationships and the properties of $TbClH_n$ are nicely modeled by band structure calculations, the essence of which is presented in Figure **5-39**. The hypothetical compound TbCl ($n = 0$) is characerized by three M–M bonding bands which, as there are 2 formula units in the cell, would be occupied by four electrons to yield weak M–M bonding. At the compositions $TbClH_{0.5}$ and $TbClH_{1.0}$, one or two of these bands respectively are drastically lowered in energy and their character becomes hydridic. The bonding situation is identical to that in discrete clusters having interstitial H atoms. Interestingly, the phase $TbClH_{1.5}$ is not stable because M–M bonding bands are available and so the excess electrons must enter an antibonding band in the d block, thus forming discrete Gd^{2+} ions. The unfavorable situation is avoided by a miscibility gap between $TbClH_{1.0}$ and $TbClH_{2.0}$.

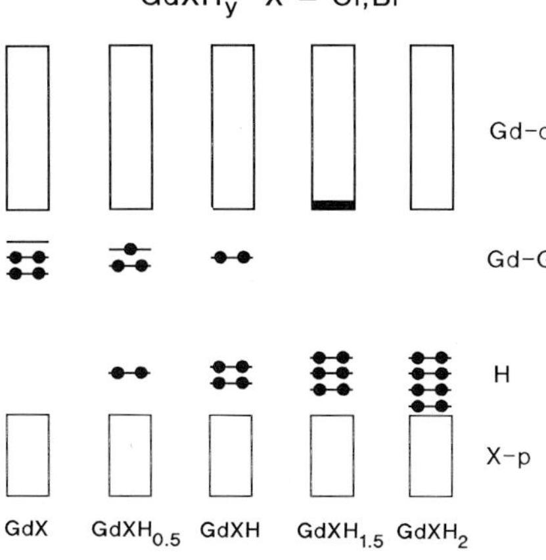

Figure 5-39. Schematic representation of the band structures for various $GdXH_y$ compositions. The bars represent bands with M–M bonding or hydridic character and do not indicate localized states.

The reversible absorption of hydrogen can be used to test the validity of the simplified ionic model for the carbide halides since the electrons can be "titrated" into bands with M–M bonding character. This has been shown for the compounds Gd_2XC (X = Cl, Br, I) and the hydrogenated phases Gd_2XCH_x. [189] When Gd_2IC is heated in H_2 at 820 K the bulky crystals split into very thin lamellae due to the insertion of H atom layers (Fig. **5-40**). The macroscopic change in the crystal together with the observed atomic shifts clearly indicate a strong electrostatic repulsion between the created H^- ions and the other anions present. As one would expect from the formulation of the host material as $[(Gd^{3+})_2X^-C^{4-}e^-]$ there is one electron per formula unit to be localized with hydrogen. The experimentally determined maximum H atom content corresponds to the exact composition Gd_2XCH, and the formation of this phase is associated with a change from the metallic to semiconducting state.

The compounds Gd_2XCH_x are again particularly well suited to illustrate the transition of a condensed cluster phase into a truly saltlike normal valence compound. Closely related phases which are based on the Gd_2X partial structure reveal another borderline.

It was mentioned earlier that the interstitial atoms in the octahedral void of a discrete cluster may also be transition metal instead of nonmetal atoms as, for example in $[Gd_6I_{12}Fe]$. [190] Octahedral clusters which are centered by a transition metal atom and condensed into chains (Pr_4I_5M) [191] and twin chains

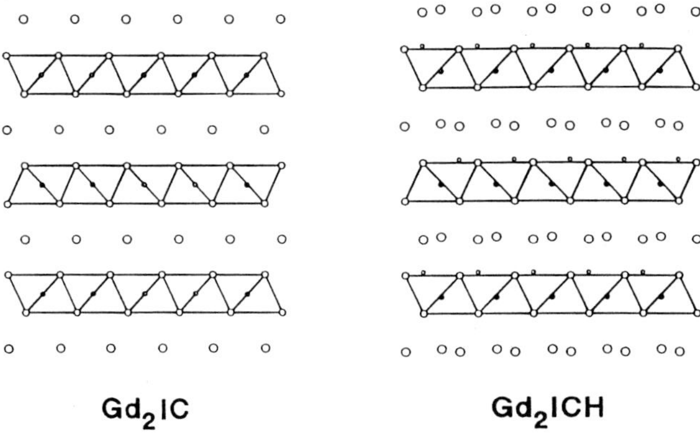

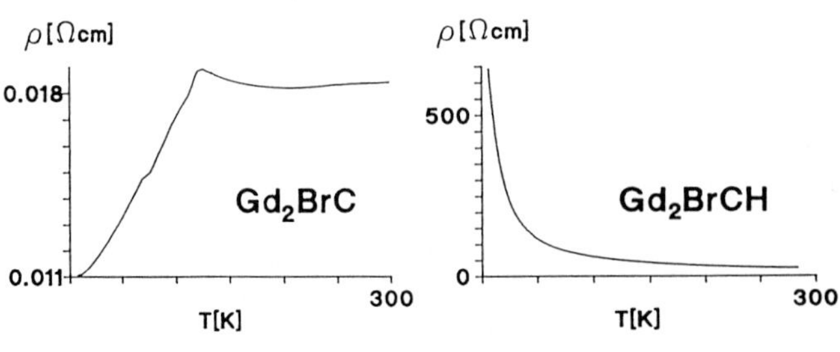

Figure 5-40. (top) Structural change of Gd_2IC (left) upon reaction with hydrogen leading to Gd_2ICH (right). The changes in composition are accompanied by changes in the electrical resistance as illustrated for the isotypic bromide phases (bottom).

($Y_6I_6Ru_2$) [192] are also known. The first examples of such compounds with layer structures have been prepared and include the phases Gd_2IFe_2 and $Y_2Br_2Fe_{2+x}$. [193]

In one layer of edge condensed octahedra there exists only one half the number of octahedral voids as there are metal atoms. Thus, all the octahedral voids are filled in Gd_2IC. Twice this number of interstitial atoms, as found in Gd_2IFe_2, can only be realized by occupying the octahedral voids by Fe_2 pairs, by occupying all the tetrahedral voids, or, last but not least, by changing the arrangement of the Gd atoms from trigonal antiprismatic into trigonal prismatic and then occupying all the voids in such a layer. Clearly, the last possibility will be avoided for strongly repulsive anions like C^{4-}. In the case of iron being the interstitial atoms, however, bonding between the Fe atoms may occur. The structure of Gd_2IFe_2 is shown in Figure **5-41**

As the calculated band structure and the COOP analysis show, a polyanionic "ferride $Fe^{2.5-}$" net forms which exhibits strong Fe–Fe bonding and essentially no M–M bonding between the electropositive Gd atoms themselves.

The structure of Gd_2IFe_2 reinforces the notion that our present state of understanding is incomplete. On the other hand, the remarkable abundance of octahedral metal clusters, whether empty or filled, justifies our focusing on this type of

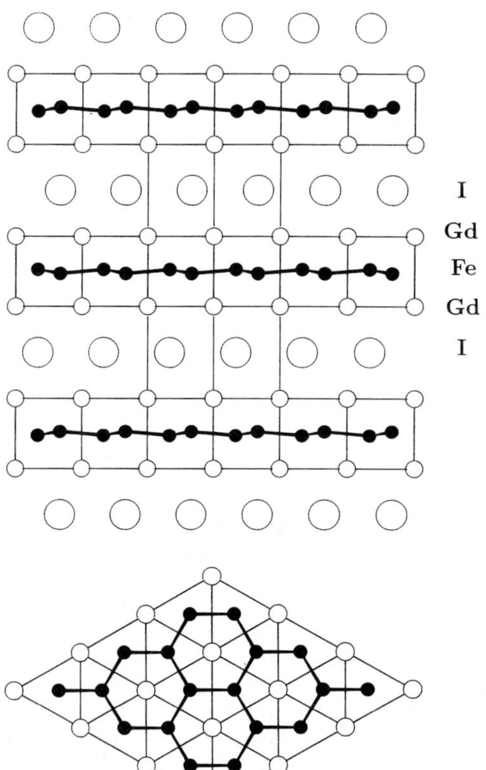

Figure 5-41. Projection of the crystal structure of Gd_2IFe_2 along a) [110] and b) [001].

unit. One should, however, at least briefly mention the multitude of other cluster types which are known as solution species and which, of course, also occur as discrete or condensed units in solid state compounds.

5.6 Condensed Non-octahedral Clusters

Numerous metal rich compounds which are composed of transition metals and p elements adopt structures that can be described in terms of condensed $[M_6X_8]$ and $[M_6X_{12}]$ type clusters, especially if distortions of such clusters are accepted. However, care must be taken for these distortions can actually introduce significant changes in the chemical bonding when compared with the discrete units. In an early analysis it was demonstrated [102] that other cluster types also play an important role within the concept of cluster condensation. These may be closely related to the octahedral clusters or may be of a completely different nature. It is not the intention here to deal with these structures in any detail, however, a short summary will be presented so as to draw attention to the breadth of the cluster condensation concept. Of particular interest are those clusters which represent fractions of the octahedral clusters, for example, the $[M_6X_8]$ type.

Figure **5-42** shows different kinds of fragments of the $[M_6X_8]$ unit. Fragments with less than three metal atoms have been omitted. In the case of the butterfly shaped M_4 unit in Figure **5-42a** an additional X atom is present. Such fragment clusters have been realized as solution species. The square pyramidal metal atom arrangement (Fig. **5-42c**) is found in the cluster ion $[Mo_5Cl_{13}]^{2-}$ [194] and the butterfly shaped M_4 unit (Fig. **5-42a**) in the cluster $[Mo_4Cl_{12}]^{4-}$. [195] The latter configuration is also observed for the oxidized species $[Mo_4Cl_{12}]^{3-}$ which, however, predominantly occurs as the rhomboid shaped planar isomer. [196] In all of these clusters the X^i_8 arrangement is complete and all X^a type halogen positions are occupied.

These M_4 and M_5 units, when condensed via their apices, are frequently observed in metal rich compounds. The rhomboidal planar cluster can be condensed into chains through corner condensation and is the main constituent in the structure of $NaMo_2O_4$ as already mentioned in the discussion on the oxomolybdates (*e. g.* $Ca_{5.45}Mo_{18}O_{32}$). The butterfly shaped cluster was already refered to as a fragment condensed to the $[M_6X_{12}]$ cluster in the $[Mo_8O_{24}]$ entity although, in this case, the O atoms were arranged above the edges. A fragment of this sort but with the nonmetal atoms positioned as they are in the $[M_6X_8]$ type cluster is the only structural element that is used in building the structure of compounds like Hf_3P_2. [197]

It is evident from Figure **5-43a** that the structure consists of parallel chains of condensed butterfly type clusters. Figure **5-43b** presents an example in which M_5 type units are condensed to form the structure (*e. g.* Rh_5Ge_3). [198] In these examples, a comparison of the bond energies of the M–X and M–M bonds indicates that the M–X bonds dominate the structures. In any case, it is interesting to find the M_6X trigonal prisms arranged in such a way so as to result in an overall arrangement which is closely related to the atomic structures of discrete clusters. One is reminded of Escher's drawings, where a regular array of one motif creates a regular array of another one such that both motifs are intrinsic parts of the drawing.

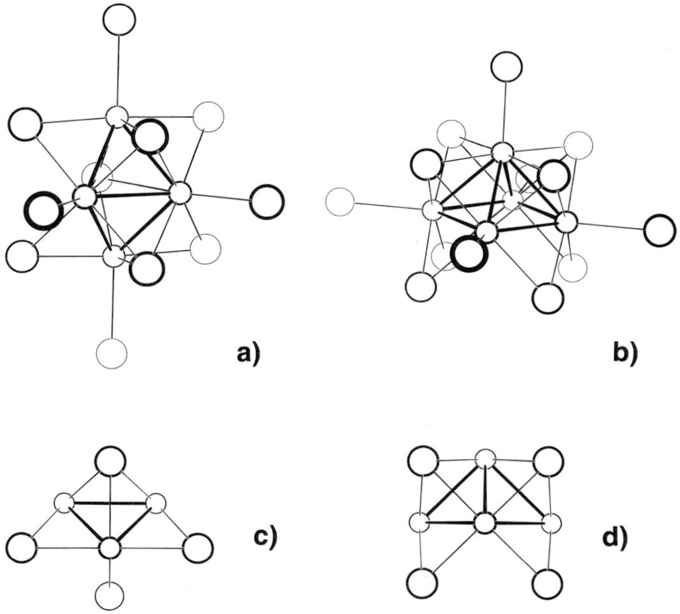

Figure 5-42. The clusters a) $[Mo_4Cl_{11}]$ and b) $[Mo_5Cl_{13}]$. c) M_4 and d) Me_5 fragments of the $[M_6X_8]$ cluster.

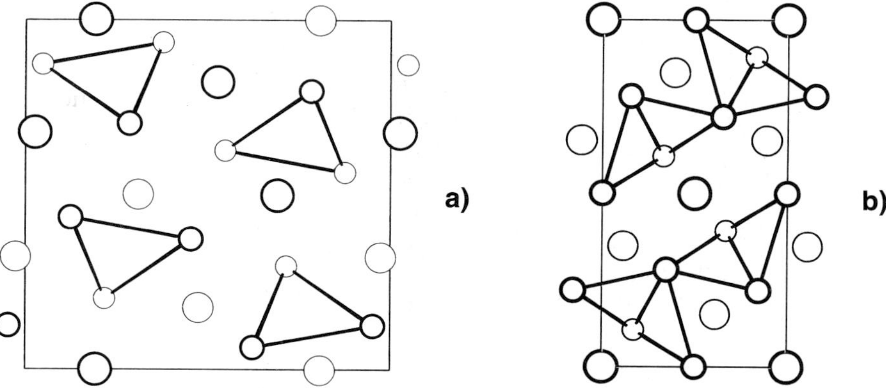

Figure 5-43. Projections of the crystal structures of a) Hf_3P_2 and b) Rh_5Ge_3 along the short axes. Metal atoms are connected by lines to indicate cluster fragments.

The concept of cluster condensation gains tremendous complexity if different cluster types are allowed to be simultaneously present in a structure. The atomic arrangement in $Zr_{14}P_9$ [199] is shown in Figure **5-44** as an example which illustrates this point. In addition to the well known M_5X_4 type chains discussed earlier, the structure also contains chains which are based on the M_4 fragments found in Hf_3P_2 as well as twin chains formed from two of these. Thus, the crystal structure of $Zr_{14}P_9$ represents an intergrowth of various structural building units. As in the case of, for example, Ti_8S_3, [200] in order for these compounds to occur their structural principles demand rather high variability. Much is left to be done in order to systematize those phases already known as well as those which have yet to be discovered, and to describe the interplay between M–X and M–M bonding in a more quantitative way than simply drawing topological lines.

The condensed cluster phases discussed so far have all involved the octahedral M_6 unit or fragments of this unit, that is, M_5 and M_4 units. Other building units may occur as, for example, the trigonal prisms discussed above. Going back to the introduction, the choice of using the octahedron as the characteristic unit in fcc metals is as arbitrary as choosing the tetrahedron. In fact, clusters built from tetrahedral M_4 cores likewise play an important role in cluster condensation. They have frequently been found both as discrete entities and in condensed ensembles. Just to mention a few examples, discrete Ru_4 tetrahedra occur in the structure of YRu_4B_4 [201] and the structure of $NdCo_4B_4$, for example, contains parallel chains of edge sharing Co_4 tetrahedra. [202] Borides having the general composition MM'_4B_4 where M = R, Th, U and M′ = Co, Rh, Ir, Ru, Os have attracted much attention due to their unusual properties at low temperature.

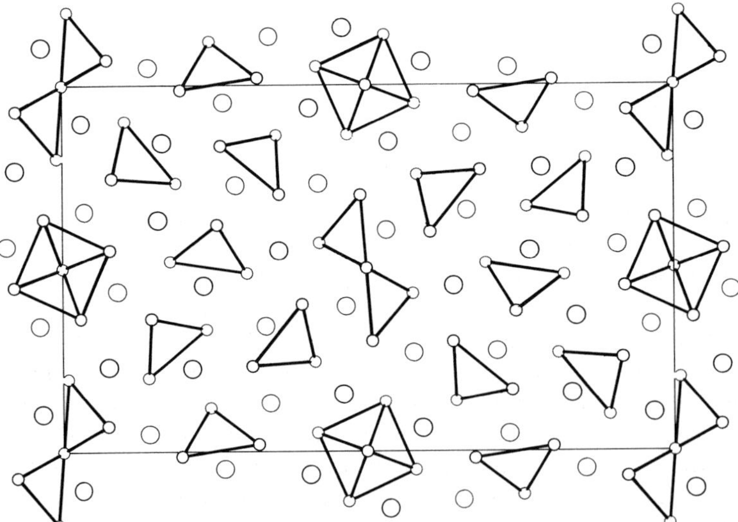

Figure 5-44. Projection of the crystal structure of $Zr_{14}P_9$ along [001]. Metal atoms are connected by lines to indicate arrangements which are reminescent of the M_6 unit and fragments of it.

[203] Reentrant superconductivity was first discovered with $ErRh_4B_4$. [204] The chain of *trans* edge condensed M_4 tetrahedra in which the coordinating atoms lie above all faces of the tetrahedra has been called a "Tetraedersternˮ by Schubert. This very common feature links the structural chemistry of condensed cluster compounds with that of the intermetallic phases. [205]

According to the Frank-Kasper concept, in the close packed tetrahedral structures of the intermetallic phases, the centered icosahedron is a frequently occurring and characteristic structural unit. It is therefore interesting to find this unit excised from such infinitely extending arrays as discrete cluster like $[Au_{13}(M_2PhP)_{10}Cl_2]^{3+}$ or as three interpenetrating icosahedra in $[Au_{13}Ag_{12}(Ph_3P)_{12}Cl_6]^{3+}$. The metal framework in the latter compound [206] represents the first step in building an infinite chain from interpenetrating icosahedra of metal atoms as it was discovered earlier in metal rich tantalum sulfides. The characteristic structural elements of Ta_6S [207, 208] and Ta_2S [209] are parallel chains of (slightly distorted) interpenetrating icosahedra or, to put it in other terms, chains of body centered pentagonal antiprisms fused together via their basal planes. The distances between the central atoms are particularly short, being 289 pm for Ta_2S and 264 pm for Ta_6S. Recently, the compound Ta_6Te_5 has been isolated and shown to contain discrete chains which are all surrounded by Te atoms according to $TaTa_{10/2}Te_{10/2}$ and bonded to one another through van der Waals bonds. [210] The structural chemistry of the tantalum chalcogenides which is based on Ta_{13} icosahedra is far from being completely known. As the structure of $Au_xTa_{15-x}S_2$- $(x < 2)$ [211] indicates, such Ta_{13} icosahedra need not result from interpenetrating Ta_6 icosahedra but may also exist as discrete units which are linked in a way reminiscent of the α structure of boron.

5.7 Bare Metal Clusters

The clusters described so far have been ordered according to their number of metal valence electrons. For elements at the center of the d series, this number is large enough so that the metal cluster can be completely surrounded by nonmetal atoms and yet have a sufficiently large number of valence electrons for M–M bonding. The "binary" $K_4[Nb_6Cl_{18}]$ and the "ternary" $[Re_6Se_4Cl_{10}]$ clusters are examples in which a total coverage of the M_6 unit by ligands occurs at both endo and exo positions. Increasing the degree of interconnection between the clusters leads to a reduction in the X/M ratio and optimizes the valence electron number for M–M bonding in the case of valence electron poor metals and/or clusters whose anions have valencies other than –1. There are two ways in which electron deficiencies are coped with and they can be realized separately or simultaneously. First, the lowering of the X/M ratio to critical values leads to cluster condensation and, second, the addition of atoms in the metal core of the cluster formally increases the electron count but rather stabilizes the arrangement by substituting weak M–M bonding by strong bonds between the M atoms and the interstitials. Clusters with no ligand spheres would be an ultimate way to cope with an extreme electron deficiency. Such clusters which contain oxygen atoms as interstitials have been found with the alkali metal suboxides. Their brief description in an article on transition metal clusters can be justified by the fact that not only is their structural chemistry an interesting extension of the discussion given so far but also because the heavy alkali metals might even act as d metals in their suboxides. [4]

The gas phase molecule $[Li_6C]$ was mentioned in the introduction. The description of chemical bonding in this compound using the ionic limit formulation as done before with metal rich solid compounds leads to the formula $[(Li^+)_6C^{4-}(e^-)_2]$ and provides a simple and easily visualized counting scheme for the number of electrons which are available for M–M bonding. In a more detailed description, all the bonding states in the octahedral molecule are occupied by the 10 valence electrons according to $(3a_{1g})^2(2t_{1u})^6(4a_{1g})^2$. The $4a_{1g}$ HOMO has Li–Li bonding character. [2]

Numerous comparable hypermetallated molecules (*e. g.* $[Li_3O]$, $[Li_4O]$, $[Li_5N]$, $[Na_3O]$, $[Na_4O]$, and $[Na_2OH]$) have been calculated to be stable against decomposition reactions in the gas phase and have been observed experimentally. [212] M–M bonded clusters are also present in solutions of the alkali metal halides. [213] The metals are entirely miscible with their molten halides, provided the temperatures are high enough. [214, 215] The clusters present in the melts are lost upon solidification due to the crystallization of the monohalides with their high lattice stabilities.

For the alkali metal clusters in the solid state, only a small island of stability exists so far, and this is with the suboxides of the heavy alkali metals Rb and Cs [4] at relatively low temperatures. Similar compounds of the lighter metals and the higher homologues of oxygen do not exist.

As shown in Figure **5-45**, the alkali metal suboxide clusters $[Rb_9O_2]$ and $[Cs_{11}O_3]$ are composed of two and three octahedral $[M_6O]$ units respectively, condensed via faces. They are the only constituents of the stoichiometric compounds Rb_9O_2 and $Cs_{11}O_3$. At low temperatures, they can occur in conjunction with additional amounts of Rb and Cs in a variety of stoichiometric phases that have even higher M/O ratios (*e. g.* Cs_4O, Cs_7O, $Cs_{11}O_3Rb_7$, $Cs_{11}O_3Rb_2$, $Cs_{11}O_3Rb$, and Rb_6O). Most of these compounds have been structurally characterized with single

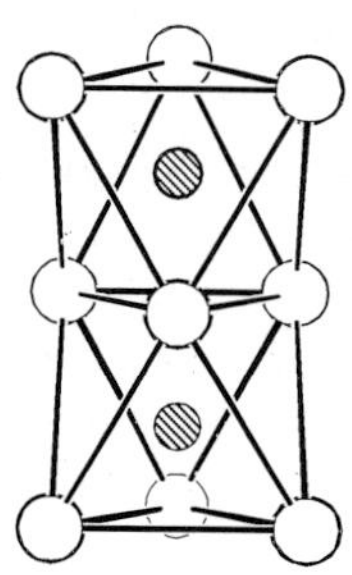

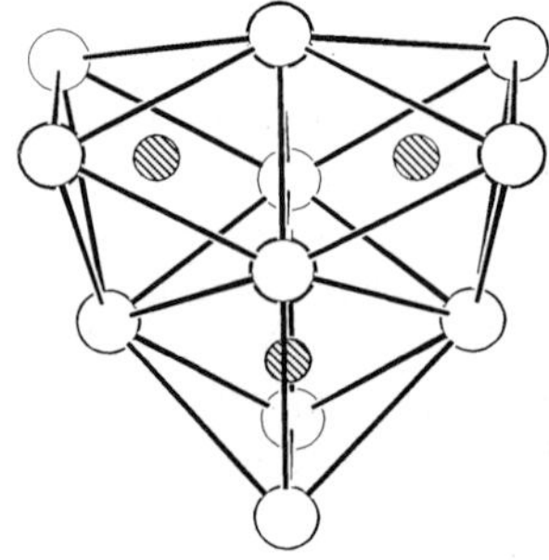

Figure 5-45. a) $[Rb_9O_2]$ and b) $[Cs_{11}O_3]$ clusters which are the characteristic structural units in alkali metal suboxides (hatched circles oxygen).

crystals. Figure **5-46** shows some of the crystal structures. The clusters are largely decomposed in melts that have the compositions of the solid phases. Cluster decomposition is known to occur from, for example, the observation that quenching a melt of composition Rb_7O at 164 K yields an amorphous product which, upon annealing at 170 K, deposits crystalline metastable $Rb_{6.33}O$ in which no $[Rb_9O_2]$ clusters exist. [216] Crystalline Rb_9O_2 is finally generated together with elemental Rb upon warming to 183 K. $[Rb_9O_2]$ and $[Cs_{11}O_3]$ clusters do not occur in the gas phase as demonstrated by mass spectroscopy of the partially oxidized metal vapors. [217]

According to Pauling's estimation of ionicity as based on differences in electronegativities, the O–M bonds in the $[Rb_9O_2]$ and $[Cs_{11}O_3]$ clusters have essentially ionic character. Hence, the formulations $[(Rb^+)_9(O^{2-})_2\ (e^-)_5]$ and $[(Cs^+)_{11}(O^{2-})_3\ (e^-)_5]$ represent a rather realistic description of the bonding in these clusters. Their stability is due to strong O–M bonds, although the formation of additional weaker M–M bonds is necessary for their existence. Model calculations using a simple Born-Mayer potential show that no stable configurations exist for the purely ionic clusters $[(Rb^+)_9(O^{2-})_2]^{5+}$ and $[(Cs^+)_{11}(O^{2-})_3]^{5+}$. [218] Such hypothetical species are unstable against loss of cations. M–M bonds, *i.e.* additional electrons in metal centered states, are necessary for a stable configuration which in the electrostatic model are taken into account as a partial shielding of the positive charges. The energy minimum is reached at a global charge of +2.3 for the $[Rb_9O_2]$ cluster. The incomplete shielding of the cationic charges by the additional electrons is probably associated with the very anisotropic charge distribution around the Rb^+ ions, which has further interesting consequences discussed later. The vibrational states of the cluster calculated for the equilibrium configuration are in good agreement with the observed Raman spectrum of Rb_9O_2. [218] Therefore, the simple model accounts for both the static and dynamic features of the cluster.

The atomic arrangement for $Cs_{11}O_3$ as shown in Figure **5-46** is reminiscent of a molecular crystal exhibiting short Cs–Cs distances (367–431 pm) as a consequence of strong ionic O–Cs bonds (275–292 pm). In contrast, the large Cs–Cs distances between the clusters, which with only one exception are longer than 500 pm, correspond to the distances found in elemental Cs. $Cs_{11}O_3$ and all other alkali metal suboxides show a metallic luster [219] and are good metallic conductors. [220] Thus, the electrons in M–M bonding states do not stay localized within the clusters but delocalize throughout the crystal due to the close contacts between the clusters.

The more metal rich suboxides may be interpreted as "intermetallic phases" of the cluster metals $[Rb_9O_2]$ and $[Cs_{11}O_3]$ with excess Rb and Cs, [4] as in $[Rb_9O_2]Rb_3$, $[Cs_{11}O_3]Cs_{10}$, and $[Cs_{11}O_3]Cs$. The molar volumes of these phases closely correspond to the sum of the molar volumes of Rb_9O_2 or $Cs_{11}O_3$ and the atomic volumes of Rb and Cs respectively. To take an example, the molar volumes of $Cs_{11}O_3Cs$ and $Cs_{11}O_3Cs_{10}$ exceed the molar volume of $Cs_{11}O_3$ by 69.9 and 696.5 $cm^3\ mol^{-1}$ respectively, and can be compared with the 69.4 $cm^3\ mol^{-1}$ volume of elemental Cs at 223 K. Clearly, the Cs atoms in these suboxides form purely metallic bonds to the $[Cs_{11}O_3]$ cluster.

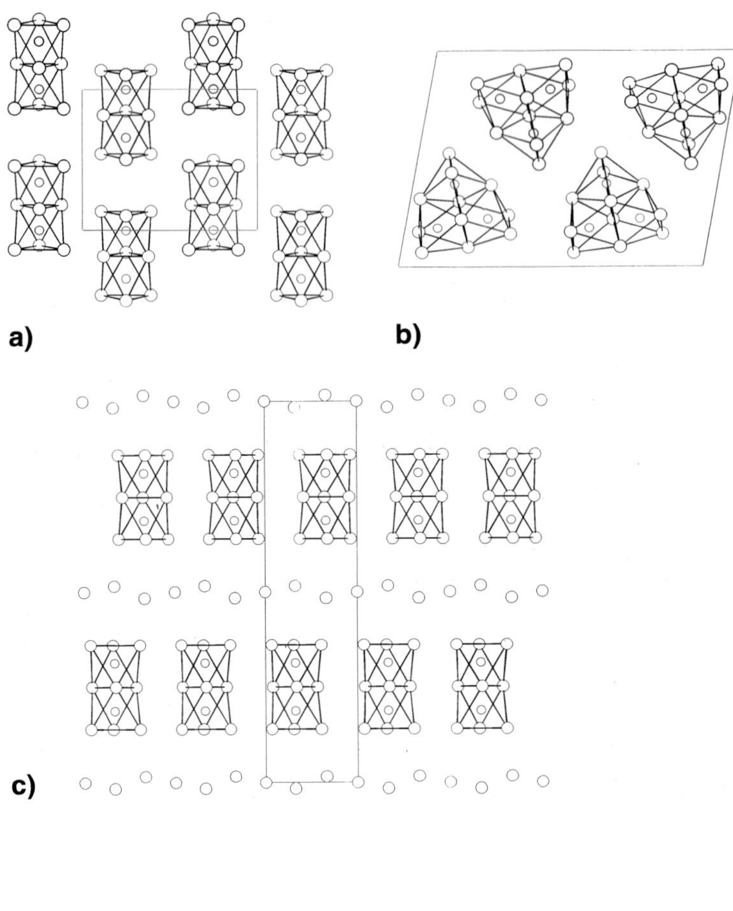

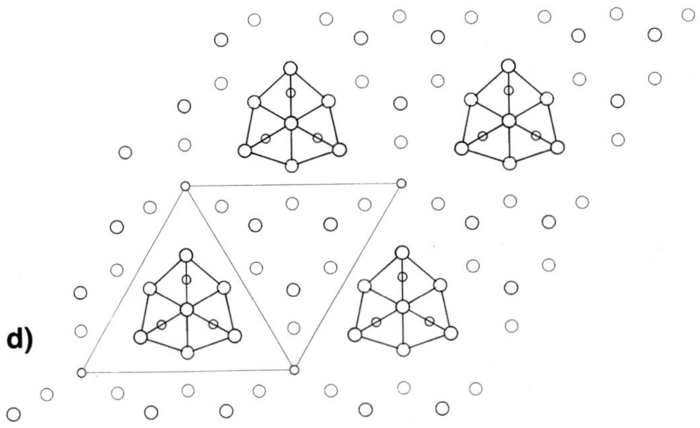

Figure 5-46. Projections of the crystal structures of the alkali metal suboxides a) Rb_9O_2, b) $Cs_{11}O_3$, c) Rb_6O (≙ $[Rb_9O_2]Rb_3$), and d) Cs_7O (≙ $[Cs_{11}O_3]Cs_{10}$). Rb_9O_2 and $Cs_{11}O_3$ clusters are outlined.

When Rb-Cs alloys of appropriate composition are partially oxidized, then the two metals which are so much alike in their aqueous chemistry exhibit an entirely different behavior. They become spatially separated in the suboxides, whereby the Cs enters into $[Cs_{11}O_3]$ clusters and the Rb is distributed in the purely metallic regions between the clusters. Examples are $Cs_{11}O_3Rb_n$ (n = 7, 2, 1). [4] There is evidence for the existence of K-Cs mixed suboxides that should, according to what has been said about the Rb-Cs suboxides, have the general composition $Cs_{11}O_3K_n$ or $Cs_{11}O_3(Cs, K)_n$. Unfortunately, they decompose at 215 K under K_2O deposition so that only preliminary X-ray and thermal analysis data have been obtained to identify their existence.

An obvious extension to the rather unique alkali metal suboxide chemistry is clear by an inspection of the periodic table which suggests a move of one step to the right with the metal and one step to the left with the nonmetal. In fact, subnitrides of the alkaline earth metals have been known for a long time and have the general formula M_2N. [221) As first determined for Ba_2N, they crystallize in the layered anti-$CdCl_2$ structure. [222] In applying the concept of condensed clusters to these one sees that the M_6 octahedra are centered by N atoms and connected via edges to form layers according to $Ba_{6/3}N$. The ionic limit of $[(M^{2+})_2N^{3-}(e^-)]$ indicates that weak M–M bonding is present both within and between the layers. Except for the two dimensional nature of the compound, Ba_2N therefore exhibits very similar bonding as that in the metal rich compounds of the neighboring elements Cs and O. Recently, a one dimensional analogue has been found. [223]

The structure of the compound $NaBa_3N$ contains parallel chains of face sharing Ba_6N octahedra having additional Na atoms which fill the voids between them as illustrated in Figure **5-47**. According to the formula $[(Ba^{2+})_3N^{3-}Na^+(e^-)_4]$, in addi-

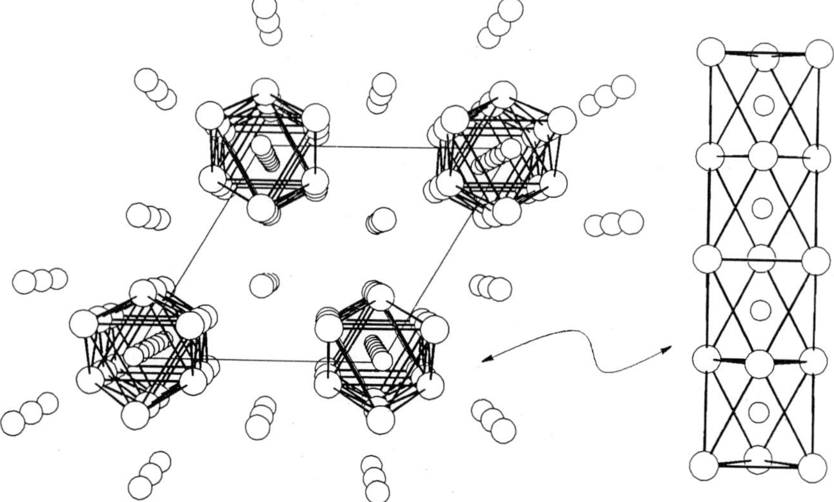

Figure 5-47. Projection of the structure of $NaBa_3N$ along [001] of the hexagonal unit cell together with the characteristic chains of face sharing Ba_6N octahedra (N, Na, and Ba atoms drawn at increasing size).

tion to the ionic bonding between the barium and nitride ions, the excess of metal valence electrons provides for M–M bonding in the periphery of the Ba_3N tubes, between the Ba_3N chains, and also to the Na atoms. The close relation between the Rb_9O_2 type cluster and the Ba_3N type chain is obvious as the latter represents the condensation of an infinite number of octahedra. Recalling the large number of phases known for the alkali metal suboxides, one is inclined to believe that more alkali/alkaline earth metal subnitrides exhibiting a variety of dimensionalities are still to come. Indeed, the new phases Na_5Ba_3N, $Na_9Ba_7N_6$ and $Na_{16}Ba_6N$ could just be isolated. [223a]

5.8 Structure Property Relations

The structures described in the preceeding sections can be generalized in a way indicated in Figure **5-48**. They belong to different categories which can be characterized as follows.

(i) Those clusters which are surrounded by nonmetal atoms yet may still approach each other to such an extent so as to get electronic delocalization between the clusters as in the case of the Chevrel type compounds. The confinement of electrons in the M–M bonding states of the single clusters is, however, more frequently observed as in numerous halide cluster compounds

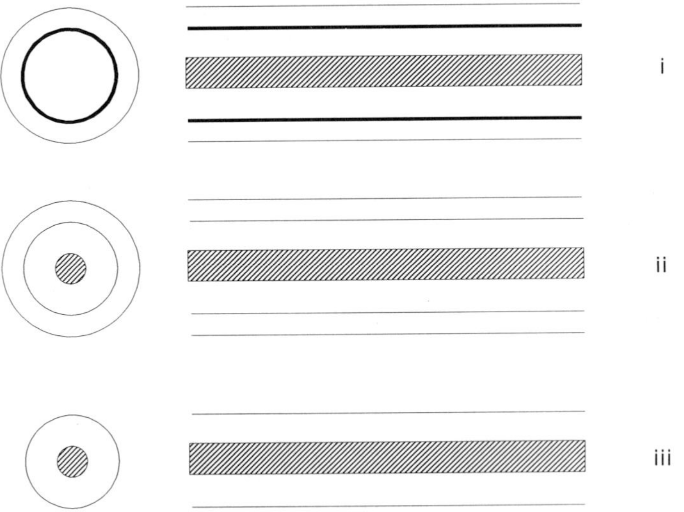

Figure 5-48. Schematized bonding in systems with discrete and condensed clusters: (i) empty metal clusters in an environment of nonmetal atoms, (ii) clusters with interstitial atoms in an environment of nonmetal atoms, and (iii) bare clusters with interstitial atoms (see text).

like Nb_6I_{11}. Condensation of such clusters leads to extended M–M bonding of low dimensionality.

(ii) For those clusters having an electron deficiency, interstitial atoms add to the stability of the clusters and condensed cluster phases. The interstitials may merely play the role of electron acceptors like interstitial hydrogen atoms. Another extreme are those interstitials which have strong covalent interaction with the surrounding metal atoms as it is found in some carbide halide systems.

(iii) A limit to cluster chemistry is reached with bare clusters stabilized by interstitial ions. Extended M–M bonding results from the close contacts between the clusters. A special situation arises from the accumulation of negative charge inside the clusters. This principle is realized with alkali metal suboxides and can be extended to the condensed cluster structures of the subnitrides.

A few chosen examples shall illustrate the remarkable physical properties which result from the generalized bonding situations given above.

5.8.1 Spin Crossover Transition in a Category (i) Cluster

Since many close lying M–M bonding orbitals characterize a metal cluster, the energy differences between them are often very sensitive to the detailed geometry of the cluster. In the case of a partial filling of these orbitals, their occupation can be critically dependant on slight geometrical variations. As a consequence, spin crossover transitions which are well known in transition metal complexes (*e. g.* Fe^{2+} complexes) should frequently be observable with metal cluster compounds. Such a spin crossover transition has been investigated in detail with the niobium iodides which contain the $[Nb_6I_8]$ and $[Nb_6I_8H]$ units. According to the formula $[(Nb^{5+})_6(I^-)_{11}(e^-)_{19}]$, Nb_6I_{11} has 19 electrons occupying M–M bonding states. This value has been verified by band structure calculations. [224]

In view of the weak electronic interaction between the clusters, the valence bands are narrow enough to describe the bonding with the MO scheme of the discrete cluster shown for the levels around the HOMO-LUMO gap in Figure **5-49**. Due to the low symmetry of the $[Nb_6I_8]$ unit in the crystal, the degeneracies of the e and t states are removed. However, the number of orbitals around the HOMO-LUMO gap clearly indicates that these orbitals originate from the t_{2u} and e_g sets of the regular cluster. In a second order phase transition near 273K, the distortion in the $[Nb_6I_8]$ cluster changes slightly due to the contortion of Nb_6 octahedron towards a trigonal prism at low temperature. [225] As a result, the one electron levels calculated on the basis of the actual structure parameters show significant changes in their relative energies. At low temperature, all but one of the 19 electrons in the M–M bonding states are spin paired. The calculated spin density distribution has also been verified by experiment in a study with polarized neutrons. [226] With increasing temperature, the energy separation between the electronic levels becomes smaller so that when the spin pairing energy falls below this energy separation, one of the electron pairs uncouples.

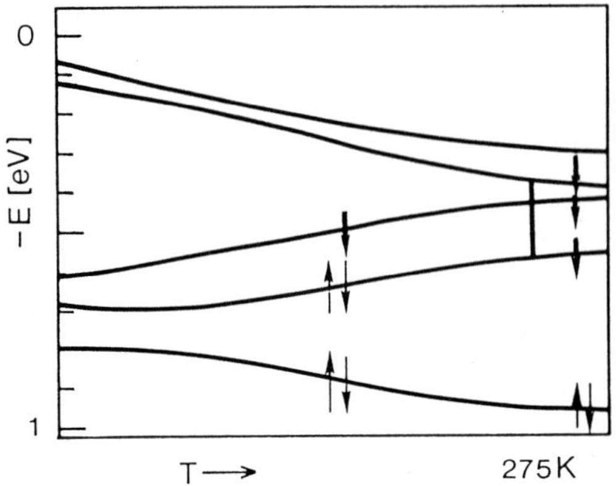

Figure 5-49. Changes in the one-electron energy levels around the HOMO-LUMO gap in Nb_6I_{11} as a function of temperature (or cluster distortion). The different electron occupations are indicated.

The phase transition [224, 227] which is primarily phonon driven, as with other spin crossover transitions, is accompanied by a change of the magnetic ground state from a doublet to a quartet and clearly seen in the respective slopes of the Curie-Weiss lines above and below the transition interval in Figure **5-50**. The compound $Nb_6I_{11}H$ undergoes a similar phase transition at slightly higher temperature. As it contains the 20 electron cluster $[Nb_6I_8H]$, the change of magnetic ground state is from singlet to triplet.

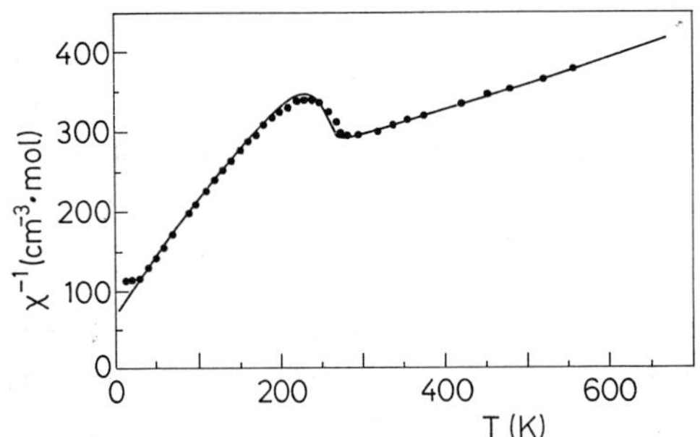

Figure 5-50. Inverse magnetic suszeptibility of Nb_6I_{11} as a function of temperature (points: experimental data, line: calculated).

A similar 20 electron cluster but free of an interstitial atom is the compound $CsNb_6I_{11}$. [34] If this is hydrogenated to $CsNb_6I_{11}H$, again an odd numbered electron count results and one finds a magnetic transition from a low spin to a high spin state with increasing temperature as in the case of Nb_6I_{11} itself. However, the transition occurs continously without a structural phase transformation. This is quite interesting in that the trigonal antiprismatic Nb_6 unit can distort towards the prism without breaking symmetry in the trigonal space group of $CsNb_6I_{11}H$.

To summarize, in the electronically discrete $[Nb_6I_8]$ cluster, the presence of a molecular orbital near the Fermi level which gets pairwise occupied upon deformation of the Nb_6 unit towards a trigonal prism leads to the magnetic phenomenon of a spin crossover transition. It is tempting to ask whether a quite similar bonding situation might be related to another physical phenomenon, superconductivity, in the case where the density of states at the Fermi level is nonzero, that is, in a metallic cluster compound.

5.8.2 Superconductivity in Chevrel Phases

The chemical and physical properties of the Chevrel compounds have been covered in books. [228] The enormous interest in these compounds was due to the fact that Mo_6S_8Pb turned out to be a superconductor ($T_c \approx 14K$), [229] with the highest upper critical fields, $H_{c2} \approx 60$ Tesla, known at the time.

Wide variations in the d electron concentration are feasible in the Chevrel phases through $(M,M')_6(X,\hat{X})_8A_x$. One possibility to change the electron balance is by substitution of the nonmetal atoms as in $Mo_6S_6Br_2$. [230] The substitution of the Mo atoms by more electron rich transition metals is also possible, and the "magic number" of 24 electrons per $(M,M')_6$ unit is reached in the semiconductors $Mo_4Ru_2Se_8$ and $Mo_2Re_4Se_8$. [231] A third possibility for varying the electron count is by the intercalation of metals A which act as electron donors. Metal atoms from nearly the entire Periodic Table have been incorporated between the clusters. The Mo–Mo distances are a sensitive probe for the number of electrons transferred from A into the cluster. At low electron concentrations, the M_6 unit is an elongated trigonal antiprism and becomes more regular as the 24 electron limit is approached. [232, 233] At the same time, the critical magnetic fields decrease and finally superconductivity is suppressed. The same holds for paramagnetic impurities. The investigations of Chevrel compounds added considerably to the knowledge about the antagonism between superconductivity and magnetism.

Large magnetic ions like those of the lanthanoids can be introduced between the clusters in a strictly ordered arrangement. The distances to the Mo atoms are large (d(Ln–Mo) > 410 pm), and their influence on the superconductivity of the phases is small. The coexistence of superconductivity and long range magnetic order could be demonstrated in these systems for the first time. [203] The Ln sublattice (*e. g.* Ln = Dy, Pr, Gd, Tb) in Mo_6S_8Ln orders antiferromagnetically without affecting the superconductivity. A ferromagnetic ordering of the Ln^{3+} ions from a strong external field is required in order to break the Cooper pairs and destroy the superconductivity. This competition between superconductivity and magnetism is the origin for the remarkable properties of Mo_6S_8Ho. It becomes a superconductor upon cooling to $T_c = 1.2K$ and a normal conductor again at the even lower temperature of 0.65K where ferromagnetic ordering of the Ho^{3+} moments occurs. Still more unusual is the behavior of $Mo_6S_{7.2}Se_{0.8}$-$Sn_{0.25}Eu_{0.75}$. [234, 235] At $T < 1K$, the superconductivity is suppressed in a weak field of approximately 1 T, but reenters at higher fields. Finally, above 22 T the phase becomes a normal conductor again. The phenomenon can be explained by assuming that an exchange field is built up by the Eu 4f electron that is opposing the external field and, therefore, compensates the latter. One may conclude from the unusual dependence of the behavior of the Chevrel phases on the magnetic impurities that the superconductivity in these compounds involves electronic bands comprising electronic states which are rather confined to the cluster substructure.

The intriguing physical properties of the Chevrel phases, as well as their comparatively simple crystal structures which are based on a primitive packing of quasi molecular $[Mo_6\hat{X}_8]$ units, triggered numerous theoretical investigations aimed at understanding the chemical bonding in these phases. [22, 23, 236, 237] They have suggested the existence for strong electronic interactions between the closely spaced clusters. In an obvious way, the transition from the local bonding picture for the "molecule" $[Mo_6\hat{X}_8]$ to the band structure of the extended crystal is shown in Figure **5-51**. In the discrete regular $[Mo_6\hat{X}_8]$ cluster, the 12 M–M bonding states lie above the M–$\hat{X}$ bonding states and are followed by the M–M anti-

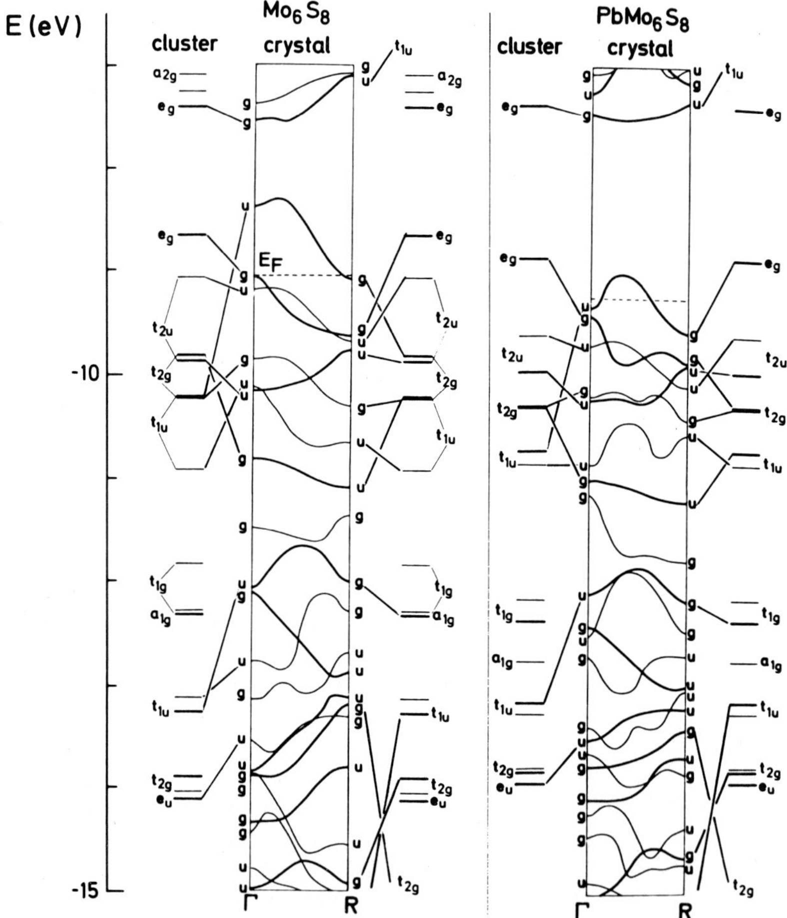

Figure 5-51. MO's around the HOMO-LUMO gap for the discrete clusters in Mo_6S_8 and Mo_6S_8Pb together with the band structures for the extended crystals. The band dispersions are shown in the direction of the threefold axis of the rhombohedral system. [64] The Fermi levels are marked by dashed lines.

bonding e_g, t_{1u}, a_{2g}, t_{1g}, and t_{2u} states. Their degeneracy is partly removed due to the deformations of the clusters with in the crystals.

The MO's for Mo_6S_8 and Pb Mo_6S_8 are drawn around the HOMO-LUMO gap in Figure **5-51**. The interaction of the cluster systems in the extended crystal leads to a broadening of all levels into bands. Their dispersion in momentum space is shown for the designated direction along the threefold axis in the rhombohedral system. The bands are filled to the Fermi level, E_F.

The special arrangements of the clusters in the $M_6\hat{X}_8$ structure (*cf.* Fig. **5-3**) allows each Mo atom to form a donor acceptor bond with an $\hat{X}$ atom of type $X^{i\text{-}a}$ between adjacent clusters. The bonding originates from an opening in the HOMO-LUMO gap. The M–M bonding e_g state is nearly unchanged in energy as there is hardly any overlap with the p orbitals on the $\hat{X}^{a\text{-}i}$ atom because the bond is of the δ type. According to the results of a population analysis, the Mo–$\hat{X}$ bonds between the clusters are virtually as strong as those within the clusters, whereas the intercluster Mo–Mo bonds have only approximately 10% of the strength of the intracluster bonds.

Self consistent band structure calculations performed for Chevrel phases having 20 to 23 d electrons in M–M bonding bands led to a result that is of significance to the understanding of their superconductivity. Obviously, the bands are not rigid but change drastically with the electron concentration in the cluster. The conduction band, in particular, changes its shape dramatically. These changes are related to the antibonding interactions between the e_u orbitals of the cluster and the p orbitals on the $\hat{X}^{a\text{-}i}$ atom. These have π character and are therefore sensitive to changes in the intercluster distances. In the case of a filled conduction band (24 electrons), these antibonding interactions demand large Mo–$\hat{X}^{a\text{-}i}$ distances. The depletion of the conduction band reduces the repulsive interaction between the clusters and, as a consequence, the intercluster distances decrease and the conduction band near Γ is raised. This change is evident from a comparison of the band structure of Mo_6S_8 (20 electrons) with that of Mo_6S_8Pb (22 electrons).

In Mo_6S_8Pb, the electrons occupy the local minimum at Γ whose position is very sensitive to any (static or dynamic) cluster deformation and to changes in the intercluster contacts. Hence, strong electron-phonon interaction involving quasi molecular electronic levels of the clusters at the Fermi level indicate a source for the pairwise attractive interactions between the conduction electrons.

5.8.3 Magnetic Order/Disorder in Category (ii) Lanthanoid Hydride Halides

In the reduced binary and ternary halides of the lanthanoids, the f electrons are well localized on the metal atoms. The coupling between the magnetic moments is weak in the case of insulators as can be seen from the low ordering temperatures of the saltlike trihalides whereby, for example, $GdCl_3$ orders ferromagnetically at 2.2K. In the presence of extended M–M bonding, the conduction electrons provide a very efficient mechanism for coupling the moments via a RKKY (*e. g.* to 292K in metallic gadolinium). Interestingly, the interaction is also enhanced if the additional electrons are not delocalized in a conduction band but localized in discrete M–M bonds. The semiconducting chain compound Gd_2Cl_3 may serve as an example [13] in that it exhibits an antiferromagnetic ordering at $T_N = 26K$.

The layered hydride (deuteride) halides $(H,D)_x$ are particularly good candidates to probe the effect of M–M bonding on the magnetic exchange interactions. The results from measurements on the systems GdBr $(H,D)_x$ and $TbBrD_x$ are in agreement with the general arguments presented above but also give a flavor for the unexpected richness of these effects and the questions still to be answered.

$GdBrH_2$ and $TbBrD_2$ are insulators and show an expected antiferromagnetic ordering at the very low temperatures of 2.5 and 5.5K respectively. A decrease in the hydrogen (deuterium) content leaves electrons which occupy bands at the Fermi level having M–M bonding character [238] (see Section 5.5 and Fig. **5-39**). As a consequence, the Neel temperature is increased to more than 30K for $GdBrD_x$ ($x \leq 1$).

Very small changes in x within the homogeneity range of the phase result in significant changes in T_N as depicted in Figure **5-52**. T_N decreases from 53K for $x = 1.0$ to 34K for $x = 0.67$. One is inclined to interpret these changes in terms of changing RKKY interactions resulting from changes in carrier concentrations. However, then $GdBrD_{0.67}$ should exhibit the highest value for T_N since the carrier concentration is highest for this composition. In fact, the situation is more com-

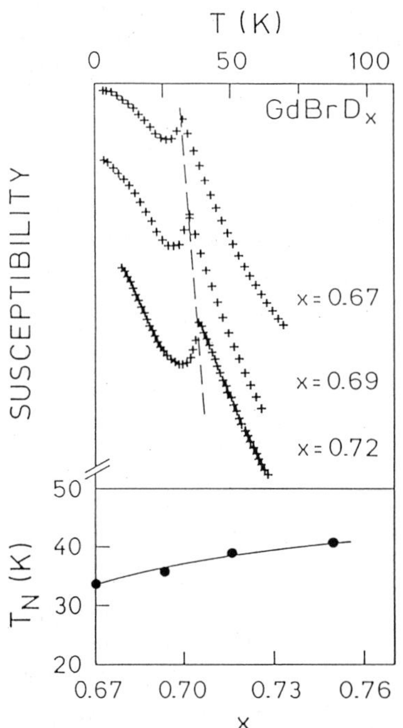

Figure 5-52. (top) Magnetic susceptibilities of $GdBrD_x$ measured around the Neel temperatures. (bottom) Dependence of the Neel temperatures on x.

plicated as indicated by the resistivity measurements shown in Figure **5-53**. Samples with the composition $GdBr(H,D)_{0.7}$ exhibit semiconducting behavior and therefore a localization of the charge carriers is observed at the lowest temperatures. The strong negative magnetoresistance of $GdBrD_{0.7}$ at low temperature indeed shows an intricate relation between antiferromagnetic order and electron localization. [239]

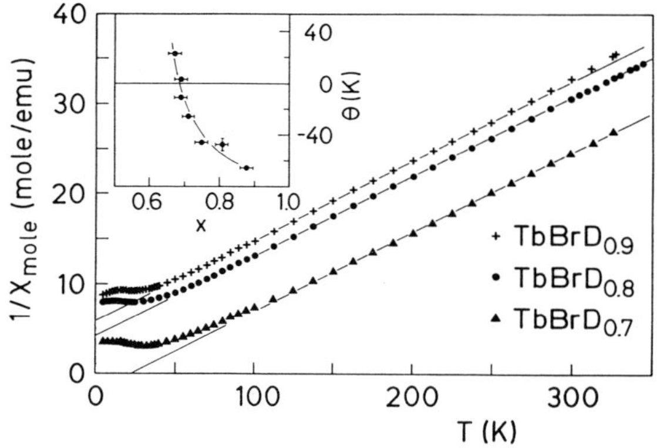

Figure 5-53. Reciprocal magnetic susceptibilities of $TbBrD_x$ and (inset) paramagnetic Curie temperatures as function of x.

Minor chemical variations can lead to entirely different physical properties. The corresponding isotypic Tb compounds $TbBr(H,D)_x$ do not exhibit electron localization and a transition to the semiconducting state at small values of x and low temperature. Obviously, crystal field effects which are absent with the S = 0 ion Gd^{3+} play an important role with the Tb^{3+} ion. The investigation on the magnetism of $TbBr(H,D)_x$ as a function of x has provided clear hints as to the importance of these effects.

The magnetic properties of $TbBrD_x$ are characterized by Curie-Weiss lines [13] which pass through the origin for the insulator $TbBrD_2$. For metallic $TbBrD_x$ ($x < 1$), the paramagnetic Curie temperature Θ shows a systematic variation with deuterium content. When $x = 0.9$ it is –66K and becomes positive, $\Theta = 23K$, for $x = 0.7$. Obviously both antiferromagnetic and ferromagnetic couplings occur. One or the other type of coupling will dominate the behavior depending on the deuterium content. Neutron diffraction studies on $TbBrD_x$ ($x \geq 0.8$) revealed that indeed the ferromagnetic ordering of the moments occurs within the layers and the antiferromagnetic ordering occurs between adjacent layers [240] as illustrated in Figure **5-54**.

With decreasing D(H) content the magnetic superstructure reflections become weaker and vanish for $x = 0.7$ despite a pronounced anomaly in the magnetic susceptibility. Obviously the competing ferro- and antiferromagnetic interactions together with the disorder in the (nonmagnetic!) sublattice of D(H) results in a spin glass formation.

More detailed investigations [241, 242] showed that all characteristics of a spin glass are present (Fig. **5-55**) Thus the samples exhibited a time dependent

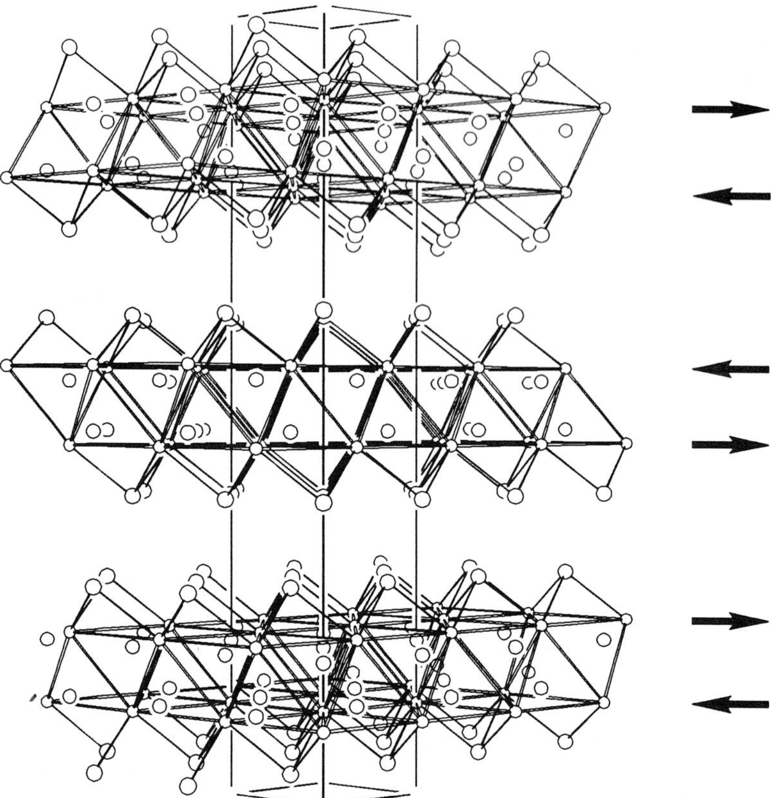

Figure 5-54. Magnetic ordering in $TbBrD_{0.9}$. The moments are parallel to the metal atom layers in [001] and antiferromagnetically coupled between adjacent layers. The magnetic cell is 3 times longer than the crystallographic cell drawn.

increase in the zerofield cooled susceptibility, drastic differences between field and zerofield cooled samples, thermal hysteresis effects, a frequency dependence in the real part of the AC susceptibility, etc. The phase $TbBrD_{0.7}$ closely approaches an ideal spin glass system in the sense that the structural disorder only occurs in the nonmagnetic sublattice, although one has to admit that this disorder also introduces some disorder in the magnetic sublattice via local shifts of the atomic positions and varying crystal fields.

5.8.4 Superconductivity in Rare Earth Carbide Halides

The structures of the layered carbide halides $RE_2X_2C_2$ are closely related to those of the metallic hydride halides discussed in the previous subsection, except for the fact that C_2 units occupy the centers of the metal atom octahedra and the halogen atoms take different positions above and below the metal atom bilayers. Regardless, the strictly two dimensional character is preserved.

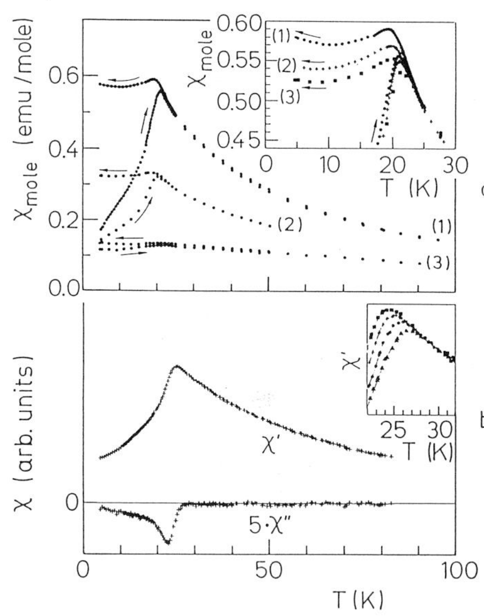

Figure 5-55. a) Field cooled (left pointing arrows) and zero field cooled (right pointing arrows) magnetic susceptibility measurements on three different $TbBrD_x$ samples: (1) $x = 0.7$ at 1.6 G, (2) $x = 0.7$ at 100 G, and (3) $x = 0.8$ at 100 G. The inset shows the field dependence for $TbBrD_{0.7}$: (1) 1.6 G, (2) 26 G, and (3) 100 G. b) Real and imaginary parts χ' and χ'' of the AC magnetic susceptibility at 11.7 Hz of $TbBrD_{0.7}$. The inset shows shift of the χ' cusp with frequency: (from left to right) 3.51, 11.7, 117 Hz, and 1.17 kHz.

According to the simplified analysis of the chemical bonding from the ionic limit formula $[(RE^{3+})_2(X^-)_2C_2^{4-}]$, semiconducting properties would be expected for the compounds since the distance d(C–C) $\approx$ 130 pm is indicative of a double bond. The compounds are, however, metallic as mentioned and explained earlier. Superconductivity has been found for compounds of the nonmagnetic rare earth metals and include $Y_2Br_2C_2$, $Y_2I_2C_2$, $La_2Br_2C_2$, and on a trace level $Lu_2Cl_2C_{2-x}$ (Fig. **5-56**). [243]

These systems provide a rather visual and local approach to the possible origin of pairwise attractive interactions between conduction electrons as a prerequisite for superconductivity. The basic idea in the case of the Chevrel phases discussed earlier and the rare earth carbide halides here is identical. In the Chevrel phases, although the electronic delocalization and hence the metallic properties are primarily due to M–$\hat{X}$ covalency, there could be a simultaneous tendency towards (pairwise) localization in an M–M bonding cluster level of quasi molecular character. Band structure calculations performed for $Gd_2Cl_2C_2$ [186] as representative of the $(RE)_2X_2C_2$ compounds clearly show the mixed d π^* character of the band states at the Fermi level (Fig. **5-36**). Again, electronic delocalization comes about through strong M–C covalency, and the quasi molecular C–C π^* orbitals at the Fermi level provide for the pairwise attractive interaction between the conduction electrons. Electron-phonon interactions will be strong for the stretching and bending vibrations of the C_2 unit as they change the position of the π^* level and extent of p–d overlap respectively. Under this general idea, it is apparently unimportant in the case of the Chevrel phases that the M–$\hat{X}$ interaction is antibonding and the M–M interaction is bonding, whereas in the carbide halides the M–C and C–C interactions are bonding and antibonding respectively.

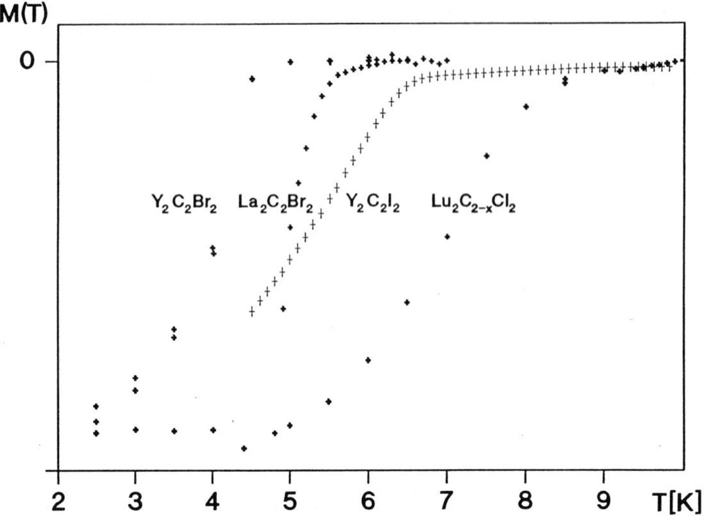

Figure 5-56. Diamagnetic shielding indicating superconductivity in $Y_2Br_2C_2$, $Y_2I_2C_2$, $La_2Br_2C_2$, and $Lu_2Cl_2C_{2-x}$.

5.8.5 Alkali Metal Suboxides as IR Emitting Photocathodes

The structures of the alkali metal suboxides have been discussed in terms of bare metal clusters stabilized by interstitial oxygen atoms. Due to the direct M–M bonding between the clusters, the suboxides are metallic and show electrical properties [220] which are similar to the free alkali metals.

Photoelectron spectra provide a more quantitative insight into the electronic properties and bonding in these compounds. All spectra in Figure **5-57** exhibit a density of states at the Fermi level (E_F = O eV) as one would expect for metallic samples. [244] In the spectrum of $Cs_{11}O_3$, the spin orbit split Cs 5p band is shifted by 0.5 eV to smaller binding energies compared to Cs, and in the spectrum of $Cs_{11}O_3Cs_{10}$ the photoemission from the 5p band is both characteristic for the $[Cs_{11}O_3]$ cluster and for pure Cs. The narrow O 2p band is evidence for the localization of electrons in the O^{2-} ion, and the measured binding energy of $E_B = 2.7$ eV, which is the lowest for all oxides, is consistent with a description of the chemical bonding in terms of O^{2-} ions that are stabilized by a weak Coulomb field from the surrounding cations.

The spectrum of $Cs_{11}O_3Cs_{10}$ could represent that of a mixture of Cs and $Cs_{11}O_3$, if it were not for the features in the spectrum that give clear evidence for a homogeneous sample. Each sample has a characteristic value for the work function, Φ, that decreases with increasing O content and shows up as a gap between the 21.2 eV excitation energy of the He I radiation and the photoemission threshold energy. The energy difference, Δ, for the satellite structures also changes systematically. These structures originate from energy losses due to surface plasmon excitation, and their energies depend on the concentration of free

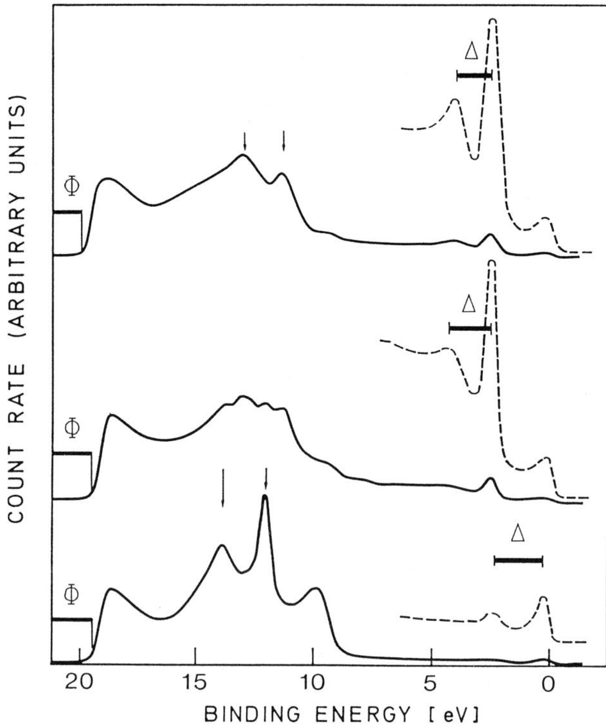

Figure 5-57. He I (21.2 eV) photoelectron spectra of (top to bottom) $Cs_{11}O_3$, Cs_7O ($\triangleq Cs_{11}O_3Cs_{10}$), and Cs. The binding energies are referenced to the Fermi level E_F = 0eV, and the arrows indicate photoemission from Cs $5p^{3/2}$ and $5p^{1/2}$ (Φ = work function, Δ = energy loss due to surface plasmon excitation). The broken lines show intensities which have been magnified 10 times.

electrons. The numbers of free electrons, 5 and 14 per formula unit respectively, as determined from the measured energy loss values for $Cs_{11}O_3$ and $Cs_{11}O_3Cs_{10}$ are in excellent agreement with those determined from the simple ionic bond model explained earlier.

Thin composite layers of Cs_2O and Cs on Ag play an important role in the IR sensitive S1 photocathodes which have been manufactured for approximately sixty years. [245, 246] The characteristic spectral response of such cathodes can be explained by allowing for the presence of alkali metal suboxides, $Cs_{11}O_3$ or "Cs_3O", since these have the appropriate electronic properties. Thus, these materials have a low work function of approximately 1 eV (compared to 2 eV for elemental Cs) and low energies of the surface plasmons (1.5 eV for $Cs_{11}O_3$). The enhancement of the photoelectric yield is due to surface plasmon decay. [247]

The small value for the plasmon energy is simply related to the low carrier concentration and intimately linked with the electron poor nature of the alkali metals. The low work function has a more interesting explanation. Although it was shown earlier that the electrical properties of the alkali metals and their suboxides are very similar, this is not quite true. In elemental Cs, the conduction electrons are allowed to take up all the space between the atoms. In contrast, in a crystal of $Cs_{11}O_3$ the conduction electrons meet a regular array of accumulated negative charge (6–) in each cluster center which is highly repulsive for the conduction electrons and becomes an avoided region for them. As such, $Cs_{11}O_3$ is a kind of microscopically foamed metal ("void metal").

As in the much investigated quantum dot and antiquantum dot structures of doped semiconductors, the confinement of electrons in space leads to quantum size effects. The estimation of such quantum size effects for $Cs_{11}O_3$ predicts a work function that is 0.9 eV lower than that for Cs. [248] More recently, metastable deexcitation spectroscopy (MDS) on Cs suboxide surfaces gave further support for this effect. [249, 250]

5.9 Closing Remark

Clusters bridge the gap between small discrete molecules and infinitely extending solids. In this article, the structures, bonding, and properties of compounds which in a narrow sense do not even contain clusters but could be called normal salts or "contaminated" metals have been included. It is hoped that by taking this approach, though it might seem arbitrary in many respects, some contribution to a unified view on matter has been made.

References

[1] P. von R. Schleyer, E.-U. Würthwein, J. A. Pople, *J. Am. Chem. Soc.* **1982**, *104*, 5839.
[2] P. von R. Schleyer in *New Horizons of Quantum Chemistry* (Eds.: P.-O. Löwdin, B. Pullman), Reidel, Dordrecht, The Netherlands, **1983**, p. 95.
[3] A. Kjekshus, T. Rakke, *Struct. Bonding (Berlin)* **1974**, *19*, 45.
[4] A. Simon, *Struct. Bonding (Berlin)* **1979**, *36*, 81.
[5] H. Schäfer, H. G. v. Schnering, *Angew. Chem.* **1964**, *76*, 833.
[6] A. Simon, *Angew. Chem.* **1988**, *100*, 163 [Int. Ed. Engl. *27*, 159].
[7] A. Perrin, C. Perrin, M. Sergent, *J. Less-Common Metals* **1988**, *137*, 241.
[8] C. Perrin, S. Ihmaïne, M. Sergent, *New J. Chem.* **1988**, *12*, 321.
[9] A. Perrin, M. Sergent, *New. J. Chem.* **1988**, *12*, 337.
[10] R. P. Ziebarth, J. D. Corbett, *Acc. Chem. Res.* **1989**, *22*, 256.
[11] R. E. McCarley in *Metal-Metal Bonds and Clusters in Chemistry and Catalysis* (Ed.: J. P. Fackler Jr.), Plenum Press, New York **1990**, pp. 91.
[12] S. C. Lee, R. H. Holm, *Angew. Chem.* **1990**, *102*, 868.
[13] A. Simon, Hj. Mattausch, G. J. Miller, W. Bauhofer, R. K. Kremer in *Handbook Physics and Chemistry of Rare Earths* (Eds.: K. A. Gschneidner Jr., L. Eyring), Elsevier Science Publ. B. V. **1991**, *15*, p. 191.
[14] J. D. Corbett in *Modern Perspectives in Inorganic Crystal Chemistry* (Ed.: E. Parthé), Kluwer Acad. Publ. **1992**, p. 27.
[15] A. Simon in *Solid State Chemistry* (Eds.: A. K. Cheetham, P. Day), Clarendon Press, Oxford **1992**, pp. 112.

[16] J. Köhler, G. Svensson, A. Simon, Angew. Chem. **1992**, *104*, 1463 [Int. Ed. Engl. *31*, 1437].

[17] C. Brosset, *Ark. Kem., Mineralog. Ser. A* **1945**, *20*, 16.

[18] P. A. Vaughan, J. H. Sturdivant, L. Pauling, *J. Amer. Chem. Soc.* **1950**, *72*, 5477.

[19] F. A. Cotton, R. E. Haas, *Inorg. Chem.* **1964**, *3*, 10.

[20] L. J. Guggenberger, A. W. Sleight, *Inorg. Chem.* **1969**, *8*, 2041.

[21] B. E. Bursten, F. A. Cotton, G. G. Stanley, *Israel J. Chem.* **1980**, *19*, 132.

[22] H. Nohl, O. K. Andersen, *Proc. 4th Conf. Supercond. d-f-Band Metals, Karlsruhe* **1982**, 161.

[23] T. Hughbanks, R. Hoffmann, *J. Am. Chem. Soc.* **1983**, *105*, 1150.

[24] J. D. Smith, J. D. Corbett, *J. Am. Chem. Soc.* **1985**, *107*, 5704.

[25] O. K. Andersen, unpublished.

[26] T. Saito, N. Yamamoto, T. Nagase, T. Suboi, K. Kobayashi, T. Yamagata, H. Imoto, K. Unoura, *Inorg. Chem.* **1990**, *29*, 764.

[27] A. Simon, H. G. v. Schnering, H. Schäfer, *Z. Anorg. Allg. Chem.* **1969**, *367*, 1.

[28] H. G. v. Schnering, *Z. Anorg. Allg. Chem.* **1971**, *385*, 75.

[29] M. Sägebarth, A. Simon, *Z. Anorg. Allg. Chem.* **1990**, *587*, 119.

[30] K. F. Herzfeld, *Phys. Rev.* **1927**, *29*, 701.

[31] N. F. Mott, *Proc. Phys. Soc.* **1949**, *A62*, 416.

[32] P. P. Edwards, M. J. Sienko, *Int. Rev. Phys.* **1983**, *3*, 83.

[33] A. Simon, H. G. v. Schnering, H. Schäfer, **1967**, *355*, 295.

[34] H. Imoto, J. D. Corbett, *Inorg. Chem.* **1980**, *19*, 1241.

[35] F. Stollmaier, A. Simon, *Inorg. Chem.* **1985**, *24*, 168.

[36] H.-J. Meyer, J. D. Corbett, *Inorg. Chem.* **1991**, *30*, 963.

[37] R. Chevrel, G. Gougeon, M. Potel, M. Sergent, *J. Solid State Chem.* **1985**, *57*, 25.

[38] R. Chevrel, M. Sergent in *Crystal Chemistry and Properties of Materials with Quasi-One-Dimensional Structures* (Ed.: J. Rouxel), Reidel, Dordrecht, Netherlands **1986**, p. 315.

[39] W. Bronger, *Pure Appl. Chem.* **1985**, *57*, 1363.

[40] W. Bronger, *Angew. Chem.* **1993**, *105*, 597.

[41] L. Leduc, A. Perrin, M. Sergent, F. LeTraon, J. C. Pilet, A. LeTraon, *Mater. Lett.* **1985**, *3*, 209.

[42] L. Leduc, A. Perrin, M. Sergent, *C. R. Acad. Sci. Paris Ser. II* **1983**, *296*, 961.

[43] H. Schäfer, H. G. von Schnering, J. Tillack, F. Kuhnen, H. Wöhrle, H. Baumann, *Z. Anorg. Allg. Chem.* **1967**, *353*, 281.

[44] C. Perrin, M. Sergent, J. Prigent, *C. R. Acad. Sci. Paris Ser. C* **1973**, *277*, 46.

[45] C. Perrin, M. Sergent, F. LeTraon, A. LeTraon, *J. Solid State Chem.* **1978**, *25*, 197.

[46] W. Bronger, H. J. Miessen, P. Müller, R. Neugroschel, *J. Less-Common Met.* **1985**, *105*, 303.

[47] R. Siepmann, H. G. von Schnering, *Z. Anorg. Allg. Chem.* **1968**, *357*, 289.

[48] W. Bronger, H.-J. Miessen, R. Neugroschel, D. Schmitz, M. Spangenberg, *Z. Anorg. Allg. Chem.* **1985**, *525*, 41.

[49] F. Klaiber, W. Petter, F. Hulliger, *J. Solid State Chem.* **1983**, *46*, 112.

[50] V. E. Federov, N. V. Podberezskaya, A. V. Mischenko, G. F. Khudorozko, I. P. Asanov, *Mater. Res. Bull.* **1986**, *21*, 1335.

[51] P. Böttcher, *Angew. Chem.* **1988**, *100*, 781 [Int. Ed. Engl. *27*, 759].

[52] L. Leduc, A. Perrin, M. Sergent, *Acta Crystallogr. C* **1983**, *39*, 1503.

[53] C. Perrin, M. Sergent, *J. Chem. Res. (S)* **1983**, 38.

[54] C. Perrin, M. Potel, M. Sergent, *Acta Crystallogr. C* **1983**, *39*, 415.

[55] P. Gougeon, M. Potel, J. Padiou, M. Sergent, *C. R. Acad. Sci. Paris Ser. 2* **1983**, *297*, 339.

[56] J. Geyer-Lippmann, A. Simon, F. Stollmaier, *Z. Anorg. Allg. Chem.* **1984**, *516*, 55.

[57] D. M. Evans, L. Katz, *J. Solid State Chem.* **1973**, *6*, 459.

[58] B. O. Marinder, *Chem. Scripta* **1977**, *11*, 97.

[59] J. Köhler, A. Simon, *Z. Anorg. Allg. Chem.* **1987**, *553*, 106.
[60] R. Burnus, J. Köhler, A. Simon, *Z. Naturforsch. B* **1987**, *42*, 536.
[61] T. Hughbanks, R. Hoffmann, *J. Am. Chem. Soc.* **1983**, *105*, 3528.
[62] R. Hoffmann, *Angew. Chem.* **1987**, *99*, 871 [Int. Ed. Engl. *26*, 846].
[63] R. Hoffmann, *Solids and Surfaces*, Verl. Chemie, Weinheim **1988**.
[64] J. Köhler, R. Tischtau, A. Simon, *J. Chem. Soc. Dalton Trans.* **1991**, 829.
[65] A. Simon, *Z. Anorg. Allg. Chem.* **1967**, *355*, 311.
[66] A. Simon, F. Stollmaier, D. Gregson, H. Fuess, *J. Chem. Soc. Dalton Trans.* **1987**, 431.
[67] B. F. G. Johnson, R. D. Johnston, J. Lewis, *Chem. Commun.* **1967**, 1057.
[68] H.-G. Fritsche, F. Dübler, H. Müller, *Z. Anorg. Allg. Chem.* **1984**, *513*, 46.
[69] A. Simon, *Ann. Chim. (Paris)* **1982**, *7*, 539.
[70] F. Dübler, H. Müller, C. Opitz, *Chem. Phys. Lett.* **1982**, *88*, 467.
[71] A. Simon, *Naturwissensch.* **1984**, *71*, 171.
[72] D. E. Eastman, J. K. Cashion, A. C. Switendick, *Phys. Rev. Lett.* **1971**, *27*, 35.
[73] C. Alefeld, J. Völkl, *Hydrogen in Metals*, Springer, Berlin, Heidelberg **1978**.
[74] O. Jepsen, R. M. Nieminen, J. Madsen, *Solid State Commun.* **1980**, *34*, 375.
[75] J. W. Lauher, *J. Am. Chem. Soc.* **1978**, *100*, 5305.
[75a] L. D. Künne, *Monatshefte Chemie* **1991**, *122*, 625
[76] S. D. Wijeyesekera, R. Hoffmann, *Organometallics* **1984**, *3*, 949.
[77] R. E. McCarley, B. G. Hughes, F. A. Cotton, F. A. Zimmermann, *Inorg. Chem.* **1965**, *4*, 1491.
[78] B. Spreckelmeyer, *Z. Anorg. Allg. Chem.* **1968**, *358*, 148.
[79] J. D. Smith, J. D. Corbett, *J. Am. Chem. Soc.* **1986**, *108*, 1927.
[80] M. W. Payne, J. D. Corbett, *Inorg. Chem.* **1990**, *29*, 2246.
[81] R. Z. Ziebarth, J. D. Corbett, *J. Solid State Chem.* **1989**, *80*, 56.
[82] H. M. Artelt, T. Schleid, G. Meyer, *Z. Anorg. Allg. Chem.* **1992**, *618*, 18.
[83] H. M. Artelt, G. Meyer, *Z. Anorg. Allg. Chem.* **1993**, *619*, 1.
[84] F. Böttcher, A. Simon, R. Kremer, H. Buchkremer-Hermanns, J. K. Cockcroft, *Z. Anorg. Allg. Chem.* **1991**, *598/599*, 25.
[85] J. Zhang, J. D. Corbett, *Inorg. Chem.* **1991**, *30*, 431.
[86] R. P. Ziebarth, J. D. Corbett, *Inorg. Chem.* **1989**, *28*, 626.
[87] J. Zhang, J. D. Corbett, *J. Less-Common Met.* **1989**, *156*, 49.
[88] J. Zhang, J. D. Corbett, *Z. Anorg. Allg. Chem.* **1991**, *598/599*, 363.
[89] R. P. Ziebarth, J. D. Corbett, *J. Am. Chem. Soc.* **1989**, *111*, 3272.
[90] T. Hughbanks, G. Rosenthal, J. D. Corbett, *J. Am. Chem. Soc.* **1986**, *108*, 8289.
[91] S. Gao, G.-X. Xu, L.-M. Li, *Inorg. Chem.* **1992**, *31*, 4829.
[91a] L. Künne, Z. Almasri, *J. Less-Common Met.* **1990**, *166*, 386
[92] A. N. Fitch, S. A. Barret, B. E. Fender, A. Simon, *J. Chem. Soc. Dalton Trans.* **1984**, 501.
[93] A. R. Miedema, R. Boom, F. R. de Boer in *Crystal Structure and Chemical Bonding in Inorganic Chemistry* (Eds.: C. J. M. Rooymans, A. Rabenau), North Holland Publ., Amsterdam, **1975**, pp. 163.
[94] A. K. Niessen, F. R. de Boer, R. Boom, P. F. Châtel, W. C. M. Mattens, A. R. Miedema, *CALPHAD – Comput. Coupling Phase Diagrams Thermochem.* **1983**, *7*, 5.
[95] D. S. Dudis, J. D. Corbett, *Inorg. Chem.* **1987**, 1933.
[96] S. Satpathy, O. K. Andersen, *Inorg. Chem.*, **1985**, *24*, 2604.
[97] A. Simon, F. Böttcher, J. K. Cockcroft, *Angew. Chem.* **1991**, *103*, 79 [Int. Ed. Engl. *30*, 101].
[98] F. A. Cotton, X. Feng, M. Shang, W. A. Wojtczak, *Angew. Chem.* **1992**, *104*, 1117 [Int. Ed. Engl. *31*, 1050].
[99] T. Braun, A. Simon, unpublished.
[100] F. Rogel, J. D. Corbett, *J. Am. Chem. Soc.* **1990**, *112*, 8199.
[101] F. Rogel, J. Zhang, M. W. Payne, J. D. Corbett, *Adv. Chem. Ser.* **1990**, *226*, 369.
[102] A. Simon, *Angew. Chem.* **1981**, *93*, 23 [Int. Ed. Engl. *20*, 1].

[103] K. R. Poeppelmeier, J. D. Corbett, *Inorg. Chem.* **1977**, *16*, 1107.
[104] F. Grønvold, A. Kjekshus, F. Raaum, *Acta Crystallogr.* **1961**, *14*, 930.
[105] P. Jensen, A. Kjekshus, *Acta Chem. Scand.* **1966**, *20*, 1309.
[106] V. Kumar, V. Heine, *Inorg. Chem.* **1984**, *23*, 1498.
[107] R. Dronskowski, *Thesis*, Stuttgart, **1990**.
[108] H. F. Franzen, *Prog. Solid State Chem.* **1978**, *12*, 1.
[109] H. F. Franzen, T. A. Beineke, B. R. Conrad, *Acta Crystallogr. B* **1968**, *24*, 412.
[110] B. Harbrecht, *Angew. Chem.* **1988**, *101*, 1696 [Int. Ed. Engl. *28*, 1660].
[111] O. K. Andersen, S. Satpathy in *Basic Properties of Binary Oxides* (Eds.: A. Dominguez, J. Castaing, R. Marquez) Serv. Publ. Univ. Sevilla, Spain, **1983**, p. 21.
[112] J. K. Burdett, T. Hughbanks, *J. Am. Chem. Soc.* **1984**, *106*, 3103.
[113] D. M. Evans, L. Katz, *J. Solid State Chem.* **1973**, *6*, 459.
[114] A. Simon, J. Köhler, R. Tischtau, G. Miller, *Angew. Chem.* **1989**, *101*, 1695 [Int. Ed. Engl. *28*, 1662].
[115] A. Magnéli, *Chem. Script.* **1989**, *26*, 535.
[116] G. Svensson, *Microsc. Microanal. Microstr.* **1990**, *1*, 343.
[117] G. Svensson, J. Köhler, M. T. Otten, A. Simon, *Z. Anorg. Allg. Chem.* **1993**, *619*, 133.
[117a] G. Svensson, J. Grins, *Acta Crystallogr. B* **1993**, *49* [in print]
[118] G. Svensson, *Chem. Commun. Univ. Stockholm* **1989**, *3*, 79.
[119] C. C. Torardi, R. E. McCarley, *J. Am. Chem. Soc.* **1979**, *101*, 3963.
[120] R. E. McCarley, *ACS Symp. Ser.* **1983**, *211*, 273.
[121] R. E. McCarley, K.-H. Lii, P. A. Edwards, L. F. Brough, *J. Solid State Chem.* **1985**, *57*, 17.
[122] J. D. Corbett, R. E. McCarley in *Crystal Chemistry and Properties of Materials with Quasi-One-Dimensional Structure* (Ed.: J. Rouxel), Reidel, Dordrecht, The Netherlands, **1986**, p. 179.
[123] K.-H. Lii, R. E. McCarley, S. Kim, R. A. Jacobson, *J. Solid State Chem.* **1986**, *64*, 359.
[124] C. C. Torardi, R. E. McCarley, *J. Less-Common Met.* **1986**, *116*, 169.
[125] C. D. Carlson, L. F. Brough, P. A. Edwards, R. E. McCarley, *J. Less-Common Met.* **1989**, *156*, 325.
[126] P. Gougeon, P. Gall, R. E. McCarley, *Acta Crystallogr. C* **1991**, *47*, 1585.
[127] P. Gougeon, P. Gall, R. E. McCarley, *Acta Crystallogr. C* **1991**, *47*, 2026.
[128] R. E. McCarley, *Polyhedron*, **1986**, *5*, 51.
[129] A. Simon, *Angew. Chem.* **1983**, *95*, 94 [Int. Ed. Engl. *22*, 95].
[130] Hj. Mattausch, A. Simon, E. M. Peters, *Inorg. Chem.* **1986**, *25*, 3428.
[131] R. A. Wheeler, R. Hoffmann, *J. Am. Chem. Soc.* **1988**, *110*, 7315.
[132] A. Simon, W. Mertin, Hj. Mattausch, R. Gruehn, *Angew. Chem.* **1986**, *98*, 831 [Int. Ed. Engl. *25*, 845].
[133] R. Ramlau, E. Fais, A. Simon, Hj. Mattausch, unpublished.
[134] R. Dronskowski, Hj. Mattausch, A. Simon, *Z. Anorg. Allg. Chem.*, **1993**, *619*, 1397
[135] S. J. Hibble, A. K. Cheetham, A. R. L. Bogle, H. R. Walwley, D. E. Cox, *J. Am. Chem. Soc.* **1988**, *110*, 3295.
[136] R. Dronskowski, A. Simon, *Acta Chem. Scand.* **1991**, *45*, 850.
[137] R. Dronskowski, A. Simon, W. Mertin, *Z. Anorg. Allg. Chem.* **1991**, *602*, 49.
[137a] B. Lindblom, R. Strandberg, *Acta Chem. Scand.* **1989**, *43*, 825.
[138] K. H. Lii, C. C. Wang, S. L. Wang, *J. Solid State Chem.* **1988**, *77*, 407.
[138a] R. Dronskowski, A. Simon, *Angew. Chem. [Int. Ed. Engl.]* **1989**, *28*, 758.
[138b] P. Gougeon, P. Gall, M. Sergent, *Acta Crystallogr. C* **1991**, *47*, 421.
[138c] P. Gougeon, M. Potel, M. Sergent, *Acta Crystallogr. C* **1990**, *46*, 1188.
[139] H. Leligny, M. Ledesert, Ph. Labbé, B. Raveau, W. H. McCarroll, *J. Solid State Chem.* **1990**, *87*, 35.
[140] P. Gougeon, R. E. McCarley, *Acta Crystallogr. C* **1991**, *47*, 241.

[141] R. Chevrel, M. Sergent in *Superconductivity in Ternary Compounds I* (Eds.: Ø. Fischer, M.B. Maple) (Top. Curr. Phys. *32*), Springer, Berlin, Heidelberg, New York, **1982**, pp. 25.
[142] H. Nohl, W. Klose, O.K. Andersen in *Superconductivity in Ternary Compounds I* (Eds.: Ø. Fischer, M.B. Maple) (Top. Curr. Phys. *32*), Springer, Berlin, Heidelberg, New York **1982**, p. 166.
[143] J.M. Tarascon, F.J. DiSalvo, J.V. Warszczak, *Solid State Commun.* **1984**, *52*, 227.
[144] R. Lepetit, P. Monceau, M. Potel, P. Gougeon, M. Sergent, *J. Low. Temp. Phys.* **1984**, *56*, 219.
[145] J.M. Tarascon, F.J. DiSalvo, C.H. Chen, P.J. Carroll, M. Walsh, L. Rupp, *J. Solid State Chem.* **1985**, *58*, 290.
[146] M. Potel, P. Gougeon, R. Chevrel, M. Sergent, *Rev. Chim. Minér.* **1984**, *21*, 509.
[147] J.M. Tarascon, *Solid State Ionics* **1986**, *18/19*, 802.
[148] J.M. Tarascon, G.W. Hull, J.V. Waszczak, *Mater. Res. Bull.* **1985**, *20*, 935.
[149] J.M. Tarascon, *Solid State Ionics* **1986**, *18/19,* 768.
[150] A.W. Struss, J.D. Corbett, *Inorg. Chem.* **1970**, *9*, 1373.
[151] A.W. Struss, J.D. Corbett, *Inorg. Chem.* **1977**, *16*, 360.
[152] S.D. Wijeyesekera, J.D. Corbett, *Inorg. Chem.* **1986**, *25*, 4709.
[153] S.D. Wijeyesekera, J.D. Corbett, *Solid State Commun.* **1985**, *54*, 657.
[154] H. Novotny, H. Boller, O. Beckmann, *J. Solid State. Chem.* **1970**, *2*, 462.
[155] A. Simon, Hj. Mattausch, N. Holzer, *Angew. Chem.* **1976**, *88*, 685 [Int. Ed. Engl. *15*, 624].
[156] K.R. Poeppelmeyer, J.D. Corbett, *Inorg. Chem.* **1977**, *16*, 694.
[157] Hj. Mattausch, J.B. Hendricks, R. Eger, J.D. Corbett, A. Simon, *Inorg. Chem.* **1980**, *19*, 2128.
[158] Hj. Mattausch, A. Simon, N. Holzer, R. Eger, *Z. Anorg. Allg. Chem.* **1980**, *466*, 7.
[159] A. Simon, *J. Solid State Chem.* **1985**, *57*, 2.
[160] S.J. Hwu, J.D. Corbett, *J. Solid State Chem.* **1986**, *64*, 331.
[161] A. Simon, Hj. Mattausch, R. Eger, *Z. Anorg. Allg. Chem.* **1987**, *550*, 50.
[162] Hj. Mattausch, R. Eger, J.D. Corbett, A. Simon, *Z. Anorg. Allg. Chem.* **1992**, *616*, 157.
[163] U. Schwanitz-Schüller, A. Simon, *Z. Naturforsch. B* **1985**, *40*, 710.
[164] W.H. Zachariasen, *Acta Crystallogr.* **1949**, *2*, 60.
[165] D.A. Lokken, J.D. Corbett, *Inorg. Chem.* **1973**, *12*, 556.
[166] A. Simon, N. Holzer, Hj. Mattausch, *Z. Anorg. Allg. Chem.* **1979**, *456*, 207.
[167] G. Ebbinghaus, A. Simon, A. Griffith, *Z. Naturforsch. A* **1982**, *37*, 568.
[168] W. Bauhofer, A. Simon, *Z. Naturforsch. A* **1982**, *37*, 568.
[169] D.W. Bullett, *Inorg. Chem.* **1985**, *24*, 3319.
[170] L.R. Morss, Hj. Mattausch, R. Kremer, A. Simon, J.D. Corbett, *Inorg. Chim. Acta* **1987**, *140*, 107.
[171] A. Simon, Hj. Mattausch, N.B. Mikheev, C. Keller, *Z. Naturforsch. B* **1987**, *42*, 666.
[172] N.B. Mikheev, A.N. Kamenskaya, I.A. Rumer, V.L. Novitschenko, A. Simon, Hj. Mattausch, *Z. Naturforsch. B* **1992**, *47*, 992.
[173] U. Schwanitz-Schüller, A. Simon, *Z. Naturforsch. B* **1985**, *40*, 705.
[174] H.-J. Meyer, N.L. Jones, J.D. Corbett, *Inorg. Chem.* **1989**, *28*, 2635.
[175] G. Meyer, H.-J. Meyer, *Chem. Mater.* **1992**, *4*, 1157.
[176] Hj. Mattausch, R.K. Kremer, A. Simon, *Z. Anorg. Allg. Chem.* **1993**, *619*, 741.
[177] A. Simon, E. Warkentin, R. Masse, *Angew. Chem.* **1981**, *93,* 1071. [Int. Ed. Engl. *20*, 1013].
[178] E. Warkentin, R. Masse, A. Simon, *Z. Anorg. Allg. Chem.* **1982**, *491*, 323.
[179] H.M. Artelt, G. Meyer, *J. Chem. Soc. Chem. Commun.* **1992**, 1320.
[180] C. Bauhofer, Hj. Mattausch, G.J. Miller, W. Bauhofer, R.K. Kremer, A. Simon, *J. Less-Common Met.* **1990**, *167*, 65.

[181] G. Argay, I. Naray-Szabo, *Acta Chim. Acad. Sci. Hung.* **1966**, *49*, 329.
[182] H. Siebert, *Z. Anorg. Allg. Chem.* **1953**, *273*, 170.
[183] C. Schwarz, *Dissertation*, Stuttgart **1987**.
[184] Hj. Mattausch, A. Simon, unpublished.
[185] A. Simon, C. Schwarz, W. Bauhofer, *J. Less-Common Met.* **1988**, *137*, 343.
[186] G. J. Miller, J. K. Burdett, C. Schwarz, A. Simon, *Inorg. Chem.* **1986**, *25*, 4437.
[187] Hj. Mattausch, A. Simon, K. Ziebeck, *J. Less-Common Met.* **1985**, *113*, 149.
[188] F. Ueno, K. Ziebeck, Hj. Mattausch, A. Simon, *Rev. Chim. Minér.* **1984**, *21*, 804.
[189] M. Ruck, A. Simon, *Z. Anorg. Allg. Chem.* **1992**, *617*, 7.
[190] T. Hughbanks, J. D. Corbett, *Inorg. Chem.* **1988**, *27*, 2022.
[191] M. W. Payne, P. K. Dorhout, J. D. Corbett, *Inorg. Chem.* **1991**, *30*, 1467.
[192] M. W. Payne, M. Ebihara, J. D. Corbett, *Angew. Chem.* **1991**, *103*, 842 [Int. Ed. Engl. *30*, 856].
[193] M. Ruck, A. Simon, *Z. Anorg. Allg. Chem.* **1993**, *619*, 327.
[194] R. Jödden, H. G. v. Schnering, H. Schäfer, *Angew. Chem.* **1975**, *87*, 595 [Int. Ed. Engl. *14*, 570].
[195] R. N. McGinnis, T. R. Ryan, R. E. McCarley, *J. Am. Chem. Soc.* **1978**, *100*, 7900.
[196] B. A. Aufdembrink, R. E. McCarley, *J. Am. Chem. Soc.* **1986**, *108*, 2474.
[197] T. Lundström, *Acta Chem. Scand.* **1968**, *22*, 2191.
[198] S. Geller, *Acta Crystallogr.* **1955**, *8*, 15.
[199] L.-E. Ergenius, B. I. Noläng, T. Lundström, *Acta Chem. Scand. A* **1981**, *35*, 693.
[200] J. P. Owens, H. F. Franzen, *Acta Crystallogr. B*, **1974**, *30*, 427.
[201] D. C. Johnston, *Solid State Commun.* **1977**, *24*, 699.
[202] Y. B. Kuzma, N. S. Bilonizhko, *Dopow. Dokl. Akad. Nauk USSR A* **1978**, *3*, 275.
[203] M. B. Maple, H. C. Hamaker, L. D. Woolf in *Superconductivity in Ternary Compounds II* (Eds.: ø. Fischer, M. B. Maple) (Top. Curr. Phys. *34*), Springer, Berlin, Heidelberg, New York **1982**, pp. 99.
[204] W. A. Fertig, D. C. Johnston, L. E. DeLong, R. W. McCallum, M. B. Maple, B. T. Matthias, *Phys. Rev. Lett.* **1977**, *38*, 987.
[205] K. Schubert, *Kristallstrukturen Zweikomponentiger Phasen*, Springer, Berlin, Göttingen, Heidelberg, **1964**.
[206] B. K. Teo, K. Keating, *J. Am. Chem. Soc.* **1984**, *106*, 2224.
[207] H. F. Franzen, J. G. Smeggil, *Acta Crystallogr. B* **1970**, *26*, 125.
[208] B. Harbrecht, *J. Less-Common Met.* **1988**, *138*, 225.
[209] H. F. Franzen, J. G. Smeggil, *Acta Crystallogr. B*, **1969**, *25*, 1736.
[210] B. Harbrecht, M. Conrad, K. Bohnen, **1992** Communication "Chemiedozententagung".
[211] B. Harbrecht, V. Wagner, **1991** Communication "AGKr".
[212] E.-U. Würthwein, P. v R. Schleyer, J. A. Pople, *J. Am. Chem. Soc.* **1984**, *106*, 6973.
[213] J. F. Jal, P. Chieux, J. Dupuy, J. P. Dupin, *J. Phys. (Les Ulis, Fr.)* **1980**, *41*, 657.
[214] M. A. Bredig, J. W. Johnson, W. T. Smith, *J. Am. Chem. Soc.* **1955**, *77*, 307.
[215] M. A. Bredig, H. R. Bronstein, W. T. Smith, *J. Am. Chem. Soc.* **1955**, *77*, 1454.
[216] H.-J. Deiseroth, A. Simon, *Z. Naturforsch. B* **1978**, *33*, 714.
[217] T. P. Martin, *Angew. Chem.* **1986**, *98*, 197 [Int. Ed. Engl. *25*, 197].
[218] T. P. Martin, H.-J. Stolz, G. Ebbinghaus, A. Simon, *J. Chem. Phys.* **1979**, *70*, 1096.
[219] G. Metsch, W. Bauhofer, A. Simon, *Z. Naturforsch. A* **1985**, *40*, 303.
[220] W. Bauhofer, A. Simon, *Z. Anorg. Allg. Chem.* **1978**, *447*, 29.
[221] S. M. Ariya, E. A. Prokofyera, *J. Gen. Chem. USSR* **1955**, *26*, 609.
[222] E. T. Keve, A. C. Skapski, *Inorg. Chem.* **1968**, *7*, 1757.
[223] P. E. Rauch, A. Simon, *Angew. Chem.* **1992**, *104*, 1505 [Int. Ed. Engl. *31*, 1519].
[223a] J. Snyder, A. Simon, unpublished **1993**.
[224] J. J. Finley, H. Nohl, E. E. Vogel, H. Imoto, R. E. Camley, V. Zevin, O. K. Andersen, A. Simon, *Phys. Rev. Lett.* **1981**, *46*, 1472.
[225] H. Imoto, A. Simon, *Inorg. Chem.* **1982**, *21*, 300.

[226] P. J. Brown, R. A. Ziebeck, A. Simon, M. Sägebarth, *J. Chem. Soc. Dalton Trans.* **1988**, 111.
[227] J. J. Finley, R. E. Camley, E. E. Vogel, V. Zevin, E. Gmelin, *Phys. Rev. B* **1981**, *24*, 1023.
[228] *Superconductivity in Ternary Compounds I, II* (Eds.: ø. Fischer, M. B. Maple) (Top. Curr. Phys. *32, 34*), Springer, Berlin, Heidelberg, New York **1982**.
[229] B. T. Matthias, M. Marezio, E. Corenzwit, A. S. Copper, H. E. Barz, *Science* **1972**, *175*, 1465.
[230] C. Perrin, R. Chevrel, M. Sergent, ø. Fischer, *Mater. Res. Bull.* **1979**, *14*, 1505.
[231] A. Perrin, M. Sergent, ø Fischer, *Mater. Res. Bull.* **1978**, *13*, 259.
[232] K. Yvon, *Curr. Top. Mater. Sci.* **1979**, *3*, 53.
[233] J. D. Corbett, *J. Solid State Chem.* **1981**, *39*, 56.
[234] N. E. Alekseevskii, M. Glinski, N. M. Dobrovolskii, V. I. Isebro, *JETP Lett. (Engl. Transl.)* **1976**, *23*, 412.
[235] T. Luhmann, D. Dew-Hughes, *J. Appl. Phys.* **1979**, *34*, 409.
[236] O. K. Andersen, W. Klose, H. Nohl, *Phys. Rev. B* **1978**, *17*, 1209.
[237] H. Nohl, W. Klose, O. K. Andersen in *Superconductivity in Ternary Compounds I* (Eds.: ø. Fischer, M. B. Maple) (Top. Curr. Phys. *32*), Springer, Berlin, Heidelberg, New York **1982**, pp. 165.
[238] J. K. Burdett, G. J. Miller, *J. Am. Chem. Soc.* **1987**, *109*, 4092.
[239] W. Bauhofer, W. Joss, R. K. Kremer, Hj. Mattausch, A. Simon, *J. Magn. Magn. Mat.* **1992**, *104–107*, 1243.
[240] J. K. Cockcroft, W. Bauhofer, Hj. Mattausch, A. Simon, *J. Less-Common Met.* **1989**, *152*, 227.
[241] R. K. Kremer, W. Bauhofer, Hj. Mattausch, W. Brill, A. Simon, *Solid State Commun.* **1990**, *73*, 281.
[242] R. K. Kremer, W. Bauhofer, Hj. Mattausch, A. Simon, *Eur. J. Solid State Inorg. Chem.* **1991**, *28*, 519.
[243] A. Simon, Hj. Mattausch, R. Eger, R. K. Kremer, *Angew. Chem.* **1991**, *103*, 1210 [Int. Ed. Engl. *30*, 1188].
[244] G. Ebbinghaus, A. Simon, *Chem. Phys.* **1979**, *43*, 117.
[245] C. H. Prescott, M. J. Kelley, *Bell System Techn. J.* **1932**, *11*, 334.
[246] A. H. Sommer, *Photoemissive Materials*, Wiley, New York **1968.**
[247] G. Ebbinghaus, W. Braun, A. Simon, K. Berresheim, *Phys. Rev. Lett.* **1976**, *37*, 1770.
[248] M. G. Burt, V. Heine, *J. Phys. C* **1978**, *11*, 961.
[249] B. Woratschek, G. Ertl, J. Küppers, W. Sesselmann, H. Haberland, *Phys. Rev. Lett.* **1986**, *57*, 1484.
[250] B. Woratschek, W. Sesselmann, J. Küppers, G. Ertl, H. Haberland, *J. Chem. Phys.* **1987**, *86*, 2411.

6 The Chemistry of Transition Metal Colloids

John S. Bradley

6.1 Introduction

6.1.1 Some Definitions

The term "colloid" – or gluelike – was originally used by Graham in 1861 [1] to describe the slow diffusion rates and noncrystallizability in apparently aqueous solutions of what were otherwise known to be insoluble materials, such as silver chloride and gold. In its original use, the term implied a suspension of one phase, either solid or liquid, in a second liquid phase, and was used for those suspensions which would not phase separate or sediment spontaneously without centrifugation. These properties indicate that the colloidal particles were in the 1 nm to 1 μm size range. However, in the context of this chapter the definition adopted will be more restricted. The solids we are concerned with here are particles of the metallic elements, principally the transition metals, and the particle size range included is that which reflects the current interest among chemists and physicists in exploring those special properties which are expected to be exhibited by small particles of solid inorganic materials, including metals, as their size approaches that of molecular clusters. That in itself is a constantly increasing lower limit as the skill of the molecular cluster chemist yields methods for the synthesis and isolation as crystalline materials of ever larger molecules. For example, the cluster $[Ni_{34}Pt_6(CO)_{48}H]^{5-}$ [2] has a metal core diameter of 11 Å, and so there is an overlap between the dimensions of the larger molecular clusters and the smallest colloidal metals. This overlap has become enlarged with the developments in the field of "giant clusters", those as yet imperfectly characterized but fascinating materials with metal atom cores containing several hundred atoms, and has caused one distinguished practitioner in that field to define the colloidal size range as beginning where giant clusters might end, at 10 nm. [3] In order to avoid the temptation of offering a definition based on any restricted size range we will include within the range of colloid chemistry those aggregates of metal atoms which have as yet no pretensions as molecules, yet might reasonably be expected to have some properties which distinguish them from the bulk. This gives us a usable range of sizes from 1–20 nm, the upper limit being as flexible as the examples with interesting properties demand.

The literature on colloidal metals is extensive and an adequate survey covering all aspects of their physical and chemical properties would require an interdisciplinary authorship of some size. This chapter reflects the predilections and limitations of the author and will concentrate on the chemical aspects of metal colloid science. Sufficient physical discussion will be given to support the descriptions of the structures and spectroscopic properties of colloidal metals in the context of their chemical applications to catalysis.

6.1.2 Historical Aspects

Since ancient times the properties of colloidal metals, most notably those of gold, and the means of preparing them, have captured the interests of scientists and laymen alike. The attraction of these materials was in their proven aesthetic and technological value as pigments and their often less well documented alchemical and therapeutic properties. Colloidal gold was used to make ruby glass and for the coloring of ceramics (applications which continue to this day), and its solutions (sometimes containing alcohol) were prescribed as tonics and elixirs, but also for more well founded treatments for arthritis. [4]

The classical age of metal colloid science can be said to have begun with the experiments of Michael Faraday on gold sols in the mid-nineteenth century. [5] Faraday demonstrated the formation of deep red solutions of colloidal gold by the reduction of chloroaurate $[AuCl_4]^-$ solutions using phosphorus as the reducing agent. Of course, Faraday had no means of determining the size of the gold particles he produced, but a recent reproduction of those preparations by J. M. Thomas, in Faraday's original laboratory at the Royal Institution in London, has demonstrated that the gold sols thus produced contain particles of 3–30 nm in diameter. [6] It is interesting to note that in those early speculative experiments Faraday laid the foundation for some of the current optical and biological applications of colloidal gold.

The turn of the century saw increased attention focus on the preparation of colloidal metal suspensions, or sols, and with particular interest in their physical chemistry. The advent of colloid science in the twentieth century is marked by Ostwald's seminal papers in 1907. [7] In the introduction the author described one aspect of colloid chemistry which is particularly pertinent to current research and to the theme of this chapter. In describing the colloidal state Ostwald began by inviting the reader to imagine a cube of a metal one centimeter on an edge. This 1 cm^3 of metal would have a surface area of 6 cm^2. If we assume that the cube was of a metal having a density of 8 g cm^{-3}, an atomic weight of 50, and an atomic diameter of 0.245 nm (iron has approximately these properties), then the cube would contain about 5×10^{24} atoms, of which only about 10^{-5} % reside on the surface. The division of the cube into one thousand smaller cubes, each 0.1 cm on an edge, results in the sample having $10^3 \times 6 \times 10^{-2}$ cm^2, or 60 cm^2 of surface area. The proportion of metal atoms at the surface has increased, but still only to 10^{-4} %, and the vast majority of the atoms in the cube will be surrounded by a shell of other atoms providing a maximum coordination number.

To reach a size scale relevant to the purposes of this review, we repeat this subdivision of the original cube into 10^{18} cubes of 10 nm (10^{-6} cm) on an edge, and the surface area of the sample becomes $10^{18} \times 6 \times 10^{-12}$ cm^2 (*i. e.* 6×10^6 cm^2), or 600 m^2. For our hypothetical iron sample the fraction of surface atoms is now 10 %. For 2.5 nm cubes this fraction is *ca.* 60 %, and if the sample were divided into cubes only 1 nm on an edge, every atom in the sample is a surface atom. Figure **6-1** shows this exercise in schematic form. This geometric phenomenon is the basis for catalysis by highly dispersed supported metals, but it has ramifications for other areas of science which are concerned with the properties of materials. The physical properties of our iron sample will clearly change from those of bulk iron for the large particle size sample to some new properties reflecting the dominance of surface atoms in the ultradispersed 1.0 nm particle sample. We thus realize, as did Ostwald, that nanoscale metal particles should display novel and interesting properties, even if only as a result of lowering the average coordination number of the atoms in the sample.

Many methods for the preparation of colloidal metals have been developed over the years since Faraday's experiments. For a survey of activity in the preparation of metal sols in the first half of the century, the reader is referred to the 1951 paper by Turkevitch, Stevenson, and Hillier [8] for detailed synthetic conditions and early electron microscopic characterizations of gold sols as prepared by the methods of Faraday (see above), Bredig (electric arc between gold electrodes

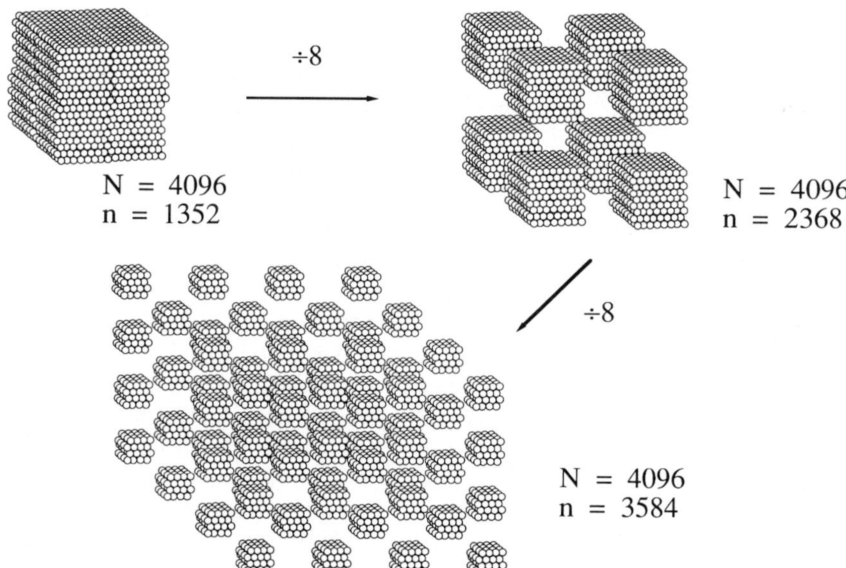

Figure 6-1. Surface statistical consequences of the subdividing of a cube with 16 atoms on a side. N = total atoms; n = surface atoms. The structural, electronic, and chemical consequences of the presence of a high proportion of surface and near-surface atoms predominate in the more highly dispersed material.

under water), and Donau (action of carbon monoxide on chloroauric acid) as well as a number of other related methods involving the action of additional reducing agents on chloroaurate solutions. These preparations introduce many of the methods used today in the current phase of preparative metal colloid chemistry. Turkevich's long study of this system yielded detailed knowledge of the nucleation, growth, and agglomeration of colloidal metals and has also allowed the reproducible preparation of gold sols having narrow particle size distributions over a range of mean sizes. [9] For example, the 20 nm sol shown in Figure **6-2** can be prepared with high reproducibility from citrate reduction of $[AuCl_4]^-$ and is widely used as a stain in electron microscopic examinations of biological samples. The particle sizes found for these and related colloids runs from *ca.* 2–100 nm, depending on the preparative conditions, with most examples falling in the large diameter range.

In the last two decades inorganic colloid chemistry has focussed more on inorganic chalcogenides [10, 11] than on the metals themselves. However, after a lengthy period in which highly dispersed metals were widely exploited in specific technological applications, such as photography and catalysis, there was a resurgence of interest in the intrinsic chemistry and physics of the many manifestations of small metal particles, and a new phase in metal colloid chemistry is now being defined.

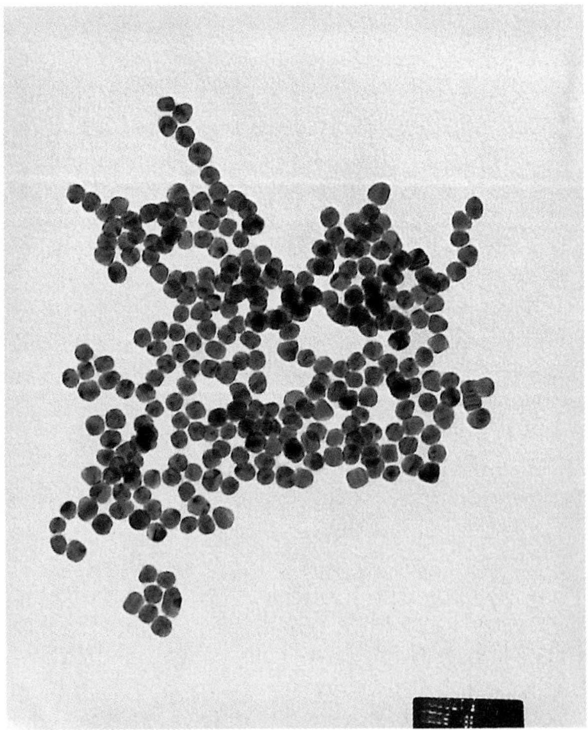

Figure 6-2. Transmission electron micrograph of a monodispersed 14 nm gold hydrosol, prepared by citrate reduction of $HAuCl_4$ (courtesy of Dr. D. Weitz, Exxon Corporate Research).

6.1.3 Current Interest

The current burst of activity in studies of small metal particles can be seen as the result of developments in inorganic (and organometallic) chemistry, characterization techniques, and solid state physics. The former two provided the means and the latter defined the goals for realizing new properties in nanoscale metal particles. The study of the chemical and physical properties of highly dispersed transition metal aggregates in the nanometer size range has become a matter of great fundamental and practical interest. From the fundamental perspective, the transition from bulk properties through the intermediate quantum size effect domain to the properties of molecular clusters has attracted the attention of chemists, physicists, and theoreticians. Physicists have predicted that particles with diameters in the range 1–10 nm will display electronic properties which reflect the incipient electronic band structure of the particles. [12] In addition, shape as well as size can play a determining role in the electronic properties of small particles. [13] Thus physical and perhaps chemical properties may result which correspond neither to those of molecular metal compounds, such as large metal carbonyl clusters, nor to those of the bulk metal. [14]

From a more practical perspective, the surface chemical and physical characteristics which derive from the potentially unique electronic properties of clusters falling in the bulk to molecular transition range are of great potential for both new technological applications of advanced materials [15] and more mature applications like catalysis. [16]

Both the advancement of our understanding in small particle science and the potential for new material science based on the chemistry and physics of nanoscale metal clusters rest on the observation and application of useful size dependent properties of these clusters. This in turn requires the preparation and isolation of these metal particles in monodispersed form and therefore a great degree of control over size, structure, and surface composition. Metals in varying levels of dispersion are found in many technological applications, but the employed methods of preparation are often defined by the intended use of the metal and not necessarily as a means of preparing physically well defined metal particles. Thus metal ceramic composites are prepared by high temperature treatment of ceramic precusors containing metal powders, and highly dispersed metal catalysts by high temperature reduction of metal precursors on oxide supports. These severe preparative techniques tend to yield metal particles with ranges of sizes and morphologies broad enough to hide the potentially unique properties which a monodispersed nanoscale sample would exhibit.

With these considerations in mind, the goals of contemporary metal colloid research, the approaches to which will be covered in this chapter, may be stated as the search for synthetic methods for metal particles which are stabilized in the 1–20 nm size range, the application of characterization methods to identify the potential novelty of the particles (structural methods, electronic properties, surface chemistry), and finally the application of these properties to various fields of chemistry and physics. Perhaps the field for which the renaissance in transition metal colloid chemistry has the most potential impact is catalysis. It has been known for many years that colloidal metals possess catalytic properties, and some of the earliest work on colloidal platinum by Bredig described the high activity of colloidal platinum in the decomposition of hydrogen peroxide. [17] It is not surprising that the extensive body of literature on the catalytic properties of supported metal particles for a wide variety of reactions has a counterpart, although less extensive, in metal colloid chemistry, and some of these aspects of colloid chemistry will be discussed in this chapter.

6.2 Preparative Methods

6.2.1 General Considerations

6.2.1.1 Generation of Metal Particles

The successful establishment of this new area of inorganic materials chemistry was dependent on being able to control the particle size, the morphology, and the composition of the metal nanoclusters. This problem was also a central one to the development of synthetic organometallic cluster chemistry. The relative ease with which the course of molecular reactions in solution can be followed by such spectroscopic methods as infrared spectroscopy and high resolution NMR allowed the development of practical synthetic protocols for preparing and isolating molecular clusters in useful yields. Even quite complex cluster compounds, such as $[(p\text{-tolyl})_3P)_{12}Au_{18}Ag_{20}Cl_{14}]$ [18] or $[Ni_{34}Pt_6(CO)_{48}H]^{5-}$, [2] can be isolated from their synthetic solutions by selective extraction and crystallization. The comparable synthetic problem with metal nanoclusters can be approached from an analogous perspective of synthetic organometallic chemistry. Thus various methods have been established over the years for the low temperature syntheses of nanoscale transition metal clusters in organic solvents based on organometallic solution synthesis techniques. There is, however, an obvious difference between molecular synthesis and molecular approaches to colloidal metals. The constant aim of the organometallic chemist to maintain the metals in complexed molecular form is reversed in the search for synthetic methods for transition metal colloids. A reaction which results in the precipitation of elemental metal in the laboratory of the molecular chemist is a failure, but it is also a potential source of colloidal metal. Such a process would normally lead to the formation of a metallic precipitate or mirror. However, in the presence of an appropriate stabilizing agent in solution, aggregation of the metal atoms can be arrested at the nanoscale stage, and the metal particles are preserved as a suspension in the organic liquid. The fact that these reactions are carried out at low temperatures, usually –100 °C to +100 °C (normal organometallic synthesis conditions), has an important collateral benefit. In this temperature range, it may be possible to isolate nonequilibrium (nonthermodynamic) structures for the metal crystallites since the particles are formed at temperatures far below their melting points and are not annealed to thermal equilibrium at the customary high temperatures used, for example, in metallurgy or classical solid state chemistry. This is an area of metal structural chemistry (which might be termed nanoscale metallurgy) which is completely unexplored.

A survey of the literature reveals that many of the various methods reported for the preparation of colloidal metal are applicable to a number of metals across the periodic table. For example, salt reduction using main group hydride reducing agents, or metal vapor routes, have been used for many metals in turn. It is not the goal of this section to provide a directory of all the reports on colloid syntheses, but only to give examples of the principal types of preparative methods which can be used.

6.2.1.2 Stabilization of Colloidal Metal Particles in Liquids

Before beginning a description of synthetic methods, we will briefly consider a general yet crucial aspect of colloid chemistry and that is the means by which the metal particles are stabilized in the dispersing medium, since small metal particles are unstable with respect to agglomeration to the bulk. At short interparticle distances, two particles would be attracted to each other by van der Waals forces which vary inversely as the sixth power of the distance between their surfaces. In the absence of any counteractive repulsive forces, an unprotected sol would coagulate. A counteraction can be achieved by two methods, *electrostatic stabilization* and *steric stabilization.* [19] In the classical gold sols prepared, for example, by the reduction of $[AuCl_4]^-$ with sodium citrate, the colloidal gold particles are surrounded by an electrical double layer arising from adsorbed citrate and chloride ions and the cations which are attracted to them. This results in a Coulombic repulsion between the particles which decays approximately exponentially with the interparticle distance, and the net result is shown schematically in Figure **6-3**. The weak minimum in the potential energy at moderate interparticle distances defines a stable arrangement for the colloidal particles which is easily disrupted by medium effects and, at normal temperatures, by the thermal motion of the particles. Thus, if the electric potential associated with the double layer is sufficently high, then electrostatic repulsion will prevent particle agglomeration. An electrostatically stabilized sol can be coagulated if the ionic strength of the dispersing medium is increased sufficiently, since this compresses the double layer and shortens the range of the repulsion. The stabilizing effect of surface ions is dependent on their concentrations. If the surface charge is reduced by the displacement of adsorbed anions by a more strongly binding neutral adsorbate, the colloidal particles can now collide and agglomerate under the influence of the van der Waals attractive forces. This phenomenon is nicely demonstrated by the addition of pyridine to a gold sol of the type mentioned above as shown in Figure **6-4**. [20] The colloidal particles aggregate into strings as a result of the residual charge on the particles, since the potential at the end of a chain of particles would be less than that along its length.

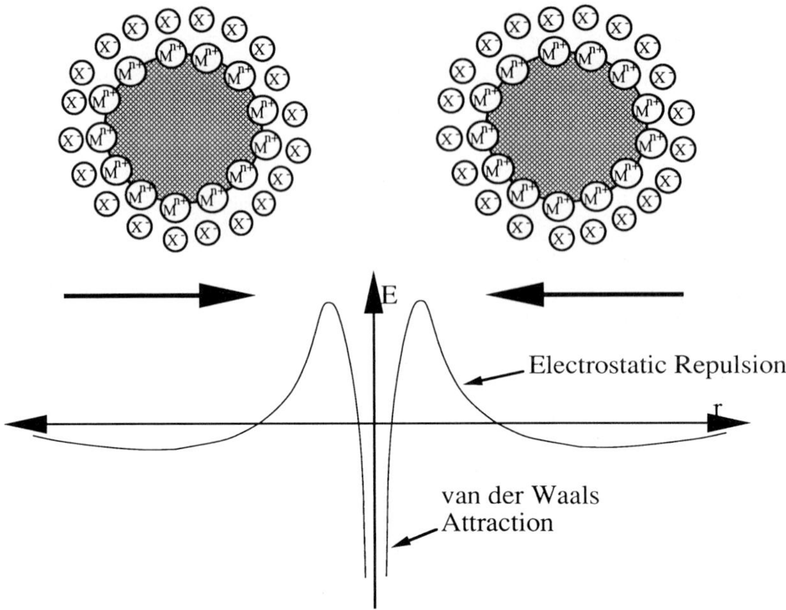

Figure 6-3. Electrostatic stabilization of metal colloid particles. Attractive van der Waals forces are outweighed by repulsive electrostatic forces between adsorbed ions and associated counterions at moderate interparticle separation.

Even in organic media where electrostatic effects would not normally be considered to be important the development of charge has been demonstrated on inorganic surfaces, including metals, in contact with organic phases such as solvents and polymers. [21] For example, the acquisition of charge by gold particles in organic liquids has been demonstrated, and the sign and magnitude of the charge was found to vary as a function of the donor properties of the liquid. [22] Thus, even for a suspension of colloidal metals in relatively nonpolar liquids, the possibility that electrostatic stabilization occurs cannot be excluded.

A second means by which colloidal particles can be prevented from aggregating is by the adsorption of large molecules, such as polymers or surfactants, at the surfaces of the particles, thus providing a protective layer. The way in which large adsorbed molecules prevent aggregation can be seen in a simplified manner by visualizing the close approach of two colloid particles, each with long chain molecules adsorbed on their surfaces as shown in Figure **6-5**. In the interparticle space, the adsorbed molecules would be restricted in motion – fewer conformations will be accessible – which causes a decrease in entropy and thus an increase in the free energy (Fig. **6-5a**). A second effect is caused by the local increase in concentration of adsorbed molecules as the protective sheaths on each particle begin to interpenetrate. This causes an osmotic repulsion as the solvent reestablishes equilibrium by diluting the polymer molecules and separating the particles (Fig. **6-5b**). It can be seen that the polymer, in order to function effectively, must

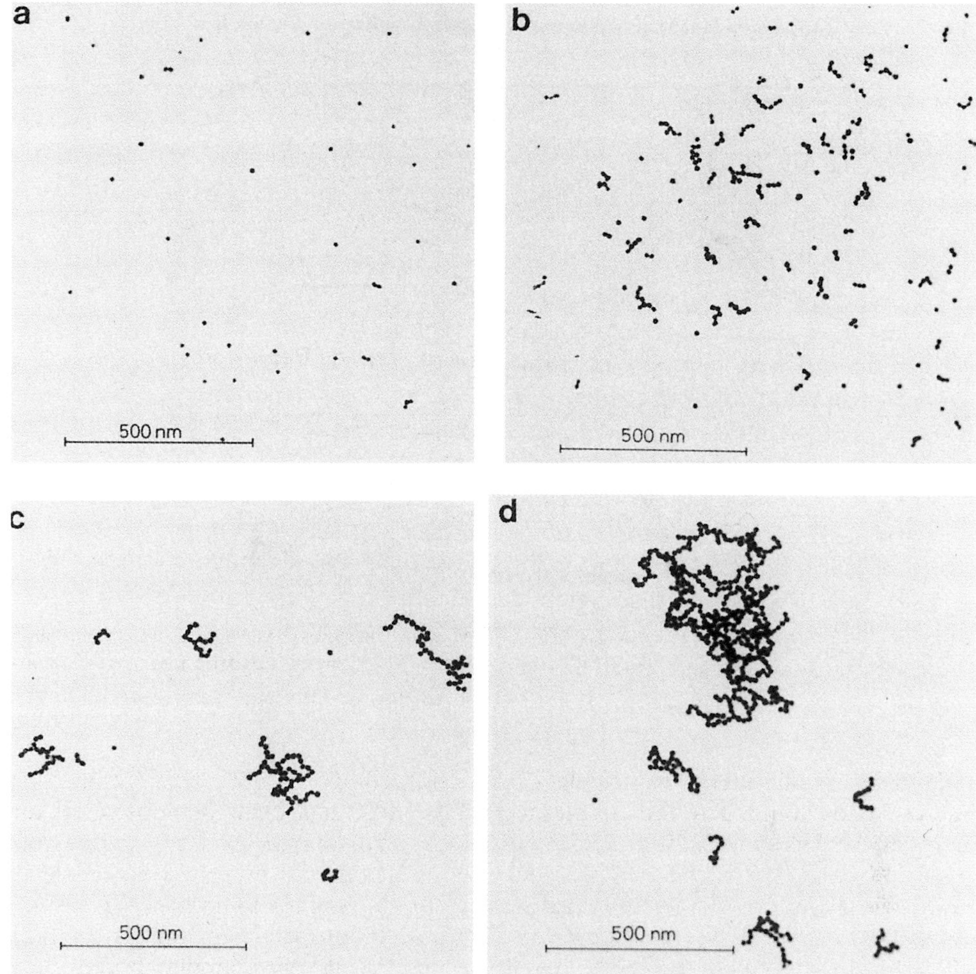

Figure 6-4. Aggregation of an electrostatically stabilized 12–18 nm gold hydrosol caused by addition of pyridine, a neutral ligand (reproduced by permission from ref. [20]).

not only coordinate to the particle surface, but also be adequately solvated by the dispersing fluid.

The choice of polymer is determined by considering the solubility of the metal colloid precursor, the solvent of choice, and the ability of the polymer to stabilize the reduced metal particles in the colloidal state. Natural polymers such as gelatin and agar were often used before the advent of synthetic polymer chemistry, and related stabilizers such as cellulose acetate, cellulose nitrate [23] and cyclodextrins [24] have been used more recently. Thiele [25] proposed the *Protective Value* as a measure of the ability of a polymer to stabilize colloidal metal. It was defined, similar to the older *Gold Number* of Zsigmondy, as the weight of the

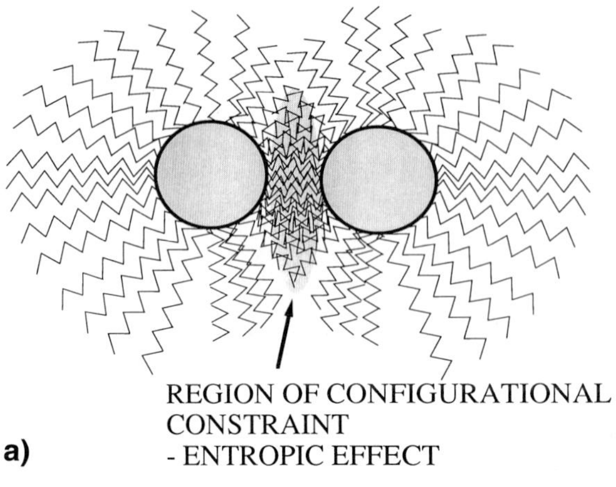

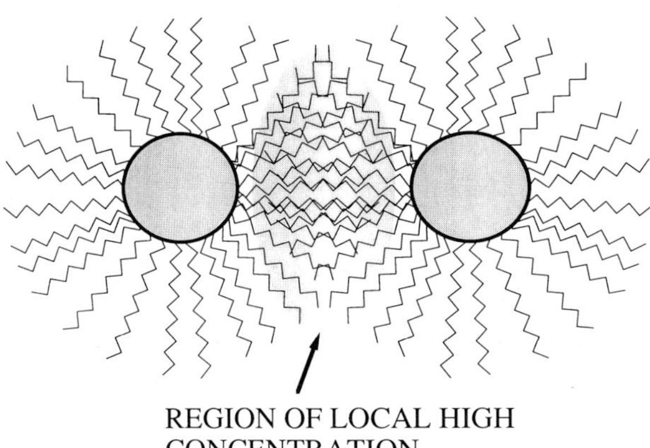

Figure 6-5. Steric stabilization of metal colloid particles by polymers or surfactant molecules: a) in the interparticle space the configurational freedom of the polymer chains of two approaching particles is restricted causing a lowering of entropy; b) the local concentration of polymer chains between the approaching particles is raised and the resulting higher local activity is osmotically counteracted by solvation.

polymer which would stabilize 1 g of a standard red gold sol containing 50 mg L^{-1} gold against the coagulating effect of 1 % sodium chloride solution. Several other studies have been performed on the relative ability of polymers to act as steric stabilizers. [26–28] Despite the fact that these rather subjective studies focussed on very specific (and quite different) sol systems, it seems that of the synthetic polymers considered (vinyl polymers with polar side groups such as poly(vinylpyrrolidone) and poly(vinyl alcohol)) are especially useful in this respect, as will become apparent in Section 6.2.2. The use of copolymers introduces another degree of variability to colloidal stabilization. For example, the use of vinylpyrrolidone-vinyl alcohol copolymers has been reported for the preparation of platinum and silver hydrosols. [29] The silver sols formed only with the copolymer, and the size of the silver particles decreased as the vinylpyrrolidone content of the copolymer increased.

Electrostatic and steric stabilization are in a sense combined in the use of long chain alkylammonium cations and surfactants, either in single phase sols or in reverse micelle synthesis of colloidal metals. Examples will be described in Section 6.2.2.5.

A new class of metal colloids has recently been established in which the surface of the particle is covered by relatively small ligand molecules such as sulfonated triphenylphosphine. [30] These materials, which will be discussed in due course below, overlap with the giant ligand-stabilized clusters mentioned in Section 6.1.3, and covered in detail elsewhere in this volume.

6.2.2 Synthetic Methods for the Preparation of Colloidal Transition Metals

The synthetic methods which have been used include modern versions of established methods for metal colloid preparation such as the mild chemical reduction of solutions transition metal salts and complexes, newer methods such as radiolytic and photochemical reduction, metal atom extrusion from labile organometallics, and metal vapor synthesis techniques. While some of these reactions have been in use for many years, others have resulted from research stimulated by the current resurgence in metal colloid chemistry. The list of preparative methods is being extended daily, and as examples of these methods are described below the reader will be made quickly aware that almost any organometallic reaction whose hitherto undesirable side effect is the facile deposition of metallic precipitates is a resource for the metal colloid chemist. The acquisition of new methods is limited only by the ingenuity of the synthetic chemist in turning a previously negative result into a synthetic possibility. Examples of each of these methods will be described below.

6.2.2.1 Salt Reduction

The reduction of transition metal salts in solution is the most widely practised method of generating colloidal suspensions of metals. In aqueous systems the reducing agent must be added or generated *in situ,* but in nonaqueous systems the solvent and the reducing agent can be one and the same. Easily oxidized solvents such as alcohols can thus function both as reducing agents and as the colloid diluent, and they have been used widely in colloid preparations. Hirai and his coworkers have made extensive use of aqueous alcohols [31, 32] and ethers [27] as reducing solvents in the preparation of colloidal precious metals stabilized by a number of organic polymers such as poly(vinylpyrrolidone), poly(vinyl alcohol) and poly(methylvinyl ether). In general, the alcohols which are useful reducing agents contain α-hydrogen (*tert*-butanol, for example, is not effective) and are oxidized to the corresponding carbonyl compound (*e. g.* methanol to formaldehyde, ethanol to acetaldehyde, *iso*-propanol to acetone) which can be detected during the reaction (Eqn. 6.1).

$$RhCl_3 + 3/2\ R_1R_2CHOH \rightarrow Rh(0) + 3/2\ R_1R_2C{=}O + 3HCl \qquad (6.1)$$

The formation of the organic carbonyl compound was ascribed to the intermediacy of an oxonium complex, followed by deprotonation to an alkoxide, and finally hydride elimination to give the carbonyl compound with subsequent elimination of HCl. The presence of a base with sufficient strength to deprotonate the oxonium intermediate is required. Thus methanol requires the addition of water or hydroxide ion in order to be effective, but higher alcohols are effective alone. The presence of acetate could also fill the requirement for base, and this is consistent with the successful use of dry methanol for the reduction of palladium acetate to colloidal palladium. [33–35] Similar alcohol reductions have been used for many years for the synthesis of low valent molecular metal complexes such as $Rh(PPh_3)_3Cl$ from $RhCl_3 \cdot 3H_2O$ in refluxing ethanol in the presence of triphenylphosphine. [36] Here the phosphine ligand stabilizes the reduced Rh(I) center by complexation at the molecular level. A similar reduction but in a polymer solution and in the absence of phosphine ligand allows the reduction to proceed further to Rh(0), which is then stabilized in the colloidal state by the dissolved polymer.

Palladium acetate has been compared with palladium chloride as a starting material for poly(vinylpyrrolidone) (PVP) stabilized colloidal palladium and found to give smaller particles and better control of particle size distribution. The effect of the Pd/PVP ratio has also been investigated. [35] Palladium organosols have been prepared in our laboratory by reduction of palladium acetate either in refluxing methanol or at a higher temperature in 2-ethoxyethanol, both in the presence of PVP, to yield *ca.* 65 Å particles. [33, 34] Organosols of platinum and palladium have been synthesized by heating salts of the metals in long chain aliphatic alcohols like 1-decanol, which not only reduce the salts to the metal but also act as stabilizers for the colloid particles. [37] The use of alcohol reduction has been extended to the synthesis of bimetallic copper-palladium colloids. [38] In this system a mixture of the acetates of palladium and copper are heated in 2-ethoxyethanol in the presence of PVP. At a reaction temperature of 135 °C, the palladium(II) is rapidly reduced to Pd(0), which in turn reduces the Cu(II) at the surface of the growing metal particle. In the reducing alcohol medium both salts are reduced to the metal, and rigorously bimetallic particles result.

The reduction of metal salts by added reducing agents in nonreducing solvents is the oldest established procedure for the preparation of aqueous suspensions of colloidal metals. A wide range of reducing agents has been used in metal colloid syntheses, and they can be frequently interchanged from metal to metal. Thus hydrazine hydrate has been used in preparations of colloidal platinum, [39] gold, [40] copper, [41] and palladium-platinum alloy. [42] A catalog of preparations will not be presented here, but examples of the use of typical reducing agents will be given.

Faraday reduced aqueous solutions of $[AuCl_4]^-$ with phosphorus vapor to produce gold hydrosols. [5] A variety of other reducing agents has subsequently been used with this gold salt, both with and without protective polymers, to give colloidal gold with particle sizes ranging from one to several hundred nanometers in diameter. Turkevitch and his coworkers have reproduced many of these methods, [8] and thereby established reliable procedures for the preparation of gold sols with quite precisely defined particle sizes. The 20 nm gold sol prepared by the reduction of $[AuCl_4]^-$ with sodium citrate has become a standard for histological staining applications. The mechanism of the metal salt reduction by citrate was also studied by the same authors. [8] They demonstrated that an induction period occurred during the reduction of $[AuCl_4]^-$ when citrate was the reducing agent but not when acetone dicarboxylate (an oxidation product of citrate) was used, when a rapid formation of colloidal gold resulted (Eqn. 6.2).

$$\begin{array}{c} CH_2\text{-}COOH \\ | \\ HOC\text{-}COOH \\ | \\ CH_2\text{-}COOH \end{array} \xrightarrow{[AuCl_4]^-} \begin{array}{c} CH_2\text{-}COOH \\ | \\ O{=}C \\ | \\ CH_2COOH \end{array} \xrightarrow{[AuCl_4]^-} CO_2 + Au(0) \quad (6.2)$$

A similar mechanism can be postulated for other reductions with citrate, and this method has been widely used for the preparation of other metal sols, such as platinum. [17, 43, 44] Citrate can also be added as an ionic stabilizer in preparations which require an additional reducing agent. The use of formate, citrate, and acetone dicarboxylate as reducing agents at various pH's was reported to give good control over particle size in the preparation of a series of platinum hydrosols. [45]

A study has been reported by Glaunsinger and coworkers [39] on the reduction of chloroplatinic acid by a variety of reducing agents including sodium borohydride, hydroxylamine hydrochloride, dimethylamine borane, sodium citrate, hydrazine monohydrate, sodium formate, trimethylamine borane, sodium trimethoxyborohydride, and formaldehyde. By careful variation of reaction temperature, protective polymer (PVP was found to be the most effective), time, pH, and reagent concentration, as well as the use of both direct and reverse addition of the reagents, it was shown that colloidal platinum could be produced as spheres, strands, or mosaic composites of platinum microcrystals.

Hydrogen has been used effectively as a reducing agent for metal salts in electrostatically stabilized metal sol syntheses. [45] Polymer stabilized hydrosols of palladium, platinum, [46–48] rhodium, [49] and iridium [50] stabilized with polyvinyl alcohol were prepared by hydrogen reduction of the metal hydroxides. Sols were produced with a broad distribution of particle sizes.

Carbon monoxide/water was used to reduce PtO_2 to colloidal platinum [51] in a reaction which effectively uses the water gas shift reaction to provide the reducing equivalents of electrons (Eqn. 6.3).

$$CO + H_2O \rightarrow CO_2 + 2H^+ + 2e^- \qquad (6.3)$$

The reduction of metal salts with borohydride or trialkylborohydride is a widely used colloid synthesis method. The preparation of platinum microcrystals having a mean diameter of 28 Å by the reduction of chloroplatinic acid with sodium borohydride has been reported as a reproducible standardized preparation. [52] PVP stabilized copper sols have been prepared by borohydride reduction of copper salts. [53, 54] In some cases, however, the formation of metal borides instead of colloidal metal has been observed. [55–57] Some recent reports in which trialkylborohydrides were used are of partucular interest. Bönnemann and coworkers [58, 59] have used tetraalkylammonium triethylborohydrides in THF solution to reduce a wide range of Group 6–11 metal halide salts (Eqn. 6.4).

$$MX_n + nNR_4\,BEt_3H \xrightarrow{THF} M_{coll} + nNR_4\,X + nBEt_3 + n/2\,H_2 \tag{6.4}$$

The reported particle sizes are generally between 1 and 5 nm, and by using appropriately substituted ($R = C_{4-20}$) tetraalkylammonium salts, the colloidal metal particles can be stabilized in the organic solvent. Organic solutions with metal concentrations of up to 1 M can be obtained by these reactions. The remarkable solubility of these materials is undoubtedly due to the adsorption of the tetraalkylammonium cations at adsorbed anion sites on the colloidal metal surface, implying the presence of an oxidized layer of metal cations at the surface.

Silanes have also found use in colloid preparation. An interesting example for the use of organosilanes in the formation of organosols of transition metals is the reduction of platinum(II) complexes, especially cyclooctadieneplatinum dichloride, to colloidal platinum by trialkoxysilanes and trialkylsilanes. This was reported by Lewis and coworkers [60] who demonstrated that the presumably homogeneous hydrosilylation catalysts formed from platinum compounds in the presence of silanes were in fact colloidal in nature, and raised the prospect for colloid activity in many homogeneous catalyst systems (see Section 6.5). The same reduction method with organosilanes has been used for the preparation of colloids of rhodium and palladium. [61]

Salt reduction has been applied to the preparation of organosols stabilized by surfactants. In a refinement of the method described above [62] a long chain tetraalkylammonium cation is added to a solution of the precursor transition metal salt with subsequent reduction by conventional means. An example of this is the treatment of palladium acetate in THF with tetraoctylammonium bromide followed by reaction with hydrogen, trialkylborane, formic acid, or simply by thermolysis, which leads to stable organosols of palladium with a lower ratio of the long chain ammonium surfactant than was necessary for the original method. A similar approach to such stabilization is exploited in the reduction of $[AuCl_4]^-$ with hydrazine [63] or of $[PtCl_6]^{2-}$ with aqueous formaldehyde, [64] after extraction of the metal salts into such organic liquids as chloroform or cyclohexane with surfactant cations like dodecylpyridinium, trioctylmethylammonium, or dioctadecyldimethylammonium. The resulting colloidal metal particles are stabilized by the surfactant cations in the essentially nonpolar organic phase.

Surfactant stabilized organosols have been prepared by extraction of the colloidal metal into an immiscible organic liquid from a preformed hydrosol. Reduction of H_2PtCl_6 by $NaBH_4$ in water containing cetyltributylphosphonium bromide followed by extraction into toluene gave a stable organosol of platinum. [65] Colloidal silver particles (8 nm) were displaced from water into cyclohexane,

n-hexane, or benzene by extraction with sodium oleate. [66] Phase transfer resulted in no change in the particle size distribution.

Water soluble sodium salts of phosphines, such as sulfonated triphenylphosphine $P(m\text{-}C_6H_4SO_3Na)_3$, and alkyltriphenylmethyltrisulfonates, $RC(p\text{-}C_6H_4SO_3Na)_3$, are examples of surfactants which function as stabilizers for hydrosols of colloidal rhodium prepared by hydrogen reduction of aqueous $RhCl_3 \cdot 3H_2O$ in the presence of the surfactant. [67, 68] The description of the colloidal material as a polyhydroxylated rhodium particle implies considerable oxidation of the rhodium surface to Rh(I), and stabilization occurs through the interaction of the anionic heads of the surfactant with the Rh(I) sites at the colloid surface. Conversely, if the rhodium sol is prepared by aqueous borohydride reduction of $RhCl_3 \cdot 3H_2O$, the colloidal particles are thought to contain only zerovalent rhodium metal and the stabilization occurs through the interaction with the hydrophobic alkyl groups on the surfactant. [69]

Just as the use of a surfactant salt of triethylborohydride introduces both the reducing agent and the stabilizer as a single reagent, [62] polymeric reducing agents can be used which also can function as a polymeric stabilizer. Examples include the use of polyethyleneimine in the preparation of colloidal silver. [70, 71]

A chemical analogue to the preparation of metal colloids by the reduction of metal salts with radiolytically produced radicals and electrons (see Section 6.2.2.2) is found in recent reports by Dye *et al.* [72, 73] They used alkalides and electrides as homogeneous reducing agents in order to reduce main group and transition metal salts in aprotic solvents. These reducing agents are crystalline salts of either alkali metal anions or trapped electrons of the form $K^+[(\text{crown})_2K]^-$ or $K^+[(\text{crown})_2e]^-$, and are prepared by reaction of the alkali metal with a crown ether. In solvents such as dimethylether or THF, these remarkable compounds reduce metal salts to the colloidal metal state in which the particle sizes fall in the nanoscale range. For example, $AuCl_3$ is reduced to an unstable colloidal gold by $K^+(15\text{-crown-}5)K^-$ in dimethylether. The familiar stable ruby red hydrosol is produced by dissolving the reaction product in water.

6.2.2.2 Thermal Decomposition and Photochemical Methods

Thermolysis

Since many organometallic compounds of the transition metals thermally decompose to their respective metals under relatively mild conditions, these compounds provide a rich source of colloidal metal precursors. It is therefore quite surprising that only relatively few examples of their use have been reported. Early work on the magnetic properties of small particles of ferromagnetic metals in organic media (so-called ferrofluids) showed that the thermal decomposition of dicobalt octacarbonyl in organic solutions containing various polymers was a versatile preparative method. [26, 74] The authors observed what has become intuitive in colloid synthesis, that is that the stabilizing polymer should not only contain polar monomer units but also be soluble in the diluting medium. Depending on the polymer used, particle sizes of from 5 nm to >100 nm were obtained. Other cobalt, [75, 76] iron, [75–77] and nickel [78] colloids have been prepared by similar organometallic decompositions. The method is widely applicable. The thermolysis of carbonyl containing complexes of rhodium, iridium, ruthenium, osmium, palladium, and platinum in polymer solutions has allowed the preparation of polymer stabilized colloidal metals having particle sizes between 1 nm and 10 nm, possibly through the decomposition of polymer bound organometallic intermediates. [75, 76, 79–81].

The syntheses of palladium and platinum organosols [82–85] by the thermolysis of such precursors as palladium acetate, palladium acetylacetonate, and platinum acetylacetonate in high boiling organic solvents like methyl-*iso*-butylketone have been reported. Likewise, bimetallic colloids of copper and palladium have been prepared from the thermolysis of mixtures of their acetates in similar solvents. [86] These preparations were performed in the absence of stabilizing polymers, and as a result, relatively broad size distributions and large particles were observed.

In view of the extensive literature covering the syntheses and structures of bimetallic metal carbonyl clusters, it is even more surprising that few attempts to use the facile decomposition of these compounds in the preparation of bimetallic colloids have been reported. Although this would be an unnecessarily complicated route to colloids of miscible metals if the salts of these metals could be coreduced to the same end, there remains the possibility of preparing bimetallic particles of immiscible metals starting from well defined molecular bimetallic clusters of immiscible metals. Such clusters are known to exist despite the immiscibility of their constituent metals in the bulk, and there is reason to believe that bulk immiscibility might not apply to very small particles due to the importance of surface effects.

Photolysis and Radiolysis

Photochemical colloid syntheses fall into two categories; the reduction of metal salts by radiolytically produced reducing agents such as solvated electrons and free radicals, and the photolysis of photolabile metal complexes. Again, the essence of these preparative procedures lies in the generation of zerovalent metal under conditions which prevent, or at least retard, the formation of bulk metal precipitates.

The photolytic method has, of course, a long and important history in the formation of photographic images from silver halide emulsions. Over the past twenty years, predominantly from the work of Henglein and of Belloni, a wide variety of colloidal metals has been prepared by this method encompassing both main group metals such as cadmium [87, 88], thallium [89, 90] and lead, [91] as well as other noble [92–98] and non-noble transition metals [99–101]. Radiolytic methods differ in the type of reducing species which is formed under irradiation, which is a function of solvent and any added solute. The radiolysis of aqueous solutions of metal ions produces solvated electrons which may either react with the dissolved metal ions directly, or with other solutes to produce secondary radicals which then reduce the metal cations. For example, irradiation of solutions of $Cu(ClO_4)_2$ containing HCOONa yields colloidal copper [101] through the reduction of Cu^{2+} by both solvated electrons and CO_2^- (Eqn. 6.5).

$$Cu^{2+} \xrightarrow{e^-(solv);\ CO_2^-} Cu(CO)_2 \xrightarrow{Cu^+} Cu_2{}^+ \xrightarrow{nCu^+} \text{nuclei} \rightarrow Cu(coll) \quad (6.5)$$

Similarly irradiation of $Pd(NH_3)_4Cl_2$ in aqueous *iso*-propanol results in the reduction of the palladium complex to colloidal palladium by both solvated electrons and 1-hydroxy-1-methylethyl radicals $(CH_3)_2COH^{\cdot}$ (formed by reaction of *iso*-propanol with the initial radiolysis products $H^{\cdot}$ and $OH^{\cdot}$). [102] Since mixtures of acetone and *iso*-propanol also yield $(CH_3)_2COH^{\cdot}$ under both radiolytic and photolytic conditions, they have been used in many studies on the radical chemistry of metal colloids. In many of these studies, the techniques of flash photolysis and pulse radiolysis were used to detect short lived intermediates, measure the kinetics, and determine the mechanisms of metal cation reduction, nucleation, and colloidal cluster formation.

An advantage of radiolytic methods for colloidal metal synthesis lies in the fact that a large number of metal nuclei are homogeneously and instantaneously produced, a condition favorable for the formation of very highly dispersed particles. In addition, the reducing agents are formed *in situ* and thus the chemical compatibility of the added reducing agents (BH_4^-, Et_3BH^-, etc.) with any other components of the colloid system is not an issue, although the possibility of photolysis of those components, as well as their reaction with primary photolysis or radiolysis products exists. This advantage has been exploited during the formation of colloidal metals in microemulsions and micelles. [103] Here the stability range of the dispersed droplets in the phase containing the metal precursor places restrictions on such reaction parameters as pH and ionic strength, thus limiting the types of chemical reducing agents which can be used (see Section 6.2.2.5).

In an interesting example of radiolytic colloid synthesis, both the colloidal metal *and* the polymer are produced simultaneously by radiolysis of the solutions of a metal salt and a monomer. Radiolysis of aqueous solutions of H_2PtCl_6 containing either acrylamide or *N*-methylolacrylamide with a ^{60}CO γ-radiation source yielded highly dispersed colloidal platinum (diameter 1.7 nm). [104]

The photolysis of metal complexes and salts is also an effective and clean method for the generation of colloidal metals, and it is amenable to all of the experimental protocols used for the stabilization of the resulting particles. Examples of this include the UV-visible irradiation of gold, silver, or platinum salts in microemulsions, surfactant solutions, or polymer solutions. [105–110] It was found from a comparative study of hydrogen reduction *vs.* photoreduction of chloroplatinic acid in the presence of surfactants that the photoreduction produced smaller platinum particles with a narrower size distribution. [105]

Since organometallic complexes are often photolabile, they could serve as sources of metal colloids under photolytic conditions. Although this method has been used for the preparation of supported metal particles in heterogeneous catalysis, the method has not found application in colloid synthesis.

6.2.2.3 Ligand Reduction and Displacement from Organometallics

Reduction of the metal can be carried out prior to colloid preparation whereby a zerovalent metal complex is the immediate colloid precursor. The thermolysis of metal carbonyls (see above) in colloid synthesis is an example of this approach. In a further refinement the supporting ligands in the zerovalent complex can be removed by ligand displacement through the addition of an excess of a weakly binding ligand or through ligand reduction. Metal particles can be generated at low temperatures by this method. The organometallic literature is replete with examples of reactions of organometallic complexes with hydrogen which give metal deposits, and some of these reactions have been applied to colloid synthesis.

The zerovalent palladium and platinum complexes with dibenzylideneacetone $Pd(dba)_2$ and $M_2(dba)_3$ (M = Pd, Pt) are known to react with either hydrogen or carbon monoxide under mild conditions to give the metal. [111] The platinum analogue has been used as a source for highly dispersed supported platinum catalyst particles by reduction of $Pt(dba)_2$ with carbon monoxide in the presence of amorphous carbon. [112] This reactivity has been applied to the preparation of palladium and platinum organosols. Reduction of $Pd(dba)_2$ with hydrogen at atmospheric pressure in dichloromethane in the presence of PVP at 25 °C takes place over a period of 30 min to several hours to yield crystalline colloidal palladium with particle sizes of *ca.* 2.5 nm. The occurrence of an induction period of unpredictable duration in this reaction implies an autocatalytic step. At 3 atm, the same reduction produces 4.0 nm palladium particles. Smaller and less well formed palladium particles result from the reaction of CO with the complex in a much faster reaction. [34] Similar reactions of both the palladium and platinum complexes in the presence of cellulose nitrate and cellulose acetate in tetrahydrofuran also yield metal organosols. [23]

Hydrogenation of olefinic ligands in zerovalent complexes is a facile method for the synthesis of metal organosols. Hydrogenation of (cyclooctadiene)(cyclooctatriene)ruthenium, Ru(COD)(COT), yields metallic ruthenium, and, in the presence of polymers such as PVP, cellulose nitrate, and cellulose acetate, ruthenium sols form. [23, 113] For the PVP stabilized sol, the particle size is very small (*ca.* 1 nm). Similar reactions of $Ni(COD)_2$, $Pt(COD)Cl_2$, or mixtures of the two have been used to prepare PVP stabilized mono- and bimetallic organosols of these metals. [114]

Colloidal copper has been prepared by an organometallic route through the reaction of (cyclopentadienyl)(*tert*-BuNC)Cu with carbon monoxide in the presence of either PVP in methylene chloride or poly(dimethylphenylene oxide), PPO, in anisole. The PPO stabilized sol contains zerovalent copper particles of 4.0 nm diameter, while the PVP stabilized colloid contains larger particles with a relatively broad size distribution. [115]

6.2.2.4 Metal Vapor Synthesis

Conceptually, if not practically, the simplest method of preparing colloidal metals is the condensation of atomic metal vapor into a dispersing medium. [116] Given the high oxidation potential of most atomic metals (for example, the oxidation potential for atomic gold is –1.5V), the use of water as the diluent phase can be ruled out, and inert organic liquids are used exclusively in this procedure. Since the activation energy for agglomeration of metal atoms is very low, the possibility for competing molecular complex formation processes which have higher activation energies can be mitigated by operating at low temperatures.

Metal Vapor Reactors

Notwithstanding priority claims in the recent literature, the use of metal vapors cocondensed with organic vapors to prepare colloidal metals in nonaqueous media was first reported by Roginski and Schalnikoff in 1927, [117] some fifty years before the recent wave of activity in metal vapor chemistry. The apparatus used is shown schematically in Figure **6-6**. The organosols were prepared at reduced pressure by the evaporation of relatively volatile metals such as cadmium, lead, and thallium, and a subsequent cocondensation of these metal vapors with the vapors of organic diluents such as benzene and toluene on a liquid air cooled cold finger. After the cocondensation was complete, a colloidal suspension of the metal was obtained by warming the frozen matrix and collecting the liquid.

Modern versions of this method employ more versatile metal evaporation techniques (*e. g.* multiple metal sources) and allow for better control of such experimental parameters as evaporation rate and pressure. The procedures reported take one of two forms. In the first, which is essentially a preparative scale matrix

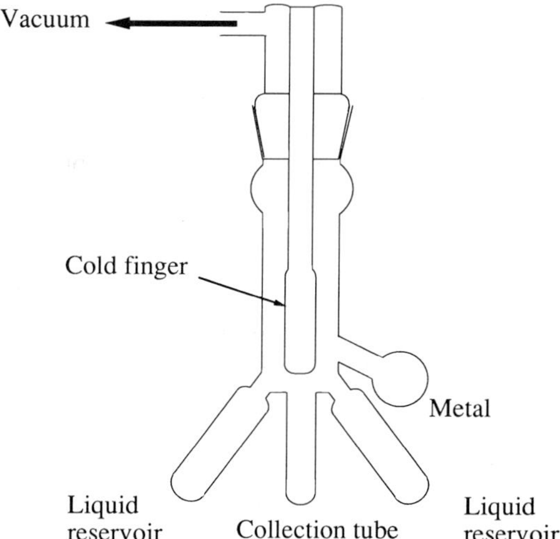

Figure 6-6. The first metal vapor synthesis reactor for preparing colloidal metals. Metal is cocondensed with organic vapors from the liquid reservoirs onto the cold finger, and the frozen matrix then allowed to melt into the collection tube (adapted from ref. [117]).

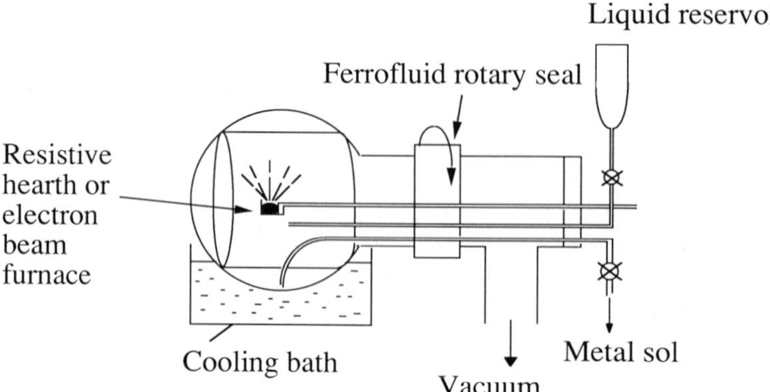

Figure 6-7. Schematic of rotary metal vapor synthesis reactor (adapted from refs. [119, 120]). Used for colloid preparations by both cocondensation and metal vapor-into-liquid processes.

isolation experiment, metal vapor is cocondensed at low temperature (usually 77 K) with the vapor of the diluent liquid in a frozen composite metal-organic glass. The colloidal metal particles nucleate and grow as the glass is warmed to melting and above. The method can be used either with a rotating reactor or a static reactor. In the former, shown schematically in Figure **6-7**, metal vapor is generated either from a resistively heated hearth or an electron beam furnace and condensed on the walls of a rotating vacuum vessel together with the vapor of an organic diluent. [118–120] In the latter, the metal vapor is cocondensed with the organic vapor onto the cooled walls of a static glass vacuum chamber of the type shown in Figure **6-8**. [121, 122] A second method involves the dissolution of the metal vapor into a well mixed liquid diluent, usually in a rotating reactor. The temperatures at which the condensations into the organic liquids

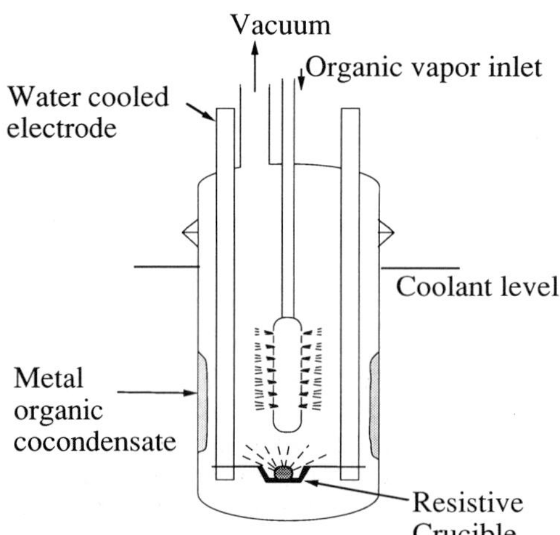

Figure 6-8. Schematic of static metal vapor reactor used in colloid preparation. Metal is evaporated from a resistively heated crucible and cocondensed with organic vapors onto the walls of a cooled evacuated vessel (adapted from ref. [121]).

must be performed is determined from a balance between the need for efficient mixing (*i. e.* relatively low viscosity) and a vapor pressure (10^{-2}–10^{-3} torr) low enough to enable efficient metal evaporation rates. If the metal is evaporated from an electron beam furnace, the temperature problem is aggravated by the maximum operating pressure of the electron gun (*ca.* 10^{-4} torr) which limits the maximum tolerable vapor pressure of the organic liquids used in the solution phase method. Add to this the fact that refractory metals such as tungsten and molybdenum require temperatures in excess of 3000 K in order to achieve useful evaporation rates at 10^{-4} torr, and it becomes clear that radiant heating of the cooled liquid and the resulting vapor pressure increase will become a problem.

A new metal vapor synthesis system for the preparative scale cocondensation of metal vapors with aerosols of organic liquids has recently been constructed in the author's laboratory. The use of an aerosol overcomes one limitation of the other methods, in which the organic diluent must either be volatile or have a liquid range which extends to a temperature low enough to reduce its vapor pressure to a useful value. The aerosol droplets, *ca.* 1 μm in diameter, are generated by feeding the organic liquid (neat liquid, polymer solutions, solutions of involatile ligands) into an ultrasonic atomizing nozzle from which they fall onto a rotating plate cooled to 77 K in a vacuum chamber as shown in Figure **6-9**. Vapors of one or more metals, obtained by simultaneous sputtering from metal or alloy targets, are then cocondensed with the aerosol. This results in the formation of a frozen organometallic matrix which is then warmed so as to allow the aggregation of the metal atoms. The resulting colloid solution is removed from the reactor under helium for characterization.

A variation on these techniques involves the generation of the metal vapor in an inert gas at moderate pressures, instead of at the highest possible vaccum.

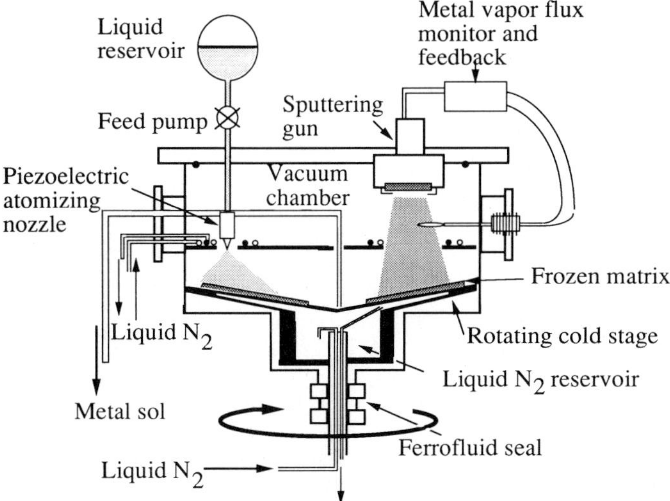

Figure 6-9. Schematic of sputtering source metal vapor/aerosol reactor for metal colloid preparation.

Originally developed for the preparation of fine metal powders, [123] this approach has been subsequently modified [124–128] in order to prepare organosols of a number of main group and transition metals. Metal vapors are thermally generated and swept by an inert gas stream into an organic liquid. The resulting colloidal suspensions contain crystalline metal particles whose sizes depend on the preparative conditions, such as inert gas atomic weight and gas pressure. The condensation of metal vapor onto the surface of a moving oil stream has also been used as a variation of this method. [129, 130]

Colloid Synthesis from Metal Vapors

The use of metal vapor routes to transition metal colloids has been reported by several workers. [15, 131–137] Klabunde has demonstrated the preparation of a variety of metal organosols by this method. [4, 138] For example, the cocondensation of gold or palladium atoms with acetone at 77 K in a static reactor followed by warming gives intensely colored colloidal suspensions of the metals with 4–9 nm particle sizes. [134, 135] The colloids are stable for several months. Simple removal of solvent results in a reversible aggregation and ultimately film formation. Condensation of gold vapor into styrene or methyl methacrylate monomer, or into cold solutions of the corresponding polymers, gave gold sols having between 7 and 15 nm diameter particles. [139] An analogous procedure in which the gold vapor was condensed into diacetylenes or their polymers gave gold colloids of 2 nm diameter. [136] Similarly, colloidal gold was prepared in fluorocarbon media [140, 141] by codepositing gold atoms into fluorocarbons like perfluorotributylamine at low temperature. Upon warming, the metal atoms aggregated to form clusters of approximately 1–2 nm diameter which could then be dispersed in polar solvents to yield brown solutions. Aging caused the clusters to aggregate into 2–10 nm colloidal particles. The dissolved uncoagulated particles could be deposited as a reflective gold film, but the film could be redissolved in organic solvents, implying the presence of substantial amounts of occluded solvent in the film. Upon warming, further agglomeration took place to form stable colloidal suspensions of metal clusters of *ca.* 5 nm diameter. Cobalt vapor has been condensed into cold toluene solutions of surfactants, which upon warming produce surfactant stabilized colloidal cobalt with particle sizes of *ca.* 5.0 nm. [137]

The condensation of vapors of palladium, platinum, copper, or nickel into methycyclohexane solutions of *iso*-butylaluminoxane, $(iso\text{-}BuAlO)_n$, in a rotating reactor was used to prepare small (<2 nm) colloidal metal particles. [133] These sols were notable in that stabilization of the metal particles was achieved in a nonpolar hydrocarbon with only Lewis acid species stabilizers, instead of the usual polar donor polymers such as PVP, etc. [142] Tri-*iso*-butylaluminum itself was also effective in stabilizing the colloidal metals and gave palladium particles with diameters less than 1.0 nm. [113]

6.2.2.5 Preparation of Colloidal Metals in Constrained Environments

Microemulsions, Micelles, Reverse Micelles, and Vesicles

The reduction of physically constrained metal precursors offers the potential for restricting the growth of the metal particles either by limiting the amount of metal particle precursor accessible to a growing nucleus or by restricting the motion of the precursor species and the growing nuclei, thus diminishing the rate of productive collisions. This tactic has found wide application, for example, in the preparation of small metal particles in zeolite cages as described in Chapter 4. In an analogous approach, liquid phase colloidal metals have been prepared in *surfactant micelles, reverse micelles, water in oil microemulsions,* and *vesicles* (Fig. **6-10**) which act as microreactors for the preparation and stabilization of metal colloids. These encapsulating organic particles are formed from the self assembly of surfactant molecules in nonpolar solvents, such that water is trapped in the intraparticle space. Although the terms micelle and microemulsion are often used interchangeably, a useful distinction between the two can be made on the basis of size – a surfactant aggregate containing *ca.* 50 molecules is a micelle, whereas one containing several hundred or more molecules is referred to as a microemulsion. Even larger particles, bilayer vesicles of 10^2–10^3 Å in diameter, can be formed, and if the surfactant molecules in the vesicle walls contain the

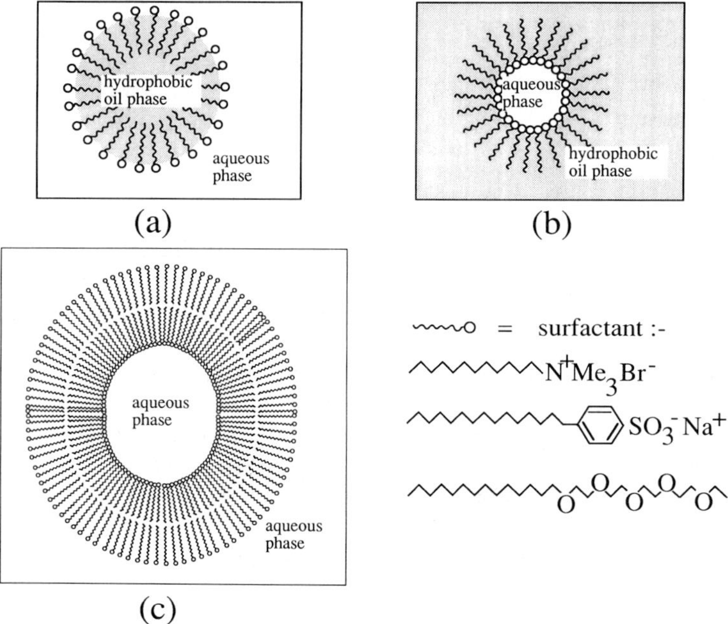

Figure 6-10. a) Surfactant oil in water micelles, b) water-in-oil reverse micelles, and c) vesicles. Metal colloids can be generated in the intraparticle space.

appropriate double bonds, they can be polymerized to form polymerized vesicles. [143] These different types of nanocapsules can contain various amounts of water, and this has allowed surfactant micelles, water-in-oil microemulsion droplets (reverse micelles), and polymerized surfactant vesicles to be used as highly dispersed reaction vessels for metal precursor solutions in metal colloid preparations. [144] The use of these media in metal colloid preparation is subject to the restrictions imposed by their stability criteria and the properties of the various systems. Microemulsions exhibit complex phase diagrams and structures [145] in addition to well defined compositional stability requirements. Their constituents often undergo rapid exchange. For example, the water pools in water-in-oil microemulsions exchange rapidly, with time constants in the milliseconds. [144, 146] Thus they are unlikely to constrain the growth of colloidal metals by limiting the number of precursor ions available to the growing metal particle. Polymerized vesicles, on the other hand, although much larger, offer greater stability and consequently better control of metal particle growth.

Aqueous suspensions of surfactant micelles were shown to be effective in stabilizing colloidal platinum produced either photochemically or by the reduction of H_2PtCl_6 by hydrogen. The former method gave smaller particles with a narrower size distribution. [105, 147] The reduction method was chosen for its compatibility with the stability criteria of the surfactant micelles. Both anionic (sodium dodecylsulfate, $Na^+[C_{12}H_{25}SO_4]^-$) and cationic (dodecyltrimethylammonium sulfate, $[C_{12}H_{25}N(CH_3)_3^+]_2\ SO_4^{2-}$) surfactants were used. Problems associated with the instability of the surfactant vesicle under conditions of high ionic strength (as demanded by the intended use of the colloid) were surmounted by using an unsaturated anionic surfactant (sodium undecenoate, $Na^+[CH_2{=}CHC_8H_{17}CO_2]^-$) to form the protective vesicle in which the platinum particles were prepared, followed by UV-polymerization of the undecenoate to produce a stable polymerized vesicle. [148] The polymerizable nonionic surfactant polyethyleneglycol undecenoate has also been used in this way. [149] Colloidal metals in polymerized surfactant vesicles have been prepared by sonication of $[PtCl_4]^{2-}$ solutions in the presence of monomeric surfactant vesicles followed by UV irradiation. [150] The vesicles were prepared from mixtures of dipalmitylphosphatidylcholine and an unsaturated surfactant such as $CH_2 = CH(CH_2)_8COO)_2NPO(OH)_2$. Under irradiation, both the reduction of $[PtCl_4]^{2-}$ to the metal sol and the polymerization of the vesicle walls result, giving a vesicle encapsulated platinum sol.

Reverse micelles [151–153] have been used in the preparation of colloidal metals. The method involves the dissolution of the metal precursor in the aqueous phase, which is then highly dispersed in an oil phase using ionic or nonionic surfactants such as cetyltrimethylammonium bromide and pentaethyleneglycol dodecyl ether. Reduction is effected either chemically, by adding a reducing agent such as hydrazine or hydrogen, or by photochemical or radiolytic methods. The sizes of the resulting colloidal metal particles were smaller than those obtained from a homogeneous aqueous reduction. It was noted in studies on the preparation of colloidal platinum, palladium, rhodium, and iridium in the aqueous phase of reverse micelles [151] that the tempting assumption that each metal particle is formed from the contents of a single reverse micelle droplet is probably not correct. Consistent with the rapid exchange rates in the water pools of the microemulsions (see above), later work on the radiolytic and photochemical production of colloidal gold from $HAuCl_4$ in water-in-oil microemulsions showed that under the conditions where each reversed micelle (with a water pool diameter of approximately 11.4 nm) contained as little as six Au(III) anions, gold particles of over 10 nm were formed. [103] Thus it is clear that equilibration between droplets of these microemulsions must occur at a rate greater than metal particle growth and at an early stage in the growth process. In addition the apparent restriction in metal particle size must result from the relative kinetics of these processes as well as the restricted diffusion of hydrated metal species in the two phase system.

Solid Polymer Matrices and Glasses

The evaporation to dryness of a suspension of a polymer stabilized metal results in a solid polymer/metal composite. All the reported electron micrographs of polymer protected colloids are, by definition, images of very thin films of the metal-polymer composite. A more elegant and potentially superior method for producing such films imvolves trapping the metal precursor compounds in a polymer matrix. This method offers another means of restricting particle mobility during nucleation and growth. Polymer films containing metal particles can be prepared by the counterdiffusion of solutions of a reducing agent and a metal salt from opposite sides of an insoluble polymer film cast from solution. Thus films of cellulose acetate or poly(vinyl alcohol) can be exposed to a $NaBH_4$ solution on one side and a solution of $RhCl_3$ on the other side, whereby metal particles are deposited as the reagents diffuse to a reaction zone inside the film. [154]

A novel approach to metal particle generation in polymers involves the synthesis of diblock copolymers in which one block comprises monomer ligand units. When cast in solid form (*e. g.* as a film), the polymer thus produced may microphase separate into microdomains, and the morphology and distribution of the microdomains can be varied in a predictable manner. Exposure of the copolymer to a metal complex or salt results in the incorporation of the metal into the poly(ligand) block of the copolymer and the resulting metal complex polymer can then be reduced to the colloidal state. [155, 156] For example, the copolymerization of *bis*(diphenylphosphino)norbornene with methyltetracyclododecene (in a ratio of 1 : 15) by ring opening metathesis polymerization gives a copolymer of this type. Equilibration with silver or gold complexes results in the incorporation of the metals into the polyphosphine blocks as shown in Scheme **6-1**. When cast as solids, these metal complex polymers phase separate, giving a lamellar morphology in the case of silver and hexagonally packed cylinders in the case of gold as shown in Figure **6-11a**. [155] Heat treatment (150 °C) of the films results in the formation of silver and gold clusters (2–10 nm and 1.5–4 nm respectively) which are restricted to the original microdomains (Fig. **6-11b**). By using a more phosphine rich copolymer and more gentle heating, single crystallites of silver could be obtained in the spherical microdomains of the copolymer. [156] In an elaboration of this method, the same authors synthesized organometallic-organic copoly-

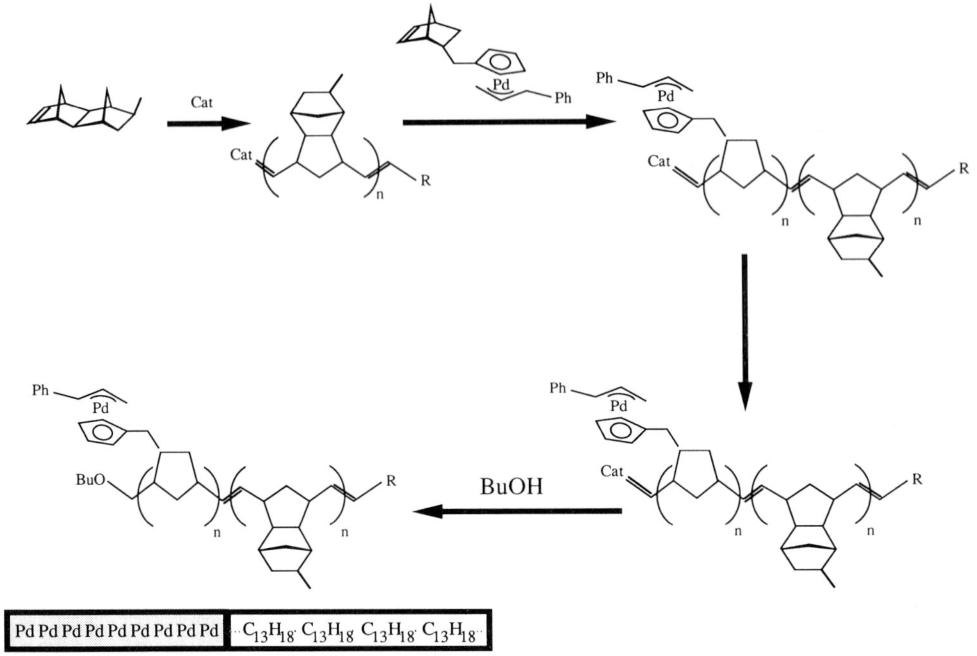

Pd Pd Pd Pd Pd Pd Pd Pd Pd | $C_{13}H_{18}$ $C_{13}H_{18}$ $C_{13}H_{18}$ $C_{13}H_{18}$

Scheme 6-1. Preparation of diblock organic/organometallic copolymers by ring opening metathesis polymerization (ROMP) of norbornenyl phosphines and norbornenyl palladium and platinum complexes (refs. [155–157]).

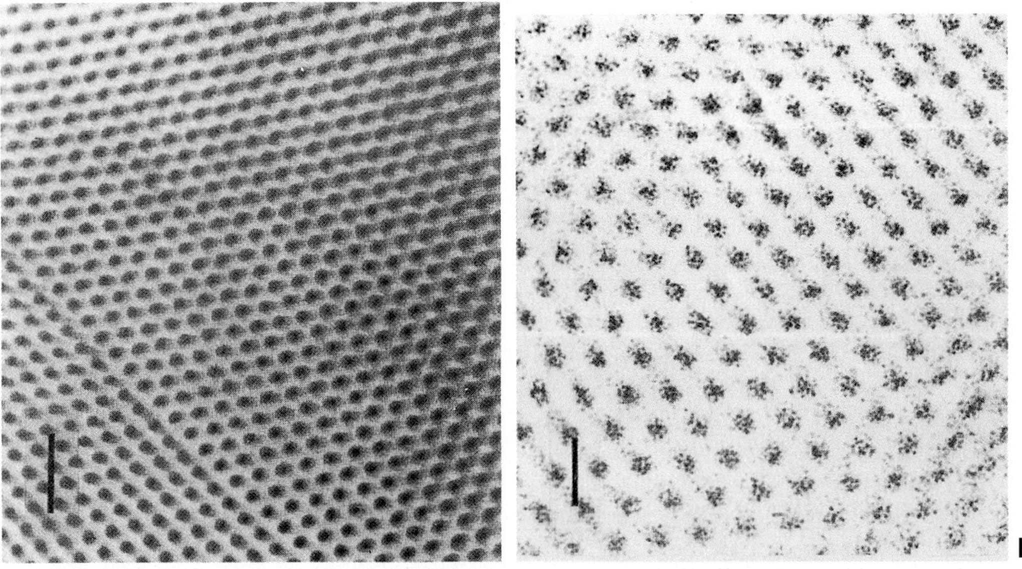

Figure 6-11. Transmission electron micrographs of gold particle preparation in hexagonal packed cylindrical microdomains of phase separated norbornenyl phosphine/gold complex copolymers: a) before and b) after gold particle formation. (Bar = 100 nm in a); 50 nm in b); reproduced by permission from ref. [155].)

mers containing palladium and platinum complexes of cyclopentadienylmethylnorbornene as the organometallic block. [157] Depending on polymer composition, phase separation into either spherical, lamellar, or cylindrical microdomains could be observed. Subsequent reduction by chemical or photochemical means not only limited the metal particle growth but also restricted the particles to the original organometallic microdomains of the polymer film due to the constraints on polymer mobility at the microdomain interfaces. This is illustrated for a lamellar morphology in Figure **6-12**. The size of the metal particles was found to vary with microdomain size in copolymer-homopolymer blends. [158] Thus, spherical, cylindrical, and lamellar concentrations of metal particles can be prepared with length scales in the 1–10 nm range.

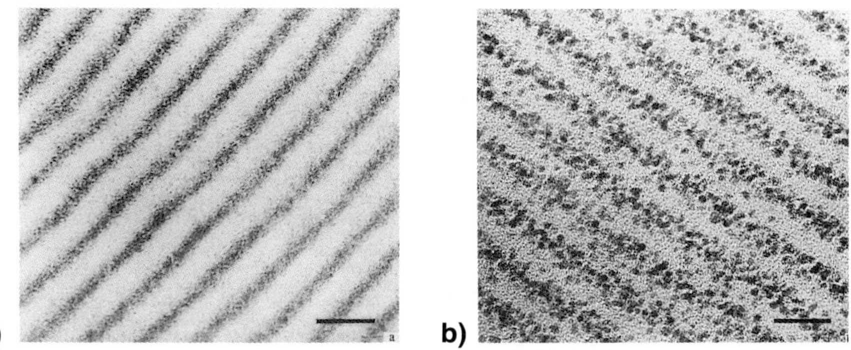

Figure 6-12. Formation of palladium particles in lamellar microdomains of phase separated poly(organopalladium)/poly(norbornene) copolymers: a) before and b) after particle formation. (Bar = 25 nm; reproduced by permission from ref. [157].)

The preparation of colloidal particles in glasses is an established area of solid state physics which arose from a desire to exploit the special electronic and optical properties of nanometer scale semiconductor particles – "quantum dots". Highly dispersed metal particles have been prepared in oxide glasses by sol-gel methods. A hydrolysable oxide precursor such as tetraethylorthosilicate, $(C_2H_5O)_4Si$, is hydrolysed to a silica sol, then a metal precursor is added, and the whole is finally gelled, calcined, and reduced to give highly dispersed metal particles entrapped in the glass matrix. [159] Alternatively triethoxysilane can be used as both the sol-gel precursor and the reducing agent for a metal salt, the resulting gel being dried to a metal containing xerogel. [160]

6.2.2.6 Bimetallic Colloids

Many of the preparative methods described above can be applied to the formation of bimetallic colloids. The simultaneous reduction of two metal salts has been reported to lead to homogeneous bimetallic particles in which a range of product compositions is possible. The synthesis of bimetallic PtAu sols by the coreduction of aqueous solutions of H_2PtCl_6 and $HAuCL_4$ with citrate has been reported. [161] Although a uniform composition of the particles was noted, there is a miscibility gap between 2 % Au and 80 % Au in the bulk.

Some Pd-Pt bimetallic colloids protected by polymers [162–165] have been prepared by the coreduction of the metal salts with aqueous alcohols. Similar bimetallic sols in nonpolar organic solvents were obtained by N_2H_4 or $NaBH_4$ reduction of palladium and platinum salt mixtures after their extraction into the organic phase (cyclohexane or chloroform) with such surfactants as trioctylphosphine oxide or distearyldimethylammonium chloride. [166]

Uniform bimetallic copper-palladium colloids (the metals are completely miscible) have been prepared by thermolysis of mixtures of the acetates in high boiling solvents such as bromobenzene, xylenes, and methyl-*iso*-butyl ketone [86] in the absence of polymer stabilizers. The resulting agglomerated bimetallic particles often contained CuO in addition to the metals. Smaller particle sizes and narrower size distributions without oxide formation were reported to result from the analogous reaction in 2-ethoxyethanol in the presence of PVP. [38] In both cases the homogeneous composition of the particles was established during the electron microprobe analysis used to determine the Cu:Pd ratio in the individual particles.

Bimetallic particles with nonuniform distributions of metals have been intentionally prepared by sequential deposition of the metals. In an early study, [45, 167] gold plated palladium colloids were prepared. A hydrosol of palladium was prepared by the citrate method, and gold(III) was then added together with the reducing agent hydroxylamine so that the gold was reduced onto the palladium seed particles. In a later elaboration on this system of colloidal miscible metals, the palladium-gold system was explored in its three possible forms: as a homogeneous alloy, as palladium plated gold, and as gold plated palladium. [168] Here again the two layer particles were prepared by prior formation of the seed sol and subsequent reduction of the second metal onto the first. Palladium-platinum colloids in their two plated forms (palladium-on-platinum and platinum-on-palladium) can likewise be prepared by reduction of a salt of the outer metal in the presence of the core metal as a preformed sol. [169] Bimetallic layered colloids of gold plated with palladium or platinum were reported to be formed from the reduction of H_2PtCl_6 or H_2PdCl_4 with hydroxylamine hydrochloride in the presence of a preformed gold sol. [170] The resulting two phase particles were isolated in their solid complexed form by addition of $[H_2N(p\text{-}C_6H_4SO_3)]^-Na^+$.

The ability of colloidal silver to act as a microelectrode by storing electrons transferred to it from organic radicals in solution (see Section 6.5.5) provides a method for plating silver colloid particles with other metals. Thus lead, indium, [171] and cadmium [172] can be deposited onto colloidal silver in the presence of radiolytically generated organic radicals. Lead can also be deposited on colloidal gold by the same method. [173] Bimetallic colloids have also been prepared by the remarkable reaction of two preformed sols in a colloid redox reaction. When colloidal suspensions of silver and lead are mixed and allowed to equilibrate, the silver particles become coated with lead, as deduced from the changes in the UV-visible spectrum of the sols. [91] The plating of submonolayer amounts of copper onto preformed 4 nm colloidal palladium stabilized by PVP also gives a bimetallic colloid in which the added copper is found to dissolve into the palladium particle. [174]

6.2.2.7 Ligand Stabilized Metal Colloids

The term "ligand stabilization" has been applied to the use of traditional organotransition metal chemistry ligands in stabilizing colloidal transition metals. Although the traditional citrate sols of gold and platinum are likewise stabilized by adsorption of citrate and other ligating anions at the particle surface, and steric stabilization by polymers must also rely on weak ligation of the polymer to the particle surface, in the context of this section we will focus on the use of phosphines as stabilizers.

Water soluble phosphines such as the sodium salt of sulfonated triphenylphosphine, $(m\text{-}C_6H_4SO_3Na)_3P$, function as stabilizers for hydrosols of colloidal rhodium, prepared by hydrogen reduction of $RhCl_3 \cdot 3H_2O$ in the presence of the surfactant. [67] The colloidal material has been described as a polyhydroxylated rhodium particle, which implies considerable oxidation of, at least, the rhodium surface to Rh(I).

The addition of solutions of water soluble phosphines such as the sodium salt of *p*-diphenylphosphinobenzenesulfonic acid, $[(C_6H_5)_2P(p\text{-}C_6H_4SO_3)]^-Na^+$, to preformed 20 nm gold hydrosols, as prepared by the citrate method, results in stable isolable gold sols. [30] The phosphines coordinate to the surfaces of the gold particles and the colloid can be isolated as a black solid by evaporation or as gold leaves by precipitation with ethanol. The solid is soluble in water giving a red solution, the original color of the gold sol. Coordination of the phosphine has no effect on the UV-visible spectrum of the sol. Electron microscopy (Figs. **6-13** and **6-14**) shows well separated 3.6 nm gold particles [175] and, in some cases, the ligand shell can even be observed. [3] Thick films of the material are metallic in appearance. Bimetallic gold-platinum and gold-palladium colloids can also be isolated using ligand stabilization. For two layer bimetallic particles (*e. g.* gold-on-platinum or palladium-on-gold), the outer metal determines the most effective ligand. Thus $[(C_6H_5)_2P(p\text{-}C_6H_4SO_3)]^-Na^+$ is preferred for outer layers of gold, and $[H_2N(p\text{-}C_6H_4SO_3)]^-Na^+$ is more effective for palladium and platinum. Exploiting this ligand preference allows the isolation of stable sols in the solid state. [170]

The use of phosphines as stabilizers for colloidal metals is reminiscent of the stabilization of low valent transition metals. Since the latter have found extensive application in homogeneous catalysis such ligand manipulation should be a powerful method for modifying colloidal metal catalysts as well.

Figure 6-13. Transmission electron micrograph of 36 nm gold colloid particles stabilized with $P(m\text{-}C_6H_4SO_3Na)_3$. The mean distance between colloid particle surfaces corresponds to two layers of the ligand. (Reproduced by permission from ref. [175].)

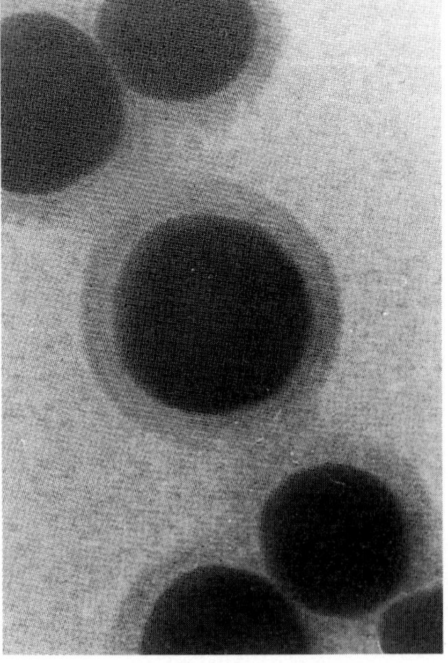

Figure 6-14. Transmission electron micrograph of phosphine stabilized gold colloid particles (44 nm) surrounded by a stabilizing ligand shell corresponding to 12 ligand molecules (7.2 nm). (Reproduced by permission from ref. [175].)

6.2.2.8 A Comment on Ligand Stabilized Giant Molecular Clusters and Colloidal Metal Particles

An important recent development in the field of molecular cluster chemistry is the synthesis by organometallic preparative techniques of very large clusters whose members comprise a family of structures with metal cores corresponding to successive close packed shells of metal atoms. [3] Each shell corresponds to the addition of $10n^2 + 2$ metal atoms to the previous cluster (the n = o shell has one atom). Thus clusters of 13, 55, 147, 309, and 561 metal atoms have been suggested to possess special stability due to their "magic numbers" of metal atoms in a closed shell. Several examples have been reported and include $Au_{55}((PPh_3)_{12}Cl_6$, [176] $Pt_{309}(phen^*)_{36}O_{30 \pm 10}$, [177] and $Pd_{561}(L)_{36}O_{200 \pm 20}$ (L = bipyridyl or phenanthroline). [178, 179] The identification of these materials rests on such data as sedimentation rates, transmission electron microscopy (TEM), and EXAFS as well as other diffraction and scattering techniques. Although individual particles of the appropriate size and structure have been imaged with atomic resolution TEM, in no case has the structure of an analytical sample been established by a single crystal X-ray diffraction study since the materials have so far resisted all attempts to grow single crystals.

A distinction might be made between these "giant molecular clusters" and colloidal metal particles purely on the basis of size. However, it might also be conversely argued that the giant clusters are more properly viewed as monodispersed colloidal metals. Even if one takes this latter point of view, there is no doubt that, on the basis of the published data, the materials reported by Schmid [3] and by Moiseev [178, 180] are remarkable for their monodispersity. On the other hand, if we adopt the former point of view, then we might justifiably regard a monodispersed 2.5 nm poly(vinylpyrrolidone) stabilized palladium colloid as a polymer ligated Pd_{561} cluster since 2.5 nm is close to the size reported by both Moiseev and Schmid for the giant cluster $Pd_{561}(L)_{36}O_{200 \pm 20}$, prepared by the reduction and subsequent air oxidation of $Pd(OAc)_2$ in the presence of ligands such as phenanthroline and bipyridyl. The difference is not merely semantic. By one definition, a molecular description demands complete homogeneity as defined by a unique electronic and molecular structure in which both are associated with a well defined energy minimum. By this definition, a monodispersed colloid of 2.5 nm $\pm$ 0.3 nm diameter could not be considered molecular. However, the following points should be considered when contemplating the nature of the giant clusters: i) since the narrow size distribution of the colloidal metal particles is within the precision limits of the TEM measurements, even a truly molecular sample would have a range of apparent diameters; ii) the mean diameter of 2.5 nm does correspond to a five shell cubic packed cluster; iii) the clusters are prepared in the absence of any severe steric constraints, such as those imposed in particle preparation in zeolite cages, and yet they do not aggregate further. All these considerations are consistent with the notion that a special stability is associated with this degree of aggregation. To describe these materials as molecules requires a rather unconventional use of the term, but perhaps no less unconventional than a description of organic polymers as molecular. They too are polydisperse and often cannot be described with the precision which a more rigorous definition of the term molecular would demand. We, however, have no hesitation in regarding them as molecules.

In reserving judgement on the nature of these clusters, their importance should not be underestimated. The physics and chemistry of small metal aggregates is under intense scrutiny, and these materials provide a unique access into this area, regardless of whether we consider them as giant molecules or extremely highly dispersed colloidal clusters. [181]

6.2.3 Summary of Synthetic Methods

It can be seen from the preceding sections that virtually any method by which metal deposits can be prepared can also be applied to the preparation of colloidal metals. Given the surge of activity in recent years in metal deposition technology, a field which has given rise to a variety of novel methods such as chemical vapor deposition and laser ablation of cluster sources for thin film deposition, the range of synthetic options is wide and increasing. This versatility allows one to choose a method which is particular to the diluent and the stabilizing agent (if any) for a colloidal metal preparation. Thus the development of new methods is limited only by the ingenuity of the inorganic chemist.

Perhaps the most constant irritant in colloid synthesis is irreproducibility. Turkevich stated quite plainly [8] that wide variations in particle sizes and size distributions can result from nucleation at adventitious impurities in colloid preparations, and that "flagrant disregard for certain common sense precautions such as cleaning vessels" can lead to failure. However the experience of many practitioners in this area, including this author, is that a chemist accustomed to the exigencies of organometallic synthesis should not encounter insuperable difficulties in reproducibility. This is not to say that we have any idea on how to control particle size through proper selection of polymers, solvents, precursors, reducing agents, or metal precursors, each of which can have an effect on particle size. The true control of particle size remains the most attractive goal for the synthetic chemist in this field.

6.3 Structural Properties of Colloidal Transition Metals

The properties of metal colloids which are of interest in the context of this chapter are those of size, structure, and composition. The questions to be answered in characterizing these particles are those which would occur both to the molecular chemist and to the solid state chemist, namely:

(i) how big are the particles and what is their size distribution?
(ii) what is their composition?

(iii) what is their structure (and is this dependent on size)?
The answers to these questions require the application of techniques drawn from solid state chemistry, metallurgy, molecular transition metal organometallic chemistry, and molecular spectroscopy.

6.3.1 Particle Size and Distribution

The first question asked about a colloidal metal suspension concerns the state of aggregation of the metal particles. How large are the particles and what is the particle size distribution? The question has been asked and answered with varying degrees of precision for more than a century. The first successful attempt to answer this question was in the application of the ultramicroscope by Siedentopf and Sigmondy in 1903 "On Making Visible and Determining the Size of Ultramicroscopic Particles with Special Attention to Gold Ruby Glasses". [182] This represented the first direct measurement of the sizes of the colloidal gold particles, which was made possible by the virtue of their light scattering properties. As the available techniques became more and more sophisticated, the characterization of metal colloid particle size became ever more precise.

Several techniques are commonly used to measure the size distributions of metal colloid particles. Electron microscopy, X-ray diffraction, and small angle X-ray scattering are the most commonly used, although classical methods such as sedimentation rates are sometimes reported. The techniques which have been extensively applied to the sizing of polymer colloids and emulsions, [183] such as light scattering and neutron scattering, have been only rarely applied to the characterization of metal sols. [103, 151, 153, 184]

There is a preponderance in the metal colloid literature of transmission electron microscopic characterizations, and this is understandable since the technique gives an immediate visualization of the particles. The high electron optical density of colloidal transition metals, especially the heavier metals, gives high contrast when the particles are dispersed in polymer films, and this renders them particularly amenable to TEM analysis. This property has led to the use of gold and platinum sols as staining agents for electron microscopic examination of biological samples.

Sample preparation for TEM analysis of colloidal metals is usually a simple procedure, involving the evaporation of a drop of suitably diluted colloid suspension onto a microscope grid. For polymer stabilized colloids, the polymer forms a thin transparent film by this procedure. Thin films can be alternatively prepared by ultramicrotomography of thick polymer/metal samples produced by evaporation of the liquid suspension.

The resolution limits of modern transmission electron microscopes are more than sufficient for the imaging of metal particles in the 1–10 nm size range of interest here. The images obtained vary in their degree of precision, whereby the information content of the image is set by the needs of the analyst. Although atomic resolution and lattice imaging are required for detailed structural analysis, a particle size distribution can be obtained at moderately low resolution for a sol with a mean diameter of 10 nm. For example, the electron micrograph of a copper-palladium organosol shown in Figure **6-15** [38] was obtained with a resolution perfectly adequate for the determination of the 4.0 nm mean particle size.

The available literature on metal colloids contains size distributions which span the range from "monodispersity" to severe agglomeration. Strictly speaking, the term monodispersed means absolutely homogeneous in size, however, in the context of colloid chemistry, a monodispersed sol contains a distribution of particle sizes with a standard deviation from the mean of less than 10% (*i. e.* a polydispersity of 1.1). Although this may seem to be a relatively generous definition of the term, it is the one in common use.

There is significant evidence that small metal particles, and especially those in polymer films, can undergo structural change and even aggregation when under the influence of the electron beam during TEM analysis. [185, 186] This is not always observed, but the recognition of this potential problem is a prerequisite in TEM analysis of colloidal metals.

Size distributions obtained by this or any other method are revealing in terms of the mechanism of particle growth. Two mechanisms can be envisioned for the growth of particles. The addition of atomic units to the growing nucleus, a process known as Ostwald ripening, or the agglomeration of particles by interparticle Brownian collisions. The former gives particle sizes characterized by a normal distribution, and the latter by a log-normal distribution. [187, 188]

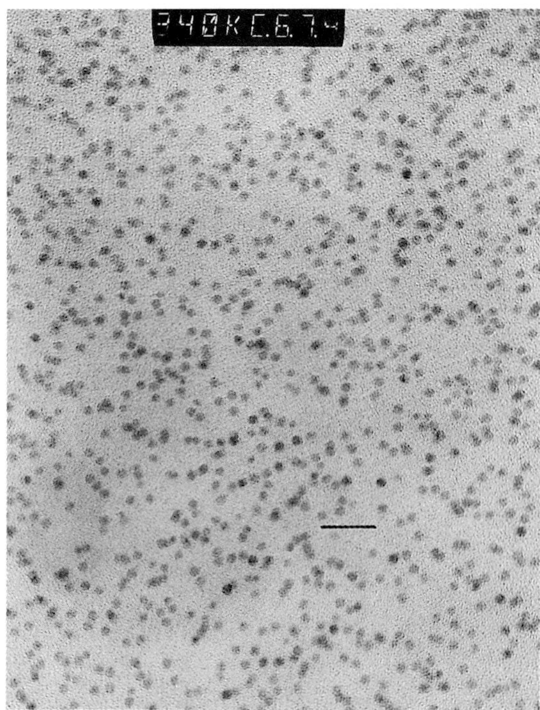

Figure 6-15. Transmission electron micrograph of 4.0 nm bimetallic copper-palladium colloid particles ($Pd_{70}Cu_{30}$) in PVP film. (Bar = 17 nm; reproduced by permission from ref. [38].)

6.3.2 Composition

The next question to be asked concerns the composition of the particles. For the case of monometallic particles this could in principal be a trivial question. If the particle is known to contain a single element, the only question which then arises concerns the oxidation states of the metal, and this can be determined by X-ray photoelectron spectroscopy. For example, the colloidal metals described in Section 6.2.2.1 and prepared by Dye and coworkers by alkalide and electride reduction of salts of gold, copper, platinum, nickel, and molybdenum (as well as several main group metals and metalloids) were analyzed by XPS [73] which showed the presence of only zerovalent metal. Oxidized metal was detected only for nickel and molybdenum (among the transition metals) and this only after exposure to an oxidizing solvent such as methanol. These results show that if a sufficiently powerful reducing agent is used in the colloid synthesis, the surface of the particles can be kept in a reduced metallic state.

However, the situation becomes more complex when one takes into account the possibility of contaminants remaining from the synthetic procedure used to prepare the colloid. The inclusion of solvent fragments during metal vapor colloid preparations, or the adsorption of unreduced metal cations and their corresponding anions during salt reduction preparations should also be considered. In the case of metal vapor methods, it was shown some years ago that certain atomic metals were reactive toward the C–H bonds in hydrocarbons [122, 189, 190] when cocondensed with vapors of these organic materials. Metal complexes containing alkylidene and hydride ligands derived from the solvent can be formed in this way. In the preparation of highly dispersed metal powders by this method, it has been shown [191] that fragments of the organic material can be incorporated in the "metal" particle during these metal vapor condensations. Klabunde has shown that in the condensation of nickel vapor with pentane, a nickel containing powder is formed which contains hydrocarbyl fragments in sufficient amounts to give a nickel:carbon ratio of up to 0.5. [192, 193] This result clearly shows that fragmentation of the organic diluent has occurred and that the organic fragments have been incorporated into the growing metal particles. Thus, the possibility for the inclusion of organic fragments in the colloidal particles when prepared by metal vapor condensation methods must always be considered.

We may also consider the composition of the surface bound species which contribute to the stability of the colloid. These include those comprising the electrical double layer in electrostatically stabilized colloids as well as other adsorbed species such as ligands, surfactants, or polymers in sterically stabilized colloids. The very fact that the metal particles are stable against flocculation demands that the surface of the particles be covered to some degree by these stabilizing agents. The nature of these surface species can often be postulated, as in a polymer stabilized colloid, but details of their structures and modes of binding to the surface are seldom analyzed and are usually represented in only schematic form (Figs. **6-3** and **6-5**). Spectroscopic approaches to determining the nature of the adsorbed species are described in Section 6.3.3.2.

When bimetallic colloids are under investigation, the question of composition becomes crucial. Just as a monodispersed system is needed in order to study particle size effects, a homogeneous composition (*i. e.* one in which all the particles have the same composition) is required if compositional effects are to be investigated. In preparing a bimetallic colloid, some synthetic insight must be used in order to avoid the formation of a sol containing particles with a range of compositions. In addition, analytical methods should be used which are sensitive to local differences in composition on the nanometer scale. Clearly, since the analysis of the bulk sol will yield only a mean composition, alternative methods must be applied. The most persuasive analytical method is energy dispersive X-ray spectroscopy (usually referred to as EDAX or EDS) and involves the selective electron bombardment of single colloid particles in the electron microscope. Each element in the irradiated particle emits X-rays at characteristic energies, and with an intensity proportional to its concentration in the particle. Thus if the analyses on a number of individual particles are identical within the statistical limits of error, it can be concluded that the sample is of homogeneous composition. Figure **6-16** shows the results of such an analysis for a bimetallic PdCu colloid

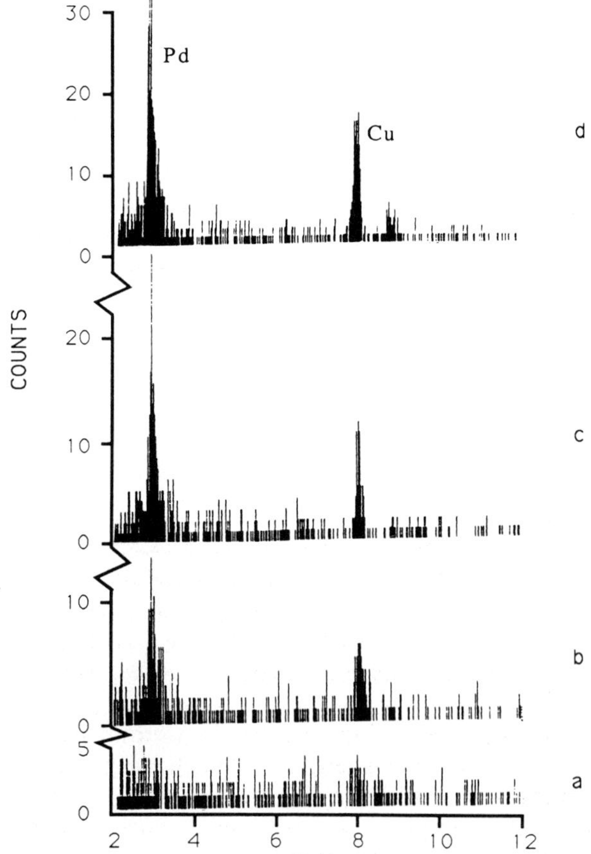

Figure 6-16. Single particle EDAX analysis of 4.0 nm bimetallic $Cu_{45}Pd_{55}$ colloid particles in PVP showing a) the absence of metal in the polymer film between the particles and b–d) a uniform composition of three particles of different sizes. (Reproduced by permission from ref. [38].)

stabilized by PVP [38] with a mean particle size of 4 nm prepared by the coreduction of copper acetate and palladium acetate. The EDAX analysis, which required placing a 1.5 nm electron probe on a sequence of particles, showed that no monometallic particles were present, that all the precursor salts had been reduced (since neither Cu nor Pd was detected in the polymer film between the particles), and that all the particles analyzed were of the same composition. This procedure is quite lengthy and requires counting times of several hundred seconds for each particle. Since the irradiated particles can move in the polymer film during this time, constant tracking is required. Despite these difficulties, it is the results from this sort of analysis which allow the properties of the colloid to be studied as a function of composition.

The excellent spatial resolution of the EDAX technique allows the composition of larger particles to be studied in even greater detail, thereby revealing compositional variation across individual particles. A series of bimetallic colloidal PdAu particles were prepared [168] using the methods described in Section 6.2.2.1 and gave either alloy particles, palladium-coated-gold, or gold-coated-palladium particles. The particles were of sufficient size (>20 nm) to allow the positioning of an electron probe at either the periphery or the center of a particle. The compositional variations as related to the probe position are shown in Figure **6-17**. Similarly, platinum-coated-gold and palladium-coated-gold particles prepared as hydrosols were analyzed by EDAX [170] which revealed their nonuniform composition. Such variations will be an important factor when we consider the structure of bimetallic colloid particles.

The question of composition can be addressed in another manner by electron microscopy. Lattice imaging techniques, which involve the reconstruction of the direct image from the diffraction pattern of a particle, can allow for the measurement of lattice spacings characteristic of the constituent phases, and thus provide constituent analysis based on the structures. [3, 194]

6.3.3 Structures

6.3.3.1 Electron Microscopy

Once the particle size distribution and composition have been determined, the question of particle structure arises. Transmission electron microscopy again plays a significant role in structure determination. However, in order to have a meaningful basis for a discussion of the structures, images at atomic resolution are necessary. For example, the images of colloidal silver [71] shown in Figure **6-18** were obtained at a resolution of 0.195 nm, and were sufficient to be used in a discussion of its structure and growth mechanism. These striking micrographs were obtained for silver particles in a sol prepared from $AgNO_3$ in acetonitrile by reduction with polyethyleneimine. It must be recognized, however, that regardless of how impressive an atomic resolution image of a single particle is, it nonetheless tells us little about the nature of the sample as a whole unless such images are typical of the whole sample, a circumstance seldom, if ever, obtained.

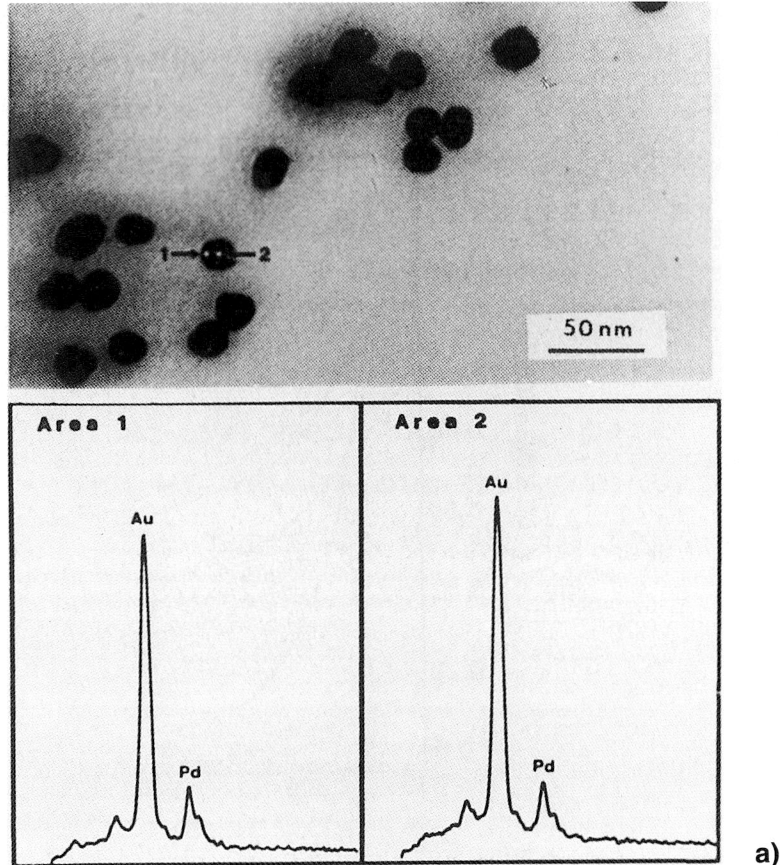

Figure 6-17. Single particle EDAX analysis showing the distribution of the metals in a) colloidal PdAu alloy, b) Au coated Pd particle (dark field image), and c) Pd coated Au particle. (Reproduced by permission from ref. [168].)

Atomic resolution electron microscopy images are obtained with particles in a specific diffraction orientation, since the single particle electron diffraction data are used for constructing the lattice image. Of course, depending on the area analyzed, electron diffraction data can contain the necessary information for structural analysis of the entire sample, or of any selected region of it. Since a sample of colloidal metal, in whatever supporting matrix, contains a large number of particles in random orientations, electron diffraction from a large area of a colloid sample results in a characteristic pattern of concentric rings which resemble the powder X-ray diffraction pattern obtained for a polycrystalline powder. Figure **6-19** shows such a pattern for a sample of colloidal palladium in PVP with a mean particle size of 6.5 nm prepared in the author's laboratory. [34] The

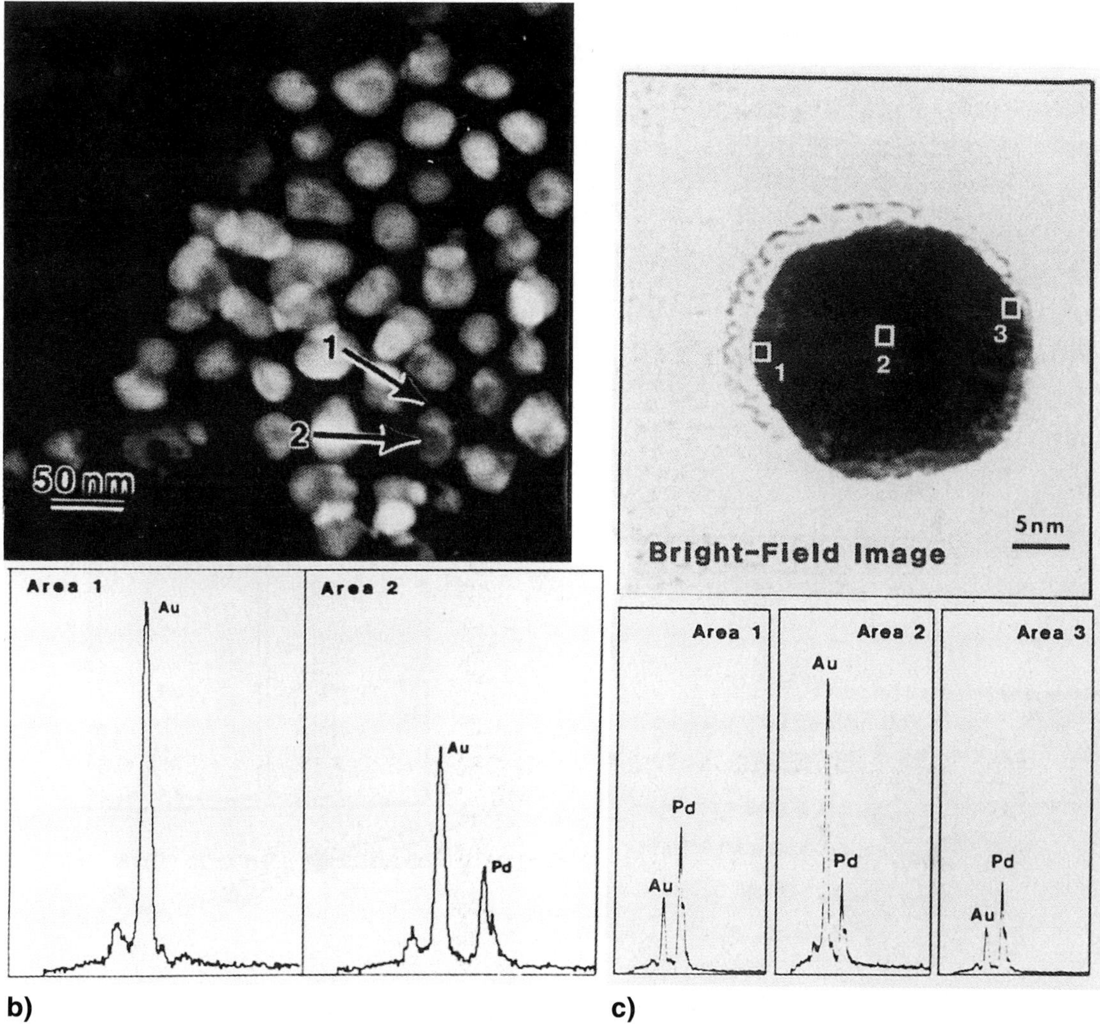

spacings correspond to the reflections expected for the face centered cubic lattice of elemental palladium. Images of individual particles in the micrograph show other evidence for crystallinity. Some particles are more highly contrasted than others, since those in an appropriate diffraction orientation will appear either lighter or darker than normal, and a clear regularity of shape is evident. Some particles are obviously polycrystalline, as shown by both optically dense and transparent regions in the same particle.

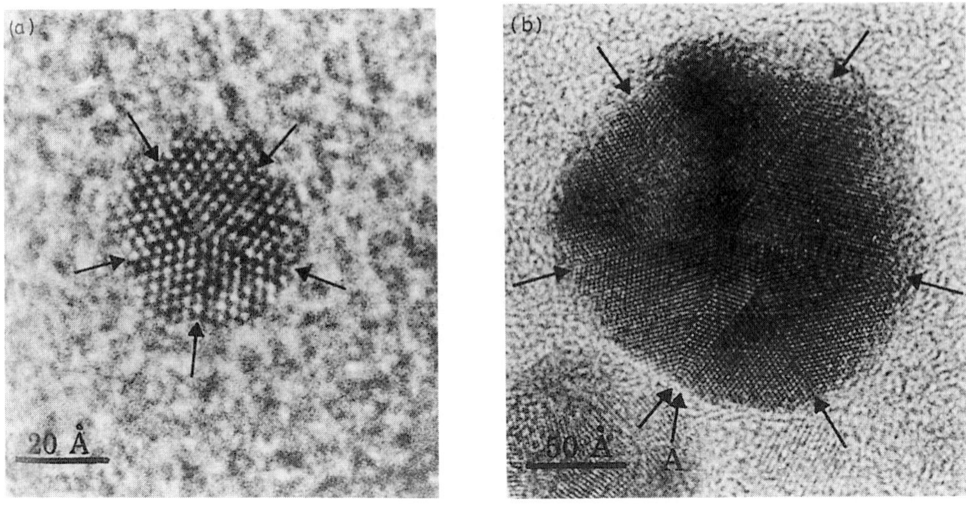

Figure 6-18. Transmission electron micrographs of decahedral silver colloid crystallites. The smaller particle a) is a perfect decahedron, the larger b) retains a residual fivefold axis but is a sevenfold twin with clear discontinuities (Reproduced by permission from ref. [71].)

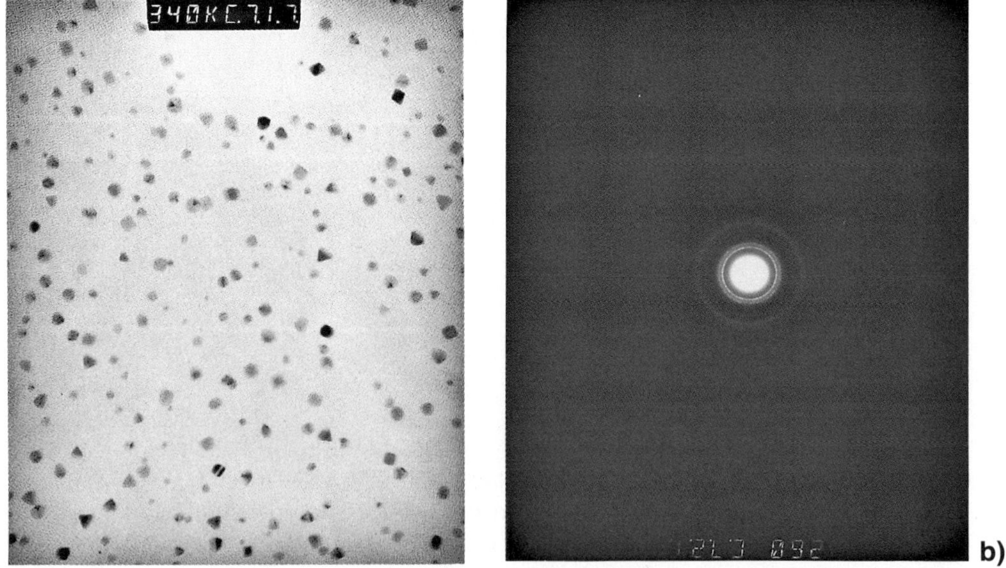

Figure 6-19. a) Transmission electron micrograph and b) electron diffraction pattern of 6.5 nm colloidal Pd in PVP showing facetting of the crystallites, contrast due to diffraction, and fcc electron diffraction pattern. (Adapted by permission from ref. [34].)

6.3.3.2 X-ray Diffraction

Similar structural data can be obtained from X-ray diffraction, although at the lower end of the range of particle sizes which are of interest (1–5 nm) broadening of the diffraction lines at usual X-ray wavelengths can make unambiguous analysis difficult. An X-ray diffraction pattern for the 6.5 nm palladium colloid described above (Fig. **6-19**), obtained using synchrotron radiation (whose high intensity is a great advantage), is shown in Figure **6-20**. The {111}, {200}, {220}, and {311} reflections from the fcc lattice can be clearly seen. However, when dealing with particles of a few tens to a few hundreds of atoms, the acquisition of useful structural information from conventional X-ray diffraction is difficult.

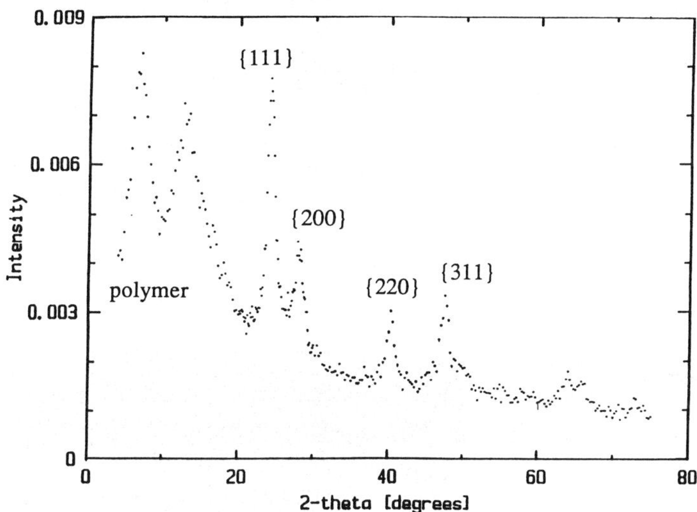

Figure 6-20. X-ray diffraction pattern of 6.5 nm Pd colloid in PVP, showing [111], [200], [220] and [311] reflections of the fcc lattice of metallic Pd. (J. S. Bradley, K. Liang, unpublished results.)

6.3.3.3 EXAFS

A powerful spectroscopic method for structural analysis which has been extensively applied to the problem of structure determination in small particles, and especially bimetallic particles, is Extended X-ray Absorption Fine Structure analysis, or EXAFS. [195] The X-ray absorption spectrum of an element contains absorption edges corresponding to the excitation of electrons from various electronic states at energies characteristic of that element. For example, K edges arise from the excitation of electrons from 1s states, and $L_{I,II,III}$ edges from excitations from 2s, $2p_{1/2}$, and $2p_{3/2}$ states. As the X-ray energy is increased above an edge, oscillations (fine structure) are observed in the absorption coefficient over an energy range of several hundred to over a thousand eV above the edge. Since these oscillations arise from the interference between waves associated with the emitted photoelectrons and photoelectrons backscattered from the neighboring atoms, they contain information on the environment of the absorbing atom. The fine structure is extracted from the X-ray absorption spectrum which, after data manipulation, is best fit to a computed EXAFS spectrum for a model structural environment for the absorbing atom. The technique is element specific and structure sensitive, and gives information on the number and identity of neighboring atoms and their distances from the absorbing atom. The information most usually sought in an EXAFS experiment comprises the number of scattering atoms of each type and their distances from the absorbing atom, that is, the composition of the mean coordination sphere about the absorbing atom. If multiple elements are present, they can be analyzed both as the absorbing atom and as the scattering atoms. A fitting technique which takes advantage of the necessary geometric relationships between the various components in a multicomponent system has been developed and is of great use in the EXAFS analysis of bimetallic particles. [196]

The use of synchrotron radiation has enormously reduced the time needed for EXAFS data acquisition, and under favorable circumstances an absorption spectrum can be obtained in less than an hour. This technique has been extremely useful in a number of fields, but especially in the study of small metal particles in heterogeneous catalysts. Here the structural entity to be analyzed is part of a complex matrix but could, in theory, contain as few as only two metal atoms, and it need not be crystalline. The method is well established in its application to small metal particles. [195] For example, the structure of a 0.5 % platinum-rhenium on alumina catalyst having metal particle sizes of 1 nm would be difficult to ascertain by either diffraction or electron microscopic methods, yet, by virtue of its element specificity and environment sensitivity, the EXAFS method can provide important details about the structure of the particles. [195] The use of this type of supported metal in fundamental studies of small metal particle chemistry has some drawbacks. Analyses are complicated by the fact that in order to prepare supported metals in this size range it is necessary to work at low metal loadings on the support so as to avoid particle aggregation and sintering during the reduction step of a classical salt impregnation/reduction catalyst preparation. At the not uncommon dilutions of less than 1 %, problems of sensitivity arise. Although the method yields information which might be impossible to obtain otherwise, a higher precision could be obtained if the concentrations were higher. It is for this reason that EXAFS analysis on certain colloidal systems is so promising for fundamental studies on nanoscale metal particles. Concentrations of up to 10 % or more can be realized in polymer stabilized metal sols by evaporating the solvent to give a dry polymer powder containing the metal particles, and much higher metal concentrations are accessible when ligand stabilized [3] or surfactant stabilized [197] colloids are used.

It must be emphasized that no information on the symmetry of the absorbing atom's environment can be derived from EXAFS experiments on metal particles, simply the number, identity, and distances of the neighboring atoms. Although interatomic distances can be determined very precisely by EXAFS analysis, the characteristic precision to which coordination numbers are established by this technique is low and usually reliable to only *ca.* 15%. However, in the application to small metal particle chemistry this is often sufficient for the determination of particle sizes and for the elucidation of element distribution in multimetallic particles. The derivation of this information follows simply from a consideration of the geometry of a particle with a close packed structure. The geometric arguments were quantified in the classic paper of van Hardeveld and Hartog [198] who calculated the statistical abundance of the various unique surface sites on particles with a wide range of morphologies, packing geometries, and particle sizes. For example let us consider an octahedrally shaped particle built from a fcc lattice. At its smallest size, the surface is dominated by sites with the lowest coordination number (CN = 4), but as the particle grows, the mean surface CN increases asymptotically to 9 and the mean overall CN to 12. A detailed analysis of a number of particle morphologies and structures has recently been performed by Benfield. [199] Thus the intuitive relationship between size and surface area is put on a useful quantitative basis, and the ability of EXAFS to provide coordination numbers despite the limited precision characteristic for the method is of immediate applicability. For example the EXAFS analysis of a dilute alloy $A_{90}B_{10}$ can reveal not only the bimetallic nature of the material (from the composition of the coordination sphere of B) and the approximate particle size (from the mean CN of A and B) but also the probable distribution of B, and therefore A, in the particle since if B has a significantly lower CN than A, it probably resides on the surface of the particle.

Thus the application of EXAFS analysis increases the precision with which structures can be determined in these systems, and when used in combination with other techniques it provides a powerful method of structural analysis. For example, from an investigation on a monometallic platinum colloid in which TEM and EXAFS analyses were combined, a lattice contraction was found for the 2.6 nm metal particles. [200]

6.3.3.4 The Structures and Compositions of Colloidal Metal Particles

What have we learned about the structure and composition of colloidal metal particles in the 1–10 nm size range from the above techniques? There is a consensus in those reports in which structure measurements were made that above a certain size, the colloid particles adopt the structure of the bulk phase. Thus palladium colloids with a 6.5 nm mean diameter were found to have the fcc structure found in bulk palladium. [113] Similar bulk-like structures were found for platinum colloids of varying sizes, [185, 200] and for colloidal platinum, palladium, rhodium, and iridium produced in reverse micelles. [151] In fact it seems that regardless of whether the colloids are prepared in aqueous or nonaqueous media, by chemical or physical means, a crystalline elemental structure can usually be identified in particles above *ca.* 2.5 nm. Below this size, at which the use of diffraction methods becomes problematic, there is evidence from reports of high resolution TEM lattice imaging investigations that crystalline particles as small as 1 nm can be observed. It should be kept in mind, however, that these studies are often records of virtuosic microscopic techniques rather than primarily structural studies on a colloidal system.

The question of structure at small size scales in metal particles can not be simply answered in a general manner. For example, a large particle of 5 nm diameter may be observed by TEM as a well facetted regular image, but closer inspection will often reveal multiple twinnings. This can be observed as the light and dark areas in the TEM image of a single particle as described in Section 6.3.3.1, or as in Figure **6-21** by a lattice image in which boundaries between crystalline domains are revealed as discontinuities in the lattice planes.

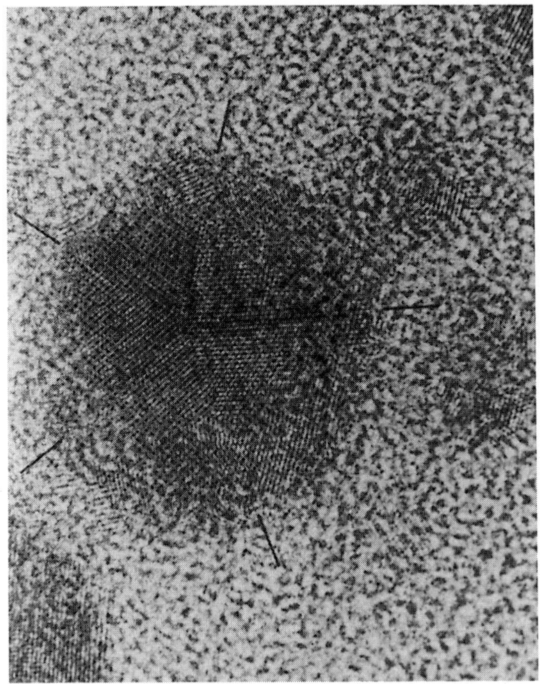

Figure 6-21. Atomic resolution transmission electron micrograph of multiply twinned colloidal Cu particle showing twin planes. (Reproduced by permission from ref. [41].)

In the smaller size range, below 4 nm, there have been several observations of structures which differ from that found in the bulk, specifically of particles with fivefold symmetry. Such a crystal cannot grow to a stable bulk lattice having a close packed structure by simple atom addition, and a number of proposals have been made to explain the transition from these nonbulk nanostructures to the bulk structures of the metals. [201] A growth mechanism has been suggested by which the fivefold symmetry crystallites could be formed, on the smallest scale, by tetrahedral capping of a tetrahedral core of four atoms. [71] Such a nucleus, in which the number of interatomic contacts are maximized for this nuclearity, would be expected to be a stable entity by analogy with the structures of a multitude of tetranuclear molecular clusters. The addition of atoms via the capping of triangular faces, starting with those of the tetrahedron, leads to a seven atom decahedron (a pentagonal bipyramid) as shown in Figure **6-22**. Subsequent addition of metal atoms can then proceed by either further capping the various triangular faces to form an icosahedron, or as a normal metallic packing to the continued development of the decahedron. Since fivefold symmetry cannot be preserved in an infinite space filled lattice, at some size the mismatch between a polytetrahedral core and an extended fcc lattice results in sufficient strain to cause the appearance of dislocations. The observation of multiple twinning as the crystal particle size increases supports this hypothesis. As shown above in Figure **6-18** where the presence of dislocations is evident, although the overall decahedral morphology is maintained, the crystal comprises no fewer than seven components in the multiply twinned particle. Such twinning has been observed for small particles of a number of metals, [201] prepared by metal evaporation and deposition, [202] as well as by colloidal methods. [71] It is often visible without the precision afforded by lattice imaging techniques, and is seen simply as light and dark areas of the same crystal since each twin, although crystalline and therefore diffracting, will be in a different orientation.

For most elemental metals, the phase diagram is extremely simple in the temperature range of interest to the synthetic inorganic chemist, and usually only one structure (fcc, bcc, or hcp) is accessible. The situation is more complex for bimetallic systems. At the first level of complexity, the question of miscibility arises. Some binary systems such as palladium-copper, show complete miscibility in the bulk over the whole range of compositions. These two metals are mutually soluble at all compositions from pure copper to pure palladium. Other binary systems are partially miscible, for example platinum-gold, where a miscibility gap exists between 98% Pt and 2% Pt. Still other binary systems are completely immiscible, for example ruthenium-copper. The question is whether or not the bulk phase diagram is valid at very small particle sizes. It might be thought that at least as far as composition and miscibility are concerned, the phase diagram would be a good guide to the compositions accessible for bimetallic colloids. This is probably generally true, although the influence of surface effects, which will be of determining importance for particles in the nanoscale size regime, might make it possible for immiscible metals to alloy in the smallest size particles. There is a report of a bimetallic colloid whose $Au_{50}Pt_{50}$ composition falls in the miscibility gap for these metals. [161] The synthesis of bimetallic nanoscale particles of immiscible metals is a worthy goal of synthetic metal colloid chemistry.

When we look beyond simple miscibility considerations, further complexity arises when the structural details of a phase diagram are considered. For example, the phase diagram of the completely miscible PdCu system is dominated above 600 °C by the α phase, a disordered fcc solid solution, and includes at lower temperatures two ordered phases at compositions near CuPd (bcc) and Cu_3Pd(fcc). Even more complexity appears as each of these phase regions is examined in detail. Again the question arises as to whether the bulk thermodynamic phase diagram is of any use in predicting the structures of bimetallic particles? The answer is once again a qualified yes, but the qualification this time is more weighty

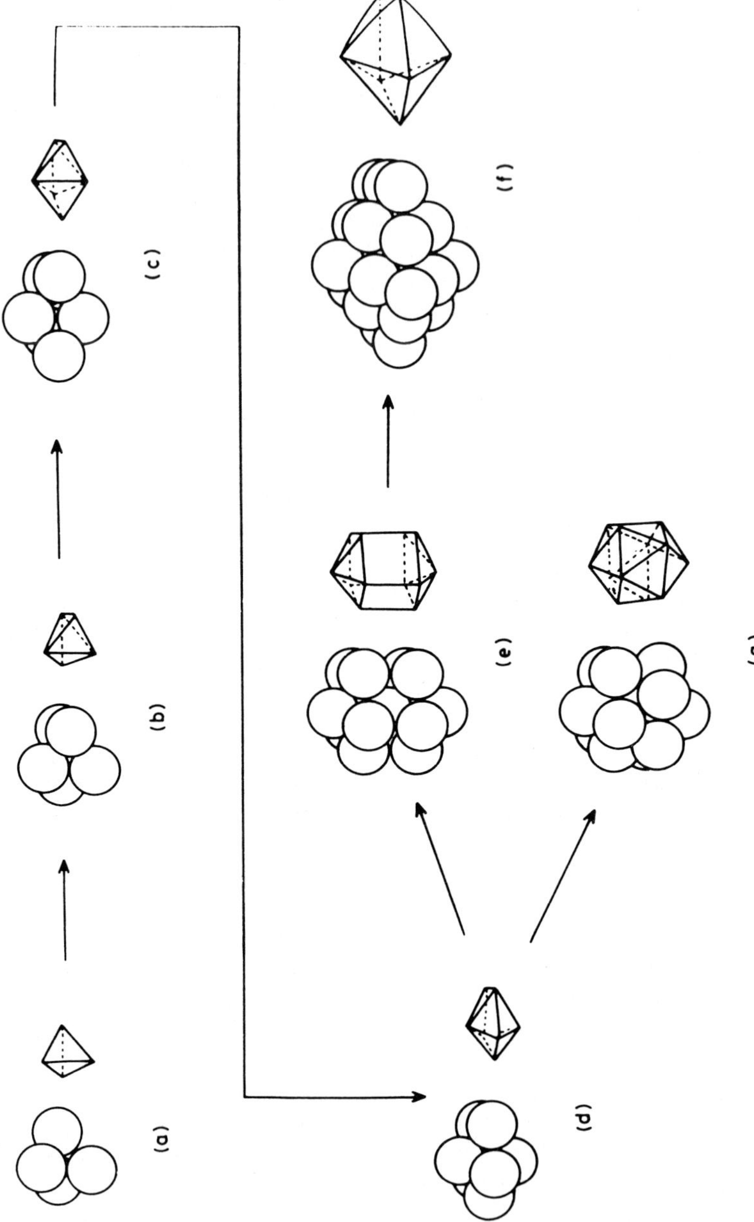

Figure 6-22. Growth sequence for decahedral metal nanocrystals based on the capping of triangular faces starting from the tetrahedron. (Reproduced by permission from ref. [71].)

and comes in the form of a recognition of the possibility of novel structures at very small sizes.

Even if the phase diagram is a valid guide for the probable structures in small particles, we run into additional problems even in a sample in which all the particles are of truly identical composition, as there is the issue of compositional variation across an individual particle. This problem arises from the influence of surface effects and takes the form of a segregation of one component over the other in the surface layer. This influence of surface proximity on composition in bulk alloys is a well studied field. [203] For an alloy of a given overall composition, one component is often found to be present in a higher concentration in a surface layer than in the bulk, whereas the next layer, the selvedge layer, is usually depleted in that component. Such oscillations in composition can extend further into the particle. The component which is enriched at the surface is that which has the lowest surface energy, but such a statement is useful in a predictive or interpretive manner only in a system for which the surface energies of the various components are known, which means under circumstances where real world complexities such as adsorbate covered surfaces are absent. The study of metal colloid particles, however, deals with surfaces which are far from clean or ideal. The surface of a particle which owes its stability to the presence of adsorbed ions or polymers in poorly defined coverages is by definition outside the realm of ideality. Nevertheless, this phenomenon of surface segregation, which can perturb an otherwise uniform composition across a bimetallic particle, will clearly be of great importance in determining the composition profile of metal particles. For particles with diameters of 1.0–3.0 nm (up to *ca.* 10 atomic diameters), most or all of the atoms in the particles are in regions which are subject to surface effects to a greater or lesser degree. This means that at this level of structural detail the questions of composition and structure for small metal colloids are not easily answered on the basis of comparison with bulk crystallographic data and phase diagrams, and so each system should be approached as a unique structural problem in its own right.

EXAFS analysis of colloidal metals has begun to shed light on this complex state of affairs. In the few examples reported so far for bimetallic colloids, a nonuniform distribution of metals has been observed. Toshima [164, 165] has studied a series of bimetallic PtPd catalysts, and concluded that the distribution of the two metals in the particles is nonuniform on the basis of differences in their respective mean coordination numbers. In our laboratory, colloidal 4.0 nm PdCu/PVP has been analyzed by this method and the distribution of the metals in the alloy particles shown to be nonuniform. However the component which is enriched at the surface appears to be palladium, in contrast to the segregation of copper to the surface of bulk copper-palladium alloys. [203] In addition, the ability of surface deposited copper to dissolve in a preformed palladium particle has been clearly demonstrated on the basis of the analysis of the coordination sphere of the copper. [204]

6.4 Spectroscopic Properties of Colloidal Metals

6.4.1 UV-Visible Spectroscopy

The first observations made on the properties of colloidal metals concerned their color, and the ruby glasses which contain inclusions of colloidal gold have already been referred to in this chapter. The color of gold sols, and the relationship of their color to particle size, was a preoccupation of Faraday and has continued to be the subject of research to the present day. Indeed, colloids of gold, silver, and copper all have characteristic colors, resulting from strong absorption in the visible region of the spectrum. For spherical particles of *ca.* 20 nm diameter, the absorption maxima in the UV-visible spectra of colloidal gold, silver, and copper occur near 520 nm, 385 nm, [205] and 560 nm [206, 207] respectively. These absorptions are a manifestation of the electronic structures of the metallic particles and are due to the excitation of plasmon resonances in the confined electron gas of the particles. These absorption spectra can be calculated using classical Mie theory, [208] in which the theoretical absorption spectrum of dilute spherical particles is related to their size and relative dielectric properties compared to the surrounding medium. The wavelength of the absorption maximum is also dependent on the size and shape of the particles and to their proximity to each other. The development of the absorption spectra with increasing particle size, and also with changing effective particle shape (for example the formation of strings of particles), has been the subject of several theoretical studies (see for example [205]). It has been calculated that the absorption spectrum for colloidal gold spheroids in water changes in a characteristic way with changes in the aspect ratio of the colloid particles. [20] As the ratio of the major to the minor axes of the particles, initially 10 nm spheres, is increased from 1 to 4, the absorption maximum splits into two peaks, one corresponding to the transverse resonance which remains at the wavelength of the spherical particles, and a second corresponding to the longitudinal resonance which shifts from 520 nm to *ca.* 770 nm. Importantly, a similar effect is even observed for spherical particles if they are allowed to approach each other to within less than one particle diameter and results in a similar splitting and shift of the longer wavelength absorption. These relationships allow the use of UV-visible absorption spectra, at least for those metals which show a pronounced absorption maximum in this region, for the estimation of changes in effective particle sizes if not for the estimation of the size itself in a given colloid. Thus the absorption spectrum for a citrate reduced gold hydrosol which is agglomerating under the influence of added pyridine (a micrograph of which is shown in Fig. **6-4**) shows a splitting in the absorption band and a distinct shift of the low frequency component to lower frequency as the effective particle size increases (Fig. **6-23b**). [209] Over a period of days, the color of the sol changes from red to blue. [20] A silver hydrosol, slowly agglomerating in the presence of pyridine, shows similar splitting and shifts (Fig. **6-23a**) whereby the color changes from yellow-brown to gray-brown. [209]

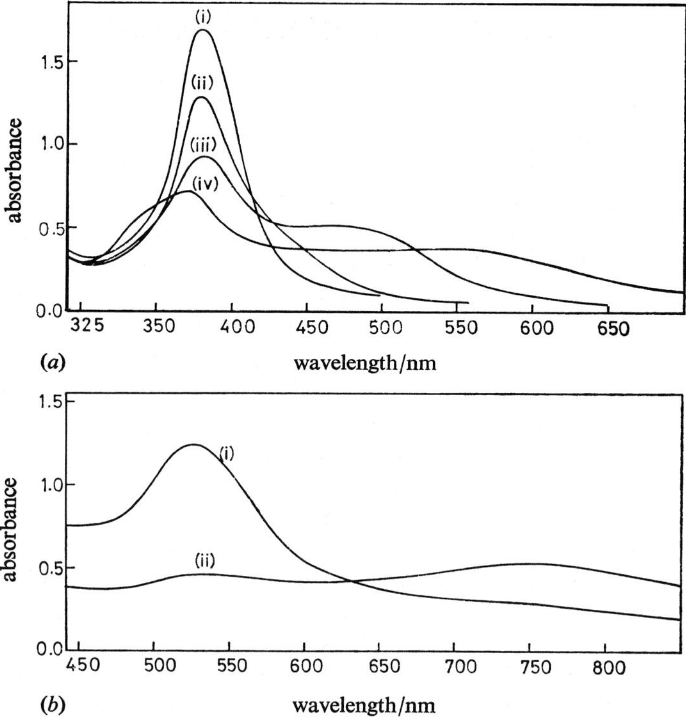

Figure 6-23. UV-visible spectra of a) a silver hydrosol and b) a gold hydrosol as they agglomerate under the influence of added pyridine. The composite spectra show the splitting of the plasmon absorption peaks and the shift to longer wavelength of the lower energy component with time. (i) As prepared, (ii) etc. after addition of pyridine. (Reproduced by permission from ref. [209].)

The surface plasmon band of a colloidal metal is sensitive to changes in the surface composition of the particle and to the presence of adsorbates. It has been observed that upon adsorption of I^-, SH^-, or $C_6H_5SH^-$ on colloidal silver, the surface plasmon band is severely damped, [210] and this is attributable to a change in the optical properties of the silver particle. Deposition of a second metal at the surface of a colloidal metal particle also has a marked effect on the UV-visible spectrum of the sol. The deposition of lead onto the surface of colloidal silver blue-shifts the plasmon absorption band from that of the pure silver sol at 380 nm, and after one or two monolayers of lead have been deposited the band is found at 340 nm. [91] Similarly the deposition of lead onto 18 nm colloidal gold [173] or of cadmium onto 4 nm colloidal silver [172] results in a shift of the plasmon absorption of the underlying metal to shorter wavelengths, and after the deposition of three monolayers, the plasmon band of the deposited metal is observed. Thus UV-visible spectroscopy provides a sensitive means of following the surface composition of colloidal metals in those cases where surface plasmon absorptions are found.

Many other colloidal metals also have absorbances in the UV-visible region of the spectrum which are similarly due to plasmon resonances or to interband transitions. Creighton has recently reported [211] the calculated UV-visible absorption spectra for fifty metallic elements as 10 nm hydrosols (and also as particles in the vacuum). In addition to copper, silver, and gold, colloids of the Group I and II metals as well as of scandium, yttrium, and tantalum show absorption bands in or near the visible region. All the other metals investigated show only broad featureless absorptions rising through the visible into the UV. This accounts for the relatively mundane colors of these metals in colloidal form, being usually brown or gray-brown in contrast to the striking colors exhibited by the colloidal coinage group metals.

6.4.2 Surface Enhanced Raman Spectroscopy (SERS)

It has been known for some time that the Raman scattering intensity of molecules is greatly enhanced when they are adsorbed onto the roughened surfaces of certain metals. For example, the Raman intensity of molecules adsorbed on electrochemically roughened silver electrodes [212] is enhanced by factors of up to 10^5. The magnitude of this enhancement is related to the roughness of the surface, that is to say, the state of division of the adsorbing surface.

Creighton and coworkers first demonstrated that these same metals in colloidal form will also exhibit this effect [209] in their study of pyridine adsorbed on gold and silver hydrosols. The addition of pyridine to a silver sol, prepared by borohydride reduction of silver nitrate in water, resulted in agglomeration of the initially spherical 20 nm particles, as described above. The initial yellow color of the sol, due to the plasma resonance at 380 nm, changed to red and then blue-gray as the particles aggregated into strings. In this aggregated state, irradiation in the yellow-green region of the spectrum gave strongly enhanced Raman scattering due to the adsorbed pyridine (Fig. **6-24a**). A similar effect was found for gold sols, as shown in Figure **6-24b**. The magnitude of the enhancement of the Raman intensity is wavelength dependent, and reaches a maximum at the wavelength of the plasma resonance for the agglomerated sol (*i. e.* the longitudinal resonance, shifted to the mid-visible region), thus establishing the connection between the SER effect and plasma excitation. [209]

Adsorbed molecules or ions formed during the various syntheses of colloidal metals have been observed by SERS. A silver hydrosol prepared by ferrous reduction of Ag^+ in the presence of citrate showed enhanced Raman intensity corresponding to adsorbed citrate, [213] and gold colloids prepared from $[Au(CN)_2]^-$ by reduction with borohydride showed SER bands due to adsorbed CN^-. [214] The Raman spectrum of $[Pt(CN)_4]^{2-}$ adsorbed on a 1.6 nm platinum colloid has been measured, [215] but it was concluded that the slight intensity increase observed (a factor of 7) for the CN mode at 2215 cm^{-1} might be due to other enhancement mechanisms. It has also been shown that ligands such as (*N*-4-dimethylaminoazobenzene-4′-sulfonyl)aspartate [216] adsorbed on colloidal silver are highly Raman active as a result of this phenomenon. Polymer films con-

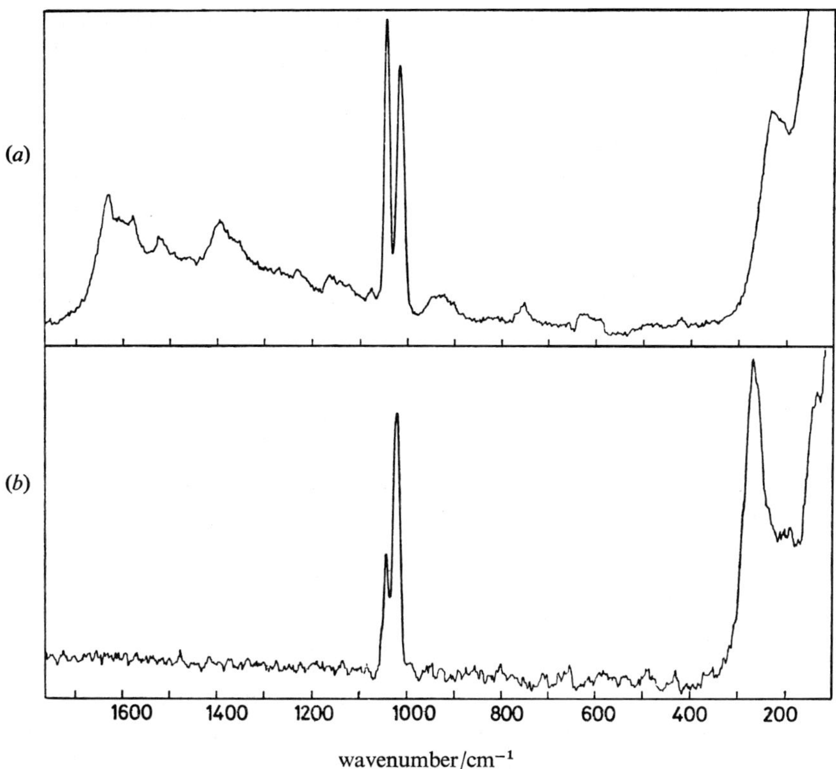

Figure 6-24. Raman spectrum of pyridine (approximately 0.005M) adsorbed from aqueous solution onto a) silver and b) gold hydrosols. Intensities of the two prominent bands near 1000 cm^{-1} are enhanced by factors of up to 100 over those for an aqueous pyridine solution. (Reproduced by permission from ref. [209].)

taining colloidal silver particles have been used as SERS sensors for CO_3^{2-} and SO_3^{2-} ions and for nucleic acid bases. [217]

SER effects have been observed for a number of adsorbates on copper hydrosols. Triphenylphosphine, diphenyl sulfide, and benzotriazole all partially displace adsorbed pyridine from colloidal copper and result in SER spectra which contain both pyridine bands and those of the coadsorbate. Diphenyl sulfide and thiophenol were found to displace pyridine. [207] A similar displacement of pyridine by thiophenol was observed with copper organosols. [218] These results exemplify the potential of this technique for investigating the surface coordination chemistry of colloidal metals. SER spectra of cytosine and of guanine and their derivatives on copper and on silver have been reported. [219, 220]

The potential for a broader application of this technique in colloid surface chemistry is limited to some degree by the dependence of the SER effect on plasmon resonance intensity. Many metals are eliminated as candidates for SERS since only very few exhibit the well defined plasmon resonance needed for this phenomenon to occur. Reports so far have concentrated overwhelmingly on the coinage metals, but it has been suggested on the basis of the calculated UV-visible absorption spectra for a wide variety of colloidal metals that several other metals merit investigation as SERS substrates. [211] The potential thus exists for the investigation of a variety of adsorbates on a number of metals. If SERS becomes more a tool in the surface chemistry of metal particles, we might anticipate being able to observe catalytic intermediates in working colloidal catalyst systems, define the surface environments in composite materials containing metal particles, and perform spectroscopic ligand analyses on ligand stabilized metal particles.

6.4.3 Infrared Spectroscopy

While Raman spectroscopy has been limited to relatively few systems in its application to the surface chemistry of metal particles, the complementary vibrational spectroscopic technique of infrared spectroscopy has been widely applied to the elucidation of the surface chemistry of small molecules adsorbed on metal surfaces in the solid state and to ligands in organometallic complexes. Of the various adsorbates which have been investigated, carbon monoxide is by far the most frequently studied. This is so for many reasons, not least among which are the facts that the intensities of CO vibrational bands increase greatly on adsorption at a metal surface or in a metal complex, and that the characteristic frequencies (1800–2100 cm^{-1}) of adsorbed or coordinated CO lie conveniently outside the range for other infrared chromophores. Thus an easily observed molecule (of some considerable catalytic importance) which has a clear spectroscopic window in the adsorbed state and also binds easily to both surfaces and metal complexes is available. Since a large data base exists for the vibrational spectra of CO in both surface chemistry and metal carbonyl chemistry, it serves as a useful probe molecule for the metal surfaces in single crystals or in the supported highly dispersed state.

The vibrational spectrum of carbon monoxide on metal surfaces has been studied from both theoretical and experimental spectroscopic perspectives. Although it is not necessary to review these results within the scope of this chapter, a summary of the most important aspects will serve as a guide for the application of the technique to colloidal metals.

6.4.3.1 Infrared Spectroscopy of CO on Metals

The surface of a metal crystallite offers a variety of binding sites for CO, as does a well defined single crystal surface. Adsorbed CO can be found in "on-top" sites (*i. e.* in a linear terminal binding mode) and twofold, threefold, or even fourfold bridging sites. On well defined single crystal surfaces, the adsorbed CO molecules are known to organize into ordered domains of various symmetries. The vibrational spectra of CO in such environments are determined by a complex interplay between the symmetry, coverage, dipolar coupling, and orientation of the CO molecules. [221]

While these phenomena define the components of the infrared spectra of adsorbed CO under the rarified circumstances of ultra high vacuum (UHV) surface physics, outside the surface physics community some approximations are usually made, based on some general observations derived from surface and molecular spectroscopy, in order to apply some useful correlations between binding geometry and vibrational spectra. These correlations are widely used in the study of supported metal particles, and are underpinned by the accumulation of three decades worth of infrared and structural data from molecular metal carbonyl chemistry. At the first (and perhaps most widely used) level of approximation, the frequencies of the vibrational transitions associated with the various binding modes of CO fall more or less into two groups identified with terminal and bridging sites. The linear terminal modes give rise to absorption bands above 2000 cm^{-1} and the various bridging modes to absorptions below this value with triply bridging CO absorbing at lower frequencies than doubly bridging CO. On ordered surfaces, the CO stretching frequencies, $\nu_{(CO)}$, increase with increasing coverage due to dipolar coupling between CO oscillators in ordered domains and islands. The assignment of specific infrared adsorption bands to the various bridging modes is not always clear cut, particularly when dealing with the relatively poorly defined binding environments for CO found on small metal crystallites supported on an amorphous oxide. In such cases inaccurate assignments can easily be made if these rules of thumb are too readily applied. However, heeding this caveat, the spectroscopic data base provides a solid resource for an investigation of CO adsorption on colloidal metals.

6.4.3.2 Infrared Spectroscopy of CO on Colloidal Metals

The infrared spectrum of CO on a platinum organosol was reported in 1986 by Lewis *et al.* in their investigation of the nature of some purportedly homogeneous organoplatinum olefin hydrosilylation catalysts. [60] The hitherto unsuspected presence of colloidal platinum in the calatytic "solutions" was confirmed by TEM, and carbon monoxide was found to adsorb onto the colloidal metal (mean particle size of 23 nm) in methylene chloride, giving rise to infrared absorptions at 2050 cm^{-1} and 1880 cm^{-1}. These bands were assigned to linear and doubly bridging CO.

Subsequently an extensive study of CO adsorption on platinum and palladium hydrosols was reported. [51, 222–226] Platinum hydrosols, prepared by the reduction of chloroplatinic acid with citrate or borohydride, adsorbed CO to give a broad infrared band centered at 2066 cm^{-1}, which rapidly broadened and shifted to lower frequency (2045 cm^{-1}) with time. [51] The reason for the shift was not determined, but possible effects from changes in surface charge or structure were suggested. Details of this system and of the analogous palladium hydrosol were subsequently elaborated on in investigations of the effects of pH, added cations and anions, and coadsorbates such as acetylene, hydrogen, and oxygen on the infrared spectra of adsorbed CO. The influence of pH for the platinum case [225] is shown in Figure **6-25** and shows a general decrease in the terminal $\nu_{(CO)}$ frequency from 2064 cm^{-1} to 2037 cm^{-1} as pH increases from acid to alkaline. For the palladium sol, the value of $\nu_{(CO)}$ (1950 cm^{-1}) for the bridged CO (the terminal CO stretch being too weak to measure precisely) was insensitive to increasing alkalinity up to pH > 11.5, at which point it decreases rapidly to 1900 cm^{-1}. These effects were interpreted in terms of hydroxide adsorption and a consequent reduction in CO coverage at high pH for the platinum sols with a concomitant decrease in vibrational frequency dipolar coupling. The absence of a change in frequency for the bridged CO on palladium sols was interpreted as being due to the formation of large islands of adsorbed CO, reduction in the size of which would not result in significant reduction in dipolar coupling and thus frequency. A similar dependence of $\nu_{(CO)}$ on the concentration of coadsorbed molecules such as hydrogen and acetylene was observed on colloidal platinum. [226] The addition of various anions resulted either in a competitive adsorption of the anion (for CN^- and SH^-) which led to a decreased or zero CO coverage, or in changes due to pH in the cases of PO_4^{3-} or CO_3^{2-}.

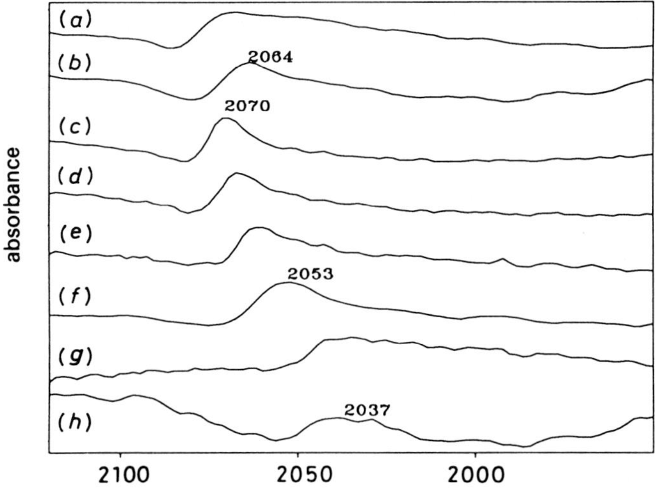

Figure 6-25. Infrared spectra of CO on a Pt hydrosol, showing the effect of changing pH. a) 2.41, b) 3.81, c) 4.86, d) 7.31, e) 9.15, f) 10.75, g) 11.66, h) 12.04. (Reproduced by permission from ref. [225].)

Particle size and structure can have a pronounced effect on the infrared spectra of adsorbed CO. A series of palladium colloids of differing sizes provided the basis for an analysis of this effect. [34, 113, 227] The PVP stabilized palladium colloids were prepared by the reduction of palladium acetate in methanol solution or by the reductive decomposition of bis(dibenzylideneacetone) palladium in the presence of PVP (see Section 6.2). The resulting colloidal palladium particles

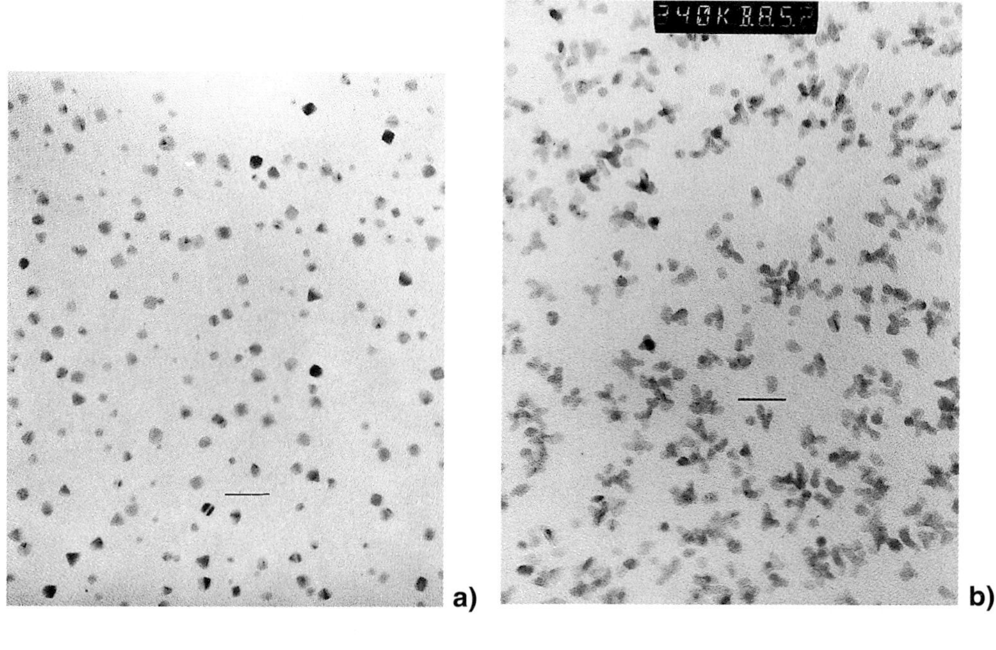

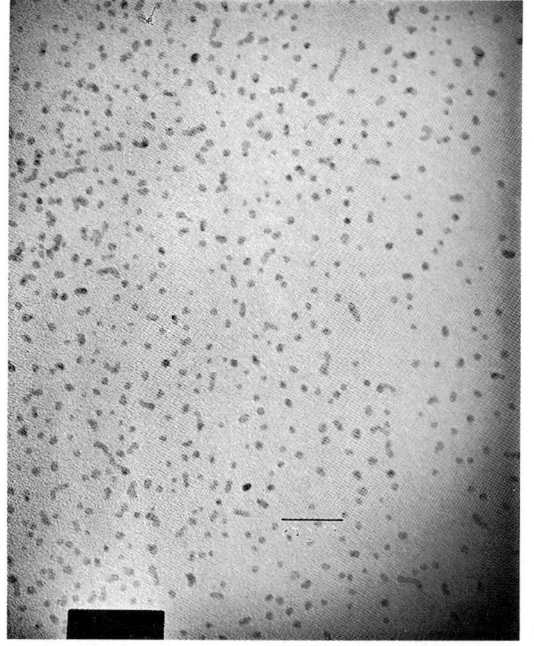

Figure 6-26. Transmission electron micrographs of three Pd/PVP colloids of differing sizes. a) 6.0 nm; b) 4.0 nm; c) 2.5 nm. (Reproduced by permission from ref. [34].)

had diameters of *ca.* 6.0 nm, *ca.* 4.0 nm, and *ca.* 2.5 nm (by TEM) as shown in Figure **6-26**, and were shown to be crystalline (fcc) palladium by electron diffraction. Carbon monoxide adsorbed on the colloidal metal was shown to bond in both terminal and bridged modes for each particle size. [34] Solution infrared spectroscopy showed terminal $\nu_{(CO)}$ at *ca.* 2050 cm^{-1} and bridging $\nu_{(CO)}$ at *ca.* 1935 cm^{-1} as shown in Figure **6-27c–e**. Small changes (± 5 cm^{-1}) in these frequencies were judged to be uninterpretable, but there was a significant size dependence for the relative occupancies of terminal and bridging CO, with the smallest particles giving the highest ratio of linear ("on top") CO to bridging CO, and the largest particles showing only bridging CO. Smaller (and probably noncrystalline) palladium sols, prepared from palladium vapor, showed a preponderance of terminal CO (Fig. **6-27a, b**), a highly unusual observation for CO on palladium. [113, 227] These results were interpreted on the basis of a preference of linear terminal CO for the palladium atoms at the edges and vertices of the colloidal metal particles, with bridged CO occupying pairs of palladium atoms on the faces. The three PVP stabilized sols all contained crystalline metal particles, and as the particle size decreased, the relative proportion of edge and vertex sites to face sites would increase. [198] The dominance of terminal CO, that is, on edge and vertex bonding sites, in the cases of the palladium vapor derived sols was seen as evidence for the irregular surfaces of these probably amorphous par-

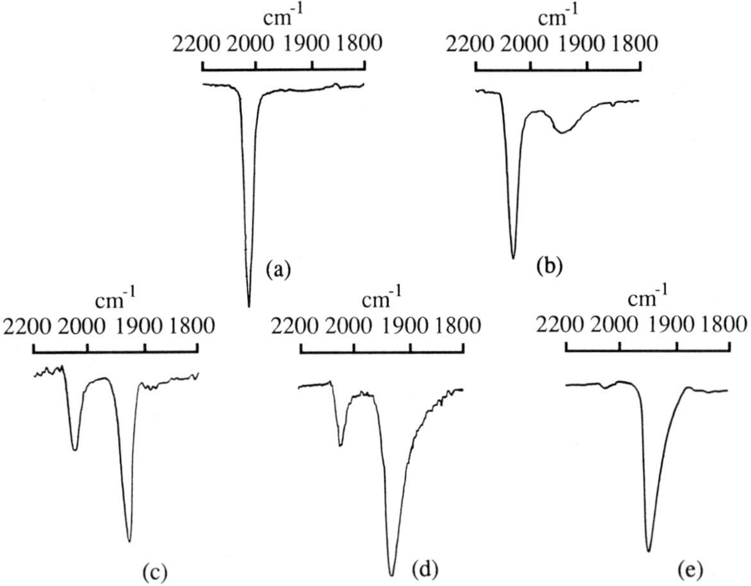

Figure 6-27. Infrared spectra of CO on colloidal Pd showing the dependence of bridging to terminal CO stretching intensities on particle size and surface structure. a) 1.0 nm, b) 1.8 nm, c) 2.5 nm, d) 4.0 nm, e) 6.0 nm. Spectra a) and b) are noncrystalline Pd colloids prepared from Pd vapor, c)–e) are crystalline Pd/PVP colloids as shown in Figure **6-26**. (Adapted from refs. [34, 113, 227].)

ticles. A similar relationship between CO binding mode and palladium crystallite size has been observed for silica supported palladium. [228]

The infrared spectrum of CO adsorbed on a vapor derived <1.0 nm platinum colloid stabilized by *iso*-butylaluminoxane in methylcyclohexane showed only a linear CO stretch at 2035 cm^{-1}. This particle size is within the range for the larger molecular platinum carbonyl cluster anions, and the colloidal platinum can in fact be hydrolysed to mixtures of $[Pt_9(CO)_{18}]^{2-}$ and $[Pt_{12}(CO)_{24}]^{2-}$ after CO adsorption. [229]

The infrared spectra of CO on colloidal platinum, palladium, and ruthenium stabilized with either cellulose acetate or nitrocellulose have been reported. [23] Although the spectra show an interesting size dependence in the case of ruthenium in nitrocellulose, interpretation is difficult since the spectra differ markedly from those reported for CO on supported ruthenium. For the platinum sols, a preference for linear CO coordination is observed, as is the case for the supported metal, while for palladium the bridging mode is preferred exclusively.

Adsorbed carbon monoxide can serve as a useful infrared probe of surface composition in bimetallic colloids if both metals bind CO. This is exemplified in the infrared spectrum of CO on a PVP stabilized colloidal alloy $Cu_{63}Pd_{37}$. [38] Carbon monoxide adsorbs readily onto these PdCu particles (*ca.* 45 Å) in dichloromethane at 25 °C, as shown by the infrared absorption spectrum in Figure **6-28**. By comparing this to the IR spectrum of CO on a pure palladium colloid of similar size [34] in Figure **6-27d**, it can be clearly seen that CO occupies both palladium and copper sites. Whereas the bands at 2046 cm^{-1} and 1936 cm^{-1} are in the frequency ranges found for linear and bridged CO on the pure palladium particles, the new band at 2089 cm^{-1} corresponds to CO on surface copper atoms, thus demonstrating that both metals are present at the surfaces of the particles.

These examples illustrate the amenability of colloidal metals to investigation by a spectroscopic method common to both molecular and surface chemistry.

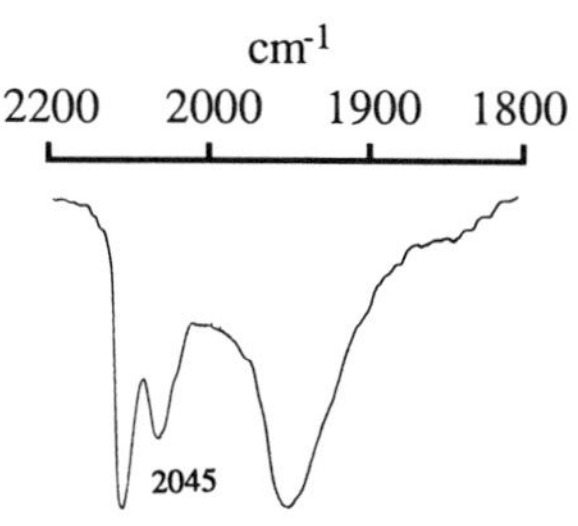

Figure 6-28. Infrared spectrum of CO on PVP stabilized colloidal $Cu_{63}Pd_{37}$. The bands at 2046 cm^{-1} and 1936 cm^{-1} are assigned to linear and bridged CO on palladium atoms, and the band at 2089 cm^{-1} corresponds to CO on surface copper atoms. (Adapted from ref. [38].)

6.4.4 NMR Spectroscopy

Two aspects of the NMR properties of colloidal metal particles will be dealt with here. One is related to the properties of the metal particle itself, and one is concerned with the NMR observation of molecules adsorbed at the colloidal metal surface. A brief summary of NMR studies on small metal particles in the solid state will serve as an introduction to both.

6.4.4.1 NMR Spectroscopy of Small Metal Particles

The resonant frequency of a metal nucleus in the bulk metal differs from that of the metal in a diamagnetic compound such as a complex or ionic solid. [230] This shift, the Knight shift, [231] is a consequence of the interaction of the metal nucleus with the conduction electrons, and is measured as a fraction of the applied field. The values of the Knight shift for the metallic elements span quite a large range [232] when it is recalled that NMR shifts in diamagnetic materials are measured in ppm. For example, the Knight shifts for ^{195}Pt and ^{105}Pd are –3.533 % and –3.0 % respectively, and those for ^{63}Cu and ^{197}Au are +0.237 % and +1.4 % respectively. [230] This means that in a field in which a standard such as $[PtI_6]^{2-}$ resonates at 100 MHz, the ^{195}Pt resonance for metallic platinum is shifted by about 3.533 MHz. Thus these shifts are a sensitive reflection of the metallic properties of the material in which the resonating metal nucleus is found. In solid state physics terminology, this "metallic nature" of the metal is expressed as the density of states at the Fermi level ("conduction electron density") and depends on the coordination number of the metal atom. Thus, surface metal atoms should show a different Knight shift than metal atoms in the interior of a metal particle, and at small enough particle sizes the two resonances should both be observable. In an example of this phenomenon, the ^{195}Pt NMR shifts of supported platinum catalyst particles were analyzed [233, 234] in terms of resolvable surface and bulk resonances.

The measurement of the NMR shift of a metal in small particulate form addresses quite directly the important question of at what size a metal particle begins to develop metallic properties. Particles of copper (less than 8 nm) have been observed to have a diminished ^{63}Cu NMR Knight shift at low temperature [235] as a result of quantum size effects (*i. e.* the loss of the metallic band structure). This will have a clear relevance to the study of colloidal metal particles.

The application of NMR to the study of molecules adsorbed on the surface of small metal particles has also been an area of considerable activity in the field of heterogeneous catalysis and is of obvious relevance to studies on colloidal metal particles. Among the adsorbates which have been studied are CO, [236–239] ethylene, [240, 241] and isoprene [242] on various highly dispersed metal catalysts. The type of metal has a strong influence on the NMR characteristics of the adsorbate, and strongly shifted resonances are common with metals which have a large Knight shift, such as platinum and palladium, due to the interaction of the adsorbate nuclei with the conduction electron density at the metal surface. Some metals such as ruthenium [237] impart negligible shifts. Although these shifts are orders of magnitude less than the Knight shift of the metal itself, some of the largest ^{13}C shifts known have been reported, for example, in the case of CO adsorbed on platinum [236] and palladium. [239] Resonances in static samples are necessarily broad, but Magic Angle Spinning (MAS) can give relatively sharp resonances [238, 239] due to the averaging of chemical shift anisotropies in the solid sample.

6.4.4.2 NMR Spectroscopy on Colloidal Metals

Although the application of NMR spectroscopy to metal colloid chemistry is only beginning and few examples of its use are known, there are some compelling reasons why the technique should be further applied to the study of colloidal metals. Since high resolution NMR relies on the use of stable homogeneous magnetic fields for its effectiveness, it is axiomatic among molecular chemists that the presence of colloidal metal in a liquid NMR sample destroys any hope of deriving useful data due to the magnetic inhomogeneity introduced by the metal particles. However, if a comparison with solid state NMR is made instead, the prospects seem much brighter. Enormous advances have been made in that field, and although the 0.1 Hz resonances of molecular high resolution spectra are not observed, a wealth of structural and physical information is available based on the relatively low (compared to liquids spectra) resolution spectra obtained in the solid state. This is especially so since the advent of MAS techniques (see above) which give relatively narrow linewidths for solid samples, and much structural information has been derived from these spectra. When seen from this perspective, the potential of liquid state NMR in application to colloidal metal solutions is great. In analyzing either surface species or bulk properties of small metal particles by NMR, the relatively free motion of the colloidal particles in the diluting liquid, analogous to freely tumbling molecules in true solutions, provides a mechanism for averaging the chemical shift anisotropy which would otherwise broaden resonances by hundreds of ppm in a static sample. This effect of Brownian motion on NMR linewidths has been estimated by calculating correlation times for the motion of colloidal particles [243] using the Stokes-Einstein equation. For particles of less than 10 nm in diameter correlation times of microseconds are found, and so only a very small contribution to the NMR linewidth results from restricted tumbling. Thus relatively narrow resonances might be expected. High-resolution NMR spectra of colloidal inorganic solids, and of species adsorbed on them, have in fact been obtained by using a conventional high-resolution NMR spectrometer. [243] The ^{31}P NMR spectrum of a colloidal suspension of calcium hydroxyapatite showed the expected line narrowing due to rotational tumbling, allowing an accurate isotropic chemical shift to be measured for the solid. Similarly, the ^{31}P NMR of a diphosphonate ligand chemisorbed on the hydroxyapatite surface was also observed in the same study. High-resolution ^{27}Al NMR spectra have been reported for colloidal suspensions of 3–10 nm aluminum trifluoride, with linewidths of several ppm being observed due to motional narrowing caused by Brownian motion of the colloidal particles in the liquid diluent. [244, 245]

These results suggest that colloidal metals could be excellent subjects for studying the NMR characteristics of small metal particles. This was recently proposed by Newmark *et al.* [246] in a report on the observation of the ^{195}Pt NMR spectrum of a platinum hydrosol. The colloid in this study was prepared by borohydride reduction of H_2PtCl_6 and had an estimated particle size of 1.6 nm. The spectrum is shown in Figure **6-29a** and shows the results of motional narrowing. The resonance is relatively sharp and symmetrical when compared to the static

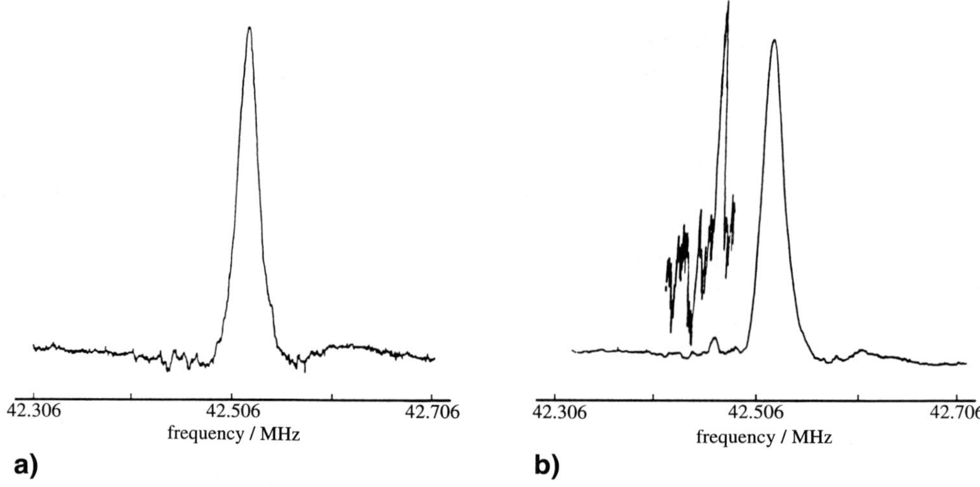

Figure 6-29. ^{195}Pt NMR of a 1.6 nm platinum hydrosol: a) as prepared and b) after addition of CO. Resonances due to the surface platinum atoms are resolved after treatment with CO. (Reproduced by permission from ref. [246].)

NMR resonance of solid platinum metal. The chemical shift of 42.5 MHz lies between those of platinum metal (41.5 MHz) and H_2PtCl_6 (43.0 MHz), and is interpreted as being a consequence of the submetallic nature of the 1.6 nm platinum particles. Thus the size issue for the transition from metallic properties to molecular properties is introduced. The symmetry of the band is surprising since more than 50 % of the platinum atoms are surface atoms at this particle size and might be expected to be resolved from the resonances for the internal platinum atoms. [234, 236] The surface atoms can, in fact, be resolved after the addition of CO to the colloid, which results in the appearance of a low intensity resonance at 42.46 MHz (Fig. **6-29b**), that is, up field (lower frequency) of the main band. This signal is assigned to the surface platinum atoms as the adsorption sites for CO. Displacement of CO by iodide broadens this resonance through the interaction with the quadrupolar iodide nucleus.

The observation of NMR resonances for molecules adsorbed on small metal particles, as opposed to resonances of the metal particles themselves, also benefits from the mobility the of colloidal particles in liquids. The ^{13}C NMR of ^{13}CO adsorbed on colloidal palladium has been reported for several systems. [33, 113, 227] When CO is adsorbed on colloidal palladium which is known to be metallic on the basis of its TEM images and X-ray diffraction pattern, it is to be expected that the resonance for the adsorbed ^{13}CO would be greatly shifted to low field, consistent with the large negative Knight shift of the metal.

In a liquid state high-resolution NMR study of adsorption on colloidal palladium, [33] it was found that under 3 atm. of 99 % ^{13}CO, the resonance which would correspond to carbon monoxide adsorbed on the 7.0 nm crystalline palladium colloid stabilized in methanol solution with PVP could not be directly observed even as far downfield as 1400 ppm. The spectrum in the 0–195 ppm range is shown in Figure **6-30** in which the feature of importance is the broad band at 185 ppm due to dissolved free ^{13}CO. The linewidth of this resonance is temperature dependent, showing reversible broadening with increasing temperature due to chemical exchange with ^{13}CO adsorbed on the colloidal metal. The resonance due to the adsorbed ^{13}CO was located by repeated application of a low power, long duration pulse over a range of frequencies from 1400 ppm to 200 ppm in 25 ppm increments, the aim of which was to saturate the ^{13}C nuclei of the ^{13}CO in the adsorbed state. The free ^{13}CO resonance at 185.5 ppm showed detectable saturation for pulses between 1100 and 500 ppm, and the maximum saturation transfer was found at 800 ppm. A plot of saturation transfer *vs.* saturation pulse frequency is shown in Figure **6-31** and suggests that the resonance corresponding to adsorbed CO has an intensity maximum at 800 ppm. This is the largest shift reported for adsorbed CO, and is consistent with the metallic nature of these large palladium colloid particles. That the maximum found in the saturation transfer curve actually corresponds to the true resonance maximum was confirmed by the observation of a band centered at 790 ppm in the solid state MAS ^{13}C spectrum of the same colloid saturated with ^{13}CO in a solid PVP matrix. [247] The value of this shift is greater than that reported for an analogous supported palladium sample containing smaller particles, consistent with the intuitive relationship between particle size and "metallic properties" (*i. e.* the development of conduction bands in the metal particle). The spin saturation transfer experiment, when carried out on ^{13}CO adsorbed on an analogous but smaller crystalline palladium colloid (2.3 nm) gave a saturation transfer maximum at *ca.*

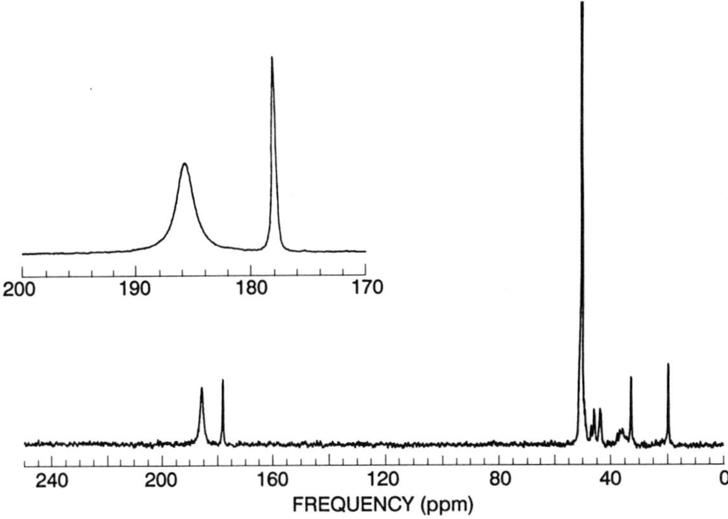

Figure 6-30. 75 MHz ^{13}C NMR of ^{13}CO in the presence of a 7.0 nm Pd/PVP colloid in methanol showing the broad resonance at 185 ppm due to dissolved ^{13}CO and the carbonyl resonance of PVP at 178 ppm. (Adapted from ref. [33].)

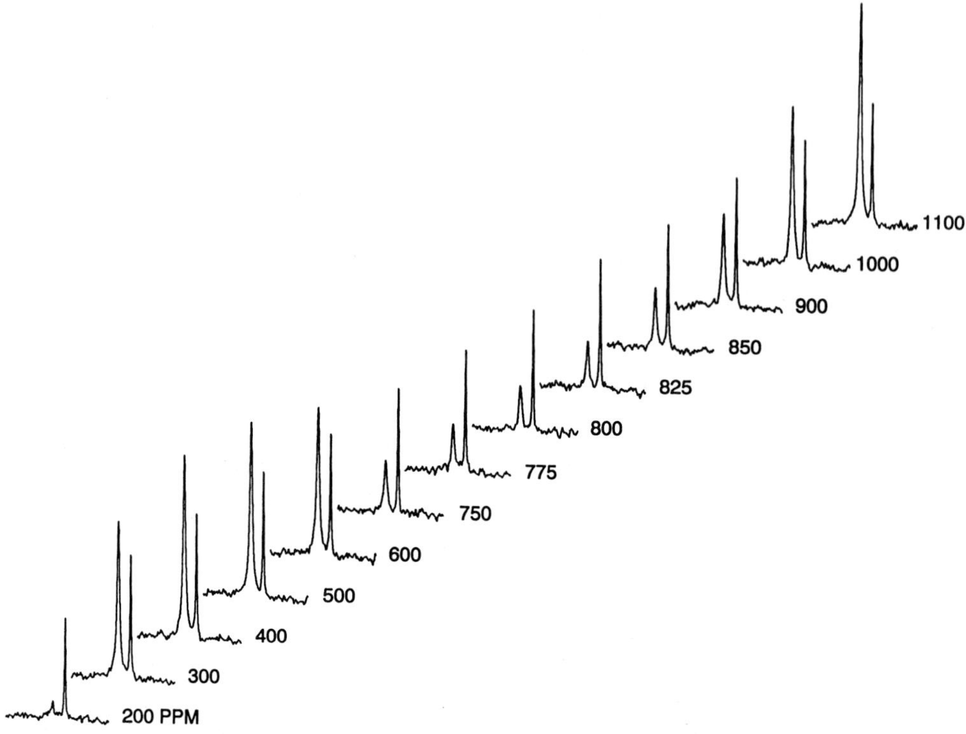

Figure 6-31. 75 MHz ^{13}C NMR of CO/Pd/PVP solution from Figure **6-30** showing the progressive saturation of the resonance due to free ^{13}CO by irradiation in the chemical shift range from 1100 to 200 ppm. Spin saturation transfer from colloid-adsorbed ^{13}CO reaches a maximum at 800 ppm. (Adapted from ref. [33].)

700 ppm, but with much greater intensity on the high field side of the maximum [113] as shown in Figure **6-32**. This is to be expected since the smaller palladium particles must have a higher proportion of edge and vertex atoms with low metal–metal coordination numbers (as shown by the relative intensities of the terminal and bridged CO bands in the infrared spectra CO of adsorbed on these particles, discussed in Section 6.4.3.2). The conduction band density of states is lower at these sites and so the metal atoms, together with their bound CO, are expected to be less Knight shifted. [248]

This structural effect on the chemical shifts of adsorbed molecules is further demonstrated in the NMR spectra of ^{13}CO on some palladium colloids derived from palladium vapor (see Section 6.2.2.4). If these are amorphous, as suspected, then their metallic properties (conduction bands) which give rise to the Knight shift should therefore not be as well developed. They adsorb CO in a predominantly terminal mode implying a disordered or rough surface (see Section 6.4.3.2), and thus might be expected to be less “metallic” in the context of this technique. In fact, the ^{13}C resonance for ^{13}CO adsorbed on such vapor derived

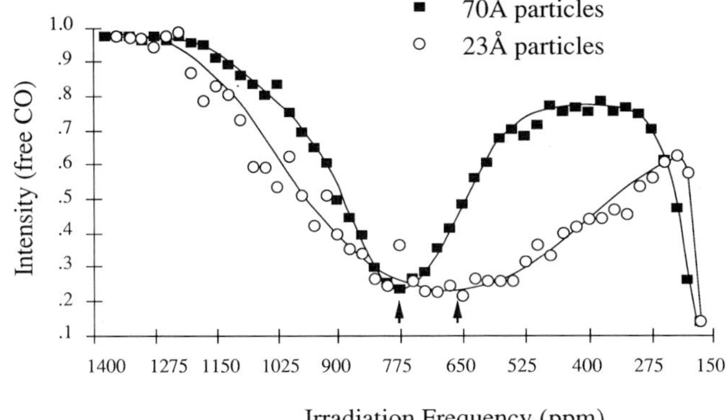

Figure 6-32. A comparison of ^{13}C NMR spin saturation transfer vs. irradiation frequency for ^{13}CO exchanging with adsorbed ^{13}CO in the presence of 7.0 nm and 2.3 nm colloidal palladium/PVP in methanol. The smaller metal particles show a greater saturation transfer at higher fields consistent with a higher incidence of low coordination number surface metal atoms (less Knight shifted) in the smaller particles. (Adapted from ref. [113].)

colloids (particle sizes 1.8 nm and 1.0 nm) shows no significant shift from the diamagnetic value of *ca.* 185 ppm. [113, 227] In the case of colloidal platinum prepared by the same method, adsorbed CO again shows no shift from the diamagnetic range, despite the very large Knight shift for the metal itself. [229] This comparison of results from crystalline and noncrystalline palladium colloids emphasizes that NMR can be used as a tool for studying metal to nonmetal transitions, not only by measuring the metal Knight shifts but also by observing the shifts associated with the adsorption of small molecules.

When ^{13}CO is adsorbed on particles of a metal having a negligible Knight shift, resonances for the various bonding modes of CO can be observed directly. In the solid state terminal and bridged CO have been resolved on small ruthenium catalyst particles supported on silica and alumina. [238] The analogous colloid experiment also revealed several binding sites for CO on a PVP stabilized ruthenium colloid prepared by hydrogenation of the zerovalent ruthenium complex Ru(cyclooctadiene)(cyclooctatriene) in methylene chloride solution (see Section 6.2.2.3). The ^{13}C NMR spectrum of ^{13}CO on this colloid is shown in Figure **6-33**, and the adsorbed CO resonances are clearly visible as broad bands at 195, 198, 242, and 286 ppm. [113] Each has a T_1 much shorter than free CO (< 0.75 s compared to > 5 s). By comparison with both the analogous solid state spectrum [238] and ^{13}C NMR data for ruthenium cluster complexes, these resonances are assigned to two terminal CO sites (195 and 198 ppm), a doubly bridging CO (242 ppm), and an as yet unassigned resonance (286 ppm) due to either a triply bridging CO or another form of carbon such as methylidene, possibly formed from a reaction of CO with surface hydrogen remaining from the colloid preparation.

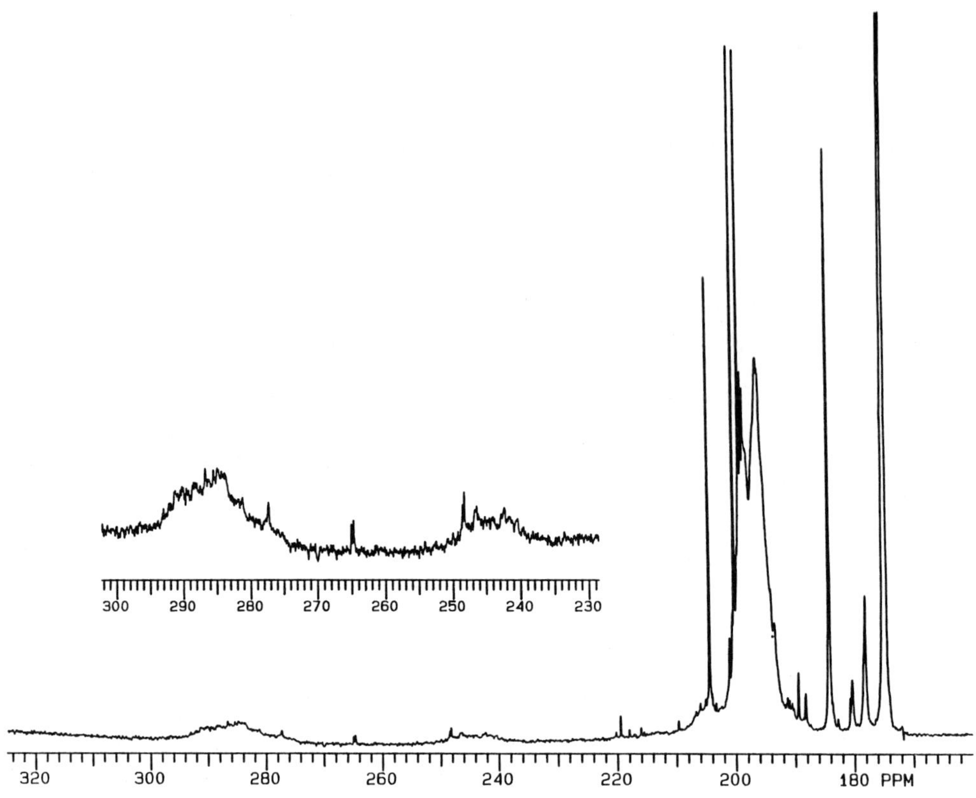

Figure 6-33. ^{13}CO NMR of ^{13}CO on colloidal Ru/PVP showing resonances assigned to terminal CO (195, 198 ppm), doubly bridging CO (242 ppm), and an unassigned resonance (286 ppm) possibly due to a triply bridging CO or another form of carbon such as methylidene. (Adapted from ref. [113].)

6.4.5 Summary of Spectroscopic Analyses of Colloidal Metals

The results described in this section show that spectroscopy in the liquid state can be applied to the analysis of both colloidal metal particles and species adsorbed on colloidal metals in a manner reminiscent of the spectroscopic investigations of molecular compounds in solution. There are established infrared and NMR databases from the molecular and solid state literature on which to base analyses of colloid spectra. The NMR data presented suggest that the study of metal particles in the important size range where transitions from the molecular to the metallic state take place will be greatly facilitated by this method. In addition, the use of NMR in observing the adsorption of small molecules promises to open the way for the development of the surface chemistry of nanoscale colloidal metal particles.

6.5 Catalysis

The extensively documented catalytic activity of small metal particles in a wide variety of reactions leaves it as surprise that colloidal metals can be catalytically active as well. The ability of metal surfaces to catalyze chemical reactions has been known since the end of the last century, and the place of colloidal transition metals in this field was established early in the history of catalysis – the vigorous catalytic decomposition of hydrogen peroxide by colloidal platinum prepared by the electric arc method was remarked upon by Bredig in 1899. This capability of colloidal metals can even become something of a problem, and the catalytic activity of adventitious colloidal metal in ostensibly homogeneous catalytic reactions has recently become an interesting (and overdue) concern in homogeneous catalysis. [249] In one case, in an implicit tribute to the reputation of metal colloids as catalysts, the dehydrogenation of aliphatic hydrocarbons by organometallic iridium complexes was considered so remarkable that the presence of colloidal metal was suspected, and considerable lengths were gone to in order to establish criteria by which it could be demonstrated that the reaction was indeed a molecularly catalyzed homogeneous process. [250]

Before introducing the areas to be covered in this section, some exclusions should be mentioned. The use of colloidal metals in such technologically important processes as electroless deposition and photography, although both involve the catalytic action of colloidal metals, is long established and will not be reviewed here. Heterogeneous supported metal catalysts are sometimes referred to as colloidal particles, [6] but in this chapter the term will be used to include only those metal particles which are synthesized and used as colloids in the liquid phase. An exception to this will be those heterogeneous catalysts which are deliberately prepared by adsorption or grafting of previously prepared colloidal metal particles onto solid supports when some unique catalyst property results from the colloidal route. Metal colloids are also implicated in the Solvated Metal Atom Dispersion method used to prepare of a number of catalytic metals from metal vapors. In this method metal atoms are condensed with the vapors of organic liquids such as toluene to give thermally labile solvated atom dispersions. Upon warming in the presence of a support, highly dispersed supported metal catalysts are produced. The preparation and use of these materials, which are best seen as novel heterogeneous catalysts, has been reviewed recently by their chief proponent. [251] Although they may be formed *via* a colloidal stage, it is not well defined and so the materials will not be reviewed here.

The catalytic uses of colloidal metals which will be described are found in a relatively small but important group of chemical reactions. These include hydrogenation, hydrosilylation, and hydration reactions of unsaturated organic molecules, as well as redox and other electron transfer processes such as the photochemical splitting of water and photocatalytic hydrogenations.

6.5.1 Novel Colloid Based Heterogeneous Catalysts

Unprotected, that is to say electrostatically stabilized, colloids of the precious metals can be attached to supports, presumably through electrostatic interactions, and there are numerous examples of heterogeneous catalysts prepared in this manner. Although the use of colloids as heterogeneous catalyst precursors will not be covered in detail here, some results should be mentioned.

One way in which the colloid routes can lead to unique catalysts is by exploiting the fine control over particle composition and microstructure which is possible in colloid synthesis prior to supporting the catalyst. An example of this can be seen in the preparation of three palladium-gold catalysts, each with a different microstructure. [168] By sequential reduction or coreduction of salts of the metals, particles with gold-on-palladium, palladium-on gold, and alloy structures can be prepared and the resulting colloid particles supported on carbon without inducing changes in the microstructures. Such control would be difficult in a traditional impregnation/reduction catalyst preparation.

Hydrogenation catalysts based on carbon supported cationic surfactant stabilized colloids of rhodium have been compared with conventional rhodium catalysts. [58, 59] These colloid based catalysts, prepared by reduction of metal salts with long chain tetraalkylammonium borohydride, as described in Section 6.2.2.1, were found to be significantly more active in the hydrogenation of butyronitrile than analogous rhodium catalysts prepared by conventional precipitation and molecular cluster precursor routes. The almost general applicability of the preparative method to the formation of very highly dispersed metal particles [62, 197] in a soluble form, which lends itself to simple addition to support materials, suggests that this could be a way of producing high loadings of highly dispersed catalysts of superior activity.

After preparation by whatever method allows for the isolation of a preferred catalyst (chosen by composition, particle size, etc.), colloidal metals can be grafted onto polymer resins under mild conditions. Colloidal platinum (3.0 nm) can be prepared by photoreduction of $[PtCl_4]^{2-}$ in solutions of a copolymer of *N*-vinylpyrrolidone and acrylamide. When exposed to a polyacrylamide gel having free aminoethyl groups under mildly acidic conditions, the colloid particles are covalently grafted to the gel. [252] Similarly, rhodium colloids (5.0 nm) protected by a copolymer of vinylpyrrolidone and methyl acrylate can be irreversibly grafted to an acrylamide gel having free aminoethyl groups over a range of pH's. [253] The immobilization mechanism was investigated in the case of a poly(vinylpyrrolidone-methyl acrylate) protected platinum sol immobilized on an aminoethyl acrylamide gel, and found to involve aminolysis of the methyl ester. As slurries in ethanol, the grafted colloid catalysts are excellent hydrogenation catalysts. For example, the rhodium catalyst is over twenty times more active for ethyl vinyl ether hydrogenation than a commercial Rh/C catalyst. That the activity is slightly lower than for the sols before immobilization implies that there are some diffusion limitations of the substrate through the polymer, and so this effect could be exploited as a selectivity enhancing tactic.

PVP stabilized rhodium or platinum sols (3.5 nm) have been grafted onto mercapto-derivatized silica. [254] The S-H groups ligate strongly to the metal particles. In this case the grafted catalyts are one to two orders of magnitude *more* active for the hydrogenation of alkenes than the unsupported colloidal catalysts in the same solvent system, and 300 times more active than commercial Rh/C and Pt/C catalysts.

6.5.2 Liquid Phase Hydrogenation of Unsaturated Organic Molecules

The use of polymer protected colloidal precious metals as hydrogenation catalysts has been known for half a century. Hydrosols of palladium, platinum, [46–48] rhodium, [49] and iridium [50] stabilized with poly(vinyl alcohol) and poly(vinyl acetate) were used as hydrogenation catalysts for a number of unsaturated organic substrates such as nitrobenzene and olefins. The colloids were prepared by hydrogen reduction of the metal hydroxides and contained relatively large particles (*ca.* 90 Å after centrifugation). At the beginning of his extensive work on colloidal transition metal catalysts, [27, 32, 54, 255–260] Hirai reported that the use of aqueous methanol as a reducing agent for $RhCl_3 \cdot 3H_2O$ produced an analogous Rh/PVA colloid. This showed a threefold activity increase [255] (per g atom Rh) for olefin hydrogenation over the prior Rh/PVA sol as a result of the reduction in size of the colloidal metal particles to *ca.* 35 Å. A further reduction in rhodium particle size to 9 Å, with a concomitant increase in hydrogenation activity, was achieved by using sodium hydroxide in methanol as the reducing agent and PVP as the stabilizing polymer. More interesting was the change in selectivity on going to the smallest rhodium particles. Internal olefins were hydrogenated more rapidly with the smaller particles, an effect attributed to a diminution of steric hindrance at the smaller catalyst particles and not, for example, to any difference in the surface composition of the colloid particles resulting from the use of sodium hydroxide in the preparation of the most active catalyst. Colloidal palladium stabilized with PVP, and prepared in the same manner, catalyzed the hydrogenation of dienes (cyclopentadiene, cyclooctadiene) to the corresponding monoenes, and dienoates (methyl linoleate) to enoates with high selectivity. [28, 32, 259]

Other forms of colloidal metals have found use in hydrogenation catalysis, and it is here that some advantages of the colloidal state begin to be seen. Organosols of platinum, possibly in reverse micelles, were formed by toluene extraction of a cetyltributylphosphonium stabilized hydrosol, and found to catalyze hydrogenation of nitrobenzene to aniline. [65] Colloidal platinum, prepared in reverse micelles [151, 152] with anionic, cationic, and nonionic surfactants, has been investigated as a catalyst for the isomerization and hydrogenation of but-1-ene. It was found that the activity of the catalyst was greatest when a nonionic surfactant was used, but that in all cases the catalysts were an order of magnitude less active than the corresponding supported platinum catalysts, probably due to the effect the surfactant has on limiting the access of the substrate to the platinum surface. A similar effect has also been observed in metal colloid catalysis of olefin hydrogenation in polymerized vesicles. [184] It was emphasized that these catalysts are suitable only for those substrates which can diffuse through the vesicle wall, and overall rates were unexceptional. These apparently negative results were in fact quite promising, since rate limitation caused by a physically based phenomenon such as diffusion can form the basis for enhanced catalytic selectivity as substrate access to the catalyst surface could be restricted if the appropriate surfactant is used.

Enhanced selectivity has in fact been accomplished in a study of micellar catalysts incorporating platinum colloids in the 2–4 nm size range. The catalysts were prepared by hydrogen reduction or (preferably) photolysis of H_2PtCl_6 in aqueous solutions of surfactants such as dodecyltrimethylammonium chloride or sodium dodecylsulfate [105] at concentrations near the critical micelle concentration. The surfactant stabilized sols were active for the hydrogenation of vinyl acetate, and the photolytically produced colloids had the highest activity (on a gram atom Pt basis) due to their smaller size. The role of the surfactant micelle was investigated in the hydrogenation of long chain unsaturated acids, such as isomeric undecenoic acids and oleic acid, catalyzed by platinum and palladium sols stabilized with the nonionic surfactant polyethyleneglycol monolaurate (Scheme **6-2**). [261] In this system the hydrophobic surface of the metal particle is protected by the hydrophobic chain of the surfactant. It was found that the rate of hydrogenation depends on the position of the double bond in the alkenyl chain of

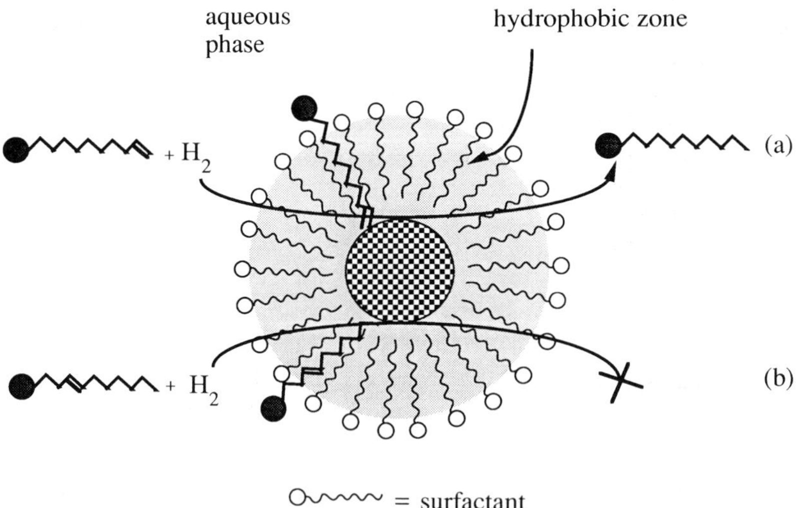

Scheme 6-2. Schematic representation of selective catalytic hydrogenation of isomeric undecenoic acids in polymerized micelle encapsulated colloidal platinum catalysts. The position of the double bond in the substrate determines the necessary orientation for access to the hydrogenation site, and this is mediated by the ordered polar micelle walls. (Adapted from ref. [261].)

the substrate. The observed hydrogenation rates were in the order undec-10-enanoic acid > oleic acid > undec-2-enoic acid. PVP stabilized sols showed no such regioselectivity, except when sodium salts of the unsaturated acids were used. (In this case, the substrate acted as a surfactant as well, and the particle should no longer be considered as a PVP stabilized colloid.)

These interesting results were explained as being due to the ordered environment provided by the micelle around the catalyst particle. A long chain unsaturated acid would presumably orientate itself in the micelle with its polar carboxylate group towards the hydrophilic layer, and its hydrophobic chain in the hydrophobic sheath of surfactant molecules which make up the micelle. Thus an internal olefinic bond would have only restricted access to the catalyst surface. This is an interesting observation in that it provides an example of selectivity modification by the stabilizing layer around the catalytic metal particle, a phenomenon virtually unique to colloidal metals and one which may provide a great advantage in colloidal catalysis of reactions of polyfunctional substrates.

In a study of colloidal rhodium hydrogenation catalysts stabilized with triphenylmethyltrisulfonates, Larpent and coworkers demonstrated an interesting surfactant effect in biphasic catalysis. Rhodium hydrosols stabilized by alkyltriphenylmethyltrisulfonates, $RC(p\text{-}C_6H_4SO_3Na)_3$ (described in Section 6.2.2.1), were shown to be active in olefin hydrogenation under biphasic conditions in which the substrate oct-1-ene was used as the organic phase. [69] An important beneficial effect of the surfactant was observed and generalized in a study of a similar rhodium sol stabilized with polyvinyl alcohol, PVA. This catalyst is similar to that reported earlier by Hirai, [255] but differs in the presence of added surfactants. [262] Using either anionic or nonionic surfactants, it was shown that the degree of interfacial tension between the aqueous (catalyst) phase and the organic (substrate) phase played a major role in determining the hydrogenation rate. With an optimum surfactant concentration in a two phase system, a rate of more than twice that in a single phase hydrogenation with the same catalyst in methanol was obtained, demonstrating that surfactants play both a colloid stabilizing role and an interfacial catalyst role in biphasic colloid catalysis.

Vapor derived colloidal catalysts in nonpolar liquids were investigated as catalysts for the hydrogenation of acenaphthylene (Eqn. 6.6). [133]

$$\text{acenaphthylene} \xrightarrow{H_2} \text{acenaphthene} \xrightarrow{2H_2} \text{tetrahydroacenaphthene} \qquad (6.6)$$

Palladium and cobalt colloids stabilized with *iso*-butylaluminoxane, $(iso\text{-}BuAlO)_n$, in methylcyclohexane catalyzed the hydrogenation of the activated double bond of acenaphthylene to acenaphthene at 25 °C, (Eqn. 6.6). In the case of the cobalt catalyst, heating to 80 °C resulted in the further hydrogenation of one of the aromatic rings in acenaphthene, a reaction which was poisoned by the addition of dibenzothiophene.

Toshima [162, 164, 165] has studied a series of PVP protected colloidal bimetallic PtPd catalysts in olefin hydrogenation catalysis. The highest activity for the hydrogenation of 1,3-cyclooctadiene to cyclooctene was observed at a Pd:Pt radio of 4:1. On the basis of EXAFS analysis [165] it was concluded that the distribution of the two metals in the particles is nonuniform, and that at the most active composition the colloid particles had a platinum core coated with palladium. A less active 1:1 Pd/Pt catalyst, with the same particle size, apparently had a more uniform distribution of metals, with both platinum and palladium atoms at the surface.

The progressive addition of cadmium to platinum in surfactant stabilized bimetallic colloidal catalysts for the hydrogenation of the unsaturated aldehyde crotonal results in a maximum in selectivity for hydrogenation of the aldehydic double bond to give crotyl alcohol at $PtCd_{0.5}$. [263] A pure platinum hydrosol is nonselective under the same conditions, hydrogenating both olefinic and aldehydic double bonds. Since overall crotonaldehyde conversion is slowed in the bimetallic system, the increased selectivity to crotyl alcohol at partial conversion is probably due to a selective poisoning of the crotyl alcohol hydrogenation step.

6.5.3 Hydrosilylation of Olefins

Addition of the Si–H bond of silanes to olefins (Eqn. 6.7) is a reaction of considerable importance in silicone technology, and several catalysts have been developed which are based on soluble platinum metal compounds. Speier's catalyst (alcoholic H_2PtCl_6) is perhaps the best known of these.

$$R_3SiH + R'CH_2{=}CH_2 \rightarrow R_3SiCH_2R' \qquad (6.7)$$

The mechanism for this ostensibly homogeneous process, the Chalk-Harrod mechanism, [264] was based on classical organometallic synthetic and mechanistic research. Its foundation lies in the oxidative addition of the silane Si–H bond to the low oxidation state metal complex catalyst, a reaction which is well established in the organometallic literature. Lewis reported in 1986 that the catalytically active solutions contained small (2.0 nm) platinum particles, and demonstrated that the most active catalyst in the system was in fact the colloidal metal. [60, 265] Subsequent studies established the relative order of catalytic activity for several precious metals to be platinum > rhodium > ruthenium = iridium ≫ osmium. [266] In addition, a dependence of the rate on colloid particle morphology for a rhodium colloid was observed. [267]

This work did much to raise awareness about the possibility that colloidal metals might be present and active as catalysts in homogeneous catalytic reactions under reducing conditions. Thus it is becoming more common for the criteria described above to be applied in determining the homogeneity or heterogeneity of catalytic processes.

6.5.4 Hydration of Unsaturated Organic Molecules

Colloidal copper, prepared by borohydride reduction of copper(II) salts in the presence of protective polymers and with particle sizes of 5.0–15.0 nm (depending on preparation details), is an active catalyst for the hydration of unsaturated nitriles to their corresponding unsaturated amides with 100 % selectivity. [53, 268] The copper particle size was unaffected by the catalytic process. The catalyst performance was optimized in a detailed study of the effects of polymer molecular weight, polymer/metal ratio, and the chemical constitution of the polymer. [268, 269]

6.5.5 Electron Transfer Reactions

A variety of inorganic redox reactions are catalyzed by platinum metal, [270] in several of which platinum sols have been observed to be active. Gold sols have also been shown to be active redox catalysts in, for example, the reaction between ferricyanide and thiosulfate ions (Eqn. 6.8). [271, 272]

$$[Fe(CN)_6]^{3-} + [S_2O_3]^{2-} \rightarrow [Fe(CN)_6]^{2-} + 1/2[S_4O_6]^{2-} \tag{6.8}$$

Here the reaction has been shown to proceed *via* electron transfer to and from the metal surface such that the metal particle acts as a highly dispersed electrode.

Organic free radical reactions are also catalyzed by metal colloids. For example, although 1-hydroxy-1-methyethyl, $(CH_3)_2COH^{\cdot}$, radicals generated by the photolysis of *iso*-propanol/acetone mixtures are unreactive towards thallium(I) ions or methylene chloride in water, the presence of colloidal silver induces them to reduce Tl^+ to colloidal thallium (a one electron process) and CH_2Cl_2 to $CHCl_3$ and Cl^- (a two electron process; Eqn. 6.9). [273]

$$Ag_n^{(m+2)-} + CH_2Cl_2 + H_2O \rightarrow Ag_n^{m-} + CHCl_3 + Cl^- + OH^- \tag{6.9}$$

Similarly the catalytic reduction of Cd^{2+}, N_2O, and $[NO_3]^-$ by $(CH_3)_2COH^{\cdot}$ radicals is also mediated by colloidal silver. [274] The importance of metal colloid chemistry to this area lies in the ability of metal sol particles to act as electron acceptors from free radicals in solution. [275] The radicals are thus diverted from the usual rapid reactions to which they are subject, such as recombination and disproportionation. Henglein has shown [276–278] that metal particles are capable of taking on many electrons from the radical species. For example, in the presence of a surfactant- or anionic polymer stabilized silver hydrosol, 1-hydroxy-1-methylethyl radicals react to liberate hydrogen at the metal surfaces. This reaction proceeds by initial electron transfer to the metal particle, followed by cathodic reduction of water by the negatively charged metal particle. (Eqns. 6.10–6.12)

$$(CH_3)_2COH^{\cdot} + Ag_n^{m^-} \rightarrow (CH_3)_2CO + H^+ + Ag_n^{(m+1)^-} \quad (6.10)$$

$$Ag_n^{(m+1)^-} + H_2O \rightarrow Ag_n^{m^-} + 1/2\ H_2 + OH^- \quad (6.11)$$

$$OH^- + H^+ \rightarrow H_2O \quad (6.12)$$

The net reaction is the liberation of hydrogen from the organic radical, however, many radicals are oxidized before hydrogen liberation begins. By measuring the pH change of the reaction during irradiation, (Eqn. 6.10), it was shown that the silver particles absorb many electrons before water is reduced. For the 7 nm silver particles used in this study, 450 stored electrons per particle were measured at the stationary state. [279] The colloidal silver particles thus act as electron pools from which electrons can be released to appropriate acceptors, in this case water.

Other metal colloids are also capable of absorbing electrons from organic radicals, and of subsequently reducing water. When this chemistry is coupled with the generation of organic radicals by visible light, we have the basis for an important application of metal colloids: the photochemical generation of hydrogen from water.

6.5.6 Photolysis of Water

The visible light induced decomposition of water into hydrogen and oxygen is a much sought after goal. One half of the process, hydrogen production, has been achieved by several systems which all share the following characteristics: absorption of photons by a photosensitizer, S, to produce an excited state, S* (Eqn. 6.13); electron transfer from the excited state S* to an electron acceptor, A, which acts as an electron relay; (Eqn. 6.14); reduction of water by the reduced electron acceptor (Eqn. 6.15). This last step is a catalyzed process in which colloidal metals, most effectively platinum sols, have played a central role in most of the systems studied. Ideally, the oxidized form of the photosensitizer would be rereduced by water to generate oxygen (Eqn. 6.16).

$$S + h\nu \rightarrow S^* \quad (6.13)$$

$$S^* + A \rightarrow S^+ + A^- \quad (6.14)$$

$$A^- + H_2O \xrightarrow{cat} 1/2\ H_2 + A + OH^- \quad (6.15)$$

$$2S^+ + H_2O \rightarrow 1/2\ O_2 + 2S + 2H^+ \quad (6.16)$$

Since reaction (6.16) is a slow multielectron transfer step, a more reactive sacrificial electron donor which can be oxidized in place of water has been added in most of the systems studied. The most frequently studied example of this process uses $[Ru(bipy)_3]^{2+}$ as the photosensitizer, *N,N'*-dimethyl-4,4'-bipyridinium (methylviologen, MV^{2+}) as an electron accepting relay, EDTA as a sacrificial electron donor and, importantly in our context, colloidal platinum as the catalyst. [280,

281] The catalyst particle acts as a repository for the electrons from the electron relay, and it is at the catalyst surface where water is reduced and hydrogen is generated. The various components of this system have been investigated in detail, but in the context of this chapter only those aspects concerning the colloidal metals will be discussed.

The goal of colloid related research in this area has been to provide the most active and stable colloidal catalysts. The model system just described, $[Ru(bipy)_3]^{2+}/MV^{2+}/EDTA/Pt$, and shown schematically in Figure **6-34**, has been used in most of these studies. Although there seems to be general agreement, based on comparative studies on the efficiency of colloidal metals, [281, 282] that platinum is the most active, details of the comparative quantum efficiencies often

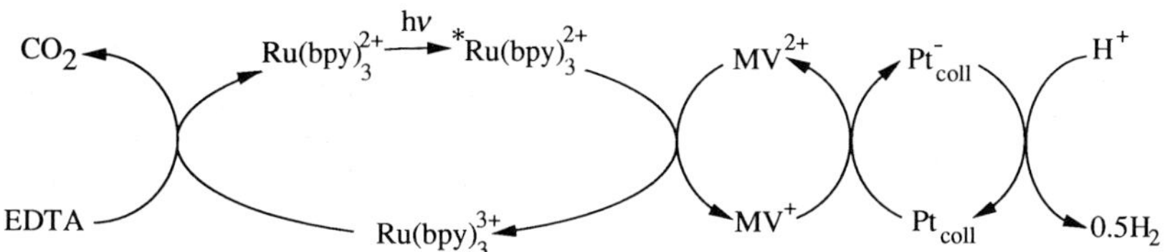

Figure 6-34. Schematic of the water splitting catalyst system $[Ru(bipy)_3]^{2+}/MV^{2+}/EDTA/Pt$. (adapted from ref. [281]).

vary from study to study. For example, although platinum was identified as the most active catalyst in two studies, the order of efficiency for Pt, Os, Ru, and Pd was found to be either Pt, Os > Ru > Pd [281] or Pt > Os, Pd > Ru. [104] The origin of this and other discrepancies clearly lies in the differing nature of the colloidal metal particles employed since the resulting particle sizes and surface compositions are dependent on the means of preparation (radiolytic, reduction of salts, identity of precursors and reducing agents). Although particle size considerations were not ignored, little control of particle size was exercised in comparing the different metals. A study on a series of various particle sized platinum colloids [283] seemed to show a maximum of activity at a diameter of 30 Å. However, since different reaction conditions were used in preparing the various platinum sols from H_2PtCl_4, the nature of the catalyst surface might not have been the same in all cases. In any case the difference in hydrogen generation activity when changing from 1.5 nm to 3.5 nm particle sizes was less than a factor of two.

The kinetics of hydrogen evolution from $MV^{+\cdot}$ radicals catalyzed by colloids of platinum and of gold are consistent with a second order dependence on catalyst concentration, and particle–particle collisions have been postulated to explain this. The likelihood of this phenomenon occuring and the inherent complexity of kinetic analysis in these systems have been critically analyzed. [284]

The nature of the stabilizing layer around the platinum catalyst particles has a significant effect on the efficiency of the catalyst. A comparison of radiolytically produced platinum/poly-*N*-methylol acrylamide sols with platinum sol (from H_2PtCl_6 chemical reduction) stabilized by a range of other polymers showed that the poly-N-methylol acrylamide sols were the most active, being *ca.* five times more efficient in hydrogen production than a polystyrene sulfonic acid stabilized sol. [104] Although the former had the smallest particle size, it was suggested that this alone was not the source of the enhanced activity. A comparative study has been made on the relative efficiencies of photolytically produced platinum sols stabilized in anionic, cationic, and neutral surfactant micelles, both monomeric and polymerized. [285] The six surfactant stabilized sols in this constant particle sized series were also compared with PVP/Pt and PAA^-/Pt (PAA = sodium polyacrylate) in the $[Ru(bipy)_3]^{2+}/MV^{2+}$/EDTA/Pt system. As shown in Table **6-1**, anionic surfactant stabilized sols were the least effective, but polymerization of the anionic surfactant raised the activity by over an order of magnitude. Polymerization of the surfactant micelle increased the activity of the colloid for all three types of surfactant, and the nonionic polymerized micelle provided the most efficient catalyst. Organic polymer stabilized sols were poor catalysts, being only slightly better than the essentially inactive anionic monomeric micelle catalyst. These results were interpreted in terms of deactivating electrostatic effects which, for cationic surfactants, reduce the accessibility of the $MV^{+\cdot}$ radical cation to the platinum surface. In the case of anionic surfactants these effects tend to immobilize both $[Ru(bipy)_3]^{3+}$ and $MV^{+\cdot}$ at the anionic interface where they can undergo reverse electron transfer to $[Ru(bipy)_3]^{2+}$ and MV^{2+}.

Although much remains to be learned about the chemistry of the catalyst particles in this important application, in light of the advancing state of the art in metal colloid characterization (see Section 6.3) there is good reason to expect

Table 6-1. Catalytic Activities of Colloidal Platinum Catalysts for Hydrogen Generation by Water Photolysis using EDTA/$Ru(bipy)_3^{2+}/MV^{2+}$/Pt(coll). (Adapted from ref. 285)

Pt sol	Protective Layer	Relative Rate
Pt/poly(ethyleneglycol)-monomethylether 10-undecenoate	nonionic polymerized micelle	306
Pt/poly(ethyleneglocol)monolaurate	nonionic monomeric micelle	302
Pt/undecenyltriethylammonium bromide	cationic polymerized micelle	267
Pt/dodecyltrimethylammonium bromide	cationic monomeric micelle	250
Pt/sodium 10-undecenoate	anionic polymerized micelle	105
Pt/sodium 10-undecenoate	anionic monomeric micelle	8
Pt/poly(vinylpyrrolidone)	nonionic linear polymer	100
Pt/poly(acrylic acid)	anionic linear polymer	60

progress. Little work has been done on the possible use of bimetallic catalysts, despite there being evidence [286] that gold-platinum bimetallic catalysts can increase hydrogen production rates. In these catalysts hydrogen adsorption at the platinum surface is inhibited, a process which can lead to deactivation by hydrogenation of the electron relay. This slow hydrogenation of MV^{2+} is one of the limiting factors in the $[Ru(bipy)_3]^{2+}/MV^{2+}$/EDTA/Pt system, and various suggestions have been made as to the nature of this reaction as well as proposals for inhibiting it. [104, 281, 287] These do not specifically concern us here, except to introduce the topic of *photocatalytic hydrogenation* by colloidal metals.

6.5.7 Photohydrogenation Catalysis

6.5.7.1 Photohydrogenation of Alkenes and Alkynes

As discussed above, the reduction of water (protons) by organic radicals is catalyzed by metal colloid particles which function as electron pools for the transfer of electrons from the radical to water. The hydrogen is presumably produced by desorption from an adsorbed atomic state at the catalyst surface. As this is the reactive state that hydrogen attains after chemisorption from the gas phase in heterogeneous hydrogenation catalysis, it is not surprising that it can add to organic substrate molecules such as alkenes or alkynes. The popular water photolysis system $[Ru(bipy)_3]^{2+}$/EDTA/metal sol, in which colloidal platinum, palladium, rhodium, ruthenium, and plated platinum-on-palladium and palladium-on-platinum act as the catalytic colloids, catalyses the photohydrogenation of alkenes and alkynes. [169] The quantum yields for ethylene hydrogenation on platinum and their pH dependence are identical to those for hydrogen production, suggesting that the two processes have the same rate determining step, the oxidation of $MV^{+\cdot}$. Thus both processes have essentially the same mechanism whereby the electrons are transferred from $MV^{+\cdot}$ to the metal colloid and the protons from water are reduced to give metal bound hydrogen atoms (Fig. **6-35**). Ethylene is then hydrogenated in a fast step. In the case of acetylene hydrogenation, the hydrogenation step is slower, but when colloidal palladium is substituted for platinum in the photocatalyst, the superior acetylene hydrogenation activity of palladium increases the quantum yield for hydrogenation despite the lower efficiency of the palladium colloids in oxidizing $MV^{+\cdot}$.

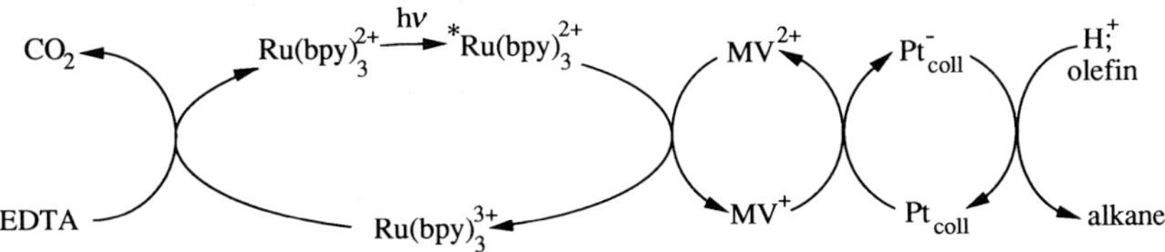

Figure 6-35. Schematic of the $[Ru(bipy)_3]^{2+}/MV^{2+}$/EDTA/Pt photohydrogenation system (adapted from ref. [169]).

Bimetallic palladium-platinum catalysts [169] show an enhanced acetylene hydrogenation activity over that obtained from a mixture of the two colloids. This has been attributed to a combination of the efficiency of platinum as a $MV^{+\cdot}$ oxidizer with that of palladium as an acetylene hydrogenation catalyst in the bimetallic particle. Similar experiments were performed in a two phase oil-water system, using palladium and platinum organosols prepared by heating the metal salts in long chain aliphatic alcohols. [37] In the biphasic system the charge relay is the double reduced MV, formed by disproportionation of $MV^{+\cdot}$. Using mixtures of preformed monometallic sols a remarkable synergism was observed between separate palladium and platinum colloids in the photohydrogenation of phenylacetylene. This was suggested to arise from interparticle collisions between the efficient $MV^{+\cdot}$ oxidizing (thus hydrogen covered) platinum particles and the catalytically efficient alkyne hydrogenating palladium particles. Aggregated bimetallic colloid formation could not be ruled out. A similar study has compared sols of ruthenium, iridium, platinum, rhodium, palladium, and silver for the photohydrogenation of unsaturated organic molecules. [282]

6.5.7.2 Photohydrogenation of Carbon Dioxide

Photoreduction of CO_2 is mediated by colloidal ruthenium and osmium in an alternative aqueous photosystem $[Ru(bipy)_3]^{2+}$ or $[Ru(bipyrazine)_3]^{2+}$/triethanolamine (sacrificial electron donor)/bipyridinium*/Ru(coll) or Os(coll), [288, 289] where bipyridinium* is one of a series of bipyridinium derivatives chosen for their varied reduction potentials. The colloids were prepared by citrate reduction of $RuCl_3$ or OsO_4. Illumination of the $[Ru(bipy)_3]^{2+}$ system under a CO_2 atmosphere resulted in the formation of methane and ethylene with simultaneous (kinetically favored) production of hydrogen. Using $[Ru(bipyrazine)_3]^{2+}$ as the photosensitizer with the ruthenium colloid yields methane, ethylene, and ethane, and no hydrogen is liberated. The reaction proceeds via electron transfer from the metal particle to activated CO_2, rather than catalyzed hydrogenation at the metal surface, and colloid adsorbed intermediates such as $=CH_2$ and $-CH_3$ are implicated. Other ruthenium(II) polypyridine photosensitizers have been investigated, from which $[Ru(phenathroline)_3]^{2+}$ has been shown to be superior to $[Ru(bipy)_3]^{2+}$. [290] Colloidal palladium stabilized by β-cyclodextrin is active in the photocatalytic reduction of $CO_2/[HCO_3]^-$ to formate, in which deazariboflavin acts as the photosensitizer and MV^{2+} as the electron relay. [291]

6.5.8 Conclusions

It is clear that colloidal metals not only possess many of the catalytic properties of supported metal catalysts in reactions carried out at temperatures mild enough to sustain colloid stability, they are also amenable to modification and application in ways more reminiscent of molecular homogeneous catalysts. The use of ligand modification with colloidal metal catalysts has hardly been mentioned in the literature, despite this being the most powerful tool for catalyst optimization in homogeneous catalysis. The precise alloying of colloidal metal surfaces has been demonstrated in spectroscopic studies, but seldom applied in catalytic applications. In addition, metal colloids potentially provide a more fundamental benefit to catalysis science. The relationship between homogeneous catalysis and heterogeneous catalysis is one which has attracted the attention of practitioners in both fields, as well as researchers seeking to understand the fundamental processes occuring at the active sites on the surfaces of metal catalysts. [292] Molecular homogeneous catalysts offer the potential advantage of being amenable to kinetic and mechanistic analyses. In homogeneous systems the spectroscopic methods which form the basis of organometallic chemistry, particularly infrared and NMR spectroscopy in solution, can be applied and their high resolution taken advantage of for precise analysis. A solution of a working homogeneous catalyst can sometimes be analyzed *in situ* (although the resulting data can also be misleading in complex systems), and the chemistry of model systems based on organometallic analogues of the purported catalyst can provide a powerful means of arriving at a convincing mechanism. The situation is more complex for heterogeneous catalysts. One of the principal goals in the field of single crystal surface chemistry has been the elucidation of reaction pathways and intermediates on catalyst surfaces under rigorously controlled conditions. To this end surface spectroscopic techniques of great sophistication and power have been developed. The precise analysis of real catalysts under working conditions is very difficult. They are often relatively ill defined metal particles supported on poorly characterized supports *via* an imperfectly understood metal support interaction. Thus heterogeneous catalysis has often turned to organometallic chemistry to provide analogues for proposed surface intermediates and reaction paths on heterogeneous catalysts. While this approach has been useful, there has been a need for more closely analogous molecular systems to give a greater confidence level to the analogy. For example, an analogy between a rhodium triphenylphosphine complex and a supported rhodium crystallite on silica is a dubious one. In response to this, the molecular chemistry community has developed the organometallic chemistry of cluster complexes in which organic fragments are bound to several adjacent metal atoms in the cluster core, and this approach has provided some useful geometric models for the binding states of a number of small organic fragments on metal surfaces.

Metal colloid chemistry provides a third alternative for analysis of surface bound reaction intermediates in metal catalyzed reactions. For some years now, a central theme in the author's laboratory has been that metal sols, especially in organic liquids, are not only subjects for fundamental research on the properties of small metal particles, but also provide the basis for an analysis of catalysis related metal particle surface chemistry by the spectroscopic techniques used in molecular organometallic chemistry. [133] Rather than being prepared by the methods of heterogeneous catalyst preparation (*i.e.* high temperature reduction of a precursor on a potentially reactive oxide support), colloidal metal particles can be prepared under mild and controlled conditions in various size ranges and in a benign isotropic environment, often being protected by a loosely bound sheath of polymer or surfactant molecules. When dispersed in a fluid medium, colloidal metal particles can be conveniently studied by solution spectroscopic techniques. When isolated as a solvent free solid, they are amenable to some of the techniques common in solid state chemistry and physics-electron and X-ray diffraction, EXAFS, solid state NMR, etc. Thus these materials can be seen to occupy a unique position between molecular organometallic chemistry and heterogeneous catalysis, and one which may provide not only insights into supported metal systems but also new understanding of the chemistry of small metal particles themselves.

6.6 Prospects in Metal Colloid Chemistry

Although catalysis is the only application of metal colloid chemistry which has been described in this chapter, the properties of nanoscale metal particles alluded to in the introduction might justify their use in a number of areas of technological importance. Composites of metal particles with polymers and glasses hold great promise as nonlinear optical materials, [15, 293] and the gold/polydiacetylene composite referred to above [136] was prepared for just such a purpose. The magnetic properties of colloidal ferromagnetic metals have long been exploited in the applications of ferrofluids, [78] and the future fabrication of nanoscale electronic and magnetic devices will benefit from new methods of preparing and assembling metal particles with precisely controlled electronic properties in various matrices. The technology of advanced materials will be an obvious beneficiary of successes in preparing composites of well defined metal particles and ceramics (cermets). The likelihood of advances in these and other areas depends on the continued aquisition of knowledge and skills in the controlled synthesis and careful characterization of nanoscale metal particles. The time is past when the simple reporting of the preparation and particle size range of a metal sol should be of particular interest. Now that techniques are readily available for detailed structural and spectroscopic analysis, the aroused interest of chemists, physicists, and materials scientists demands that this field be the subject of interdisciplinary research, the possibility of which is more amenable today than it has ever been.

References

[1] T. Graham, *Phil. Trans. Roy. Soc.* **1861,** *151,* 183–190.

[2] A. Ceriotti, F. Demartin, G. Longoni, M. Manassero, G. Marchionna, G. Piva, M. Sansoni, *Angew. Chem. Int. Ed. Engl.* **1985,** *24,* 696–698.

[3] G. Schmid, *Chem. Rev.* **1992,** *92,* 1709–1727.

[4] M. T. Franklin, K. J. Klabunde in *High Energy Process in Organometallic Chemistry* (Ed.: K. S. Suslick), American Chemical Society, Washington, D. C. **1987,** pp. 246–259.

[5] M. Faraday, *Phil. Trans. Roy. Soc.* **1857,** *147,* 145–153.

[6] J. M. Thomas, *Pure and Appl. Chem.* **1988,** *60,* 1517–1528.

[7] W. Ostwald, *Colloid-Zeitschrift* **1907,** *1,* 291–331.

[8] J. Turkevich, P. C. Stevenson, J. Hillier, *Disc. Faraday Soc.* **1951,** *11,* 55–75.

[9] J. Turkevich, *Gold Bulletin* **1985,** *18,* 86–91.

[10] M. L. Steigerwald, L. E. Brus, *Ann. Rev. Mater. Sci.* **1989,** *19,* 471–495.

[11] A. Henglein, *Top. Curr. Chem.* **1988,** *143,* 113–180.

[12] W. P. Halperin, *Rev. Mod. Phys.* **1986,** *58,* 533–606.

[13] K. Kimura, *Z. Phys. D.* **1989,** *11,* 327–332.

[14] G. Schmid, *Mater. Chem. Phys.* **1991,** *29,* 133–142.

[15] M. P. Andrews, G. A. Ozin, *Chem. Mater.* **1989,** *1,* 174–187.

[16] J. F. Hamilton, R. C. Baetzold, *Science* **1979,** *205,* 1213–1220.

[17] J. Turkevich, R. S. J. Miner, I. Okura, S. Namba, *Proc. Swedish. Symp. Catal.* **1981,** 111–126.

[18] B. K. Teo, M. Hong, H. Zhang, D. Huang, X. Shi, *J. Chem. Soc., Chem. Commun.* **1988,** 204–206.

[19] J. T. G. Overbeek, in *Colloidal Dispersions,* J. W. Goodwin, ed. Royal Society of Chemistry, London **1981,** pp. 1–23.

[20] C. G. Blatchford, J. R. Campbell, J. A. Creighton, *Surface Science* **1982,** *120,* 435–455.

[21] M. E. Labib, *Colloids and Surfaces* **1988,** *29,* 293–304.

[22] M. E. Labib, R. Williams, *J. Colloid Interface Sci.* **1984,** *97,* 356–366.

[23] A. Duteil, R. Quéau, B. M. Chaudret, C. Roucau, J. S. Bradley *Chem. Mater.* **1993,** *5,* 341–347.

[24] M. Komiyama, H. Hirai, *Bull. Chem. Soc. Jpn.* **1983,** *56,* 2833–2834.

[25] v. H. Thiele, H. S. von Levern, *J. Colloid Sci.* **1965,** *20,* 679–694.

[26] P. H. Hess, P. H. Parker, *J. Appl. Polymer. Sci.* **1966,** *10,* 1915–1927.

[27] H. Hirai, Y. Nakao, N. Toshima, *J. Macromol. Sci., Chem.* **1979,** *A13,* 727–750.

[28] H. Hirai, *Makromol. Chem. Suppl.* **1985,** *14,* 55–69.

[29] K. Meguro, Y. Nakamura, Y. Hayashi, M. Torizuka, K. Esumi, *Bull. Chem. Soc. Jpn.* **1988,** *61,* 347–350.

[30] G. Schmid, A. Lehnert, *Angew. Chem. Int. Ed. Eng.* **1989,** *28,* 780–781.

[31] H. Hirai, Y. Nakao, N. Toshima, *J. Macromol. Sci.-Chem.* **1978,** *A12,* 1117–1141.

[32] H. Hirai, *J. Macromol. Sci.-Chem.* **1979,** *A13,* 633–649.

[33] J. S. Bradley, J. M. Millar, E. W. Hill, *J. Am. Chem. Soc.* **1991,** *113,* 4016–4017.

[34] J. S. Bradley, E. W. Hill, S. Behal, C. Klein, B. Chaudret, A. Duteil, *Chem. Mater.* **1992,** *4,* 1234–1239.

[35] F. Porta, F. Ragaini, S. Cenini, G. Scari, *Gazz. Chim. Ital.* **1992,** *122,* 361–363.

[36] F. A. Cotton, G. Wilkinson, *Advanced Inorganic Chemistry,* 5th ed **1988,** Wiley, New York, p. 905.

[37] D. Mandler, I. Willner, *J. Phys. Chem.* **1987,** *91,* 3600–3605.

[38] J. S. Bradley, E. W. Hill, C. Klein, B. Chaudret, A. Duteil, *Chem. Mater.* **1993,** *5,* 254–256.

[39] P. R. van Rheenen, M. J. McKelvey, W. S. Glaunsinger, *J. Solid State Chem.* **1987,** *67,* 151–169.

[40] H. Ishizuka, T. Tano, K. Torigoe, K. Esumi, K. Meguro, *Colloids Surf* **1992,** *63,* 337–340.
[41] A. C. Curtis, D. G. Duff, P. P. Edwards, D. A. Jefferson, B. F. G. Johnson, A. I. Kirkland, A. S. Wallace, *J. Phys. Chem.* **1988,** *92,* 2270–2275.
[42] R. Touroude, P. Girard, G. Maire, J. Kizling, K. M. Boutonnet, P. Stenius, *Surfactant Sci. Ser.* **1992,** *44,* 357.
[43] D. N. Furlong, A. Launikonis, W. H. F. Sasse, J. V. Saunders, *J. Chem. Soc., Faraday Trans.* **1984,** *80,* 571–588.
[44] A. Harriman, G. R. Millward, P. Neta, M. C. Richoux, J. M. Thomas, *J. Phys. Chem.* **1988,** *92,* 1286–1290.
[45] J. Turkevitch, G. Kim, *Science* **1970,** *169,* 873–879.
[46] L. D. Rampino, F. F. Nord, *J. Am. Chem. Soc.* **1941,** *63,* 2745–2749.
[47] L. D. Rampino, F. F. Nord, *J. Am. Chem. Soc.* **1941,** *63,* 3268.
[48] K. E. Kavanagh, F. F. Nord, *J. Am. Chem. Soc.* **1943,** *65,* 2121–2125.
[49] L. Hernandez, F. F. Nord, *J. Colloid Sci.* **1948,** *3,* 363–375.
[50] W. P. Dunsworth, F. F. Nord, *J. Am. Chem. Soc.* **1950,** *72,* 4197–4198.
[51] M. R. Mucalo, R. P. Cooney, *J. Chem. Soc., Chem. Commun.* **1989,** 94–95.
[52] P. R. van Rheenen, M. J. McKelvey, R. Marzke, W. S. Glaunsinger, *Inorg. Synthesis* **1983,** *24,* 238–242.
[53] H. Hirai, H. Wakabayashi, M. Komiyama, *Chem. Lett.* **1983,** 1047–1050.
[54] H. Hirai, H. Wakabayashi, M. Komiyama, *Bull. Chem. Soc. Jpn.* **1986,** *59,* 367–372.
[55] G. N. Glavee, K. J. Klabunde, C. M. Sorensen, G. C. Hadjipanayis, *Langmuir* **1993,** *9,* 162–169.
[56] G. N. Glavee, K. J. Klabunde, C. M. Sorensen, G. C. Hadjipanayis, *Inorg. Chem.* **1993,** *32,* 474–477.
[57] I. Ravet, A. Gourgue, Z. Gabelica, J. B. Nagy, *Proc. VIII International Congress on Catalysis* **1984,** vol. 4, 871–878.
[58] H. Bönnemann, W. Brijoux, R. Brinkmann, E. Dinjus, T. Joussen, B. Korall, *Angew. Chem., Int. Ed. Engl.* **1991,** *30,* 1312–1314.
[59] H. Bönnemann, W. Brijoux, R. Brinkmann, E. Dinjus, R. Fretzen, T. Joussen, B. Korall, *J. Mol. Catal.* **1992,** *74,* 323–333.
[60] L. N. Lewis, N. Lewis, *J. Am. Chem. Soc.* **1986,** *108,* 7228–7231.
[61] L. N. Lewis, N. Lewis, *Chem. Mater.* **1989,** *1,* 106–114.
[62] H. Bönnemann, R. Brinkmann, R. Köppler, P. Neiteler, J. Richter, *Advanced Materials* **1993,** in press.
[63] K. Esumi, N. Sato, K. Torigoe, K. Meguro, *J. Colloid Interface Sci.* **1992,** *149,* 295–298.
[64] K. Meguro, M. Torizuka, K. Esumi, *Bull. Chem. Soc. Jpn.* **1988,** *61,* 341–345.
[65] V. M. Deshpande, P. Singh, C. S. Narasimhan, *J. Chem. Soc., Chem. Commun.* **1990,** 1181–1182.
[66] H. Hirai, H. Aizawa, H. Shiozaki, *Chem. Lett.* **1992,** 1527–1530.
[67] C. Larpent, H. Patin, *J. Mol. Catal.* **1988,** *44,* 191–195.
[68] C. Larpent, F. Brisse-Le Menn, H. Patin, *New J. Chem.* **1991,** *15,* 361–366.
[69] C. Larpent, F. Brisse-Le Menn, H. Patin, *J. Mol. Catal.* **1991,** *65,* 35–40.
[70] v. H. Thiele, J. Kowallik, *Kolloid-Zeitschrift* **1965,** *234,* 1017–1029.
[71] D. G. Duff, A. C. Curtis, P. P. Edwards, D. A. Jefferson, B. F. G. Johnson, D. E. Logan, *J. Chem. Soc., Chem. Commun.* **1987,** 1264–1266.
[72] J. L. Dye, K. L. Tsai, *Faraday Discuss. Chem. Soc.* **1991,** *92,* 45–55.
[73] K.-L. Tsai, J. L. Dye, *Chem. Mater.* **1993,** *5,* 540–546.
[74] J. R. Thomas, *J. Appl. Phys.* **1966,** *37,* 2914–2915.
[75] T. W. Smith, *US Patent 4252674* **1981.**
[76] T. W. Smith, *US Patent 4252673* **1981.**
[77] C. H. Griffiths, H. P. O'Horo, T. W. Smith, *J. Appl. Phys.* **1979,** *50,* 7108.
[78] M. Kilner, G. J. Russell, S. R. Hoon, B. K. Tanner, *J. Magn. Mag. Mater.* **1983,** *39,* 107–110.

[79] T. W. Smith, *U.S. 4252671* **1981**.
[80] T. W. Smith, *US Patent 4252672* **1981**.
[81] T. W. Smith, *US Patent 4252678* **1981**.
[82] T. Tano, K. Esumi, K. Meguro, *J. Colloid Interface Sci.* **1989,** *133,* 530–533.
[83] K. Esumi, M. Suzuki, T. Tano, K. Torigoe, K. Meguro, *Colloids Surf.* **1991,** *55,* 9–14.
[84] K. Esumi, O. Sadakane, K. Torigoe, K. Meguro, *Colloids Surf* **1992,** *62,* 255–257.
[85] K. Esumi, T. Tano, K. Meguro, *Langmuir* **1989,** *5,* 268–270.
[86] K. Esumi, T. Tano, K. Torigoe, K. Meguro, *Chem. Mater.* **1990,** *2,* 564–567.
[87] A. Henglein, J. Lille, *J. Phys. Chem.* **1981,** *85,* 1246–1251.
[88] A. Henglein, M. Guttierez, E. Janata, B. G. Ershov, *J. Phys. Chem.* **1992,** *96,* 4598–4602.
[89] J. Butler, A. Henglein, *Radiat. Phys. Chem.* **1980,** *15,* 603–612.
[90] B. G. Ershov, A. Henglein, *J. Phys. Chem.* **1993,** *97,* 3434–3436.
[91] A. Henglein, A. Holzwarth, P. Mulvaney, *J. Phys. Chem.* **1992,** *96,* 8700–8702.
[92] A. Henglein, *Ber. Bunsenges. Phys. Chem.* **1977,** *81,* 556–561.
[93] A. Henglein, R. Tausch-Treml, *J. Coll. Interfac. Sci.* **1981,** *80,* 84–93.
[94] S. Mosseri, A. Henglein, E. Janata, *J. Phys. Chem.* **1989,** *93,* 6791–6795.
[95] G. Mills, A. Henglein, *Radiat. Phys. Chem.* **1985,** *26,* 385–390.
[96] M.-O. Delcourt, J. Belloni, J.-L. Marignier, C. Mory, C. Colliex, *Radiat. Phys. Chem.* **1984,** *23,* 485–487.
[97] J. Belloni, M. D. Delcourt, C. Leclere, *Nouv. J. Chim.* **1982,** *6,* 507–509.
[98] M. O. Delcourt, N. Keghouche, J. Belloni, *Nouv. J. Chim.* **1983,** *7,* 131–136.
[99] B. G. Ershov, E. Janata, A. Henglein, *Radiat. Phys. Chem.* **1992,** *39,* 123–126.
[100] J. L. Marignier, J. Belloni, M. O. Delcourt, J. P. Chevalier, *Nature* **1985,** *317,* 344–345.
[101] B. G. Ershov, E. Janata, M. Michaelis, A. Henglein, *J. Phys. Chem.* **1991,** *95,* 8996–8999.
[102] M. Michaelis, A. Henglein, *J. Phys. Chem.* **1992,** *96,* 4719–4724.
[103] K. Kurihara, J. Kizling, P. Stenius, J. H. Fendler, *J. Amer. Chem. Soc.* **1983,** *105,* 2574–2579.
[104] R. Rafaeloff, Y. Haruvy, J. Binenboym, G. Baruch, L. A. Rajbenbach, *J. Mol. Catal.* **1983,** *22,* 219–233.
[105] N. Toshima, T. Takahashi, H. Hirai, *Chem. Lett.* **1985,** 1245–1248.
[106] Y. Yonezawa, T. Sato, M. Ohno, H. Hada, *J. Chem. Soc., Faraday Trans.* **1987,** *83,* 1559–1567.
[107] P. Barnickel, A. Wokaun, *Mol. Phys.* **1990,** *69,* 1–9.
[108] K. Torigoe, K. Esumi, *Langmuir* **1992,** *8,* 59–63.
[109] Y. Yonezawa, T. Sato, S. Kuroda, K. Kuge, *J. Chem. Soc., Faraday Trans.* **1991,** *87,* 1905–1910.
[110] T. Sato, S. Kuroda, A. Takami, Y. Yonezawa, H. Hada, *Appl. Organomet. Chem.* **1991,** *5,* 261–268.
[111] Y. Takahashi, T. Ito, S. Sakai, Y. Ishii, *J. Chem. Soc. Chem. Commun.* **1970,** 1065–1066.
[112] P. Gallezot, D. Richard, G. Bergeret in *Novel Materials in Heterogeneous Catalysis,* ACS Symposium Series No. 437; (Eds.: R. T. K. Baker, L. L. Murrell), American Chemical Society, Washington, D. C. **1990,** pp. 150–159.
[113] J. S. Bradley, J. M. Millar, E. W. Hill, S. Behal, B. Chaudret, A. Duteil, *Faraday Disc.* **1991,** *92,* 255–268.
[114] J. S. Bradley, E. W. Hill, C. Klein, *unpublished results* **1993**.
[115] V. Agelou, A. Duteil, B. Chaudret, R. Mazel, C. Roucau, J. S. Bradley, *unpublished results,* **1993**.
[116] J. R. Blackborow, D. Young, *Metal Vapor Synthesis,* Springer Verlag, New York, **1979**.
[117] S. Roginsky, A. Schalnikoff, *Kolloid Z.* **1927,** *43,* 67–70.
[118] F. W. S. Benfield, M. L. H. Green, J. S. Ogden, D. Young, *J. Chem. Soc., Chem. Commun.* **1973,** 866–867.

[119] P. L. Timms in *Cryochemistry* (Eds.: M. Moskovits, G. A. Ozin), **1976,** Wiley, New York, pp. 61–136.
[120] M. Andrews, G. A. Ozin, C. G. Francis, *Inorg. Syn.* **1981,** *22,* 116–123.
[121] K. J. Klabunde, P. Timms, P. S. Skell, S. D. Ittel, *Inorg. Syn.* **1979,** *19,* 59–69.
[122] M. L. H. Green, D. O'Hare in *High Energy Processes in Organometallic Chemistry* (Ed.: K. S. Suslick), **1987,** American Chemical Society, Washington, D.C., pp. 260–277.
[123] N. Wada, *Jpn. J. Appl. Phys.* **1967,** *6,* 553–558.
[124] K. Kimura, S. Bandow, *Bull. Chem. Soc. Japan* **1983,** *56,* 3578–3584.
[125] K. Kimura, *Bull. Chem. Soc. Japan* **1984,** *57,* 1683–1684.
[126] N. Satoh, K. Kimura, *Bull. Chem. Soc. Japan* **1989,** *62,* 1758–1763.
[127] N. Satoh, S. Bandow, K. Kimura, *J. Colloid Interfac. Sci.* **1989,** *131,* 161–165.
[128] N. Wada, M. Ichikawa, *Jpn. J. Appl. Phys.* **1976,** *15,* 755–756.
[129] S. Yatsuya, T. Ohno, S. Yatsuya, R. Uyeda, *Japan. J. Appl. Phys.* **1977,** *16,* 705–717.
[130] S. Yatsuya, T. Hayashi, H. Akoh, E. Nakamura, A. Tasaki, *Japan. J. Appl. Phys.* **1978,** *17,* 355–359.
[131] M. P. Andrews, G. Ozin, *J. Phys. Chem.* **1986,** *90,* 2922–2928.
[132] M. P. Andrews, M. E. Galvin, S. A. Heffner, *Proc. Mat. Res. Soc. Smyp.* **1989,** *131,* 21–33.
[133] J. S. Bradley, E. W. Hill, M. E. Leonowicz, H. Witzke, *J. Mol. Catal.* **1987,** *41,* 59–74.
[134] S. T. Lin, M. T. Franklin, K. J. Klabunde, *Langmuir* **1986,** *2,* 259–260.
[135] G. Cardenas-Trevino, K. J. Klabunde, E. B. Dale, *Langmuir* **1987,** *3,* 986–992.
[136] A. Olsen, Z. H. Kafafi, *J. Am. Chem. Soc.* **1991,** *113,* 7758–7760.
[137] M. Kilner, N. Mason, D. Lambrick, P. D. Hooker, P. L. Timms, *J. Chem. Soc. Chem. Commun.* **1987,** 356–357.
[138] K. J. Klabunde, *Platinum Met. Rev.* **1992,** *36,* 80–84.
[139] K. J. Klabunde, J. Habdas, G. Càrdenas-Treviño, *Chem. Mater.* **1989,** *1,* 481–483.
[140] E. B. Zuckerman, K. J. Klabunde, B. J. Olivier, C. M. Sorensen, *Chem. Mater.* **1989,** *1,* 12–14.
[141] K. J. Klabunde, G. Youngers, E. J. Zuckerman, B. J. Tan, S. Antrim, P. M. Sherwood, *Eur. J. Solid State Inorg. Chem.* **1992,** *29,* 227–260.
[142] J. S. Bradley, E. W. Hill, *US Patent 4857492* **1989.**
[143] J. H. Fendler, P. Tundo, *Acc. Chem. Res.* **1984,** *17,* 3–8.
[144] J. Fendler, K. Kurihara in *Metal Containing Polymeric Systems* (Eds.: J. E. Sheats, C. E. Carraher, Jr., C. U. Pittman, Jr.), Plenum Press **1985,** pp. 341–353.
[145] D. Langevin, *Accts. Chem. Res.* **1988,** *21,* 255–260.
[146] H. Eicke F., *Topics in Curr. Chem.* **1980,** *87,* 85–145.
[147] N. Toshima, T. Takahashi, *Bull. Chem. Soc. Japan* **1992,** *65,* 400–409.
[148] N. Toshima, T. Takahashi, H. Hirai, *Chem. Lett.* **1986,** 35–38.
[149] N. Toshima, T. Takahashi, H. Hirai, *Chem. Lett.* **1987,** 1031–1034.
[150] K. Kurihara, J. H. Fendler, *J. Am. Chem. Soc.* **1983,** *105,* 6152–6153.
[151] M. Boutonnet, J. Kizling, P. Stenius, G. Maire, *Colloids and Surf.* **1982,** *5,* 209–225.
[152] M. Boutonnet, J. Kizling, R. Touroude, G. Maire, P. Stenius, *Applied Catal.* **1986,** *20,* 163–177.
[153] J. P. Wilcoxon, R. L. Williamson, *Proc. Mat. Res. Soc. Symp.* vol 177, Materials Research Society, Pittsburgh **1990,** pp. 269–274.
[154] M. Nishida, J.-I. Ishiyama, Y. Kurokawa, *Reactive Polymers* **1991,** *14,* 205–211.
[155] Y. Ng Cheong Chan, R. R. Schrock, R. E. Cohen, *Chem. Mater.* **1992,** *4,* 24–27.
[156] Y. Ng Cheong Chan, R. R. Schrock, R. E. Cohen, *J. Am. Chem. Soc.* **1992,** *114,* 7295–7296.
[157] Y. Ng Cheong Chan, G. S. W. Craig, R. R. Schrock, R. E. Cohen, *Chem. Mater.* **1992,** *4,* 885–894.
[158] Y. Ng Cheong Chan, R. R. Schrock, *Chem. Mater.* **1993,** *5,* 566–570.
[159] B. Breitscheidel, J. Zieder, U. Schubert, *Chem. Mater.* **1991,** *3,* 559–566.

[160] J. M. Tour, S. L. Pendalwar, J. P. Cooper, *Chem. Mater.* **1990,** *2,* 647–649.
[161] P. A. Sermon, J. M. Thomas, K. Keryou, G. R. Millward, *Angew. Chem. Int. Ed. Engl.* **1987,** *26,* 918–919.
[162] N. Toshima, K. Kushihashi, T. Yonezawa, H. Hirai, *Chem. Lett.* **1989,** 1769–1772.
[163] B. Zhao, N. Toshima, *Chem. Express* **1990,** *5*, 721.
[164] N. Toshima, T. Yonezawa, M. Harada, K. Asakura, Y. Iwasawa, *Chem. Lett.* **1990,** 815–818.
[165] N. Toshima, M. Harada, T. Yonezawa, K. Kushihashi, K. Asakawa, *J. Phys. Chem.* **1991,** *95,* 7448–7453.
[166] K. Esumi, M. Shiratori, H. Ishizuka, T. Tano, K. Torigoe, K. Meguro, *Langmuir* **1991,** *7,* 457–459.
[167] R. S. J. Miner, S. Namba, J. Turkevich, *Stud. Surf. Sci. Catal.* **1981,** *7,* 160–172.
[168] J. B. Michel, J. T. Schwartz in *Catalyst Preparation Science, IV* (Eds.: B. Delmon, P. Grange, P. A. Jacobs, G. Poncelet), Elsevier Science Publishers, New York, 1987, pp. 669–687.
[169] Y. Degani, I. Willner, *J. Chem. Soc. Perkin Trans. II* **1986,** 37–41.
[170] G. Schmid, A. Lehnert, J.-M. Malm, J.-O. Bovin, *Angew. Chem. Int. Ed. Engl.* **1991,** *30,* 874–876.
[171] A. Henglein, P. Mulvaney, A. Holzwarth, T. E. Sosebee, A. Fojtik, *Ber. Bunsenges. Phys. Chem.* **1992,** *96,* 754–759.
[172] A. Henglein, P. Mulvaney, T. Linnert, A. Holzwarth, *J. Phys. Chem.* **1992,** *96,* 2411–2414.
[173] P. Mulvaney, M. Giersig, A. Henglein, *J. Phys. Chem.* **1992,** *96,* 10419–10424.
[174] J. S. Bradley, G. Via, L. Bonneviot, *unpublished results.*
[175] G. Schmid, A. Lehnert, U. Kreibig, Z. Adamczyk, P. Belouschek, *Z. Naturforsch.* **1990,** *45b,* 989–994.
[176] G. Schmid, U. Giebel, W. Huster, A. Schwenk, *Inorg. Chim. Acta.* **1984,** *85,* 97–102.
[177] G. Schmid, B. Morun, J.-O. Malm, *Angew. Chem. Int. Ed. Engl.* **1989,** *28,* 778–779.
[178] N. M. Vargaftik, I. I. Moiseev, D. I. Kochubey, K. I. Zamaraev, *Faraday Discuss. Chem. Soc.* **1991,** *92,* 13–29.
[179] G. Schmid, *Polyhedron* **1988,** *7,* 2321–2329.
[180] M. N. Vargaftik, V. P. Zagorodnikov, I. P. Stolyarov, I. L. Moiseev, V. A. Likholobov, D. I. Kochubey, A. L. Chuvilin, V. I. Zaikovsky, K. I. Zamaraev, G. I. Timofeeva, *J. Chem. Soc., Chem. Commun.* **1985,** 937–939.
[181] G. Schmid, *Aspects of Homogeneous Catalysis* **1990,** *7,* 1–36.
[182] H. Siedentopf, R. Zsigmondy, *Annalen der Physik* **1903,** *10,* 1–8.
[183] *Colloidal Dispersions* (Ed.: J. W. Goodwin) **1982,** Royal Society of Chemistry, London.
[184] K. Kurihara, J. H. Fendler, I. Ravet, J. Nagy, *J. Mol. Catal.* **1986,** *34,* 325–335.
[185] N. J. Long, R. F. Marzke, M. McKelvy, W. S. Glaunsinger, *Ultramicroscopy* **1986,** *20,* 15–20.
[186] H. Poppa, F. Rumpf, R. D. Moorhead, C. Henry, *Proc. Mat. Res. Soc. Symp.* Vol. 111, Materials Research Society **1988,** pp. 1–11.
[187] C. G. Granqvist, R. A. Buhrman, *J. Appl. Phys.* **1976,** *47,* 2200–2219.
[188] C. G. Granqvist, R. A. Buhrman, *J. Catal.* **1976,** *42,* 477–479.
[189] F. G. N. Cloke, A. E. Derome, M. L. H. Green, D. O'Hare, *J. Chem. Soc., Chem. Commun.* **1983,** 1312–1313.
[190] J. A. Bandy, F. G. N. Cloke, M. L. H. Green, D. O'Hare, K. Prout, *J. Chem. Soc., Chem. Commun.* **1984,** 240–242.
[191] B. A. Scott, R. M. Plecenik, G. S. Cargill 3rd, T. R. McGuire, S. R. Herd, *Inorg. Chem.* **1980,** *19,* 1252–1259.
[192] S. C. Davis, S. Severson, K. Klabunde, *J. Amer. Chem. Soc.* **1981,** *103,* 3024–3029.
[193] S. C. Davis, K. J. Klabunde, *J. Am. Chem. Soc.* **1978,** *100,* 5973–5974.

[194] C.J. Baddeley, D.A. Jefferson, R.J. Lambert, R.M. Ormerod, T. Rayment, G. Schmid, A.P. Walker in *Proc. Mat. Res. Soc. Symp.* (Eds.: P.D. Persans, J.S. Bradley, R.R. Chianelli, G. Schmid), Materials Research Society **1992,** pp. 85–90.
[195] J.H. Sinfelt, G.H. Via, F.W. Lytle, *Catal. Rev.-Sci. Eng.* **1984,** *26,* 81–140.
[196] G.H. Via, K.F. Drake Jr., G. Meitzner, F.W. Lytle, J.H. Sinfelt, *Catal. Lett.* **1990,** *5,* 25–34.
[197] H. Bönnemann, W. Brijoux, E. Dinjus, T. Joussen, *Angew. Chem. Int. Ed. Engl.* **1990,** *29,* 273–275.
[198] R. van Hardeveld, F. Hartog, *Surf. Sci.* **169,** *15,* 189–230.
[199] R.E. Benfield, *J. Chem. Soc., Faraday Trans* **1992,** *88,* 1107–1110.
[200] D.G. Duff, P.P. Edwards, J. Evans, J.T. Gauntlett, D.A. Jefferson, B.F.G. Johnson, A.J. Kirkland, D.J. Smith, *Angew. Chem. Int. Ed. Engl.* **1989,** *28,* 590–593.
[201] P.-A. Buffat, M. Flüeli, R. Spycher, P. Stadelmann, J.-P. Borel, *Faraday Discuss. Chem. Soc.* **1991,** *92,* 173–187.
[202] L.D. Marks, A. Howie, *Nature (London)* **1979,** *282,* 196–198.
[203] *Surface Segregation Phenomena* (Eds.: P.A. Dowben, A. Miller) **1990,** CRC Press: Boca Raton.
[204] G.H. Via, L. Bonneviot, J.S. Bradley, *unpublished results.*
[205] A.J. Creighton in *Surface Enhanced Raman Scattering* (Eds.: R.K. Chang, T.E. Furtak), Plenum, New York **1982,** 315–337.
[206] G.C. Papavassiliou, T. Kokkinakis, *J. Phys. F: Metal Phys.* **1974,** *4,* L67–68.
[207] J.A. Creighton, M.S. Alvarez, D.A. Weitz, S. Garoff, M.W. Kim, *J. Phys. Chem.* **1983,** *87,* 4793–4799.
[208] M. Kerker, *The Scattering of Light and Other Electromagnetic Radiation.* Academic Press, New York **1969.**
[209] J.A. Creighton, C.G. Blatchford, M.G. Albrecht, *J. Chem. Soc. Faraday II* **1979,** *75,* 790–798.
[210] T. Linnert, P. Mulvaney, A. Henglein, *J. Phys. Chem.* **1993,** *97,* 679–682.
[211] J.A. Creighton, D.G. Eadon, *J. Chem. Soc., Faraday Trans.* **1993,** *87,* 3881–3891.
[212] R.P. van Duyne in *Chemical and Biological Applications of Lasers* (Ed.: C.B. Moore), Academic Press, New York **1978,** p. 101.
[213] M. Kerker, O. Siiman, L.A. Bumm, W.-S. Wang, *Appl. Opt.* **1980,** *19,* 3253–3255.
[214] K.U. von Raben, R.K. Chang, B.L. Laube, *Chem. Phys. Lett.* **1981,** *79,* 465–469.
[215] R.E. Benner, K.U. von Raben, K.C. Lee, J.F. Owen, R.K. Chang, B.L. Laube, *Chem. Phys. Lett.* **1983,** *96,* 65–69.
[216] O. Siiman, A. Lepp, M. Kerker, *Chem. Phys. Lett.* **1983,** *100,* 163–168.
[217] Y. Kurokawa, Y. Imai, Y. Sasaki, T. Maeda, *Anal. Biochem.* **1993,** *209,* 247–250.
[218] J. Pate, A. Leiden, B.J. Bozlee, R.L. Garrell, *J. Raman Spectrosc.* **1991,** *22,* 477–480.
[219] S. Sanchez-Cortes, J.V. Garcia-Ramos, *J. Raman Spectrosc.* **1992,** *23,* 61–66.
[220] S. Sanchez-Cortes, J.V. Garcia-Ramos, *Vib. Spectrosc.* **1993,** *4,* 185–192.
[221] N. Sheppard, T.T. Nguyen in *Advances in Infrared and Raman Spectroscopy* (Eds.: R.J.H. Clarke, R.E. Hester) **1978,** 106.
[222] M.R. Mucalo, R.P. Cooney, *J. Chem. Soc., Faraday Trans* **1991,** *87,* 3779–3785.
[223] M.R. Mucalo, R.P. Cooney, J.B. Metson, *Colloids Surf* **1991,** *60,* 175–197.
[224] M.R. Mucalo, R.P. Cooney, *Can. J. Chem.* **1991,** *69,* 1649–1655.
[225] M.R. Mucalo, R.P. Cooney, *J. Chem. Soc., Faraday Trans* **1991,** *87,* 1221–1227.
[226] M.R. Mucalo, R.P. Cooney, *J. Colloid Interface Sci.* **1992,** *150,* 486–491.
[227] J.S. Bradley, J.M. Millar, E.W. Hill, M. Melchior, *J. Chem. Soc., Chem. Commun.* **1990,** 705–706.
[228] L.L. Shieu, Z. Karpinski, W.M.H. Sachtler, *J. Phys. Chem.* **1989,** *93,* 4890–4894.
[229] J.S. Bradley, J.M. Millar, E.W. Hill, S. Behal, *J. Catal.* **1991,** *129,* 530–539.
[230] C. Kittel, *Introduction to Solid State Physics,* Wiley, New York **1976,** p. 514.
[231] W.D. Knight, *Phys. Rev.* **1949,** *76,* 1259–1270.

[232] G. C. Carter, L. H. Bennett, D. J. Kahan, “Metallic Shifts in NMR, Part 1“, *Progress in Materials Science,* Vol. 20 **1977,** Pergamon, Oxford.
[233] H. E. Rhodes, P.-K. Wang, H. T. Stokes, C. P. Slichter, J. H. Sinfelt, *Phys. Rev. B.* **1982,** *26,* 3559–3568.
[234] J. P. Bucher, J. Buttet, J. J. van der Klink, M. Graetzel, E. Newson, T. B. Truong, *J. Mol. Catal.* **1987,** *43,* 213–220.
[235] M. J. Williams, P. P. Edwards, D. P. Tunstall, *Faraday Disc.* **1991,** *92,* 199–215.
[236] C. P. Slichter, *Ann. Rev. Phys. Chem. 1986, 37,* 25–51.
[237] T. M. Duncan, T. W. Root, *J. Phys. Chem.* **1988,** *92,* 4426–4432.
[238] T. M. Duncan, K. W. Zilm, D. M. Hamilton, T. W. Root, *J. Phys. Chem.* **1989,** *93,* 2583–2590.
[239] K. W. Zilm, L. Bonneviot, D. M. Hamilton, G. G. Webb, G. L. Haller, *J. Phys. Chem.* **1990,** *94,* 1463–1472.
[240] M. Pruski, J. C. Kelzenberg, B. C. Gerstein, T. S. King, *J. Am. Chem. Soc.* **1990,** *112,* 4232–4240.
[241] I. D. Gay, *J. Catal.* **1987,** *108,* 15–23.
[242] M. Rigole, C. Depecker, G. Wrobel, P. Legrand, M. Guelton, J. P. Bonnelle, *J. Phys. Chem.* **1990,** *94,* 6743–6748.
[243] J. P. Yesinowski, *J. Amer. Chem. Soc.* **1981,** *103,* 6266–6267.
[244] K. Kimura, N. Satoh, *Chem. Lett.* **1989,** 271–274.
[245] N. Satoh, K. Kimura, *J. Am. Chem. Soc.* **1990,** *112,* 4688–4692.
[246] R. D. Newmark, M. Fleischmann, B. S. Pons, *J. Electroanal. Chem. Interfacial Electrochem.* **1988,** *255,* 325–330.
[247] J. M. Millar, J. S. Bradley, *unpublished results.*
[248] K. W. Zilm, L. Bonneviot, G. L. Haller, O. H. Han, M. Kermarec, *J. Phys. Chem.* **1990,** *94,* 8495–8498.
[249] R. H. Crabtree, *Chem. Rev.* **1985,** *85,* 245–269.
[250] D. R. Anton, R. H. Crabtree, *Organometallics* **1983,** *2,* 855–859.
[251] K. J. Klabunde, Y.-X. Li, B.-J. Tan, *Chem. Mater.* **1991,** *3,* 30–39.
[252] H. Hirai, M. Ohtaki, M. Komiyama, *Chem. Lett.* **1986,** 269–272.
[253] H. Hirai, M. Ohtaki, M. Komiyama, *Chem. Lett.* **1987,** 149–152.
[254] Y. Wang, H. Liu, Y. Jiang, *J. Chem. Soc. Chem. Comm.* **1989,** 1878–1879.
[255] H. Hirai, Y. Nakao, N. Toshima, K. Adachi, *Chem. Lett.* **1976,** 905–910.
[256] H. Hirai, Y. Nakao, N. Toshima, *Chem. Lett.* **1978,** 545–548.
[257] H. Hirai, H. Chawanya, N. Toshima, *Makromol. Chem. Rapid Commun.* **1981,** *2,* 99–103.
[258] H. Hirai, S. Komatsuzaki, N. Toshima, *Bull. Chem. Soc. Jpn.* **1984,** *57,* 488–494.
[259] H. Hirai, H. Chawanya, N. Toshima, *Bull. Chem. Soc. Jpn.* **1985,** *58,* 682–687.
[260] H. Hirai, H. Wakabayashi, M. Komiyama, *Bull. Chem. Soc. Jpn.* **1986,** *59,* 545–550.
[261] M. Ohtaki, N. Toshima, M. Komiyama, H. Hirai, *Bull. Chem. Soc. Jpn.* **1990,** *63,* 1433–1440.
[262] C. Larpent, E. Bernard, *J. Chem. Soc. Chem. Comm.* **1992,** 535–536.
[263] V. M. Deshpande, C. S. Narasimhan, P. Singh, *J. Mol. Catal.* **1990,** *63,* L5–10.
[264] A. J. Chalk, J. F. Harrod, *J. Am. Chem. Soc.* **1967,** *89,* 1640–1647.
[265] L. N. Lewis, N. Lewis, R. J. Uriarte, *Adv. Chem. Ser.* **1992,** *230,* 541–549.
[266] L. N. Lewis, R. J. Uriarte, *Organometallics* **1990,** *9,* 621–625.
[267] L. N. Lewis, R. J. Uriarte, N. Lewis, *J. Mol. Catal.* **1991,** *66,* 105–113.
[268] H. Hirai, H. Wakabayashi, M. Komiyama, *Bull. Chem. Soc. Jpn.* **1986,** *59,* 545–550.
[269] H. Hirai, H. Wakabayashi, M. Komiyama, *Makromol. Chem., Rapid Commun.* **1984,** *5,* 381–384.
[270] M. Spiro, P. W. Griffin, *Chem. Comm.* **1969,** 262–263.
[271] P. L. Freund, M. Spiro, *J. Phys. Chem.* **1985,** *89,* 1074–1077.
[272] P. L. Freund, M. Spiro, *J. Chem. Soc., Faraday Trans 1* **1986,** *82,* 2277–2282.
[273] A. Henglein, *J. Phys. Chem.* **1979,** *83,* 2858–2862.

[274] A. Henglein, *Ber. Bunsenges. Phys. Chem.* **1980,** *84,* 253–259.
[275] A. Henglein, *Chem. Rev.* **1989,** *89,* 1861–1873.
[276] A. Henglein, *Angew. Chem. Int. Ed. Eng.* **1979,** *18,* 418.
[277] A. Henglein, *J. Phys. Chem.* **1979,** *83,* 2209–2216.
[278] A. Henglein, B. Lindig, J. Westerhausen, *J. Phys. Chem.* **1981,** *85,* 1627–1628.
[279] A. Henglein, J. Lillie, *J. Am. Chem. Soc.* **1981,** *103,* 1059–1066.
[280] A. Moradpour, E. Amouyal, P. Keller, H. Kagan, *Nouv. J. Chim.* **1978,** *2,* 547–549.
[281] E. Amouyal, P. Koffi, *J. Photochem.* **1985,** *29,* 227–242.
[282] C.-K. Tan, V. Newberry, T. R. Webb, C. A. McAuliffe, *J. Chem. Soc., Dalton Trans.* **1987,** 1299–1303.
[283] N. Toshima, M. Kuriyama, Y. Yamada, H. Hirai, *Chem. Lett.* **1981,** 793–796.
[284] M. S. Matheson, P. C. Lee, D. Meisel, E. Pellizetti, *J. Phys. Chem.* **1983,** *87,* 394–399.
[285] N. Toshima, T. Takahashi, H. Hirai, *J. Macromol. Sci.-Chem.* **1988,** *A25,* 669–686.
[286] A. Harriman, *J. Chem. Soc. Chem. Com.* **1990,** 24–26.
[287] M.-O. Delcourt, E. Amouyal, M. Georopolis, *New J. Chem.* **1986,** *11,* 607–611.
[288] R. Maidan, I. Willner, *J. Am. Chem. Soc.* **1986,** *108,* 8100–8101.
[289] I. Willner, R. Maidan, D. Mandler, H. Dürr, G. Dörr, K. Zengerle, *J. Am. Chem. Soc.* **1987,** *109,* 6080–6086.
[290] H. Durr, H. P. Trierweiler, I. Willner, R. Maidan, *New J. Chem.* **1990,** *14,* 317–320.
[291] D. Mandler, I. Willner, *J. Am. Chem. Soc.* **1987,** *109,* 7884–7885.
[292] See, for example, *Proc. VI Int. Symp. on Relations Between Homog. and Heterog. Catal.,* published as *J. Mol. Catal.* **1989,** *55 and 56,* 1989.
[293] A. W. Olsen, Z. H. Kafafi, *Proc. Mat. Res. Soc. Symp. 206* Materials Research Society **1991,** pp. 175–181.

7 Perspectives

Günter Schmid

"The world of the neglected dimensions" (W. Ostwald, 1915), known in principle since Michael Faradays synthesis of the ruby red gold colloids, is presently experiencing a tremendous renaissance. During the first half of our century, it was mainly physical chemists who were interested in small metal particles because of their special behavior in solution where they occupied a position between pure solutions and emulsions. With the discovery of the metal carbonyls, a new era of complex chemistry began. Molecules like [$Co_4(CO)_{12}$] or [$Rh_6(CO)_{16}$] in which metal–metal interactions occur between several atoms nowadays represent some of the first members of a continuous series of particles which finally ends with the formation of small metal-like particles, the colloids. Studies on these materials, which reach from molecular clusters consisting of only a few metal atoms up to metal colloids with thousands of atoms, has meanwhile grown to an independent interdisciplinary scientific field. During the past 10–15 years it has been recognized that small particles possess size-dependent properties. For that reason, metal clusters and colloids became not only interesting objects from an academic viewpoint but also from a practical one. Our present knowledge about these particles promises the discovery of novel properties in materials which might have revolutionary character for numerous areas in science and technology.

In the present book, we have attempted to describe the actual state of our knowledge in the most important areas of cluster and colloid science on a level which satisfies the demands of the scientific community. On the other hand, it should also be appropriate to offer a basis to those who want to familiarize themselves with this exciting science. Although the results described in the various chapters are still mainly in the realm of basic research, there are already numerous starting points for promising applications.

According to the estimations from numerous scientists, the future of metal clusters and colloids will be focussed in the following fields:

- Condensed clusters separated by nonmetal atoms, or ligand stabilized clusters in crystal or pellet forms, and composed of appropriate elements, may lead to important new magnetic materials.
- Clusters at the initial (or terminal) ends of the metallic state in which the last freely mobile electrons are applied as quantum dots, may contribute to a new revolution in microelectronics. Single electron transitions (SET) between clusters may lead to the final step in the miniaturization of electronic devices. Novel generations of computers and minilasers become evident.

- The semiconductor properties of metal clusters can help to develop new and more effective photovoltaic elements.
- Mixed metallic clusters and colloids can be used as a basis for unknown alloys.
- Transition metal clusters in combination with main group elements like P, As, S, Se can act as ideal precursors for novel semiconducting materials.
- The use of defined clusters and colloids in catalysis is still in its beginning stages. The present results are promising.

These prospecting aspects require such a lot of innovation that the future of cluster and colloid science can be viewed with great optimism. It is further hoped that this field of science will lead to a continued cooperation between physicists, chemists, and material scientists. Hardly another discipline is as well suited for this as cluster and colloid science.

Index